Optical Nano and Micro Actuator Technology

Optical Nano and Micro Actuator Technology

Edited by
George K. Knopf
Yukitoshi Otani

CRC Press
Taylor & Francis Group
Boca Raton London New York

CRC Press is an imprint of the
Taylor & Francis Group, an **informa** business

CRC Press
Taylor & Francis Group
6000 Broken Sound Parkway NW, Suite 300
Boca Raton, FL 33487-2742

© 2013 by Taylor & Francis Group, LLC
CRC Press is an imprint of Taylor & Francis Group, an Informa business

No claim to original U.S. Government works

Printed in the United States of America on acid-free paper
Version Date: 2012924

International Standard Book Number: 978-1-4398-4053-5 (Hardback)

Visit the Taylor & Francis Web site at
http://www.taylorandfrancis.com

and the CRC Press Web site at
http://www.crcpress.com

This book is dedicated to the spirit, fortitude, and resilience of the people of Japan.

Contents

PART I   Introduction to Optical Actuation

PART II   Photoresponsive Materials

PART III   Harnessing Light and Optical Forces

PART IV   Optically Driven Systems

PART V   Applications of Optical Actuation

Preface

Do not Bodies and Light act mutually upon one another; that is to say, Bodies upon Light in emitting, reflecting, refracting and inflecting it, and Light upon Bodies for heating them, and putting their parts into a vibrating motion wherein heat consists?

Sir Isaac Newton, Opticks (1704), Book 3, Query 5, 133

Our quest to understand the fundamental physics of how light interacts with solid objects and affects the inherent thermal, mechanical, and electrical properties of the material is a scientific journey that is more than 300 years old. Recent advances in precision instrumentation and a deeper understanding of material science have opened doors not envisioned by Newton and his contemporaries. Modern tools for scientific investigation have enabled researchers around the globe to explore the micron and submicron worlds, where electromagnetic radiation can play a dominant role in changing the behavior of certain materials—in other words, in the world of the *very very small*, light does matter. Through material design and clever part fabrication, the properties of light (radiation pressure, intensity, wavelength, phase) can be transformed into small, yet meaningful, displacements and forces.

From an engineering perspective, actuators are the devices within machines and complex systems that perform mechanical work in response to a command or control signal. The device can be separated into two parts: the *actuator shell* and the *method of actuation*. The shell is the basic structure of the actuator and, often, contains deformable or moving parts. Examples of deformable microactuator shells include cantilever beams, microbridges, diaphragms, and torsional mirrors. The main function of any shell design is to provide a mechanism for the desired actuation method to produce useful work. The actuation method is the means by which a control signal is converted to a force that is applied to the actuator shell and creates physical movement. The output of the overall system is the desired response given as a small displacement, force, or pressure value. The different methods of nano- and microactuation take advantage of mechanical, electrostatic, piezoelectric, magnetic, thermal, fluidic, acoustic, chemical, biological, or optical principles.

Although optically activated nano- and microactuators are probably the least developed of all force-generating structures, they offer several interesting design features. All-optical circuits and devices have advantages over conventional electronic components because they are activated by photons instead of currents and voltages. In many of these designs, the photons provide both the energy into the system and control signal to initiate the desired response. Furthermore, optical systems are free from current losses, resistive heat dissipation, and mechanical friction forces that greatly diminish the performance and efficiency of conventional electronic or electromechanical systems. The negative effects of current leakage and power loss are greatly amplified as design engineers strive for product miniaturization through the exploitation of micro- and nanotechnology. Even the radiation pressure arising from a beam of photons becomes a viable force for driving mechanisms that have a picogram mass or exist in a nanometer size. Optical actuators are also ideal components for smart structures because they are immune from electromagnetic interference, safe in hazardous or explosive environments, and exhibit low signal attenuation.

The fundamental and unique characteristics of light-activated optical actuators are explored in this book. The more commonly studied photoactivated and light-driven systems that utilize off-the-shelf photocells to generate a voltage to directly drive an electromagnetic motor are not discussed. Here, the primary means of actuation is to project light onto a nano- or microactuator shell in an effort to generate mechanical deformation that, in turn, produces the desired displacement or force.

Light is used to both initiate movement (i.e., power) and control the actuation mechanism to perform work. The opportunities for product innovation, by applying the basic mechanisms of optical actuation, are illustrated by several novel system designs.

This book provides a multidisciplinary, multiple author perspective of this rapidly evolving field. The book begins with the essential background necessary to understand light-driven systems and then advances to innovative optical actuator technologies for realizing solutions in a wide range of applications. The chapters include introductory topics such as the nature of light and the interaction between light and NEMS/MEMS devices. This book also explores several photoresponsive materials that enable the design of a variety of optically driven structures and mechanisms, light-driven technologies that permit the manipulation of nano- and microscale objects, and applications in optofluidics, BIOMEMS and biophotonics, medical device design, and micromachine control.

The editors and chapter contributors sincerely hope that this book will serve as a tool to help researchers, scholars, and students advance light-driven technologies that improve the quality of human life in the twenty-first century. The book, therefore, has three major objectives. The first is to present the scientific language and fundamental principles of this emerging interdisciplinary technology. The second is to provide readers with a holistic view of optical nano- and microactuator systems, thereby enabling them to begin the process of understanding how light can be used to physically manipulate material and mechanical structures. The third goal is to help readers realize the potential and practical applications of light-driven systems. By achieving these goals, the authors hope to advance the underlying technology and provide an inspiration for the next generation of scientists and engineers as they move forward to develop innovative solutions far beyond anything that we can imagine today.

The book is a collection of 22 chapters organized in 5 parts covering the fundamentals of photoresponsive materials, issues related to harnessing the forces of light, design of optically driven systems, and applications. The chapters are written by an international group of leading experts from academia and industry, representing Asia, Australia, Europe, and North America.

PART I: INTRODUCTION TO OPTICAL ACTUATION

Part I consists of two chapters that provide the reader with key definitions and an overview of the fundamental characteristics of light and its interaction with various materials. The discussion focuses on different perspectives of optical actuation and how these photon-driven nano- and micro-deformable structures can be incorporated into a variety of technologies. The contributing authors provide a diverse perspective on the topics drawing from material science, engineering design, electronics, and applied physics. Topics include an introduction to optical actuation, the nature of light, and the transfer of optical energy to material surfaces.

PART II: PHOTORESPONSIVE MATERIALS

Future advances in optically driven actuators are dependent on rapidly evolving photoresponsive materials that enable designers to create very small optically driven structures, mechanisms, and machines that have never been envisioned previously. Part II consists of five chapters that explore a variety of light-responsive materials that can generate measureable displacements and forces. This includes an introduction to the principles of molecular photoactuation, a close examination of reversible light-switchable molecules, and a new perspective on photostrictive actuators. New concepts in material design have also enabled researchers to create light-responsive mechanical actuation. A detailed discussion on the photomechanical actuation based on carbon nanotubes and an in-depth look at light-sensitive hydrogels are presented.

PART III: HARNESSING LIGHT AND OPTICAL FORCES

The ability to manipulate and interact with particles at micro- and nanoscales has been the key to many recent advances in science and technology. Optical manipulation of particles offers an attractive choice due to its inherent flexible and often noninvasive nature. In the field of optical manipulation, optical tweezers (OT) and optoelectronic tweezers have emerged as the dominant methods. Optical tweezing can be used to trap particles and minute objects by exploiting the optical gradient forces that arise from a tightly focused laser beam. Optoelectronic tweezers (OET), on the other hand, exploit light-induced dielectrophoresis (LIDEP). In this technique, the optical field interacts with a photoconductive material to create virtual electrodes that form electric field gradient landscapes that trap the particles. Part III presents four chapters that explore how the optical forces generated from OT and OET can be used to manipulate particles and activate small mechanisms that initiate mechanical movement in microfluidic channels. Many current applications require highly controllable OT and OET systems that can manipulate fragile living cells, fine industrial particles, droplets, and aerosols. Precision control for sensitive biological applications can be realized by incorporating new concepts such as holographic optical tweezers (HOT).

PART IV: OPTICALLY DRIVEN SYSTEMS

Optical actuators can either *directly* or *indirectly* transform light energy into a force. These transducers can be designed to operate under different properties of light such as intensity, wavelength, phase, and polarization. Part IV consists of five chapters that explore how light-driven technologies can be used to drive mechanisms, manipulate liquids, control optofluidic systems, and improve the performance of NEMS and MEMS devices. Recent progress toward developing monolithic optically driven devices fabricated using ultrashort pulsed femtosecond lasers and the integration of soft polymer micro- and nanostructures into optical MEMS are also examined.

PART V: APPLICATIONS OF OPTICAL ACTUATION

Although optical actuation technologies have been studied over the past several decades, it was not until the introduction of new photoresponsive materials and nanofabrication technologies that more effective and efficient optically driven systems have been developed. Part V consists of six chapters that discuss the application of optical nano- and microactuator technologies to solving real-world problems. Chapters include a discussion on how optical actuation plays a critical role in BIOMEMS and biophotonics, the development of an all-optical MEMS endoscope, and investigation of a variety of light-activated and powered shape memory actuators. The design of optically driven microrobotics and light propulsion systems for spacecraft is also explored. The book concludes with a chapter that gives the contributing authors an opportunity to provide their viewpoints as to the future prospects of optical nano- and microactuation technology and innovative applications. It is hoped that this more speculative discussion will spur discussion and provide young researchers with insight on future directions. This presentation may also be of interest to individuals who wish to see a snapshot of views of a rapidly changing technology at the early stages of its development.

Acknowledgments

We would like to express our sincere and heartfelt thanks to all the contributors to this book for their time and effort in preparing the chapters. Their excellent work is very much appreciated. We would also like to express our sincere gratitude to Ashley Gasque, press editor, and Kari Budyk, project coordinator, from Taylor & Francis Group/CRC Press for their assistance, advice, and patience during the editing phase of the book.

Indeed we are very much indebted to our families and wives, Eirin and Ikuko, who have generously supported this project at each step by letting us use family time during evenings, weekends, and holidays. The Knopf (Erik and Karl) and Otani (Akine and Gen) children have always understood and offered humor to ease the task. Yukitoshi also wishes to acknowledge the unwavering support of his mother, Matsuko, during difficult times.

George K. Knopf
Yukitoshi Otani

Editors

George K. Knopf is a professor in the Department of Mechanical and Materials Engineering at the University of Western Ontario (London, Ontario, Canada). His areas of research interest include bioelectronics, laser microfabrication technologies, micro-optics, optical sensors and microactivators, intelligent CAD, and interactive data visualization. Professor Knopf has contributed to the development of several intelligent systems for engineering design, including the efficient packing of 3D parts for layered manufacturing and adaptive reconstruction of complex freeform surfaces. In recent years, the focus of his research has expanded in the areas of laser microfabrication, micro-optics, and light-driven technologies. This work includes a unique approach to surface geometry measurement using an unconstrained range-sensor head [U.S. patent 6,542,249], micro-optic element design for large area light guides and curtains, nonlithographic fabrication of metallic micromold masters by laser machining and welding, laser micropolishing and development of several bioelectronic devices that exploit the photoelectric signals generated by dried bacteriorhodopsin (bR) films. These light-activated transducers represent a new sensor technology that can be fabricated on flexible polymer substrates for creating novel imaging and biosensor systems [U.S. patent 7,573,024].

Professor Knopf has coedited a CRC Press volume entitled *Smart Biosensor Technology* and several SPIE Proceedings on *Optomechatronics Systems*. He has also acted as a technical reviewer for numerous academic journals, conferences, and granting agencies and has cochaired several international conferences.

Yukitoshi Otani is a professor in the Center for Optical Research and Education (CORE) at Utsunomiya University (Tochigi, Japan). He received his master's degree from Tokyo University of Agriculture and Technology in 1990 and his doctorate from the University of Tokyo in 1995. After working for a brief period at HOYA Corporation, he joined Tokyo University of Agriculture and Technology in 1991 and moved to Utsunomiya University in 2010. He was a visiting professor at the College of Optical Sciences, the University of Arizona, from 2004 to 2005. Dr. Otani is an SPIE fellow, where he was honored in 2010 for his pioneering research in the interdisciplinary field of optics and mechanics (optomechatronics).

Dr. Otani current research interests include optomechatronics, optical actuator and manipulator, varifocus lens, 3D profilometry by Moiré topography and interferometry, scatterometry, birefringence mapping, polarization, and polarimetry. Over the past two decades, Dr. Otani has contributed to the development of novel measurement techniques in polarization engineering based on birefringence, Stokes parameters, and Mueller matrix. In addition, he has developed several optically driven microactuators. One technique creates mechanical movement by producing a photothermal effect on the tip of a polyvinylidine difluoride (PVDF) microcantilever. The microactuator was successfully used for microrobotic and microassembly applications.

Dr. Otani has authored and coauthored more than 195 technical papers in major refereed journals and conference proceedings, including in *Optical Engineering*, *Optical Review*, *Optics Communications*, *Optics Letters*, and the *Proceedings of SPIE*. In addition, he has published 32 Japanese language papers and presented 8 invited talks at major international conferences. In recent years, he has coedited three SPIE Proceedings on *Optomechatronic Actuators and Manipulation* and has written a CRC Press book chapter on "Birefringence Measurement" (T. Yoshizawa ed., *Handbook of Optical Metrology: Principles and Applications*, 2009). Dr. Otani is currently an associate editor for the *International Journal of Optomechatronics* and has been chairing organized sessions on mechano-photonics in the Japanese Society of Precision Engineering since 2003.

Contributors

Khaled Al-Aribe
Department of Mechanical and Materials
Engineering
The University of Western Ontario
London, Ontario, Canada

Theodor Asavei
School of Physical Sciences
Centre for Biophotonics and Laser Science
University of Queensland
Brisbane, Queensland, Australia

Philippe Bado
Translume Inc.
Ann Arbor, Michigan

Christopher J. Barrett
Department of Chemistry
McGill University
Quebec, Montreal, Canada

Yves Bellouard
Department of Mechanical Engineering
Eindhoven University of Technology
Eindhoven, the Netherlands

Roman Boulatov
University of Illinois at Urbana-Champaign
University of Illinois
Urbana, Illinois

Bernd Dachwald
Department of Aerospace Engineering
FH Aachen University of Applied Sciences
Aachen, Germany

Mark Dugan
Translume Inc
Ann Arbor, Michigan

Norman Heckenberg
School of Physical Sciences
Centre for Biophotonics and Laser Science
University of Queensland
Brisbane, Queensland, Australia

Nien-Tsu Huang
Department of Mechanical Engineering
University of Michigan
Ann Arbor, Michigan

Toshiaki Iwai
Department of Bio-Applications and Systems
Engineering
and
Department of Electrical and Electronic
Engineering
Graduate School of Bio-Applications and
Systems Engineering
Tokyo University of Agriculture and
Technology
Fuchū, Tokyo, Japan

Arash Jamshidi
Department of Electrical Engineering and
Computer Sciences
University of California, Berkeley
Berkeley, California

Peipei Jia
Biomedical Engineering program
Faculty of Engineering
The University of Western Ontario
London, Ontario, Canada

George K. Knopf
Department of Mechanical and Materials
Engineering
The University of Western Ontario
London, Ontario, Canada

Kah How Koh
Department of Electrical and Computer
Engineering
National University of Singapore
Singapore, Singapore

Timothy J. Kucharski
Department of Chemistry
University of Illinois at Urbana-Champaign
Urbana, Illinois

Katsuo Kurabayashi
Department of Mechanical Engineering
and
Department of Electrical Engineering and
    Computer Science
University of Michigan
Ann Arbor, Michigan

Chengkuo Lee
Department of Electrical and Computer
    Engineering
National University of Singapore
Singapore, Singapore

Vince Loke
School of Physical Sciences
Centre for Biophotonics and Laser Science
University of Queensland
Brisbane, Queensland, Australia

Zahid Mahimwalla
Department of Chemistry
McGill University
Quebec, Montreal, Canada

Jun-ichi Mamiya
Chemical Resources Laboratories
Tokyo Institute of Technology
Yokohama, Japan

Shoji Maruo
Department of Mechanical Engineering
Graduate School of Engineering
Yokohama National University
Yokohama, Japan

Steven L. Neale
Biomedical Engineering Research Division
School of Engineering
University of Glasgow
Glasgow, United Kingdom

Timo A. Nieminen
Centre for Biophotonics and Laser Science
School of Physical Sciences
University of Queensland
Brisbane, Queensland, Australia

Suwas K. Nikumb
Centre for Automotive Materials and
    Manufacturing
National Research Council of Canada
London, Ontario, Canada

Hideki Okamura
Department of Physics
International Christian University
Tokyo, Japan

Yukitoshi Otani
Center for Optical Research and Education
Utsunomiya University
Tochigi, Japan

Balaji Panchapakesan
Department of Mechanical Engineering
University of Louisville
Louisville, Kentucky

Halina Rubinsztein-Dunlop
School of Physical Sciences
Centre for Biophotonics and Laser Science
University of Queensland
Brisbane, Queensland, Australia

Ali A. Said
Translume Inc.
Ann Arbor, Michigan

M. Chandra Sekhar
Department of Physics and Astronomy
The University of Western Ontario
London, Ontario, Canada

Atsushi Shishido
Chemical Resources Laboratories
Tokyo Institute of Technology
Yokohama, Japan

Alex Stilgoe
School of Physical Sciences
Centre for Biophotonics and Laser Science
University of Queensland
Brisbane, Queensland, Australia

Atsushi Suzuki
Department of Materials Science & Research
Institute of Environment and Information
    Sciences
Yokohama National University
Yokohama, Japan

Hiroshi Toshiyoshi
Research Center for Advanced Science and
    Technology
Institute of Industrial Science
The University of Tokyo
Tokyo, Japan

Yi-Chung Tung
Research Center for Applied Sciences
Academia Sinica
Taipei, Taiwan

Kenji Uchino
Department of Electrical Engineering
The Pennsylvania State University
University Park, Pennsylvania
and
Office of Naval Research
ONR Global-Asia
Tokyo, Japan

Robert Vogel
School of Mathematics and Physics,
The University of Queensland
Brisbane, Queensland, Australia

Ming C. Wu
Department of Electrical Engineering and
    Computer Sciences
Berkeley Sensor and Actuator Center
University of California-Berkeley
Berkeley, California

Kevin G. Yager
Brookhaven Laboratories
Upton, New York

Johtaro Yamamoto
Faculty of Advanced Life Sciences
Hokkaido University
Sapporo, Japan

Jun Yang
Department of Mechanical and Materials
    Engineering
The University of Western Ontario
London, Ontario, Canada

Part I

Introduction to Optical Actuation

Actuators perform useful work in response to external or internal stimuli. The amount of work they can perform, the functions they fulfill, and the energy expenditures they require to do the desired work depend drastically on the method of actuation and energy input. Some of the fundamental and unique characteristics of light-activated nano- and microactuators are explored in Chapter 1. Optical actuation can be divided into direct and indirect optical methods. Direct optical actuation uses light to interact with the active parts of the actuator and cause actuation, while indirect optical methods take advantage of the secondary effects of light such as heating, photon-generated electricity, or photoconductivity of photoresponsive materials. Comparing these two types of optical actuation, direct optical processes can be much faster than indirect ones. Due to the optical nature of photomechanical actuation technologies, they offer many advantages over traditional actuation technologies. Besides the capability of remote energy transfer and coupling into the small systems, optical actuation also enables remote controllability, easy system design and construction, electrical-mechanical decoupling, free of electromagnetic noise, elimination of electrical circuits, better scaling capability, and ability to work in harsh environments. Such merits of optical actuation could potentially lead to many applications in the actuation and sensing fields. Potential applications of optical actuators include an alternative mechanism for converting solar energy directly into mechanical motion for planetary exploration, direct corrective control in adaptive optics/ interferometer, optical micro- and nanopositioning and control, solar tracking actuator/shutter for self-alignment of the spacecraft to the sun for optimal power generation using solar sails, optically controlled valves for space applications, optically controlled microrobots, nanorobots, MOMS, and photophones. Potential sensing applications of optical actuators include a variety of tunable sensors for incident radiation (UV, visible) based on the detection of incident radiation intensity and indirect micro- and nanochemical and biological sensors based on photodetection when the device is loaded with foreign materials such as chemicals, particles, and cells. In developing the aforementioned applications, photomechanical actuation can be incorporated with advanced optical components and systems such as fiber optics, waveguides, and integrated optoelectronic and optical systems to produce integrated "smart" systems.

In Chapter 2, Drs. Suwas K. Nikumb and M. Chandra Sekhar provide an overview of electromagnetic radiation and light with a focus on light as waves and streams of photons. The nature of light is central to the field of optics and is the starting point for understanding how light interacts with matter. Electromagnetic radiation is first introduced with a brief discussion on the history of the wave–particle duality of light and the optical wavelengths. This is followed by a look at the generation of laser light and the various physical characteristics of low- and high-power lasers that benefit engineering and technology. The transfer of energy from laser source to material surface is examined in terms of the physics of illumination, light absorption, surface modification (melting, ablation, vaporization), and plasma effects. The chapter concludes with a brief overview on laser material removal, introducing relevant surface modification processes based on continuous wave or long-pulse, nanosecond duration, and femtosecond time scale laser interactions, relevant applications, and its effect on surface profile formation of laser-ablated materials.

1 Light-Driven and Optically Actuated Technologies

George K. Knopf

CONTENTS

1.1 INTRODUCTION

Nano- and microactuators are material structures and microscopic devices that perform a mechanical action in response to a specific command or control signal. Many of these devices are no more than a few molecules (10^{-9} m) to several hundred microns (10^{-6} m) in size. The mechanical action may produce a physical displacement or induce a force that effects its immediate environment. From an engineering perspective, the microscopic mechanism can be separated into two fundamental parts: the actuator shell and the method of actuation (Tabib-Azar 1998). The *shell* is the basic physical structure of the actuator and, often, contains deformable or moving parts. The main function of any shell design is to provide an effective and efficient mechanism for the desired actuation method to produce useful work (Knopf 2006). Examples of deformable microactuator shells include cantilever beams, microbridges, flexible diaphragms, torsional mirrors, and expansive polymer gels. The physical form and mechanical principles exploited by these very tiny microactuators may be similar to analogous systems found in the larger macroworld or merely the expansive and contractive properties of materials to environmental stimuli. Under certain conditions, the actuator shell may actually transcend scale and become physically massive such as solar sails driven by tiny quantities of radiation pressure.

In contrast, the physical structure of the nanoactuator shell no longer looks familiar to many engineers and behaves in a nonrigid fashion. Nanoactuator shells often involve groups of interconnected molecules moving in unison under an external energy source. Rather than describing actuator components and integrated systems at this very minute scale, many researchers have focused on developing a nanomotor as the primary molecular device for converting energy into movement. A typical molecular nanomotor is capable of producing forces in the order of pico-Newtons (pN) (Li and Tan 2002). One branch of this research involves transforming molecular motor proteins found in living cells into molecular motors and integrating them in artificial devices and systems (Kang et al. 2009). The motor protein is able to transport a cargo within the nano/micro device in a manner that is similar to how kinesin in eukaryotic cells move various molecules along the tracks of microtubules inside cells (Setou et al. 2000, Vale 2003). Other types of motor proteins include myosin, which enables the contraction of muscle fibers in animals and dynein found in flagella.

For both nano- and microsystems, the *actuation method* is the means by which a control signal, in the form of energy, is converted to a force that can be applied to the actuator shell and induce physical movement. Different methods of actuation take advantage of mechanical, electrostatic, piezoelectric, magnetic, thermal, fluidic, acoustic, chemical, biological, or optical principles.

Although optically driven actuators are one of the least developed of all force generating mechanisms, they provide for several interesting and beneficial design opportunities when developing technologies at very small physical scales. These optical transducers can be designed to operate under different properties of light such as intensity, wavelength, phase, and polarization characteristics. Optical actuators can either *directly* or *indirectly* transform light energy into the desired structural displacement (Tabib-Azar 1998, Knopf 2006). Direct optical methods use photons to interact with the photosensitive properties of the material used to construct the actuator shell and, thereby, directly cause mechanical deformation. An example of direct optical microactuation occurs when a light source is used to alter electrostatic forces that move a silicon microcantilever beam (Tabib-Azar 1998). On the other hand, indirect optical methods exploit the ability of light to generate heat when it strikes the surface of a material and, thereby, influence the thermal properties of gases, fluids, and solids located in the immediate environment. One possibility is to use a light source to heat a gas that expands sufficiently to deform a very thin flexible diaphragm. The remote heating of liquids in Lab-on-a-Chip (LoC) devices by optical methods also represents new approaches for driving fluid flow along microchannels or mixing streams of liquids prior to analysis and chemical synthesis.

Indirect optical methods often have simpler designs and generate more actuation power than direct optical methods. However, these methods of actuation utilize thermal and phase transformation properties of fluids and solids, and can be comparatively slow. In contrast, direct optical microactuators are relatively fast but produce very small forces. However, as the physical scale of technology shrinks from a macro to a micro world, the forces and displacements required are reduced proportionally and the methods of direct optical actuation become more relevant and appealing to system design engineers. Direct optical actuators have become, therefore, crucial in the development of sophisticated micromachines that are constrained by spatial restrictions, low noise requirements, and minimal power consumption.

At the microscale, optically driven devices and circuits have an important advantage over conventional electronic circuitry because these mechanisms are activated by streams of photons instead of electrical currents and voltages. In many of these optically driven technologies, the photons provide both the energy into the system and control signal used to initiate the desired actuator response. Optically driven systems are also free from undesirable side-effects that include electrical current losses, resistive heat dissipation, and friction forces that may significantly reduce the nano- and microsystem performance and efficiency. The negative effects of small current leakage, resistive heat, and power loss may not be limiting design condition in every-day consumer electronics and mechatronic products, but the negative effects of unshielded electric fields and unwanted temperature gradients can be detrimental to the performance of products that depend on reliable nano- and microscale components. These perceived nuisances are often ignored for the sake of manufacturing efficiency and reduced product cost but will become serious problems as engineers strive for product miniaturization through the enhanced exploitation of materials engineering and nanotechnology. The nondependence on electric current also makes optical actuators ideal components for smart structures that must be immune from electromagnetic interference, exhibit low signal attenuation, and function safely in hazardous or explosive environments.

Some of the fundamental aspects and unique characteristics of light activated nano- and microactuators are explored in this chapter. The more commonly studied systems constructed from off-the-shelf photocells used to generate a voltage potential that drives an electromagnetic motor or solenoid will not be discussed. Instead, the primary means of optical actuation is to focus a light beam onto a nano- or microactuator shell in an effort to generate the required mechanical deformation or displacement that, in turn, produces the desired movement or force. In this context, light is used to both initiate movement (i.e., power) and control the actuation mechanism used to perform useful work.

The following is a brief overview of optically driven nano and microactuators. The discussion begins in Section 1.2 with a look at how light can be used to exert pressure on mechanical structures, and how the optical gradient forces that arise from the evanescent bonding between light waves can manipulate micro-objects and biological cells. Section 1.3 examines several other direct methods of optical actuation that take advantage of novel properties of photoresponsive materials. Some of these polymer materials can alter their shape, bend, or experience a change in their volume when exposed to light as discussed in Section 1.3.1. One approach is to use photoisomerizable molecules such as azobenzene to develop polymers that contract under certain illumination conditions. Liquid crystal elastomers (LCEs), for example, can significantly alter their shape by up to 400% in response to changes in the chemical environment, temperature, and light. Azobenzene can also be used to create a light-driven single-molecule DNA nanomotors. Other researchers have shown that the light-induced reactions of a photoresponsive dithienylethene can be used to reversibly trigger paralysis in living organisms. Although the latter research does not involve developing mechanical micromachines, it does represent a different philosophy where the movement of biological organisms can be regulated by light. Materials such as lanthanum-doped lead zirconate titanate (PLZT) ceramic have the ability to convert photon energy into mechanical motion by photostrictive effects arising from the superposition of the photovoltaic effect and the converse-piezoelectric effect are introduced in Section 1.3.2. The photomechanical effect consistent with the surface piezoelectric phenomenon can also be used to produce photo-stimulated vibrations in microcantilever beams as discussed in Section 1.3.2.2. Section 1.3.2.3 describes an alternative optomechanical effect that has been observed in certain amorphous semiconducting materials such as chalcogenide glasses (ChG) when absorbing polarized light. These materials are used to create microcantilever structures that can move between two extreme positions by merely altering the angle of polarized light relative to the device. Recent advances in carbon nanotube technology have given rise to a new class of photomechanical actuators. Light-induced elastic responses from single-walled carbon nanotubes and fibrous networks are briefly discussed in Section 1.3.3. Other direct optical methods that induce mechanical strain and deformation by using photon-generated electrons to alter the electrostatic pressure on a cantilever beam that forms a parallel plate capacitor with a ground plate are summarized in Section 1.4. These electron-mediated optical microactuators are faster than approaches that depend upon thermal radiation.

Indirect optical actuators often use photothermal interactions arising from light striking a material surface or fluid to activate and drive actuation. The methods discussed in Section 1.5 are often simpler than direct optical actuators but require some understanding of how light transfers energy to the medium. A brief introduction to the transfer of energy from a beam of light to fluids and solids is given in Section 1.5.1. A simple optothermal expansion of a fluid is first described in Section 1.5.2. The remote heating of liquids in LoC devices by optical methods also represents new approaches for driving fluid flow along microchannels or mixing streams of liquids prior to analysis. The concept of light-driven microflows is presented in Section 1.5.3. It is even possible to develop a gas-bubble piston to pump fluid in a microchannel where the piston is set into motion by light-induced thermo-capillary forces. The temperature gradient caused by light striking a material surface can also be used to induce mechanical changes to shape-memory alloys (SMA) as described in Section 1.5.4. These solids experience a discontinuous change in their physical structure near a crystalline phase transformation temperature. The dimensional change is significantly greater than thermal expansion and can be used to pull on structures with sufficient forces to perform mechanical work. Finally, the photothermal vibration of specially contoured fiber optic strands is introduced in Section 1.5.5 as a means of creating controlled movement. The displacements are sufficient to enable the construction of miniature walking platforms, robotic manipulators, and optical "chop sticks."

1.2 HARNESSING THE FORCES OF LIGHT

In the mid-1800s, the highly imaginative French author Jules Verne envisioned a fantastic world in his classic novel "From Earth to the Moon" where spaceships could be powered by the force of light (Chapter XIX, Verne 1993). A remarkable fact of this fictional story is that it was originally published in 1865 several years before the scientist James Maxwell predicted in 1871 (Maxwell 1873) that it could even be theoretically possible to generate any measurable force with a light source and decades before Pyotr Lebedev (Lebedev 1901) experimentally confirmed Maxwell's prediction in the early 1900s. For many years Lebedev's observations were nothing more than scientific curiosities with little impact on serious engineering or the development of practical technologies. The possibilities of using light to drive technology and various modes of transportation, however, continued to inspire the imagination of countless science fiction writers throughout the twentieth century and well into the next.

1.2.1 LIGHT PRESSURE

In a very broad sense, electromagnetic radiation exerts a very small pressure or force upon any exposed surface. If absorbed by the surface, the *radiation pressure* can be described as the power flux density divided by the speed of light. Conversely, if the radiation is completely reflected by the surface, then the total pressure experienced by the surface is doubled. In terms of focused light from a laser, the pressure is equal to beam power divided by the universal constant for the speed of light in a vacuum c ($\sim 3 \times 10^8$ m/s). Unfortunately, this results in a *very very weak* force. For example, a typical 1 mW diode laser pointer will generate a force in the range of a pico-Newton (Tang 2009). Therefore, to move or suspend a small solid object such as a coin in the air, it would be necessary to use 10^9 similar lasers all pointed at the same spot on the object's surface. Increasing the power of the light source 10^9 times will result in an equivalent kW laser with sufficient energy to ablate or vaporize the coin surface (depending on beam spot size).

The driving force generated by light pressure is based on the transfer of photons. Photons have no mass but carry energy at the speed of light. The momentum, p_{ph}, of each photon in a light beam is the result of this energy (Steen 1998, Ashkin 2000) and can be given by

$$p_{ph} = \frac{h\upsilon}{c} \tag{1.1}$$

where
  h is Planck's constant $(6.63 \times 10^{-34} \text{ Js})$
  c is the speed of light in a vacuum $(\sim 2.99 \times 10^8 \text{ m/s})$
  and υ is the optical frequency related to the wavelength (λ) of the light source by

$$\upsilon = \frac{c}{\lambda} \qquad (1.2)$$

The optical forces arising from light-material interaction result from the exchange of momentum between the incident photons and irradiated object (Jonáš and Zemánek 2008). A focused light beam with power of P generates $P/h\upsilon$ photons per second. If the stream of photons strikes a mirror surface with no absorption, then the photons are assumed to be reflected straight back and produce a total change-in-momentum per second,

$$\left(\frac{dp_{Total}}{dt} \right) = \left(\frac{2P}{h\upsilon} \right)\left(\frac{h\upsilon}{c} \right) = 2\frac{P}{c} \qquad (1.3)$$

The principle of conservation of momentum (Ashkin 2000) implies that the highly reflective mirror acquires an equal (dp_{Total}/dt), or force, in the direction of the light. In other words,

$$F_{mirror} = \left(\frac{dp_{Total}}{dt} \right) = 2\frac{P}{c} \qquad (1.4)$$

Alternatively, if the stream of photons strikes a surface that is 100% absorption, then the corresponding force on the structure is reduced to one-half because the recoil momentum will be nonexistent.

Consider a 1 W laser with a small beam diameter being projected onto a totally reflecting mirror. The force due to the photons striking the surface is

$$F_{mirror} = 2\frac{P}{c} = 2\frac{1\,\text{N} \times \text{m/s}}{2.99 \times 10^8\,\text{m/s}} = 0.668 \times 10^{-8}\,\text{N} = 6.7\,\text{nN} \qquad (1.5)$$

This is the maximum force that can be expected from the photon momentum at a light power of 1 W. Increasing the laser power will proportionally increase the force in absolute terms, but the value will still remain very small.

Although such tiny forces are not able to move a large spacecraft like NASA's shuttle or even a cup of coffee across the table, this perceived weakness can be turned into an engineering strength if the objective is to manipulate liquids and solids objects that exist in the nano and microworlds and weigh only few picograms. At this scale, optics can be used to shape and redirect light beams to strike the object surface or actuator shell at precise locations. Assume that the 1 W laser is focused to a very small spot size, then the light pressure on the surface or structure can be controlled. For example, the pressure, P_{mirror}, due to a 1 μm diameter beam on the reflective mirror surface is

$$P_{mirror} = \frac{F_{mirror}}{\text{Area}_{spot}} = \frac{4(6.7 \times 10^{-9}\,\text{N})}{\pi(1 \times 10^{-6}\,\text{m})^2} = 8509\,\frac{\text{N}}{\text{m}^2} \qquad (1.6)$$

Even though the spot size is only in the micro range, the pressure at this point is 8.5 kPa, which is significant if the mechanism being pushed optical forces is also in the microscale.

Furthermore, leading-edge fabrication technologies exist that enable a wide variety of nano- and micro-optical components and waveguides to be accurately reproduced on different types of rigid and mechanically flexible substrates. A major drawback in developing optically driven microactuators and micromachines is, however, that the forces generated by radiation pressure push against the actuator shell in only one direction. This is problematic for designing viable actuation systems and motors because the design goal is most often to develop a mechanism that can either push or pull on command.

1.2.2 OPTICAL GRADIENT FORCES

An alternative approach to light pressure for inducing micro-object movement is to take advantage of the *optical gradient forces* that arise from the evanescent bonding between light waves. The gradient optical force was first explored by Ashkin (1970) more than four decades ago to develop an optical trap and "tweezer" that could manipulate molecules through a specially designed microscope (Dholakia and Reece 2006). In his initial work, Ashkin was able to experimentally observe the forced acceleration of freely suspended particles by radiation pressure from continuous wave (CW) visible laser beam. Based on this observation, he was able to demonstrate that it was possible to lift a glass sphere off a glass plate and levitate it above the surface (Ashkin 1970). A later experiment with his colleague Dziedzic (Ashkin and Dziedzic 1971, Ashkin and Dziedzic 1980) showed that it was possible to trap a glass sphere in a vertical laser.

The physicists used a 0.25 W TEM_{00} 0.515 μm laser beam to support a 20 μm sphere in mid-air. The sphere was observed to sit stably for hours in an on-axis position because the transverse forces, caused by the difference between the refractive indices of glass and air, pushed it toward the maximum light intensity creating a potential well. The maneuverability of suspended spheres, too transparent and too small for direct viewing, were observed by scattering the light of the supporting beam or an ancillary, very low power laser. In 1980, Ashkin and Dziedzic (Ashkin and Dziedzic 1980) showed that a refinement on this micromanipulation enabled two laser beams to assemble aggregates of two, three, and four spheres in mid-air.

In the late 1990s, Higurashi et al. (1997) demonstrated how carefully shaped fluorinated polyimide micro-objects with a 6–7.5 mm cross-sectional radius can be rotated using Ashkin's method of optical trapping and tweezing. In this series of experiments, radiation forces near the focal point were used to position and rotate the micro-object about the laser beam axis. These micro-objects were surrounded by a *medium* with either high or low relative-refractive index, and optically trapped by exerting radiation pressure through their center openings using a strongly focused trapping laser beam. If the micro-object had a low relative refractive index, then the pressure was exerted on the inner walls of the object. Conversely, if the micro-object has a high-relative refractive index, then the pressure is exerted on the outer wall as shown in Figure 1.1. The micro-objects were both trapped and rotated by radiation pressure when the horizontal cross-sections of these objects show rotational symmetry. In addition, the rotation speed versus optical power and the axial position of the laser focal point were investigated for high relative-refractive index micro-objects. The rotational speed, with respect to optical power, was found to be in the range of 0.44–0.73 rpm/mW (Figure 1.1c).

In recent years, applications of the gradient optical force have driven advances in optical trapping and tweezing (Dholakia and Reece 2006, Nieminen et al. 2006, Kröner et al. 2007, Dienerowitz et al. 2008). Research has been performed in various areas including cell biology, colloidal dynamics, statistical mechanics, hydrodynamics, Brownian motion, and nanomanipulation. Researchers at MIT (Tang 2009) were some of the first to exploit gradient optical forces on a microchip for nano-manipulation. Based on Maxwell's equations and applying fundamental principles, the MIT researchers were able to generate a gradient force in the pico-Newtons range that was sufficient to activate a nanoscale oscillator. The experimental device consisted of two parallel optical waveguides and a light source at a known frequency. Although the two waveguides maintained a fixed separation distance of the parallel beams and exhibited very little optical loss, the optical fields generated an electromagnetic bond that could be measured and observed (Li et al. 2008, 2009, Tang 2009). This same principle was later exploited by Hong X. Tang (Tang 2009) and his team at Yale to produce oscillations using one single-mode waveguide rather than two. In this single waveguide design, the

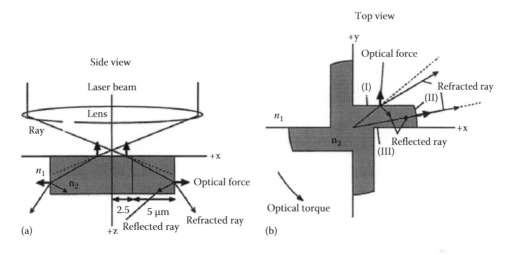

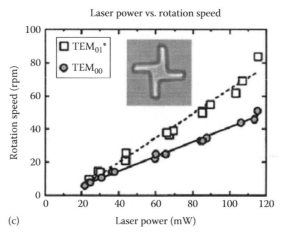

FIGURE 1.1 The basic principle of optical trapping and tweezing. (a) Side view and (b) top view of the optical forces acting on an optically transparent *high relative–refractive index* micro-object (n_1, n_2) in the Gaussian light beam. (c) The dependence of rotation speed on laser-power and laser mode (TEM00- and TEM01*). (Reprinted with permission from Higurashi, E., Ohguchi O., Tamamura T., Ukita H., and Sawada R., Optically induced rotation of dissymmetrically shaped fluorinated polyimide micro-objects in optical traps, *J. Appl. Phys.*, 82(6), 2773–2779. Copyright 1997, American Institute of Physics.)

optical field around the waveguide is asymmetrical in order to create an imbalance that is necessary to exert a net force.

It was only recently that researchers like Tang (Li et al. 2008, Tang 2009) were able to realize that these incredibly tiny forces could be used to drive mechanical structures at the sub-micron level. The seemingly small transverse gradient force generated at the sub-micron scale does not require reflective surfaces like radiation pressure and, therefore, provides a more versatile approach to planar actuator shells. Tang and his colleagues also demonstrated that it is possible to detect the transverse forces in an integrated silicon photonic circuit and exploit this force to drive nanoscale mechanical devices. The minute transverse force on the free-standing single-mode waveguide occurs because guided light is evanescently coupled to the dielectric substrate. The net optical force is determined by integrating the Maxwell stress tensor over the surfaces of the waveguide. For a specified optical input power, both the *effective refractive index* and the force on the waveguide depend on the separation distance between the waveguide and substrate (Figure 1.2). The calculations

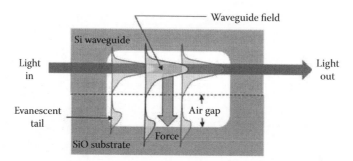

FIGURE 1.2 The gradient optical force exerted in a waveguide suspended in an air gap. According to Maxwell's equations (Maxwell 1873), the asymmetry between the open air on top of the waveguide and the thin air gap underneath should distort the optical field creating a downward force.

by Li et al. (2008) show that as the separation is reduced from 500 to 50 nm, the magnitude of the optical force increases from 0.1 to 8pN $\mu m^{-1} mW^{-1}$. Although these transverse gradient optical forces are very small, this principle has been used to throw switches in silicon optical circuits (Tang 2009) and to develop nanomechanical beam resonators embedded in a photonic circuit with an on-chip interferometer for displacement sensing (Li et al. 2008, 2009).

1.3 PHOTORESPONSIVE MATERIALS

Over the past two decades, numerous other researchers from around the globe have studied a variety of materials that exhibit optical-to-mechanical energy conversion properties and, therefore, represent exciting opportunities for creating novel light-driven nano- and microactuators. These material systems include light-induced shape-changing polymers (Jiang et al. 2006, Lendlein et al. 2005, Mitzutani et al. 2008), single-DNA nanomotors (Li and Tan 2002, Kang et al. 2009), molecular photoswitches (Al-Atar et al. 2009, Carling et al. 2009, Boyer et al. 2010), shape-changing LCEs (Warner and Terentjev 1996, Finkelmann et al. 2001, Warner and Terentjev 2003, Yu et al. 2003, Tabiryan et al. 2005, van Oosten et al. 2007, 2009), photostrictive materials that take advantage of photovoltaic and inverse piezoelectric effects (Uchino 1990, Morikawa and Nakada 1997, Poosanass et al. 2000), photomechanical actuators with charge-induced surfaces (Lagowski et al. 1975, Suski et al. 1990, Datskos et al. 1998), ChG influenced by mechanical polarization effects (Krecmer et al. 1997, Stuchlik et al. 2001), and photomechanical actuation of carbon nanotubes (Zhang and Iijima 1999, Verissimo-Alves et al. 2001, Piegari et al. 2002, Kroerner et al. 2004, Lui et al. 2009). Each of these optically driven material systems are briefly summarized next.

1.3.1 LIGHT-INDUCED SHAPE-CHANGING POLYMERS

A variety of stimulus-responsive polymers have been developed that exhibit large macroscopic directional movement or changes in shape when exposed to external stimuli such as temperature, electric fields, pH, or light. The shape or volume change in these polymers and polymeric gels is usually reversible once the stimulus is removed. For example, a freestanding photoinduced polymer cantilever beam will bend under irradiation and return to its original shape once the light source is removed. Some researchers have described how photo-responsive shape-memory polymers (SMP) containing cinnamic groups (Jiang et al. 2006) can be deformed and fixed into fairly complicated predetermined shapes by exposing the material to ultraviolet (UV) radiation. The original shapes can then be recovered at ambient temperatures under UV illumination at a different wavelength. These photoresponsive SMP have advantages over thermally induced shape memory materials because they can recover the original shapes at ambient temperatures and not react prematurely to environmental temperature fluctuations (e.g., biomedical implant that is to be activated in the body).

In order to respond to light, polymers need to posses photosensitive functional groups or fillers (Jiang et al. 2006). The most commonly used photosensitive molecules can be grouped as photoisomerizable molecules such as azobenzenes, triphenylmethane leuco derivatives that undergo photoinduced ionic dissociation, and photoreactive molecules such as cinnamates (Figure 1.3). Azobenzenes can exist in either *cis* or *trans* conformation states and under exposure to a particular wavelength of light these molecules can be made to switch reversibly from one state to the other. If the azobenzene groups are linked to macromolecules, then the switch can cause relatively large changes in the polymeric material. In contrast, triphenylmethane leuco derivatives will dissociate

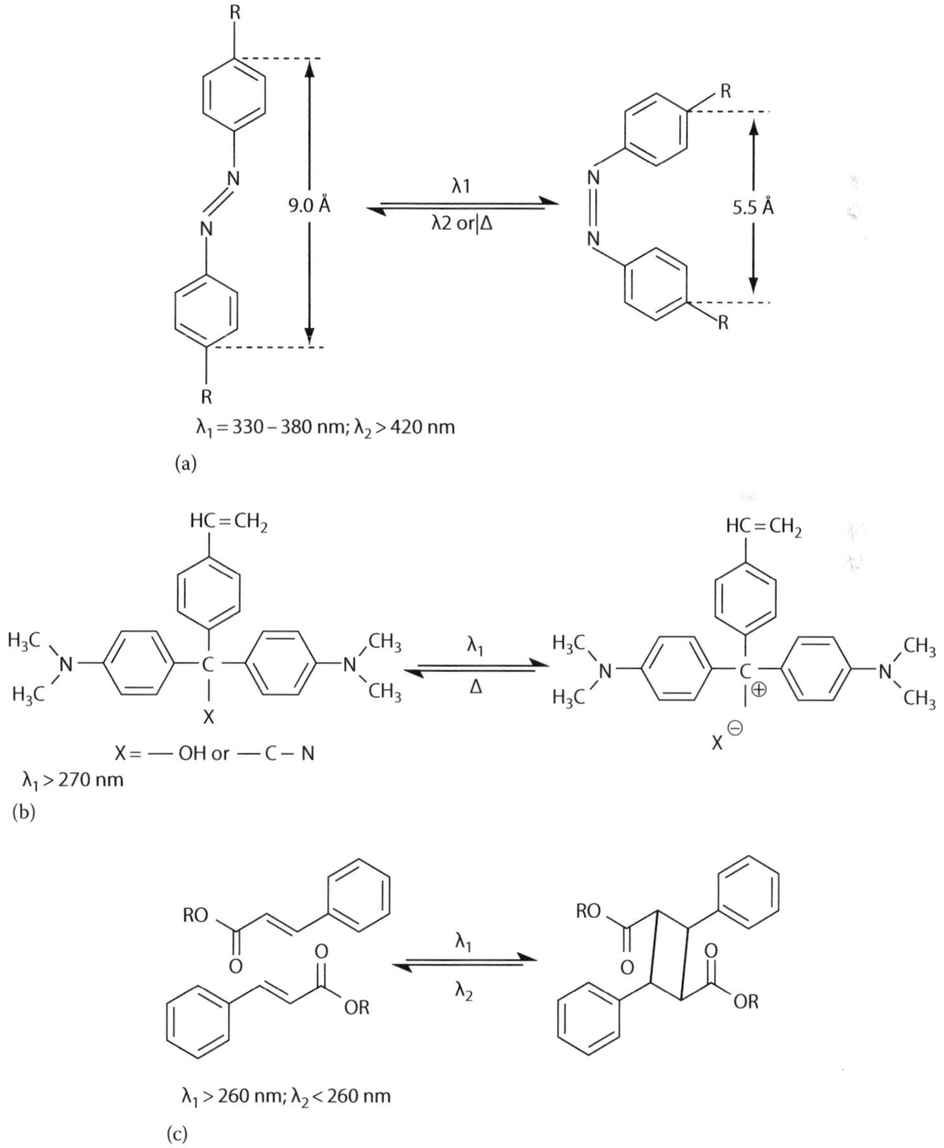

FIGURE 1.3 The light-induced isomerization and photochemical reactions for (a) *trans–cis* photoisomerization of azobenzene groups, (b) photoinduced ionic dissociation of triphenylmethane leuco derivatives, and (c) photodimerization of the cinnamic acid (CA) group. (Jiang, H., Kelch, S., and Lendlein, A.: Polymers move in response to light. *Adv. Mater.* 2006. 18. 1471–1475. Copyright Wiley-VCH Verlag GmbH & Co. KGaA. Reprinted with permission).

into ion pairs under UV illumination. The reverse or back reaction that recombines the ion pairs will occur under dark conditions. If the triphenylmethane leuco derivatives are incorporated in a polymer, then the reversible variation of electrostatic repulsion between the photogenerated charges will induce the solid-state polymer or gel to expand and contract. Cinnamate-type groups are photochemically reactive molecules that are able to form photoreversible covalent cross-links in polymers. These molecules can be used to create light-responsive SMP.

The light-induced contraction of polymers containing azobenzene has been extensively studied (Warner and Terentjev 1996, Finkelmann et al. 2001, Jiang et al. 2006). For nematic LCEs containing azobenzene groups, the light-induced reaction from *trans → cis* isomerization of the azobenzene units produces the movement of the liquid crystal domains and subsequently causes the collapse of the alignment order producing a significant contraction. Studies have shown (Finkelmann et al. 2001, Yu et al. 2003) that nematic LCEs exhibit large-strain reversible actuation properties over a narrow temperature interval at their nematic-isotropic (N-I) transition temperature. The bending direction of certain LCE films can be controlled by applying linearly polarized light (Yu et al. 2003, Tabiryan et al. 2005). In these cases, the incident light is largely absorbed by the surface layer of the polymer film because of the strong absorption by the azobenzene moieties. In other words, the *trans → cis* isomerization occurs only at the surface layer causing significant volume contraction at the surface and forcing the whole film to bend.

Hydrogels based on polyacrylamide, or poly(*N*-iso-propyl acrylamide) (PNIPAM), with incorporated triphenyl-methane leuco derivatives will swell when exposed to UV radiation and shrink when the UV light is removed (Hirasa 1993). The hydrogel's swelling is caused by an increase of the osmotic pressure within the gels due to ionization reactions and ion-pair formation initiated by the UV irradiation. It is possible to create visible light-sensitive gels by introducing trisodium salt of copper chlorophyllin into the thermosensitive PNIPAM gels. An increase in temperature arising from light absorption will then alter the swelling behavior of the gel. Some PNIPAM gels have also been observed to undergo volume change under radiation even if no photosensitive molecules are combined with the macromolecules (Hirasa 1993).

In contrast to the two-directional movement of the azobenzene-containing polymers, light-induced SMP enable temporary fixation of a user-defined shape. The original shape can be recovered upon irradiating the sample with light of a suitable wavelength. Recent research has shown that certain photoresponsive SMP (photoisomerization of the cinnamic acid [CA] group) can undergo shape recovery at ambient temperature in response to UV radiation (Jiang et al. 2006). The photoresponsive SMP consist of two components on the molecular level: molecular switches and permanent "netpoints." The molecular switches are the result of photoreactive CA and cinnamylidene acetic acid (CAA) functional groups. These groups are able to fix and later release the deformed shape by forming photo-reversible covalent cross-links when exposed to alternating wavelengths of light. The netpoints are cross-linking polymer-chain segments that determine the permanent shape. When this type of polymer is stretched, the randomly coiled segments of the amorphous polymer chains between the two netpoints are elongated (Jiang et al. 2006). For example, when exposed to an UV light source with a wavelength $\lambda > 260$ nm, the elongated segments of the chains are partially fixed because of the formation of the photo-reversible cross-links. This results in a new elongated temporary shape. The photo-reversible cross-links are split by UV irradiation at $\lambda < 260$ nm, driving the film to recover its original permanent shape.

There are several factors that influence the shape-recovery speed and the extent of recovery for these SMP. These include wavelength and intensity of the light source, and irradiation time. By increasing the light intensity or decreasing the thickness of the polymer film, it is possible to improve the photochemical reactions and, therefore, response times. Yu et al. (2003) describe a fast light response in azobenzene-containing LCEs for a film as thin as 7 μm.

An alternative material with high potential for creating light-driven actuators and microrobotic systems is poly(vinylidene difluoride) (PVDF) (Mizutani et al. 2008). PVDF is a ferroelectric polymer that possesses both pyroelectric and piezoelectric properties. When the temperature of the PVDF material is increased through light irradiation, it causes a piezoelectric effect to occur and the material mechanically deforms. Mizutani et al. (2008) demonstrate how a strip of this material can be used to create a cantilever and leg for a small microrobot. One surface of the PVDF film is coated with a thin silver (Ag) electrode. When irradiated with a He–Ne laser, the PVDF cantilever generates an electric field by means of the pyroelectric effect. The thickness of the PVDF cantilever is 28 μm and the Ag electrode is 6 μm thick. The effect causes conduction electrons to be generated and dispersed on the Ag electrode. The electric field in the cross-section of the PVDF film is, however, inhomogeneous, causing the film to bend along the cross-sectional direction as a result of the inverse piezoelectric effect (i.e., bend toward the light). In the experiments, a 10 mW irradiation is used to move the cantilever 250 mm in 0.5 ms.

1.3.1.1 Liquid Crystal Polymers

Liquid crystal polymer actuators have been shown to produce large and reversible shape deformations in response to applied stimuli such as changes in chemical environment, heat, and light. The strain response can be significant with 300% in one dimension and can be controlled by varying the magnitude of the input stimulus. Consequently, polymer microactuators have significant potential for MEMS, microfluidic and LOC systems.

LCEs obtain their unique shape-changing properties from the interrelationship between the elastic properties of the polymer networks and the ordering of the mesogenic liquid crystalline moieties. As a result, nematic LCEs are able to change their shape by up to 400% over a relatively narrow temperature interval straddling their N-I transition temperature (Finkelmann et al. 2001). A uniaxially aligned monodomain nematic LCE will exhibit spontaneous contraction along its *director axis* when heated to its N-I transition temperature (Finkelmann et al. 2001). This occurs because of the coupling between the average polymer chain anisotropy and the nematic order (Warner and Terentjev 1996). Significant changes to the material shape can arise from small variations in microscopic ordering.

Light can be used to alter the underlying nematic order parameter in the elastomer network and initiate the deformation. Liquid-crystal molecules containing photoisomerizable, such as azobenzene, can experience a reduction in their nematic order when exposed to ultraviolet (UV) radiation. The change induces a phase transformation in the liquid-crystal molecules from nematic to isotropic state (Figure 1.4). The transformation occurs because of the UV-induced *cis–trans* isomerization of the azo (N=N) bond where the photosensitive mesogenic molecules change from rod-like shape to a bent or kinked shape upon irradiation ($\lambda = 365$ nm) (Hogan et al. 2002). The rod-like shape stabilizes the liquid crystal phase while the "kinked" shape acts as an impurity and destabilizes the nematic phase, reducing its order parameter (Ikeda et al. 1990). The reverse *trans* → cis reaction occurs on heating or irradiation ($\lambda = 465$ nm) (Finkelmann et al. 2001, Hogan et al. 2002).

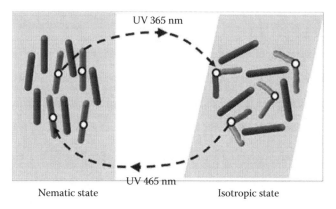

Nematic state   Isotropic state

FIGURE 1.4 An illustration showing the nematic-isotropic phase transformation in a liquid crystal elastomer when the material is exposed to UV light.

Many experimental studies have been conducted on evaluating the optomechanical response of the nematic elastomers. Finkelmann et al. (2001) found that the fractional contraction of elastomers achieved a large value of 22% at a temperature of 313 K, which is all the mechanical response that the elastomer would experience on heating from 313 K to the isotropic state. Yu et al. (2003) demonstrated that the large bending of a single film of liquid crystal network containing an azobenzene chromophore can be induced by UV light. Tabiryan et al. (2005) reported a reversible bi-directional bending of the azo LC polymer by switching the polarization of the light beam between orthogonal directions where not only the magnitude but also the sign of photoinduced deformation can be controlled by the polarization state of the light beam. At the nanoscale, a single molecule level optomechanical cycle has been realized by optically lengthening or contracting individual polymers through switching the azobenzenes between their *trans* and *cis* configurations (Hugel et al. 2002). Photomechanical actuation of nematic elastomers is slow, nonelastic, and requires polarized light at specific wavelength ranges.

Yamada et al. (2008) performed a series of tests on a laminated structure composed of a LCE layer and a flexible plastic sheet in an effort to study the mechanical forces generated on the composite file when exposed to UV light radiation (Figure 1.5). In these tests, both ends of the film were rigidly clamped and the film was then loaded with a force of 44 mN at 30°C. The internal stresses generated by the film for different UV intensities are shown in Figure 1.5b. The study showed that although a single-layer LCE layer was brittle and cracked at high intensities due to low mechanical strength, the composite file was able to generate significant forces without breakage.

The authors used this concept to create a light-driven motor that directly converted light energy into rotation (Figure 1.6). The laminated LCE film was used to construct a plastic belt attached to a pulley system (Figure 1.6a). By irradiating the belt with UV light from the top while irradiating the bottom with visible light, rotation was induced in the pulley system. Yamada et al. (2008) suggest a local contraction force occurs at the irradiated section of the LCE laminated film near the right pulley. The contraction force is along the alignment direction of the azobenzene mesogens, which is parallel to the long axis of the plastic belt. Consequently, these forces act on the right pulley causing it to rotate in the counterclockwise direction. The simultaneous irradiation of the belt on the top

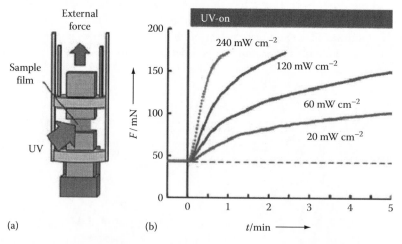

(a)   (b)

FIGURE 1.5 Experimental setup (a) to determine the mechanical forces (b) generated in LCE laminated films when exposed to UV irradiation. The liquid crystalline elastomers were coated on the both sides of the flexible polyethylene film to make the initial sample flat. The dimensions of the test samples were 2.5 mm × 5 mm. The intensity of the 366 nm UV light source was set at 20, 60, 120, and 240 mW/cm². (Yamada, M., Kondo, M., Miyasato, R., Mamiya, J., Yu, Y., Kinoshita, M., Barrett, C.J., and Ikeda, T.: Photomobile polymer materials: Towards light-driven plastic motors. *Angew. Chem. Inst. Ed.* 2008. 47. 4986–4988. Copyright Wiley-VCH Verlag GmbH & Co. KGaA. Reprinted with permission.)

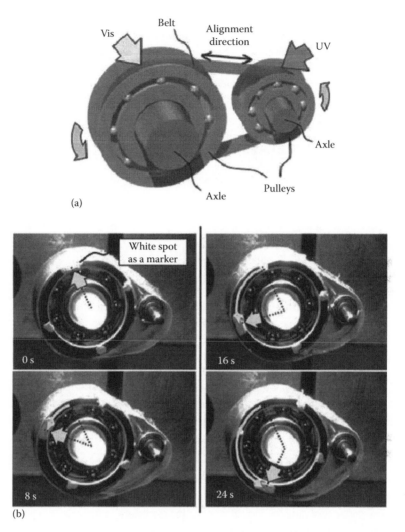

FIGURE 1.6 A light-driven motor with a plastic belt constructed from LCE laminated film. The basic motor design and relationship between the rotation direction and UV irradiation positions are illustrated in (a). The diameter of the larger left pulley is 10 mm and the smaller right is 3 mm. The LCE plastic belt is 36 mm × 5.5 mm. The time series of images in (b) show the rotation profiles of the LCE laminated film induced by 366 nm, 240 mW/cm² UV irradiation and >500 nm, 120 mW/cm² visible light. (Yamada, M., Kondo, M., Miyasato, R., Mamiya, J., Yu, Y., Kinoshita, M., Barrett, C.J., and Ikeda, T.: Photomobile polymer materials: Towards light-driven plastic motors. *Angew. Chem. Inst. Ed.* 2008. 47. 4986–4988. Copyright Wiley-VCH Verlag GmbH & Co. KGaA. Reprinted with permission.)

left of the figure generates a local expansion force at the exposed region, producing a counterclockwise rotation of the left pulley in the motor. The simultaneous contraction and expansion forces at different regions along the long axis of the belt induce rotation in the pulleys and move the belt in the same direction. Rotation transfers new portions of the belt into the radiation fields enabling the motor system to rotate continuously.

Van Oosten et al. (2009) demonstrated how an all-polymer microdevice could be fabricated using *inkjet printing technology* and that this manufacturing method would introduce new design opportunities. Specifically, the researchers took advantage of the self-assembly properties of the liquid crystal to create large strain gradients. By applying multiple inks, they were able to create light-driven microactuators with different subunits that could be selectively addressed by changing the wavelength (or color) of the light source. The actuators could be used to create flow and mixing in wet environments similar to the cilia found in natural microorganisms such as paramecia. Inkjet printing was selected for microfabrication because this process enables producing variations of the material composition in the plane substrate in a single manufacturing step. The authors use a commercial inkjet printer with a monomeric liquid crystal mix containing one of the two dyes deposited on the substrate. For the experiments, the reactive monomeric mix had a crystalline-nematic transition just above room temperature and, therefore, a solvent was used to print the mix.

There are several azobenzene dyes that can induce shape deformations when included in a liquid crystal network or liquid crystal rubber (van Oosten et al. 2007, 2008, 2009). Two such dyes are A3MA with a *trans*-absorption peak of 358 nm (UV light) and DR1A with a peak at 490 nm (green visible). The optimal dye concentration will depend upon the thickness of the film. For example, a 10 μm-thick film requires an azobenzene concentration of 4 wt% for the A3MA dye and 1 wt% for the DR1A dye. The concentrations are sufficient to initiate a response but not too high to prevent light from penetrating deep into the film.

Van Oosten et al. (2008) investigated the reversible nonlinear response behavior of photostimulated bending of light-driven LC microactuators. The study looked at the bending action of a planar uniaxially aligned film with internal composition gradient of poly(C6M/C61BP/A3MA), photopolymerized over the range of 0.5–10 mW/cm^2. When exposed to laser light (351 nm) from one direction, the LC film would rapidly bend toward the light source. After a period of prolonged exposure, the film relaxed but over time it curved in the opposite direction as shown in Figure 1.7a. The bending radius of the film over time is shown in Figure 1.7b. An interesting observation made by the authors is that as the LC film uncurls the speed of the transformation increases because the unbending sample is exposed to more direct light.

One goal of van Oosten's work (van Oosten et al. 2009) was to create light-driven artificial cilia that could be used as a microfluidic mixer or pump. The artificial cilia were similar to the natural cilia found in paramecia in that it may produce a flapping, asymmetrical motion causing the surrounding liquid to flow. The asymmetric motion was the result of a backward stroke different from the forward stroke which had been introduced into the liquid crystal azobenzene artificial cilia by temporally varying the intensity and wavelength of the light over the microactuator surface.

1.3.1.2 Single-DNA Molecular Nanomotors

Nanomotors are molecular devices that directly convert energy into some type of movement. Molecular motors comprised of single protein or DNA molecules are commonly found in biology. DNA nanomotors are able to switch between intramolecular tetraplex and intermolecular duplex states (Li and Tan 2002) by alternating through DNA hybridization and strand exchange reactions causing the motor to shrink and expand like an inchworm (Li and Tan 2002, Kang et al. 2009). These molecular scale nanomotors have potential applications in gene therapy, drug delivery, biochip design, and even nanoscale manufacturing (Bishop and Klavins 2007). The DNA nanomotors require the addition and removal of fuel and motor strands in order to function properly. Although a variety of energy sources have been proposed in the literature including electromagnetic fields (Kang et al. 2009), photon-driven single DNA nanomotors appear to be particularly useful at the nanoscale.

Kang et al. (2009) describe a single-molecule DNA hairpin-structured nanomotor driven by light. This particular nanomotor is a DNA hairpin-structured molecule incorporated with azobenzene moiety. The azobenzene enables the molecules to undergo a reversible photocontrollable switching operation. The azobenzene-DNA molecular motor eliminates the need for multiple DNA strands as required by other proposed nanomotor designs. When exposed to a repeated UV and visible light irradiation, it is possible to cause an opening and closing structural change of the DNA as shown in Figure 1.8. Movement is created by regulating the dehybridization (open state) and hybridization (closed state) of a hairpin structure by controlling the azobenzene moieties integrated on the DNA bases in the hairpin's duplex stem segment. The structural change of DNA from contraction

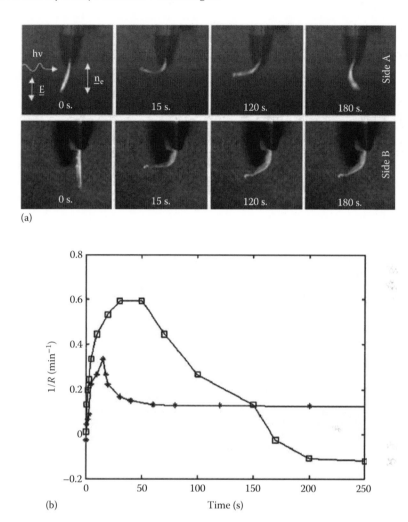

(a)

(b)

FIGURE 1.7 Photostimulated bending of light-driven LC actuators driven by the isomerization of azobenzene is illustrated in (a). The top row of images shows the response of the liquid crystal (LC) film oriented with Side A toward the 351 nm, 150 mW/cm² laser light. The bottom row is the same LC film actuator with Side B oriented toward the laser light. (b) The bending radius (1/R) of the sample as it is exposed to light for both Side A (□) and Side B (*). (Reprinted with permission from Van Oosten, C.L., Corbett, D., Davies, D., Warner, M., Bastiaansen, C.W.M., and Broer, D.J., Bending dynamics and directionality reversal in liquid crystal network photoactuators, *Macromolecules* 41, 8592–8596. Copyright 2008, American Chemical Society.)

(closed state) when the tethered azobenzene moiety (Azo-) takes the *trans* conformation under visible light and the corresponding expansion (Open) state when Azo- takes on the *cis* conformation under UV irradiation. Since the open-close cycle of the hairpin molecule exhibits reversible extension and contraction behavior, it can be identified as a nanomotor. The functionality of the single-DNA molecule motor is determined by a predominant intramolecular interaction within the molecule instead of disjunctive strand exchange as found in earlier DNA-based nanomotors. The photo regulation of this simple system is concentration-independent and, therefore, suitable for fabricating high-density molecular motors.

1.3.1.3 Molecular Photoswitch: Turning Organisms On-and-Off

Molecular photoswitches exploit light to move between states that exhibit different colors, absorption, or emission properties. Recently, Neil Branda and colleagues at Simon Fraser University

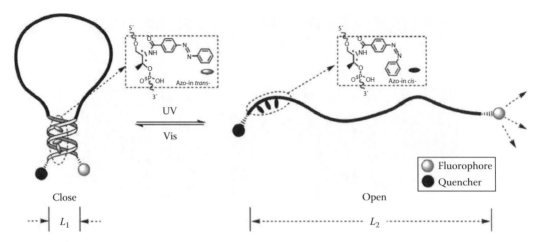

FIGURE 1.8 Illustration of the photo-switchable single-DNA nanomotor driven by photon energy. The main components of the nanomotor include a hairpin backbone, azobenzene moiety, and fluorophore/quencher pair for signaling motor movement. The average size of hairpin structure is L_1, and the average size of extended molecules based on persistence length of a single-DNA strand is L_2. The single-DNA nanomotor exhibits a CLOSE state when exposed to visible irradiation and an OPEN state with irradiated by UV light. (Reprinted with permission from Kang, H., Liu, H., Phillips, J.A., Cao, Z., Kim, Y., Chen, Y., Yang, Z., Li, J., and Tan, W., Single-DNA molecule nanomotor regulated by photons, *Nano Lett.* 9(7), 2690–2696. Copyright 2009, American Chemical Society.)

(Al-Atar et al. 2009, Carling et al. 2009, Boyer et al. 2010) demonstrated that the light-induced reactions of a photoresponsive dithienylethene can be used reversibly trigger paralysis in a living organism. Transparent nematode worms, *Caenorhabditis elegans*, were incubated in a mixture of the bipyridinium dithienylethene and a buffer containing 10% dimethlsulfoxide (DMSO). The *C. elegans* nematodes were chosen for experimentation because it is simple organism with 302 neurons and is optically transparent so that the color change can be monitored (Al-Atar et al. 2009). Exposing the colorless nematodes, which were fed ring-open form of the mixture, to a UV light (365 nm) for 2 min resulted in an immediate change in color to blue, suggesting the dithienylethene photoswitch has undergone the characteristic ring-closing reaction. Visible light (>490 nm) triggered the reverse reaction producing the ring-open isomer causing the nematodes to become colorless. *C. elegans* worms that were exposed to ring-closed form of the mixture would become immediately blue in color and could be made colorless by exposure to the visible light (490 nm) for 20 min (Figure 1.9). Depending upon the type of isomer present, the mobility of the nematodes could be controlled. The nematodes with ring-closed isomers became immobile and could be reactivated with exposure to visible light.

Although Branda and his fellow researchers are not powering or controlling a traditional "artificial" machine, they are exploiting light to regulate the movement of biological organism, which can then be used to perform a directed task (although a simple one). This study and its possible extension to in vivo drug delivery are of particular interest to researchers who are developing new actuation technologies for biomedical applications. The authors believe that photoswitches offer potential to advance photodynamic therapy as a noninvasive alternative to surgical treatments (Al-Atar et al. 2009, Carling et al. 2009).

1.3.2 OPTOMECHANICAL STRESSES AND STRAINS

1.3.2.1 Photostrictive Materials

Photostrictive materials produce mechanical strain when irradiated by light and, thereby, directly convert photon energy into mechanical motion. The photostrictive behavior of the material is the

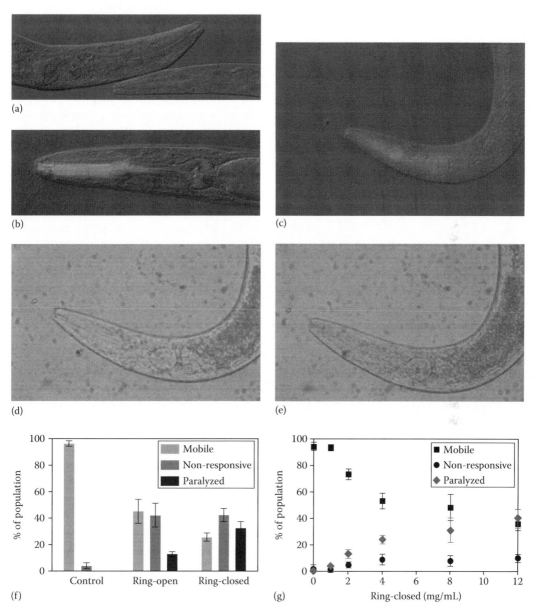

FIGURE 1.9 (**See color insert.**) Photocontrolled molecular switch used to induce paralysis in *C. elegans* nematodes. (a) Fluorescence microscopy image of *C. elegans* incubated with only 10% dimethylsulfoxide. Images of the nematode fed the ring-closed isomer of the photoswitch at (b) 10 min and (c) 60 min. Microscopy images also show the photoswitch being converted between colorless ring-open isomer (d) to ring-closed blue isomer (e) when exposed to visible wavelengths greater than 490 nm and UV light (365 nm) for 20 and 2 min, respectively. The number of mobile, nonresponsive, and paralyzed nematodes for samples that have been treated with the ring-open or the ring-closed form of the photoswitch compared to controls after 60 min incubation is shown in (f). Finally, the samples exposed to varying amounts of ring-open isomer are presented in (g). (Reprinted with permission from Al-Atar, U., Fernandes, R., Johnsen, B., Baillie, D., and Branda, N.R., A photocontrolled molecular switch regulates paralysis in a living organism, *J. Am. Chem. Soc.*, 131, 15966–15967 Copyright 2009 American Chemical Society.)

combined effect of both photovoltaic and the converse-piezoelectric properties (Poosanaas et al. 2000). The photovoltaic property produces relatively large voltages when the material is irradiated while the converse-piezoelectric effect causes the same material to expand or contract when the voltage is applied.

The photostrictive phenomenon can be observed in a number of different ferroelectric materials (Uchino 1990). For example, polarized ferroelectric ceramics will generate a relatively high voltage that exceeds the band gap energy when exposed to a uniform light source that has the same wavelength as the absorption edge of the material (Poosanaas et al. 2000). This photovoltage can be in the order of kV/cm for some materials. Furthermore, the photovoltage generated across the material sample induces mechanical deformation because of the simultaneous existence of an inverse piezoelectric effect. The *figure of merit* for the photostrictive material is often expressed numerically as the product of photovoltage E_{ph} and the piezoelectric constant d_{33}. This quantity can be used to characterize the performance of the ferroelectric material. In other words, to improve the photostrictive effect for a particular sample, it is necessary to optimize both of these parameters.

The photostrictive effect has been largely observed in ferroelectric polycrystalline materials such as PLZT ceramic. The PLZT ceramics exhibit large photostriction properties almost instantaneously when exposed to light and are, therefore, useful for developing rapid response optical microactuators, relays, and photon-driven micromachines (Uchino 1990). Certain ferroelectric materials, such as $(Pb,La)(Zr,Ti)O_3$ (PLZT) doped with WO_3, exhibit a more pronounced photostriction effect when exposed to a uniform near-UV light.

Morikawa and Nakada (1997) created an optical microactuator from ceramic polycrystalline material $(Pb,La)(Zr,Ti)O_3$ that had a bimorph structure and was capable of generating a displacement of up to several hundred micrometers. The bimorph-type optical actuator consisted of a pair of adhesively bonded PLZT ceramic elements oriented in opposite polarized directions. When a 365 nm UV light source irradiates the upper PLZT element in the cantilever device, the element stretches in the polarized direction due to the piezoelectric effect. The lower PLZT element is not illuminated and therefore does not expand. Since the two PLZT elements are connected to common electrodes and the lower nonexpanding element is oriented in the opposite polarized direction of the top, the lower element will experience a negative voltage and correspondingly contract due to the same piezoelectric phenomena. The combined effect of the two PLZT ceramic elements is a significant downward motion of the microcantilever beam. Subsequent UV illumination of the lower element will then cause the optically driven beam to bend upward.

1.3.2.2 Charge-Induced Surface Photo-Mechanical Actuators

A photomechanical effect consistent with the surface piezoelectric phenomenon has also been used to produce photo-stimulated vibrations in silicon (Si) microcantilever beams that are covered in a thin polycrystalline ZnO film. Early studies (Lagowski and Gatos 1972, Gatos and Lagowski 1973, Lagowski et al. 1975, Suski et al. 1990) have shown that thin crystals of polar (non-centrosymmetric) semiconductors can exhibit piezoelectric properties in response to light-induced electronic transitions. The photomechanical response arises from the depopulation and population of surface states by sub-bandgap illumination while the overall number of bulk-free carriers remains the same (Suski et al. 1990). The barrier height of the depleted layer can also change under these illumination conditions resulting in surface stress variations. When the light source is modulated, the barrier height will vary according to the fluctuations in photon intensity producing surface stresses with the same frequency. Consequently, it is possible to create resonant vibrations in the Si microcantilever beam by optically altering the frequency of the surface barrier so that it closely matches the natural frequency of the beam (Lagowski et al. 1975, Suski et al. 1990). The observed photomechanical effect is consistent with the surface piezoelectric phenomenon where the external stress applied to polar semiconductors leads to a modification of the surface barrier height and causes pronounced changes in contact potential difference (Lagowski et al. 1975).

This effect was demonstrated in the early 1990s by Suski et al. (1990). A 10 mm × 1.5 mm × 50 μm Si/SiO$_2$/ZnO microcantilever beam (ZnO film ~5 μm thick) was fabricated and activated using an argon laser ($\lambda \sim$ 520 nm). The experimental device resonated at a frequency of ~350 Hz with a maximum deflection of 160 nm under 130 μW of light power. The photomechanical effect was also reported in bulk semiconductor materials where the photon energy irradiating on the Si microcantilever was above the band gap. As the photons become absorbed in the semiconductor material, the free electrons are excited and move from the valance band into conduction band leaving holes in the lattice. The movement of electrons and creates local mechanical strain in the material.

The photo-generation of free charge carriers (electrons and holes) in a semiconductor can also produce mechanical strain (Datskos et al. 1998). The phenomenon is related to the temperature changes and thermal stress formed on the solid structure when it absorbs the photons. The heating and subsequent pressure build up produces an acoustic wave at the frequency of the modulated light signal. The photoinduced stress is sufficient to cause a silicon (Si) microcantilever beam to deflect. The photoinduced stress is in the opposite direction and four times larger than the stress resulting from only thermal excitation.

Datskos et al. (1998) studied the photon-induced stress in a micro silicon cantilever of 100 μm × 20 μm × 0.5 μm in size. Upon absorption 9 nW infrared light at 780 nm, the cantilever deflected ~1 nm. The experiments also showed that the photoinduced stress is in the opposite direction of the stress arising from only thermal excitation, and that the photoinduced stress is approximately four times as large. The ability to control the resonant frequency by light suggests a viable mechanism for actuating microstructures, however, only in a sub-micron or nanoscale range. Datskos et al. (2001) would later exploit this simple microcantilever effect in developing a new class of chemical microsensors.

1.3.2.3 Photoinduced Optical Anisotropy in ChG Films

An optomechanical effect has been also observed in certain amorphous semiconducting materials, such as ChG, when absorbing polarized light (Stuchlik et al. 2004). Reversible photoinduced anisotropy (PA) was the first reported by Krecmer et al. (1997) who showed that a thin amorphous film of As$_{50}$Se$_{50}$ deposited on a clamped AFM cantilever exposed to polarized irradiation would exhibit reversible nanocontraction. Upon irradiation with polarized light, a very small movement in the chalcogenide glass was measured parallel to the direction of the electric field of the light and minute expansion along the axis orthogonal to the electric field of the light was observed. To demonstrate the effect, the authors performed measurements on a 200 μm cantilever beam with a thickness of 0.6 μm. The surface of the beam was covered with a thin 250 nm As$_{50}$Se$_{50}$ film. When exposed to polarized light, the beam bent approximately ±1 μm (Stuchlik et al. 2001).

The optomechanical effect observed in ChG is linked to photoinduced optical anisotropy. In this context, the polarized light causes preferential absorption and reflection of the linearly polarized inducing light in a previously isotropic chalcogenide bulk or film sample (Krecmer et al. 1997). When exposed to polarized illumination, either parallel or perpendicular to the main axis of the cantilever, the beam bends to one of two "extreme" positions. Further experiments by Stuchlik et al. (2001) revealed that it is possible to drive the cantilevers to intermediate positions between these two "extremes" simply by changing the angle of polarized incident light relative to the cantilever. This means that a step-like actuation scheme could be obtained between the "extremes" by step-like tuning the polarization angles. The ability to regulate the cantilever position enables precise position control at the nanoscale by simply controlling the rotation angles of the polarized light sources, which could be a useful approach in optical nano-manipulation applications. However, the strain response is both small and slow. Furthermore, the structural origin of the mechanical polarization is not clear but Stuchlik et al. (2001) suggest that the effect arising from some anisotropic structural elements are aligned by the linearly polarized light. The structural elements, however, still remain to be identified.

1.3.3 Photomechanical Actuation Using Carbon Nanotubes

The extraordinary mechanical properties and unique electrical properties of carbon nanotubes (CNTs) have stimulated extensive research activities worldwide (Sun and Li 2007). These CNTs are a class of new, one-dimensional carbon nanomaterials discovered by Iijima in 1991. Driven by a wide variety of applications including nanocomposite materials, nanoelectrode materials, field emitters, nanoelectronics, and nanoscale sensors, significant progress has been made in CNT synthesis and characterization in the past decade.

Structures comprised of one cylindrical tube are called single-walled carbon nanotubes (SWCNTs). SWCNTs have a relatively small diameter (~0.4 nm) and based on their structure can be either metallic or semiconducting. The SWCNT structure (Figure 1.10) is characterized by a vector (m,n) designating the orientation of the graphene layer relative to the tube axis: armchair $(n=m)$, zigzag $(n=0$ or $m=0)$, or chiral (any other n and m). Armchair SWNTs are always metals. SWCNTs with $n-m=3k$, where k is a nonzero integer, are semiconductors with a tiny band gap. All other types of SWCNTs are semiconductors with a band gap that inversely depends on the nanotube diameter (Sun and Li 2007). While semiconducting SWCNTs (S-SWCNTs) can be used to build high-performance field-effect transistors and sensors, metallic SWCNTs (M-SWCNTs) appear to be useful for creating interconnects.

Multi-walled carbon nanotubes (MWCNTs) are structures that contain a concentric set of cylinders with a constant interlayer separation of 0.34 nm (Figure 1.11). MWCNTs have relatively large diameter, ranging from a few nanometers to several tens of nanometers, and are conducting materials. The electronic properties of perfect MWCNTs are rather similar to those of perfect SWCNTs, because the coupling between the cylinders is weak in MWCNTs.

Individual CNTs exhibit various extraordinary properties including mechanical, electrical, thermal, and chemical properties. The Young's modulus of a CNT is over 1 TPa and the tensile strength

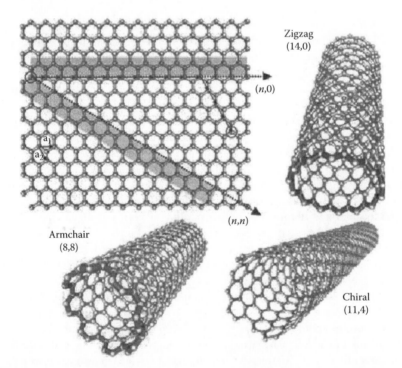

FIGURE 1.10 Roll-up of a graphene sheet leading to the three different types of carbon nanotube (CNT). (Balasubramanian, K. and Burghard, M.: Chemically functionalized carbon nanotubes. *Small*. 2005. 1(2). 180–192. Copyright Wiley-VCH Verlag GmbH & Co. KGaA. Reprinted with permission.)

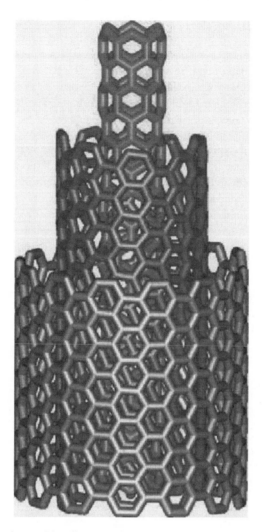

FIGURE 1.11 Structure of a multi-walled carbon nanotube (MWCNT) made up of three cylindrical shells. These types of MWCNT can have diameters up to 100 nm. (Balasubramanian, K. and Burghard, M.: Chemically functionalized carbon nanotubes. *Small*. 2005. 1(2). 180–192. Copyright Wiley-VCH Verlag GmbH & Co. KGaA. Reprinted with permission.)

is about 200 GPa. The thermal conductivity can be as high as 3000 W/mK. CNTs have very high chemical stability and can be chemically functionalized: i.e., it is possible to attach a variety of atomic and molecular groups to their ends (Sun and Li 2007).

From a fabrication perspective, SWCNTs are often not created as isolated tubes or filaments rather they formed as bundles. The electronic conductance and thermoelectric properties of the SWCNTs change when they are in bundle form. This modification is explained by considering the effects of intertube or inter-rope contacts or tubule effects (Kaiser et al. 1998). Studies in the late 1990s suggested that the structure distortion caused by van der Waals force can modify the electronic structure of the CNT and influence both the optical and mechanical properties of SWCNT bundles.

Light-induced elastic responses from SWCNT bundles and fibrous networks were first reported by Zhang and Iijima (1999). These early researchers observed the elastic movement of bundles of ~20 – 50 µm long SWCNT filaments when exposed to visible light. The electrostatic interaction of the SWCNT bundles was believed to be the cause of the elastic filament behavior. The authors concluded that the effect was the result of photovoltaic or light-induced thermoelectric effect related to the modification of electronic structure during the bundle formation. Since the introduction of light-induced CNTs, other researchers have shown theoretically that the lattice strain and conformational distortions of carbon nanotubes are optically induced by polaron (electron–hole pair) generations when exposed to light (Verissimo-Alves et al. 2001, Piegari et al. 2002).

The orientation and alignment of highly anisotropic carbon nanotubes is also an important factor in determining the photomechanical properties. To understand the effects of nanotube alignment on optical actuation, the photoconductivity of both unaligned and partially aligned CNT thin films were studied (Lui et al. 2009). The study showed that a high degree of nanotube orientation can improve the power conversion efficiency by about 10%. In addition, the film constructed from partially oriented CNTs produced faster response times and achieved a higher internal photon to electron power conversion efficiency than the film made up of nonoriented nanotubes.

In the past decade, a number of SMP photomechanical actuators that exploit CNTs (Lu and Panchapakesan 2006, 2007) have been proposed (Figure 1.12). Shape memory (recovery) is the ability of a material to reversibly recover inelastic strain energy when exposed to particular environmental stimuli. The strain energy is captured in the material through a reversible morphology change induced by shape deformation or by a suppression of molecular relaxation. Examples of morphological change by deformation include the martensitic transformation of SMA and thermal- or strain-induced crystallization for SMP. In terms of molecular relaxation, this would occur when quenching through a glass transition or crystallization temperature for SMPs. For most cases, the material is able to recover the original shape when the temperature is raised above the critical thermal transition. Polymer carbon nanotube and nanocomposites have been receiving a great deal of attention in recent years because of enhanced mechanical properties and unique electrical and thermal properties.

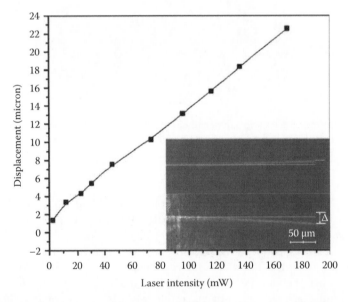

FIGURE 1.12 The displacement of a micro-optomechanical cantilever actuator constructed from a carbon nanotube film (CNF) and a photoresist SU-8 epoxy resin. The bending of the 30 μm × 300 μm × 7 μm microcantilever is a function of laser intensity (mW). The near-linear response has a maximum displacement of ~23 μm when exposed to 808 nm, 170 mW light source. The cross-sectional view of actuation under laser light stimulus is shown by the inset. (Reprinted with permission from Lu, S. and Panchapakesan, B., Nanotube micro-optomechanical actuators, *Appl. Phys. Lett.* 88(25), 253107-1–253107-3. Copyright 2006, American Institute of Physics.)

One of the earliest material-based systems was a carbon nanotube filled thermoplastic elastomer (Morthane) nanocomposite introduced by Kroerner et al. (2004). Morthane is a linear, hydroxyl terminated polyester polyurethane that exhibits a low glass-transition temperature ($T_g = -45°C$), near-ambient melting temperature of the soft-segment crystallites ($T_{m,s} = 48°C$), and exhibits huge strain-induced deformations (~700%). Significant deformations at room temperature ($T_g < T_{room} < T_{m,s}$) cause the flexible polymer segments to crystallize. The crystallization process forms physical cross-links, which prevent the polymer from undergoing strain recovery when the applied stress is removed. Subsequent heating and melting of the strain-induced soft-segment crystallites releases the constrained polymer chains forcing the material to revert back to its stress-free shape.

CNTs can be introduced to the thermoplastic elastomer to regulate the heating process and, thereby, control the shape memory effect. Kroerner et al. (2004) demonstrated that a uniform dispersion of 1–5 vol.% of carbon nanotubes in the thermoplastic elastomer yields nanocomposites that could store and release up to 50% more recovery stress than the pristine Morthane resin. The anisotropic nanotubes not only increased the rubbery modulus of the elastomer by a factor of 2–5, but also improved the strain-induced crystallization necessary for fixing the material shape. When heated by an infrared radiation source, the CNTs raised the internal temperature of the composite Morthane PCN melting the strain-induced polymer crystallites and triggered the release of the stored strain energy altering the material shape. The authors were able to deform the PCN material by 300%, exerting ~19 J to lift 60 g weight more than 3 cm with a force of approximately 588 N.

1.4 ELECTRON-MEDIATED OPTICAL MICROACTUATORS

Several other direct optical methods use photons to interact with the light-sensitive properties of the shell material and, thereby, induce mechanical strain and deformation. One approach uses photo-generated electrons to alter the electrostatic pressure on a microcantilever beam that forms a parallel plate capacitor (Tabib-Azar 1990, Tabib-Azar and Leane 1990, Tabib-Azar et al. 1992). There are a number of benefits from using electron-mediated direct optical actuation over thermal radiation processes including faster speed, lower power consumption, versatility in design, and ability to tailor the device to a particular application using conventional semiconductor doping and etching (Tabib-Azar 1998).

Tabib-Azar (1990, 1998) describes in detail one "warped capacitor" design that uses photo-generated electrons to change the electrostatic pressure on a thin Si microcantilever beam that forms a parallel plate capacitor with a conductive ground plate (Figure 1.13). In this electrostatic

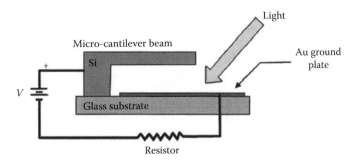

FIGURE 1.13   Side view of an electrostatic microactuator showing the basic electrical connections and current paths. The optical generation of electron–hole pairs in the semi-insulating layer between the Si cantilever beam and Au ground plate leads to cantilever actuation. (Adapted from Tabib-Azar, M., *Nanotechnology*, 1, 81, 1990).

microactuator, a thin P$^+$ silicon (Si) cantilever is attached to an insulating glass substrate and suspended over a gold (Au) ground plate forming a parallel plate capacitor with a surface area, A, and separation distance, d. The capacitance of the flat parallel plates is given by the expression

$$C = \frac{k\varepsilon_0 A}{d} = \frac{\varepsilon_0 L b}{d} \tag{1.7}$$

where

  k is relative permittivity of the dielectric material between the plates ($k \approx 1$ for air)
  ε_0 is the free-space permittivity (8.854×10^{-12} F/m)
  b and L are the width and length of the cantilever beam, respectively
  d is the distance between the cantilever beam and the ground plane

From the definition of capacitance, the unit Farad (F) is equal to Coulomb/volt.

A separate power source, V_0, is then connected across the capacitor through a resistor forming the bias current. The stored charge, Q, necessary for the electrostatic pressure to deform the cantilever is

$$Q = CV_0 \tag{1.8}$$

When exposed to a light source with sufficient energy, photo-generated electrons are formed on the Au ground plate. The photoelectrons will migrate through the air gap to the microcantilever beam reducing the charge on the capacitor and causing the cantilever to bend.

If the change in capacitance as the cantilever beam deforms is neglected, then the steady-state deflection at the end of the beam can be described in terms of the charge on the capacitor, $\delta(Q)$,

$$\delta(Q) = \frac{3L^2}{4\varepsilon Y b^3 w^2} Q^2 \tag{1.9}$$

where

  Y is the Young's modulus
  b is the thickness of the cantilever beam

Since this is a nonlinear relationship, the cantilever beam will bend in a stable manner up to the threshold of spontaneous collapse (Tabib-Azar 1998) given by

$$\delta_{threshold} \approx \frac{6\varepsilon V^2}{Yb^3} \frac{L^4}{8d^2} \tag{1.10}$$

The fastest time that it can smoothly traverse this distance is approximately equal to the period of the fundamental mode of free vibration.

Tabib-Azar (1998) describes a microactuator that uses a 600 μm × 50 μm × 1 μm cantilever beam with a gap of 12 μm. A bias voltage of 6 V and optical power less than 0.1 mW/cm^2 was used to move the cantilever 4 μm in approximately 0.1 ms. Continuously charging the capacitor with a current $i \leq (i_{max}/2)$, where i_{max} is determined by the battery circuit, allows light-controlled actuation in either direction. A continuous photon flux, $\Phi < (i/\eta)$, where η is the quantum efficiency, short circuits the capacitor more slowly than the battery charges it, causing a charge build-up which closes the plates. A photon flux $\Phi > (i/\eta)$ causes an opposing photocurrent greater than the charging current. The net charge then decreases, and the capacitor plates relax open. Both the static and dynamic behavior of the light-controlled actuators is investigated and described in detail by the author.

1.5 OPTICALLY DRIVEN PHOTOTHERMAL EFFECTS

The photothermal interaction arising from light striking a solid material or liquid can also be used to drive actuation. By introducing localized heat energy, it is possible to deform an actuating shell to create the desired displacement or induce thermo viscous effects that enable the small amounts of fluid to be optically driven through narrow microchannels. An example of the first case is when a focused light beam is used to heat a confined liquid to induce a transformation to a pressurized gas. The phase change causes the gas to expand and deform a thin flexible membrane or diaphragm. The movement and forces generated by the change in shape of the diaphragm produces the desired actuation. In contrast, the second case involves using the energy from light to directly alter a property, such as viscosity, of the target solution enabling the fluid to be optically driven along microscale channels. The transfer of energy from a light source to a solid material or fluid for actuation is summarized later. The optical heating of gases, liquids, and solids can be used to create diaphragm microactuators, mixing of microfluidic streams, driving and pumping fluid through microchannels, inducing phase transformation in SMA, and inducing photothermal vibrations in "specially" contoured optical fibers.

1.5.1 LASER MATERIAL INTERACTIONS

Indirect optical actuation methods often exploit the ability of light to generate heat and, thereby, directly influence the thermal properties of gases, liquids, and solids. These methods have simple designs and often generate more actuation power than direct optical methods based on photoresponsive materials as discussed in Section 1.3. However, these actuation methods that utilize thermal and phase transformation properties of fluids and solids can be comparatively slow. In all situations, the underlying mechanisms of indirect optical actuation depend on the transfer of energy from the light source to the surface of the actuator shell. Most engineering applications utilize lasers as the light source because the light rays from a CW or pulsed laser travel in the same direction, are essentially the same wavelength (monochromatic), and are in-phase (coherent). Furthermore, the laser beam does not diverge significantly as it moves through the air and maintains a high energy density (Steen 1998). The energy from a laser can be delivered to the surface through an optical waveguide or via free space.

Consider the situation when a stream of photons, from a focused laser, is projected onto a material's surface. The laser–surface interaction may result in illumination, light reflection, energy absorption, thermal and thermodynamic effects, melting, vaporization, or plasma effects. The precise nature of the interaction depends on the optical power of the laser, duration of exposure, reflective and absorption properties of the target surface, thermal properties of the exposed material, and local environmental conditions. The absorption properties of the target surface determine how efficiently the photon energy is transferred into the material. This is dependent on the material's absorption coefficient (relates to the amount of optical energy transferred per unit depth), reflectivity (if 100% reflective then no light is absorbed), and material surface finish (smooth or rough).

When the laser light strikes the surface, it produces a measureable localized thermal change to the material. Irradiance (E) is a measure of the incident laser power per unit area (W/cm^2 or W/m^2),

$$E = \frac{\Phi}{A} \tag{1.11}$$

where
Φ is the power of the laser source (W)
A is the area of the beam spot in (m^2)

The term E is often referred to as the power density in the literature but this not technically true. To appreciate the relationship between beam size and irradiance, consider a 2 kW laser focused to a 0.2 mm diameter beam (Steen 1998). The irradiance is

$$E = \frac{\Phi}{\pi r^2} = \frac{2000 \text{ W}}{\pi (0.1 \times 10^{-3})^2} = 6.3 \times 10^{10} \text{ W/m}^2$$

In other words, a kW laser with a small beam size will vaporize an element in a fraction of a milli-second and, therefore, a concentrated (focused) light beam can be used as a *powerful* "heat source."

Using laser light to perform the heating of a material surface also requires a basic understanding of the thermal properties of the material (Steen 1998, Steen and Mazumder 2010). Thermally conductive, insulating, and semiconducting materials behave differently when exposed to a focused, concentrated light beam. Since no energy band gap, or forbidden band, exists between the valence and *conduction bands* (CB) of a thermal conductive metal, a large amount of the electrons in the CB can easily absorb the photon energy. These CB electrons transfer their energy to the material through electron–lattice collisions. With *very low energy* from a laser, the photons are easily absorbed by metals and the acquired energy is turned to heat. In contrast, the large band gaps cause the insulator materials to have essentially no CB electrons and, therefore, exhibit no thermal conduction. A large energy band gap in the material will require a significant amount of energy from the laser for photon absorption. The thermal behavior of semiconductors is between conductors (e.g., metals) and insulators. A very small band gap exists between the conduction and valence bands for semiconductor materials and the energy can be transferred fairly easily between the CB and the lattice structure. However, the amount of energy necessary from the laser for photons to be absorbed by the material is greater than conductive metals.

Thermal properties reflect how the heat energy from the laser beam striking the surface flows into the material. Key engineering properties include *thermal conductivity* and *thermal diffusivity*. Thermal conductivity describes how fast the heat flows through the material while thermal diffusivity reflects how fast the material will conduct the thermal energy. The *heat flow* through a material depends on the thermal conductivity (k) and on the specific heat (c_h) of a material. Thermal diffusivity is defined as

$$\alpha = \frac{k}{\rho c_p} \tag{1.12}$$

where
    ρ is the material density
    k is thermal conductivity
    c_p is specific heat of the material

Thermal diffusivity (α) tells us how fast materials will accept/conduct thermal energy and can be used to approximate the depth that heat will travel per pulse (with time t) through the material:

$$\text{Depth} = \sqrt{(4\alpha t)} \tag{1.13}$$

Materials with a low value for thermal diffusivity, such as stainless steel and some nickel alloys, will limit the penetration depth into the material. Note that heat flow is dependent on the *specific heat* of a material as it is used to determine the rate of change of temperature.

Laser intensity and pulse duration also influence heat penetration in a material. For some metals like stainless steel that have *low thermal diffusivity*, a lower-powered laser with a long pulse is used. In contrast, for metals such as copper that have a *high thermal diffusivity*, a higher power laser with shorter pulses to overcome the losses can be used.

The specific heat capacity (c_p) changes with respect to temperature and is defined as

$$c_p = \frac{dQ}{mdT} \tag{1.14}$$

where c_p is the ratio of the heat (dQ) to the product of the mass m and temperature change (dT).

In addition to conductivity and diffusivity, the *thermodynamic properties* relate to the amount of energy required to heat, melt, or vaporize the material. This depends upon the target material's density, heat capacity, melting and vaporization temperatures, and the latent heats of fusion and vaporization (Steen 1998). Heat capacity is a measureable physical quantity related to heat, mass, and change in temperature.

Some laser micromachining operations such as drilling and cutting require the material to be vaporized. Metallic materials rapidly reach their vaporization temperatures when exposed to high irradiance sources. The time required to reach vaporization is

$$t_v = \left(\frac{\pi}{4} \right) \left(\frac{k\rho c_p}{E_a^2} \right) (T_v - T_o)^2 \tag{1.15}$$

where

k is thermal conductivity
ρ is material density
c_p is the material's heat capacity
T_v is the vaporization temperature
T_o is ambient temperature
E_a is the absorbed irradiance

During laser drilling, pulse durations typically range from 100 μs to 10 ms.

Intuitively, one would think that a higher level of laser power will remove more material, but this is not always the case. Laser beams with high irradiance tend to vaporize the material near the beginning of the pulse, and then the remaining energy in the laser pulse will slightly thermally ionize the vaporized material and heat it to a high temperature because it is now absorbing the incident laser energy. This effect cascades until a hot opaque ionized plasma is formed. At this stage, the plasma is absorbing most of the incident energy and not interacting as much with the target material and the process of vaporization ceases. The plasma effect leads to a phenomenon known as *laser supported absorption* (LSA) wave (Willmott and Huber 2000). This wave is created right above the target surface and rises back toward the laser beam path accompanied with a bright flash of light and a loud noise. This effect could even propagate a shockwave into the material which can be used for shock hardening certain alloys (Steen 1998, Steen and Mazumder 2010).

1.5.2 Optothermal Expansion of Fluid

Many indirect optical methods for mechanical actuation take advantage of the heat generated by the light source to create the desired force or pressure to move the actuator shell (Hale et al. 1988, 1990, Hockaday and Waters 1990, McKenzie and Clark 1992, McKenzie et al. 1995). When a simple gas is heated, it expands according to the ideal gas law

$$P_g V_g = nRT \tag{1.16}$$

where

P_g is the gas pressure
V_g is the gas volume
n is the number of moles
R is the gas constant (0.0821 L atm/mol °K)
T is the absolute temperature

The optically actuated silicon diaphragm device shown in Figure 1.14 exploits this simple principle to deflect a flexible membrane, called a *diaphragm*, in order to perform mechanical work. The cavity

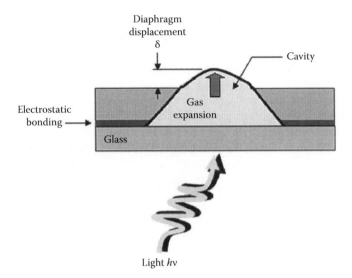

FIGURE 1.14 Light-driven silicon diaphragm device based on optical heating of enclosed gas. The basic concept has been used for a variety of engineering applications. (Adapted from Hale, K.F. et al., *IEE Proc.*, 135(5), 348, 1988; Hale, K.F. et al., *Sens. Actuators*, A21–A23, 207, 1990).

is filled with a gas or oil that expands when heated from the light source. As the diaphragm expands under pressure, it produces the desired deflection, δ.

The displacement produced by a diaphragm actuator (Tabib-Azar 1998), δ, at the center from its equilibrium position, can be described as

$$P_g = \frac{4a_1 b}{L^2}\sigma_0 \delta + \frac{16 a_2 f(\upsilon)b}{L^4}\left(\frac{Y}{1-\upsilon}\right)\delta^3 \qquad (1.17)$$

where
  L is length
  σ_0 is the residual stress
  $(Y/(1-\upsilon))$ is the bi-axial modulus
  b is the thickness of the diaphragm

The dimensionless parameters a_1, a_2, and $f(\upsilon)$ depend on the geometry of the diaphragm. Tabib-Azar (1998) describes a square diaphragm given as $a_1 = 3.04$, $a_2 = 1.37$, and $f(\upsilon) = 1.075 - 0.292\upsilon$. This type of microfabricated flow controllers has been shown to have speeds of 21 ms in air-flow and 67 ms in oil-flow, with sensitivities of 304 and 75 Pa/mW, respectively.

Mizoguchi et al. (1992) used this same simple concept to create a micropump that included an array of five closed diaphragm-actuated devices called microcells (Figure 1.15). Each microcell consisted of a pre-deflected 800 μm × 800 μm square membrane that was micromachined in 0.25 mm³ of silicon and filled with Freon 113, a liquid with a boiling point of approximately 47.5°C. A carbon-wool absorber was placed inside the cell to convert the incident light from the optic fiber into heat. The microcell exhibited a relatively large deflection, approximately 35 μm, when the cell's contents were heated and the Freon 113 undergone a phase change from liquid to gas. The fluid that is being transported by the pump is fed into a flow channel between the glass plate and deflecting membrane using very small harmonic movements. The harmonic order of the membrane's deflection determines the fluid flow rate and direction. The small quantities of Freon in each cell allowed relatively low optical powers to be used to change the phase of the liquid to gas, giving the large membrane deflections needed to operate the pump. The microcell was fabricated and operated by a

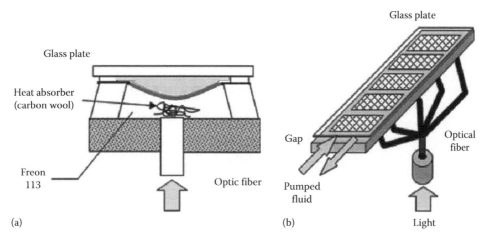

FIGURE 1.15 A simple design for a micropump comprised of a sequence of closed light-activated diaphragm devices originally proposed by Mizoguchi et al. (Adapted from Mizoguchi, H. et al., Design and fabrication of light driven pump. *Micro Electro Mechanical Systems'92*, Travemunde, Germany, pp. 31–36, 1992.)

laser source with no more than 10 mW. The micropump achieved a head pressure of approximately 30 mmag and flow rate of 30 nL/cycle.

1.5.3 LIGHT-DRIVEN MICROFLOWS

The miniaturization and integration of various analytical laboratory processes on a single platform is a key design requirement for many microfluidic and LOC devices used for medicine and environmental monitoring (Eijkel and van den Berg 2005, Dittrich and Manz 2006). The primary building blocks of these integrated microsystems include micromixers, micropumps, separators, filters, reaction chambers, and waste disposal chambers. The reduction in physical size of these functional components have increased the speed of analysis, lowered operating costs due to the consumption of small quantities of reagents, and enabled novel system designs that avoid biological sample cross-contamination.

One of the most critical components on the microsystem platform is the microfluidic mixer. Efficient mixing of chemicals and biological substances in order to create the desired reactions is an essential step in preparing the sample for analysis. The typical mechanism for mixing fluids in the macroworld is to create turbulent flows at high Reynolds numbers. In the microdomain, however, mixing liquids becomes more difficult because laminar flow dominates the process and slow process of molecular diffusion becomes the primary mixing mechanism (Hessel et al. 2005, Nguyen and Wu 2005). To improve the mixing rate over shorter length microchannels, researchers have developed a variety of micromixer designs. These microfluidic mixers can be categorized into passive or active mixing, depending on whether an external source of energy is being used to assist the mixing operation (Nguyen and Wereley 2002).

Passive mixers do not require external power sources and, typically, use specially designed microfluidic channel microstructures to improve the overall mixing rate. The simplest passive micromixers designs for combining two adjacent streams of liquid are T- and Y-micromixers (Kamholz et al. 1999, Kamholz and Yagar 2002). Unfortunately, a relatively long microchannel is required because the mixing mechanism is completely depended on the molecular diffusion rate. To improve the mixing rate, Kamholz et al. (1999) use a parallel lamination micromixer to split the streams into a number of substreams. These substreams are then recombined to improve mixing over shorter channel lengths. The higher mixing rate of the split–join approach is due to the increased contact surfaces between fluids.

Instead of splitting both streams to increase the contact surface between the adjacent fluids, injection mixers use a nozzle array to inject one stream into another. Alternatively, chaotic advection mixers with specially design features to generate splitting, stretching, folding, and breaking of the microflows (Wang et al. 2002) have also been created to improve the mixing operation. Three-dimensional complex microstructures such as the connected out-of-plane L-shape microchannels (Chen and Meiners 2004) and twisted microchannels have also been reported to enhance the chaotic advection mixing.

The advantages of passive mixers have been their relative operating simplicity, reliability, and robustness over the designed microflow range. However, to be efficient, these mechanisms require lengthy microchannels or complex 3D microchannel structures. The additional material and fabrication steps for 3D structures can significantly increase the cost of any LOC device or micrototal analysis system (μTAS) that must be produced in large quantities.

In contrast, active micromixers use an external energy source to create disturbances in liquid streams to improve mixing rates. Fujii et al. (2003) describe an external micropump used to generate a pressure disturbance that improves mixing. Electrohydrodynamic and electrokinetic disturbances are also reported to improve mixing rates (El Moctar et al. 2003). Suzuki and Ho (2002) describe a micromixer integrated with electrical conductors that generate magnetic fields, which then move 1–10 μm diameter magnetic beads to create a disturbance in the flow stream. Liu et al. (2002) reported using the acoustic streaming technique to induce air bubbles in streams where the bubbles disturb the flows to improve fluid mixing. Tsai and Lin (2002) reported utilizing a microheater embedded in microchannels to thermally generate air bubbles that disturb the streams.

In all cases, active micromixers require peripheral hardware to introduce additional energy into the flow stream or to activate a moving mechanism. This approach results in higher mixing rates but may become unreliable because of increased opportunities for mechanism failure. Furthermore, the higher unit cost of fabricating LOC devices with embedded active components has discouraged industry from commercializing microfluidic devices. The following section describes an active micromixer driven solely by a focused laser beam. The motivation of this study is that the diffusion coefficient for nonreactive fluids can be altered by raising localized temperature of the adjacent fluid streams. A variation of this principle has been described by Weinert and Braun (2008) who developed a light pump approach that moves liquid through LOC devices. The methodology enables fluid to be transported without the need for channel walls to define the fluid motion path. An alternative approach for using light to drive fluid through a microchannel is a gas-bubble piston concept. Bezuglyi and Ivanova's (2007) gas-bubble piston is set in motion by light-induced thermo-capillary forces. These bubbles conform to the shape of the channel and do not leave "stagnant" amounts of liquid behind. Unfortunately, all these methods require the liquid to be heated and the heat may affect the quality of the sample and, when raised to a high temperature, the movement through the very narrow channels may be difficult to control because of the conductive dissipation of heat in the channel walls.

1.5.3.1   Laser-Assisted Micromixing

Optical techniques have proven to be useful for manipulating solids and fluids at a remote distance. The precise control of fluid flow fields or the concentration of solutes is essential for investigating chemical, biological, and cellular processes (Weinert and Braun 2009) in LOC technology. These microfluidic devices perform very fast analysis on small volume samples over mixing and reactive distances of only a few millimeters. Another advantage of this technology is that it is possible to simultaneously perform multi-parallel analyses. Unfortunately, controlling these fluids at these small scales can be problematic. The fluid flow is laminar, thereby the transport, mixing, and flow control require complex LOC designs with multiple active and passive mixers, channels, and valves.

Molecular diffusion is the primary mixing mechanism for parallel liquid streams undergoing laminar flow with a small Reynolds number, $Re < 100$. Under this condition, the mixing action can be represented by Fick's law (Nguyen and Wereley 2002)

$$\tau = \frac{d^2}{2 \cdot D} \qquad (1.18)$$

where

τ is the time required to complete the molecular diffusion process (s)
d is the channel width (μm)
D is the diffusion coefficient(μm^2/s)

The microchannel length (μm) required for mixing to be completed can be approximated as

$$L = U \cdot \tau \qquad (1.19)$$

where U is the average flow velocity (μm/s). The higher average flow rate U implies that a longer channel is required to complete the mixing.

The diffusion coefficient D in Equation 1.18 is defined as

$$D = \frac{k_B \cdot T}{6 \cdot \pi \cdot \mu \cdot r_A} \qquad (1.20)$$

where

k_B is the Boltzmann constant (1.38×10^{-23} J/K)
T is the system temperature (K)
μ is the dynamic viscosity of the pure solvent
r_A is the radius of the solute particle (Bird et al. 2002)

The diffusion coefficient, D, is directly proportional to the temperature. As the temperature increases, the amount of energy available to the particles in the flow stream also increases, causing them to move faster and, thereby, increasing the diffusion rate. A higher diffusion rate implies the mixing of the streams can be completed in a shorter period of time τ or over a shorter microchannel length.

Therefore, a focused laser beam can function as an active micromixer and shorten the required length of the mixing channels (Shiu et al. 2010). This very simple idea increases the rate of molecular diffusion for nonreactive fluids by elevating the temperature. Preliminary experiments on a Y-channel micromixer (Figure 1.16) were conducted using a 1 mW, 670 nm laser. The laser beam was focused on the microchannel using a 100 mm focal length objective lens. The laser-assisted mixing of the test fluids showed a 36.4% increase in the average diffusion coefficient value with 1–10 μL/min flow rates (Figure 1.16b). The maximum percentage difference of diffusion distances had increased by approximately 7.85% over the nonlaser-assisted conditions.

The laser's wavelength for these experiments did not correspond to the highest absorption wavelength for water and, therefore, the liquid was not overheated because the elevated temperatures may destroy the biological samples. For some applications, engineers can select a wavelength that is invisible to the biological samples in an effort to avoid laser radiation damage yet increase the buffer solution temperature to appropriately mixing the solution in the shortest time possible. Another problem that may arise is an unsymmetrical temperature distribution because darker colored particles suspended in the liquid tend to absorb more heat energy than the lighter, often transparent, water. However, the primary advantage of using the laser-assisted technique is that it requires simple channel structures without the need of complicated multiple split-join microchannels to improve mixing rate or the specially designed 3D microchannel structures to create the chaotic advection type of passive mixing.

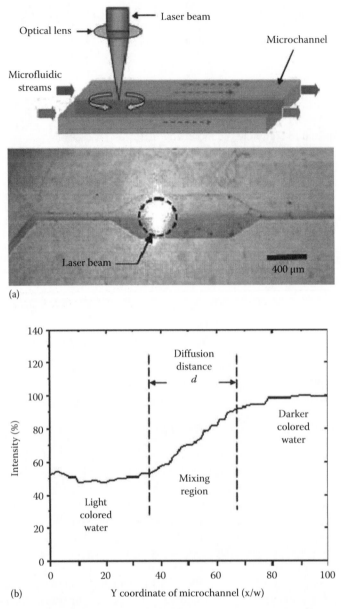

FIGURE 1.16 An optically assisted active micromixer (a) where a focused laser beam delivers a small amount of energy that is absorbed by a microfluidic stream and alters the localized diffusion coefficient (b) enabling increased mixing over a shorter channel distance.

1.5.3.2 Microflows Using Thermoviscous Expansion

A variation on laser-assisted mixing is the light pump approach introduced by Weinert and Braun (2008). The authors show how small volumes of fluid could be optically pumped along a predefined path of a moving warm spot that was created using an infrared laser. The repetitive motion of the laser beam is used to remotely drive 2D microflows of liquid with a resolution of 2 μm. The experiments produced pumping speeds of 150 μm/s with a maximum 10 K temperature increase at the localized light spot. The study also confirmed that the fluid motion was the

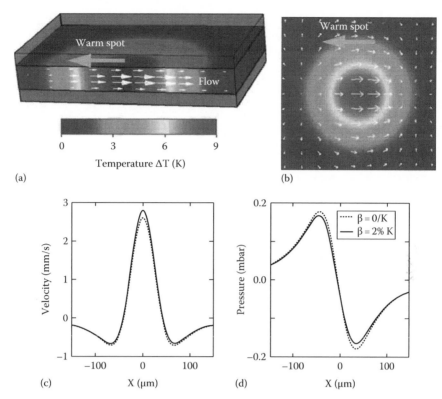

FIGURE 1.17 (**See color insert.**) Finite element simulation of the fluid velocity for thermo viscous pumping around the laser spot in both (a) 3D and (b) 2D. The color-coded temperature spot is moving to the left producing fluid expansion in its front (left side) and contraction in its wake (right side). The velocity and pressure profiles created by the laser spot, along the x-direction, are shown in (c) and (d), respectively. The pressure profile is positive in front of the spot, as a result of the thermal expansion, and negative in its wake. (Reprinted with permission from Weinert, F.M. and Braun, D., Optically driven fluid flow along arbitrary microscale patterns using thermoviscous expansion, *J. Appl. Phys.*, 104, 104701–104701-10. Copyright 2008 American Institute of Physics.)

result of the dynamic thermal expansion in the gradient of liquid viscosity. The viscosity of the liquid at the light spot is reduced by an increase in local temperature, resulting in a broken symmetry between the thermal expansion and thermal contraction in the front and wake of the spot (Figure 1.17). Consequently, the fluid is observed to move in an opposite direction to the movement of the heated spot because of this asymmetric thermal expansion at the spot front with respect to the corresponding asymmetric thermal contraction in the wake. The authors use the Navier–Stokes equations to predict the fluid speed. Experiments using water show that this nonlinearity can produce fluid steps in the range of 100 nm for each passage of the spot. However, the spot movement can be repeated in kHz range resulting in fluid speeds that exceed 100 μm/s. The authors further demonstrated the concept by pumping nanoparticles over millimeters through a gel.

Weinert and Braun (2008) also developed a mathematical model to predict the pump velocity. Since the flow is laminar, it is possible to neglect the inertia terms. In addition, the dimensions parallel to the surface are much larger than the width of the water layer, so it will be possible to perform a thin-film approximation in order to describe the 2D flow. The pump velocity, v_{pump}, can be approximated as

$$\upsilon_{pump} = -\frac{3\sqrt{\pi}}{4} f\alpha\beta b\Delta T^2 \tag{1.21}$$

where
   f is the repetition frequency of the laser
   b is the spot width
   α is the expansion coefficient
   β is the temperature dependence of the viscosity
   ΔT is the amplitude of the temperature spot

The temperature was obtained by imaging a temperature-sensitive fluorescent dye dissolved in the water. The shape of the moving warm spot could be measured by stroboscopic illumination.

The theoretical model was verified by experiments in the low frequency range of the laser ($f \ll 1$ kHz). Furthermore, the analytical model predicts a linear response of the pump velocity to both the thermal expansion $\alpha\Delta T_o$ and the change in temperature-dependent viscosity $\beta\Delta T_o$ for a similar shaped temperature spot. If the spot temperature is increased with a higher laser power, then the pump velocity increases proportionally to ΔT^2.

Light-driven pumping can reach speeds of up to 150 mm/s for moderate heating (Weinert and Braun 2008). In comparison, pressure-driven microfluidics will operate at significantly faster flow speeds. However, to achieve these flow speeds, it is necessary to incorporate millimeter-sized interfacial connections to outside pumps. The light pump approach enables highly localized fluid actuation. Furthermore, since the pump pattern is programmable in real time, it can readily be adapted to a variety of geometric patterns. This type of thermoviscous pumping could selectively enhance the transport of molecules from one cell to another.

Interestingly, the light pump based on thermoviscous expansion can be used to move fluids through an unstructured environment (Weinert and Braun 2009, Weinert et al. 2009). In this work, an infrared laser was used to melt liquid channels into a sheet of ice. Since the entire channel was not melted at once, the liquid would refreeze behind the laser spot. The repetition of the laser spot motion with high frequencies enabled the water to undergo a series of melting and freezing cycles, thereby increasing the velocity of the pumping action. As the warm spot formed by the laser was moved through a thin ice sheet, the ice was observed to thaw at the front of the spot while simultaneously freezing in the wake (Figure 1.18). The density change during the phase transition induced divergent microflows due to mass conservation, with a sink (div $\upsilon < 0$) at the front of the source and (div $\upsilon > 0$) at the back. The solid ice boundaries confine the liquid flow such that the movement is from the back of the molten spot to the front.

The pumping action was performed with a repetition rate of $f = 650$ Hz and a chamber temperature of $T_o = -10°C$ (Weinert et al. 2009). With densities $\rho_{water} = 1000$ kg/m^3 and $\rho_{ice} = 917$ kg/m^3, the pump velocity is $\upsilon_{pump} = 9.5$ mm/s. Experimentally, the authors were able to measure the pump velocity at 11 mm/s. The length of the molten spot depends on the temperature of the ice sheet. At low ice temperatures, only a short spot molten. At higher ice temperatures, the molten spot can reach lengths beyond 500 μm with the pump velocity exceeding 50 mm/s.

From a design perspective, the light pump approach implemented on "ice sheets" as introduced by Weinert and Braun (2009) does not require separate valves to switch between pump paths, thereby significantly reducing the hardware necessary for the LOC devices. Further, the approach to fluid transport does not require channel walls to define the fluid motion path or permit external pressure control to drive flow. In other words, it is possible to directly and locally drive fluids without structural changes to the substrate, eliminating some of the steps in microfabrication.

1.5.3.3 Micropumping by Optically Driven Bubbles

Bezuglyi and Ivanova (2007) introduced another method for optically pumping a fluid through a narrow microchannel using a gas-bubble piston concept. The gas-bubble piston was set in motion by

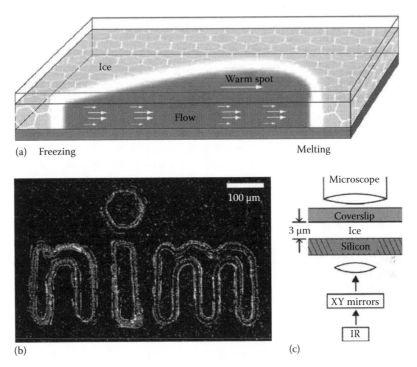

FIGURE 1.18   An example of moving an infrared laser spot through a micrometer thin ice sheet. (a) The ice thaws at the front of the spot (right side) and freezes in the wake (left side). (b) The heating laser spot traces the letters "nim" at 10 Hz. The temporarily molten microchannels formed by the thermo viscous pumping of ice are 25 μm wide and 3 μm thick. An overlay picturing the differences of successive images of 1 μm fluorescent polystyrene spheres suspended in ice is shown in the photograph. (c) The laser spot is focused on the ice sheet from the bottom as illustrated in the experimental setup. (Reprinted with permission from Weinert, F.M. et al. Light driven microflow in ice, *Appl. Phys. Lett.* 94, 113901-1–113901-3. Copyright 2009, American Institute of Physics.)

light-induced thermo-capillary force. Bubble methods for pumping small amounts of fluid have been an attractive approach for a variety of microfluidic applications because these bubbles can be readily produced by external microheaters. As these bubbles move, they tend to conform to the shape of the channel and, thereby, do not leave "stagnant" amounts of liquid behind the flow (Figure 1.19). Unfortunately, these bubble pumps require the fluid that is to be pumped to be heated to a high temperature which may be difficult to control due to the conductive dissipation of heat into the microchannel walls.

In an effort to minimize the heat dissipation problem, Bezuglyi and Ivanova (2007) introduced the notion of exploiting thermocapillary forces induced by a focused light beam to move bubbles

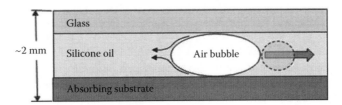

FIGURE 1.19   Schematic showing the experimental setup used to optically pump a fluid through a narrow channel using the gas-bubble piston concept originally introduced by Bezuglyi and Ivanova. (Adapted from Bezuglyi, B.A. and Ivanova, N.A., *Fluid Dyn.*, 42(1), 91, 2007.)

suspended in a solution. To demonstrate the optically-driven bubble piston pumping system, the authors created a prototype that consisted of two reservoirs connected by two microchannels. The controlling light source is a NVO-100 arc lamp with a 0.6 mm diameter beam and a focal plane power of about 100 mW. Using a bubble that has a length (L_w) approximately equal to the channel width (l) as a piston, the authors were able to produce a pumping action that took 15 s for a bubble-piston velocity of 0.5 mm/s. The volume of fluid pumped per piston stroke was 370 nL. Since the volume of the fluid column in the channel was determined to be 460 nL, the efficiency of the stroke is approximately 80% (Bezuglyi and Ivanova 2007).

The method of optically driven microfluidic pumping has several important advantages over "bubble" methods that use electrodynamic microheaters. First, the air bubble in an optically driven system is initially at the temperature of the fluid making it possible to pump the solution at small temperature differences (>10 K). Second, the authors did not observe any significant in the microflow during the pumping action. This occurs with other microheaters because it is often necessary to provide a high-power pulse to the bubble in the solution and then switch off the heater to prevent the bubble from collapsing. Third, the use of optical radiation energy is efficient because the light radiation is released directly in the fluid volume and converted into thermal energy for $\sim 10^{-12}$ s. Furthermore, the energy supply for the optically driven system does not require conductors with high electrical conductivity of a given cross-section. The problem is that the cross-section of the electrical conductors cannot be arbitrarily small as is the case of resistive methods. The diameter and wavelength of the laser beam can be controlled to provide an optimal solution for the desired applications. The performance can be further improved by producing smooth microchannel walls, improving the thermal insulation properties of the channel material, choosing a light radiation wavelength that can be efficiently absorbed by the fluid, and choosing a fluid with a high-temperature surface tension coefficient.

1.5.4 Temperature-Induced Phase Transformation of Solids

The photothermal effect arising from light striking a solid actuator shell can also be used to generate measureable forces and controlled movements (Yoshizawa et al. 2001, Okamura et al. 2009). SMA, such as 50/50 nickel–titanium (Ni–Ti) and 50/50 gold–cadmium (Au–Cd), are a group of metal alloys that can directly transform thermal energy into mechanical work. These solids experience a discontinuous change in their physical structure near a crystalline phase transformation temperature. The dimensional change is significantly larger than the linear volume change that occurs under normal thermal expansion. The corresponding phase transformation temperature can be adjusted over a wide range of temperatures by varying the composition of the alloy. The mechanism for actuation (Figure 1.20) is the martensite-to-austenite phase transformation that occurs as the material is heated and its subsequent reversal during cooling. The hysteresis effect implies that the temperature at which the material undergoes a phase change during heating is different from the temperature that causes the same material to return to the martensite state during cooling. The hysteresis effect for Ni–Ti is typically in the order of 20°C for SMA material (Gilbertson 1993).

The alloy can be formed into a wire or strip at high temperatures when it resides in an austenitic condition. Increasing the temperature applied to a pre-loaded Ni–Ti wire, originally at ambient temperature, will cause the material to undergo a phase transformation and move the position of the attached load a distance of approximately 4% of the overall wire length (Figure 1.20b). In essence, the small force created during the contraction period can be used to perform mechanical work (Gilbertson 1993). The reduction in the wire length can be recovered by cooling the material back to ambient temperature. The number of times the Ni–Ti material can exhibit the shape memory effect is dependent upon the amount of strain, and consequently, the total distance through which the wire is displaced. The amount of wire deflection is also a function of the initial force applied. The amount of pull-force generated for the applied current is significant for the size of the wire. Thicker

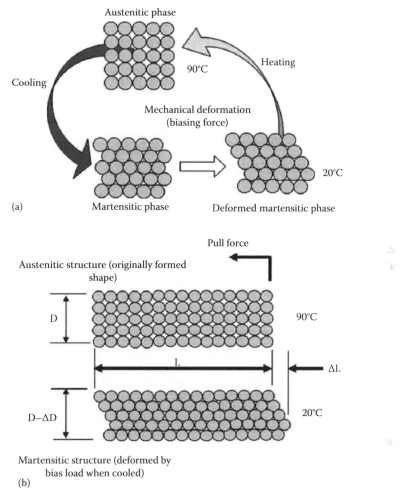

Austenitic phase

Cooling

90°C

Heating

Mechanical deformation
(biasing force)

20°C

(a)   Martensitic phase   Deformed martensitic phase

Pull force

Austenitic structure (originally formed
shape)

D

90°C

L

ΔL

D−ΔD

20°C

Martensitic structure (deformed by
bias load when cooled)
(b)

FIGURE 1.20 Schematic showing the (a) crystalline structure transformation of NiTi alloy as it is heated and cooled and (b) the change in NiTi wire length during transformation.

wires will generate greater forces but require larger currents and longer cooling time. For example, a 200 μm NI–Ti wire produces 4× the force (~5.8 N) than a 100 μm wire but takes 5.5× as long (~2.2 s) to cool down once heating has ceased (Gilbertson 1993).

Although SMA materials exhibit unique and useful design characteristics such as large power/ weight ratio, small size, cleanness and silent actuation, the successful application of the material has been limited to a small number of actuation devices that produce small linear displacements.

An example of an optically driven walking machine that employs SMA is shown in Figure 1.21 (Yoshizawa et al. 2001). The miniaturized machine consists of two parts: a body made of SMA and springs, and feet made of magnets and temperature-sensitive ferrites. The feet stick to the carbon steel floor due to magnetic force balance caused by the incident light beam, and the body repeats stretching and shrinking using the deformation of SMA caused by the switching ON and OFF of the projected light beam.

1.5.5 PHOTOTHERMAL VIBRATION

Inaba et al. (1995) described how the photothermal vibration of the quartz core of an optical fiber by laser light can be used to construct a vibration-type transducer. The microcantilever beam in this design

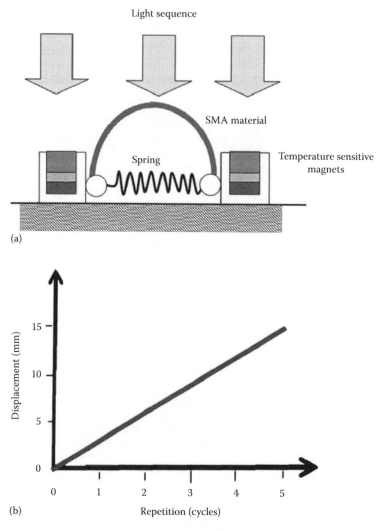

FIGURE 1.21 An optically driven walking machine initially proposed by Yoshizawa et al. (a) The machine consists of two basic parts: a body made of SMA and springs, and feet made of magnets and temperature-sensitive ferrites. The authors were able to control the movement sequentially heating the magnets and SMA actuator. (b) Linear relationship between the displacement and the number of cycles. (Adapted from Yoshizawa, T. et al., A walking machine driven by a light beam, in *Optomechatronic Systems II*, Cho, H.-Y., Ed., *Proceedings of the SPIE*, Vol. 4564, pp. 229–236, 2001.)

was the quartz core of the fiber and fabricated by etching the clad layer from the optical fiber tip. The resonance frequency depended largely upon the physical qualities of the cantilever such as size, density, and Young's modulus. The effect is also partially dependent upon the density of the gas or liquid which surrounds the cantilever because the *resonance shapeness* of the beam is a function of the viscosity coefficient for the gas or liquid. The resonance frequency for the microcantilever was observed to decrease from 16.69 to 16.59 kHz with an increase in pressure from 1 to 100 Pa, and a reduction in the resonance sharpness with an increase in pressure from 100 Pa to 10 kPa (Inaba et al. 1995).

Based on the concept of photothermal vibration, Otani et al. (2001) proposed a dynamic optical actuator that was driven solely by light. The device is a walking miniature robot constructed from three optical fibers, which represent legs, attached to a base as shown in Figure 1.22. Each fiber was cut for a bevel and the surface was painted black so that it could absorb light and convert it

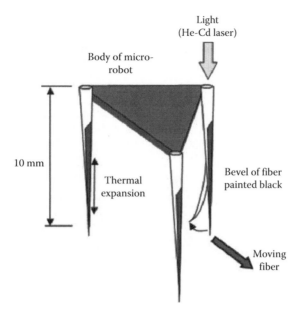

FIGURE 1.22 Illustration of the fiber optic walking platform proposed by Otani et al. Each leg of the microrobot bent due to thermal expansion when the light was turned on. The optically actuated robot was able to move 25 μm/s using a pulsating white light source. (Adapted from Otani, Y. et al., Photothermal actuator composed of optical fibers, in *Optomechatronic Systems II*, Cho, H.-Y., Ed., *Proceedings of the SPIE*, Vol. 4564, pp. 216–219, 2001.)

to heat. The photothermal effect occurred in response to a flashed incident beam, with a constant cycle time, onto one side of the optical fiber leg. The flashing light source produced a stretch vibration on the tip of the fiber that enabled it to operate like a flat spring. The authors experimentally demonstrated that the diameter of the fiber has an influence on the amount of deformation. It was observed that a 10 mm long fiber with a diameter of 250 μm would deform by 30 μm, while a 1000 μm diameter fiber of the same length would deform by as much as 50 μm. Furthermore, Otani et al. (2001) studied the effect of fiber length on the amount of displacement generated. A 1 mm long fiber with a diameter of 250 μm was found to deform 10 μm, while a 15 mm fiber of the same diameter deformed 90 μm.

In the microrobot design, the optical fiber became bent due to the thermal expansion that occurred when the light was turned on. If the light is turned off, it returns to its original shape. The switching frequency of the 12.1 mW helium cadmium laser (442 nm) was 4 Hz. The light was delivered to an individual optical fiber by a microscopic object lens. An acousto-optic modulator was used to switch intensity on and off. A mirror and object lens were mounted on a moving stage that followed the movement of the photothermal actuator. The movement of the photothermal actuator also depended on the balance of the device. Otani et al. (2001) adjusted the balance using a small weight on the base of the optical fiber. The size of the optical actuated walking robot was 3 mm × 10 mm and moved 2.3 mm at 25 μm/s using a pulsating light source.

A micromanipulator based on the photothermal bending effect experienced by a beveled optical fiber was described in a paper by Jankovic et al. (2004). The micromanipulator design incorporated four fingers, two bendable fibers for actively grasping small objects and two stationary fibers to provide structural support while holding the object (Figure 1.23). Each finger was a 1 mm diameter acrylic optic fiber with a 25 mm beveled edge near the tip. The beveled edge as coated with a thin layer of black paint where the thickness has a measurable impact on the amount of tip deflection. A light beam, from a 150 W halogen illuminator, was directed into the fixed end of the sculpted optic fiber causing the tip at the free end to deflect by approximately 50 μm. Several experiments

(a)   (b)

FIGURE 1.23 A micromanipulator based on the photothermal bending effect of optical fibers.). The device (a) consists of two bendable fingers and two stationary fingers. (b) A sequence of images demonstrating the controlled release of a 20 g steel ball. Note that the ball is released when the central finger tip is fully illuminated. (Adapted from Jankovic, N. et al., Light actuated micromanipulator based on photothermal effect of optical fibers, in *Optomechatronic Sensors, Actuators and Control*, Moon, K., Ed., *Proceedings of the SPIE*, Vol. 5602, pp. 63–72, 2004.)

were conducted to demonstrate that this simple microgripper is able to grasp, hold, and release a variety of small metal screws and ball bearings. Finite element analysis is used to further investigate the physical properties of the optical actuator. The theoretical deflections were slightly greater than the experimentally observed values. The FEM analysis is also used to estimate the maximum force ($\sim$0.7 mN) generated at the actuator tip during deflection.

1.6 SUMMARY

The fundamental aspects and unique characteristics of light-activated nano- and microscale actuators were summarized in this chapter. The primary means of optical actuation was to focus a laser beam onto a nano- or microactuator shell in an effort to generate mechanical deformation, displacement, or force. In this manner, light was used to both initiate movement (i.e., power) and control the actuation mechanism used to perform useful work. These optically driven systems harness radiation pressure, take advantage of photoresponsive materials, enable electrons to flow in electrostatic devices, or exploit optically driven photothermal effects. The physical size and complexity of the actuator design increase as driving mechanism moves from direct radiation pressure and optical gradient forces, to photothermal vibrations in sculpted optical fibers. Light-induced shape-changing polymers exploit the behavior of photoisomerizable molecules, such as azobenzene, to bend or expand when exposed to specific wavelengths. Azobenzene has also been used to create light-driven single-molecule DNA nanomotors. Other researchers have shown that the light-induced reactions of a photoresponsive compounds like dithienylethene can be used to reversibly trigger paralysis in living organisms. The recent introduction of carbon nanotube technologies has enabled the development of a new class of photomechanical actuators. Indirect optical actuators exploiting the photothermal interactions arising from light striking a material surface or fluid to activate and drive actuation have been used to create a variety of microscale switches, valves, and micropumps based on expanding mechanical diaphragms. Simple optothermal expansion of a fluid can also be used to drive fluid flow along microchannels or mixing streams of liquids prior to analysis. Light can even induce thermo-capillary forces that drive bubbles which push fluid through microchannels. Not only fluids but certain solids such as SMA can undergo measureable dimensional changes when heated by an optical source. The dimensional changes are significantly greater than mere thermal expansion, and have been used to pull to create machines that perform mechanical work. Alternatively, the photothermal vibration occurring at a tip of an optical fiber can be optically activated and controlled, enabling the design of innovative miniature walking platforms, robotic manipulators, and optical chop sticks.

REFERENCES

Al-Atar, U., Fernandes, R., Johnsen, B., Baillie, D., and Branda, N.R. 2009. A photocontrolled molecular switch regulates paralysis in a living organism. *J. Am. Chem. Soc.* 131: 15966–15967.

Ashkin, A. 1970. Acceleration and trapping of particles by radiation pressure. *Phys. Rev. Lett.* 24: 156–159.

Ashkin, A. 2000. History of optical trapping and manipulation of small-neutral particle, atoms and molecules. *IEEE J. Sel. Top. Quantum Electron.* 6(6): 841–856.

Ashkin, A. and Dziedzic, J.M. 1971. Optical levitation by radiation pressure. *Appl. Phys. Lett.* 19: 283–285.

Ashkin, A. and Dziedzic, J.M. 1980. Observation of light scattering using optical levitation. *Appl. Phys.* 19: 660–668.

Balasubramanian, K. and Burghard, M. 2005. Chemically functionalized carbon nanotubes. *Small* 1(2): 180–192.

Bezuglyi, B.A. and Ivanova, N.A. 2007. Pumping of a fluid through a microchannel by means of a bubble driven by a light beam. *Fluid Dyn.* 42(1): 91–96.

Bird, R.B., Stewart, W.E., and Lightfoot, N.E. 2002. *Transport Phenomena*, 2nd edn. New York: John Wiley & Sons, Inc.

Bishop, J.D., and Klavins, E. 2007. An improved autonomous DNA nanomotor. *Nano Lett.* 7(9): 2574–2579.

Boyer, J.-C., Carling, C.-J., Gates, B.D., and Branda, N.R. 2010. Two-way photoswitching using one type of near-infrared light, upconverting nanoparticles, and chaining only light intensity. *J. Am. Chem. Soc.* 132: 15766–15772.

Carling, C.-J., Boyer, J.-C., and Branda, N.R. 2009. Remote-control photoswitching using NIR light. *J. Am. Chem. Soc.* 131: 10838–10839.

Chen, H. and Meiners, J.C. 2004. Topologic mixing on a microfluidic chip. *Appl. Phys. Lett.* 84: 2193–2195.

Datskos, P.G., Rajic, S., and Datskou, I. 1998. Photoinduced and thermal stress in silicon microcantilevers. *Appl. Phys. Lett.* 73(16): 2319–2321.

Datskos, P.G., Sepaniak, M.J., Tipple, C.A., and Lavrik, N. 2001. Photomechanical chemical microsensors. *Sens. Actuators B* 76: 393–402.

Dholakia, K. and Reece, P. 2006. Optical micromanipulation takes hold. *Nanotoday* 1(1): 18–27.

Dienerowitz, M., Mazilu, M., and Dholakia, K. 2008. Optical manipulation of nanoparticles: A review. *J. Nanophotonics* 2: 021875 (32pp.).

Dittrich, P.S. and Manz, A. 2006. Lab-on-a-chip: Microfluidics in drug discovery. *Nat. Rev.* 5: 210–218.

Eijkel, J.C.T. and van den Berg, A. 2005. Nanofluidics: What is it and what can we expect from it. *Microfluid Nanofluid* 1: 249–267.

El Moctar, A.O., Aubry, N., and Batton, J. 2003. Electro-hydrodynamic micro-fluidic mixer. *Lab Chip* 3: 273–280.

Finkelmann, H., Nishikawa, E., Pereira, G.G., and Warner, M. 2001. A new opto-mechanical effect in solids. *Phys. Rev. Lett.* 87(1): 015501-1–015501-4.

Fujii, T., Sando, Y., Higashino, K., and Fujii, Y. 2003. A plug and play microfluidic device. *Lab Chip* 3: 193–197.

Gatos, H.C. and Lagowski, J. 1973. Surface photovoltage spectroscopy—A new approach to the study of high-gap semiconductor surfaces. *J. Vac. Sci. Technol.* 10(1): 130–135.

Gilbertson, R.G. 1993. *Muscle Wires Project Book.* San Rafael, CA: Mondo-Tronics.

Hale, K.F., Clark, C., Duggan, R.F., and Jones, B.E. 1988. High-sensitivity optopneumatic converter. *IEE Proc.* 135(5): 348–352.

Hale, K.F., Clark, C., Duggan, R.F., and Jones, B.E. 1990. Incremental control of a valve actuator employing optopneumatic conversion. *Sens. Actuators* A21–A23: 207–210.

Hessel, V., Lowe, H., and Schonfeld, F. 2005. Micromixers—A review on passive and active mixing principles. *Chem. Eng. Sci.* 60: 2479–2501.

Higurashi, E., Ohguchi, O., Tamamura, T., Ukita, H., and Sawada, R. 1997. Optically induced rotation of dissymmetrically shaped fluorinated polyimide micro-objects in optical traps. *J. Appl. Phys.* 82(6): 2773–2779.

Hirasa, O. 1993. Research trends of stimuli-responsive polymer hydrogels in Japan. *J. Intell. Mater. Syst. Struct.* 4: 538–542.

Hockaday, B.D. and Waters, J.P. 1990. Direct optical-to-mechanical actuation. *Appl. Opt.* 29(31): 4629–4632.

Hogan, P.M., Tajbakhsh, A.R., and Terentjev, E.M. 2002. UV manipulation of order and macroscopic shape in nematic elastomers. *Phys. Rev. E* 65: 041720-1–041720-10.

Hugel, T., Holland, N.B., Cattani, A., Moroder, L., Seitz, M., and Gaub, H.E. 2002. Single-molecule optomechanical cycle. *Science* 296: 1103–1106.

Iijima, S. 1991. Helical microtubules of graphitic carbon. *Nature* 354: 56–58.

Ikeda, T., Horiuchi, S., Karanjit, D.B., Kurihara, S., and Tazuke, S. 1990. Photochemically induced isothermal phase transition in polymer liquid crystals with mesogenic phenyl benzoate side chains 1. Calorimetric studies and order parameters. *Macromolecules* 23: 36–42.

Inaba, S., Kumazaki, H., and Hane, K. 1995. Photothermal vibration of fiber core for vibration-type sensor. *Jpn. J. Appl. Phys.* 34: 2018–2021.

Jankovic, N., Zeman, M., Cai, N., Igwe, P., and Knopf, G.K. 2004. Light actuated micromanipulator based on photothermal effect of optical fibers. In *Optomechatronic Sensors, Actuators and Control*, ed. K. Moon, *Proceedings of the SPIE*, Vol. 5602, pp. 63–72.

Jiang, H., Kelch, S., and Lendlein, A. 2006. Polymers move in response to light. *Adv. Mater.* 18: 1471–1475.

Jonáš, A. and Zemánek, P. 2008. Light at work: The use of optical forces for particle manipulation, sorting and analysis. *Electrophoresis* 29: 4813–4851.

Kaiser, A.B., Düsburg, G., and Roth, S. 1998. Heterogeneous model for conduction in carbon nanotubes. *Phys. Rev. B* 57(3): 1418–1421.

Kamholz, A.E., Weigl, B.H., Finlayson, B.A., and Yager, P. 1999. Quantitative analysis of molecular interactive in microfluidic channel: The T-sensor. *Anal. Chem.* 71: 5340–5347.

Kamholz, A.E. and Yager, P. 2002. Molecular diffusive scaling laws in pressure-driven microfluidic channels: Deviation from one-dimensional Einstein approximations. *Sens. Actuators B* 82: 117–121.

Kang, H., Liu, H., Phillips, J.A., Cao, Z., Kim, Y., Chen, Y., Yang, Z., Li, J., and Tan, W. 2009. Single-DNA molecule nanomotor regulated by photons. *Nano Lett.* 9(7): 2690–2696.

Knopf, G.K. 2006. Optical actuation and control. In *Handbook of Optoelectronics*, eds. J. Dakin and R.G.W. Brown, pp. 1453–1479, Boca Raton, FL: CRC Press.

Krecmer, P., Moulin, A.M., Stephenson, R.J., Rayment, T., Welland, M.E., and Elliott, S.R. 1997. Reversible nanocontraction and dilatation in a solid induced by polarized light. *Science* 277: 1799–1802.

Kroerner, H., Price, G., Pearce, N.A., Alexander, M., and Vaia, R.A. 2004. Remotely actuated polymer nanocomposites—Stress recovery of carbon nanotube filled thermoplastic elastomers. *Nat. Mater.* 3: 115–120.

Kröner, G., Parkin, S., Nieminen, T.A., Loke, V.L.Y., Hechenberg, N.R., and Rubinsztein-Dunlop, H. 2007. Integrated optomechanical microelements. *Opt. Express* 15(9): 5521–5530.

Lagowski, J. and Gatos, H.C. 1972. Photomechanical effect in noncentrosymmetric semiconductors-CdS. *Appl. Phys. Lett.* 20(1): 14–16.

Lagowski, J., Gatos, H.C., and Sproles, E.S. 1975. Surface stress and normal mode of vibration of thin crystals: GaAs. *Appl. Phys. Lett.* 26(9): 493–495.

Lebedev, P. 1901. Untersuchungen über die druckkräfte des lichtes. *Ann. Phys.* 311: 433–458.

Lendlein, A., Jiang, H., Jünger, O., and Langer, R. 2005. Light-induced shape-memory polymers. *Nature* 434: 879–882.

Li, M., Pernice, W.H., and Tang, H.X. 2009. Broadband all-photonic transduction of nanocantilevers. *Nat. Nanotechnol.* 4: 377–382.

Li, M., Pernice, W.H., Xiong, C., Baehr-Jones, T., Hochberg, M., and Tang, H.X. 2008. Harnessing optical forces in integrated photonic circuits. *Nature* 456(27): 480–484.

Li, J.J. and Tan W. 2002. A single DNA molecule nanomotor. *Nano Lett.* 2(4): 315–318.

Liu, Y., Lu, S., and Panchapakesan, B. 2009. Alignment enhanced photoconductivity in single walled carbon nanotube films. *Nanotechnology* 20: 1–7.

Liu, R.H., Yang, J., Pindera, M.Z., Athavale, M., and Grodzinski, P. 2002. Bubble-induced acoustic micromixing. *Lab Chip* 2: 151–157.

Lu, S. and Panchapakesan, B. 2006. Nanotube micro-optomechanical actuators. *Appl. Phys. Lett.* 88: 253107-1–253107-3.

Lu, S. and Panchapakesan, B. 2007. Photomechanical responses of carbon nanotube/polymer actuators. *Nanotechnology* 18: 305502 (8pp.).

Maxwell, J.C. 1873. *A Treatise on Electricity and Magnetism.* Oxford, U.K.: Clarendon Press.

McKenzie, J.S. and Clark, C. 1992. Highly sensitive micromachined optical-to-fluid pressure converter for use in an optical actuation scheme. *J. Micromech. Microeng.* 2: 245–249.

McKenzie, J.S., Hale, K.F., and Jones, B.E. 1995. Optical actuators. In *Advances in Actuators*, eds. A.P. Dorey and J.H. Moore, pp. 82–111. Bristol, U.K.: IOP Publishing Ltd.

Mizoguchi, H., Ando, M., Mizuno, T., Takagi, T., and Nakajima, N. 1992. Design and fabrication of light driven pump. *Micro Electro Mechanical Systems'92*, Travemunde, Germany, pp. 31–36.

Mizutani, Y., Otani, Y., and Umeda, N. 2008. Optically driven actuators using poly(vinylidene difluoride). *Opt. Rev.* 15(3): 162–165.

Morikawa, Y. and Nakada, T. 1997. Position control of PLZT bimorph-type optical actuator by on-off control. *IECON Proc.* 3: 1403–1408.

Nguyen, N.T. and Wereley, S.T. 2002. *Fundamentals and Applications of Microfluidics Fabrication Techniques for Microfluidics.* Norwood, MA: Artech House.

Nguyen, N.T. and Wu, Z. 2005. Micromixers—A review. *J. Micormech. Microeng.* 15: R1–R16.

Nieminen, T.A., Higuet, J., Kröner, G., Loke, V.L.Y., Parkin, S., Singer, W., Hechenberg, N.R., and Rubinsztein-Dunlop, H. 2006. Optically driven micromachines: Progress and prospects. *Proc. SPIE* 6038: 237–245.

Okamura, H., Yamaguchi, K., and Ono, R. 2009. Light-driven actuator with shape memory alloy for manipulation of macroscopic objects. *Int. J. Optomechatronics* 3: 277–288.

Otani, Y., Matsuba, Y., and Yoshizawa, T. 2001. Photothermal actuator composed of optical fibers. In *Optomechatronic Systems II*, ed. H.-Y. Cho. *Proceedings of the SPIE*, Vol. 4564, pp. 216–219.

Piegari, E., Cataudella, V., Marigliano Ramaglia, V., and Iadonisi, G. 2002. Comment on "polarons in carbon nanotubes". *Phys. Rev. Lett.* 89(4): 049701.

Poosanaas, P., Tonooka, K., and Uchino, K. 2000. Photostrictive actuators. *Mechatronics* 10: 467–487.

Setou, M., Nakagawa, T., Seog, D.-H., and Hirokawa, N. 2000. Kinesin superfamily motor protein KIF17 and mLin-10 in NMDA receptor-containing vesicle transport. *Science* 288: 1796–1802.

Shiu, P.P., Knopf, G.K., Ostojic, M., and Nikumb, S. 2010. Laser-assisted active microfluidic mixer. In Optomechatronic Technologies (ISOT), Toronto, ON, Oct. 25–27, (5pp), Doi. 10.1109/ISOT.2010.5687346.

Steen, W.M. 1998. *Laser Material Processing*, 2nd edn. London, U.K.: Springer-Verlag.

Steen, W.M. and Mazumder, J. 2010. *Laser Material Processing*, 4th edn. London, U.K.: Springer-Verlag.

Stuchlik, M., Krecmer, P., and Elliott, S.R. 2001. Opto-mechanical effect in chalcogenide glasses. *J. Optoelectron. Adv. Mater.* 3(2): 361–366.

Stuchlik, M., Krecmer, P., and Elliott, S.R. 2004. Micro-nano actuators driven by polarised light. *IEE Proc. Sci. Meas. Technol.* 151(2): 131–136.

Sun, X. and Li, C.-Z. 2007. Fundamental aspects and applications of nanotubes and nanowires for biosensors. In *Smart Biosensor Technology*, eds. G.K. Knopf and A.S. Bassi, pp. 291–330, Boca Raton, FL: CRC Press.

Suski, J., Largeau, D., Steyer, A., Van de Pol, F.C.M., and Blom, F.R. 1990. Optically activated $ZnO/SiO_2/Si$ cantilever beams. *Sens. Actuators A* 24: 221–225.

Suzuki, H. and Ho, C.M. 2002. A magnetic force driven chaotic micro-mixer. In *Proceedings of MEMS'02, 15th IEEE International Workshop on Micro Electromechanical System*, Las Vegas, NV, pp. 40–43.

Tabib-Azar, M. 1990. Optically controlled silicon microactuators. *Nanotechnology* 1: 81–92.

Tabib-Azar, M. 1998. *Microactuators: Electrical, Magnetic, Thermal, Optical, Mechanical, Chemical, and Smart Structures.* Norwell, MA: Kluwer Academic.

Tabib-Azar, M. and Leane, J.S. 1990. Direct optical control for a silicon microactuator. *Sens. Actuators A21–A23*: 229–235.

Tabib-Azar, M., Wong, K., and Ko, W. 1992. Aging phenomena in heavily doped (p⁺) micromachined silicon cantilever beams. *Sens. Actuators A* 33: 199–206.

Tabiryan, N., Serak, S., Dai X.-M., and Bunning, T. 2005. Polymer film with optically controlled form and actuation. *Opt. Express* 13(9): 7442–7448.

Tang, H.X. 2009. May the force of light be with you. *IEEE Spectr.* 46(10): 46–51.

Tsai, J.H. and Lin, L. 2002. Active microfluidic mixer and gas bubble filter driven by thermal bubble pump. *Sens. Actuators A* 97–98: 665–671.

Uchino, K. 1990. Photostrictive actuator. *IEEE Ultrasonics Symposium*, Honolulu, HI, pp. 721–723.

Vale, R.D. 2003. The molecular motor toolbox for intracellular transport. *Cell* 112: 467–480.

Van Oosten, C.L., Bastiaansen, C.W.M., and Broer, D.J. 2009. Printed artificial cilia from liquid-crystal network actuators modularly driven by light. *Nat. Mater.* 8: 677–682.

Van Oosten, C.L., Corbett, D., Davies, D., Warner, M., Bastiaansen, C.W.M., and Broer, D.J. 2008. Bending dynamics and directionality reversal in liquid crystal network photoactuators. *Macromolecules* 41: 8592–8596.

Van Oosten, C.L., Harris, K.D., Bastiaansen, C.W.M., and Broer, D.J. 2007. Glassy photomechanical liquid-crystal network actuators for microscale devices. *Eur. Phys. J. E* 23: 329–336.

Verissimo-Alves, M., Capaz, R.B., Koiller, B., Artacho, E., and Chacham, H. 2001. Polarons in carbon nanotubes. *Phys. Rev. Lett.* 86(15): 3372–3375.

Verne, J. 1993. From Earth to the Moon. Translated by Lowell Bair. New York: Bantam Books. Original book published in 1865.

Wang, H., Iovenitti, P., Harvey, E., and Masood, S. 2002. Optimizing layout of obstacles for enhanced mixing in microchannels. *Smart Mater. Struct.* 11: 662–667.

Warner, M. and Terentjev, E.M. 1996. Nematic elastomers—A new state of matter? *Prog. Polym. Sci.* 21: 853–891.

Warner, M. and Terentjev, E.M. 2003. *Nematic Liquid Crystal Elastomers*. Oxford, U.K.: Clarendon Press.

Weinert, F.M. and Braun, D. 2008. Optically driven fluid flow along arbitrary microscale patterns using thermoviscous expansion. *J. Appl. Phys.* 104: 104701-1–104701-10.

Weinert, F. and Braun, D. 2009. Light driven microfluidics-pumping water optically by thermoviscous expansion. *International Symposium on Optomechatronic Technologies*, Istanbul, Turkey, September 21–23, 2009, pp. 383–386.

Weinert, F.M., Wühr, M., and Braun, D. 2009. Light driven microflow in ice. *Appl. Phys. Lett.* 94: 113901-1–113901-3.

Willmott, P.R. and Huber, J.R. 2000. Pulsed laser vaporization and deposition. *Rev. Mod. Phys.* 72(1): 315–328.

Yamada, M., Kondo, M., Miyasato, R., Mamiya, J., Yu, Y., Kinoshita, M., Barrett, C.J., and Ikeda, T. 2008. Photomobile polymer materials: Towards light-driven plastic motors. *Angew. Chem. Int. Ed.* 47: 4986–4988.

Yoshizawa, T., Hayashi, D., Yamamoto, M., and Otani, Y. 2001. A walking machine driven by a light beam. In *Optomechatronic Systems II*, ed. H.-Y. Cho. *Proceedings of the SPIE*, Vol. 4564, pp. 229–236.

Yu, Y., Nakano, M., and Ikeda, T. 2003. Directed bending of a polymer film by light. *Nature* 425: 145.

Zhang, Y. and Iijima, S. 1999. Elastic response of carbon nanotube bundles to visible light. *Phys. Rev. Lett.* 82(17): 3472–3475.

2 Nature of Light

Suwas K. Nikumb and M. Chandra Sekhar

CONTENTS

2.1   INTRODUCTION

Light is a wonderful gift of the world we live in. It is vital to almost all living beings on the planet of earth that senses, feels, and sees it. Witness the whole panorama of existence where we can see rich colors emanating from different objects that are relevant to our environment, e.g., scenery of rainbow colors, green forests, flowery gardens, colorful birds, or even a lighting firefly. The nature of light is central to the field of optics over many centuries and is regarded as a key point in understanding its interaction with matter as well as any modern technologies involving light. The principles revealed through the exploration to unveil the nature of light are crucial for the renaissance of optics with the development of lasers, fiber optics, opto-electronics, and semiconductor photonic devices. To gain deeper knowledge of the physical phenomenon involving light, it is important to know its basic properties; for instance, reflection, refraction, dispersion, polarization, interference, and diffraction which essentially contribute to underlying mechanisms. In this chapter, we will discuss the nature of light and some important concepts, processes, and optical technologies that are critical to the light-driven systems and leading-edge, modern optical actuator technologies used for realizing solutions in a wide variety of applications.

To begin with, we introduce electromagnetic radiation and light, which includes a brief history on the wave–particle duality of light, wavelength distribution in the electromagnetic spectrum, sources of natural and artificial light, e.g., incandescent light, gas lamps, semiconductor light sources, and lasers, with brief description on polarization properties, and coherent and incoherent sources of light. Subsequent sections focus on topics related to the absorption and emission of radiation, population inversion, generation of laser light, and various types of continuous wave (CW) and pulsed lasers including recently developed fiber laser, disk laser, and the bio laser. Next we discuss some of the important laser beam characteristics such as linewidth, mode and beam quality, divergence, and the shaping of laser beams in the context of the core subject matter of this text. Leading further, the discussion hinges on the energy transfer during interactions between laser and material surfaces to cover topics such as physics of illumination, light absorption, melting, vaporization, and plasma effects. The chapter concludes with a brief overview on laser material removal introducing relevant surface modification processes based on continuous wave or long-pulse, nanosecond duration and femtosecond time scale laser interactions, relevant applications, and its effect on surface profile formation of laser ablated materials.

2.2   ELECTROMAGNETIC RADIATION AND LIGHT

The physical nature of light has puzzled mankind since ancient times. The revelation of its nature witnessed many dramatic developments in the history. The science of optics, which embraces significant pool of knowledge accumulated over roughly 3000 years of human history, has been subjected to debate many fundamental questions such as, What is light made of? What is a photon? How it behaves? One particular question, on the description of light as particles or waves, plagued human beings since the sixteenth century. The Greeks believed that light consisted of tiny particles (*corpuscles*) emitted by a source of light. By 1900 most of the physical principles underlying different optical phenomena were formulated. However, the discovery of quantum physics, nearly a century ago, revolutionized the fundamentals of the nature of light. Therefore, it is imperative to comprehend the nature of light to reveal the interaction mechanism between light and matter, which is essential to explore the dynamics of the universe and many modern days' applications.

2.2.1 Wave–Particle Nature of Light

The Dutch astronomer Christian Huygens (1629–1695) [1] introduced the notion that light as a wave motion, which spreads out from a light source in all directions as a series of waves and propagates through all-pervasive elastic medium called ether. He showed that wave theory could explain refraction and reflection. As waves are unaffected by gravity, it was assumed that they slow down upon entering a denser medium. In the same era, Newton [2,3] (1642–1727) strongly advocated the particle theory of light, in which he regarded rays of light as streams of very small particles emitted from a source traveling in straight lines in space. In 1801, Thomas Young [2,4] in his famous double-slit experiment demonstrated that under appropriate conditions light rays interfere with each other, strongly supporting to the wave theory. Augustine Fresnel [5] also upheld the wave character in his Fresnel equations on light reflection and transmission at a plane interface in 1821. Later, James Clerk Maxwell [6] in 1865 proposed that electromagnetic waves travel with the speed of light and at this time it seemed that wave theory of light was firmly established. Interestingly, neither the particle theory nor the wave nature could explain all the basic optical phenomena and subsequent experiments on photoelectric effect. It was noted that light behaves like waves in the propagation during interference and diffraction experiments, while it behaves as particles in its interaction with matter. The principles of quantum mechanics emerged in the beginning of the nineteenth century, revealed by the revolutionary work of Max Plank, Albert Einstein, Niels Bohr, and other prominent figures of the time. The quantization model showed that atoms emit light in discrete energy packets or chunks rather than in a continuous manner and the energy of light wave is present in a stream of particles called *photons* whose energy is proportional to the frequency of electromagnetic wave specified by the following Planck's equation [7]

$$E = hn \tag{2.1}$$

where
The energy E is a quantum of electromagnetic radiation
h is Planck's constant of 6.63×10^{-34} J-s
v is the frequency of radiation (c/λ, where c being the velocity of light, i.e., 2.99×10^8 ms^{-1}, and λ is the wavelength of light in meters)

The quantum electrodynamics field developed at a later stage revealed a notable finding, that *photons* were neither waves nor particles, but a bit more complex than either. However, with simple intuition, it is possible to choose physical models as waves or particles to treat many optical phenomena, in which wavelike characteristics dictate in some instances while particle-like attributes are evident in some other cases. This characteristic duality of light also varies with the wavelength or energy of the photons [8]. For instance, at longer wavelengths from radio to the visible blue region, the wave theory explains most of the optical phenomena where as high-energy photons of shorter wavelengths in the x-rays and gamma ray region, the particle theory dominates.

In 1873, James Clerk Maxwell [6] studied the electromagnetic radiation and light in more detail. He discovered that self-propagating electromagnetic waves travel through space at a constant speed. This similarity with the earlier measured speed of light was also confirmed by Hertz in 1887 [9]. This fact led him to conclude that light was indeed a form of electromagnetic radiation. He then provided a mathematical description of the behavior of electric and magnetic fields in the form of Maxwell's equations. The experiments performed by Albert Michelson and Edward Morley in 1887 [10] indicated that ether did not exist. Then onwards, light is accepted

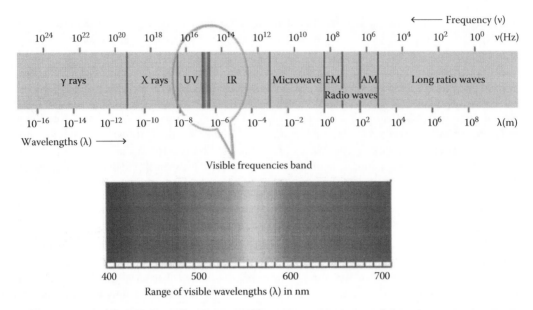

FIGURE 2.1 **(See color insert.)** Electromagnetic Spectrum. (Adopted from: http://upload.wikimedia.org/wikipedia/commons/f/f1/EM_spectrum.svg)

to be an electromagnetic wave, which requires no medium for its propagation and all light wavelengths were viewed and located as a portion of the electromagnetic radiation spectrum, as illustrated in Figure 2.1.

2.2.2 ELECTROMAGNETIC SPECTRUM

The electromagnetic waves are generated by oscillating electric charges. All electromagnetic radiation exhibits wavelike properties except differing in wavelength. The waves consist of oscillating electric and magnetic fields at right angles to each other and to the direction of wave propagation (Figure 2.2). Radiated waves can be detected at considerable distances. Furthermore, electromagnetic waves carry energy and momentum and therefore can exert pressure on a

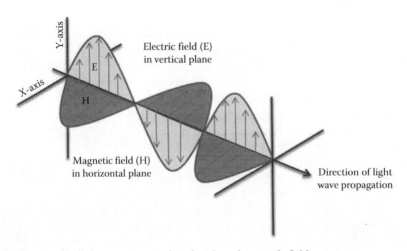

FIGURE 2.2 Propagating light wave representing electric and magnetic fields.

given surface. The electromagnetic waves range from ultra long (e.g., electric power line), to very short (e.g., gamma ray radiation), wavelengths covering the entire spectrum. Of significant interest to many photonics applications the region from infrared to ultraviolet. Figure 2.2 shows the range of optical wavelengths containing three specific regions, i.e., infrared (0.76–300 μm), visible (390–760 nm), and ultraviolet (10–390 nm), which correspond to the optical frequencies ranging from 1 THz in the far-infrared region to 3×10^{16} Hz in the extreme ultraviolet region [11]. Also, the continuous shift in the optical wavelength from 390 to 760 nm corresponds to the colors of the visible spectrum of light from violet to red.

2.2.3 Sources of Natural and Artificial Light

There are many different kinds of sources of light. Typically they can be classified as either natural or artificial [12]. Examples of natural sources are the sun, stars in the sky, and thunderstorm lightning. Examples of artificial sources of radiation are incandescent lamp filaments, fluorescent lights, black-body radiation, hot coal in a campfire, a candle flame, heater coils, lasers, masers, radio and television antennas, radars, and x-ray tubes. Almost all materials with temperatures above absolute zero emit electromagnetic radiation because of thermal motion of molecules. Whereas *luminescent light* emission is observed, when excited electrons fall to lower energy levels [12]. Light produced during the thunderstorm is because of electric discharges through ionized gases. Certain chemicals produce visible radiation by chemiluminescence. In living insects, e.g., fireflies, this process is referred to as bioluminescence. Some substances, when illuminated by more energetic radiation, produce fluorescent light, while some other materials emit light some time after excitation by more energetic radiation, a process known as phosphorescence. Deceleration of a free charged particle, e.g., an electron, can produce visible radiation such as synchrotron radiation and bremsstrahlung radiation. Also, particles travelling at a speed faster than the phase velocity of light, in a medium, can cause visible Cherenkov radiation [13,14]. Atoms emit and absorb light at characteristic energies and produces emission lines in the spectrum. Every atom and molecule has its own distinctive set of spectral lines. Earlier observations of the line spectra led to significant understanding of the nature of atoms. Hot, dense materials emit continuous spectrum containing bands of frequencies. Mainly two types of spectra are prominent in photonics: the emission spectra and the absorption spectra. An absorption spectrum is from the light that has passed through an absorbing medium. An emission spectrum is emitted by a light source. Such emission can be spontaneous, as in commercial neon signs, fluorescent materials, gas discharge, or flash lamps. Emission can also be stimulated as in laser or microwave maser. The importance of the light was further exemplified with the remarkable discovery of laser [15]. The word "laser" is an acronym that stands for "*l*ight *a*mplification by *s*timulated *e*mission of *r*adiation." It is a monochromatic, meaning a single color, light source of electromagnetic radiation which can travel farther distances with remarkably little spreading. It was a significant breakthrough that enhanced the field of optics and photonics with a multitude of technological innovations and applications, encompassing many diverse fields of science and technology.

2.2.4 Polarization

In earlier sections, we have seen that light wave could interfere just like sound waves and that light could be polarized, if it were a transverse wave. The relationship between light and electromagnetism was first established by Michael Faraday in 1845 when he discovered that the plane of polarization of linearly polarized light is rotated when light rays travel along the magnetic field direction in the presence of a transparent dielectric, and the effect known as Faraday rotation [16,17]. Further in 1873, James Maxwell explained the complex, time-varying behavior of electric and magnetic fields associated with the light wave. Later, Hertz's experiments showed electromagnetic waves could be reflected, refracted, and diffracted. He also demonstrated that the oscillating nature of electromagnetic waves radiated by a transmitter. In other words, light waves are composed of mutually perpendicular electric and magnetic fields with wave propagation at right angles to both fields as shown earlier in Figure 2.2. It carries light energy as it propagates and the amount of energy per second across a unit area perpendicular to the direction of travel is known as irradiance or flux density. Polarization arises from the direction of electric field vector with respect to the direction of wave propagation. Since the light wave's electric field component vibrates in a direction perpendicular to its propagation motion, it is a transverse wave and therefore polarizable [10,18]. If the electric vector is aligned only in one direction, then the light is *linearly polarized or plane-polarized*, i.e., the electric field primarily resides in the plane of vibration, although its amplitude and sign varies in time. If the electric vector component has two directions, with equal intensity and at right angles to each other, then the light is said to be *circularly* polarized. If one of the vectors is stronger than the other, then the resultant light would be *elliptically* polarized. Finally, light is said to be *unpolarized* if it is composed of vibrations in many different directions with no preferred orientation or to render any single resultant polarization [17].

2.2.5 COHERENT AND INCOHERENT SOURCES

Conventional light sources such as flashlights, fluorescent tubes, and tungsten bulbs are incoherent sources. In such devices, the emitted photons from the filament of the bulb carry many wave frequencies, oscillating in random chaotic directions, and the photons exhibit frequent and abrupt changes in phase within the beam at all time [19]. There is a total absence of a constant phase relationship. Incoherent light waves produced by such sources do not constitute an orderly pattern. On the other hand, two waves are said to be coherent if they have a constant relative phase and produce constructive or destructive interference when added together or subtracted from each other. Coherence is a property of waves that enable stationary interference. Coherent light wave exhibits orderly emission of photons, i.e., the change of phase within the beam occurs for all the emitted photons at the same time, very similar to marching soldiers in unison. More generally, coherence describes all the measurable properties of the correlation between waves such as amplitude, phase, and frequency. Light produced by lasers is coherent, intense, monochromatic, and highly directional. Other examples of coherent sources can be multiple loud speakers driven by the same amplifier at a concert hall or several radio antennas powered by the same transmitter.

2.2.6 ABSORPTION AND EMISSION OF RADIATION

To better understand the physics behind the generation of laser light, let us look at the absorption, emission of radiation and transition processes between atomic levels. From the atomic physics we know that the atoms and molecules, sometimes referred as species, can exist in one of the discrete, allowed or highly probable levels or states. In the energy domain, these species can make upward or downward transitions between two allowed states, e.g., E_0 and E_1, by absorbing or emitting a photon of light. Albert Einstein in 1917 first recognized three main processes by which atoms can interact with an electromagnetic field. Therefore, for a simple two-level atomic system, following conditions can occur [20]:

1. *Absorption*: An atom in lower level E_0 may absorb photon of frequency $(E_1 - E_0)/h$ (where h is Planck's constant) and make an upward transition to the higher energy level E_1. The rate of absorption depends on the number of atoms present in the lower level and the energy density of radiation present in the system.
2. *Spontaneous emission*: An atom in an excited level E_1 can make a downward transition spontaneously due to inherent interactions of atomic structure by emitting a photon corresponding to the energy difference between the two levels. Often, this type of transition takes place because of a large number of atoms present in the excited level. In this type of emission, the photons radiate in random directions without any polarization contributing to the field.

3. *Stimulated emission*: Opposite to absorption process, an atom in an excited state can also make a downward transition to a lower state in the presence of an external electromagnetic radiation of frequency $(E_1 - E_0)/h$. The rate of stimulated or induced emission depends on the energy density or the strength of the external radiation and the number of atoms present in the excited state. An unusual feature of this process is that the emitted photon is in phase with, has the polarization of, and propagates in the same direction as the stimulated wave that induced the atom to undergo this kind of transition. This means, if an atom is stimulated to emit light energy by a propagating wave, the additional quantum of energy that is released in the process tends to add to or amplify the incident wave, increasing its amplitude or flux density. This type of radiation possesses coherent characteristics and it is a key to the operation of the laser [20].

2.2.7 POPULATION INVERSION AND LASING REQUIREMENTS

For a laser medium to be effective, its electron energy level structure has to be such that there is an upper excited level having a long spontaneous lifetime and a lower excited level that decays rapidly back to the ground state. Most importantly, another mechanism must be present by which the long life of upper level can be preferentially populated. A pumping source is normally required to provide energy to the atoms and molecules in the lasing medium to produce the excited states. The energy can be fed via highly energetic electrons, energetic heavy particles such as protons, neutrons, or in the form of optical radiation. Depending on the nature of the lasing medium common sources such as electrical voltage for gaseous and semiconductor lasers and optical methods, e.g., flash lamps or another laser wavelength as pump source for crystalline and liquid dye medium are used.

Population inversion is another requirement for light amplification within a laser resonator. Amplification occurs when amplifying medium populates with more atoms having electrons at a higher energy level than at the lower energy level for a specific pair of allowed transition. In other words, the number of atoms in the excited state or upper level exceeds the number of atoms in the lower level. This condition is known as population inversion [15,19]. Under this condition, if an electromagnetic wave passes through the resonator cavity, more photons will be stimulated, contributing to an overall increase in the photon flux or an amplification of the output beam.

2.3 GENERATION OF LASER LIGHT

A typical laser unit consists of an energized active medium that is capable of providing optical amplification and a resonator or an optical cavity with mirror arrangement that supply the necessary optical feedback of the amplified beam into the gain medium [11] as illustrated in Figure 2.3. The active medium consists of atoms, molecules, ions in solid, liquid or gaseous form. The resonator cavity consists of two highly reflecting plane or spherical mirrors at opposite ends, aligned suitably to confine the optical energy allowing multiple reflections between the mirrors. The light waves

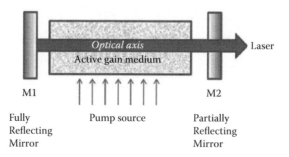

FIGURE 2.3 Design of a laser resonator cavity.

travel from one end of the optical cavity to the other, interacting with atoms and molecules in the medium, increasing the intensity of light by stimulated emission, to produce a highly directional, intense, coherent and monochromatic beam of light. The laser beam is obtained by allowing a small portion of light to escape passing through one of the end mirrors that is partially transmitting or by using a mirror with a small hole in it. The wavelength, λ, of a typical laser is determined by the energy difference between the excited states and the lower energy level when the transition takes place. The lasers produce output wavelengths or frequencies that cover a wide range of the electromagnetic spectrum, spanning from the microwave and infrared region to the visible, ultraviolet, vacuum ultraviolet, and the soft-x-ray spectral regions. Some lasers are tunable, meaning they can produce several wavelengths, adjustable one at a time, which are particularly attractive for spectroscopy and communication applications. Depending on the engineering design and application, the resonator sizes can vary from just a few tens of microns, e.g., microchip or semiconductor lasers to larger dimensional high-power multi-kilowatt lasers, toward very large fusion laser systems which can occupy a very large building space. Lasers that are designed to provide a continuous stream of light are called *continuous wave* (CW) lasers or they can be designed to provide pulsed light beam of varying frequencies for specific applications. Laser output powers can differ from a few nanowatts to multi-terawatts ($>10^{12}$ W) regime depending on resonator type, design considerations, and other specifications. An example of a compact, mobile, 5 TW, 70 fs pulse duration laser system is shown in Figure 2.4 [21]. In the pulsed mode, lasers can generate short pulses or bursts of light from millisecond (10^{-3} s) duration to more recently investigated femtosecond (10^{-15} s) and attosecond (10^{-18} s) durations, providing extremely high peak intensities for probing many innovative applications in science and technology. In general, the average powers of ultrashort pulsed lasers are much lower in comparison to multi-kilowatt class CW lasers. Laser light, unlike ordinary light, is unique in that it usually consists of monochromatic light. All laser resonators have a finite wavelength width over which gain can occur [21].

In recent years, lasers have greatly transformed the world. Lasers operating at wavelengths from the infrared to the vacuum-ultraviolet range have become an indispensable part of numerous

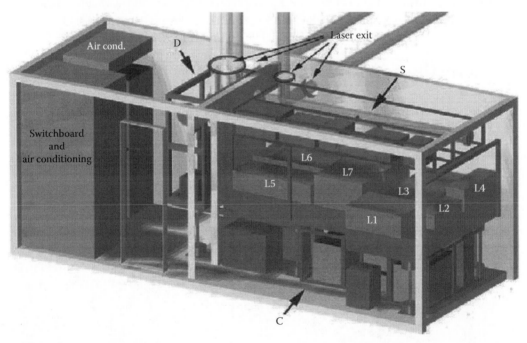

FIGURE 2.4 A compact five terawatt femtosecond pulse laser system. (From Wille, H. et al., *Eur. Phys. J. Appl. Phys.*, 20, 183, 2002.)

applications in Medicine, Industry, Entertainment and Defense. Conventional tools are being replaced with powerful lasers for a variety of manufacturing applications such as cutting, drilling, welding, brazing, alloying, texturing, heat treating, cladding, and micromachining of materials in both the industrial and medical world. Recent experimental evidence suggests that lasers could sustain continuous output powers beyond 100 kW for several minutes [22], indicating the potential of using lasers for military applications. Other applications involving lasers are communication, remote sensing, metrology, surveying and products, e.g., laser pointers, fiber optic sensors, compact disk players, surgical devices, laser scanners, and communication devices to name a few.

2.4   TYPES OF LASERS

Lasers are generally classified according to the type of lasing medium used that causes the lasing action. Most common types are gas lasers, solid state lasers, liquid lasers, metal vapor lasers, semiconductor lasers, free-electron lasers, and more recently developed fiber lasers. Several text books are available in the literature describing design details, energy-level diagrams, specific lasing transition, life time of an allowed and participating meta-stable states, mathematical representations, models, etc., in detail. The intent here is simply to provide a brief introduction on some of the laser types only. Readers may refer to suggested reading list at the end of this chapter for further understanding of the specifics of the subject matter.

2.4.1   SOLID-STATE LASERS

Solid-state lasers use transparent, grown, end-surface polished crystals or glass material as the host matrix which are normally doped with ionic species of laser atoms. Some typical materials include aluminum oxides (sapphire), yttrium aluminum garnets (YAG), yttrium lithium fluoride (YLF), yttrium vanadate (YVO4), yttrium aluminum perovskite (YAP; YAlO3), or phosphate or silica glass, etc. Doping species are rare earth ions or transition ions with outer shells unfilled or defect center's called color centers. Typical dopants are neodymium (Nd^{3+}), ytterbium (Yb^{3+}), erbium (Er^{3+}), chromium (Cr^{3+}), titanium (Ti^{3+}), etc. For color center lasers the host materials are typically an alkali-halide crystal and an electron trapping crystal defect works as laser species. The pumping source is usually a flash lamp or another efficient laser light wavelength. A short description of various types of solid-state lasers is presented later.

2.4.1.1   Ruby Laser

The ruby laser, which emits red visible wavelength of 694.3 nm, was the very first laser demonstrated in 1960 by Maiman [23]. A helical Xenon flash lamp, enclosed in elliptical reflector, was used as a pumping source to excite the laser levels of chromium doped (~0.05% by weight) in crystalline sapphire (Al_2O_3) host matrix material, as the lasing medium. Here, the energy levels of the Cr ions take part in the lasing action, absorbing the pump radiation wavelengths typically from 400 to 550 nm. The excited Cr ions rapidly undergo a nonradiative transition in 10^{-8}–10^{-9} s time scale to the metastable upper level, from which the lasing transition occurs to the ground level, emitting bright red color radiation at 694.3 nm wavelength. In the past, pulsed ruby laser were often used for applications such as spot welding and hole drilling.

2.4.1.2   Neodymium Laser

The doped neodymium ions are efficient energy absorbers into host substances such as yttrium-aluminum-garnet (YAG) or glass [24]. After absorption at an appropriate band, the excited ions decay down to a metastable state from which a lasing transition occurs to a terminal level, before reaching the ground level. The neodymium laser emits 1060 nm wavelength for YAG and 1050 nm wavelength for glass materials. The resonator is constructed with an end polished cylindrical YAG crystal which is pumped optically either by a flash lamp or a semiconductor diode laser of appropriate frequency. The crystal and flash lamps are mounted at different foci of a highly polished, gold-coated, elliptical cavity made of metal or ceramic material. The cooling is usually provided with de-ionized water circulating around the YAG crystal and the flash lamp. An important property of Nd:YAG material is its high thermal conductivity, which lessens the probability of thermal damages or cracking tendency during operation and permits using crystal geometries, e.g., cylindrical rods, discs, and slabs for fabrication. High average-power outputs can be achieved using either an oscillator–amplifier system or by combining several beams together. The neodymium lasers, in the pulsed or CW mode, are ideal candidates for a wide range of material-processing applications including heat treating, marking, engraving, annealing, soldering, welding, cutting, scribing, perforating, structuring, drilling, ablation, and for other applications such as processing of materials e.g., plastics, metals, textiles, leather, cardboard, glass, ceramics, semiconductor wafers, and thin films or foils of metals [25,26].

2.4.1.3   Disk Lasers

Another remarkable innovation in the field of solid-state lasers is the development of disk laser. An example illustrating the construction of a thin-disk laser design is presented in Figure 2.5 [30]. In this case, the active medium is shaped like a disk of Yb:YAG crystal which is irradiated from the front surface using a high-power InGaAs diode laser as pump source [27,28]. The back surface is fused to an efficient heat sink for cooling purposes. The pump light penetrates the laser disk active medium several times until fully absorbed. The emitted beam is coupled out of the resonator using specially designed optics. A small disc, approximately 0.3 mm thick and 7 mm in diameter, can produce >500 W of laser power at 1030 nm wavelength. There are many significant advantages resulting from this geometry and the cavity design [29]. First, superior cooling of the laser active material, a reduced thermal lensing effect, and therefore improved beam quality in comparison to commonly available laser systems. Secondly, higher dopant concentrations are manageable with the disk geometry. Outstanding beam quality, minimal divergence, and with laser powers exceeding 16 kW at the work piece, these lasers are perfect for many production floor applications [30]. Disk lasers are ideal for welding and cutting of metals. Moreover, the disk laser's beam quality enables enough power density for large working distances and applications using narrow focusing optics. Using laser light cables, they can be easily integrated into production lines adding processing flexibility and can be easily combined with industrial robots. Some applications are seam welding of pipes or tubing, and thick metal plates cutting. Currently, many industrial sectors such as automotive, transportation, aerospace, and heavy industries are benefiting using these laser devices for production of parts in their manufacturing environment.

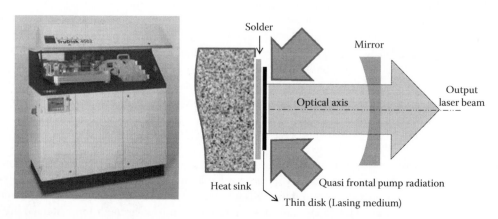

FIGURE 2.5   An example of disk laser design and commercial disk laser unit. (*Photograph courtesy of TRUMPF Laser Technology Center, MI 48170.*)

2.4.2 FIBER LASERS

Fiber lasers are relatively recent developments in the field and these lasers are doped plastic or glass fibers that are either end-pumped or side-pumped by means of diode lasers [31]. The fiber diameters can be as small as 100 μm, which permits oscillations to be contained within a very narrow volume, providing an excellent mode quality of the beam which significantly exceeds the conventional flash lamp pumped Nd:YAG laser sources. High-power, high-quality, multi-kilowatt CW or pulsed fiber lasers operating at 1085 nm are readily available [32,33]. A 2 kW fiber laser focused at a spot size of 50 μm can provide a power density of >100 MW/cm^2. These lasers are highly efficient and long-lasting for many thousands of hours of maintenance-free operation. Fiber lasers with power levels up to 50 kW [34,35] have already been tested in the field, for applications from detonating land mines and unexploded ordnance to short-range rocket defense. In manufacturing, high-power fiber lasers are finding many application avenues, e.g., weld steel from several meters distance, a process known as remote welding, cutting of >50 mm thick steel plates, and high-speed scribing of thin-film solar panels.

2.4.3 SEMICONDUCTOR OR DIODE LASERS

These lasers are based on semiconductor gainmedia, where optical gain is usually achieved by stimulated emission at an interband transition under conditions of a high carrier density in the conduction band [36]. The most commonly known applications of these lasers are in communication devices and in compact disk players to detect the digitally coded information. They are also used in high-speed printing systems and laser pointers. Semiconductor lasers are tiny, a few tens of microns in dimensions, and are very efficient devices. The emitted wavelength ranges from ~600 nm to 1.8 μm and typically provides continuous output powers with possible extension to other wavelength regions such as blue and green. Most semiconductor lasers are laser diodes, which are pumped with an electrical current in a region where an n-doped and a p-doped semiconductor material meet. However, there are also optically pumped semiconductor lasers, where carriers are generated by absorbed pump light, and quantum cascade lasers, where intraband transitions are utilized. Some of the common materials used for semiconductor lasers are GaAs (gallium arsenide), GaP (gallium phosphide), InGaP (indium gallium phosphide), GaN (gallium nitride), InGaAs (indium gallium arsenide), GaInNAs (indium gallium arsenide nitride), InP (indium phosphide), and GaInP (gallium indium phosphide). The gain medium essentially forms a $p-n$ junction, which is typically a thin slice of a semiconductor crystal cleaved in short segment of these materials [37,38]. The inherently reflective cleaved end surfaces often serve as mirrors of the elongated gain region to generate enough amplification for laser emission. The laser operation occurs in the longitudinal mode due to its short cavity length and produces output beam with large angular divergence. Mass quantities of semiconductor lasers are usually produced by depositing various layers of p- and n-doped material, different insulating layers, and other metal materials as contacts. Because of their low threshold current and low power consumption, heterostructured quantum-well semiconductor lasers are most common in their use.

2.4.4 FREE-ELECTRON LASER

The free-electron lasers [39–43] are distinctly different, in that the output beam does not occur from discrete transitions from excited states of atoms or molecules as in the case of gaseous, liquid, or solid-state lasers. In this case, a high-energy electron beam passes through a varying magnetic field, which causes electrons to oscillate in transverse direction. These oscillations cause electrons to radiate and to stimulate other electrons to oscillate, thereby radiate at same frequency, and in phase with the original oscillating electrons. Reflecting mirrors at both ends provide optical amplification producing an intense beam of light, also known as synchrotron radiation. Essentially, the laser consists of a circuit where relativistic electrons speeding around a circular path within a magnetic field, emitting photons on each turn in the same direction as the traveling electrons. The emitted radiation offers high-power output from deep infrared to x-ray radiation.

2.4.5 LIQUID (DYE) LASERS

Dye lasers are wavelength tunable lasers and fall in this category [44–46]. They can operate at high power with pulse lengths varying from CW to femtoseconds. The gain medium typically consists of a host solvent liquid, e.g., alcohol or water, in which an organic dye, e.g., Rhodamine 6G, is dissolved. Dyes have different spectral bandwidth which allows emission of laser from UV to infrared wavelengths. By changing a dye concentration in the solution and adjusting the flow pressure, wavelengths from 320 to 1500 nm could be selected. The dye solution typically absorbs light from a pump source, e.g., a UV nitrogen laser, argon–ion laser, green copper vapor laser, or a flash lamp. The wavelength tuning is accomplished by using a coated, reflective diffraction grating or a prism as one of the laser mirrors. Using a mechanical drive the grating or prism can be rotated to select the specific laser wavelength over the spectrum of the typical dye. Each dye has a tunable gain bandwidth of ~30 to 40 nm with a linewidth as narrow as 10 GHz or less. Dye lasers are very versatile. In addition to their wavelength agility, these lasers can offer very large pulsed energies or very high-average powers. Flashlamp-pumped dye lasers have been shown to yield hundreds of Joules per pulse and copper-laser-pumped dye lasers can yield average powers in the kilowatt regime. Dye lasers are used in many applications including astronomy, isotope separation [47], and spectroscopy [48]. In medical applications, the wide range of tunable wavelengths enables a good matching with the absorption lines of certain tissues, such as melanin or hemoglobin, while the narrow bandwidth helps reduce the possibility of damage to the surrounding tissue [49]. They are used to treat blood vessel disorders such as scars, port-wine stains, and kidney stones as well as tattoo removal and a number of other applications.

2.4.6 GAS LASERS

The most commonly known gas lasers are the helium-neon laser, the argon and krypton ion laser, the carbon dioxide laser, and the rare-gas-halide excimer laser. Other metal vapor lasers, e.g., copper vapor, helium-cadmium, and the hydrogen-fluoride chemical lasers, also fit in the same category; however, only a few common lasers are discussed later. These lasers can be operated continuously or in the pulsed mode to provide a range of power outputs from modest to multi-kilowatt levels suitable for industrial applications.

2.4.6.1 Helium-Neon Laser

The helium-neon laser or HeNe laser [15,17,19] was the first gas laser developed in 1961 by Ali Javan, which is the forerunner of the whole gas laser family [50]. The laser gain medium consists of a mixture of helium and neon inside a small bore capillary discharge tube, sealed with mirrors at both ends, and usually excited by a DC electrical energy. Initially, the helium metastable atoms are excited by electron collisions with helium ground-state atoms. This energy is then transferred to the desired neon excited energy levels, thereby producing the required population inversion with lower-lying helium energy levels. Some applications require externally mounted mirrors and a delicate alignment procedure. In such resonators, the ends of the laser tube are made of pyrex or quartz set at a Brewster's angle to the axis of the laser, to produce polarized light output. The most widely used laser wavelength is in the red region of the spectrum at 632.8 nm, which has the highest gain. HeNe lasers can also be operated at the 543.5 nm green and several infrared wavelengths, e.g., 1.15 and 3.39 μm, by choosing the resonator mirror coatings with their peak reflectance at these other wavelengths. Since visible transitions at wavelengths other than 633 nm have slightly lower gain, these lasers produce lower output powers ranging from 1 to 100 mW in CW mode. In addition to industrial and scientific laboratories, this laser is typically used in surveying, construction, supermarket scanners, printers, and many other applications [51,52].

2.4.6.2   Argon Ion and Krypton Ion Lasers

The argon ion laser and the krypton ion lasers were discovered shortly after the HeNe lasers [53,54]. These lasers offer a wide range of visible (in the blue and green spectral region) and ultraviolet laser wavelengths. Both argon and krypton ion lasers are similar in their basic principles of operation. They are rare gas ion lasers and use similar basic resonator hardware configuration and power requirements. These lasers primarily differ in gas fill of the plasma tube and the optics (mirrors or prisms, etc.) for selecting the output wavelength. Applying a DC current through a 50–200 cm long, narrow-bore plasma discharge tube filled with low-pressure (0.1 torr) argon or krypton gas produces the population inversion. The argon ion laser can be operated as a continuous gas laser at several different wavelengths ranging from 408.9 and 686.1 nm; however, the most prominent transitions are in the green at 488 and 514.5 nm, respectively. The lasers can operate at much higher powers providing 20–100 W of continuous power. Also, because of higher input electrical power requirements, these lasers are relatively inefficient, but still extremely useful for certain applications because of their short wavelengths. Argon and krypton ion lasers find applications in many diverse fields including, printing, scanning, lithography, photo plotting, forensic medicine, general and ophthalmic surgery, large-scale light shows holography, electro optics, and as an optical "pumping" source for other lasers.

2.4.6.3   Helium-Cadmium Laser

The helium-cadmium or HeCd laser operates in continuous mode emitting primarily the blue at 441.6 nm and ultraviolet at 325 nm wavelengths, respectively [54]. The construction of these lasers is far more complex compared to the HeNe-type lasers. The laser resonators are about 40–100 cm in length for output powers ranging from 20 to 200 mW. The lasing element is cadmium, which is a metal at room temperature. The laser tube contains a reservoir for cadmium and a heater to vaporize it. For population inversion, the metallic cadmium must be evaporated, and then the vapor must be distributed uniformly down the narrow-bore quartz discharge tube and running an electrical discharge current of up to 100 mA through the tube. The cadmium vapor diffuses into the excited helium gas within the gain region, where it is ionized and the cataphoresis force on the cadmium ions propels them towards the cathode, thereby causing uniform distribution of the metal, producing uniform gain profile in the discharge region. The resonator also needs to sustain a higher level of internal pressure for the vaporized cadmium to remain in the tube. The operating lifetime of the laser depends on the amount of cadmium available in the reservoir. Other limiting factors being the control of vapor and the loss of helium via diffusion through the glass tube walls. Because of their excellent beam quality, HeCd lasers find wide applications in several areas, e.g., fabrication of CD master plates, 3-D stereo lithography, fluorescence information analysis, confocal microscopy, holography, fabrication of diffraction grating, printing spools, inspection and measurement, nondestructive testing, and microlithography [55].

2.4.6.4   Excimer Lasers

Excimer or rare-gas halide excimer lasers emit pulses of light in the ultraviolet region of the spectrum consisting of mixtures of noble gas atoms such as argon, krypton, and xenon with reactive gases (halide molecules) such as fluorine or chlorine operating in a high-pressure gaseous discharge [56]. The term excimer is short form of "excited dimer," which indicates the excited-state nature of the lasing molecules. They are the most powerful and efficient lasers in the UV. While a lot of different excimer laser transitions have been used to generate light pulses at various wavelengths between 126 and 660 nm, the most commonly used lasers are krypton fluoride (KrF, 248 nm), argon fluoride (ArF, 193 nm), xenon fluoride (XeF, 351 nm), and xenon chloride (XeCl, 308 nm). In addition, the deep UV wavelength of the fluorine laser (F_2, 157 nm) also found applications in industry. Typically, these lasers produce pulsed output with pulse energy ranging from tens of millijoules to several joules with pulse durations of 10–50 ns and repetition rates up to 1 kHz. The excitation occurs when electrons within the discharge colliding with and ionizing the rare gas atoms while dissociating the halide molecules into either F or Cl atoms to form negative halogen ions. These negative ions and the positive rare gas ions readily combine in an excited state to produce the laser species. Applications include deep-ultraviolet photolithography [57,58], laser eye surgery, dental procedures and manufacturing of semiconductor microelectronic devices [59,60].

2.4.6.5 Carbon Dioxide Laser

The carbon dioxide laser (CO_2), first invented by Kumar Patel in 1964 [61], is among the most power-ful laser category operating primarily in the middle infrared spectral region at a wavelength of 10.6 µm, producing CW powers of over 150 kW and pulse energies more than 10 kJ [22,41]. Their ability to produce high-power beams along with parameter control flexibility earned them the title, "indus-trial work horse," for a multitude of material-processing applications in manufacturing industries. Different versions of low-power CO_2 lasers are also available in smaller sizes with a wide range of power levels. The most common versions are sealed off, slow axial flow, fast axial flow, transverse flow, and transverse electric atmospheric (TEA) pulsed lasers [62,63]. The active medium of the lasers consists of mixture of helium, nitrogen, and CO_2 gases in certain proportion. However, CO_2 is the active molecule that emits the 10.6 µm radiation from a transition between two of its vibrational modes. Excitation is normally accomplished by a direct electrical discharge within the gas mixture. Electron collisions with the nitrogen molecules within the discharge produce metastable energy levels. The energy contained in those levels is subsequently transferred by collisions to the CO_2 molecule, where the population inversion is produced. This is one of the most powerful and efficient lasers, with wall-plug conversion efficiency achievable up to 30% or more. The carbon dioxide laser finds many useful applications in industry, in particular, welding, drilling, heat treating and cutting of sheet metals and other engineering materials [20].

2.4.7 Bio Laser

In the past half-century, laser action has been demonstrated in a wide variety of materials. However, in an interesting recent development, lasing action was observed for the first time in living cells [64–66]. In this example, human embryonic kidney cells were genetically engineered to produce green fluorescent protein. This protein when used as a gain medium, light amplification can occur (Figure 2.6). The experiment was performed by placing the protein cells between two small mirrors, some 20 µm apart forming the laser cavity, in which light could bounce several times through the cell. When the cells were exposed to blue light, the laser cavity emitted an intense, directional green laser light at 516 nm with a pump-energy threshold of ~1 nJ. The ability to induce lasing action in living cells with no harm to the cells interestingly opens up new possibilities for bioimaging applications. Since the emitted mode patterns depend on the refractive index, revealing 3D struc-tures of the cell interior could be a promising application. In addition, increasing the speed of flow cytometry measurements or using the bright green fluorescence for therapeutic studies, photody-namic therapies, and providing an interface between electronics and living organisms are other possible applications [67].

2.5 LASER BEAM PROPERTIES

Laser beams exhibit several unique characteristics which make them suitable for diverse applications. Apart from the power output of a laser, properties such as wavelength, coher-ence, polarization, brightness, power distribution or mode, beam divergence, and beam focusing capabilities need special considerations during manufacturing of specific laser system design. Shaping of a particular laser beam with advanced optical techniques can also provide unique tools for materials-processing application. Although we have touched upon some of the basic properties in earlier sections, in the context of the subject matter presented in this book some key properties are discussed in this section.

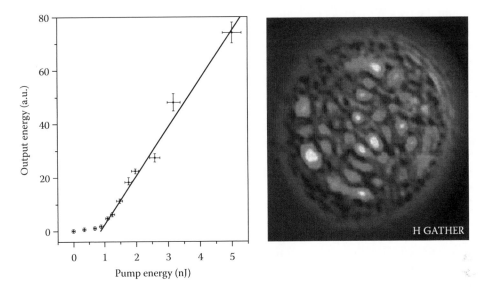

FIGURE 2.6 A living cell doped with a fluorescent protein emits laser light at 516 nm with a pump-energy threshold at ~1 nJ and the microscope image of the lasing cell. (From Gather, M.C. and Yun, S.H., *Nat. Photonics*, 5, 406, 2011.)

2.5.1 WAVELENGTH

The wavelength of a laser depends on the transitions taking place by stimulated emission. Often laser undergoes simultaneously on distinct transition lines or wavelengths. When application demands a highly monochromatic, spectrally pure beam, the resonator must be designed to oscillate only on one single transition. Practically, this can be achieved by inserting wavelength sensitive reflective and or transmissive elements (e.g., prism or dispersive grating as rare mirror) into the laser optical cavity. On passing through the prism, the individual beam wavelengths undergo dispersion. The wavelength whose ray is incident normal to the mirror is reflected back into the cavity, while other wavelengths suffer losses, resulting in the oscillation of a single wavelength within the optical cavity. The wavelength tuning of the laser can thus be achieved by rotating the prism or diffraction grating.

2.5.2 MODE AND LASER BEAM DIAMETER

Normally the output of a laser resonator consists of oscillating single or multi-modes or discrete frequency components. The mode quality depends on the boundary conditions, mirror shape and their separation. Two types of modes exist in the laser cavity. The longitudinal or axial modes which correspond to separate light beams travelling along a distinct path between the cavity mirrors and having an exact integral number of wavelengths along the path, and the transverse modes which are represented by slightly different optical paths as they travel through the resonator cavity. Therefore, each transverse mode could consist of several longitudinal modes oscillating along its path. The directional property of the output beam is reduced as the number of transverse mode increases. Also, the spectral purity is reduced as the number of longitudinal modes increases because the beam contains more discrete frequency components. Depending on the application, appropriate modes have to be chosen while suppressing the unwanted ones for a specific application. For low beam divergence and fine focus spot diameter, the Gaussian or Transverse Electromagnetic Mode (TEM_{00}) is often specified [68]. Consequently, with higher-order modes, since the energy is concentrated away from the resonator axis, the beam diameter becomes wider. In some applications, it is possible to avoid the higher-order modes by introducing a diaphragm, normal to the axis, within the optical cavity. For a highly monochromatic output beam, the longitudinal mode option is preferred. To reduce the number of longitudinal modes that oscillate within a laser cavity, various techniques are available, e.g., altering the cavity length or using Fabry-Pérot etalon [69]. It is important to note that the intensity of laser light is not the same throughout the cross-section of the beam diameter. This is because the cavity also controls the transverse modes, or intensity cross-sections. Therefore for TEM_{00} beam which has an ideal symmetrical round cross-section, the distribution of intensity minimal at the edges and gradually maximizes at the center. The laser beam diameter is defined as the diameter of a circular beam at a certain point where the intensity drops to half, i.e., full width at half maximum (FWHM), $1/e$ (0.368) and $1/e^2$ (0.135) of its maximum value. In other words, beam diameter is the diameter of the laser beam cross-section between points near the outer edge of the beam where its intensity is only 50%, 63%, and about 86% of the intensity at the beam center.

2.5.3 MONOCHROMATICITY AND LINEWIDTH

The light from a laser typically comes from an atomic transition with a single precise wavelength. So we expect the laser light has a single spectral color and is almost the purest monochromatic light [69]. However, the laser light in general is not exactly monochromatic. It produces a very narrow finite bandwidth around a single, central wavelength. This is due to the Doppler broadening effect of moving atoms or molecules from which the spectral emission originates. The degree of monochromaticity can be described as a measure of the bandwidth which is often referred to as linewidth in terms of wavelength as $\Delta\lambda$ or in frequency as Δv. The narrower is the linewidth, the more monochromatic is the beam [70]. Typically, for a He–Ne laser at a wavelength of 632.8 nm, $\Delta\lambda$ is about 0.01 nm. On the other hand, for a typical diode laser operating at a wavelength of 900 nm, $\Delta\lambda$ is about 1 nm. Conventional LED light has a linewidth of approximately from 30 to 60 nm. The Nd:YAG laser operating at 1.064 nm exhibits a typical $\Delta\lambda$ of 0.00045 nm. The spectrally pure output is critical for a multitude of applications, including remote sensing for specific chemical constituents, high-resolution spectroscopy, and high signal-to-noise ratio (SNR) in communications.

2.5.4 DIRECTIONALITY AND BEAM DIVERGENCE

A prominent property of a laser beam, mentioned earlier, is its directionality. The mirrors at opposite ends of the resonator cavity facilitate back and forth beam reflections amplifying the gain by the stimulated emission and produce a well-collimated beam [71]. The high degree of collimation primarily arises from the resonator cavity that has nearly parallel front and back mirrors, which confines laser's output beam perpendicular to those mirrors. Therefore, in principle, a perfectly collimated beam would be parallel to the optical axis with zero degree divergence angle. In reality, the diffraction phenomena play a pivotal role in determining the size of the laser spot projected at a given distance. The narrow light beam from the resonator cavity subsequently diverges at an angle depending on the resonator configuration, the size of the output aperture, and the beam diffraction along its path. The diffraction imposes a limit on the minimum achievable beam diameter after passing through an optical beam delivery system. The beam divergence of a laser beam is a measure for how fast the beam expands far from the beam waist. For a given laser, the beam emerging from the output mirror (or the output aperture) and the diffraction effects on the beam will limit the minimum divergence and spot size of the beam. For a perfect Gaussian beam at wavelength λ, the divergence θ_o (half angle) is given by

$$\theta_o = \left(\frac{1}{\pi} \left[\frac{\lambda}{w_o} \right] \right) \tag{2.2}$$

where w_0 is the beam radius (with $1/e^2$ intensity) at the beam waist. Note that the beam divergence increases with wavelength, and decreases as the beam (or output lens aperture) diameter increases. Beam divergence measurements can be performed by measuring the beam radius at different distances, using standard beam profilers [72]. A large value of beam divergence for a given beam radius signifies poor beam quality. Beams with much smaller divergence propagating over significantly larger distances are considered as near collimated laser beams for practical purposes. A low beam divergence angle can be critical for pointing applications as well as optical communication.

2.5.5   BEAM QUALITY AND M^2 FACTOR

Many material-processing applications requiring higher target intensities utilize tightly focused laser beams. Hence, the quality of a laser beam is a key parameter for the design of optical beam delivery for commercial laser systems [71]. We know that an unmodified laser beam diverges along its path by diffraction effects. The beam is diffraction-limited when its diameter is comparable to its initial waist value. Such a laser beam can be focused more tightly to obtain larger depth of focus, the distance over which the focused beam has more or less the same intensity. In addition, such a beam can be guided using smaller size beam-delivery optics. In order to denote the quantified representation of the beam propagation characteristics, the concept of a dimensionless beam propagation parameter, M^2, was developed in 1970 for all types of lasers. M^2 is a quantitative measure of the quality of a laser beam and according to ISO standard # 11146 [73], and is defined as the beam parameter product (BPP) divided by λ/π. The BPP is the product of a laser beam divergence and the diameter of the beam at its narrowest point or at the beam waist. M^2 beam quality factor provides an accurate representation of the propagation characteristics of the beam. It limits the degree to which a laser beam can be focused for a given beam divergence, which is limited by the numerical aperture of the focusing lens. The value of $M^2 = 1$ for a single-mode, diffraction-limited TEM_{00} Gaussian beam. The M^2 factor varies significantly for different lasers and affects their characteristics and, therefore, cannot be neglected in optical designs. The HeNe lasers typically have M^2 factor < 1.1, but for ion lasers, it is typically between 1.1 and 1.3. Similarly, for collimated TEM_{00} diode laser, this value ranges from 1.1 to 1.7, whereas for high-energy multimode lasers, the M^2 factor can be as high as 3 or 4, respectively. High-power lasers exhibit much higher M^2 value of 100 or greater. Thermal distortions in the active gain media, use of poor optical quality of components, diffraction effects at apertures, etc., are the main reasons for the poor beam quality. In general, the beam quality deteriorates when the laser operates at higher-order cavity modes.

2.5.6   FOCUSABILITY AND DEPTH OF FOCUS

Several manufacturing applications require laser beam to be focused to a very small spot size. Therefore, the beam diameter at the focal spot becomes an important parameter for processing of materials. The smallest spot size that can be obtained from a single TEM00 mode laser which is roughly the dimension of wavelength of the laser and is given by the equation [74]

$$d_{min} = \frac{2.44 f \lambda}{D} \tag{2.3}$$

where
   f is the lens focal length
   D is the unfocussed beam diameter
   λ is wavelength of light

Note that other factors that may influence the focal spot size are the lens spherical aberration and the thermal lensing effects. During transmission of the beam through focusing optics, laser power fluctuations can often cause shape changes of the lenses, affecting the focal spot size. In general, high-power lasers can be focused to spot diameters in the order 100–200 μm; however, extra precautions are needed in terms of beam parameters, mode quality, and focusing optics to obtain tightly focused spot diameters.

Another parameter of interest for material processing is the depth of focus (DOF) [75]. On passing through a spherical lens the laser beam initially converges at the lens focal plane and then diverges to wider beam diameter. DOF is defined as the distance over which the focused beam has roughly the same intensity or it is the distance over which the focal spot size varies within ~5%. The equation for DOF is

$$\text{DOF} = 2.44\lambda \left(\frac{f}{D} \right)^2 \tag{2.4}$$

The superb quality beam focusing achievable from a laser with Gaussian TEM00 mode is extremely useful for optical trapping applications [76,77]. An optical trap is formed by tightly focusing a laser beam with an objective lens of high numerical aperture (NA). In this case, a dielectric particle near the focus will experience a force due to the transfer of momentum from the scattering of incident photons. The resultant optical force is composed of a scattering force in the direction of light propagation and a gradient force in the direction of spatial light gradient. Such optical force could levitate micron-size dielectric particles in a medium such as water or air. Specific properties of the laser, e.g., power stability, pointing stability, thermal drift, wavelength, and mode quality, are important considerations. In general, trapping forces ~1 pN per 10 mW of power can be achieved with micron-scale beads. Wavelength is an important consideration when biological material is trapped, particularly for in vivo trapping of cells or small organisms. Solid-state lasers, e.g., Nd:YAG and Nd:YLF, are suitable for working with a variety of sample materials. Diode-pumped lasers offer superior amplitude and pointing stability; however, they suffer from mode instabilities and noncircular beam dimensions. The Ti:sapphire lasers operating at power range ~250 mW and wavelengths from 830 to 970 nm are adequate sources for optical-trapping applications [77].

2.6   SHAPING OF LASER BEAMS

Laser beam shaping, essentially the art of redistributing the irradiance and phase profile of a laser's output radiation, is an enabling technology used in many diverse applications [78–84]. For example, laser material processing, laser-matter interaction studies, welding, cutting, hole drilling, as well as medical procedures, e.g., corneal surgery and cosmetic skin treatments. Other applications include electronic component trimming, data-image processing, lithography, semiconductor manufacture, graphic arts, optical data processing, and military uses. The shape of the beam is defined by its irradiance distribution. However, the phase of the shaped beam is a major factor in determining the propagation properties of the beam profile. Also, a large beam with a uniform phase front will maintain its shape over a considerable propagation distance. The shaped beam cross-section can be arbitrary including rectangular, line, triangular, circular, and hexagonal or doughnut shape. In Figure 2.7, a commercially available beam shaper [80] is shown with the profiles of intensity distribution at the focal point. Beam-shaping technology can be applied to both coherent and incoherent beams. The predominant methods for beam shaping include use of homogenizers [83], amplitude masks, refractive elements [84], and diffractive optical elements (DOE) [85]. More advanced optical devices include spatial light modulators, deformable mirrors, and tunable acoustic gradient index (TAG) lenses which allow real-time modulation of beam's intensity profile on the surface [86–88].

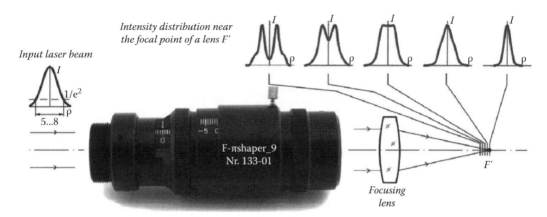

FIGURE 2.7 Commercial beam shaper. (From Laskin, A.V. and Laskin, V., Variable beam shaping with using the same field mapping refractive beam shaper. *LASE SPIE Photonic West Conference*, San Francisco, CA, Paper # 8236–13, pp. 21–26, 2012.)

In general, the laser beam shaping techniques can be categorized as follows:

1. *Aperture shaping*: In this case, the beam is expanded and an aperture is used to select a suitable flat section of the beam. Though sometimes useful, this technique experiences significant loss at the output plane.

2. *Field mapping method*: This technique transforms the input field into a beam with desired field. Here, a single-mode Gaussian input beam can be transformed in such a way that rays are bent in a plane and are uniformly distributed in the output plane. The ray bending defines a wave front that can be associated with an optical phase element. Field mappers in general are loss-free and this beam-shaping approach is usually suitable for well-defined single-mode lasers.

3. *Beam integrators or beam homogenizers*: In this case, the input beam is divided into a number of beamlets by lenslet array (or small mirrors for reflective beam integrator) and superimposed in the output plane using a primary lens. Here, the output pattern is the sum of diffraction patterns determined by the individual lenslet apertures. This method is well suited for multimode lasers, e.g., excimer lasers and multi-kilowatt CO_2 lasers, and incur no loss of optical power in its use.

4. *Diffractive diffuser beam shapers*: These can be viewed as beam integrators with lenslet aperture size approaching zero with the phase varying from one lenslet to other lenslet. They are designed to diffract the incident beam into the desired irradiance distribution with a built-in speckle (or random) pattern. These types of beam shapers are more tolerant to alignment errors in comparison to the conventional field mappers.

2.7 ENERGY TRANSFER: PHYSICS OF ILLUMINATION AND SURFACE INTERACTIONS

One of the major advantages of the laser as a tool for material-processing applications is its ability to accurately deposit a controlled amount of energy on a predefined region of the material surface. This controlled delivery of laser energy is achieved by selecting the right processing parameters to modify specific surface properties of the material under consideration. Often, the desired result could be in the form of change of surface chemistry, surface texture, crystal structure, or surface morphology without affecting the rest of the bulk material [79,89]. For example, by altering the texture or by imprinting desired microfeatures with a laser beam, the surface appearance and absorption properties can be controlled precisely. In another instance, simply heat treating the component surface with a defocused laser beam, its wear, friction, or other tribological characteristics could be modified [90]. Therefore, in order to develop a good understanding of these surface modification processes, it is necessary to realize the underlying physical phenomena that take place when the laser beam interacts with the targeted material surface.

Laser-induced changes in material chemistry, crystal structure, and surface morphology have been well documented in the literature [69,90–94]. When a laser beam irradiates the material's surface, many complex phenomena take place, depending on the laser power density, wavelength, and the material characteristics [20]. The incident light energy is partially reflected, partially absorbed, and partially transmitted depending on the type of the material, e.g., metal, glass, composite, and ceramics. Of the total light energy irradiating the surface, the fraction that is absorbed by the material is of particular interest in different material-processing applications. The main reason being in this interaction the temperature distribution within the material as a result of these heat transfer processes depends on several thermo-physical properties of the sample material [95,96]. These properties include density, emissivity, thermal conductivity, specific heat, latent heat of fusion, thermal diffusivity, geometrical dimensions and the sample thickness, and the laser processing parameters such as absorbed energy, beam cross-section, or spot size. In general, the interaction begins with the absorption of the light photons in the form of electronic and vibration excitation of the atoms in material, and subsequent energy transfer to the lattice phonons converting it into heat, which then dissipates to surrounding atoms. As more and more photons are absorbed, the material temperature increases and the process triggers a knock-on effect which causes a further rise in temperature rapidly, within a very short time span, e.g., a millisecond or so in the case of welding of materials. The rate of temperature rise depends on the balance between absorption of energy and its dissipation into the material. The magnitude or the achievable temperature rise due to heating governs the different physical effects in the material such as melting, sublimation, vaporization, dissociation, plasma formation, and ablation that is responsible for material removal. Note that the absorptivity is influenced by the orientation of the material surface with respect to the beam direction and reaches a maximum value for angles near normal incidence.

Even though the basic laser–material interaction is similar, certain unique aspects among types of materials such as metals, ceramics, glasses, and plastics must be taken into consideration. In contrast to metals, ceramics and glasses absorb well at both ends of the spectrum [97]. Due to poor thermal characteristics and high melting points, ceramics are difficult to process in comparison to metals. Glasses possess poor thermal conductivity and can melt easily by absorbing only a small fraction of the incident energy at infrared wavelengths. Plastics are even better absorbers of laser energy, especially in the UV and CO_2 regions. UV laser wavelengths are highly effective in breaking certain bonds in the plastic molecule and could be used to alter a material's surface properties [94]. Furthermore, it is also possible to modify material properties under the surface for transparent materials. The following discussion addresses these ideas in more specific details.

2.7.1 Surface Reflectivity

Among material properties, the surface reflectivity is of significance during laser-matter interactions. It is defined as the fraction of the incident radiation reflected by a surface. In general, surface reflectivity is treated as a directional property that is a function of the reflected and the incident direction, and the incident wavelength of light. For metals, its color depends on the behavior of reflectivity that varies as a function of wavelength [79,95]. Some metal reflectivity values are relatively high, >90% at normal incidence in the visible spectrum of light. Further, at longer infrared wavelengths, all metals exhibit higher reflectivity. Metals with higher values of electrical conductivity possess higher values of surface reflectivity. Therefore, for such metals, the shorter wavelengths will give better processing results [98]. For other materials such as glass, plastics, and ceramics, the reflectivity values also differ. Due to the changes in reflectivity, some materials will appear shiny or dull in color. For diffuse surfaces, e.g., matte finish, white painted, the reflectivity is uniform, meaning that the radiation is reflected in all angles somewhat equally. Such surfaces are called Lambertian [71]. In specular surfaces, like glass or polished metal, the reflectivity will be almost zero, which means the reflected radiation will follow a different path from incident radiation for all cases other than radiation normal to the surface. In general, most real objects contain some mixture of diffuse and specular reflective properties.

The absorptivity or the amount of light absorbed by a metallic surface is proportional to $1 - R$, where R is the reflectivity [20,95]. If absorptivity decreases, only a small fraction of the light incident on the surface is absorbed and available for heating the work piece. In addition, the importance of R value is more pronounced at longer wavelengths. For, shorter wavelengths, the absorptivity $1 - R$ is much higher compared to longer infrared wavelengths since more photons are absorbed by the greater number of bound electrons. At higher temperatures, the reflectivity falls and absorptivity increases for some metallic materials such as copper, aluminum, and steels. Since reflectivity is a surface phenomenon, the presence of films, e.g., oxide and oil, on the surface will have a large impact on absorption of radiation. The variation of the surface reflectivity with angle of incidence has a significant effect on the absorptivity of a material since at certain angles the electrons may be limited from vibrating [20]. Surface roughness also has a considerable effect on absorption due to multiple reflections in the undulations.

2.7.2 ABSORPTION AND DEPTH OF HEAT PENETRATION

Inside the material surface, the absorbed intensity of the light starts decaying as a function of depth at a rate determined by the material's absorption coefficient β. The absorption coefficient is a material property that describes the fractional amount of optical energy that is deposited per unit depth in the material. Here, β is a function of wavelength and temperature, but for a constant value of β, intensity I decays exponentially with depth z according to the Beer–Lambert law [99],

$$I(z) = I_o e^{-\beta z} \tag{2.5}$$

where
  $I(z)$ is the light intensity at a depth z
  I_o is the intensity just below the surface neglecting the reflection losses

As stated, the energy deposition is maximum at the surface and decreases slowly underneath the surface with most energy deposition takes place within a depth $\delta = 1/\beta$, which is the depth at which the intensity of light transmitted, drops to $1/e$ (~68%) of its original value at the surface. This depth δ is called the optical absorption depth. Figure 2.8 illustrates the optical absorption depth as a function of wavelength, for a number of metals and semiconductor materials [79]. In general, the absorption depths are much shorter in comparison to the material dimensions. Also, there is an obvious advantage of selecting wavelengths with shorter absorption depths which can enable modification of local surface properties without significantly altering the bulk of the material. Particularly for wavelengths in the UV region, the absorption depth is as short as 10 nm. The forgoing discussion considers only linear optical phenomena and may not be applicable to all materials. For example, materials such as glasses exhibit strong nonlinearities in their index of refraction and often lead to self-focusing effects [100]. Therefore for CW or nanosecond pulse lasers, the absorption is assumed to be due to single photon interactions. However, with pico and femtosecond pulses, the interaction between the laser and materials occurs in a nonlinear multiphoton absorption processes because of the availability of extremely high-peak intensities and shorter time frames. Such strong field interactions can decrease the absorption depth [101]. The absorption process is so fast that the beam essentially plucks the atoms from the surface of the material without interacting or disturbing the

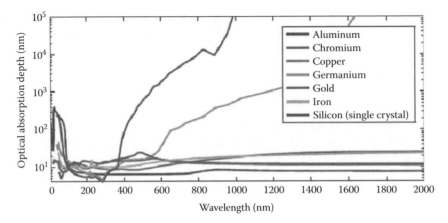

FIGURE 2.8 (**See color insert.**) Optical absorption depth versus wavelength for different materials. (From Brown, M.S. and Arnold, C.B., Fundamentals of laser-material interaction and application to multiscale surface modification, in *Laser Precision Microfabrication*, Sugioka, K. et al. (Eds.). Springer-Verlag, Berlin, Germany, Chapter 4, 2010.)

neighboring atoms. These ultrashort pulse lasers are ideal for applications involving micro-scale material removal processes, e.g., microtexturing or micromachining, mainly because they produce an exceptionally clean post-machined surface without any heat affected zone (HAZ) present into the bulk of the material.

The material response to the absorption of laser energy depends on the flow of heat inside the material. Also, the temporal and spatial evolution of the temperature field is governed by the heat equation, which includes thermal properties of the material such as the mass density ρ, specific heat at constant pressure c_p, and the thermal conductivity K, of the substrate. The heating rate is inversely proportional to the specific heat per unit volume ρc_p and the critical factor K for heat flow is $K/\rho c_p$, which has the dimensions of cm^2/s, and the diffusion coefficient is known as thermal diffusivity. The thermal diffusivity of a material determines how fast the material will receive and conduct heat energy. Lower value of thermal diffusivity represents a restriction on heat penetration into the material. Similarly, higher thermal conductivity allows larger penetration of the fusion front with no thermal cracking in the material. The depth of penetration of the heat inside the material at time τ (the thermal diffusion length) is given by the equation [99]

$$D = \left(4K\tau\right)^{1/2} \tag{2.6}$$

where
   D is the depth of penetration characterizes the distance over which temperature changes propagate in time τ
   K is the thermal diffusivity

If the depth of penetration is longer than the optical absorption length, temperature rise at the laser spot is limited. On the other hand, if it is shorter than the optical absorption length, there will be a rapid rise in temperature, which would lead to melting or possible vaporization of the material. For practical applications, the thermal time constant plays a crucial role and it is defined as the length of time required for heat to penetrate to a specific depth. Thus for a metal plate of thickness x, the thermal time constant would be $x^2/4K$. For processes such as heat treating, cladding, scribing, cutting, welding, drilling, marking, or micromachining, the laser parameters, e.g., wavelength, power, pulse repetition rate, and pulse-width, must be selected carefully.

For appropriate selection of process parameters, a judicious threshold can be determined by equating the optical absorption length to the depth of penetration. Since the optical absorption lengths for most metals are of the same order, the difference in time scales results from the differences in the depth of penetration (Figure 2.8). Therefore, metals such as stainless steel can be machined by much longer pulse-widths and silicon requires shorter pulse durations to cause ablation largely due to the variation in their respective conductivity properties [102]. Also, for processing of materials, there must be an adequate amount of energy contained within each pulse to heat up a useful volume of the material. In addition, for a given pulse energy, as the pulse duration shortens, the heat increasingly confines near the focal region and eventually leads to heating, melting, ablation, and vaporization.

2.7.3 Light Interaction with Metals

Metals are characterized by an abundance of the conduction band electrons. They also conduct heat remarkably well and since their vibration frequencies are not quantified because of their being bound to the atoms, they absorb most of the wavelengths in visible and infrared spectrum of light. Metals are nontransparent, their optical absorption depths are very small, and they behave opaquely to the incident radiation [103]. Upon irradiation the energy will be partially absorbed and partially reflected from a solid metal target. The absorption mechanism in metals is largely dominated by the free electrons in an electron gas through inverse bremsstrahlung effect [20,104], the process in which photons are absorbed by electrons as opposed to their emission by excited electrons. The absorption is followed by a fast energy relaxation within the electronic subsystem. Energy transfer to the lattice takes place by diffusion, collisions, and the electron–phonon coupling. The time required for the excited electronic states to transfer energy to phonons and subsequent lattice heating depends on the specific material characteristics and the mechanism of light absorption. For most metals, this thermalization time is of the order of 10^{-12}–10^{-10} s [79] due to significant variation in the absorption mechanism. Therefore, interaction with pulse-widths in the nanosecond range, the temperature equilibrium is established during the initial phases of the pulse. For the remainder of the pulse, the energy in excess of the diffused heat melts and evaporates the material. The surface profile is determined by the vapor pressure and, the pressure and flow in the melt pool [105], which yields a rough ablated surface. For pulses in the picosecond ranges, these effects are somewhat diminished but the ablation is still quite adequately described by the melting and evaporation processes.

For the ultrashort pulse-widths, in the femtosecond range, although the heat diffusion into the lattice is not completely absent, its effect is diminished considerably. In this pulse-width range, the electron–phonon coupling effects become significant [106]. Since the relaxation time is too long, no meaningful temperature can be assigned, particularly to the lattice subsystem. Thus, the ablation process in the most part is described by the energy absorbed by the free electrons, which is then transferred to the lattice by collisions, breaking the molecular and atomic bonds. Finally, the material is expelled by the electromagnetic repulsive forces. Since little melt is present and most of the material is removed directly, the surface closely conforms to the beam shape. If the intensity profile of the beam is adequately shaped, a cleaner cut results. If the energy is sufficiently high, the ablated plasma has enough energy to escape reducing the debris and dross. Energy is transferred to the lattice by the phonons [107]. Some energy is transferred to the lattice by its collisions with the plasma, which continues even after the pulse disappears. This yields a small amount of melt, resulting in some roughness of the surface and formation of debris. However, an adequately shaped plasma plume, which results from filamentation, reduces these effects. At still higher energies, the melt is created during the initial phases of the pulse. Further energy transfer from the pulse and subsequent energy transfer by the collisions with plasma increase the melt and evaporation conditions, rendering the processes similar to the long pulse ablation. Therefore, cleaner fabricated features can be obtained with the energy in an intermediate, filamentation range.

Sometimes the microstructural defects and material impurities can slow down the thermalization time. For thicker metals with low thermal diffusivity, the thermal time constant x becomes too large for heat penetration. However, for metals with higher thermal diffusivity values, e.g., copper, silver, and aluminum, this value is in the range of a few milliseconds and heat can penetrate through during laser interaction. The thermal time constant thus provides a convenient means for estimating the time that a laser beam must interact with a surface in order to enable full penetration of heat through the material. Note that in all materials including metals, semiconductors, and insulators the effects produced by the electromagnetic irradiation are caused primarily by the electric field component of the electromagnetic radiation rather than the magnetic component, which is negligible [108]. Metals are also able reflectors because the electrons oscillate at the frequency of the incoming light but do not convert much of this energy into heat. At the metal surface, the re-radiated light interferes with itself to form a reflected beam similar to a dielectric surface, but with more intensity [74]. When the absorbed laser energy is directly transformed into heat, the process is termed as photothermal or pyrolytic [79]. Laser processing of metallic materials with nanosecond or wider pulse duration is typically characterized by photothermal mechanism. On the other hand, when the excitation energies are sufficient enough to break the bonds structure, e.g., irradiation of polymers with ultrashort pulse laser, the process is typically a nonthermal photochemical or photolytic. The femtosecond laser enables both photochemical processing of metals and semiconductors as well as it can cause thermal modification of material through lattice phonons, in which case the mechanism is referred to as photophysical [79].

2.7.4 Light Interaction with Semiconductors

A material is characterized as metals, semiconductors, and insulators according to their band structure. Semiconductors are materials which include a band of electronic states, filled at absolute zero, separated from another band, empty at absolute zero, by a region of forbidden energies called a band gap [109]. In semiconductors, the absorption of laser light predominantly occurs through resonant excitations such as transitions of valence band electrons to the conduction band or within bands. The free carriers, composed of an electron and a hole pair, are created in the inter-band transitions. The holes tend to be mobile and behave like electrons but with the opposite charge. As a consequence, a single photon collision is enough to ionize a molecule or atom by lifting an electron from the bound state to the conduction band or even removing it. Thus during the initial phases of a pulse, significant number of electrons are generated in the conduction band for the material to behave as metal. Further absorption of photons in semiconductors then leads to the creation of electron–hole pairs with some kinetic energy. These energized carriers heat up colliding among each other and at certain critical temperature transfer their energy to the lattice via recombination and photon generation, resulting in lattice heating and melting [110]. Semiconductors are produced from silicon or gallium arsenide type of materials by doping small quantities of impurities to increase the number of free electrons (n-type) or to reduce the number of bound electrons (p-type). The optical behavior of the semiconductor is influenced by the band gap. At longer wavelengths when photon energies are less than the band gap, the material will not interact with the conduction electrons and behave like a dielectric. However, shorter wavelengths are sufficiently energetic to interact and, as a consequence, free carriers contribute to the high reflectance of many semiconductors in the visible range. Near the band gap, the absorptivity of the material depends strongly on its temperature and level of doping.

2.7.5 Light Interaction with Dielectrics

Light interaction process in the dielectrics differs fundamentally from the metals and semiconductors [111,112]. The dielectrics are characterized by an almost complete absence of the free electrons and a large electronic band gap. Therefore, photo-ionization requires multi-photon absorption, which is a low-probability phenomenon. Consequently, ablation cannot rely on direct bond breaking by the photons. Instead, the energy is absorbed by the few free electrons available by the inverse brems-strahlung phenomenon, which is diffused into the electron subsystem by collisions. When some of the free electrons acquire energy equal to or greater than the band gap, the lattice subsystem can ionize by impact ionization. The seed electrons are sometimes provided by added impurities; otherwise the initial multi-photon ionization remains their only source. After sufficient conduction band electrons are generated, the energy is transferred by similar to metallic materials and semiconductors. For longer pulses, the photo-energy absorbed by the conduction band electrons is transferred to the lattice in the form of heat, which can melt the material, and the dielectric acquires somewhat metal-like properties. Energy transfer by the collisions still continues. For shorter pulses, no thermal equilibrium establishes for the duration of the pulse. After the photo-energy transfer ceases, the energy diffusion among the conduction band electrons and to the lattice by impact continues. During this period, thermal equilibrium develops and generates heat in the lattice. Next, during the ablation phase, the energy is transferred to the lattice by the phonons. Due to the diffusion of energy within the electronic subsystem and to the lattice, ablation process continues after the pulse no longer transfers the energy to the free electrons. In fact, most of the material removal in the dielectrics occurs during the post-pulse period [113]. For shorter pulses, the energized phonons transfer energy to the lattice. This can cause damage to the material directly and by energizing the electrons in the bound states, which in turn are removed by impact ionization. Dielectrics damage is lesser for the short pulses in comparison to the long pulses. However, damage has been observed in the dielectrics for all pulse-widths. Most of the damage to the dielectrics for short pulses results from high density of the free electrons. Avalanche effect can generate the free electrons at an enormous rate. As their number density reaches the threshold, which is at about 10^{21} per unit volume, they cause melt and other damage, which result from the impact ionization and the phonon energy. Melt results from the fact that at such high density of free electrons, the dielectric behaves as metal. At these energies, ablation takes place mostly by melting and vaporization. Therefore, mainly two mechanisms play a role in the ablation of dielectrics by short pulse lasers: Coulomb explosion or gentle ablation and thermal vaporization or strong ablation [114]. Both mechanisms coexist to some extent in all short pulse ablation of the dielectrics. During the initial stages, Coulomb explosion dominates and, after reaching the critical density, strong ablation dominates. Thus, a cleaner fabricated surface results at the fluences in an intermediate range, which is high enough for an effective ablation but sufficiently low so that the critical density is not reached and strong ablation does not occur [115].

2.8 MELTING AND VAPORIZATION

Melting and vaporization play a predominant role in most of the material-processing applications such as welding, cutting, and drilling with high-intensity laser beams [20]. When a high-intensity laser beam strikes the target material, generally four phases of the material—solid, liquid, vapor, and plasma—appear simultaneously since the intense electromagnetic field of the laser beam drives the temperature up past the material's critical point [116]. The absorbed heat energy penetrates into the solid surface through thermal conduction and quickly reaches its melting temperature. At this juncture, a liquid interface sets in and propagates further into the material. If the incident laser irradiance is too high, the surface begins to vaporize immediately before a significant depth of molten material or a hole is produced. For example, effective surface melting and metal welding depends on propagation of a fusion front through the sample during the laser interaction period, while avoiding the vaporization. However to achieve this, a careful balance between penetration depth and surface vaporization is necessary. Often laser energy can be controlled by means of pulse-stretching or beam-shaping techniques for desirable effects for monitoring specific processes. For a practical application involving deep penetration, welding of thick metals can be achieved through the keyhole formation using multi-kilowatt lasers, while avoiding the vapor–plasma effects. Although surface vaporization is not suitable for welding applications, it is advantageous for applications such as cutting and hole drilling [74].

In laser drilling of metals, melt ejection can be a significant mechanism. Here, the material is removed via vaporization and physical ejection of the molten material. The melt ejection process in single-pulse laser drilling can be described as follows [117]. Immediately after the start of the laser pulse, the substrate surface starts to heat up. After a time, the surface temperature reaches the melting point and a molten layer is formed. Vaporization produces a recoil pressure that acts on the molten layer, removing molten material from the region ahead of the ablation front. The recoil pressure initially overcomes the threshold required for melt ejection sometime shortly after the start of the pulse [118]. At the initiation of melt ejection, the thermal gradients in the material and the vaporization rate at the surface, and hence the recoil pressure, will be at their highest optimum and the molten layer will be at its thinnest. A large number of small droplets are therefore ejected at a relatively high velocity. As the pulse progresses further, the molten layer thickens, resulting in somewhat larger size droplets. The recoil pressure decreases, reducing the ejection velocity, but continues to remove material from the ablation front. The molten material moves along the hole walls relatively smoothly, breaking up into discrete droplets under the influence of surface tension on exiting the hole. Surface tension effects may also increase the angle of ejection as the molten material follows the curve of the hole entrance on exiting it. The extent to which melt ejection occurs depends both on material properties and the laser parameters [119]. For example, a threshold beam intensity must be reached before melt ejection occurs. As the pulse intensity increases, more pressure is generated and thereby causes an increase in the flow velocity of the molten material. Note that the amount of material removed increases with further increase in pulse intensity. Low-intensity pulses tend to generate relatively thick molten layers, whereas high-intensity pulses eject melt more efficiently. The relationship between melt ejection and pulse intensity can be exploited to control melt ejection by temporally shaping the laser beam or utilizing pulse trains [81,120].

Commonly used lasers in hole-drilling process operate with pulse durations in a few tens of microseconds to approximately 10 ms regime. This time scale is sufficient to interact with a surface to heat through to melting and vaporization temperature. However, the vaporization occurs at a continually retreating surface, with the laser energy delivered to the bottom of the developing hole. Therefore, the time t_B to reach the vaporization temperature is given by [99]

$$t_B = \left(\frac{\pi}{4}\right)\left(\frac{K\rho c}{F^2}\right)(T_B - T_0)^2 \tag{2.7}$$

where
   K, ρ, and c are the thermal conductivity, density, and heat capacity of the material, respectively
   T_B is the vaporization temperature
   T_0 is the ambient temperature
   F is the absorbed irradiance

Also, note that smaller focal spot requires higher irradiance to cause vaporization. Under most of the conditions, the heat of vaporization of the work piece is the dominant variable in determining how much material is removed. For high values of laser irradiance, only a small amount of incident energy is lost via thermal conduction out of the irradiated spot and the depth D of the hole may be given by [99]

$$D = \frac{Ft_p}{L\rho A} \tag{2.8}$$

where
   t_p is the pulse duration
   L is the specific latent heat of vaporization
   A is the focal area

It cleary suggests that the amount of material vaporized depends on the exact conditions of the laser irradiation.

2.9 PLASMA GENERATION AND SHIELDING EFFECTS

The laser–material interaction initiated by the absorption of photons raises the surface temperature of the material. At higher fluence, the laser beam suffers the self-focusing Kerr effect, transitioning through a gaseous medium, which increases its intensity further causing gas breakdown. Additional photoelectric and thermoelectric processes occurring near the surface also increase the electron density [105,121]. Incident radiation may also contribute atoms, thermal electrons, and ions near the surface by vaporizing surface asperities. If the density of the plasma in the vicinity of the focal point exceeds certain critical value, it begins to distort the beam profile and consequently deforms the surface. All these processes serve to increase electron, ion, and neutral atom density (plasma) near the surface and absorb more laser radiation complimenting to the heat input. The plasma formation is essentially a gas composed of a mixture of free positive and negative charges, metal atoms, ions, electrons, and components of the ambient gas atmosphere. Shock waves generated by rapid material ablation also heat up the surrounding medium to transform it in ionized plasma and propagate along the laser beam axis. Almost two orders of magnitude increase in plasma density induced by a CO_2 laser beams have been observed in various gases, attributed to shock waves. Plasma also acts as an energy source supplying heat to the material beyond the pulse duration, which can result in melting of the material. As the plasma continues to absorb more energy, it begins to move into the surrounding region via radiation. Note that, although the plasma generation may distort the laser beam, high-energy particles in plasma, which participate in the material ablation, have a positive effect on achieving better processing results. Thus, material processing with high-intensity pulse lasers involves complex phenomena with both favorable and unfavorable impact on the surface quality. The tendency to form plasma is determined by the atomic/molecular weight of the gas, its thermal conductivity, and its ionization energy. Molecular gases also consume dissociation energy before becoming ionized. Low molecular weight increases the recombination rate between metal ions and electrons of the plasma, so that the plasma becomes suppressed or less dense. High thermal conductivity gas shroud increases heat transfer from the plasma to the surroundings and thereby decreases the temperature of the plasma and hence its density.

Occurrence of unacceptable plasma effects is a common concern, particularly for high-power laser-welding applications [75,122,123]. During conduction welding, the heat is transferred from the surface into the material by thermal conduction. This type of welding is typically performed with relatively low-power (<500 W) lasers, in which case the power density is not sufficient to create a keyhole (see Figure 2.9). The weld is usually wide and is differentiated with a shallow profile. High-power laser welding, on the other hand, is characterized by keyhole welding or deep penetration welding. In this case, the laser beam energy is transferred deep into the material via a cavity filled with metal vapor. The process begins with laser power densities $>10^5$ W/mm^2 which melts and partly vaporizes the metal and the vapor pressure displaces the molten metal so that a cavity is formed into the bulk, known as keyhole. Inside the keyhole, the absorption rate of laser radiation increases due to multiple reflections from the wall surface.

When the beam hits the wall surface of the keyhole, a part of the beam energy is absorbed by the material. During welding, the temperature in the keyhole becomes so intense that plasma formation occurs. The ionized metal vapor temperature exceeds far more than 10,000 K. During keyhole formation, the plasma acts as a shield in the energy transfer process and reduces the power efficiency. Further, the evaporation pressure in the keyhole causes the plasma to expand above the keyhole and form a cloud. This causes the incoming laser beam to defocus and scatter around by the plasma cloud, resulting in increased spot size with subsequent changes in the focal position and effective power density. The extended plasma cloud also causes the penetration depth of the weld to decrease. The quality of the weld formation depends on the existence of the appropriate column of plasma. In general, the weld profile is a nail-head shape due to the energy absorption in the cloud. If plasma formation is extensive, the welding process may even be interrupted. The keyhole welding enables very deep and narrow welds into the metallic material.

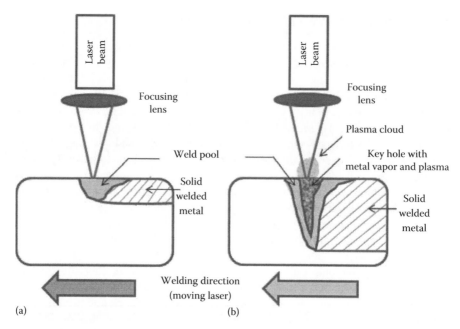

FIGURE 2.9   (a) Conduction welding and (b) Key-hole welding.

Shielding gases play a critical role in protecting the metal oxidation, oxide inclusions, and porosity that can affect weld quality [123,124]. The density of shielding gas, e.g., is important for proper protection of the weld area. Low-density gases such as He do not displace air as easily and escape quickly from the weld zone. Therefore, either a high-speed or a high-flow rate gas supply is necessary for protecting weld zone. Argon has high density and can be a better alternative to replace air more effectively. A mixture of helium/argon provides combined advantage, higher ionization potential of helium, and higher density of Argon, to achieve better weld zone protection in CO_2 laser welding. The shielding effect is different when CO_2 and Nd:YAG lasers are used mainly due to their wavelength difference and mode of operation. Often specially designed nozzles are used to achieve good shielding against oxidation without disturbing melt flow around the keyhole. For example, the welding gas can be directed in the beam path through a coaxial nozzle. A helium gas jet is useful when plasma formation becomes a serious issue, particularly in welding of thicker metal plates with CO_2 laser. Also, a stream of gas with high velocity can displace the plasma cloud from above the keyhole to provide better shielding of the weld pool. When a jet nozzle is used, the focusing optics, i.e., mirrors or lenses, must be protected from fume deposition and spatter. In this case, a high-velocity gas stream across the laser beam can keep away the fumes and spatter. The welding gas increases the welding speed as well as improves the mechanical properties of the joint.

2.10   LASER MATERIAL REMOVAL AND SURFACE MODIFICATION

In general, laser material removal using a CW or long-pulsed laser, the key mechanism involves melting of solid piece of the material, followed by an expulsion of the molten metal by an assist gas jet. The removal of material thus accomplished by laser–material interaction leads to an assortment of useful processes such as hole drilling, cutting, texturing, grooving, marking, micromachining, and scribing. These processes involve a diverse range of lasers operating in different temporal modes, from CW to pulsed mode with pulse duration ranging from millisecond (ms), nanosecond (ns) to ultrashort picoseconds (ps) and femtosecond (fs). Therefore, it is necessary to develop a good understanding of the benefits and limitations of the physical processes involved during the laser–material interaction. As discussed earlier, upon impingement of the beam on the surface of the material, the laser energy is first absorbed by the free electrons. The absorbed energy then propagates through the electron subsystem and is ultimately transferred to the lattice. There are three distinct characteristic time scales that play an important role during the laser–matter interaction [125]: electron cooling time, which is roughly of the order of a picosecond, the lattice heating time, and the laser pulse duration (pulse-width). Since the electron cooling time is very much shorter than the lattice heating time, different conditions occur when the incident laser pulse duration varies from a millisecond or infinite (CW) to nanosecond to femtosecond regime.

For longer pulse durations (e.g., >1 ms) or infinite as in the case of CW laser interaction, the typical time scale is much larger than the electron–lattice energy coupling time. Here, the removal mechanism is melting of the material and its subsequent ejection. Laser cutting of steel, nonferrous metals, and nonmetals are common process examples in this time scale category. However, in the industrial setting, an oxygen-assist gas is frequently used for laser cutting of steel materials [126]. The assist gas provides exothermic energy and helps increase the throughput cutting speed. Also, in oxygen-assisted laser cutting formation of the striation marks on the cut surface are mainly due to the internal hydrodynamic instabilities of the cutting process coupled with cyclic oxidation phenomena.

Interactions involving laser pulses with nanosecond duration, the by the energy absorbed free electron has enough time to be transferred to the lattice, during this period the electron and the lattice can reach thermal equilibrium, and the main energy loss is the heat conduction by the solid target material. Therefore, the material initially melts and then quickly vaporizes from the liquid state at higher laser fluences. In this case, the primary material removal mechanism is ablation but substantial melting is still present if a target material is metal. Typical process examples in this regime are laser drilling, grooving, marking, microfabrication and scribing of materials. In addition, the observed HAZ is much smaller [127], normally in the order of a few microns, as compared to the CW laser processing. In this time regime, the machining processes are performed with Q-switched, diode pumped solid state (DPSS) Nd:YAG lasers and its harmonic wavelengths.

For material removal using femtoseconds laser pulses [128], the interaction mechanism changes dramatically. In this time scale, the laser pulse duration is much shorter than the electron cooling time. Therefore, when the laser beam interacts with the material, excited free electrons instantly heat up to high temperatures by the absorption of photons and through collisions with ions. Subsequent electron–photon interaction allows transfer of the absorbed energy to their positive lattice ions, within about a picosecond time frame for most materials. If the energy intensity is sufficiently strong, which is often in the case of ultrashort pulsed lasers, they break off immediately without having adequate time to transfer their energy to their neighboring lattice ions and a direct solid–vapor transition occurs. In short, for melt-free ablation, ultrashort pulse duration and high enough pulse energy is necessary. As stated earlier, since there is not enough time for energy transfer to the lattice, the material can be ablated very precisely with little or no collateral damage. The material removal process is also independent of laser wavelength and the HAZs are minimal, resulting into clean, burr-free machining. The typical lasers used are Ti:sapphire laser or femtosecond fiber laser with fundamental wavelength at 780 nm and a fluence ranging from 0.1 to 10 J/cm^2. High-power densities and minimal HAZ are major benefits associated with ultrashort-pulsed laser processing. Since direct solid–vapor transition is achieved, all processes including cutting, drilling, texturing, grooving, marking, scribing, structuring, and micromachining facilitate this time scale and precise material removal is feasible without changes in the bulk material properties. For these reasons, ultrashort pulse laser processing of materials is becoming a preferred technique for various applications such as drilling, cutting, microfabrication with complex features, and surface texturing/modification. Processes are well established to carve features significantly smaller than the diffraction limit, i.e., sub-micron features inside transparent bulk materials [114], nanoscale features on the surface of thin metal films, and scribing of photovoltaic film interconnects [129] on polymeric and metallic substrates, by taking an advantage of the threshold effect. Almost all materials, e.g., metals, alloys, polymers, glass, and ceramics, can be machined with high-quality results [130].

2.10.1 Surface Profile Formation in Laser Ablated Materials

With ultrashort pulse laser interaction, the ablation proceeds with photo-ionization creating an imbalance of the electromagnetic forces in the lattice, which essentially expels the material [131]. Higher intensity beam transfers sufficient energy during the pulse for material removal. The consequent plasma emanating from the material also acquires high energy by collision transfer, preventing it from re-settling and thus resulting in the reduction of debris. Therefore, cleaner fabricated features with less debris can be expected. However, this requires a well-shaped beam as the fabricated surface profile conforms closely to the incident intensity distribution. At sufficiently high intensities, the laser radiation propagating through a gaseous medium such as air, He, and Ne suffers nonlinear, multiphoton effects such as self-focusing of the beam. This further enhances the intensity, resulting in gaseous breakdown. The scattering effects due to the generated plasma compensate for self-focusing to some extent, but they also distort the beam profile with consequent distortion of the surface features. Intensity is still sufficiently high to ablate and generate material plasma. Addition of plasma increases the distortion to the beam profile. Plasma also acts as an energy source supplying heat to the material beyond the pulse duration, which may result in melting, offsetting the benefits further. Thus, material processing with high-power ultrashort pulse lasers involves complex phenomena with both favorable and unfavorable impact on the machined surface quality.

2.11 CONCLUDING REMARKS

In this chapter, we have presented the nature of light starting with a brief history on the wave–particle duality of light, electromagnetic spectrum, sources of natural and artificial light, and an introductory review of various types of lasers along with brief description on their basic properties in the context of the core subject of this book. Subsequent sections cover specific topics related to important laser beam characteristics such as linewidth, mode and beam quality, significance of M^2 factor and the shaping of laser beams and discussion on the energy transfer during the interactions between laser and material surface, physics of illumination, light absorption, melting, vaporization, and plasma effects. The chapter concludes with a brief summary on laser material removal introducing relevant surface modification processes based on CW, nanosecond and femtosecond time scale interactions, and their effect on surface profile formation of laser ablated materials. Over all, we have attempted to introduce many versatile capabilities of laser processing to modify the surface properties of materials and to enhance their performance in several applications. The amazing "LIGHT" has thus transformed our world while opening many avenues to process emerging new materials and also to find unique solutions to novel applications.

REFERENCES

1. Rouse Ball, W. W. 1908. *A Short Account of the History of Mathematics*, 4th edn. MacMillan, New York. http://www-history.mcs.st-andrews.ac.uk/Mathematicians/Huygens
2. Robinson, A. 2006. *Last Man Who Knew Everything*. Pearson Education, Inc., New York.
3. Bernard, C. I. 1980. *The Newtonian Revolution,* Cambridge University Press, Cambridge, New York.
4. Carnal, O. and Mlynek, J. 1991. Young's double-slit experiment with atoms: A simple atom interferometer, *Phys. Rev. Lett.* 66: 2689–2692.
5. Darling, D., 1999. The worlds of David Darling, *Encyclopedia of Science*.
6. Clerk, M. J. 1865. A dynamical theory of the electromagnetic field, *Roy. Soc. Trans.* CLV, 459.
7. Kuhn, T. S. 1978. *Black Body Theory and Quantum Discontinuity (1894–1912),* 1st edn. Oxford University Press, Inc., Oxford, U.K.
8. Born, M. and Wolf, E. 1999. *Principles of Optics*, 7th edn. Cambridge University Press, Cambridge, U.K.
9. Caes, C. J. 2001. *How Do We Know the Speed of Light?,* 1st edn. The Rosen Publishing Group, Inc. New York.
10. Michelson, A. A. and Morley, E. W. 1887. On the relative motion of the earth and the luminiferous ether, *Am. J. Sci.* 34: 333–345.

11. Vandergriff, L. J. 2008. Nature and properties of light (Module 1.1), in *Fundamentals of Photonics*, Roychoudhuri, C. (Ed.). Tutorial Texts TT79, SPIE Press. Also see: Meyers, R. A. (Ed.). 1990. *Encyclopedia of Modern Physics*. Academic Press, Inc., New York.

12. Valeur, B. and Berberan-Santos, M. N. A. 2011. Brief history of fluorescence and phosphorescence before the emergence of quantum theory, *J. Chem. Educ.* 88(6): 731–738.

13. Jelley, J. V. 1958. *Cerenkov Radiation and Its Applications*. Pergamon Press, London, U.K.

14. Landau, L. D., Liftshitz, E. M., and Pitaevskii, L. P. 1984. *Electrodynamics of Continuous Media*. Pergamon Press, New York.

15. O'shea, D. C., Callen, R. W., and Rhodes, W. T. 1978. *Introduction to Lasers and Their Applications*. Addison-Wesley Publishing, Boston, MA.

16. Goldstein, D. 2003. *Polarized Light*, 2nd edn. Marcel Dekker, New York.

17. Hecht, E. 2002. *Optics*, 4th edn. Addison Wesley, Reading, MA.

18. Zvezdin, A. K. and Kotov, V. A. (Eds.). 1997. *Modern Magneto Optics and Magneto Optical Materials*. IOP, Bristol, U.K.

19. Silfvast, W. T. 2004. *Laser Fundamentals*, 2nd edn. Cambridge University Press, Cambridge, U.K.

20. Steen, W. M. and Mazumder, J. 2010. *Laser Material Processing*, 4th edn. Springer-Verlag, London, U.K.

21. Wille, H. et al. 2002. Teramobile: A mobile femtosecond-terawatt laser and detection system. *Eur. Phys. J. Appl. Phys.* 20: 183–190.

22. Hecht, J. 2010. Laser weapon lasts for six hours, *New Scientist*, December 8.

23. Maiman, T. H. 1960. Stimulated optical radiation in ruby. *Nature* 187(4736): 493–494.

24. Geusic, J. E., Marcos, H. M., and Van Uitert, L. G. 1964. Laser oscillations in nd-doped yttrium aluminum, yttrium gallium and gadolinium garnets. *Appl. Phys. Lett.* 4(10): 182.

25. Yariv, A. 1989. *Quantum Electronics*, 3rd edn. Wiley, New York, pp. 208–211.

26. Koechner, W. 1965. *Solid-State Laser Engineering*. Springer-Verlag, New York.

27. Giesen, A. et al. 1994. Scalable concept for diode-pumped high-power solid-state lasers. *Appl. Phys. B* 58: 365–372.

28. Stewen, C. et al. 2000. A 1-kW CW thin disc laser. *IEEE J. Sel. Top. Quantum Electron.* 6(4): 650–657.

29. Ricaud, S. et al. 2011. Yb:CaGdAlO$_4$ thin-disk laser. *Opt. Lett.* 36(21): 4134.

30. Giesen, A. and Speiser, J. 2007. Fifteen years of work on thin-disk lasers: Results and scaling laws. *IEEE J. Sel. Top. Quantum Electron.* 13(3): 598.

31. Popov, S. 2009. Fiber laser overview and medical applications, in *Tunable Laser Applications*, 2nd edn., Duarte, F. J. (Ed.). CRC Press, New York.

32. Ueda, K. 1999. Scaling physics of disk-type fiber lasers for kW output. *Lasers Electro-Opt. Soc.* 2: 788–789.

33. Overton, G. 2009. IPG-photonics-offers-worlds-first-10-kw-single-mode-production-laser, *Laser Focus World*, June.

34. Gapontsev, V. P. et al. 2007. Large mode area fiber for low loss transmission and amplification of single mode lasers. U.S. Patent # 7,283,714B1.

35. Hecht J., 2009. Photonic frontiers-fiber-lasers: Fiber-lasers-ramp-up-the-power, *Laser Focus World*, December.

36. Patil, D. S. (Ed-in-chief), 2012. Semiconductor Laser Diode Technology and Applications, InTech Publications.

37. Chow, W. W. and Koch, S. W. 1999. *Semiconductor-Laser Fundamentals*. Springer-Verlag, Berlin, Germany.

38. Saleh, B. E. A. and Teich, M. C. 1991. *Fundamentals of Photonics*. John Wiley & Sons, Inc., New York.

39. Madey, J. 1974. Stimulated emission of radiation in periodically deflected electron beam. U.S. Patent # 3822410.

40. Emma, P. et al. 2010. First lasing and operation of an angstrom-wavelength free-electron laser. *Nat. Photonics* 4: 641–647.

41. Duarte, F. J. (Ed.). 1995. *Tunable Lasers Handbook*. Academic, New York, Chapter. 9.

42. Feldhaus et al. 2005. X-ray free-electron lasers. *J. Phys. B At. Mol. Opt. Phys.* 38(2005): S799–S819.

43. Freund, H. P. and Antonsen, T.M. Jr. 1996. *Principles of Free electron Lasers*, 2nd edn. Chapman & Hall, London, U.K.

44. Schafer, F. P. 1990. *Dye Lasers*. Springer-Verlag, Berlin, Germany.

45. Fork, R. L., Greene, B. I., and Shank, C. V. 1981. Generation of optical pulses shorter than 0.1 psec by colliding pulse mode locking. *Appl. Phys. Lett.* 38: 671–672.

46. Hänsch, T. W. 1972. Repetitively pulsed tunable dye laser for high resolution spectroscopy. *Appl. Opt.* 11: 895–898.

47. Akerman, M. A. 1990. Dye laser isotope separation, in *Dye Laser Principles*, Duarte, F. J. and Hillman, L. W. (Eds.). Academic, New York, Chapter. 9.

48. Demtröder, W. 2003. *Laser Spectroscopy*, 3rd edn. Springer-Verlag, Berlin, Germany.
49. Costela, A. et al. 2009. Medical applications of dye lasers, in *Tunable Laser Applications*, Duarte, F. J. (Ed.), 2nd edn. CRC Press, New York, Chapter. 8.
50. Javan, A., Bennett, W. R., and Herriott, D. R. 1961. Population inversion and continuous optical maser oscillation in a gas discharge containing a He-Ne mixture. *Phys. Rev. Lett.* 63: 106–110.
51. White, A. D. 2011. Recollections of the first continuous visible laser. *Opt. Photonics News* 22: 34–39.
52. Willet, C. S. 1974. *An Introduction to Gas Lasers*. Pergamon Press, Oxford, U.K.
53. Svelto, O. 1998. *Principles of Lasers*, Plenum Press, New York.
54. Hecht, J. 1986. *The Laser Guidebook*, McGraw-Hill Book Company, New York.
55. Pawley, J. B. 2006. Laser sources for confocal microscopy, in *Handbook of Biological Confocal Microscopy*, Springer, Berlin, Germany, Chapter 5, p. 103.
56. Basting, D. and Marowsky, G. (Eds.). 2005. *Excimer Laser Technology*. Springer-Verlag, Berlin, Germany.
57. Jain, K. et al. 1982. Ultrafast deep-UV lithography with excimer lasers. *IEEE Electron. Device Lett.* 3(3), 53–55.
58. Lin, B. J. 2009. *Optical Lithography*. SPIE Press, Bellingham, WA.
59. Blum, S. E., Srinivasan, R. and Wynne, J. J. Far ultraviolet surgical and dental procedures. U.S. Patent #4784135. Issued 1988-10-15.
60. Linsker, R. et al. 1984. Far-ultraviolet laser ablation of atherosclerotic lesions. *Lasers Surg. Med.* 4(1): 201–206.
61. Patel, C. K. N. 1964. Continuous-wave laser action on vibrational-rotational transitions of CO_2. *Phys. Rev.* 136(5A): A1187–A1193.
62. Beaulieu, A. J. 1970. Transversely excited atmospheric pressure CO_2 lasers. *Appl. Phys. Lett.* 16(12): 504–505.
63. Pearson, P. R. and Lamberton, H. M. 1972. Atmospheric pressure CO_2 lasers giving high output energy per unit volume. *IEEE J Quantum Electron.* 8(2): 145–149.
64. Palmer, J. 2011. Laser is produced by a living cell. *BBC News*, http://www.bbc.co.uk/news/science-environment-13725719
65. Gather, M. C. and Yun, S. H. 2011. Single-cell biological lasers. *Nat. Photonics* 5: 406–410.
66. Pikas, D. J. et al. 2002. Nonlinear saturation and lasing characteristics of green fluorescent protein. *J. Phys. Chem. B* 106: 4831–4837.
67. Hecht, J. 2011. Organic lasers: Living fluorescent-protein-doped cell lases. *Laser Focus World*, August 2011.
68. Siegman, E. 1993. Defining, measuring and optimizing laser beam quality. *Proc. SPIE* 1868: 2.
69. Siegman, A. E. 1986. *Lasers*, University Science Books, Sausalito, CA.
70. Clark R. S. 1987. *The Photonics Design and Application Handbook,* Lauren Publishing, Pittsfield, MA.
71. Miller, F. P., Vandome, A. F., and McBrewster, J. 2010. *Lambertian Reflectance,* VDM Verlag Dr: Mueller e.K., Saarbrücken, Germany, 2010.
72. Bolton, R., 2002. Give your laser beam a checkup, *Photon Spectra Mag.* 36: 107.
73. ISO Standard 11146. 2005. Lasers and laser-related equipment-test methods for laser beam widths, divergence angles and beam propagation ratios.
74. Charschan, S. S. (Ed.). 1986. *Lasers in Industry*. Laser Institute of America, Orlando, FL.
75. Kannatey-Asibu, E. Jr. 2009. *Principles of Lasers Materials Processing*. John Wiley & Sons, Hoboken, NJ.
76. Ashkin, A. 1970. Acceleration and trapping of particles by radiation pressure. *Phys. Rev. Lett.* 24: 156–159.
77. Neuman, K. C. and Block, S. M. 2004. Optical trapping. *Rev. Sci. Instrum.* 75(9): 2787–2809.
78. Dickey, F. M. and Holswade, S. C. 2000. *Laser Beam Shaping: Theory and Techniques*. Marcel Dekker, Inc., New York.
79. Brown, M. S. and Arnold, C. B. 2010. Fundamentals of laser-material interaction and application to multiscale surface modification, in *Laser Precision Microfabrication*, Sugioka, K. et al. (Eds.). Springer-Verlag, Berlin, Germany, Chapter. 4.
80. Laskin, A. V. and Laskin, V. 2012. Variable beam shaping with using the same field mapping refractive beam shaper. *LASE SPIE Photonic West Conference*, San Francisco, CA, Paper # 8236–13, pp. 21–26.
81. Dickey, F. M. and Holswade, S. C. 1996. Gaussian laser beam profile shaping. *Opt. Eng.* 35: 3285–3295.
82. Weichman, L. S. et al. 2000. Beam shaping element for compact fiber injection system. *Proc. SPIE* 3929: 176.
83. Zhang, S. Y. et al. 2003. Ultrashort laser pulse beam shaping. *Appl. Opt.* 42(4): 715.
84. Romero, L. A. and Dickey, F. M. 1996. Lossless laser beam shaping. *J. Opt. Soc. Am. A* 13: 751–760.
85. Gale, M. T. 1997. Replication techniques for diffractive optical elements. *Microelectron. Eng.* 34: 321–339.

86. Sanner, N. et al. 2005. Programmable focal spot shaping of amplified femtosecond laser pulses. *Opt. Lett.* 30(12): 1479.

87. Nemoto, K. et al. 1997. Optimum control of the laser beam intensity profile with a deformable mirror. *Appl. Opt.* 36(30): 7689.

88. McLeod, E. et al. 2006. Multiscale Bessel beams generated by a tunable acoustic gradient index of refraction lens. *Opt. Lett.* 31(21): 3155.

89. Gregson, V. 1984. *Laser Materials Processing.* Holland Publishing Co., Amsterdam, the Netherlands.

90. Dahotre, N. 1998. *Lasers in Surface Engineering, Surface Engineering Series*, Vol. 1. ASM International, Materials Park, OH.

91. von Allmen, M. and Blatter, A. 1995. *Laser-Beam Interactions with Materials: Physical Principles and Applications.* Springer Series in Materials Science, Springer, Berlin, Germany.

92. Ion, J. C. 2005. *Laser Processing of Engineering Materials: Principles, Procedure and Industrial Applications.* Elsevier Butterworth-Heinemann, Oxford, U.K.

93. Majumdar, J. D. and Manna, I. 2003. Laser processing of materials. *Sadhana* 28: 495–562.

94. Elliott, D. J. 1995. *Ultraviolet Laser Technology and Applications.* Academic Press, Inc., London, U.K.

95. Bass, M. (Ed.). 2009. *Handbook of Optics,* Vol. 1, McGraw-Hill, New York, ISBN: 0071498893.

96. MiGliore, L. 1976. *Lasers Materials Processing.* Marcel Dekker, Inc., New York.

97. Ready, J. F. and Farson, D. F. (Eds.). 2001. *LIA Handbook of Laser Materials Processing.* Laser Institute of America, Magnolia Publishing, Inc., Destin, FL.

98. Ready, J. F. 1982. Material processing—An overview. *Proc. IEEE* 70: 533.

99. Charschan, S. S. (Ed.). 1993. *Guide to Laser Material Processing.* Laser Institute of America, Orlando, FL.

100. Slusher, R. E. and Eggleton, B. J. 2004. *Nonlinear Photonic Crystals*, 1st edn. Springer-Verlag, Berlin, Germany.

101. Ghofraniha, N. et al. 2007. Shocks in nonlocal media. *Phys. Rev. Lett.* 99(4): 043903.

102. Staudt, W., Borneis, S., and Pippert, K. D. 1998. TFT Annealing with Excimer Laser - Technology and Market Outlook. *Phys. Status Solidi A Appl. Res.* 166(2): 743.

103. Melinger, J. S. et al. 1998. Pulsed laser induced single event upset and change collection measurements as a function of optical penetration depth. *J. Appl. Phys.* 84: 690–703.

104. Haug, E. and Nakel, W. 2004. *The Elementary Process of Bremsstrahlung,* Vol. 73, World Scientific Lecture Notes in Physics. World Scientific, River Edge, NJ.

105. Vatsya, S. R., Li, C., and Nikumb, S. K. 2005. Surface profile of material ablated with high-power lasers in ambient air medium. *J. Appl. Phys.* 97: 034912.

106. Vatsya, S. R. and Virk, K. S. 2003. Solution of two-temperature thermal diffusion model of laser–metal interactions. *J. Laser Appl.* 15(4): 273.

107. Vatsya, S. R. and Nikumb, S. K. 2003. Modeling of fluid dynamical processes during pulsed-laser texturing of material surfaces. *Phys. Rev. B* 68: 035410-1–035410-5.

108. von Allmen, M. F. 1983. Coupling of laser radiation to metals and semiconductors. in *Physical Processes in Laser–Materials Interactions*, Bertolloti, M. (Ed.). Plenum, New York, p. 50.

109. Laser Matter Interaction: Up-to-date research reference, July 8, 2003 Selected Topics in Laser-Matter Interaction. The Stefan University Press Series on Frontiers in Interdisciplinary Physics.

110. Vatsya, S. R. and Nikumb, S. K. 2002. Modeling of laser-induced avalanche in dielectrics. *J. Appl. Phys.* 91: 344–351.

111. Vatsya, S. R. et al. 2009. High power femtosecond laser machining of metals in ambient medium, in *High-Power and Femtosecond Lasers—Properties, Materials and Applications*, Barret, P.-H. and Palmer, M. (Eds.). Nova Science Publishers, Inc., New York, Chapter. 6.

112. Apostolova, T. and Hahn, Y. 2000. Modeling of laser-induced breakdown in dielectrics with subpicosecond pulses. *J. Appl. Phys.* 88: 1024–1034.

113. Stuart, B. C. et al. 1996. Nanosecond-to-femtosecond laser-induced breakdown in dielectrics. *Phys. Rev. B* 53: 1749–1761.

114. Gamaly, E. G. 2006. Laser-matter interaction in the bulk of a transparent solid: Confined microexplosion and void formation. *Phys. Rev. B* 73: 214101.

115. Jiang, L. and Tsai, H. L. 2003. Femtosecond lasers ablation: Challenges and opportunities. *Aerospace Eng.* 51–53.

116. Hyungson, K. 2010. On vaporization in laser material interaction. *J. Appl. Phys.* 107: 104908-1.

117. Voisey, K. T. et al. 2003. Melt ejection during laser drilling of metals. *Mater. Sci. Eng.* 356: 414–424.

118. Ki, H., Mohanty, P. S., and Mazumder, J. 2001. Modelling of high-density laser-material interaction using fast level set method. *J. Phys. D Appl. Phys.* 34: 364–372.

119. Modest, M. F. 1996. Three-dimensional, transient model for laser machining of ablating/decomposing materials. *Int. J. Heat Mass Transfer* 39: 221–234.

120. Dixon, R. D. and Lewis, G. K. 1983. The influence of plasma during laser welding. *Proc. ICALEO LIA* 38: 44–49.

121. Zhang, X. 2003. Modeling and application of plasma charge current in deep penetration laser welding. *J. Appl. Phys.* 93: 8842.

122. Li, L. et al. 1996. Plasma charge sensor for in-process, non-contact monitoring of the laser welding process. *Meas. Sci. Technol.* 7: 615.

123. Berkmanns, J. and Mark, F. *Laser Welding*. Linde AG, Linde Gas Division, Pullach, Germany.

124. Dawes, C. 1992. *Laser Welding: A Practical Guide*. Woodhead Publishing, Cambridge, U.K.

125. Yao, Y. L. et al. 2005. Time scale effects in laser material removal: A review. *Int. J. Adv. Manuf. Technol.* 26: 598–608.

126. Chichkov, B. N. 1996. Femtosecond, picosecond, and nanosecond laser ablation of solids. *Appl. Phys. A Solids Surf.* 63: 109–115

127. Bordatchev, E., Nikumb, S., and Hsu, W. 2004. Laser micromachining of the miniature functional mechanisms. *Proc. SPIE* 5578: 579–588.

128. Shirk, M. D. and Molian, P. A. 1998. A review of ultrashort pulsed laser ablation of materials. *J. Laser Appl.* 10(1): 18–28.

129. Rekow, M. et al. 2011. CdTe P1, P2, P3 Scribe optimization using a pulse programmable industrial fiber laser. *Proceedings of the 26th EU PVSEC*, Hamburg, Germany.

130. Nikumb, S. K. et al. 2005. Precision glass machining, drilling and profile cutting by short pulse lasers. *Thin Solid Films* 477(2005): 216–221.

131. Perrière, J., Millon, E., and Fogarassy, E. 2006. *Recent Advances in Laser Processing of Materials*. Elsevier Publications, Amsterdam, the Netherlands.

SUGGESTED FURTHER READING

Arata, Y. 1986. *Plasma, Electrons and Laser Beam Technology—Development and Use in Material Processing*. American Society for Metals, Carnes Publication Services, Inc., Novelty, OH.

Barret, P.-H. and Palmer, M. (Eds.). 2009. *High-Power and Femtosecond Lasers—Properties, Materials and Applications*. Nova Science Publishers, Inc., New York.

Bertolotti, M. 1999. *The History of the Laser*. Institute of Physics, Bristol, PA.

Born, M. and Wolf, E. 1980. *Principles of Optics*. Pergamon Press, New York.

Bracewell, R. N. 1978. *Fourier Transform and its Applications*. McGraw-Hill, New York.

Csele, M. 2004. *Fundamentals of Light Sources and Lasers*. Wiley, Hoboken, NJ.

Davis, J. R. 2003. *Handbook of Materials for Medical Devices*. ASM International, Novelty, OH.

Duarte, F. J. (Ed.). 1991. *High Power Dye Lasers*. Springer-Verlag, Berlin, Germany.

Duley, W. W. 1976. *CO₂ Lasers: Effects and Applications*. Academic Press, Inc., New York.

Duley, W. W. 1983. *Laser Processing and Analysis of Materials*. Plenum, New York.

Hecht, J. and Teresi, D. 1982. *Laser: Supertool of the 1980s*. Ticknor & Fields, New York.

Hecht, E. and Zajac, A. 1979. *Optics*. Addison-Wesley Publishing, Boston, MA.

Koechner, W. 1992. *Solid-State Laser Engineering*, 3rd edn. Springer-Verlag, Berlin, Germany.

Madou, M. J. 2002. *Fundamentals of Microfabrication—The Science of Miniaturization*, 2nd edn. CRC Press LLC, New York.

Meyers, R. A. (Ed.). 1990. *Encyclopedia of Modern Physics*. Academic Press, New York.

Ready, J. F. 1978. *Industrial Applications of Lasers*. Academic Press, Inc., New York.

Rhodes, C. K. (Ed.). 1979. *Excimer Lasers*. Springer-Verlag, Berlin, Germany. Collection of review articles.

Schaaf, P. 2010. *Laser Processing of Materials: Fundamentals, Applications and Developments*. Springer-Verlag, Berlin, Germany.

Silfvast, W. T. 1996. *Laser Fundamentals*. Cambridge University Press, Cambridge, U.K.

Steen, W. M. 2003. *Laser Material Processing*. Springer-Verlag, Berlin, Germany.

Sugioka, K., Meunier, M., and Pique, A. 2010. *Laser Precision Microfabrication*. Springer-Verlag, Berlin, Germany.

Svelto, O. 1998. *Principles of Lasers*, 4th edn. (trans. David Hanna), Springer-Verlag, Berlin, Germany.

Taylor, N. 2000. *LASER: The Inventor, the Nobel Laureate, and the Thirty-Year Patent War*. Simon & Schuster, New York.

Wilson, J. and Hawkes, J. F. B. 1987. *Lasers: Principles and Applications*. Prentice Hall International Series in Optoelectronics, Prentice Hall, Upper Saddle River, NJ.

Part II

Photoresponsive Materials

This part focuses on how light-responsive materials can generate very small mechanical displacements and forces. Chapter 3 provides a look into some materials that are able to convert photons into directional movement by reversibly changing their shape or volume when irradiated with light. Roman Boulatov and Timothy Kucharski from the University of Illinois-Urbana introduce the notion of photochemical actuation, where the aspect ratio of an irradiated material changes during the photoisomerization of its consistent molecular components (e.g., photoactive monomers or dopants). These photochemical actuating materials are ensembles of molecular chromophores that exist in at least two isomers of significantly different molecular shapes. Dimensional changes at the macroscale are a cumulative effect of a large number of independent molecular-scale structural rearrangements induced by the absorption of photons. The authors present a chemist's view of the operation of molecular photoactuation and discuss the challenges in optimizing their performance, with an emphasis on how molecular thermal and photochemical processes dictate the fundamental limits of a material's operating capabilities. The remainder of the chapter presents the molecular basis for photochemical actuation and then discusses the general design criteria for creating photoactuating structures.

In Chapter 4, Christopher Barrett and his colleagues in North America and Japan describe how some photo reversible molecules can use light to induce shape changes in the host material. This class of reversible light-switchable molecules includes photoresponsive molecules that photodimerize, such as coumarins and anthracenes; those that allow intra molecular photo-induced bond formation, such as fulgides, spiro-pyrans, and diarylethenes; and those that exhibit photoisomerization, such as stilbenes, crowded alkenes, and azobenzene. The application of reversible changes in shape that can be induced with various material systems incorporating azobenzene to effect significant reversible mechanical actuation, is presented. This photomechanical effect can be defined as the reversible change in shape inducible in some molecules by the adsorption of light, which results in a significant macroscopic mechanical deformation of the host material.

A new perspective on photostrictive actuators is provided by Professor Kenji Uchino in Chapter 5. Photostrictive materials exhibit light-induced strains that arise from a superposition of the "bulk" photovoltaic effect, i.e., generation of large voltage from the irradiation of light, and the converse-piezoelectric effect, i.e., expansion or contraction under the voltage applied. (Pb,La)(Zr,Ti)O_3 (PLZT) ceramics doped with WO_3 exhibit large photostriction under uniform illumination of near-ultraviolet light. Using a bimorph configuration, a photo-driven relay and a microwalking device have been demonstrated. However, for the fabrication of these devices, higher response speed must be achieved. The chapter first reviews the theoretical background for the photostrictive effect and then describes enhanced performance through composition modification and sample preparation technique (thickness and surface characteristics of the sample). Its potential future applications are also briefly described.

In Chapter 6, Balaji Panchapakesan from the University of Louisville reviews the current state of the art in carbon nanotube–based photomechanical actuators and their composites. The chapter presents key points in the photomechanical responses of carbon nanotubes and shows how orientation, alignment, and anisotropic optical properties of carbon nanotubes influence photomechanical actuation. The current state of the field of carbon nanotube polymer photomechanical actuators is discussed. The photomechanical responses of nanotube–polymer systems depend on the structure of the actuator, nanotube alignment, entanglement, and the presence of pre-strains in the sample. While the mechanism of photomechanical actuation is not fully understood, Panchapakesan describes some possible mechanisms that contribute to the overall photomechanical responses of carbon nanotubes and their polymer composites. It is expected that the interplay between elastic, electrostatic, polaronic, and thermal interactions gives rise to the overall photomechanical responses of carbon nanotubes. The author also provides some insights into how the photomechanical responses of carbon nanotubes and their polymers may be coupled to the band structure of carbon nanotubes. Finally, the applications of photomechanical actuation of carbon nanotube in micro-electro-mechanical systems (MEMS) and nanotechnology are reviewed.

Professor Atushi Suzuki from Yokohama National University describes the volume phase transition in hydrogels induced by irradiation of visible light in Chapter 7. The gels consist of covalently crosslinked copolymer network of thermoresponsive polymer and chromophore. Without light illumination, the gel volume changes sharply (continuously or discontinuously) at a characteristic temperature when the temperature is varied. Upon illumination of visible light, the characteristic temperature is lowered. Even in the case of a gel that shows a continuous change, the volume change becomes discontinuous above a certain threshold of light intensity. In this case, a discontinuous volume transition is observed when the light intensity is gradually changed at a fixed temperature of appropriate value. The phase transition can be understood in terms of temperature increment at the immediate vicinity of polymer chains due to the heating by light. The results are qualitatively described by the equation of state of gels. By making use of the hysteresis phenomenon in response to the changes in external stimuli, the phase of gel can be successfully activated by visible light (switched on) and deactivated (switched off) by altering the local environment conditions. This switching phenomenon is of technological importance in developing various optical applications such as actuators, sensors, and display units.

3 Fundamentals of Molecular Photoactuation

Timothy J. Kucharski and Roman Boulatov

CONTENTS

3.1 INTRODUCTION

Photoactuating materials convert photon energy into directional translation at meso- to macroscales by reversibly changing their shape or volume when irradiated with light. Because they can be powered wirelessly and controlled remotely and are insensitive to magnetic noise and potentially compatible with rapid (>100 Hz) work cycles, particularly when compared to materials that require diffusive mass transport, photoactuating materials are of growing interest in chemistry, materials science, engineering, and soft matter physics. Here, we define photophysical actuation as that due to heating, pyroelectric and piezoelectric effects (Mizutani et al. 2008) resulting from photon absorption, radiation forces (Juodkazis et al. 2000), or their combination. This chapter is devoted to photochemical actuation, in which aspect ratio(s) of an irradiated material changes due to photoisomerization of its constituent molecular components (e.g., photoactive monomers or dopants), and we will use the terms *photoactuation* and *photochemical actuation* interchangeably. Photochemical actuating materials are ensembles of molecular chromophores that exist in at least two isomers of significantly different molecular shape (Figure 3.1). Dimensional change at the macroscale is a cumulative effect of a large number of often independent molecular-scale structural rearrangements, each induced by the absorption of a photon. Photochemical actuation provides perhaps the most direct link between chemical reactivity at the molecular level and useful properties of the bulk material. Consequently, the classical chemical concepts, including molecular design and reaction dynamics, may be particularly impactful in the development of new photoactuating materials and understanding the behavior of the existing ones at the molecular level.

Photoactuation has been demonstrated in molecular crystals, liquid-crystalline films, and polymers. Amorphous photoactuating polymers typically contain ≤10% of photoisomerizable groups,

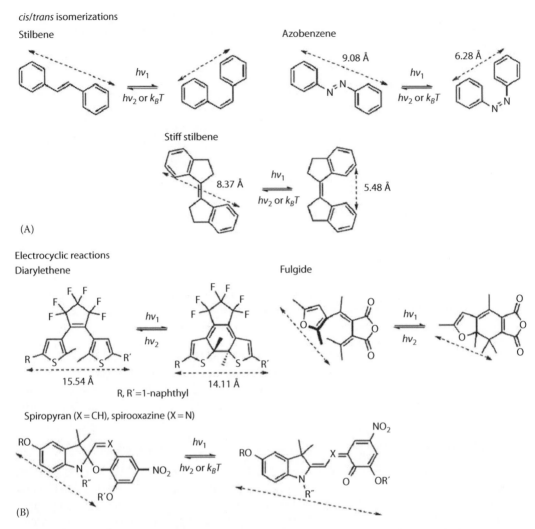

cis/trans isomerizations

Stilbene

Azobenzene

Stiff stilbene

(A)

Electrocyclic reactions

Diarylethene

Fulgide

R, R′=1-naphthyl

Spiropyran (X = CH), spirooxazine (X = N)

(B)

FIGURE 3.1 Representative molecules of different classes of photochromic systems suitable for photoactuation, categorized as (A) *cis/trans* isomerizations or (B) electrocyclic reactions. Selected interatomic distances that change between the two forms and could serve as points of attachment for extended structures are indicated by the dashed arrows. Distances for azobenzene and stiff stilbene were calculated at the B3LYP/6-31G(d) level of density functional theory in the gas phase; distances for the diarylethene shown were determined crystalographically (From Morimoto, M. and Irie, M., *J. Am. Chem. Soc.*, 132, 14172, 2010.). h, k_B, and T are Plank's constant, Boltzmann's constant, and absolute temperature, respectively.

often in cross-links, and can change their dimensions by a few percent upon irradiation (Ikeda et al. 2007). This dimensional change is called photoinduced strain (photostrain), although unconstrained samples are strain-free before and after irradiation. Larger photostrains appear accessible in molecular crystals (Kobatake et al. 2007) and cocrystals (Morimoto and Irie 2010) of photoisomerizable diarylethenes (Figure 3.1B), which undergo single-crystal–single-crystal transitions when irradiated with either UV or visible light, resulting in reversible macroscopic shape changes. Liquid-crystal (LC) films doped with photoisomerizable dyes (Eelkema et al. 2006) and liquid-crystal elastomers (LCEs) with photoisomerizable mesogens (Corbett 2009, Ikeda et al. 2007, Warner and Terentjev 2003a,b) have been shown to move micro- and macroscopic objects when irradiated with light. Several aspects of photoactuating materials, including their syntheses and proof-of-concept demonstrations of potential applications have been previously reviewed (Barrett et al. 2007, Ikeda et al. 2007, Irie 2008, Jiang et al. 2006, Koerner et al. 2008, Russew and Hecht 2010, Yu and Ikeda 2008) and are not repeated in this chapter.

The design of photoactuating materials presents several conceptual and technical challenges. The key operating parameters of such materials—strain, stress, energy conversion efficiency, power output, and operating frequency—cannot be adjusted independently, and their coupling is complex and little understood. Particularly, little studied is the effect of bulk stresses on the kinetics and thermodynamics of the molecular processes, both productive (e.g., photoisomerization) and parasitic (e.g., side reactions and thermal relaxations of strained molecules without changes in bulk aspect ratios). In this chapter, we will briefly describe a chemist's view of the operation of photoactuating materials and the challenges in optimizing their performance, focusing on how molecular thermal and photochemical processes dictate the fundamental limits of a material's operating capabilities.

The remainder of the chapter is organized as follows: First, the molecular basis for photochemical actuation will be examined. Second, general design criteria for photoactuating materials will be discussed, followed by an examination of the outstanding questions that remain, in both cases noting the existing molecular approaches for understanding these issues. Then, we will briefly introduce the principles of the chemomechanical formalism and will subsequently discuss the force-dependent kinetics of photoactuating monomers within this conceptual framework. Finally, we will examine how the kinetics and thermodynamics of actuating reactions impose the fundamental limits on the achievable performance characteristics of photoactuating polymers and will conclude with our speculations regarding future prospects for advancing the field.

3.2 MOLECULAR BASIS FOR PHOTOCHEMICAL ACTUATION

3.2.1 EFFECTING CHANGES IN MOLECULAR SHAPE

In photochemical actuation, irradiation-induced changes in the aspect ratio(s) of bulk material are the end result of a photochemical process, isomerization of a chromophore upon photon absorption. Chromophores that exist as multiple isomers are typically photochromic, i.e., the isomers have different absorption spectra. Photochromism has long been studied, and because many physical properties of photochromic isomers are different, such molecules are used in a variety of applications (Bouas-Laurent and Dürr 2001, Kobatake and Irie 2003, Tamai and Miyasaka 2000). The most important properties of photochromes for photoactuating applications are differences in (1) molecular dimensions, (2) absorption spectra of the two isomers, and (3) high quantum efficiency of photoisomerization with minimal side reactions. Of the numerous photochromic molecules known, those that have been considered for photoactuating applications include (Figure 3.1) stilbenes (Waldeck 1991), azobenzenes (Barrett et al. 2007, Yu and Ikeda 2008), stiff stilbenes (Huang and Boulatov 2010), diarylethenes (Irie 2000, Morimoto and Irie 2005), fulgides (Yokoyama 2000, 2001), and spiropyrans/spirooxazines (Minkin 2004). All these chromophores have at least one internuclear distance that differs significantly in the two isomers.

Upon photon absorption, a molecule is excited to a new electronic state. In a photochrome, the stable nuclear configuration of the ground electronic state differs significantly from that of the excited states. Consequently, upon excitation the molecule undergoes rapid structural rearrangement that brings it to a conical intersection (seam), connecting the ground and excited energy surfaces, followed by further evolution of the structure to an energy that is minimum at the ground state. (Domcke et al. 2004, Levine and Martínez 2007). The whole process is extremely rapid, occurring on the ps timescale (Tamai and Miyasaka 2000). Even in sterically congested systems, ps-timescale C=C photoisomerizations enable directed rotation at MHz frequencies (Klok et al. 2008).

Azobenzene is the most popular photochrome for use in photoactuating materials and the mechanism of its photoisomerization has been studied extensively. In the gas phase (Bandara et al. 2010,

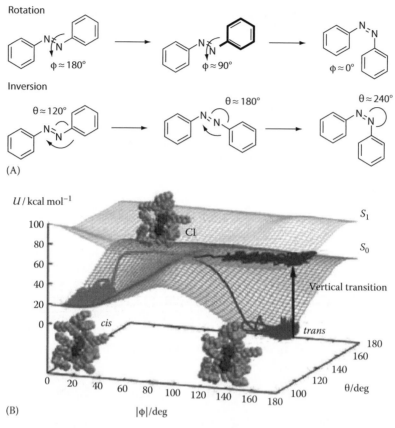

(A)

(B)

FIGURE 3.2   **(See color insert.)** (A) $E \rightarrow Z$ isomerization by rotation and inversion in azobenzene, showing the changes in the C–N=N–C dihedral angle (φ) and the C–N=N angle (θ). (B) Molecular dynamics simulation in n-hexane (light gray molecules) of $S_0 \rightarrow S_1$ photoisomerization of *trans*-azobenzene to *cis*-azobenzene (dark gray, blue, and red molecule), proceeding through the conical intersection (CI) connecting the first excited singlet (S_1) and ground state singlet (S_0) potential energy surfaces; U is internal energy. (Reproduced from Tiberio, G. et al. *Chem. Phys. Chem.*, 11, 1018, 2010.)

Tiberio et al. 2010) light absorption in the singlet ground state (S_0) of the E isomer can excite the molecule to either the first excited singlet state ($S_0 \rightarrow S_1$, n–π^* transition, ~450 nm light) or the second excited singlet state ($S_0 \rightarrow S_2$, π–π^* transition, ~320 nm light). The S_1 state generated directly from the S_0 state has a lifetime of 2.6 ps, during which the molecule rotates around the N–C bond (Figure 3.2A). The S_2 state rapidly relaxes to the S_1 state, which then isomerizes by a concerted-inversion mechanism with a lifetime of only 500 fs. In condensed phases, the steric restrictions of the environment (Creatini et al. 2008, Tiberio et al. 2010) may result in a mixed rotation–inversion mechanism of $E \rightarrow Z$ photoisomerization (Figure 3.2B).

Typically, isomerization of C=C or N=N (Figure 3.1A) results in larger structural differences than electrocyclic ring opening/closing (Figure 3.1B). The shorter Z isomers of photochromic molecules containing a photoisomerizable C=C or N=N are less thermodynamically stable than the longer E isomers, with free energy differences of up to ~11 kcal/mol for azobenzenes. Constraining an internuclear distance that is different in the two isomers perturbs the relative stability of the two isomers up to the point of making the Z isomer the global energy minimum (Funke and Grützmacher 1987, Norikane et al. 2008). Large structural differences between the two isomers have also been used to photochemically generate highly strained molecules not easily accessible by conventional synthetic routes (Huang and Boulatov 2010).

3.2.2 TRANSMITTING MOLECULAR SHAPE CHANGES TO LARGER LENGTH SCALES

Photoisomerization of a chromophore connected to the polymer matrix at two atoms whose separation differs significantly in the two isomers (Figure 3.1) transiently creates a strained isomer because molecular isomerization is too fast for the rest of the polymer matrix to adjust to the new local structure (arrows, Figure 3.3). This local strain slowly dissipates by the rearrangement of polymer chains. Irradiation-induced changes in the aspect ratio(s) of amorphous photoactuating polymers reflect primarily the total difference in molecular dimensions of individual chromophores projected onto a specific macroscopic axis. Consequently, the relative magnitude of accessible photostrain is limited to the relative elongation (contraction) of a monomer upon isomerization times the fraction of the material made up of the monomers. The net change is usually modest (Ikeda et al. 2007) and often occurs on the ms timescale (Irie 2008) because the macromolecular rearrangement can be inhibited by chain entanglement.

In highly ordered materials with long-range interactions, such as LCEs or single crystals, photoisomerization can induce a phase change resulting from the disruption of long-range order (Ichimura 2000), rapidly generating photostrains that may exceed 100%.

Photoinduced single-crystal–single-crystal transitions, which yield macroscopic dimensional changes, have been observed in single crystals of diarylethenes (Irie 2008, Kobatake et al. 2007). Optically thick (the irradiation is completely absorbed within a fraction of the material) single-crystal cantilevers exhibit macroscopic bending due to photoisomerization of the constituent molecules only near the transiently irradiated surface (Figure 3.4). Expansion of only those unit cells that are within the penetration depth of incident irradiation creates internal stresses in the crystal that are relieved by its reversible deformation. Large-scale photoactuation in such systems can be very rapid, with $<5\,\mu s$ response times for 8 ns pulses, even at 4.7 K (Morimoto and Irie 2010). Such rapid responses are possible because each individual unit cell expands or contracts independently of other cells (Figure 3.4, compared to the rearrangement of individual macromolecules in polymer films, Figure 3.3). In the example shown in Figure 3.4, in situ x-ray diffraction studies revealed that photoactuation was accompanied by the elongation of the crystal axis *orthogonal* to the molecular axis of the chromophore that undergoes the largest dimensional change upon isomerization (Morimoto and Irie 2010). The contraction along the *a* axis (i.e., parallel to the diarylethene axis indicated in Figure 3.1) did not cause the macroscopic response. This example shows that designing photoactuating molecular crystals is a problem of both crystal engineering (the relationship between crystal packing and molecular shape) and molecular design (the difference in molecular shapes of the two isomers).

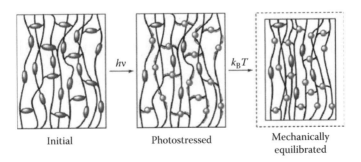

| Initial | Photostressed | Mechanically equilibrated |

FIGURE 3.3 Molecular mechanism for photoactuation in a nanoscopic region of a bulk polymer sample. In the initial state, photoisomerizable comonomers are in their elongated state (ovals, e.g., (*E*)-azobenzene). Irradiation leads to photoisomerization of the photoactive units to their shorter state (circles, e.g., (Z)-azobenzene), which creates local strains in their vicinity (arrows). Relaxation of these strains requires relatively slow, thermally activated, rearrangement of macromolecular chains, which results in contraction of the bulk material.

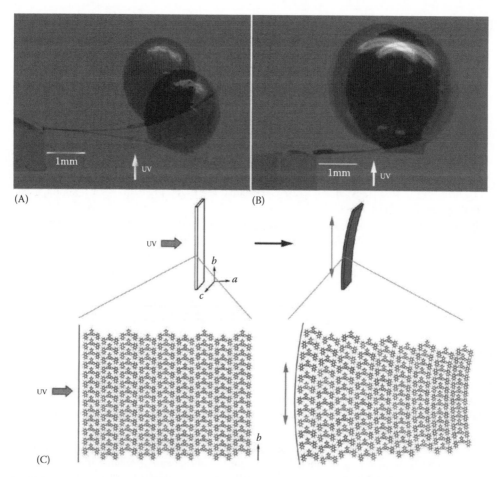

FIGURE 3.4 Superimposed photographs before and after irradiation with UV light (direction indicated) of a molecular crystal cantilever holding (A) a 2 mm lead ball, 46.77 mg (0.17 mg cantilever) or (B) a 3 mm steel ball, 110.45 mg (0.18 mg cantilever); (C) the proposed mechanism of photoactuation; molecules closer to the irradiated crystal face are predominately the ring-*closed* form of the diarylethene following photoisomerization, which have a unit cell that is *longer* along the *b* axis. (Modified From Morimoto, M. and Irie, M., *J. Am. Chem. Soc.*, 132, 14172, 2010.)

Photoisomerization has also been used to control long-range order of LC films, both monomeric and polymeric (Eelkema et al. 2006). Alignment in LC films can be directed either by photoactive mesogens dispersed throughout the film or by a layer of photoactive molecules on the contacting surface. With such photoactive "command surfaces," the geometric effects of photoisomerization can be amplified as much as $\sim 10^4$-fold due to the strong dependence of the long-range order of a LC on the properties of the contacting surface (Ichimura 2000). In films that exhibit textured surfaces, such as cholesteric LCs with helix axes aligned parallel to the surface, photoisomerization of dopants can modify the texture of the surface, which can yield directional microscopic translation. In one example, photoisomerization of the central C=C in the dopant and its subsequent thermal rearrangement inverted the handedness of molecule's helicity (Figure 3.5A). The cholesteric LC film containing $\sim 1\%$ by weight of this dopant responded to irradiation by bulk reordering, causing a rotation of the surface texture and with it the rotation of a µm-sized glass rod resting on the surface (Figure 3.5C; Eelkema et al. 2006).

LCEs are free-standing materials composed of cross-linked polymer chains that incorporate mesogens in the polymer backbones, side chains, and/or the cross-links. These materials have attracted increasing interest for photochemical actuation because the phase transition triggered by photoisomerization of mesogens can greatly amplify the difference in the molecular structure of the two isomers (Corbett 2009, Ikeda et al. 2007, Warner and Terentjev 2003a,b). The mechanism for photoactuation in LCEs based on a phototriggered nematic-isotropic transition is illustrated in Figure 3.6. Nematic mesogens (e.g., (*E*)-azobenzene, Figure 3.1) are rod-shaped molecules that induce long-range ordering by aligning their long axes (nematic directors) to maximize π–π interactions in aromatic systems and allow internal flow. Polymers containing a few mole percent of nematic mesogens can form long-range structures in which the polymer backbones are aligned preferentially along a single axis (nematic director, typically the long molecular axis of the mesogen, Figure 3.6A). Regions of the material in which molecules have the same orientation of their nematic directors compose domains, and bulk LCEs may be mono- or polydomain. Upon irradiation, the photoisomerization of a critical fraction of the mesogens to the nonmesogenic (bent) geometry diminishes the enthalpic gain of optimal π–π stacking below the entropic cost of maintaining a long-range order and triggers a nematic $\rightarrow$ isotropic phase change. In the isotropic phase, the polymer molecule adopts a spherical shape distribution. This shape change corresponds to a contraction along the direction of the (now eliminated) nematic director and an expansion perpendicular to that

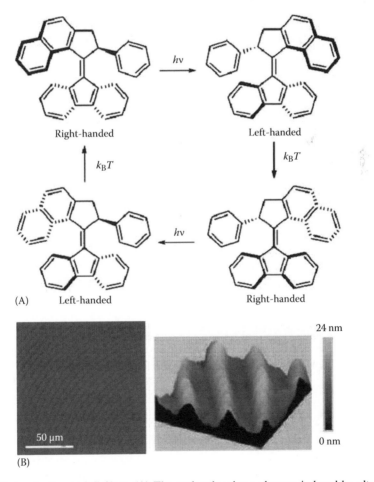

FIGURE 3.5 Photoactuation in LC films. (A) The molecular shape changes induced by alternating steps of photoisomerization (*hv*) and thermal rearrangement ($k_B T$). The handedness of the molecular helix is indicated for each conformation. (B) The surface relief of a cholesteric LC film containing 1% by weight of the molecule in (A).

(*continued*)

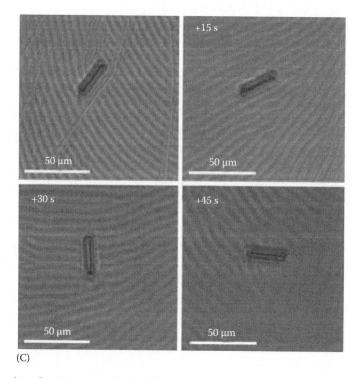

(C)

FIGURE 3.5 (continued)   Photoactuation in LC films. (C) A series of images taken at 15 s intervals show-ing the rotation of a glass rod as the surface texture of the LC film rotates under UV irradiation. The glass rod is rotated by 28°, 141°, and 226° relative to the first frame. (Modified from Eelkema, R. et al., *Nature*, 440, 163, 2006.)

direction. If the material is constrained to the initial shape by an external load, photoisomerization creates stress, whose relaxation can drive translation of the load (mechanical work). If the nonme-sogenic isomer is thermodynamically metastable, its thermal isomerization of the mesogenic form restores the long-range nematic order. Alternatively, the mesogens can be regenerated by photoi-somerization, completing an actuation cycle.

Single-domain, optically thick LCE cantilevers bend when irradiated on one face, because only a thin layer of the film undergoes a nematic–isotropic phase transition, similar to the molecular crystal cantilevers described earlier. The response can be rapid, indicating a more facile macro-molecular rearrangement in the LCEs compared to that in amorphous photoactuating polymers. In one example, a monodomain LCE cantilever containing azobenzene-based mesogens could be made to oscillate by >110° at 270 Hz (Serak et al. 2010, White et al. 2008) in the absence of an external load. For comparison, a hummingbird beats its wings at 20–80 Hz (White et al. 2008). Applications of such LCE cantilevers as actuators in autonomous microaerial devices have been suggested. However, it remains to be established whether such oscillations can support loads as the wings of a hummingbird do.

Amorphous elastomers, LCEs, molecular crystals, and monomeric LC films generate large-scale actuation by transmitting the effects of molecular shape changes to large length scales. LC films and LCEs also magnify the molecular-level shape changes to achieve strains >100% (Corbett 2009, Ichimura 2000) even in response to subtle variations in the molecular geometry or orienta-tion (White et al. 2009). Macroscopic changes in shape result from elastic mechanical equilibra-tion of local strains created by photoisomerization of individual chromophores. Maximum internal stresses that can be sustained are determined by the elastic moduli of the material, which range from <1 MPa for LC photoelastomers (Harris et al. 2005) to >11 GPa for molecular crystals and

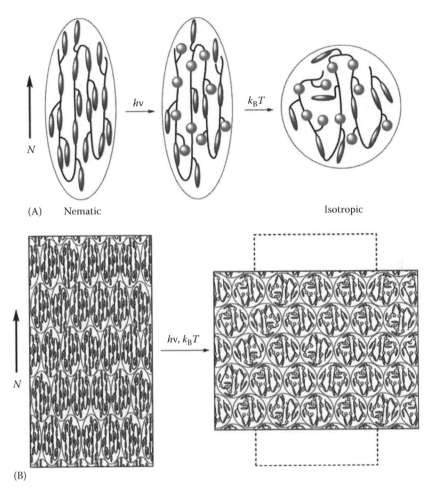

FIGURE 3.6 Molecular mechanism for photoactuation in LCEs. (A) The shape distributions of polymer chains in the nematic (with director N) and isotropic states with photoisomerizable mesogens in the backbone and in side chains, shown in their rod-like (ovals, e.g., (E)-azobenzene) and bent forms (circles, e.g., (Z)-azobenzene). Irradiation leads to photoisomerization, which disrupts the nematic ordering of the mesogens, leading to thermal rearrangement of the polymer chain to an isotropic state. (B) The isothermal nematic → isotropic phase change induced by photoisomerization leads to contractile strain parallel to the nematic director of a region of an LCE and expansion perpendicular to it (assuming sample volume is conserved). Cross-links between polymer chains are omitted for clarity.

glassy LC networks (LCNs; Morimoto and Irie 2010). Consequently, the maximum stresses that a photoactuating material can operate against vary from <2 kPa for LCEs to >2 MPa for LCNs.

3.3 GENERAL DESIGN CRITERIA FOR PHOTOACTUATING SYSTEMS

Designing a photoactuating material for specific applications may require the optimization of several coupled characteristics, such as maximum strain, stall stress, energy conversion efficiency, power output, and operating frequency. The relative importance of these parameters is application specific. For example, applications in a high-volume pump require large strains but modest cycle frequencies and stall stresses can be acceptable. Alternatively, high-pressure pumps require shape changes against large stresses. Photoactuators intended to rapidly move rigid levers (e.g., microaerial vehicles) would require the generation of large stresses and high cycle frequencies, and those that served as cantilevers themselves may also require large strains. In this section, we will examine how the thermodynamics and kinetics of the photochemical reaction limits the achievable combination of these performance characteristics with the goal of defining molecular-design criteria in terms of the engineering properties of the resultant material.

3.3.1 Importance of Anisotropy for Photoactuation

Isotropic photoactuation has been demonstrated in some materials, for example, the reversible expansion and contraction of solvent-swollen polymer gels by changing the hydrophobicity/hydrophilicity of dyes via their photoisomerization (van der Veen and Prins 1971, van der Veen et al. 1974). Because of the coupling between macroscopic response and molecular-level changes in ordered materials (e.g., LC systems), anisotropy at the micro- or nanoscale usually results in anisotropic responses at the macroscale. However, not only is anisotropic photoactuation desirable for applications such as flapping cantilevers (Serak et al. 2010) or bending films that swim (Camacho-Lopez et al. 2004), for the vast majority of photoactuating materials not dependent on mass transport, anisotropy in either the material or the irradiation conditions is essential for bulk photoactuation itself, regardless of the specific application intended.

The necessity for anisotropy is best explained with the example of photoactuating nematic LCEs (Corbett 2009, Warner 2003a,b). Nematic mesogens (Figure 3.1) are anisotropic molecules, which typically have distinct molar absorptivities along their molecular axes (Harris and Bertolucci 1989). Consequently, polarized light can be used to photoisomerize a subset of mesogens with appropriate alignment relative to the material's axes. In polydomain LCEs, the orientation of the nematic director of individual domains (typically μm-size) is random. When such a material is irradiated with unpolarized light (a uniform distribution of electric field vectors), photoactive mesogens in all of the domains absorb light, resulting in contraction of each domain along one axis and expansion along the other two. Because the domains have random orientation, the polydomain sample does not change its aspect ratio upon irradiation (Corbett 2009). In other words, the microscopic anisotropy (i.e., within each domain) is defeated by macroscopic isotropy (i.e., bulk material and unpolarized light). However, if plane-polarized light is used, only domains comprising locally aligned mesogens with an optical axis parallel with the incident light absorb light and change dimensions. Hence, bulk photoactuation is achieved with a macroscopically isotropic material.

It was established early in the development of photoactuating LCEs that the actuation induced by irradiation of monodomain samples was the same as that obtained by heating the samples above their nematic–isotropic phase transition temperatures (Corbett 2009, Warner 2003a,b). The photochemical and photothermal actuation mechanisms were differentiated by irradiating polydomain samples with polarized light (Harvey et al. 2007). If actuation were due to irradiative heating, no bulk actuation would be observed (as is indeed the case for thermally induced nematic → isotropic transitions) because heat generated by optical absorption in the properly aligned domains would rapidly diffuse to nearby domains, leading to stresses in conflicting directions. Because actuation was observed, only a subset of domains changed dimensions upon irradiation, suggesting primarily photochemical, rather than photothermal, actuation (Corbett 2009, Harvey et al. 2007). Thermal effects may be important in photochemical actuation using intense light sources (Corbett and Warner 2008a).

Photoactuation can also be achieved with unpolarized light and monodomain LCEs (Finkelmann et al. 2001) or molecular crystals (Morimoto and Irie 2010), which are macroscopically anisotropic. The enabling breakthrough in preparing monodomain LCEs was the development of a two-step cross-linking process (Küpfer and Finkelmann 1991) in which a lightly cross-linked polydomain LC polymer sample is stretched until all domains are aligned with their directors along the strain axis, which causes the sample to become transparent. The stretched polymer then undergoes further cross-linking due to the presence of slowly reacting moieties in the polydomain sample, preserving the long-range ordering once the applied stress is removed. This method was used to prepare the first example of photoactuating polymers capable of significant (20%) photostrains (Finkelmann et al. 2001). Monodomain LCEs can also be prepared by cross-linking LC polymers aligned by boundary conditions, for example, by photocross-linking in a thin cell coated with polyimide layers, which have been physically aligned by rubbing (Li et al. 2009). This method is the only one that yields highly cross-linked (glassy) LC polymers, but it has also been applied for the preparation of photoactuating LCEs (Serak 2010, White et al. 2008, 2009).

The photoactuating properties of the material are also affected by the location of the photoactive mesogens: one study indicated that LCEs containing azobenzene in cross-linkers were more effective photoactuators than LCEs with azobenzene in side chains (Kondo et al. 2010).

3.3.2 PHOTOPHYSICAL CONSIDERATIONS FOR PHOTOACTUATING MATERIALS

As mentioned earlier, different material responses are observed for both optically thick and thin samples. Optical thickness refers to the amount of light absorbed as it passes through the material, in the sense that optically thick samples absorb more light than optically thin ones. The optical thickness is determined not only by the physical thickness of the material but also by the concentration of photoactive constituents present and the intensity of the irradiation source. In optically thin samples, photoactuation is manifested as contractile stresses and strains, because the light transmitted through the material is sufficient to achieve the same degree of actuation throughout. Optically thick samples exhibit bending/curling when irradiated because of the non-uniform distribution of photostresses along the depth of the sample; this bending can be reversed either by thermal relaxation or by irradiating the opposite side of the material to symmetrize the distribution of photostresses with respect to the midplane of the sample (Morimoto and Irie 2010, Serak et al. 2010).

If bending/curling is inhibited by macroscopic constraints (e.g., a strain gauge), larger net contractile stresses are obtained for samples with lower concentrations of photoisomerizable molecules (Kondo et al. 2010). In LCEs, once a domain has undergone its isothermal nematic → isotropic phase change, photoisomerization of more mesogens by the absorption of additional light does not increase the photostress generated. It may, however, increase the penetration depth of the incident irradiation, with more material experiencing internal stresses. The maximally efficient use of light energy in terms of photostress generation would, therefore, occur in materials that have the minimum concentration of photoactive mesogens needed to induce nematic → isotropic phase changes when irradiated. This concentration, however, may also depend on the imposed load: for a more highly strained LCE sample, a larger fraction of its mesogens need to be eliminated by photoisomerization to induce the isothermal nematic → isotropic phase transition (Ichimura 2000, Warner and Terentjev 2003b). The effect of concentration of covalently bound photoactive mesogens on the performance characteristics of the material has been relatively little studied.

Experimentally observed bending/curling of optically thick samples of photoelastomers can be modeled by the distribution of photostresses (Dunn 2007). The effect of light polarization on the photoactuation of polydomain LCEs has also been modeled (Corbett and Warner 2006, 2008b, Dunn 2007). In addition to the nonlinear responses with respect to light intensity in those studies, nonlinear penetration of the light (i.e., deviations of from Beer's law) in optically thick samples irradiated with intense beams has also been considered (Corbett et al. 2008, Corbett and Warner 2008a).

3.3.3 STRESS, STRAIN, AND ENERGY CONVERSION EFFICIENCY

Stress and strain in a material are related by its modulus of elasticity. Soft materials exhibit large strains at small stresses, and stiff materials require large stresses to achieve small strains.

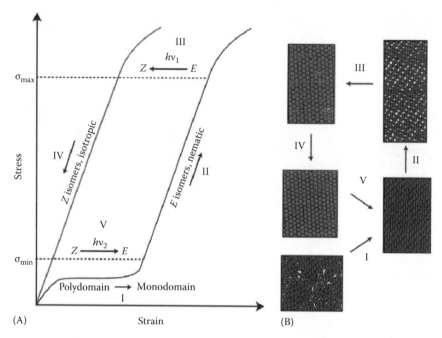

FIGURE 3.7 **(See color insert.)** Model work cycle for a photoactuating polydomain LCE; the E isomers of the photoactive mesogens are rod-like, and the Z isomers are bent. (A) The work cycle depicted on a stress–strain plot. The descriptions of the steps (I–V) in the work cycle are described in the main text. (B) Illustrations of the material's microstructural changes in Steps I–V, depicting the shape distributions of polymer chains in the nematic (blue ovals) and isotropic (red circles) states.

The amount of energy required to deform a material is given by the area under its stress–strain curve $\sigma(\epsilon)$, where σ is stress and ϵ is strain. Integrating $\sigma(\epsilon)$ from $\epsilon = 0$ to the yield strain defines the material's modulus of resilience and thus its capacity to absorb mechanical energy without undergoing irreversible (plastic) deformation. During an actuating cycle, the elastic modulus of the material may change by as much as 500% for skeletal muscle and polypyrrole electromechanical actuators (Spinks and Truong 2005).

The simplest photoactuating work cycle is shown in Figure 3.7, using the example of a photoactuating LCE. (The related single-molecule cycle is presented in the following section.) Starting with an unstrained polydomain nematic LCE containing mostly the E isomers of photoactive mesogens, the material is axially loaded in the dark to stress σ_{max}. At stresses below σ_{min}, the material undergoes a strain-induced polydomain–monodomain transition (Path I, Figure 3.7; Corbett 2009, Warner and Terenjev 2003b). Above σ_{min} the (now monodomain) sample behaves elastically (Path II). Irradiation the material with light of frequency ν_1 to cause $E \rightarrow Z$ photoisomerization of the photoactive mesogens while maintaining constant stress results in its axial contraction (Path III) due to a decrease in the nematic order. This contraction performs work on the external load. Slowly relaxing the external load down to σ_{min} allows further contraction of the material (Path IV) and generates more useful work. Irradiation of the isotropic phase at stress σ_{min} with light of frequency ν_2 to cause $Z \rightarrow E$ photoisomerization of the photoactive mesogens returns them to their rod-like state and induces an isothermal isotropic $\rightarrow$ nematic phase change, during which the sample elongates (Path V).

The reversible work w_{rev} done by the material in one cycle is given by the area inside the curves $II \rightarrow III \rightarrow IV \rightarrow V$ if each step is performed infinitely slowly, i.e., w_{rev} is the thermodynamic limit of conversion of light to mechanical work by the material. The corresponding maximum external energy conversion efficiency is, therefore,

$$\eta_{max} = \frac{w_{rev}}{h\nu_1\Phi_1 + h\nu_2\Phi_2} \tag{3.1}$$

where

h is Plank's constant

Φ_1 and Φ_2 are the total absorbed photon fluxes during the first and second irradiation steps, respectively

The maximum work achievable depends on the elastic moduli of the nematic and isotropic phases. The elastic moduli of LCEs have been shown to depend on the degree of molecular alignment and the cross-linking density (Harris et al. 2005) and are somewhat tunable through molecular design. The maximum energy conversion efficiency additionally depends on the number of photons that must be absorbed by the material to induce the phase transition and the frequencies of light used, which depends on the quantum yields and absorption spectra of the chromophore and is also amenable to molecular design. For example, the effect of substituents on the wavelengths of E/Z photoisomerization of C=C and N=N in stilbenes, azobenzenes, and, more recently, isoindigos has been studied extensively computationally (Blevins and Blanchard 2004, Jacquemin et al. 2007, Perpète et al. 2006) and experimentally (Forber et al. 1985, Görner and Kuhn 1995, Lunák et al. 2010, Nishimura et al. 1976, Ross and Blanc 1971).

3.3.4 OUTSTANDING QUESTIONS IN MOLECULAR PHOTOACTUATOR DESIGN

The general rules for the design of photoactuating materials and monomeric chromophores remain to be formulated, but several trends have emerged. For example, available photostresses could be increased by working with highly cross-linked polymers with high elastic moduli (Spillmann et al. 2006) or microstructurally ordered glassy LC polymers (Harris et al. 2005, van Oosten 2007) in which the nematic director twists by 90° through the sample, similar to the helical director ordering in cholesteric LC phases. Improving the long-range order of LCEs is thought to increase accessible photostrains. Potential strategies for maximizing work capacity and energy conversion efficiencies of photoactuating materials are less clear. From a molecular-design perspective, this goal requires optimizing work-generating molecular parameters by maximizing quantum yields for photoisomerization, the difference in the extinction coefficients of the two isomers and isomerization wavelength, while minimizing dissipative processes, including photochemical and thermal side reactions and unproductive relaxation of the metastable isomer, particularly at high bulk stresses. Other parameters, including the concentration of the chromophores, their localization in the polymer matrix and molecular compliances of nonchromophoric parts of the material may also be critical for achieving acceptable energy conversion efficiencies and work densities.

It may be more efficient to understand the chemistry of photoactuation in individual macromolecules before moving to the less tractable bulk materials. The advances in micromanipulation techniques in the past few decades have allowed individual macromolecules to be stretched controllably (Neuman and Nagy 2008). Photochemical actuation by individual chains of oligoazobenzene was demonstrated in 2002 (Holland et al. 2003, Hugel et al. 2002). An azobenzene-containing oligomer (Figure 3.8A) was bound at its termini to a glass slide and an Au-coated AFM tip, allowing it to be stretched by separating the force probes while recording the force required to maintain the probe separation (Figure 3.8B). The oligomer was stretched to 80 pN and irradiated at 420 nm to increase the fraction of its azobenzenes in the E form. The oligomer was then stretched to 200 pN (Path I, Figure 3.8) and subsequently irradiated with 365 nm light (Path II) to photoisomerize azobenzene monomers to the shorter Z form. The stretching force was gradually reduced to 85 pN (Path III) allowing the oligomer to contract. Irradiation with 420 nm light increases the fraction of E isomers, completing the optomechanical cycle (Path IV).

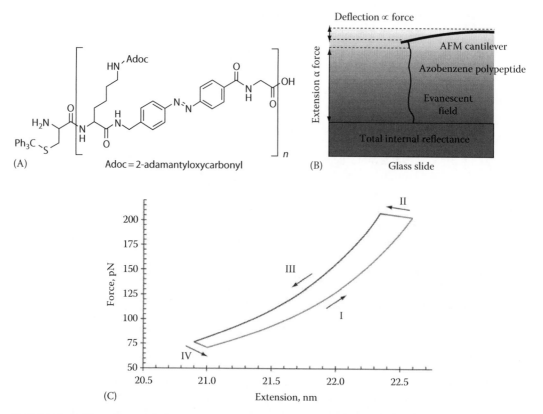

FIGURE 3.8 Experimental demonstration of a single molecule optomechanical work cycle. (A) Chemical structure of the azobenzene-containing oligomer used; $n \approx 47$. (B) Experimental setup for stretching a single oligomer between the glass slide and atomic force microscope (AFM) cantilever with photoisomerization of the N=N being induced by the evanescent field due to total internal reflection of the excitation irradiation in the glass slide. (C) Wormlike chain fits (Paths I and III) to an experimentally demonstrated optomechanical cycle plotted in terms of molecular restoring force versus extension. The molecule contains a greater fraction of (Z)- azobenzene in the black trace than it does in the grey trace. Paths I–IV are described in the main text. (From Hugel, T., *Science*, 296, 1103, 2002.)

This and similar experiments can be understood quantitatively within a general single-chain work cycle (Figure 3.9), comprising loading (stretching the polymer by applying an increasing tensile force to its termini), irradiation, unloading, and recovery (Steps I–IV, respectively). The single-chain cycle is comparable to that for a bulk material (Figure 3.7) except that stress (volumetric energy density) is replaced with force (linear energy density), because a macromolecule is effectively one-dimensional. The relationship between the average applied tensile force and the average separation of the polymer's termini divided by the number of chromophores in the polymer is described by Equation 3.2 where s_0 and λ are the contour length and stretching compliance per chromophore of the all-E strain-free polymer, Δs and $\Delta \lambda$ are the differences in these variables between the all-E and all-Z forms, χ_E is the molar fraction of the E isomers of the chromophores in the polymer, b is the Kuhn length derived from a modified isomeric state model for the all-E polymer (Huang and Boulatov 2011), α is an empirical parameter that accounts for the loss of entropy per chromophore upon photoisomerization, and β is the inverse thermal energy (i.e., $\beta = (k_B T)^{-1}$):

$$s(F) = \left(s_0 + \chi_E \Delta s + F\left(\lambda + \chi_E \Delta \lambda\right)\right)\left(\coth\left(F\left(b + \chi_E\left(\Delta s + \alpha\right)\right)\beta\right) - \frac{1}{F\left(b + \chi_E\left(\Delta s + \alpha\right)\right)\beta}\right) \quad (3.2)$$

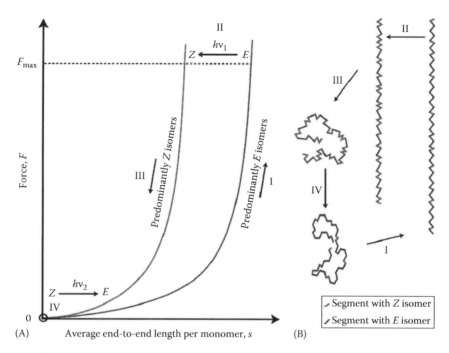

FIGURE 3.9 **(See color insert.)** Model work cycle for a single molecule of a photoactuating polymer. (A) The work cycle depicted on a force–extension plot. The descriptions of the steps (I–IV) in the work cycle are described in the main text. (B) Illustrations of the polymer shape changes in Steps I–IV, indicating the E (blue) and Z (red) isomers in each state.

Similarly to the macroscopic cycle, the reversible work per monomer per cycle is given by the area bound by the two solid force-extension curves and the broken line, describing the isotensional photochemical contraction. This cycle can be performed reversibly only at 0 K because at higher temperatures the thermal relaxation of the metastable Z isomer to the E analog would collapse the two force-extension curves.

The structural and kinetic parameters of Equation 3.2 (s, Δs, λ, $\Delta\lambda$, and χ_E) all depend on the force experienced by the macromolecule. For example, thermal $Z \rightarrow E$ isomerization would be accelerated in an overstretched polymer relative to the strain-free analog and the quantum yields of generating the more strained Z isomer will be suppressed. Such effects maybe neglected for polymers operating against stresses <1 MPa, as is typical for demonstrations of photoactuation by LCEs, but must be taken into account in designing polymers for operation against larger stresses. There is evidence of such effects in the experiment described in Figure 3.8 (Hugel et al. 2002), where maximum contraction of the stretched oligoazobenzene corresponded to conversion of only ~28% (mol) of azobenzene monomers into the Z isomer, compared to ~55% in the absence of strain. Part of the difference may be attributed to the nonuniform irradiation of the macromolecule in these experiments. This comparison and other demonstrations of the effects of macroscopic stresses on thermal and photochemical kinetics (Lee et al. 2010) highlight the need for a general kinetic model incorporating stress as a kinetic variable.

3.4 MULTISCALE REACTION DYNAMICS

The (typical) acceleration of chemical reactions in bulk strained materials is often called mechanochemistry. The effects can be quite dramatic, with up to 10^{15}-fold accelerations in reaction rates reported. Mechanochemical phenomenology, its biological and technological significance, and the experimental and computational methods of study have been reviewed previously

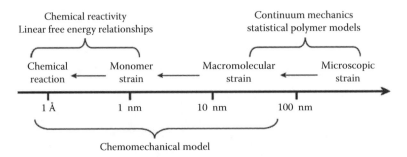

FIGURE 3.10 Hierarchy of length scales involved in multiscale phenomena and the regimes of the established models that describe phenomena at various length scales (top). Because the length scales of continuum mechanics and chemical kinetics do not overlap, a new model integrating the two formalisms is required to arrive at quantitative descriptions of mechanochemical phenomena. (Reproduced from Kucharski, T.J. and Boulatov, R., *J. Mater. Chem.*, 21, 8237, 2011.)

(Black et al. 2011, Caruso et al. 2009, Huang and Boulatov 2011, Kucharski and Boulatov 2011). Until our work, the quantitative predictions of kinetic and/or thermodynamic stabilities of reactive sites in stretched polymers have been prevented by the failure of traditional chemical and physical models to bridge the "formidable gap" of the length scale spanning ~50 nm and ~1 nm, where coupling between large-scale strains and chemical reactions is important (Figure 3.10; Boulatov 2011, Kucharski and Boulatov 2011). Chemomechanics is a conceptual framework to allow such predictions.

3.4.1 BASIC PRINCIPLES OF THE CHEMOMECHANICAL FORMALISM

Qualitatively, mechanochemistry can be viewed as the effect of molecular strain on chemical reactivity, a topic that chemists have studied for the past 100 years. However, mechanochemical phenomena are not amenable to a quantitative description within the formalism of linear free energy relationships (LFERs) that underlie the kinetic models of molecular strain. The reason is that materials and molecular strain are different concepts. Materials strain is rigorously defined and quantitative (Tauchert 1974), but molecular strain is qualitative (Kucharski and Boulatov 2011). The quantitative discussion of strain-induced chemistry relies on the concept of strain energy, which is defined formally as the difference in the enthalpy of formation of a strained molecule and its hypothetical strain-free analog. Within the formalism of the LFERs, strain energy allows semi-quantitative estimates of strain-induced differences in activation barriers of reactions without any information about the transition state. This is possible in conventional chemical reactions because the kinetically relevant changes in strain energy are confined to small and well-defined volumes of space (strained reactant and product) and are fairly easy to estimate. In contrast, in reactions in bulk anisotropically strained environments, strain energies of the initial and final states are difficult, if not impossible, to define, and the strain-energy change that contributes to the reaction barrier is only a tiny fraction of the total strain energy.

A quantitative relationship between large-scale strain and localized reaction kinetics requires an intensive quantifier of strain, such as force (Boulatov 2011, Huang and Boulatov 2011, Kucharski and Boulatov 2011). Force is the gradient of strain energy along a constrained degree of freedom. A macroscopically constrained degree of freedom, such as an axis of anistropic material, is often called the mechanical coordinate. The simplest example is a reaction within an inert polymer that is stretched, with the mechanical coordinate being the separation of the polymer termini, τ (Figure 3.11A). Either τ or its conjugate variable, mechanical force, F_τ, is the control parameter. The two parameters for isolated macromolecules are related by several models (e.g., Figure 3.11B).

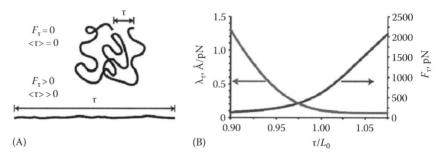

FIGURE 3.11 (A) Depiction of the mechanical coordinate τ in single polymer chains; $\langle\rangle$ denotes ensemble average values. (B) The compliance (λ_τ) and restoring force (F_τ) of a single molecule of polyethylene as a function of its strain τ/L_0; the contour length, L_0, is 150 nm. Note considerable restoring force at the polymer end-to-end distances below its contour length, reflecting the considerable reduction of macromolecular entropy as the polymer termini are constrained (Flory 1969, Rubinstein 2003). Calculated with data and methods in Hugel et al. (2005). (Modified from Kucharski, T.J. and Boulatov, R., *J. Mater. Chem.*, 21, 8237, 2011.)

If the mechanical coordinate remains in thermal equilibrium with its environment during an elementary localized reaction, the corresponding rate constant of a long flexible polymer stretched to average restoring force F_τ, $k(F_\tau)$, relative to the same reaction in the strain-free reactant, k_0, is given by Equation 3.3, where λ is the total stretching compliance of mechanical coordinate τ in the ground state, and τ_{ts} and $\Delta\lambda$ are the differences in the strain-free end-to-end separation and in the compliances of τ between the ground and transition states. The famous Eyring–Bell–Evans ansatz (Bell 1978, Evans 2001, Evans and Ritchie 1997, Kauzmann and Eyring 1940) postulating exponential dependence of the reaction rate with restoring force is a limiting case of Equation 3.3:

$$\beta^{-1}\ln\frac{k\left(F_\tau\right)}{k_0} = \frac{\lambda}{\lambda+\Delta\lambda}\left(F_\tau\tau_{ts}+\frac{F_\tau^2}{2}\Delta\lambda\right) \tag{3.3}$$

Macroscopic stretching compliances, λ, of many polymers have been measured by single-molecule force spectroscopy. However, τ_{ts} and $\Delta\lambda$, which characterize strain-free macromolecular *transition states*, are not available for reactions of interest to materials chemists. To allow practical applications of Equation 3.3, for example, for predicting the changes in the kinetic or thermodynamic stability of a reactive site as the polymer containing it is stretched, we formulated and computationally validated the local approximation of chemomechanics. In its simplest formulation, which applies to reactions in stretched polymers, the local approximation postulates the existence of a molecular coordinate of the reactive site, q, such that

1. The strain-free difference in the mechanical coordinate of the macromolecular reactant between its ground and transition states approximately equals the difference in coordinate q between the strain-free ground and transition states of the minimal reactant, i.e., $\tau_{ts}\approx\Delta q$
2. The mechanical force (i.e., the restoring force of the mechanical coordinate) equals the molecular restoring force of the local coordinate q (i.e., $F_\tau\approx F_q$)

With these assumptions, Equation 3.3 can be rewritten in terms of readily available structural and kinetic parameters of the minimal reactant, Δq and $\Delta\lambda_q$, i.e., the strain-free difference in the dimensions and compliance of a local molecular coordinate between the ground and transition states of the reaction (Equation 3.4; Boulatov 2011, Huang and Boulatov 2011, Kucharski and Boulatov 2011). Unlike the more familiar molecular force constants, molecular compliances are independent of the coordinate system, so that in the absence of electronic effects, the compliance of a localized degree of freedom in a large molecule (e.g., a polymer) is the same as the compliance of the same degree of freedom in an isolated site (Baker 2006, Jones and Swanson 1976):

$$\beta^{-1} \ln \frac{k(F_\tau)}{k_o} \approx \beta^{-1} \ln \frac{k(F_q)}{k_o} = \frac{\lambda}{\lambda + \Delta\lambda_q}\left(F_q\Delta q + \frac{F_q^2}{2}\Delta\lambda_q\right) \tag{3.4}$$

3.4.2 FORCE-DEPENDENT KINETICS OF PHOTOACTUATING MONOMERS AND MESOGENS

The local approximation cannot be validated experimentally by studying reactions in stretched polymers because the existing methods lack the resolution and reaction scope to yield the necessary data. Therefore, we developed an alternative to traditional micromanipulation techniques to measure reaction rates as a function of the restoring force of a molecular degree of freedom: molecular force probes (Boulatov 2011, Huang and Boulatov 2010, 2011, Kucharski and Boulatov 2011).

The method relies on synthesizing a series of increasingly strained macrocycles in which a reactive site of interest is incorporated in a 5–10 atom-long inert strap constraining the E isomer of stiff stilbene (red, Figure 3.12A). Quantum-chemical computations confirmed that in this molecular architecture the reactive site experiences approximately the same pattern of strain as it does in a stretched polymer. The magnitude of the restoring force of different molecular degrees of freedom of the reactive site is controlled by the length and conformational flexibility of the strap (X + reactive moiety + Y, Figure 3.12B) so that a series of ~10–12 macrocycles of 15–20 endocyclic atoms can reproduce the range of the restoring forces accessible in a typical single-molecule force experiment (<100 to >650 pN) with small (<50 pN) increments between individual macrocycles. We have used molecular force probes to validate the chemomechanical formalism for force-dependent kinetics of broad reaction classes, including electrocyclic ring opening (Huang et al. 2009, Yang et al. 2009) and nucleophilic displacement (Kucharski et al. 2009, 2010).

The same approach is useful in measuring the effect of axial strain on the kinetics of thermal and photochemical $Z \rightarrow E$ isomerization of stiff stilbene or another chromophore. This approach showed that both barriers for thermal and quantum yields of photochemical isomerization of C=C decrease with force. Consequently, the composition of the photostationary state is particularly sensitive to applied force, varying by >1000-fold over 600 pN in the case of stiff stilbene.

The sensitivity of different chromophores to imposed constraints is determined by the difference in the dimension of the constrained degree of freedom between the two isomers. This trend suggests that coupling chromophores to the polymeric network through rigid moieties such as fused aromatic rings or acetylene units would increase the maximum dimensional change upon irradiation of the material at the expense of decreasing both the quantum yields of generating the strained (typically Z) isomer and its thermal stability against relaxation to the less-strained (typically E) analog. The latter may limit the work density, the energy conversion efficiency, and the maximum stress that can be practically generated by the material.

The stresses at which these effects would become significant can be estimated as follows. Neutron diffraction data suggest that in well-aligned polymeric materials, individual chains can be approximated as filling cylinders with a radius of 2 nm (Li et al. 1993). Assuming a hexagonal close packing of polymer molecules (area fraction of 0.9069), a tensile stress of 44 MPa (comparable to the stresses generated by photoactuating molecular crystals; Morimoto and Irie 2010) corresponds to an average tensile force of $F_\tau \approx 610$ pN per macromolecule. Using Equation 3.4 and neglecting the second-order effects ($\Delta\lambda = 0$), the contraction of the constrained molecular degree of freedom between the ground and transition states of $\Delta q \approx 2.85$ Å (e.g., that for thermal isomerization of stiff stilbene connected through its C6,C6′ positions or for azobenzene connected at the *para* positions) could correspond to a strain-induced decrease of the barrier for thermal relaxation of the Z isomer to its less strained E analog of $F_q\Delta q \approx 1735$ pN Å = 25 kcal/mol at room temperature. The value is more than half of that for strain-free $Z \rightarrow E$ isomerization of stiff stilbene (38 kcal/mol) and is greater than that of azobenzene (23 kcal/mol)! Clearly, operation at high stresses would require correspondingly large barriers to thermal isomerization to avoid (nearly) barrierless thermal relaxation kinetics. Several strategies exist to increase the activation energies of isomerization by substitution, but such increases are typically coupled to (undesirable) increases in the energy of photoisomerization (Figure 3.13). The correlation between the activation energy for thermal $Z \rightarrow E$ isomerization and the energy of the absorption band for the π system that leads to photoisomerization is striking, particularly because the data are for three different classes of compounds (alkenes, stilbenes, and azobenzenes) measured in a variety of media.

Alternatively, in LCEs in which the photoisomerizable mesogens are in non-load-bearing fragments (i.e., side chains, rather than cross-links or the main chain), F_q is probably determined solely by interaction with the environment and presently is difficult to estimate. Rather than relying on the translation of F_q into macroscopic stresses, photoactuating LC systems operate by photochemically inducing an isothermal phase change. If F_q for nonload-bearing photoisomerizable mesogens is significantly lower than that for their load-bearing analogs, their thermal relaxation kinetics may

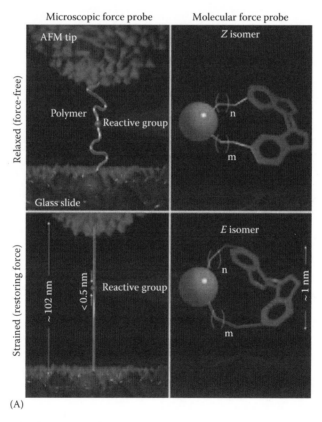

(A)

FIGURE 3.12 (**See color insert.**) (A) Comparison of methods for measuring force-dependent kinetics of localized reactions. On the left, conventional single-molecule force spectroscopy requires the incorporation of the reactive moiety (blue sphere) into a long flexible polymer, attaching the polymer to a pair of microscopic force probes (here, the tip of the atomic force microscope cantilever and a glass slide on a piezoelectric stage) and stretching it by separating the probes. The size of the reactive moiety is typically less than the surface roughness of the probes or the magnitude of their thermal fluctuations, which significantly limits the accuracy of the measurements and the scope of reactions amenable to such studies. The right panels show a molecular force probe containing the same reactive moiety. (Reproduced from Yang, Q.Z. et al., *Nat. Nanotechnol.*, 4, 302, 2009.)

(continued)

Minimal reactant

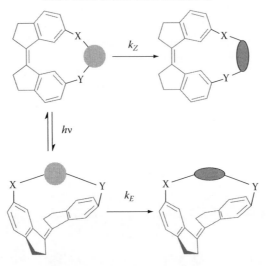

Strain-free *Z* isomers serve as reference

k_Z

$h\nu$

k_E

(B)

Strained *E* isomers mimic reactive moieties
in streched polymers

FIGURE 3.12 (continued)   (See color insert.) (B) The general method for measuring force-dependent kinetics with molecular force probes. The strained *E* isomers are obtained by photoisomerization of strain-free *Z* analogs, which are synthesized using conventional chemistry. (Modified from Kucharski, T.J. et al., *J. Phys. Chem. Lett.*, 1, 2820, 2010.)

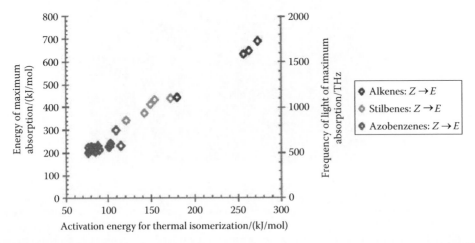

FIGURE 3.13   (See color insert.) Apparent relationship between the activation energy for thermal $Z \rightarrow E$ isomerization and the energy of the maximum absorption band leading to photoisomerization of the *Z* isomer. Absorption energies for the *Z* isomers of azobenzenes were estimated as the absorbance energy in the *E* isomer +the absorbance energy of (*Z*)-azobenzene – the absorbance energy of (*E*)-azobenzene. Alkene data from Bouman (1985), Huh (1990), Walker (1986), Gary (1954), Molina (1999), and Kistiakowsky (1934); stilbene data from Görner (1995) and Ross (1971), and azobenzene data from Nishimura (1976).

be far less perturbed by the macromolecular stresses present. However, increasing macroscopic stresses may instead reduce the effectiveness of such photoisomerizations, but this can be counteracted by increasing the concentration of such mesogens in the material, as mentioned previously. Thus, if both isomers of a photoactive mesogen are sufficiently thermally stable in the absence of force, an LCE with such mesogens in side chains may be able to maintain high photostresses and photostrains for long periods of time. This may be one reason for the success of photoactuating LCE materials.

3.5 CONCLUSIONS

In this chapter, we have reviewed the physical chemistry of photoactuation in polymers, molecular crystals, LCs, and LCEs. The two key events in photochemical actuation are isomerization following photon absorption, which results in structural changes at the molecular level, and the transmission of these changes to larger length scales. Bringing the full potential of chemistry to bear to yield new photoactuating materials requires a general, physically sound, and quantitative model to describe dynamic coupling between the molecular and mesoscopic scales. By allowing useful estimates of the kinetics and thermodynamics of productive and parasitic reactions in materials under external loads, such a model would help guide the design of new chromophores and polymer architectures capable of achieving practically useful energy conversion efficiencies and work densities.

Chemomechanics is one such model, and within the local approximation it provides the only available approach to estimate the fundamental limits of photomechanical energy conversion using molecular isomerization. Bulk strains can significantly perturb the kinetics of actuating reactions, suppressing quantum efficiencies of generating and accelerating undesirable relaxation of metastable strained isomers.

REFERENCES

Baker, J. 2006. A critical assessment of the use of compliance constants as bond strength descriptors for weak interatomic interactions. *J. Chem. Phys.* 125: 014103/1–6.

Bandara, H. M. D., T. R. Friss, M. M. Enriquez et al. 2010. Proof for the concerted inversion mechanism in the *trans→cis* isomerization of azobenzene using hydrogen bonding to induce isomer locking. *J. Org. Chem.* 75: 4817–4827.

Barrett, C. J., J.-i. Mamiya, K. G. Yager, and T. Ikeda. 2007. Photo-mechanical effects in azobenzene-containing soft materials. *Soft Matter* 3: 1249–1261.

Bell, G. I. 1978. Models for the specific adhesion of cells to cells. *Science* 200: 618–627.

Black, A. L., J. M. Lenhardt, and S. L. Craig. 2011. From molecular mechanochemistry to stress-responsive materials. *J. Mater. Chem.* 21: 1655–1663.

Blevins, A. A. and G. J. Blanchard. 2004. Effect of positional substitution on the optical response of symmetrically disubstituted azobenzene derivatives. *J. Phys. Chem. B* 108: 4962–4968.

Bouas-Laurent, H. and H. Dürr. 2001. Organic photochromism. *Pure Appl. Chem.* 73: 639–665.

Boulatov, R. 2011. Reaction dynamics in the formidable gap. *Pure Appl. Chem.* 83: 25–41.

Bouman, T. D. and A. E. Hansen. 1985. Electronic spectra of mono-olefins. RPA calculations on ethylene, propene, and *cis-* and *trans-*2-butene. *Chem. Phys. Lett.* 117: 461–467.

Camacho-Lopez, M., H. Finkelmann, P. Palffy-Muhoray, and M. Shelley. 2004. Fast liquid-crystal elastomer swims into the dark. *Nat. Mater.* 3: 307–310.

Caruso, M. M., D. A. Davis, Q. Shen et al. 2009. Mechanically-induced chemical changes in polymeric materials. *Chem. Rev.* 109: 5755–5798.

Corbett, D., C. L. van Oosten, and M. Warner. 2008. Nonlinear dynamics of optical absorption of intense beams. *Phys. Rev. A* 78: 013823/1–4.

Corbett, D. and M. Warner. 2006. Nonlinear photoresponse of disordered elastomers. *Phys. Rev. Lett.* 96: 237802/1–4.

Corbett, D. and M. Warner. 2008a. Bleaching and stimulated recovery of dyes and of photocantilevers. *Phys. Rev. E* 77: 051710/1–11.

Corbett, D. and M. Warner. 2008b. Polarization dependence of optically driven polydomain elastomer mechanics. *Phys. Rev. E* 78: 061701/1–13.

Corbett, D. and M. Warner. 2009. Changing liquid crystal elastomer ordering with light—A route to opto-mechanically responsive materials. *Liq. Cryst.* 36: 1263–1280.

Creatini, L., T. Cusati, G. Granucci, and M. Persico. 2008. Photodynamics of azobenzene in a hindering environment. *Chem. Phys.* 347: 492–502.

Domcke, W., D. R. Yarkony, and H. Köppel (eds.). 2004. *Conical Intersections: Electronic Structure, Dynamics and Spectroscopy*. Hackensack, NJ: World Scientific.

Dunn, M. L. 2007. Photomechanics of mono- and polydomain liquid crystal elastomer films. *J. Appl. Phys.* 102: 013506/1–7.

Eelkema, R., M. M. Pollard, J. Vicario et al. 2006. Nanomotor rotates microscale objects. *Nature* 440: 163.

Evans, E. 2001. Probing the relation between force—lifetime—and chemistry in single molecular bonds. *Annu. Rev. Biophys. Biomol. Struct.* 30: 105–128.

Evans, E. and K. Ritchie. 1997. Dynamic strength of molecular adhesion bonds. *Biophys. J.* 72: 1541–1555.

Finkelmann, H., E. Nishikawa, G. G. Pereira, and M. Warner. 2001. A new opto-mechanical effect in solids. *Phys. Rev. Lett.* 87: 015501/1–4.

Flory, P. J. 1969. *Statistical Mechanics of Chain Molecules*. New York: Interscience Publishers.

Forber, C. L., E. C. Kelusky, N. J. Bunce, and M. C. Zerner. 1985. Electronic spectra of *cis-* and *trans-*azobenzenes: Consequences of ortho substitution. *J. Am. Chem. Soc.* 107: 5884–5890.

Funke, U. and H.-F. Grützmacher. 1987. Dithia-diaza[n.2]paracyclophane-enes. *Tetrahedron* 43: 3787–3795.

Gary, J. T. and L. W. Pickett. 1954. The far ultraviolet absorption spectra of the isomeric butenes. *J. Chem. Phys.* 22: 599–602.

Görner, H. and H. J. Kuhn. 1995. *Cis–Trans* photoisomerization of stilbenes and stilbene-like molecules. In *Advances in Photochemistry*. Neckers, D. C., D. H. Volman, and G. v. Bünau. (eds.), pp. 1–117. New York: John Wiley & Sons, Inc.

Harris, D. C. and M. D. Bertolucci. 1989. *Symmetry and Spectroscopy: An Introduction to Vibrational and Electronic Spectroscopy*. New York: Dover Publications, Inc.

Harris, K. D., R. Cuypers, P. Scheibe et al. 2005. Large amplitude light-induced motion in high elastic modulus polymer actuators. *J. Mater. Chem.* 15: 5043–5048.

Harvey, C. L. M. and E. M. Terentjev. 2007. Role of polarization and alignment in photoactuation of nematic elastomers. *Eur. Phys. J. E* 23: 185–189.

Holland, N. B., T. Hugel, G. Neuert et al. 2003. Single molecule force spectroscopy of azobenzene polymers: Switching elasticity of single photochromic macromolecules. *Macromolecules* 36: 2015–2023.

Huang, Z. and R. Boulatov. 2010. Chemomechanics with molecular force probes. *Pure Appl. Chem.* 82: 931–951.

Huang, Z. and R. Boulatov. 2011. Chemomechanics: Reaction kinetics in multiscale phenomena. *Chem. Soc. Rev.* 40: 2359–2384.

Huang, Z., Q.-Z. Yang, D. Khvostichenko, T. J. Kucharski, J. Chen, and R. Boulatov. 2009. Method to derive restoring forces of strained molecules from kinetic measurements. *J. Am. Chem. Soc.* 131: 1407–1409.

Hugel, T., N. B. Holland, A. Cattani, L. Moroder, M. Seitz, and H. E. Gaub. 2002. Single-molecule optomechanical cycle. *Science* 296: 1103–1106.

Hugel, T., M. Rief, M. Seitz, H. E. Gaub, and R. R. Netz. 2005. Highly stretched single polymers: Atomic-force-microscope experiments versus *ab-initio* theory. *Phys. Rev. Lett.* 94: 048301/1–4.

Huh, D. S., J. Y. Um, S. J. Yun, K. Y. Choo, and K.-H. Jung. 1990. Gas phase thermal cis-trans isomerization reaction of 1-bromopropene. *Bull. Korean Chem. Soc.* 11: 391–395.

Ichimura, K. 2000. Photoalignment of liquid-crystal systems. *Chem. Rev.* 100: 1847–1874.

Ikeda, T., J. i. Mamiya and Y. Yu. 2007. Photomechanics of liquid-crystalline elastomers and other polymers. *Angew. Chem. Int. Ed.* 46: 506–528.

Irie, M. 2000. Diarylethenes for memories and switches. *Chem. Rev.* 100: 1685–1716.

Irie, M. 2008. Photochromism and molecular mechanical devices. *Bull. Chem. Soc. Jpn.* 81: 917–926.

Jacquemin, D., E. A. Perpete, G. E. Scuseria, I. Ciofini, and C. Adamo. 2007. TD-DFT performance for the visible absorption spectra of organic dyes: Conventional versus long-range hybrids. *J. Chem. Theory Comput.* 4: 123–135.

Jiang, H. Y., S. Kelch, and A. Lendlein. 2006. Polymers move in response to light. *Adv. Mater.* 18: 1471–1475.

Jones, L. H. and B. I. Swanson. 1976. Interpretation of potential constants: Application to study of bonding forces in metal cyanide complexes and metal carbonyls. *Acc. Chem. Res.* 9: 128–134.

Juodkazis, S., N. Mukai, R. Wakaki, A. Yamaguchi, S. Matsuo, and H. Misawa. 2000. Reversible phase transitions in polymer gels induced by radiation forces. *Nature* 408: 178–181.

Kauzmann, W. and H. Eyring. 1940. The viscous flow of large molecules. *J. Am. Chem. Soc.* 62: 3113–3125.

Kistiakowsky, G. B. and W. R. Smith. 1934. Kinetics of thermal cis-trans isomerization. III. *J. Am. Chem. Soc.* 56: 638–642.

Klok, M., N. Boyle, M. T. Pryce, A. Meetsma, W. R. Browne, and B. L. Feringa. 2008. MHz unidirectional rotation of molecular rotary motors. *J. Am. Chem. Soc.* 130: 10484–10485.

Kobatake, S. and M. Irie. 2003. Photochromism. *Annu. Rep. Prog. Chem. Sect. C Phys. Chem.* 99: 277–313.

Kobatake, S., S. Takami, H. Muto, T. Ishikawa, and M. Irie. 2007. Rapid and reversible shape changes of molecular crystals on photoirradiation. *Nature* 446: 778–781.

Koerner, H., T. J. White, N. V. Tabiryan, T. J. Bunning, and R. A. Vaia. 2008. Photogenerating work from polymers. *Mater. Today* 11: 34–42.

Kondo, M., M. Sugimoto, M. Yamada et al. 2010. Effect of concentration of photoactive chromophores on photomechanical properties of crosslinked azobenzene liquid-crystalline polymers. *J. Mater. Chem.* 20: 117–122.

Kucharski, T. J. and R. Boulatov. 2011. The physical chemistry of mechanoresponsive polymers. *J. Mater. Chem.* 21: 8237–8255.

Kucharski, T. J., Z. Huang, Q.-Z. Yang et al. 2009. Kinetics of thiol/disulfide exchange correlate weakly with the restoring force in the disulfide moiety. *Angew. Chem. Int. Ed.* 48: 7040–7043.

Kucharski, T. J., Q.-Z. Yang, Y. Tian, and R. Boulatov. 2010. Strain-dependent acceleration of a paradigmatic S_N2 reaction accurately predicted by the force formalism. *J. Phys. Chem. Lett.* 1: 2820–2825.

Küpfer, J. and H. Finkelmann. 1991. Nematic liquid single crystal elastomers. *Makromol. Chem. Rapid Commun.* 12: 717–726.

Lee, C. K., D. A. Davis, S. R. White, J. S. Moore, N. R. Sottos, and P. V. Braun. 2010. Force-induced redistribution of a chemical equilibrium. *J. Am. Chem. Soc.* 132: 16107–16111.

Levine, B. G. and T. J. Martínez. 2007. Isomerization through conical intersections. *Annu. Rev. Phys. Chem.* 58: 613–634.

Li, M. H., A. Brûlet, P. Davidson, P. Keller, and J. P. Cotton. 1993. Observation of hairpin defects in a nematic main-chain polyester. *Phys. Rev. Lett.* 70: 2297–2300.

Li, C., C. W. Lo, D. Zhu, C. Li, Y. Liu, and H. Jiang. 2009. Synthesis of a photoresponsive liquid-crystalline polymer containing azobenzene. *Macromol. Rapid Commun.* 30: 1928–1935.

Lunák, S., Jr., P. Horáková, and A. Lycka. 2010. Absorption and fluorescence of arylmethylidenoxindoles and isoindigo. *Dyes Pigments* 85: 171–176.

Minkin, V. I. 2004. Photo-, thermo-, solvato-, and electrochromic spiroheterocyclic compounds. *Chem. Rev.* 104: 2751–2776.

Mizutani, Y., Y. Otani, and N. Umeda. 2008. Optically controlled bimorph cantilever of poly(vinylidene difluoride). *Appl. Phys. Express* 1: 041601/1–3.

Molina, V., M. Merchán, and B. O. Roos. 1999. A theoretical study of the electronic spectrum of cis-stilbene. *Spectrochim. Acta A–M.* 55: 433–446.

Morimoto, M. and M. Irie. 2005. Photochromism of diarylethene single crystals: Crystal structures and photochromic performance. *Chem. Commun.* 3895–3905.

Morimoto, M. and M. Irie. 2010. A diarylethene cocrystal that converts light into mechanical work. *J. Am. Chem. Soc.* 132: 14172–14178.

Neuman, K. C. and A. Nagy. 2008. Single-molecule force spectroscopy: Optical tweezers, magnetic tweezers and atomic force microscopy. *Nat. Methods* 5: 491–505.

Nishimura, N., T. Sueyoshi, H. Yamanaka, E. Imai, S. Yamamoto, and S. Hasegawa. 1976. Thermal *Cis*-to-*Trans* isomerization of substituted azobenzenes II. Substituent and solvent effects. *Bull. Chem. Soc. Jpn.* 49: 1381–1387.

Norikane, Y., R. Katoh, and N. Tamaoki. 2008. Unconventional thermodynamically stable *cis* isomer and *trans* to *cis* thermal isomerization in reversibly photoresponsive [0.0](3,3′)-azobenzenophane. *Chem. Commun.* 1898–1900.

Perpète, E. A., J. Preat, J.-M. André and D. Jacquemin. 2006. An ab initio study of the absorption spectra of indirubin, isoindigo, and related derivatives. *J. Phys. Chem. A* 110: 5629–5635.

Ross, D. L. and J. Blanc. 1971. Photochromism by *cis–trans* isomerization. In *Photochromism*. G. H. Brown. (ed.). pp. 471–556. New York: John Wiley & Sons, Inc.

Rubinstein, M. and R. H. Colby. 2003. *Polymer Physics*. New York: Oxford University Press.

Russew, M.-M. and S. Hecht. 2010. Photoswitches: From molecules to materials. *Adv. Mater.* 22: 3348–3360.

Serak, S., N. Tabiryan, R. Vergara, T. J. White, R. A. Vaia, and T. J. Bunning. 2010. Liquid crystalline polymer cantilever oscillators fueled by light. *Soft Matter* 6: 779–783.

Spillmann, C. M., J. Naciri, M.-S. Chen, A. Srinivasan, and B. R. Ratna. 2006. Tuning the physical properties of a nematic liquid crystal elastomer actuator. *Liq. Cryst.* 33: 373–380.

Spinks, G. M. and V.-T. Truong. 2005. Work-per-cycle analysis for electromechanical actuators. *Sens. Actuators A Phys.* 119: 455–461.

Tamai, N. and H. Miyasaka. 2000. Ultrafast dynamics of photochromic systems. *Chem. Rev.* 100: 1875–1890.

Tauchert, T. R. 1974. *Energy Principles in Structural Mechanics*. New York: McGraw-Hill.

Tiberio, G., L. Muccioli, R. Berardi, and C. Zannoni. 2010. How does the *trans–cis* photoisomerization of azobenzene take place in organic solvents? *Chem. Phys. Chem.* 11: 1018–1028.

van Oosten, C., K. Harris, C. Bastiaansen, and D. Broer. 2007. Glassy photomechanical liquid-crystal network actuators for microscale devices. *Eur. Phys. J. E* 23: 329–336.

van der Veen, G., R. Hoguet, and W. Prins. 1974. Photoregulation of polymer conformation by photochromic moieties–II. Cationic and neutral moieties on an anionic polymer. *Photochem. Photobiol.* 19: 197–204.

van der Veen, G. and W. Prins. 1971. Photomechanical energy conversion in a polymer membrane. *Nature Phys. Sci.* 230: 70–72.

Waldeck, D. H. 1991. Photoisomerization dynamics of stilbenes. *Chem. Rev.* 91: 415–436.

Walker, I. C., T. M. Abuain, M. H. Palmer, and A. J. Beveridge. 1986. The electronic states of propene studied by electron impact spectroscopy and ab initio configuration interaction calculations. *Chem. Phys.* 109: 269–275.

Warner, M. and E. Terentjev. 2003a. Thermal and photo-actuation in nematic elastomers. *Macromol. Symp.* 200: 81–92.

Warner, M. and E. M. Terentjev. 2003b. *Liquid Crystal Elastomers*. New York: Oxford University Press.

White, T. J., S. V. Serak, N. V. Tabiryan, R. A. Vaia, and T. J. Bunning. 2009. Polarization-controlled, photo-driven bending in monodomain liquid crystal elastomer cantilevers. *J. Mater. Chem.* 19: 1080–1085.

White, T. J., N. V. Tabiryan, S. V. Serak et al. 2008. A high frequency photodriven polymer oscillator. *Soft Matter* 4: 1796–1798.

Yang, Q.-Z., Z. Huang, T. J. Kucharski, D. Khvostichenko, J. Chen, and R. Boulatov. 2009. A Molecular force probe. *Nat. Nanotechnol.* 4: 302–306.

Yokoyama, Y. 2000. Fulgides for memories and switches. *Chem. Rev.* 100: 1717–1740.

Yokoyama, Y. 2001. Molecular switches with photochromic fulgides. In *Molecular Switches*. pp. 107–121. Ben, L. F., Weinheim, Germany: Wiley-VCH Verlag GmbH.

Yu, Y. and T. Ikeda. 2008. Photodeformable materials and photomechanical effects based on azobenzene-containing polymers and liquid crystals. In *Smart Light-Responsive Materials*. pp. 95–144Y. Zhao and T. Ikeda. (eds.). Hoboken, NJ: John Wiley & Sons, Inc.

4 Photo-Mechanical Azo Polymers for Light-Powered Actuation and Artificial Muscles

Zahid Mahimwalla, Kevin G. Yager, Jun-ichi Mamiya,
Atsushi Shishido, and Christopher J. Barrett

CONTENTS

4.1 INTRODUCTION

The change in shape inducible in some photo-reversible molecules using light can effect powerful changes to a variety of properties of a host material. This class of reversible light-switchable molecules includes photo-responsive molecules that photodimerize, such as coumarins and anthracenes; those that allow intramolecular photo-induced bond formation, such as fulgides, spiro-pyrans, and diarylethenes; and those that exhibit photoisomerization, such as stilbenes, crowded alkenes, and azobenzene. The most ubiquitous natural molecule for reversible shape change however, and perhaps the inspiration for all artificial biomimics, is the rhodopsin/retinal protein system that enables vision, which is perhaps the quintessential reversible photoswitch for performance and robustness. Here, the small retinal molecule embedded in a cage of rhodopsin helices isomerizes from a *cis* geometry to a *trans* geometry around a C=C double bond with the absorption of just a single photon. The modest shape change of just a few angstroms is quickly amplified however, and sets off a cascade of larger shape and chemical changes, eventually culminating in an electrical signal to the brain of a vision event, the energy of the input photon amplified many thousands of times in the process. Complicated biochemical pathways then revert the *trans* isomer back to *cis* and set the system back up for another cascade upon subsequent absorption. The reversibility is complete, and many subsequent cycles are possible. The reversion mechanism back to the initial *cis* state is complex and enzymatic however, so direct application of the retinal/rhodopsin photoswitch to engineering systems is difficult. Perhaps the best artificial mimic of this strong photo-switching effect however, for reversibility, speed, and simplicity of incorporation, is azobenzene. *Trans* and *cis* states can be switched in microseconds with low power light, reversibility of 10^5 and 10^6 cycles is routine before chemical fatigue, and a wide variety of molecular architectures is available to the synthetic materials chemist permitting facile anchoring and compatibility, as well as chemical and physical amplification of the simple geometric change.

This chapter focuses on the study and application of reversible changes in shape that can be induced with various material systems incorporating azobenzene to effect significant reversible mechanical actuation. This photo-mechanical effect can be defined as the reversible change in shape inducible in some molecules by the adsorption of light, which results in a significant macroscopic mechanical deformation of the host material. Thus, it does not include simple thermal expansion effects, nor does it include reversible but nonmechanical photoswitching or photochemistry, nor any of the wide range of optical and electro-optical switching effects for which good reviews exist elsewhere. These azobenzenes are similarly of great interest for light energy harvesting applications across much of the solar spectrum, yet this emerging field is still in an early enough stage of research output as to not yet warrant review.

4.2 PHOTO-MECHANICAL AZO POLYMERS

Azobenzene, with two phenyl rings separated by an azo (–N=N–) bond, serves as the parent molecule for a broad class of aromatic azo compounds. These chromophores are versatile molecules, and have received much attention in research areas both fundamental and applied. The strong electronic absorption maximum can be tailored by ring substitution to fall anywhere from the ultraviolet to red-visible regions, allowing chemical fine-tuning of color. This, combined with the fact that these azo groups are relatively robust and chemically stable, has prompted extensive study of azobenzene-based structures as dyes and colorants. The rigid mesogenic shape of the molecule is well suited to spontaneous organization into liquid crystalline phases, and hence polymers doped or functionalized with azobenzene-based chromophores (azo polymers) are common as liquid crystalline media. With appropriate electron donor/acceptor ring substitution, the π electron delocalization of the extended aromatic structure can yield high optical nonlinearity, and azo chromophores have seen extensive study for nonlinear optical applications as well. One of the most interesting properties of these chromophores however, and the main subject of this chapter, is the readily induced and reversible isomerization about the azo bond between the *trans* and the *cis* geometric isomers and the geometric changes that result when azo chromophores are incorporated into polymers and other materials. This light-induced interconversion allows systems incorporating azobenzenes to be used as photoswitches, effecting rapid and reversible control over a variety of chemical, mechanical, electronic, and optical properties.

Examples of such photocontrol have been demonstrated in photo-switchable phase changes [1], phase separation [2], or reversal of phase separation [3], solubility changes [4,5], and crystallization [6]. These suggest a highly promising route toward novel functional materials: the incorporation of photo-physical effects into self-assembling systems. The inherent amplification of molecular order to macroscopic material properties can be coupled with molecular-scale photoswitching. For instance, in amphiphilic polypeptide systems, self-assembled micelles were stable in the dark but could be disaggregated with light irradiation [7]. This construct can act as a transmembrane structure, where the reversible formation and disruption of the aggregate enabled photo-switchable ion transport [8]. In another example, cyclic peptide rings connected by a *trans*-azo unit would hydrogen-bond with their neighbors, forming extended chains. The *cis*-azo analog, formed upon irradiation, participates in intramolecular hydrogen bonding, forming discrete units and thereby disrupting the higher-order network [9,10]. A system of hydrogen-bonding azobenzene rosettes was also found to spontaneously organize into columns, and these columns assemble into fibers. Upon UV irradiation, this extended ordering was disrupted [11], converting a solid organogel into a fluid. Similarly, large changes in viscosity can be elicited by irradiating a solution of azo polyacrylate associated with the protein bovine serum albumin [12]. In a liquid crystal system, light could be used to induce a glass-to-LC phase transition [13]. A wide variety of applications (such as microfluidics) is possible for functional materials that change phase upon light stimulus.

The primary and secondary shapes of azo-containing self-assembled structures in solution can also be controlled with light. Azo block-copolymers can be used to create photo-responsive micelles [14–18], and vesicles [19]. Since illumination can be used to disrupt vesicle encapsulation, this has been suggested as a pulsatile drug delivery system [20]. The change in azo dipole moment during isomerization plays a critical role in determining the difference between the aggregation in the two states and can be optimized to produce a highly efficient photo-functional vesicle system [21]. The use of azo photoisomerization to disrupt self-assembled systems may be particularly valuable when coupled with biological systems. With biomaterials, one can exploit the powerful and efficient biochemistry of natural systems, yet impose the control of photoactivation. The azobenzene unit in particular has been applied to photo-biological experiments with considerable success [22]. Order–disorder transitions can also be photoinduced in biopolymers. Azo-modified polypeptides may undergo transitions from ordered chiral helices to disordered solutions [23–25], or even undergo reversible α-helix to β-sheet conversions [26]. In many cases, catalytic activity can be regulated due to the presence of the azo group. A cylcodextrin with a histidine and azobenzene pendant, for example, was normally inactive because the *trans* azo would bind inside the cyclodextrin pocket, whereas the photo-generated *cis* version liberated the catalytic site [27]. The activity of papain [28,29] and the catalytic efficiency of lysozyme [30] were similarly modulated by photo-induced disruption of protein structure. Instead of modifying the protein structure itself, one can also embed the protein in a photo-functional matrix [29,31,32], or azo derivatives can be used as small-molecule inhibitors [33]. Azobenzene can also be coupled with DNA in novel ways. In one system, the duplex formation of an azo-incorporating DNA sequence could be reversibly switched [34], since the *trans* azobenzene intercalates between base pairs, stabilizing the binding of the two strands, whereas the *cis* azobenzene disrupts the duplex [35]. The incorporation of an azobenzene unit into the promoter region of an otherwise natural DNA sequence allowed photocontrol of gene expression [36], since the polymerase enzyme has different interaction strengths with the *trans* and *cis* azo isomers. The ability to create biomaterials whose biological function is activated or inhibited on demand via light is of interest for fundamental biological studies, and, possibly, for dynamic biomedical implants.

Perhaps of a range as wide as the interesting phenomena displayed by azo aromatic compounds is the variety of molecular systems into which these chromophores can be incorporated. In addition to liquid crystalline (LC) media and amorphous glasses, azobenzenes can be incorporated into self-assembled monolayers and superlattices, sol–gel silica glasses, and various biomaterials. The photochromic or photo-switchable nature of azobenzenes can also be used to control the properties of novel small molecules, using an attached aromatic azo group. This chapter focuses on the study and application of reversible changes in shape that can be achieved with various systems incorporating azobenzene through the photo-mechanical effect, defined here as the reversible change in molecular shape inducible in some molecules by the adsorption of light, which results in a significant macroscopic mechanical deformation or actuation of the host material.

A comprehensive review will be presented here of the underlying photochemical and photophysical nature of chromophores in host polymers, the geometric and orientational consequences of this isomerization, and some of the interesting ways in which these phenomena have been exploited recently for various photo-mobile applications ranging from one-dimensional (1D) motion on flat surfaces, two-dimensional (2D) transport, micro- and macroscale motion in three-dimensions (3D), through to full actuation applications in robotics.

4.2.1 Azobenzene Chromophores

In this chapter, as in most on the subject, we use "azobenzene" and "azo" in a general way: to refer to the class of compounds that exhibit the core azobenzene structure, with different ring substitution patterns (even though, strictly, these compounds should be referred to as "diazenes"). There are many properties common to nearly all azobenzene molecules. The most obvious is the strong electronic absorption of the conjugated π system. The absorption spectrum can be tailored, via the ring substitution pattern, to lie anywhere from the ultraviolet to the visible-red region. It is not surprising that azobenzenes were originally used as dyes and colorants, and up to 70% of the world's commercial dyes are still based on azobenzene [37,38]. The geometrically rigid structure and large aspect ratio of azobenzene molecules makes them ideal mesogens: Azobenzene small molecules and polymers functionalized with azobenzene can exhibit liquid crystalline phases [39,40]. The most startling and intriguing characteristic of the azobenzenes is their highly efficient and fully reversible photoisomerization. Azobenzenes have two stable geometric isomer states: a thermally stable elongated *trans* configuration and a metastable bent *cis* form. Remarkably, the azo chromophore can interconvert between these two isomers upon absorption of just a single photon, as the quantum yield in many systems approaches unity. For most azobenzenes, the molecule can be optically isomerized from *trans* to *cis* with light anywhere within the broad absorption band in the near UV and visible, and the molecule will subsequently thermally relax back to the *trans* state on a timescale dictated by the substitution pattern. This clean photochemistry is central to azobenzene's potential use as a tool for nanopatterning, and the efficient and tuneable and low energy absorption range is especially attractive for sunlight-driven applications and solar energy harvesting.

Azobenzenes can be separated usefully into three spectroscopic classes, well described first by Rau [41]: azobenzene-type molecules, aminoazobenzene-type molecules, and pseudo-stilbenes (refer to Figure 4.1, for examples). The energies and intensities of their absorption spectra

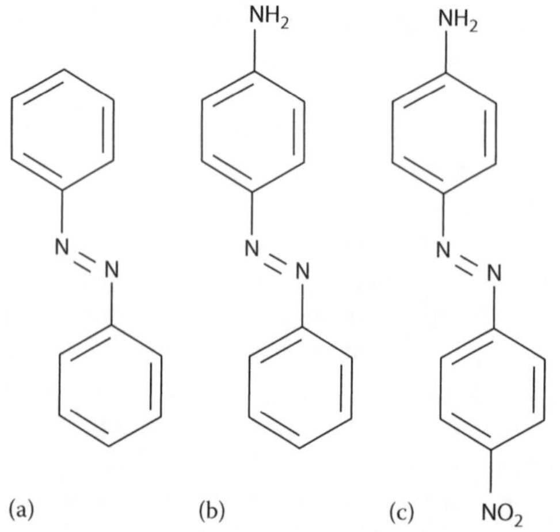

(a)   (b)   (c)   NO$_2$

FIGURE 4.1 Examples of azo molecules classified as (a) azobenzenes, (b) aminoazobenzenes, and (c) pseudo-stilbenes.

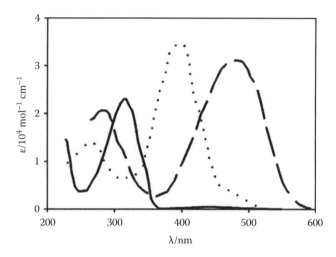

FIGURE 4.2 Schematic of typical absorbance spectra for *trans*-azobenzenes. The azobenzene-type molecules (solid line) have a strong absorption in the UV and a low-intensity band in the visible (barely visible in the graph). The aminoazobenzenes (dotted line) and pseudo-stilbenes (dashed line) typically have strong overlapped absorptions in the visible region.

(shown in Figure 4.2) give rise to their prominent and characteristic class colors: yellow, orange, and red, respectively. Most smaller azo molecules exhibit absorption characteristics similar to the unsubstituted azobenzene archetype, where the molecules exhibit a low-intensity $n \to \pi^*$ band in the visible region, and a much stronger $\pi \to \pi^*$ band in the UV. Although the $n \to \pi^*$ is symmetry-forbidden for *trans*-azobenzene (C_{2h}), vibrational coupling and some extent of non-planarity make it nevertheless observable [42].

Adding substituents of increasing electronic interaction to the azobenzene rings leads to increasing changes in spectroscopic character. Of particular interest is ortho- or para-substitution with an electron-donating group (usually an amino, $-NH_2$), which results in a new class of compounds. In these aminoazobenzenes, the $n \to \pi^*$ and $\pi \to \pi^*$ bands are much closer, and in fact the $n \to \pi^*$ absorption band may be completely buried beneath the intense $\pi \to \pi^*$. Whereas azobenzenes are fairly insensitive to solvent polarity, aminoazobenzene absorption bands shift to higher energy in nonpolar solvents, and shift to lower energy in polar solvents. Further substituting azobenzene at the 4 and 4′ positions with an electron donor and an electron acceptor (such as an amino and a nitro, $-NO_2$, group) leads to a strongly asymmetric electron distribution (often referred to as a "push/pull" substitution pattern). This shifts the $\pi \to \pi^*$ absorption to lower energy toward the red and then well past the $n \to \pi^*$ band. This reversed ordering of the absorption bands defines the third spectroscopic class of the pseudo-stilbenes (in analogy to stilbene, phenyl–C=C–phenyl). The pseudo-stilbenes are very sensitive to the local environment and are highly solvato- and enviro-chromic, which can be useful in some applications.

Especially in condensed phases, the azos are also sensitive to packing and aggregation. The π–π stacking gives rise to shifts of the absorption spectrum. If the azo dipoles adopt a parallel (head-to-head) alignment, called J-aggregates, they give rise to a red-shift of the spectrum (bathochromic) as compared to the isolated chromophore. If the dipoles self-assemble antiparallel (head-to-tail), they are called H-aggregates, and a blue-shift (hypsochromic) is observed. Fluorescence is seen in some aminoazobenzenes and many pseudo-stilbenes, but not in azobenzenes, whereas phosphorescence is absent in all three classes. By altering the electron density, the substitution pattern also necessarily affects the dipole moment in addition to the absorbance and in fact all the higher-order multipole moments. This becomes a significant tool to tailor many nonlinear optical (NLO) properties, such as orientation extent in an applied electric field (poling), and the higher-order moments define the extent molecule's second- and third-order nonlinear response [43], and the strongly asymmetric distribution of the delocalized electrons that results from push/pull substitution results in some superb NLO chromophores.

4.2.2 AZOBENZENE PHOTOCHEMISTRY

Key to some of the most intriguing results and interesting applications of azobenzenes is the facile and reversible photoisomerization about the azo bond, converting between the *trans* (E) and *cis* (Z) geometric isomers (Figure 4.3). This photoisomerization is completely reversible and free from side reactions, prompting Rau to characterize it as "one of the cleanest photoreactions known [41]." The *trans* isomer is more stable by approximately 50–100 kJ/mol [44,45], and the energy barrier to the photo-excited state (barrier to isomerization) is on the order of 200 kJ/mol [46]. In the dark, azobenzene molecules will be found initially in the *trans* form. Upon absorption of a photon (with a wavelength in the *trans* absorption band), the azobenzene will convert, with high efficiency, into the *cis* isomer. A second wavelength of light (corresponding to the *cis* absorption band) can cause the back-conversion, and both these forward and reverse photoisomerizations typically exhibit picosecond timescales [47,48]. Alternately, azos will thermally reconvert from the *cis* to *trans* state, with a timescale ranging from milliseconds to hours, or even days, depending on the substitution pattern and local environment. More specifically, the lifetimes for azobenzenes, aminoazobenzenes, and pseudo-stilbenes are usually on the order of hours, minutes, and seconds, respectively, another useful predictive outcome of Rau's classification scheme. The energy barrier for thermal isomerization is on the order of 100–150 kJ/mol [49,50]. Considerable work has gone into elongating the *cis* lifetime with the goal of creating truly bistable photo-switchable systems, such as by attaching bulky ring substituents to hinder the thermal back reaction. Polyurethane main-chain azos achieved a lifetime of more than 4 days (thermal rate-constant of $k = 2.8 \times 10^{-6}$ s^{-1}, at 3°C) [51], and one azobenzene para-substituted with bulky pendants had a lifetime of even 60 days ($k < 2 \times 10^{-7}$ s^{-1}, at room temperature) [52], yet the back re-conversion could not be entirely arrested at environmental temperatures. The conformational strain of macrocyclic azo compounds can be used however to metastably "lock" the *cis* state, where lifetimes of 20 days ($k = 5.9 \times 10^{-7}$ s^{-1}) [53], 1 year (half-life 400 days, $k = 2 \times 10^{-8}$ s^{-1}) [54,55], and even 6 years ($k = 4.9 \times 10^{-9}$ s^{-1}) [56] were observed. Similarly, using the hydrogen-bonding of a peptide segment to generate a cyclic structure, a *cis* lifetime of ~40 days ($k = 2.9 \times 10^{-7}$ s^{-1}) was demonstrated [10]. Of course, one can also generate a system that starts in the *cis* state and where isomerization (in either direction) is completely hindered. For instance, attachment to a surface [57], direct synthesis of ring-like azo molecules [58], and crystallization of

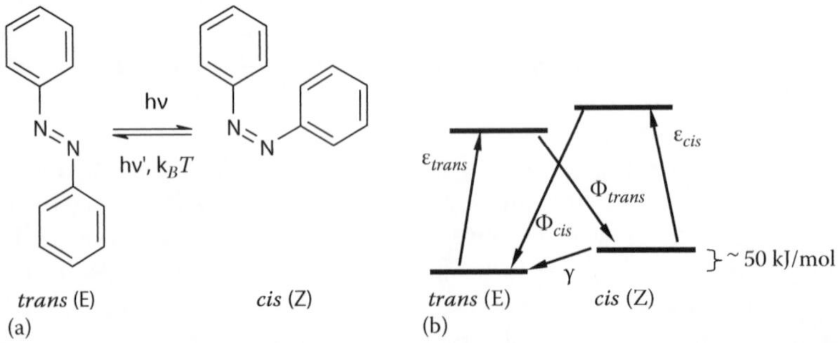

FIGURE 4.3 (a) Azobenzene can convert between *trans* and *cis* states photochemically and relaxes to the more stable *trans* state thermally. (b) Simplified state model for azobenzenes. The *trans* and *cis* extinction coefficients are denoted ε_{trans} and ε_{cis}. The Φ refer to quantum yields of photoisomerization, and γ is the thermal relaxation rate constant.

the *cis* form [59,60] can be used to maintain one state, but such systems are obviously not bistable photoswitches, nor are they reversible.

A bulk azo sample or solution under illumination will achieve a photostationary state, with a steady-state *trans/cis* composition based on the competing effects of photoisomerization into the *cis* state, thermal relaxation back to the *trans* state, and *cis* reconversion upon light absorption. The steady-state composition is unique to each system, as it depends on the quantum yields for the two processes (Φ_{trans} and Φ_{cis}) and the thermal relaxation rate constant. The composition also thus depends upon irradiation intensity, wavelength, temperature, and the matrix (gas phase, solution, liquid crystal, sol–gel, monolayer, polymer matrix, etc.). Azos are photochromic (their color changes upon illumination), since the effective absorption spectrum (a combination of the *trans* and *cis* spectra) changes with light intensity. Thus, absorption spectroscopy can be conveniently used to measure the *cis* fraction in the steady state [61,62], and the subsequent thermal relaxation to an all-*trans* state [63–66]. NMR spectroscopy can also be used [67]. Under moderate irradiation, the composition of the photostationary state is predominantly *cis* for azobenzenes, mixed for amino-azobenzenes, and predominantly *trans* for pseudo-stilbenes. In the dark, the *cis* fraction is below most detection limits, and the sample can be considered to be in an all-*trans* state. Isomerization is induced by irradiating with a wavelength within the azo's absorption spectrum, preferably close to λ_{max}. Modern experiments typically use laser excitation with polarization control, delivering on the order of 1–100 mW/cm² of power to the sample. Various lasers cover the spectral range of interest, from the UV (Ar$^+$ line at 350 nm), through blue (Ar$^+$ at 488 nm), green (Ar$^+$ at 514 nm, YAG at 532 nm, HeNe at 545 nm), and into the red (HeNe at 633 nm, GaAs at 675 nm). The ring substitution pattern affects both the *trans* and the *cis* absorption spectra, and for certain patterns, the absorption spectra of the two isomers overlap significantly (notably for the pseudo-stilbenes). In these cases, a single wavelength of light activates both the forward and reverse reactions, leading to a mixed stationary state, and continual interconversion of the molecules. For some interesting azobenzene photomotions, this rapid and efficient cycling of chromophores is advantageous, whereas in cases where the azo chromophore is being used as a switch, it is clearly undesirable.

The mechanism of isomerization has undergone considerable debate. Isomerization takes place either through a rotation about the activated N–N single bond, with rupture of the π bond, or through inversion, with a semilinear and hybridized transition state, where the π bond remains intact (refer to Figure 4.4). The thermal back-relaxation is generally agreed to proceed via rotation, whereas for the photochemical isomerization, both mechanisms appear viable [68]. Historically the rotation mechanism (as necessarily occurs in stilbene) was favored for photoisomerization, with some early hints that inversion may be contributing [69]. More recent experiments, based on matrix or molecular constraints to the azobenzene isomerization, strongly support inversion [70–73]. Studies using picosecond Raman and femtosecond fluorescence imply a double bond (N=N) in the excited state, confirming the inversion mechanism [74,75]. By contrast, Ho et al. [76] found evidence that the pathway is compound specific: a nitro-substituted azobenzene photoisomerized via the rotation pathway. Furthermore, ab initio and density functional theory calculations indicate that both pathways are energetically accessible, although inversion is preferred [77,78]. Thus, both mechanisms may be competing, with a different one dominating depending on the particular chromophore and environment. The emerging consensus nevertheless appears to be that inversion is the dominant pathway for most azobenzenes [79]. The availability of the inversion mechanism explains how azos are able to isomerize easily even in rigid matrices, such as glassy polymers, since the inversion mechanism usually has a much smaller free volume requirement than the rotation.

The thermal back-relaxation is generally first order, although a glassy polymer matrix can lead to anomalously fast decay components [80–83], attributed to a distribution of chromophores in highly strained configurations. Higher matrix crystallinity increases the rate of decay [84]. The decay rate itself can act as a probe of local environment and molecular conformation [85,86]. The back-relaxation of azobenzene is acid catalyzed [87], although strongly acidic conditions will lead to side reactions [60]. For the parent azobenzene molecule, quantum yields (which can be indirectly

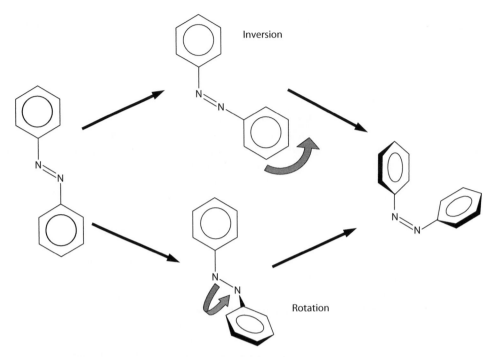

FIGURE 4.4 Mechanism of azobenzene isomerization proceeds either via rotation or inversion. The *cis* state has the phenyl rings tilted at 90° with respect to the CNNC plane.

measured spectroscopically [80,88,89]) are on the order of 0.6 for the *trans* → *cis* photoconversion, and 0.25 for the back photoreaction. Solvent has a small effect, increasing the *trans* → *cis* and decreasing the *cis* → *trans* yield as polarity increases [90]. Aminoazobenzenes and pseudo-stilbenes isomerize very quickly and can have quantum yields as high as 0.7–0.8.

4.2.3 CLASSES OF AZOBENZENE SYSTEMS

Azobenzenes are robust and versatile chromophores, and have been extensively investigated and applied as small dye molecules, pendants on other molecular structures, or incorporated (doped or covalently bound) into a wide variety of amorphous, crystalline, or liquid crystalline polymeric host systems. Noteworthy examples include self-assembled monolayers and superlattices [91], sol–gel silica glasses [92], and biomaterials [22,93,94]. A variety of small structural molecules incorporating azobenzene have been synthesized, including crown-ethers [95], cyclodextrins [96,97], proteins such as bacteriorhodopsin [98], and 3D polycyclics such as cubane [99] and adamantane [100]. Typically, however, azo chromophores are embedded into a solid matrix for studies and real devices. As a result, matrix effects are inescapable: The behavior of the chromophore is altered due to the matrix, and in turn the chromophore alters the matrix [101]. Although either could be viewed as a hindrance, both can in fact be quite useful: The chromophore can be used as a delicate probe of the matrix (free volume, polarizability, mobility, morphology, viscoelasticity, etc.), and when the matrix couples to chromophore motion, molecular motions can be translated to larger lengthscales, for example using nanometer "command surfaces" of azo chromophores to re-orient surrounding micron-sized layers of inert liquid crystals [101]. Thus, the incorporation strategy can be extremely valuable for transferring and amplifying azobenzene's photo-reversible effects.

Doping small molecules azobenzenes into polymer matrices is perhaps the convenient inclusion technique [102,103]. These "guest–host" systems can be cast or spin-coated from solution mixtures of polymer and azo small molecules, where the azo content in the thin film is easily adjusted

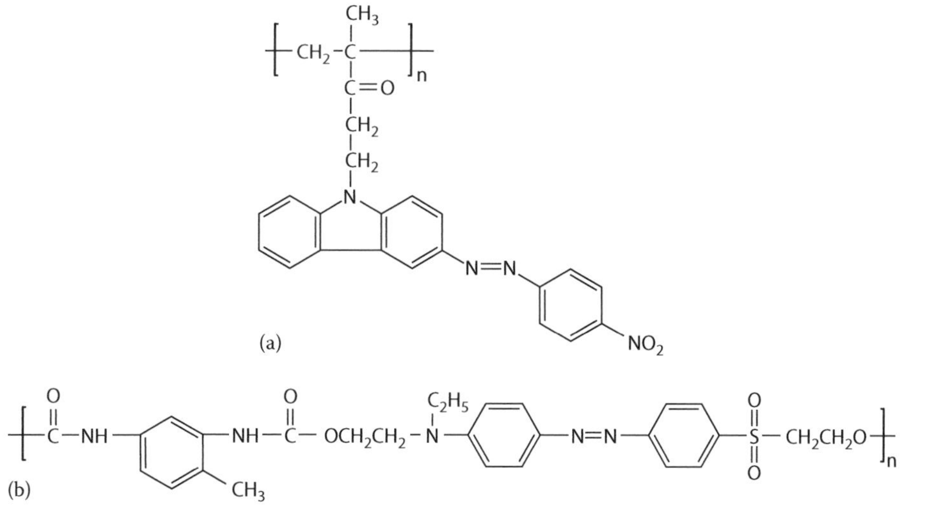

FIGURE 4.5 Examples of azo polymer structures, showing that both (a) side-chain and (b) main-chain architectures are possible.

via concentration. Although doping leaves the azo chromophores free to undergo photo-induced motion unhindered, it has been found that many interesting photo-mechanical effects do not couple to the matrix in these systems. Furthermore, the azo mobility often leads to instabilities, such as phase separation or microcrystallization. Thus, one of the most reliable, robust, and effective methodologies for incorporating azobenzene into functional materials is by covalent attachment to polymers. The resulting materials benefit from the inherent stability, rigidity, and processability of polymers, in addition to maximizing the photo-responsive behavior of the azo chromophores. Both side-chain and main-chain azobenzene polymers have been prepared (Figure 4.5) [104]. Many different backbones have been used as scaffolds for azo moieties, including imides [105], esters [106], urethanes [107], ethers [108], organometallic ferrocene polymers [109], even conjugated polydiacetylenes [110], polyacetylenes [111], and main-chain azobenzenes [112,113]. The most common azo polymers are acrylates [114], methacrylates [115], and isocyanates [116].

Another clever and unique strategy that allows for the simplicity of doping while retaining the stability of covalent polymers is to engineer complementary noncovalent attachment of the azo dyes to the polymer backbone. In particular, ionic attachment can lead to a homogenous and stable matrix (when dry) [117]. The use of surfactomesogens (molecules with ionic and liquid crystalline properties) also enables a simple and programmatic way of generating new materials [118]. It has been demonstrated that azobenzenes can be solubilized by guest–host interactions with cyclodextrin [119], and it is thus possible that similar strategies could be fruitfully applied to the creation of bulk materials.

Considerable research has also been conducted on azobenzene dendrimers [120,121] and molecular glasses [122]. These inherently monodisperse materials offer the possibility of high stability, excellent sample homogeneity (crucial for high-quality optical films), and excellent spatial control (with regard to lithography, for instance) without sacrificing the useful features of amorphous linear polymers. The synthetic control afforded with such systems allows one to carefully tune solubility, aggregation, thermal stability, and crystallinity [123–125]. The unique structure of dendrimers can be used to exploit azobenzene's photochemistry [126–128]. For instance, the dendrimer structure can act as an antenna, with light-harvesting groups at the periphery, making energy available via intramolecular energy transfer to the dendrimer core [129,130]. Thus a dendrimer with an azo core could be photoisomerized using a wavelength outside of its native absorption band. The dendrimer architecture can also be used to amplify the molecular motion of azo isomerization. For instance, a dendrimer with three azobenzene arms exhibited different physical properties for all the various isomerization combinations, and the isomers could be separated by thin-layer chromatography on this basis [131]. Thin films of azo material are typically prepared with spin-coating, where a polymer solution is dropped onto a rotating substrate. This technique is fast and simple, and generally yields high-quality films that are homogeneous over a wide area. Films can also be prepared via solvent evaporation, the Langmuir–Blodgett technique [132–135], self-assembled monolayers [136], or layer-by-layer electrostatic self-assembly to produce azo polyelectrolyte multilayers (PEMs) [137–141].

Azobenzenes are also ideal candidates to act as mesogens: molecules of a shape that readily forms liquid crystalline (LC) mesophases. Many examples of small-molecule high aspect-ratio azobenzene liquid crystals have been studied, and some azo polymers also form LC phases (refer to Figure 4.6 for a typical structure). For side-chain azobenzenes, a certain amount of mobility is required for LC phases to be present; as a rule, if the tether between the chromophore and the backbone is less than 6 alkyl units long, the polymer should exhibit an amorphous and isotropic solid-state phase, whereas if the spacer is longer than 6 units, LC phases typically form. The photoisomerization of azobenzene leads to modification of the phase and alignment (director) in LC systems [101,142]. The director of a liquid crystal phase can be modified by orienting chromophores doped into the phase [143,144], by using an azobenzene-modified "command surface [145–147]," using azo copolymers [148], and, of course, in pure azobenzene LC phases [149,150]. One can force the LC phase to adopt an in-plane order (director parallel to surface), homeotropic alignment (director perpendicular to surface), tilted or even biaxial orientation [151]. These changes are fast and reversible. While the *trans* azobenzenes are excellent mesogens, the *cis* azos typically are not. If even a small number of azo molecules are distributed in an LC phase, *trans* → *cis* isomerization can destabilize the phase by lowering the nematic-to-isotropic phase transition temperature [152]. This enables fast isothermal photocontrol of phase transitions [79,153–155].

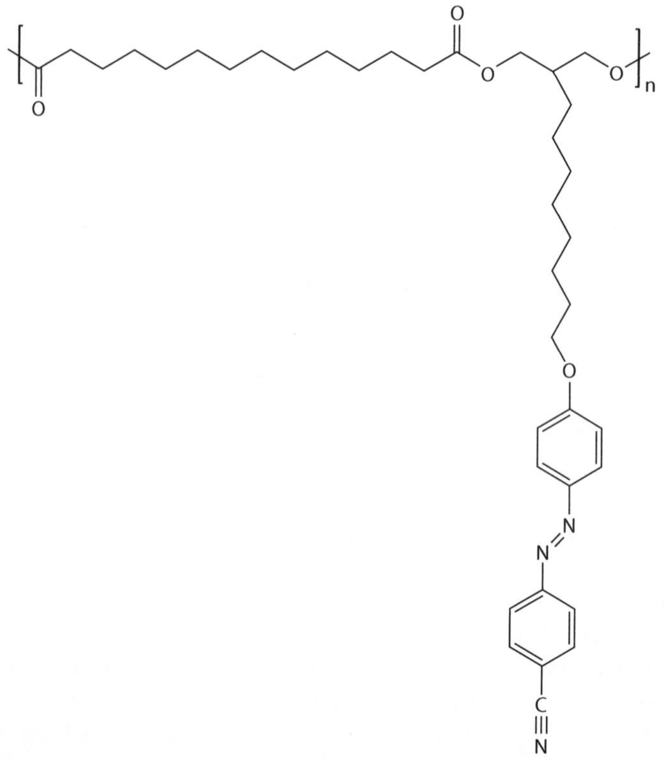

FIGURE 4.6 Typical liquid crystalline side-chain azobenzene polymer.

4.3 PHOTO-INDUCED MOTIONS AND MOVEMENTS

Irradiation with light produces molecular geometric changes in azobenzenes, and under appropriate conditions, these changes can translate into larger-scale motions and even macroscopic movements of material properties. As classified first by Natansohn and Rochon [156], we will describe these motions roughly in order of increasing size scale. However, since the motion on any size-scale invariably affects (and is affected by) other scales, clear divisions do not exist, and these effects are often concurrent and interdependent. The most relevant motion to eventual actuation applications is expansion and contraction due to the photo-mechanical effect, so this effect will be highlighted for this chapter—the photo-mechanical effect in amorphous and LC azo polymers, though the related effect of all-optical surface patterning ("photo-morphing") which occurs when an azo polymer film is exposed to a light intensity gradient will also be introduced and discussed briefly [43,156–158].

4.3.1 MOLECULAR MOTION

The root fundamental molecular photomotion in azobenzenes is the geometrical change that occurs upon absorption of light. In *cis*-azobenzene, the phenyl rings are twisted at 90° relative to the C–N=N–C plane [71,159]. Isomerization reduces the distance between the 4 and 4' positions from 0.99 nm in the *trans* state to 0.55 nm for the *cis* state [160–162]. This geometric change creates or increases the dipole moment: whereas the *trans* form has no dipole moment in parent azobenzene, the *cis* form has a dipole moment of 3.1 D [59]. The free volume requirement of the *cis* can be much larger than the *trans* [163], and the free volume required to *cycle between* these two states is still larger. It has been estimated that the minimum free volume pocket required to allow isomerization to proceed through a transition state via the inversion pathway is 0.12 nm^3 [71,81] and via the rotation pathway is approximately 0.38 nm^3 [51]. The effects of matrix free volume constraints on photochemical reactions in general have been considered [164]. The geometrical changes in azobenzene are very large, by molecular standards, and it is thus no surprise that isomerization modifies a wide host of material properties. More recent measurements via high-pressure spectroscopy (10^4–10^5 atm) on the force applied and energy exerted through this isomerization suggest that azobenzene is indeed an extremely powerful little artificial muscle, and application optimization for actuators depends largely on clever engineering of the mechanical advantage provided and is not inherently materials limited.

This molecular displacement generates a significant nanoscale force, which has been measured in single-molecule force spectroscopy experiments [165,166] and compared well to theory [167]. In these experiments, illumination causes contraction of an azobenzene polymer, showing that each chromophore can exert pN to nN molecular forces on demand. This force can be further engineered into an on–off or "ratchet motion" switch bridging the gap between force and simple machinery, such as that demonstrated by an ingenious two-state pseudo-rotaxane that could be reversibly threaded-dethreaded using light that Stoddard and coworkers called an "artificial molecular-level machine [168,169]." The ability to activate and power molecular-level devices using light is of course attractive in many applications, since it circumvents the limitations inherent to diffusion or wiring, and permits a remote (or even quite distant) power supply. The fast response and lack of waste products in azo isomerization are also advantageous. Coupling these molecular-scale motions to do actual human-scale useful work is of course the next challenging step. Encouraging progress in this direction is evident, however, from a wide variety of molecular switches that have been synthesized. For example, an azo linking two porphyrin rings enabled photocontrol of electron transfer [170], and in another example, dramatically different hydrogen-bonding networks (intermolecular and intramolecular) could be favored based on the isomeric state of the azo group linking two cyclic peptides [9,10]. Other recently reported examples include osmotic pressure pumps [171], created by the photo-controlled solubility of azobenzene, analytical columns that increase the effluent rate of developing solvents [172], reversible light controlled conductance switching [173], and photo-responsive gold nanoparticle solvation [174], and network formation [175].

4.3.2 MACROSCOPIC MOTION

While it is important to study the nm-scale azobenzene molecular conformational changes that give rise to macroscopic phenomena, by far the most useful applications to actuation are the reversible changes that can result in changes to bulk phenomena, or to macroscopic motion over the μm to cm size-scale. The first consideration is perhaps whether the host material can expand or contract to an appreciable extent. In floating monolayers at a liquid surface, it is well established that the larger molecular size of the *cis* isomer leads to a corresponding lateral expansion of many 10s of % [176], which can modify other bulk properties. For instance, this allows photomodulation of a monolayer's water contact angle [177] or surface potential [178]. Using fluorinated azo polymers, good photocontrol was demonstrated over photopatterning [179,180], and wettability has been demonstrated [181–183]. A monolayer of azo-modified calixarene, when irradiated with a light gradient, produced a gradient in surface energy sufficient to move a macroscopic oil droplet [101]. In more recent work, surfactants of azobenzene were used to create a liquid–liquid interface between oleic acid droplets in an aqueous solution [184]. Photoisomerization of the azobenzene surfactant created a wavelength-dependant interfacial tension capable of inducing interfacial flow, and this interfacial flow then generated large-scale droplet motion in a direction opposite to the gradient. The photo-controlled droplet motion was thus used to direct droplets into the trajectories of various shapes and letters. It also suggests possible applications of the aforementioned materials to microfluidics. Modest photo-induced contact angle changes for thin polymer films have also been reported [84]. Recently an azobenzene copolymer assembled into polyelectrolyte multilayer showed a modest 2° change in contact angle with UV light irradiation. However, when the same copolymer was assembled onto a patterned substrate, the change in contact angle upon irradiation was enhanced to 70° [183]. That surface roughness plays a role in contact angle is well established, and shows that many systems can be optimized to give rise to a large change in surface properties through clever amplification.

In layered inorganic systems with intercalated azobenzenes, reversible photochanges in the basal spacing (on the order of 4%) can be achieved [185,186]. In polymer films, there is evidence that the film thickness increases, as measured by in situ single wavelength ellipsometry, atomic force microscopy (AFM), and in situ neutron reflectometry [146]. Photocontraction for semicrystalline main-chain azos has been measured [105,187–191], where the extent and direction (expansion or contraction) of photo-mechanical change can be tuned by using ambient temperature as demonstrated by variable-temperature neutron reflectometry experiments. The experiments demonstrated unambiguously that both photoexpansion and photocontraction could be optimized in a single azo material merely by varying the dominance of these two competing effects with low and high extremes of temperature, respectively.

The most convincing demonstration of macroscopic motion due to azo isomerization is the mechanical bending and unbending of a free-standing LC polymer film [192,193]. The macroscopic bending direction may be selected either with polarized light, or by aligning the chromophores with rubbing. Bending occurs in these relatively thick films because the free surface (which absorbs light) contracts, whereas the interior of the film (which is not irradiated owing to the strong absorption of the upper part of the film) does not contract. Because the direction of bending can be controlled with polarized light, the materials enable full directional photo-mechanical control [194]. Other examples include expansion-bending of cantilevers coated with an amorphous azobenzene thin film [195,196], and macroscopic contraction-bending of fibers and cantilevers made of azobenzene liquid crystalline elastomers [197–203]. One can also invert the coupling of mechanical and optical effects: By stretching an elastomeric azo film containing a grating, one can affect its wavelength-selection properties and orient chromophores [204]. Much of this bending and related motion at the macroscale are invariably related to or have possible applications in actuation and will be further discussed in Sections 4.4 and 4.5.

4.3.3 PHOTOTRANSPORT ACROSS SURFACES

In 1995, an unexpected and unprecedented optical effect was discovered in polymer thin films containing the azo chromophore Disperse Red 1 (DR1), shown in Figure 4.7. The Natansohn/Rochon research team [205] and the Tripathy/Kumar collaboration [206] simultaneously and independently discovered a large-scale surface mass transport when the films were irradiated with a light interference pattern. In a typical experiment, two coherent laser beams, with a wavelength in the azo absorption band, are intersected at the sample surface (refer to Figure 4.8). The sample usually consists of a thin spin-cast film (10–1000 nm) of an amorphous azo polymer on a transparent substrate. The sinusoidal light interference pattern at the sample surface leads to a sinusoidal surface patterning, i.e., a relief grating often referred to in the literature as a surface relief grating (SRG), though the effect is not limited to just gratings, and might more accurately and generally be called photopatterning, phototransport, or photomorphing. These gratings were found to be extremely large, up to hundreds of nanometers, as confirmed by AFM (Figure 4.9), which means that the light induced motion of many 100s of nm of the polymer chains to "walk" across the substrate surface. The SRGs diffract very cleanly and efficiently, and in retrospect it is clear that many reports of large diffraction efficiency prior to 1995, attributed to birefringence, were in fact due to surface gratings unbeknownst to the experimenters. The process occurs readily at room temperature (well below the T_g of the amorphous polymers used) with moderate irradiation (1–100 mW cm^{-2}) over seconds to minutes. The phenomenon is a reversible mass transport, not irreversible material ablation, since a flat film with the original thickness is recovered upon heating above T_g. Critically, it requires the presence and isomerization of azobenzene chromophores, as other absorbing but nonisomerizing chromophores do not produce SRGs. Many other systems can exhibit optical surface patterning [207], but the amplitude of the modification is much smaller, does not involve mass transport, and usually requires additional processing steps. The all-optical patterning unique to azobenzenes has been studied intensively since its discovery and many reviews of the remarkable body of experimental results are available [43,104,156,158].

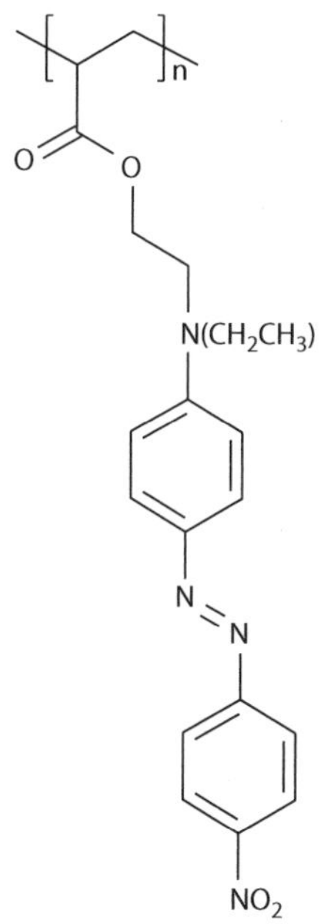

FIGURE 4.7 Chemical structure of poly(disperse red 1) acrylate, pdr1a, a pseudo-stilbene side-chain azo polymer that generates high-quality surface-relief structures.

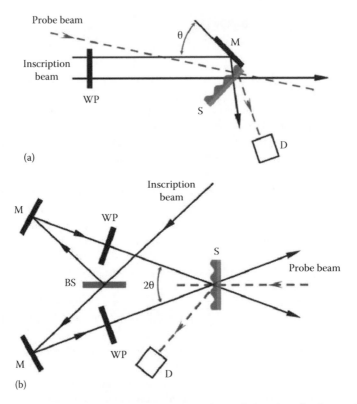

FIGURE 4.8 Experimental setup for the inscription of a surface relief grating: S refers to the sample, M are the mirrors, D is a detector for the diffraction of the probe beam, WP is a waveplate (or generally a combination of polarizing elements), and BS is a 50% beam splitter. The probe beam is usually a HeNe (633 nm) and the inscription beam is chosen based on the chromophore absorption band (often Ar⁺ 488 nm). (a) A simple one-beam inscription involves reflecting half of the incident beam off a mirror adjacent to the sample. (b) A two-beam interference setup enables independent manipulation of the polarization state of the two incident beams.

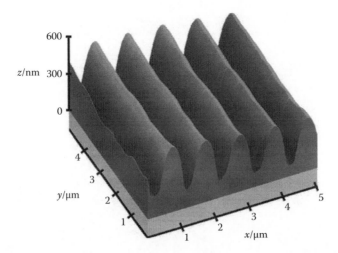

FIGURE 4.9 AFM image of a typical surface relief grating (SRG) optically inscribed into an azo polymer film. Grating amplitudes of hundreds of nanometers, on the order of the original film thickness, are easily obtained. In this image, the approximate location of the film-substrate interface has been set to $z = 0$, based on knowledge of the film thickness.

4.3.3.1   Photopatterning and Photomorphing

In a typical photo-patterning grating inscription experiment, a sinusoidally varying light pattern is generated at the sample surface. What results is a sinusoidal surface profile: a surface relief grating (SRG). This is the pattern most often reported in the literature, because it is most conveniently generated (by intersecting two coherent beams) and most easily monitored (by recording the diffraction intensity at a nonabsorbing wavelength, usually using a HeNe laser at 633 nm). However, it must be emphasized that the azo surface mass transport can produce arbitrary patterns. Essentially, the film encodes the impinging light pattern as a topography pattern (as a Fourier transform), encoding holographically both the spatial intensity and the polarization patterns of the incident light. What appears to be essential is a *gradient* in the intensity and/or polarization of the incident light field. For instance, a single focused Gaussian laser spot will lead to a localized pit depression, a Gaussian line will lead to an elongated trench, etc. [208]. In principle, any arbitrary pattern could be generated through an appropriate mask, interference/holographic setup, or scanning of a laser spot [156].

Concomitant with the inscription of a surface relief is a photo-orientation of the azo chromophores, which depends on the polarization of the incident beam(s). The orientation of chromophores in SRG experiments has been measured using polarized Raman confocal microspectrometry [209–211], and the strong surface orientation has been confirmed by photoelectron spectroscopy [212]. What is found is that the chromophores orient perpendicular to the local polarization vector of the impinging interference pattern. Thus, for a $(+45°, -45°)$ two-beam interference: In the valleys $(x = 0)$, the electric field is aligned in the y-direction, so the chromophores orient in the x-direction; in the peaks $(x = \Lambda/2)$, the chromophores orient in the y-direction; in the slope regions $(x = \Lambda/4)$, the electric field is circularly polarized and thus the chromophores are nearly isotropic. For a (p, p) two-beam interference, it is observed that the chromophores are primarily oriented in the y-direction everywhere, since the impinging light pattern is always linearly polarized in the x-direction. Mass transport may lead to perturbations in the orientational distribution, but photo-orientation remains the dominant effect.

The anisotropy grating that is submerged below a surface relief grating apparently also leads to the formation of a density grating under appropriate conditions. It was found that upon annealing a SRG, which erases the surface grating and restores a flat film surface, a density grating began growing beneath the surface (and into the film bulk) [213,214]. This density grating only develops where the SRG was originally inscribed, and it appears that the photo-orientation and mass transport leads to the nucleation of liquid crystalline "seeding aggregates" that are thermally grown into larger-scale density variations. The thermal erasure of the SRG, with concomitant growth of the density grating, has been both measured [215] and modeled [216]. Separating the components due to the surface relief and the density grating is described in a later section, but briefly: The diffraction of a visible-light laser primarily probes the surface relief, whereas a simultaneous x-ray diffraction experiment probes the density grating. The formation of a density grating is similar to, and consistent with, the production of surface topography [217] and surface density patterns [218], as observed by tapping mode AFM on an azo film exposed to an optical near field. In these experiments, it was found that volume is not strictly conserved during surface deformation [219], consistent with changes in density.

4.3.3.2   Dependence of Phototransport on Material Properties

For all-optical surface motion and patterning to occur, one necessarily requires azobenzene chromophores in some form. There are, however, a wide variety of azo materials that have exhibited surface mass patterning. This makes the process much more attractive from an applied standpoint: It is not merely a curiosity restricted to a single system but rather a fundamental phenomenon that can be engineered into a wide variety of materials. It was recognized early on that the gratings do not form in systems of small molecules (for instance, comparing unreacted monomers to their corresponding polymers). The polymer molecular weight (MW), however, must not be too large [220].

Presumably a large MW eventually introduces entanglements that act as cross-links, hindering polymer motion. Thus intermediate molecular weight polymers (MW $\sim 10^3$, arguably oligomers) are optimal. That having been said, there are many noteworthy counterexamples. Weak SRGs can be formed in polyelectrolyte multilayers, which are essentially cross-linked-polymer systems [211,221–223]. Efficient grating formation has also been demonstrated using an azo cellulose with ultra-high molecular weight (MW $\sim 10^7$) [224,225]. In a high molecular weight polypeptide (MW $\sim 10^5$), gratings could be formed where the grating amplitude was dependant on the polymer conformation [226], and restricted conformations (α-helices and β-sheets) hindered SRG formation.

The opposite extreme has also been investigated: Molecular glasses (amorphous nonpolymeric azos with bulky pendants) exhibited significant SRG formation [227–229]. In fact, the molecular version formed gratings more quickly than its corresponding polymer [230]. Another set of experiments compared the formation of gratings in two related arrangements: (1) a thin film of polymer and small-molecule azo mixed together and (2) a layered system, where a layer of the small-molecule azo was deposited on top of the pure polymer [231]. The SRG was negligible in the layered case. Whereas the authors suggest that "layering" inhibits SRG formation, it may be interpreted that coupling to a host polymer matrix enhances mass transport, perhaps by providing rigidity necessary for fixation of the pattern. A copolymer study did in fact indicate that strong coupling of the mesogen to the polymer enhanced SRG formation [232], and molecular glasses with hindered structures also enhanced grating formation [233].

Gratings have also been formed in liquid crystalline systems [234,235]. In some systems, it was found that adding stoichiometric quantities of a nonazo LC guest greatly improved the grating inscription [236,237]. This suggests that SRG formation may be an inherently cooperative process, related to the mesogenic nature of the azo chromophore. The inscription sometimes requires higher power (>1 W/cm^2) than in amorphous systems [238], and in dendrimer systems, the quality of the SRG depends on the generation number [239]. Maximizing the content of azo chromophore usually enhances SRG formation [240], although some studies have found that intermediate functionalization (50%–80%) created the largest SRG [241,242]. Some attempts have been made to probe the effect of free volume. By attaching substituents to the azo ring, its steric bulk is increased, which presumably increases the free volume requirement for isomerization. However, substitution also invariably affects the isomerization rate constants, quantum yield, refractive index, etc. This makes any universal analysis ambiguous. At least in the case of photo-orientation, the rate of inscription appears slower for bulkier chromophores, although the net orientation is similar [243,244]. For grating formation, it would appear that chromophore bulk is of secondary importance to many other inscription parameters. The mass transport occurs readily at room temperature, which is well below the glass-to-rubber transition temperature (T_g) of the amorphous polymers typically used. Increasing polymer, T_g [245], has been suggested to favor grating formation and modulation depth, and gratings can even be formed in polymers with exceptionally high T_g [246], sometimes higher than 370°C [247]. These gratings, however, can sometimes be difficult to completely erase reversibly via annealing [248].

4.3.3.3 Photo-Pressure Mechanism of Azo Photomotion

Several mechanisms have been described to account for the microscopic origin of the driving force in azobenzene optical patterning. Arguments have appealed to thermal gradients, diffusion considerations, isomerization-induced pressure gradients, and interactions between azo dipoles and the electric field of the incident light. At present, the isomerization-induced pressure gradient mechanism appears to provide the most complete and satisfactory explanation consistent with most known observations, and quantitative modeling (though some strong polarization dependence is still unexplained). Subsequent viscoelastic modeling of the movement process has also been quite successful, correctly reproducing nearly all experimentally observed surface patterns, without needing to directly describe the microscopic nature of the driving force. Fluid mechanics models provided suitable agreement with observations [249] and were later extended to take into account a depth dependence and a velocity distribution in the film [250,251], which reproduces the thickness dependence of SRG inscription. A further elaboration took into account induced anisotropy in the film and associated anisotropic polymer film deformation (expansion or contraction in the electric field direction) [252]. The assumption of an anisotropic deformation is very much consistent with experimental observations [253]. Such an analysis, remarkably, was able to reproduce most of the polarization dependence, predicted phase-inverted behavior at high power, and even demonstrated double-period (interdigitated) gratings. A nonlinear stress-relaxation analysis could account for the nonlinear response during intermittent (pulse-like) exposure [254]. Finite-element linear viscoelastic modeling enabled the inclusion of finite compressibility [255]. This allowed the nonlinear intermittent-exposure results, and, critically, the formation of density gratings, to be correctly predicted. This analysis also demonstrated, as expected, that surface tension acts as a restoring force that limits grating amplitude (which explains the eventual saturation). Finally, the kinetics of grating formation (and erasure) have been captured in a lattice Monte Carlo simulation that takes into account isomerization kinetics and angular redistribution of chromophores [256–258]. Thus, the nonlinear viscoelastic flow and deformation (compression and expansion) of polymer material appear to be well understood. What remains to be fully elucidated is the origin of the force inside the material. More specifically, the connection between the azobenzene isomerization and the apparent force must be explained.

One of the first mechanisms to be presented was the suggestion by Barrett et al. of pressure gradients inside the polymer film [220,249]. The assumption is that azobenzene isomerization generates pressure due to the greater free volume requirement of the *cis* and due to the volume requirement of the isomerization process itself. Isomerization of the bulky chromophores leads to pressure that is proportional to light intensity. The light intensity gradient thus generates a pressure gradient, which leads to plastic material flow modeled well by traditional fluid mechanics. Order-of-magnitude estimates were used to suggest that the mechanical force of isomerization would be greater than the yield point of the polymer, enabling flow. Plastic flow is predicted to drive material out of the light, consistent with observations in amorphous systems. At first it would seem that this mechanism cannot be reconciled with the polarization dependence, since the pressure is presumably proportional to light intensity, irrespective of its polarization state. However, one must more fully take orientational effects into account. Linearly polarized light addresses fewer chromophores than circularly polarized light, and would thus lead to lower pressure. Thus, pure polarization patterns can still lead to pressure gradients. Combined with the fact that the polarized light is orienting (and in a certain sense photobleaching and spectral hole burning), this can explain some aspects of the polarization data. The agreement, however, is still not perfect. For instance, the (*s,s*) and (*p,p*) combinations lead to very different gratings in experiments. It is possible that some missing detail related to polarization will help explain this discrepancy, such as orientation due to shear thinning during the resultant flow, where shear flow orients the azo mesogens *away* from the light vector axis in the case of (*s,s*), diminishing the driving force, yet aligns the azo mesogens *back into* the light axis in the case of (*p,p*), enhancing the driving force.

Considering the balance of experimental results from the literature, it now appears the simple mechanical argument of a pressure mechanism may be most correct, though other contributing coeffects such as photosoftening may assist the process. In one key experiment, irradiation of a transferred Langmuir–Blodgett film reversibly generated ~5 nm "hills," attributed to nanoscale buckling that relieves the stress induced by lateral expansion [259]. This result is conspicuously similar to the spontaneous polarization-dependant formation of hexagonally arranged ~500 nm "hills" seen on an amorphous azo polymer sample irradiated homogeneously [260,261]. In fact, homogeneous illumination of azo surfaces has caused roughening [262] and homogeneous optical erasure of SRGs leads to similar pattern formation [263]. The early stages of SRG formation, imaged by AFM, again show the formation of nano-sized hills [264]. Taken together, these seem to suggest that irradiation of an azo film leads to spontaneous lateral expansion, which induces a stress that can be relieved by buckling of the surface, thereby generating surface structures. In the case of a light gradient, the buckling is relieved by mass transport coincident with the light field that generated the pressure inside the film. In an experiment on main-chain versus side-chain azo polymers, the polarization behavior of photodeformation was opposite [219]. This may be explained by postulating that the main-chain polymer contracts upon isomerization, whereas the side-chain polymer architecture leads to net expansion. Similarly, the opposite phase behavior in amorphous and LC systems may be due to the fact that the former photoexpand and the latter photocontract [253]. Lastly, many large surface structures were observed in an azo-dye-doped elastomer film irradiated at high power (4 W/cm^2) [265]. The formation of structures both parallel and perpendicular to the grating direction could be attributed to photoaggregation of the azo dye molecules and/or buckling of the elastomeric surface. Thus a purely photo-mechanical explanation may be able to describe surface mass transport in azo systems. Further investigations into reconciling this model with the polarization dependence of inscription are in order, and presumably photo-orientation will play a key role, as may a photo-softening effect of lower viscosity under irradiation.

4.4 NANO-FABRICATION APPLICATIONS OF AZO PHOTOMOTION

The rapid, facile, reversible, and single-step all-optical surface patterning effect discovered in a wide variety of azobenzene systems has, of course, been suggested as the basis for numerous applications. Azobenzene is versatile, amenable to incorporation in a wide variety of materials, and the photopatterning is low power, single step, and reversible, which offers attractive advantages. On the other hand, one may use a system where cross-linking enables permanent fixation of the surface patterns [266]. Many proposed applications are optical and fit well with azobenzene's already extensive list of optical capabilities. The gratings have been demonstrated as optical polarizers [267], angular or wavelength filters [268,269], and couplers for optical devices [270]. They have also been suggested as photonic band gap materials [271] and have been used to create lasers where emission wavelength is tunable via grating pitch [272,273]. The process has, of course, been suggested as an optical data storage mechanism [274]. The high-speed and single-step holographic recording has been suggested to enable instant holography [275], with obvious applications for industry or end consumers. Since the hologram is topographical, it can easily be used as a master to create replicas via molding. The surface patterning also allows multiple holograms to be superimposed into hierarchical structures. This has been used to create multilayered structures [276], with phase correlation between layers of the active azobenzene and an alternating spacer layer, to form 3D linear, tetragonal and hexagonal relief gratings with a hierarchical structure. Another novel suggestion is to use the holographic patterning for rapid prototyping of optical elements [277]. Optical elements could be generated or modified quickly and during device operation. They could thereafter be replaced with permanent components, if required.

The physical structure of the surface relief can be exploited to organize other systems. For instance, it can act as a command layer, aligning neighboring liquid crystals phases [278–282]. The grating can be formed after the LC cell has been assembled, and can be erased and rewritten. Holographic films can use the structure of the surface relief to organize fluorophores into various 2D micropatterns [283,284]. Colloids can also be arranged into the grooves of an SRG, thereby templating higher-order structures [285,286], and these lines of colloids can then be sintered to form wires [287]. The surface topography inscription process is clearly amenable to a variety of optical-lithography patterning schemes. These optical patterns are amenable to soft-lithographic approaches of replica molding using PDMS stamps to reproduce the gratings on a variety of substrates [288] and have been used to fabricate analyte sensors. These sensors were based upon the observed change in the diffraction efficiency of a grating upon analyte absorption. In a recent example [289], the diffraction grating of an azobenzene-based material was transferred onto a stimuli responsive hydrogel functionalized with glucose oxidase and has been used to demonstrate glucose sensors capable of quantitative and continuous measurements in solution (see Figure 4.10).

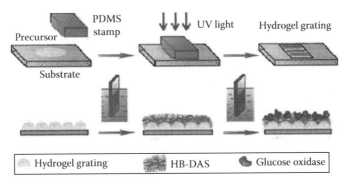

FIGURE 4.10   Schematic of the fabrication of the glucose-sensing hydrogel gratings. (Reproduced with permission from Ye, G. et al., *Chem. Commun.*, 46, 3872, 2010.)

Another advantage of holographic patterning is that there is guaranteed registry between features over macroscopic distances. This is especially attractive as technologies move toward wiring nanometer-sized components. One example in this direction involved evaporating metal onto an SRG, and then annealing. This formed a large number of very long (several mm) but extremely thin (200 nm) parallel metal wires [290]. Of interest for next-generation patterning techniques is the fact that the azo surface modification is amenable to near-field patterning, which enables high-resolution nanopatterning by circumventing the usual diffraction limit of far-field optical systems. Proof of principle was demonstrated by irradiating through polystyrene spheres assembled on the surface of an azo film. This results in a polarization-dependent surface topography pattern [217] and a corresponding surface density pattern [218]. Using this technique, resolution on the order of 20 nm was achieved [291]. This process appears to be enhanced by the presence of gold nano-islands [292]. It was also shown that volume is not strictly conserved in these surface deformations [219]. In addition to being useful as a sub-diffraction limit patterning technique, it should be noted that this is also a useful technique for imaging the near-field of various optical interactions [293]. The (as of yet not fully explained) fact that sub-diffraction limit double-frequency surface relief gratings can be inscribed via far-field illumination [210,211,294,295] further suggests the azo polymers as versatile high-resolution patterning materials.

4.5   PHOTO-INDUCED MECHANICAL RESPONSE AND ACTUATION

If an actuator is defined as an energy transducer converting an input energy into mechanical motion, then azobenzene-based systems are excellent candidates for photo-mechanical actuation for many niche applications involving small size, localized actuation, remoteness of the power source, and freedom from the encumbrance of batteries, electrons, and internal moving parts, where advantageous. The most convincing demonstration of macroscopic motion due to azo isomerization is the mechanical bending and unbending of a free-standing polymer thin films [192,193]. As described thoroughly earlier, the film bending direction is tunable by chromophore alignment or polarized light. Bending occurs in these films through surface contraction, while the thick inner layer does not contract as it is not irradiated. As the direction of bending can be controlled via the polarization of the light, the materials enable full directional photo-mechanical control [194] and have been used to drive macroscopic motion of a floating film [296]. The contraction of these materials (as opposed to expansion) appears again to be related to the main-chain azo groups, and may also be related to the LC nature of the cross-linked gels. For a thin film floating on a water surface, a contraction in the direction of polarized light was seen for LC materials, whereas an expansion was seen for amorphous materials [253]. A related amplification of azo motion to macroscopic motion is the photo-induced bending of a microcantilever coated with an azobenzene monolayer [195].

Other examples include macroscopic bending and 3D control of fibers made of azobenzene liquid crystalline elastomers [197–199], light driven micro valves [297], and full plastic motors [298]. In this section a survey summary of various manifestations of the photo-mechanical effect leading to macroscale actuation with various azobenzene-based materials will be described.

4.5.1 PHOTOACTUATION IN MONOLAYERS AND INTERFACIAL FILMS

Monolayers of azobenzene polymers are easily prepared at the air/water interface, and much of the earliest work focused on these simple systems. In the monolayer state, changes in the molecular shape and orientation can be directly related to the film properties such as a film area and a surface pressure, providing further ease of direct molecular interpretation of results. Thus, monolayer films of azobenzene are ideal for understanding macroscopic deformations in terms of molecular level processes. When monolayers of the azobenzene polymers are prepared at interfaces, the motion of azobenzene moieties occurring at a molecular level is transferred directly and efficiently, and can be readily amplified to a macroscopic material. Photo-mechanical effects of a monolayer consisting of polyamides with azobenzene moieties in the main chain were first reported by Blair et al. in 1980 [121,122]. At the air/water interface, a decrease in stress was observed upon UV light irradiation of the monolayer, indicating a contraction of the monolayer. In the dark, the stress increased again, and the cycle could be repeated many times. For these main-chain type monolayers, the azobenzene moieties were considered to lie flat on the water surface. The photo-mechanical effects were then simply attributed to the *trans-cis* isomerization of the azobenzene moieties, which occupy a larger area at the interface when they are in the more linear *trans* form than in the *cis* form.

Higuchi et al. prepared a polypeptide monolayer composed of two α-helical poly(α-methyl L-glutamate) rods linked by an azobenzene moiety [123]. The *trans-cis* photoisomerization and the consequent change in geometry of the azobenzene produced a bending of the main chain of the molecule and a decrease in the limiting area per molecule. It was estimated that the bending angle between the two α-helical rods, produced by irradiation with UV light, was about 140°, and the photo-induced bent structure resulted in a reduction of the molecular area at the air/water interface owing to a decrease in the distance between the ends of the molecule. An important finding here was that the photo-induced changes in the area of the monolayer occurred more slowly than the spectral changes of the azobenzene moieties and that the photo-induced changes in the surface area may arise from the rearrangement of the bent molecules, induced by photoisomerization of the azobenzene moieties in the main chain. The intermolecular interaction in the condensed monolayer may have served to slow down the rate of their rearrangement process.

In monolayers of side-chain type polymers, photo-mechanical effects of related azobenzene-containing polypeptides were also investigated by Menzel et al. in 1992 [124]. They prepared poly(L-glutamate)s with azobenzene groups in the side chains coupled to the backbone via alkyl spacers. The resulting monolayers showed a photoresponsive behavior that was opposite to the aforementioned systems, however, as they *expanded* when exposed to UV light, and *shrank* when exposed to visible light. This was perhaps the first observation of curious opposite expansion/contraction behavior from the same class of chromophores. The *trans-cis* photoisomerization of the azobenzene moiety upon UV light irradiation in this work led to a large increase in the dipole moment of this unit however, and this gain in affinity to a water surface was proposed to be responsible for the net contraction [125]. In perhaps the first set of studies into quantifying the effect generally, and optimizing some photo-mechanical systems, Seki et al. prepared poly(vinyl alcohol) containing azobenzene side chains and observed photo-induced changes in areas on a water surface in an excellent series of papers beginning in 1993 [125–134]. These monolayers at the air/water interface exhibited a three-fold expansion in area upon UV light irradiation and reversibly shrunk by visible light irradiation. The mechanism of the photo-induced changes in area was interpreted in terms of the change in polarity of the azobenzene moiety: The *trans-cis* photoisomerization led to an increase in dipole moment, bringing about a higher affinity of the *cis*-azobenzene to the water surface and the expansion of the monolayers. *Cis-trans* back isomerization by visible light irradiation then gave rise to a recovery of the monolayers to the initial structure. By analyzing the XRD data, it was shown that the thickness of the monolayer becomes larger for the *trans* form than the *cis* form. The resulting change in the thickness by 0.2–0.3 nm due to the *trans-cis* isomerization in the hydrophobic side chain was then directly observed in situ on the water surface [131]. These results with azo monolayers indicate that the photo-induced deformations of the azobenzene-containing monolayers can depend strongly on the location of the azobenzene moieties in the dark: When the azobenzenes are on or in the water subphase, the structural response of the monolayers is determined by the geometrical change of the photochromic units. On the other hand, the change in polarity of the azobenzene moieties is more important when they are away from the water subphase in the dark. The potential of azobenzene monolayers for actuation-based applications has been demonstrated by Ji et al. [195] through the amplification of azo motion in monolayers to macroscopic motion. A monolayer of thiol terminated azobenzene derivative was deposited onto a gold-coated microcantilever, and exposure to UV light resulted in the reversible deflection of the microcantilever due to molecular repulsion in the monolayer.

4.5.2 PHOTOACTUATION IN AMORPHOUS THIN FILMS

Azo polymers offer advantages over azo monolayers as superior materials in view of higher processability, the ability to form free-standing films with a variety of thicknesses from nanometer to centimeter scales, flexibility in molecular design, and precisely controlled synthesis. Hence, azo polymers have emerged as the azo material of choice for most applications. From this point of view, polymer actuators capable of responding to external stimuli and deforming are most desirable for practical applications, either amorphous or organized (such as liquid crystalline). Various chemical and physical stimuli have been applied such as temperature [135], electric field [136,137], and solvent composition [138] to induce deformation of polymer actuators.

The use of structural changes of photo-isomerizable chromophores for a macroscopic change in size of polymers was first proposed by Merian in 1966 [139], when he observed that a nylon filament fabric dyed with an azobenzene derivative shrank upon photoirradiation. This effect was postulated to involve the photochemical structural change of the azobenzene group absorbed on the nylon fibers, yet these fibrous systems were sufficiently complex that the real mechanism could only be speculated upon. The observed shrinkage was also quite small, only about 0.1%, which made it further difficult to draw firm conclusions. Following this interesting work, however, much effort was made to find new photo-mechanical systems with an enhanced efficiency [140,141]. Eisenbach, for example, investigated in 1980 the photo-mechanical effect of poly(ethyl acrylate) networks cross-linked with azobenzene moieties and observed that the polymer network contracted upon exposure to UV light due to the *trans-cis* isomerization of the azobenzene cross-links and expanded by irradiation with Vis light due to *cis-trans* back isomerization [142]. This photo-mechanical effect was mainly attributed to the conformational change of the azobenzene cross-links by the *trans-cis* isomerization of the azobenzene chromophore. It should also be noted that the degree of deformation was also very small in these systems, around 0.2%.

Matejka et al. also synthesized several types of photochromic polymers based on a copolymer of maleic anhydride with styrene containing azobenzene moieties both in the side chains and in the cross-links of the polymer network [143–145]. The photo-mechanical effect observed here was enhanced with an increase in the content of photochromic groups, and for a polymer with 5.4 mol% of the azobenzene moieties, a photo-induced contraction of the sample of 1% was achieved. Most recently, the photo-induced expansion of thin films of acrylate polymers containing azobenzene chromophores was tracked directly in real time by Barrett et al., using a variety of techniques including in situ single wavelength ellipsometry, atomic force microscopy (AFM), and in situ neutron reflectometry [146]. An initial expansion of the azobenzene polymer films was found to be irreversible with an extent of relative expansion observed of 1.5%–4% in films of thickness ranging from 25 to 140 nm, and then a subsequent and reversible expansion was observed with repeated irradiation cycles, achieving a relative extent of expansion of 0.6%–1.6%. The extent and direction (expansion or contraction) of photo-mechanical change could be tuned for the first time just by using ambient temperature, suggesting that competing dynamic effects exist during isomerization. These variable-temperature neutron reflectometry experiments demonstrated unambiguously that both photo-expansion and photocontraction could be optimized in a single azo material merely by varying the dominance of these two competing effects with low and high extremes of temperature respectively. This implicates a fundamental competition of mechanisms and helps unify both the photo-contraction and photo-expansion literature. In particular, it now appears that most azo materials exhibit photoexpansion below a well-defined crossover temperature and photocontraction above this temperature. Highly mobile materials will thus be above their crossover temperature at ambient conditions, whereas rigid materials will be below.

As another technique to measure the photo-mechanical effect directly, recent developments of single-molecule force spectroscopy by AFM have enabled one quite successfully to measure mechanical force produced at a molecular level. Gaub et al., for example, synthesized a polymer with azobenzene moieties in its main chain [147,148] and then coupled the ends of the polymer covalently to the AFM tip and a supporting glass substrate by heterobifunctional methods to ensure stable attachment and investigated the force (pN) and extension (nm) produced in a single polymer in total internal reflection geometry using the slide glass as a wave guide. This clever excitation geometry proved very useful to avoid thermo-mechanical effects on the cantilever. They were thus able to photochemically lengthen and contract individual polymer chains by switching the azobenzene moieties between their *trans* and *cis* forms by irradiation with UV (365 nm) and visible (420 nm) light, respectively. The mechanical work executed by the azobenzene polymer strand by *trans-cis* photoisomerization could then be estimated directly as $W \approx 4.5 \times 10^{-20}$ J. This mechanical work observed at the molecular level resulted from a macroscopic photoexcitation, and the real quantum efficiency of the photo-mechanical work for the given cycle in their AFM setup was only on the order of 10^{-18}. However, a theoretical maximum efficiency of the photo-mechanical energy conversion at a molecular level can be estimated as 0.1, if it is assumed that each switching of a single azobenzene unit is initiated by a single photon carrying an energy of 5.5×10^{-19} J [147,148].

Photo-induced reversible changes in elasticity of semi-interpenetrating network films bearing azobenzene moieties were achieved recently by UV and Vis light irradiation [149]. These network films were prepared by cationic copolymerization of azobenzene-containing vinyl ethers in a linear polycarbonate matrix. The network film showed reversible deformation by switching the UV light on and off, and the photo-mechanical effect was attributed to a reversible change between the highly aggregated and dissociated state of the azobenzene groups [149–151]. In other studies, similar films of azobenzene-containing vinyl ethers films with polycaprolactone have achieved rapid (0.1 min) anisotropic deformation and recovery. The films, placed under constant tensile stress, were stretched perpendicular and parallel to the tensile stress before irradiation. Photoisomerization of these films resulted in film contraction for stretching parallel to the tensile stress and film elongation for stretching perpendicular to the tensile stress. The photo-mechanical response was observed to increase with film stretching and speculated to arise from anisotropic responses caused by the isomerization induced vibration of azobenzene molecules that decreases the modulus of the deformed amorphous area [299]. Other polymer films that exhibit high bending intensity and large bending angles (90°) have also been reported [300].

The photo-mechanical expansion of azobenzene has been used to create a simple UV sensor [301,302] and has been proposed for applications in mechanically tunable filters and switching devices. The sensor, based upon a fiber Bragg grating coated with an azobenzene polymer, measured UV light intensity by monitoring the center wavelength shift in the fiber Bragg grating. Upon photoimerization (proportional to incoming UV light) the encapsulating azobenzene material applied a photo-mechanical axial strain upon the fiber Bragg grating proportionally shifting its center wavelength. Another interesting and similar mode of deformation of polymer colloidal particles by light was reported by Wang et al. [14] and by Liu et al. [303–306]. The former observed that spherical polymer particles containing azobenzene moieties changed their shape from a sphere to an ellipsoid upon exposure to interfering linearly polarized laser beams, and the elongation of the particles was induced along the polarization direction of the incident laser beam. The latter reported the deformation of the micellar structure between spherical and rod-like particles under alternating UV and visible light irradiation. Gels of polymer films containing azobenzenes are also potential materials for applications, however, in general the gels reported have a disadvantage in that the response is slow, and the degree of deformation of the polymer films is too small to be practically utilized. It is generally agreed now that it is crucial to develop only photo-mechanical systems that can undergo fast and large deformations.

4.5.3 PHOTOACTUATION IN LIQUID CRYSTALLINE AZO POLYMERS

The previous monolayer, gel, and amorphous polymer films described are generally without microscopic or macroscopic order, so the photo-mechanical deformations mostly occur in an isotropic and uniform way, i.e., there is no preferential direction for deformation. If materials with anisotropic physical properties are instead used however, the mechanical power produced can increase significantly, and more control can be realized. Liquid crystalline elastomers (LCEs) are materials that have advantageous properties of both LCs and elastomers arising from polymer networks. Due to the LC properties, mesogens in LCEs show alignment, and this alignment of mesogens can be coupled with polymer network structures. This coupling gives rise to many characteristic properties of LCEs, and depending on the mode of alignment of mesogens in LCEs, they are classified as nematic LCEs, smectic LCEs, cholesteric LCEs, etc. If one heats nematic LCE films toward the nematic–isotropic phase transition temperature, the nematic order will decrease, and when the phase transition temperature is exceeded, one observes a disordered state of mesogens. Through this phase transition, the LCE films show a general contraction along the alignment direction of the mesogens, and if the temperature is lowered back below the phase transition temperature, the LCE films revert back to their original size by expanding. This anisotropic deformation of the LCE films can be very large, and along with good mechanical properties this provides the LCE materials with promising properties as artificial muscles [307–310]. By incorporating photochromic moieties into LCEs, which can induce a reduction in the nematic order and in an extreme case a nematic–isotropic phase transition of LCs, a contraction of LCE films has been observed upon exposure to UV light to cause a photochemical reaction of the photochromic moiety [311–313]. Most recently, a 2D movement, bending, of LCE films has been reported by Ikeda et al. after incorporation of the photochromic moieties into LCEs [192,193]. Light-driven actuators based on LCE materials are a topic of recent intensive studies, and a variety of actuation modes have been proposed and developed. LCEs are usually lightly cross-linked networks, and it is known that the cross-linking density has a great influence on the macroscopic properties and the phase structures [307,314–317]. The mobility of chain segments is reduced with an increase of cross-linking points, and consequently the mobility of mesogens in the vicinity of a cross-link is suppressed. The film modulus also increases with cross-linking [316]. A cross-link is recognized as a defect in the LC structure and an increase in the cross-linking density produces an increasing number of defects. Therefore, LC polymers with a high cross-linking density are referred to as LC thermosetting polymers (duromers) distinguished from LCEs.

Cooperative motion of LCs may be most advantageous in changing the alignment of LC molecules by external stimuli. If a small portion of LC molecules changes its alignment in response to an external stimulus, the other LC molecules also change their alignment. This means that only a small amount of energy is needed to change the alignment of whole LC films: Such a small amount of energy as to induce an alignment change of only 1 mole% of the LC molecules is enough to bring about the alignment change of the whole system. This means that a huge force or energy amplification is possible in LC systems. When a small amount of azobenzene is incorporated into LC molecules and the resulting guest/host mixtures are irradiated to cause photochemical conversion of the photochromic guest molecules, a LC to isotropic phase transition of the mixtures can be induced isothermally. The *trans* form of the azobenzenes, for instance, has a rod-like shape, which stabilizes the phase structure of the LC phase, while the *cis* form is bent and tends to destabilize the phase structure of the mixture. As a result, the LC-isotropic phase transition temperature (Tc) of the mixture with the *cis* form (Tcc) is much lower than that with the *trans* form (Tct). If the temperature of the sample (T) is set at a temperature between Tct and Tcc, and the sample is irradiated to cause *trans-cis* photoisomerization of the azobenzene guest molecules, then Tc decreases with an accumulation of the *cis* form, and when Tc becomes lower than the irradiation temperature T, an LC-isotropic phase transition of the sample is induced. Photochromic reactions are usually reversible, and with *cis-trans* back isomerization the sample reverts to the initial LC phase. This means that phase transitions of LC systems can be induced isothermally and reversibly by photochemical reactions of photoresponsive guest molecules. Tazuke et al. reported the first explicit example of the nematic–isotropic phase transition induced by *trans-cis* photoisomerization of an azobenzene guest molecule dispersed in a nematic LC in 1987 [318].

Ikeda et al. reported the first example of a photochemical phase transition in LC polymers; they demonstrated that by irradiation of LC polymers doped with low-molecular-weight azobenzene molecules with UV light to cause *trans-cis* isomerization, the LC polymers underwent a nematic–isotropic phase transition, and with *cis-trans* back isomerization, the LC polymers reverted to the initial nematic phase [153,319,320]. However, it soon became apparent that LC copolymers are superior to the doped systems because in the doped systems phase separation was observed when the concentration of the photochromic molecules was high. A variety of LC copolymers was prepared and examined for their photochemical phase transition behavior [153,321–323]. The effects of nano confinement and macromolecular geometry on the orientation and photo-mechanical volume change of LC has also been examined [324,325].

One of the important factors of the photoresponsive LCs is their response rate to optical excitation. In this respect, the response time of the photochemical phase transition has been explored by time-resolved measurements [322,326,327]. The nematic–isotropic phase transition of the LC polymer was induced after a sufficient amount of the *cis*-form had been produced with a single pulse of the laser and the isothermal phase transition of the LC polymers occurred in a time range of ~200 ns, which is comparable to that of low-molecular-weight LCs [322,326,327]. LCEs show good thermoelastic properties: Across the nematic–isotropic phase transition, they contract along the alignment direction of the mesogens and by cooling below the phase transition temperature they show expansion. By a combination of this property of LCEs with a photochemical phase transition (or photochemically induced reduction of nematic order), one can induce deformation of LCEs by light quite efficiently [311–313]. In fact, Finkelmann et al. have succeeded in inducing a contraction of 20% in an azobenzene-containing LCE upon exposure to UV light to cause the *trans-cis* isomerization of the azobenzene moiety [311]. They synthesized monodomain nematic LCEs containing a polysiloxane main chain and azobenzene chromophores at the cross-links. From the viewpoint of the photo-mechanical effect, the subtle variation in nematic order by *trans-cis* isomerization causes a significant uniaxial deformation of LCs along the director axis, if the LC molecules are strongly associated by covalent cross-linking to form a 3D polymer network. Terentjev et al. have incorporated a wide range of azobenzene derivatives into LCEs as photoresponsive drivers, and examined their deformation behavior upon exposure to UV light, and analyzed in detail these photo-mechanical effects [312,313].

More recently, Keller et al. synthesized monodomain nematic azobenzene side-on (mesogens parallel to the long axis of the film) elastomers by photopolymerization using a near-infrared photoinitiator [328]. The photopolymerization was performed on aligned nematic azobenzene monomers in conventional LC cells, and thin films of these LCEs showed fast (less than 1 min) photochemical contraction, up to 18%, by irradiation with UV light and a slow thermal back

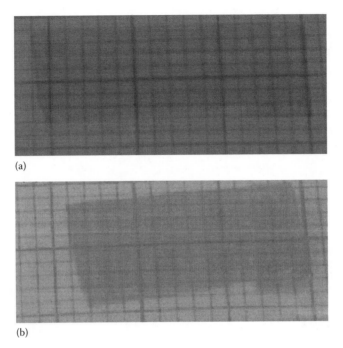

(a)

(b)

FIGURE 4.11   Photographs of the photodeformation of Keller's azobenzene CLCP before UV light irradiation (a) and under UV light irradiation (b). (From Li, M.H. et al., *Adv. Mater.,* 15, 569, 2003.)

reaction in the dark (Figure 4.11). Two-dimensional movements of LCE films have since been demonstrated, and many three-dimensional examples have been envisaged and are discussed later. Ikeda et al. was the first to report photo-induced bending behavior of macroscopic LC gel systems [193] and LCEs containing azobenzenes [317,192,193,329]. In comparison with a 1D contraction or expansion, the bending mode, a full 2D movement, could be advantageous for a variety of real manipulation applications. Figure 4.12 [330] depicts the bending and unbending processes induced by irradiation of UV and visible light, respectively. It was observed that the monodomain LCE film bent toward the irradiation direction of the incident UV light along the rubbing direction, and the bent film reverted to the initial flat state after exposure to Vis light. This bending and unbending behavior was reversible just by changing the wavelength of the incident light. In addition, after the film was rotated by 90°, the bending was again observed along the rubbing direction. Importantly, these results demonstrated that the bending can be anisotropically induced along the rubbing direction of the alignment layers.

One great challenge to optimizing these systems is the extinction coefficient of the azobenzene moieties at ~360 nm, which is usually so large that more than 99% of the incident photons are absorbed by the near-surface region within 1 μm. Since the thickness of the films used is typically 20 μm, the reduction in nematic order occurs only in the surface region facing the incident light, but in the bulk of the film the *trans*-azobenzene moieties remain unchanged. As a result, the volume contraction is generated only in the surface layer, causing the bending toward the irradiation direction of the incident light, yet far from optimal efficiency. Furthermore, the azobenzene moieties are preferentially aligned along the rubbing direction of the alignment layers, and the decrease in alignment order of the azobenzene moieties is thus produced just along this direction, contributing to the anisotropic bending behavior. Monodomain LCE films with different cross-linking densities were prepared by copolymerization [317]. The films showed the same bending behavior, but the maximum bending extents were different among the films with different cross-linking densities. Because the film with a higher cross-linking density holds a higher order parameter, the reduction

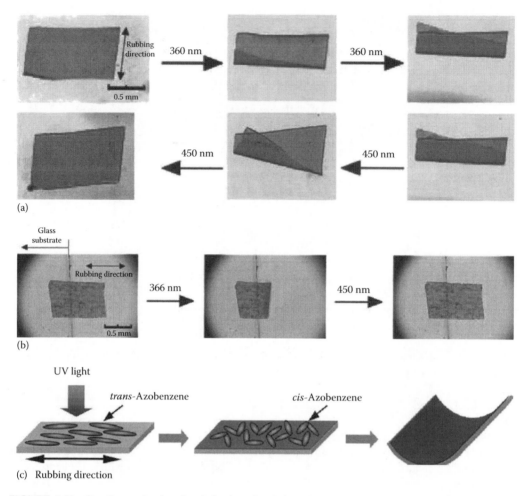

FIGURE 4.12 Bending and unbending behavior of an LC gel in toluene (a) and a CLCP film in air (b). (c) Plausible mechanism of the photo-induced bending of CLCP films. (From Yu, Y.L. et al., *Angew. Chem. Int. Ed.*, 46, 881, 2007.)

in the alignment order of the azobenzene moieties gives rise to a larger volume contraction along the rubbing direction, contributing to a larger bending extent of the film along this direction. By means of the selective absorption of linearly polarized light in the polydomain LCE films, Ikeda et al. succeeded in realizing a photo-induced direction-controllable bending in that a single polydomain LCE film can be bent repeatedly and precisely along any chosen direction (Figure 4.13) [192]. The film bent toward the irradiation direction of the incident light, with significant bending occurring parallel to the direction of the light polarization.

In a related system, Palffy-Muhoray et al. demonstrated that by dissolving azobenzene dyes into a LCE host sample, its mechanical deformation in response to nonuniform illumination by visible light becomes very large (more than 60° bending) [296]. When a laser beam from above is shone on such a dye-doped LCE sample floating on water, the LCE "swims" away from the laser beam, with an action resembling that of flatfish (Figure 4.14). A similar azobenzene LCE film with extraordinarily strong and fast mechanical response to the influence of a laser beam was developed [331], where the direction of the photo-induced bending or twisting of LCE could be reversed by changing the polarization of the laser beam. The phenomenon is a result of photo-induced reorientation of azobenzene moieties in the LCE. Broer et al. prepared LCE films with a densely cross-linked, twisted configuration of azobenzene moieties [332]. They have shown a large amplitude bending and

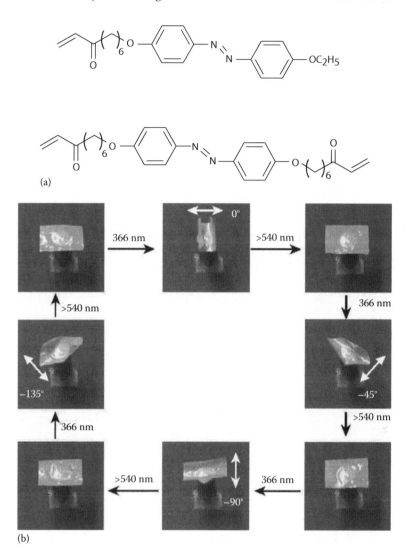

FIGURE 4.13 Precise control of the bending direction of a film by linearly polarized light. Chemical structures of the LC monomer (3a) and cross-linker (3b) used for preparation of the film and photographic frames of the film in different directions in response to irradiation by linearly polarized light of different angles of polarization (white arrows) at 366 nm and bending flattened again by Vis light longer than 540 nm. (From Yu, Y.L. et al., *Angew. Chem. Int. Ed.*, 46, 881, 2007.)

coiling motion upon exposure to UV light, which arises from the 90° twisted LC alignment configuration. The alignment of the azobenzene mesogens in the LCE films was examined for how it affects the photo-induced bending behavior. Homeotropically aligned films were prepared and exposed to UV light, and it was found that the homeotropic LCE films showed a completely different bending; upon exposure to UV light they bent *away* from the actinic light source [333]. Additionally, LCE films with varying chromophore concentration and location were prepared [334]. In films with a low azo content and thickness, under continuous UV irradiation, film bending was observed before a relaxation back to its initial shape. This bending and unbending motion was attributed to the penetration of light through the film resulting in chromophore isomerization on the opposite side of the film. This photocontraction causes the film to revert to its initial shape. The largest mechanical force generated by photoirradiation of the various films was measured as 2.6 MPa [334].

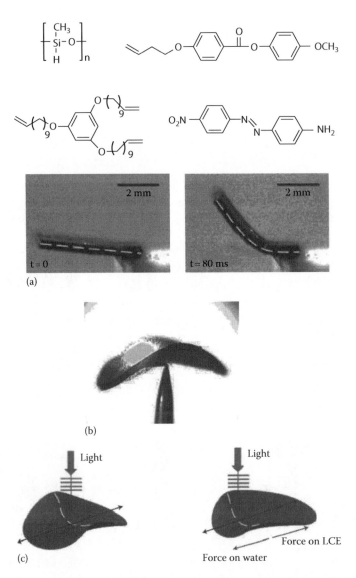

FIGURE 4.14 (a) Photo-mechanical response of the "swimming" CLCP sample. (b) The shape deformation of a CLCP sample upon exposure to 514 nm light. (c) Schematic illustration of the mechanism underlying the locomotion of the dye-doped CLCP sample. (From Camacho-Lopez, M. et al., *Nat. Mater.,* 3, 307, 2004.)

Ferroelectric LCE films with a high LC order and a low T_g were also prepared [330], where irradiation with 366 nm light induced the films to bend at room temperature toward the irradiation direction of the incident light along the direction with a tilt to the rubbing direction of the alignment layer. The bending process was completed within 500 ms upon irradiation, and the mechanical force generated by photoirradiation was measured as 220 kPa, similar to the contraction force of human muscles (~300 kPa). More recently, Ikeda and coworkers [335] and others [199–202] have prepared artificial muscle-like fibers. The fibers of the former report were composed of cross-linked liquid crystalline polymers (CLCPs) capable of 3D movement controlled by irradiation intensity and direction, with a mechanical force generated by the fibers under photoirradiation measured as 210 kPa. A solution liquid crystalline azobenzene polymers can also enable microparticle actuation as recently reported by Kurihara et al. [336] The fast and repeatable translation motion of polystyrene (PS) microspheres was observed when placed in a liquid crystalline azobenzene solution irradiated with light. The PS microspheres moved toward the UV light source and away from the visible light source when placed into the azobenzene solution and irradiated with UV or Vis light. The direction of PS microsphere motion was thus controllable by the manipulation of the light source while the speed was controllable by the intensity of irradiation or the concentration of azobenzene doped into the film.

A hierarchical self-assembled film of liquid crystalline polymer brushes containing azobenzene into films has been reported [337]. In the absence of chemical cross-linking, the resulting bi-morphic film is capable of photo-mechanical bending due to the amplification of the azobenzene photoisomerization across the hierarchical film structure. In this hierarchical assembly the monomer is polymerized into polymer brushes that assemble into cylinders. These cylinders form rectangular 2D lattices and can undergo large scale macroscopic alignment in the outer layers by using uniaxially stretched teflon sheets. Bending only occurs when the lattices on both sides of the bimorph film are parallel to each other demonstrating the need for hierarchical amplification of the azobenzene photoisomerization in the absence of chemical cross-linking.

4.5.4 PHOTOACTUATION IN AZOBENZENE CRYSTALS

While most azobenzene photo-mechanical systems are based upon amorphous or liquid crystalline polymers, there are also some very recent reports of photo-mechanical crystals of azobenzene in the literature [338–342]. There has been reports of numerous solid state reactions in molecular crystals [343,344], and of these, crystalline photoreactions are especially interesting as they are often accompanied by molecular motion and morphological changes at the crystal surfaces [345–348]. Irie et al. were among the first to report on these crystalline photoreactions in diarylethene microcrystals accompanied by a rapid, reversible shape change of the crystal under alternating UV and visible light irradiation [349,350].

In contrast to the diarylethene derivatives, the photoisomerization of azobenzene, requiring a larger free volume, is hindered in the bulk crystal. An early AFM study demonstrated the reversible alteration of the layered structure of an azobenzene crystal under UV and visible light suggesting that the topmost bilayers of the azobenzene crystal are capable of isomerization [351]. Conclusive evidence of reversible photoisomerization in azobenzene crystals has only been recently reported through a reversible 3.5% reduction in particle size of azobenzene crystals dispersed in water [341], and the fraction of the *cis* isomer was determined to level off at 30% in the photostationary state. In further work, photo-induced particle size deformation of crystalline azobenzene and silica nanohybrids fabricated by dry grinding was also reported [340].

In other examples, photoisomerization in crystalline azobenzene was demonstrated by the formation of a surface relief grating upon single crystal azobenzene derivatives [352] and the observation of photo-induced vitrification near the surfaces of the single crystals of azobenzene-based molecular materials possessing a glass-forming ability [342]. In further work [339], the reversible mechanical bending of plate-like microcrystals of azobenzene derivatives has been reported. Here, photoisomerization of the *trans* azobenzene molecules on the (001) crystal surface elongates the unit cell length near the (001) surface giving rise to uneven features. As the inner unit cells do not undergo photoisomerization their dimensions remain constant and thus, result in crystal bending.

More recently, Kyu and coworkers have observed variously the "swimming," sinking, and stationary floating of azobenzene crystals in a triacrylate solution (TA) (Figure 4.15) [338]. The authors explain such motion through the creation of concentration/surface tension gradients formed around the liquid crystal interface by the rejection of TA solvent from the growing crystal fronts. When these gradients act on different facets in an unbalanced manner the crystal is propelled forward and "swims." Solvent rejection in the vertical direction causes crystal flotation while balanced forces on all facets results in stationary crystal growth. Additionally, stationary rhomboidal crystals of azobenzene were shown to swim (move away from the UV light source)

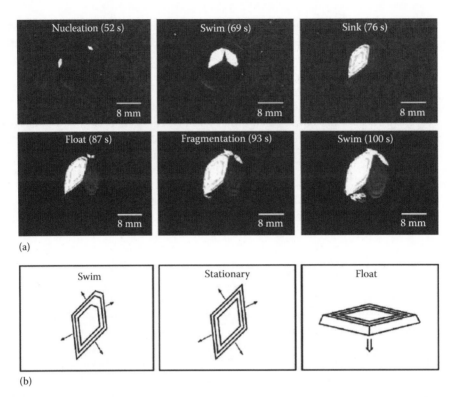

(a)

(b)

FIGURE 4.15 (a) Self-motions of azobenzene crystals in 35 wt% solution, showing swimming, sinking, floating, and birth of baby single crystals. (b) The sketch on the left is a conjecture of self-motion due to the unbalanced forces of the rejected solvent creating a concentration/surface tension gradient from the lateral crystal growth fronts, propelling the rhomboidal crystal to swim on the surface; the drawing in the middle shows the stationary crystal growth as the forces and the surface tension gradients on each facet are balanced, and the sketch on the right represents the solvent rejection in the vertical direction causing the pyramid crystal to float. (Reproduced from Milam, K. et al., *J. Phys. Chem. B,* 114, 7791, 2010.)

upon irradiation. This has been attributed by the authors to the generation of a mechanical torque within the crystal by higher isomerization rates in the sections closer to the UV light. Additionally, isomerization induced changes in the polarity and thus solubility of the azobenzene crystals could result in system instability, driving phase segregation and greater solvent rejection rates from the crystal front closer to the UV light.

4.6 APPLICATIONS IN ROBOTICS

While there has admittedly been far more research into the materials, mechanisms, and measurements of the photo-mechanical effect with azo chromophore materials, there has been some preliminary proof-of-principle applications of note toward real actuation. Of course, the vast majority of existing robotic applications employs traditional electro-mechanical machines, but a wider definition can encompass photo-actuated materials as artificial muscles as well, and in some niche applications they could be competitive or even advantageous. In bio-mimetic actuation (muscle-like movement) and in many micromachines, traditional electronic metallic components can suffer from fundamental drawbacks such as low flexibility, bioincompatibility, ready corrosion, and low strength-to-weight ratios, as compared to polymeric materials. Additionally, polymeric materials have established advantages such as high processability, easy fabrication and relatively low weight density, and low cost and environmental impact [353,354]. Traditional electronic robots also require an integral or attached electric power source, and a variety of related components to successfully operate [355,356]. Thus, simple soft materials driven by light could play an important role as efficient energy conversion systems for bio-mimetic actuation and micromachines. Scaling down is also more easily envisaged for photo-driven polymeric systems, where one can then avoid the growing problem of nanoscale electrical connection and be free of the "nano-batteries" that would otherwise be required for nanoelectrodevices. Small-scale actuation of photopolymers with no internal "moving" parts also circumvents the fundamental problem of friction, adhesion, and wear at small lengthscales, as tradition robotic engineering motifs based on axels, pulleys, wheels, and gears grinds to a halt as size is reduced through to the nanoscale.

While there have indeed been good advances achieved toward conductive and electrostrictive polymers, it may be just as valid a strategy to explore changing the input power source for actuation from electrons to photons, as nature has always done, to take advantage of the wider range of polymers permitted by azo chromophore incorporation. Additionally, this opens up an exciting class of materials that can harvest sunlight directly into mechanical work, without wasteful energy interconversion, and permits one to power devices and robotics completely remotely—even at large distances: through transparent barriers, through space on even astronomical length scales, in liquids, or even inside living biological tissue with low invasion. There are also quite clearly many applications and materials (such as most bio-materials) that are simply not compatible with batteries and moving electrons. One might observe that essentially all locomotion, actuation, and movement in biology is nonelectric, so bio-inspired engineering and bio-mimic approaches have a natural place in investigating photoactuation for artificial muscles. As a final observation for motivation, one might also observe the enormous potential for sunlight-driven applications as a "free" and sustainable energy source, and that even natural "chemical energy" muscle devices in biology can trace their energetic source back to photosynthesis. Indeed, by this metric, one can consider all mechanical energy in the natural world to have been produced by photons at some earlier stage of origin.

Azobenzene-based materials are ideal candidates for such photo-robotics applications as they are capable of strong and efficient mechanical actuation powered by light energy without the need for additional components such as batteries or wires. The photo-induced deformations (expansion/contraction and bending) can be translated with appropriate engineering into rotational and other motions capable of producing applicable work. The first example of such engineering was demonstrated by Ikeda, Yamada, Barrett, and coworkers [298], who translated the photo-induced deformations of an LCE film into rotational motion by joining two ends of an LCE film to create a continuous ring. The azobenzene mesogens in this light actuated motor were aligned in the circular direction of the ring. The azobenzene film laminated with a thin polyethylene sheet was then mounted onto a pulley system. Irradiating the belt with simultaneous UV and visible light on the downside right and upside left, respectively, caused film rolling and caused the pulleys to be driven through belt rotation in a counter-clockwise direction, demonstrating a first light-powered motor. The azobenzene mesogens were aligned parallel to the long axis. Thus, irradiation near the right pulley of the belt results in a contraction force while the visible light near the upper left pulley causes a local expansion force causing a counter-clockwise rotation in the left and right pulleys. The rotation then exposes new sections of the belt to irradiation continuing the photo contraction and expansion of the belt and thus a continuous rotation of the pulleys (see Figure 4.16).

Other examples of robotic actuation by this same group [335,357–361] include an "inchworm" locomotion (Figure 4.17) achieved by a macroscale sheet of CLCPs on flexible PE substrates with asymmetric sliding friction [357]. In this application, the film undergoes photo-mechanical contraction while the asymmetric end shapes on the PE films act as a ratchet, directing film motion. Robotic arm-like actuation of flexible PE sheets was also demonstrated by using the azobenzene moieties as hinges (Figure 4.18). Different sections of a flexible PE film were laminated with azo CLCPs enabling specific control (expansion or contraction) at various positions of the film as each

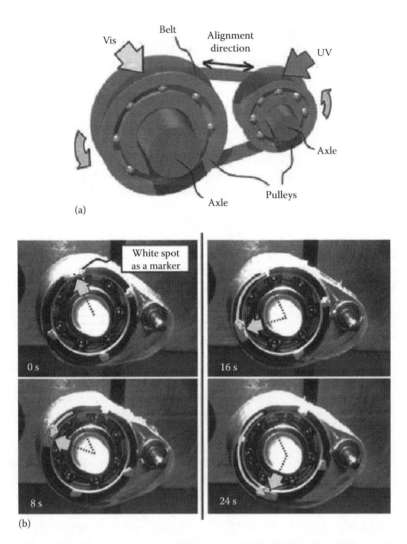

FIGURE 4.16 **(See color insert.)** Light-driven plastic motor with the LCE laminated film. (a) Schematic illustration of a light-driven plastic motor system, showing the relationship between light irradiation positions and a rotation direction. (b) Series of photographs showing time profiles of the rotation of the light-driven plastic motor with the LCE laminated film induced by simultaneous irradiation with UV (366 nm, 240 mW cm^{-2}) and visible light (>500 nm, 120 mW cm^{-2}) at room temperature. Diameter of pulleys: 10 mm (left), 3 mm (right). Size of the belt: 36 mm × 5.5 mm. Thickness of the layers of the belt: PE, 50 μm; LCE, 18 μm. (Reproduced from Yamada, M. et al., *Angew. Chem. Int. Ed.,* 47, 4986, 2008.)

of the sections was individually addressable optically. The laminated sections of azobenzene mesogens thus act as hinge joints enabling various 3D motions of the entire film, acting as arms with remote-control over elbows and wrists [357]. More recent advancements using e-beam cross-linking have improved film durability [362], as compared to the previously laminated films composed of an adhesive layer. It has also made possible fabrication of controlled, large area, adhesive free, photomobile polymer materials.

Most recently Yu et al. have described the design and fabrication of a full sunlight responsive robotic arm capable of lifting up and moving an object weighting 10 mg (10× the weight of the robotic arm; Figure 4.19) [203,363,364]. This robot consisted of several azobenzene containing

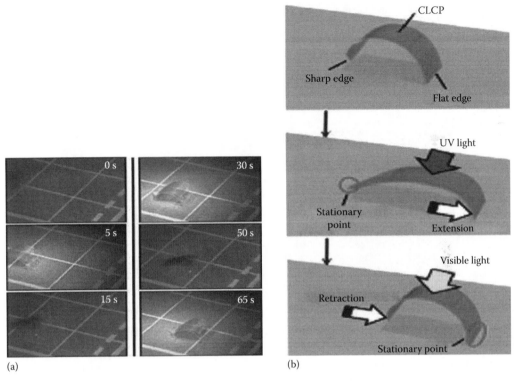

(a)   (b)

FIGURE 4.17   (a) Series of photographs showing time profiles of the photo-induced inchworm walk of the CLCP laminated film by alternate irradiation with UV (366 nm, 240 mW cm^{-2}) and visible light (>540 nm, 120 mW cm^{-2}) at room temperature. The film moved on the plate with 1 cm × 1 cm grid. (b) Schematic illustrations showing a plausible mechanism of the photo-induced inchworm walk of the CLCP laminated film. Upon exposure to UV light, the film extends forward because the sharp edge acts as a stationary point (the second frame), and the film retracts from the rear side by irradiation with visible light because the flat edge acts as a stationary point (the third frame). Size of the film: 11 mm × 5 mm; the CLCP laminated part: 6 mm × 4 mm. Thickness of the layers of the film: PE, 50 μm; CLCP, 18 μm. (Reproduced from Yamada, M. et al., *J. Mater. Chem.*, 19, 60, 2009.)

CLCP films on PE substrates connected by joints to mimic the arm, wrist, hand, and even fingers of the human arm. Thus, the robotic arm could be bent and manipulated to perform complex actions by individually addressing various sections or films of azobenzene, i.e., an object could be picked up or dropped by addressing the fingers with light, while the entire robotic arm could be moved by addressing the arm with light at different locations. Smaller localized movements are possible by light-contracting the wrist, etc.

Further work by the same authors has also shown a similar adaptation of the CLCP films for the design and fabrication of microvalves [297] and micropumps [365]. The microvalves were created by fitting a CLCP film over an inlet valve in a sealed valve chamber, where the film in this state completely blocks the inlet preventing flow [366]. Upon irradiation, film bending results in an unblocking of the inlet valve as well as a concave cavity under the bent film that allows solution to flow from the inlet to the nearby outlet. In the case of micropumps [365], the CLCP film is placed on the outside of a membrane covering a sealed cavity. Upon irradiation the CLCP film bends, forcing the membrane downward, reducing cavity volume and increasing the cavity pressure. Thus, fluid flows out through the outlet valve. Upon film contraction, the membrane is pulled upward increasing the cavity volume, decreasing cavity pressure and forcing fluid

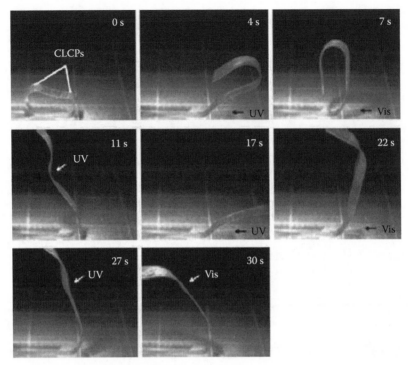

FIGURE 4.18   (See the color insert.) Series of photographs showing time profiles of the flexible robotic arm motion of the CLCP laminated film induced by irradiation with UV (366 nm, 240 mW cm^{-2}) and visible light (>540 nm, 120 mW cm^{-2}) at room temperature. Arrows indicate the direction of light irradiation. Spot size of the UV light irradiation is about 60 mm^2. Size of the film: 34 mm×4 mm; the CLCP laminated parts: 8 mm×3 mm and 5 mm×3 mm. Thickness of the layers of the film: PE, 50 μm; CLCP layers, 16 μm. (Reproduced from Yamada, M. et al., *J. Mater. Chem.*, 19, 60, 2009.)

inflow through the inlet valve. Related to possible microfluidic applications, van Oosten et al. [367] have reported the design and construction of bio-inspired artificial cilia for microfluidic pumping and mixing applications. Using commercial inkjet printing technology droplets of reactive azo LC monomers were deposited onto a film of polyvinyl alcohol (PVA) and a thin layer of rubbed polyamide for LC alignment. After self-assembly and cross-linking of the LC monomers another layer of the same or different azobenzene monomer-based ink is added to create mono or bicomponent cilia capable of responding to different wavelengths of light. Dissolving the PVA releases the cilia, which are capable of intensity-dependant upward bending when irradiated with UV light from above. The bicomponent cilia were capable of different bending properties due to their separately addressable sections, and the activation of these two components in sequence with different wavelengths of light would thus imply a nonreciprocal motion, permitting the cilia to pump fluids [368].

Ingenious high frequency photo-driven oscillators have also been designed and reported by Bunning, White, and coworkers [197–199]. The oscillators were cantilevers made of azo functionalized liquid crystal polymer networks capable of achieving oscillation frequencies of up to 270 Hz and an efficiency of 0.1% under a focused laser beam, with a range of motion nearing the maximal 180° achievable (Figure 4.20). The cantilevers possessed a storage modulus ranging from 1.3 to 1.7 GPa and were shown to bend faster and attain larger bending angles with monodomain orientation, increasing azobenzene concentration, and reduced thickness. The bending angle was also dependent upon the polarization of incoming light as well as atmospheric pressure. Remarkably

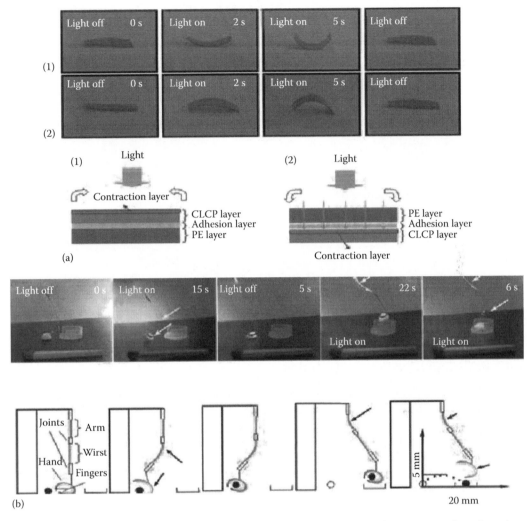

FIGURE 4.19 **(See color insert.)** (a) Photographs of the bilayer films used to construct the microrobot. When visible light (470 nm, 30 mW cm^{-2}) is irradiated on the azobenzene-based cross-linked liquid crystalline polymer (CLCP) layer the films can either bend upward (1) or downward (2) depending upon the position of the CLCP layer. A schematic illustration of the bending is also included in the bottom with the number corresponding to the photographs. Size of the film: 7 mm×4 mm×30 μm. (b) (Top) Photographs showing the microrobot picking, lifting, moving, and placing the object to a nearby container by turning on and off the light (470 nm, 30 mW cm^{-2}). Length of the match in the pictures: 30 mm. Thickness of PE and CLCP films: 12 μm. Object weight: 10 mg. (Bottom) Schematic illustrations of the states of the microrobot during the process of manipulating the object. The insert coordinate indicates the moving distance of the object in vertical and horizontal directions. White and black arrows denote the parts irradiated with visible light. (Reproduced from Cheng, F. et al., *Soft Matter*, 6, 3447, 2010.)

these azo polymer cantilevers were also shown to oscillate under a focused beam of sunlight [199] and thus offer the potential for remotely triggered photoactuation (using sunlight or a focused laser), adaptive optics and most importantly energy harvesting. Such a high frequency oscillator could thus power a micro-optomechanical system as it is a single unit containing both the force generation component (azobenzene) and a kinematic structure (cantilever) capable of amplification or transmission of the work.

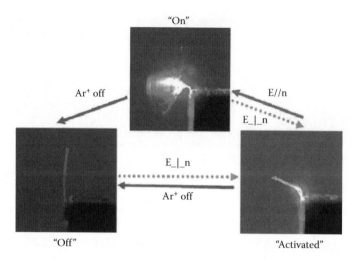

FIGURE 4.20 Optical protocol for activating the light-powered oscillation of a cantilever. The nematic director (n) is positioned parallel to the long axis of the polymer cantilever of dimension 5 mm × 1 mm × 50 μm. When exposed to light polarized orthogonal to n (E ⊥ n) bending occurs toward the laser source. Cycling the Ar+ laser from E ⊥ n to E // n can turn oscillation "on," while blocking the Ar+ or returning the polarization of the laser beam to E ⊥ n turns the oscillation "off." (Reproduced from White, T.J. et al., *Soft Matter*, 4, 1796, 2008.)

4.7 CONCLUSIONS AND OUTLOOK

The azobenzene chromophore is a unique and powerful molecular switch, exhibiting a clean and reversible photoisomerization that induces a reversible change in geometry. This motion can be exploited directly as a photoswitch and can also be further amplified so that larger-scale material properties are switched or altered in response to light. Thus, azo materials offer a promising potential as photo-mechanical materials. Light is an efficient power source for many of these applications, a direct transfer of photonic energy into mechanical motion with no moving parts, and light is also an ideal triggering mechanism, since it can be localized (in time and space), is selective, is nondamaging, and allows for remote activation and remote delivery of energy to a system. Thus, for sensing, actuation, and motion, photo-functional materials are of great interest. Azo materials have demonstrated a wide variety of switching behavior, from altering optical properties, to surface energy changes, to even eliciting bulk material phase changes. Azobenzene is the leader amongst the small class of photo-reversible molecules, and soft azo polymers can be considered promising materials for next-generation photo-mechanical applications because of their ease of incorporation and efficient and robust photochemistry. This chapter described the photo-mechanical effects observed in monolayers, thin films, gels, crystals, amorphous polymers, and LCEs containing azobenzene. In various systems, full macroscopic light-driven actuation has been achieved; however, the mechanical forces produced thus far and the efficiency for light energy conversion are still far from optimal. LCEs in particular are promising materials for artificial muscles and motors driven by light, and in these systems not only 2D but 3D motions have now been achieved, which are competitive and promising for many applications as soft actuators. Many problems also still remain unsolved however, such as fatigue resistance and biocompatibility of these materials, which need further intensive investigation.

Overall, azobenzene materials might still be viewed more as "solutions in need of a problem to solve," as material development has far outpaced application. For the field to progress now, it requires creative and inspired engineering, continuing on from this body of excellent and successful science, to identify the major unique niches in actuation where azobenzene-based materials and photoactuation in general are capable of becoming a competitive solution. This chapter has identified various strengths, properties and possibilities that azobenzene-based systems are capable of as well as the ability to incorporate azobenzene into various materials and systems. It still, however, lacks unifying problems or application areas where it can display its inherent advantages and potential. The few recent "proof-of-principle" applications described in the last section have provided much encouragement and confidence, however, toward the ability of azobenzene-based materials to fabricate real macro- and micro-scale robots amenable to remote operation and control, as well as the advantages offered in design simplification and scale-down afforded by the replacement of electrons by photons. Driving actuation with light by this powerful emerging class of photo-energy harvesting materials can offer important and significant advantages that warrant much further study of these materials into their full potential.

REFERENCES

1. K. Aoki, M. Nakagawa, K. Ichimura, *Journal of the American Chemical Society* 2000, *122*, 10997.
2. S. Kadota, K. Aoki, S. Nagano, T. Seki, *Journal of the American Chemical Society* 2005, *127*, 8266.
3. J. J. Effing, J. C. T. Kwak, *Angewandte Chemie, International Edition in English* 1995, *34*, 88.
4. H. Yamamoto, A. Nishida, T. Takimoto, A. Nagai, *Journal of Polymer Science, Part A: Polymer Chemistry* 1990, *28*, 67.
5. K. Arai, Y. Kawabata, *Macromolecular Rapid Communications* 1995, *16*, 875.
6. T. D. Ebralidze, A. N. Mumladze, *Applied Optics* 1990, *29*, 446.
7. M. Higuchi, N. Minoura, T. Kinoshita, *Chemistry Letters* 1994, *2*, 227.
8. M. Higuchi, N. Minoura, T. Kinoshita, *Macromolecules* 1995, *28*, 4981.
9. C. Steinem, A. Janshoff, M. S. Vollmer, M. R. Ghadiri, *Langmuir* 1999, *15*, 3956.
10. M. S. Vollmer, T. D. Clark, C. Steinem, M. R. Ghadiri, *Angewandte Chemie, International Edition* 1999, *38*, 1598.
11. S. Yagai, T. Nakajima, K. Kishikawa, S. Kohmoto, T. Karatsu, A. Kitamura, *Journal of the American Chemical Society* 2005, *127*, 11134.
12. G. Pouliquen, C. Tribet, *Macromolecules* 2005, *39*, 373.
13. P. Camorani, M. P. Fontana, *Physical Review E (Statistical, Nonlinear, and Soft Matter Physics)* 2006, *73*, 011703.
14. G. Wang, X. Tong, Y. Zhao, *Macromolecules* 2004, *37*, 8911.
15. P. Ravi, S. L. Sin, L. H. Gan, Y. Y. Gan, K. C. Tam, X. L. Xia, X. Hu, *Polymer* 2005, *46*, 137.
16. S. L. Sin, L. H. Gan, X. Hu, K. C. Tam, Y. Y. Gan, *Macromolecules* 2005, *38*, 3943.
17. E. Yoshida, M. Ohta, *Colloid and Polymer Science* 2005, *283*, 872.
18. E. Yoshida, M. Ohta, *Colloid and Polymer Science* 2005, *283*, 521.
19. H. Sakai, A. Matsumura, T. Saji, M. Abe, *Studies in Surface Science and Catalysis* 2001, *132*, 505.
20. X.-M. Liu, B. Yang, Y.-L. Wang, J.-Y. Wang, *Chemistry of Materials* 2005, *17*, 2792.
21. X. Tong, G. Wang, A. Soldera, Y. Zhao, *Journal of Physical Chemistry B* 2005, *109*, 20281.
22. I. Willner, S. Rubin, *Angewandte Chemie, International Edition in English* 1996, *35*, 367.
23. G. Montagnoli, O. Pieroni, S. Suzuki, *Polymer Photochemistry* 1983, *3*, 279.
24. H. Yamamoto, A. Nishida, *Polymer International* 1991, *24*, 145.
25. A. Fissi, O. Pieroni, E. Balestreri, C. Amato, *Macromolecules* 1996, *29*, 4680.
26. A. Fissi, O. Pieroni, F. Ciardelli, *Biopolymers* 1987, *26*, 1993.
27. W.-S. Lee, A. Ueno, *Macromolecular Rapid Communications* 2001, *22*, 448.
28. I. Willner, S. Rubin, A. Riklin, *Journal of the American Chemical Society* 1991, *113*, 3321.
29. I. Willner, S. Rubin, *Reactive Polymers* 1993, *21*, 177.
30. T. Inada, T. Terabayashi, Y. Yamaguchi, K. Kato, K. Kikuchi, *Journal of Photochemistry and Photobiology, A: Chemistry* 2005, *175*, 100.
31. I. Willner, S. Rubin, R. Shatzmiller, T. Zor, *Journal of the American Chemical Society* 1993, *115*, 8690.
32. I. Willner, S. Rubin, T. Zor, *Journal of the American Chemical Society* 1991, *113*, 4013.
33. K. Komori, K. Yatagai, T. Tatsuma, *Journal of Biotechnology* 2004, *108*, 11.
34. H. Asanuma, X. Liang, T. Yoshida, M. Komiyama, *ChemBioChem* 2001, *2*, 39.
35. X. Liang, H. Asanuma, H. Kashida, A. Takasu, T. Sakamoto, G. Kawai, M. Komiyama, *Journal of the American Chemical Society* 2003, *125*, 16408.
36. M. Liu, H. Asanuma, M. Komiyama, *Journal of the American Chemical Society* 2005, *128*, 1009.
37. H. Zollinger, *Azo and Diazo Chemistry*, Interscience, New York, 1961.

38. H. Zollinger, *Colour Chemistry, Synthesis, Properties, and Applications of Organic Dyes*, VCH, Weinheim, Germany, 1987.

39. S. Kwolek, P. Morgan, J. Schaefgen, *Encyclopedia of Polymer Science and Engineering*, John-Wiley, New York, 1985.

40. G. Möhlmann, C. van der Vorst, *Side Chain Liquid Crystal Polymers*, C. Mcardle, Ed., Plenum and Hall, Glasgow, U.K., 1989, 92–93, 361.

41. H. Rau, Photoisomerization of azobenzenes, in *Photochemistry and Photophysics*, J. Rebek, Ed., CRC Press, Boca Raton, FL, 1990, p. 119.

42. H. Rau, *Berichte der Bunsen-Gesellschaft* 1968, *72*, 408.

43. J. A. Delaire, K. Nakatani, *Chemical Reviews* 2000, *100*, 1817.

44. F. W. Schulze, H. J. Petrick, H. K. Cammenga, H. Klinge, *Zeitschrift fuer Physikalische Chemie (Muenchen, Germany)* 1977, *107*, 1.

45. I. Mita, K. Horie, K. Hirao, *Macromolecules* 1989, *22*, 558.

46. S. Monti, G. Orlandi, P. Palmieri, *Chemical Physics* 1982, *71*, 87.

47. T. Kobayashi, E. O. Degenkolb, P. M. Rentzepis, *Journal of Physical Chemistry* 1979, *83*, 2431.

48. I. K. Lednev, T.-Q. Ye, R. E. Hester, J. N. Moore, *Journal of Physical Chemistry* 1996, *100*, 13338.

49. E. V. Brown, G. R. Granneman, *Journal of the American Chemical Society* 1975, *97*, 621.

50. P. Haberfield, P. M. Block, M. S. Lux, *Journal of the American Chemical Society* 1975, *97*, 5804.

51. L. Lamarre, C. S. P. Sung, *Macromolecules* 1983, *16*, 1729.

52. Y. Shirota, K. Moriwaki, S. Yoshikawa, T. Ujike, H. Nakano, *Journal of Materials Chemistry* 1998, *8*, 2579.

53. Y. Norikane, K. Kitamoto, N. Tamaoki, *Journal of Organic Chemistry* 2003, *68*, 8291.

54. H. Rau, D. Roettger, *Molecular Crystals and Liquid Crystals Science and Technology, Section A: Molecular Crystals and Liquid Crystals* 1994, *246*, 143.

55. D. Rottger, H. Rau, *Journal of Photochemistry and Photobiology A: Chemistry* 1996, *101*, 205.

56. S. A. Nagamani, Y. Norikane, N. Tamaoki, *Journal of Organic Chemistry* 2005, *70*, 9304.

57. B. K. Kerzhner, V. I. Kofanov, T. L. Vrubel, *Zhurnal Obshchei Khimii* 1983, *53*, 2303.

58. U. Funke, H. F. Gruetzmacher, *Tetrahedron* 1987, *43*, 3787.

59. G. S. Hartley, *Nature (London)* 1937, *140*, 281.

60. G. S. Hartley, *Journal of the Chemical Society, Abstracts* 1938, 633.

61. E. Fischer, *Journal of Physical Chemistry* 1967, *71*, 3704.

62. H. Rau, G. Greiner, G. Gauglitz, H. Meier, *Journal of Physical Chemistry* 1990, *94*, 6523.

63. G. Gabor, E. Fischer, *Journal of Physical Chemistry* 1971, *75*, 581.

64. C. D. Eisenbach, *Macromolecular Rapid Communications* 1980, *1*, 287.

65. S. R. Hair, G. A. Taylor, L. W. Schultz, *Journal of Chemical Education* 1990, *67*, 709.

66. P. L. Beltrame, E. D. Paglia, A. Castelli, G. F. Tantardini, A. Seves, B. Marcandalli, *Journal of Applied Polymer Science* 1993, *49*, 2235.

67. S. W. Magennis, F. S. Mackay, A. C. Jones, K. M. Tait, P. J. Sadler, *Chemistry of Materials* 2005, *17*, 2059.

68. S. Xie, A. Natansohn, P. Rochon, *Chemistry of Materials* 1993, *5*, 403.

69. D. Gegiou, K. A. Muszkat, E. Fischer, *Journal of the American Chemical Society* 1968, *90*, 3907.

70. H. Rau, E. Lueddecke, *Journal of the American Chemical Society* 1982, *104*, 1616.

71. T. Naito, K. Horie, I. Mita, *Macromolecules* 1991, *24*, 2907.

72. Z. F. Liu, K. Morigaki, T. Enomoto, K. Hashimoto, A. Fujishima, *Journal of Physical Chemistry* 1992, *96*, 1875.

73. A. Altomare, F. Ciardelli, N. Tirelli, R. Solaro, *Macromolecules* 1997, *30*, 1298.

74. T. Fujino, T. Tahara, *Journal of Physical Chemistry A* 2000, *104*, 4203.

75. T. Fujino, S. Y. Arzhantsev, T. Tahara, *Journal of Physical Chemistry A* 2001, *105*, 8123.

76. C.-H. Ho, K.-N. Yang, S.-N. Lee, *Journal of Polymer Science, Part A: Polymer Chemistry* 2001, *39*, 2296.

77. C. Angeli, R. Cimiraglia, H.-J. Hofmann, *Chemical Physics Letters* 1996, *259*, 276.

78. B. S. Jursic, *Chemical Physics Letters* 1996, *261*, 13.

79. T. Ikeda, O. Tsutsumi, *Science (Washington, DC)* 1995, *268*, 1873.

80. W. J. Priest, M. M. Sifain, *Journal of Polymer Science, Part A: Polymer Chemistry* 1971, *9*, 3161.

81. C. S. Paik, H. Morawetz, *Macromolecules* 1972, *5*, 171.

82. C. Barrett, A. Natansohn, P. Rochon, *Macromolecules* 1994, *27*, 4781.

83. C. Barrett, A. Natansohn, P. Rochon, *Chemistry of Materials* 1995, *7*, 899.

84. N. Sarkar, A. Sarkar, S. Sivaram, *Journal of Applied Polymer Science* 2001, *81*, 2923.

85. L. L. Norman, C. J. Barrett, *Journal of Physical Chemistry B* 2002, *106*, 8499.

86. K. Tanaka, Y. Tateishi, T. Nagamura, *Macromolecules* 2004, *37*, 8188.
87. H. Rau, A. D. Crosby, A. Schaufler, R. Frank, *Zeitschrift fuer Naturforschung, A: Physical Sciences* 1981, *36*, 1180.
88. S. Malkin, E. Fischer, *Journal of Physical Chemistry* 1962, *66*, 2482.
89. Y. Q. Shen, H. Rau, *Macromolecular Chemistry and Physics* 1991, *192*, 945.
90. P. Bortolus, S. Monti, *Journal of Physical Chemistry* 1979, *83*, 648.
91. S. Yitzchaik, T. J. Marks, *Accounts of Chemical Research* 1996, *29*, 197.
92. D. Levy, L. Esquivias, *Advanced Materials (Weinheim, Germany)* 1995, *7*, 120.
93. M. Sisido, Y. Ishikawa, M. Harada, K. Itoh, *Macromolecules* 1991, *24*, 3999.
94. B. Gallot, M. Fafiotte, A. Fissi, O. Pieroni, *Macromolecular Rapid Communications* 1996, *17*, 493.
95. S. Shinkai, T. Minami, Y. Kusano, O. Manabe, *Journal of the American Chemical Society* 1983, *105*, 1851.
96. J. H. Jung, C. Takehisa, Y. Sakata, T. Kaneda, *Chemistry Letters* 1996, *2*, 147.
97. H. Yamamura, H. Kawai, T. Yotsuya, T. Higuchi, Y. Butsugan, S. Araki, M. Kawai, K. Fujita, *Chemistry Letters* 1996, *9*, 799.
98. A. K. Singh, J. Das, N. Majumdar, *Journal of the American Chemical Society* 1996, *118*, 6185.
99. S. H. Chen, J. C. Mastrangelo, H. Shi, T. N. Blanton, A. Bashir-Hashemi, *Macromolecules* 1997, *30*, 93.
100. S. H. Chen, J. C. Mastrangelo, H. Shi, A. Bashir-Hashemi, J. Li, N. Gelber, *Macromolecules* 1995, *28*, 7775.
101. K. Ichimura, *Chemical Reviews* 2000, *100*, 1847.
102. R. Birabassov, N. Landraud, T. V. Galstyan, A. Ritcey, C. G. Bazuin, T. Rahem, *Applied Optics* 1998, *37*, 8264.
103. F. L. Labarthet, T. Buffeteau, C. Sourisseau, *Journal of Physical Chemistry B* 1998, *102*, 2654.
104. N. K. Viswanathan, D. Y. Kim, S. Bian, J. Williams, W. Liu, L. Li, L. Samuelson, J. Kumar, S. K. Tripathy, *Journal of Materials Chemistry* 1999, *9*, 1941.
105. F. Agolini, F. P. Gay, *Macromolecules* 1970, *3*, 349.
106. K. Anderle, R. Birenheide, M. Eich, J. H. Wendorff, *Makromolekulare Chemie, Rapid Communications* 1989, *10*, 477.
107. J. Furukawa, S. Takamori, S. Yamashita, *Angewandte Makromolekulare Chemie* 1967, *1*, 92.
108. M. C. Bignozzi, S. A. Angeloni, M. Laus, L. Incicco, O. Francescangeli, D. Wolff, G. Galli, E. Chiellini, *Polymer Journal (Tokyo)* 1999, *31*, 913.
109. X.-H. Liu, D. W. Bruce, I. Manners, *Chemical Communications (Cambridge)* 1997, *3*, 289.
110. M. Sukwattanasinitt, X. Wang, L. Li, X. Jiang, J. Kumar, S. K. Tripathy, D. J. Sandman, *Chemistry of Materials* 1998, *10*, 27.
111. M. Teraguchi, T. Masuda, *Macromolecules* 2000, *33*, 240.
112. A. Izumi, M. Teraguchi, R. Nomura, T. Masuda, *Journal of Polymer Science, Part A: Polymer Chemistry* 2000, *38*, 1057.
113. A. Izumi, M. Teraguchi, R. Nomura, T. Masuda, *Macromolecules* 2000, *33*, 5347.
114. S. y. Morino, A. Kaiho, K. Ichimura, *Applied Physics Letters* 1998, *73*, 1317.
115. A. Altomare, F. Ciardelli, B. Gallot, M. Mader, R. Solaro, N. Tirelli, *Journal of Polymer Science, Part A: Polymer Chemistry* 2001, *39*, 2957.
116. N. Tsutsumi, S. Yoshizaki, W. Sakai, T. Kiyotsukuri, *MCLC S&T, Section B: Nonlinear Optics* 1996, *15*, 387.
117. A. Priimagi, S. Cattaneo, R. H. A. Ras, S. Valkama, O. Ikkala, M. Kauranen, *Chemistry of Materials* 2005, *17*, 5798.
118. C. M. Tibirna, C. G. Bazuin, *Journal of Polymer Science, Part B: Polymer Physics* 2005, *43*, 3421.
119. S. A. Haque, J. S. Park, M. Srinivasarao, J. R. Durrant, *Advanced Materials (Weinheim, Germany)* 2004, *16*, 1177.
120. H. B. Mekelburger, K. Rissanen, F. Voegtle, *Chemische Berichte* 1993, *126*, 1161.
121. D. M. Junge, D. V. McGrath, *Chemical Communications (Cambridge)* 1997, *9*, 857.
122. V. A. Mallia, N. Tamaoki, *Journal of Materials Chemistry* 2003, *13*, 219.
123. K. Naito, A. Miura, *Journal of Physical Chemistry* 1993, *97*, 6240.
124. H. Ma, S. Liu, J. Luo, S. Suresh, L. Liu, S. H. Kang, M. Haller, T. Sassa, L. R. Dalton, A. K.-Y. Jen, *Advanced Functional Materials* 2002, *12*, 565.
125. V. E. Campbell, I. In, D. J. McGee, N. Woodward, A. Caruso, P. Gopalan, *Macromolecules* 2006, *39*, 957, ASAP, doi 10.1021/ma051772o.
126. O. Villavicencio, D. V. McGrath, Azobenzene-containing dendrimers, in *Advances in Dendritic Macromolecules*, G. R. Newkome, Ed., JAI Press, Oxford, U.K., 2002, p. 1.

127. A. Momotake, T. Arai, *Polymer* 2004, *45*, 5369.

128. A. Momotake, T. Arai, *Journal of Photochemistry and Photobiology, C: Photochemistry Reviews* 2004, *5*, 1.

129. D.-L. Jiang, T. Aida, *Nature (London)* 1997, *388*, 454.

130. T. Aida, D.-L. Jiang, E. Yashima, Y. Okamoto, *Thin Solid Films* 1998, *331*, 254.

131. D. M. Junge, D. V. McGrath, *Journal of the American Chemical Society* 1999, *121*, 4912.

132. T. Seki, M. Sakuragi, Y. Kawanishi, T. Tamaki, R. Fukuda, K. Ichimura, Y. Suzuki, *Langmuir* 1993, *9*, 211.

133. G. Jianhua, L. Hua, L. Lingyun, L. Bingjie, C. Yiwen, L. Zuhong, *Supramolecular Science* 1998, *5*, 675.

134. J. Razna, P. Hodge, D. West, S. Kucharski, *Journal of Materials Chemistry* 1999, *9*, 1693.

135. J. R. Silva, F. F. Dall'Agnol, O. N. Oliveira Jr., J. A. Giacometti, *Polymer* 2002, *43*, 3753.

136. S. D. Evans, S. R. Johnson, H. Ringsdorf, L. M. Williams, H. Wolf, *Langmuir* 1998, *14*, 6436.

137. R. C. Advincula, E. Fells, M.-K. Park, *Chemistry of Materials* 2001, *13*, 2870.

138. S. Balasubramanian, X. Wang, H. C. Wang, K. Yang, J. Kumar, S. K. Tripathy, L. Li, *Chemistry of Materials* 1998, *10*, 1554.

139. F. Saremi, B. Tieke, *Advanced Materials (Weinheim, Germany)* 1998, *10*, 389.

140. K. E. Van Cott, M. Guzy, P. Neyman, C. Brands, J. R. Heflin, H. W. Gibson, R. M. Davis, *Angewandte Chemie, International Edition* 2002, *41*, 3236.

141. O. Mermut, C. J. Barrett, *Journal of Physical Chemistry B* 2003, *107*, 2525.

142. V. Shibaev, A. Bobrovsky, N. Boiko, *Progress in Polymer Science* 2003, *28*, 729.

143. S. T. Sun, W. M. Gibbons, P. J. Shannon, *Liquid Crystals* 1992, *12*, 869.

144. K. Anderle, R. Birenheide, M. J. A. Werner, J. H. Wendorff, *Liquid Crystals* 1991, *9*, 691.

145. W. M. Gibbons, P. J. Shannon, S.-T. Sun, B. J. Swetlin, *Nature (London)* 1991, *351*, 49.

146. K. Ichimura, Y. Hayashi, H. Akiyama, T. Ikeda, N. Ishizuki, *Applied Physics Letters* 1993, *63*, 449.

147. A. G. Chen, D. J. Brady, *Applied Physics Letters* 1993, *62*, 2920.

148. U. Wiesner, N. Reynolds, C. Boeffel, H. W. Spiess, *Makromolekulare Chemie, Rapid Communications* 1991, *12*, 457.

149. J. Stumpe, L. Mueller, D. Kreysig, G. Hauck, H. D. Koswig, R. Ruhmann, J. Ruebner, *Makromolekulare Chemie, Rapid Communications* 1991, *12*, 81.

150. S. Hvilsted, F. Andruzzi, C. Kulinna, H. W. Siesler, P. S. Ramanujam, *Macromolecules* 1995, *28*, 2172.

151. O. Yaroschuk, T. Sergan, J. Lindau, S. N. Lee, J. Kelly, L.-C. Chien, *Journal of Chemical Physics* 2001, *114*, 5330.

152. M. Eich, J. Wendorff, *Journal of the Optical Society of America B: Optical Physics* 1990, *7*, 1428.

153. T. Ikeda, S. Horiuchi, D. B. Karanjit, S. Kurihara, S. Tazuke, *Macromolecules* 1990, *23*, 42.

154. T. Hayashi, H. Kawakami, Y. Doke, A. Tsuchida, Y. Onogi, M. Yamamoto, *European Polymer Journal* 1995, *31*, 23.

155. T. Kato, N. Hirota, A. Fujishima, J. M. J. Frechet, *Journal of Polymer Science, Part A: Polymer Chemistry* 1996, *34*, 57.

156. A. Natansohn, P. Rochon, *Chemical Reviews* 2002, *102*, 4139.

157. N. K. Viswanathan, S. Balasubramanian, L. Li, S. K. Tripathy, J. Kumar, *Japanese Journal of Applied Physics, Part 1: Regular Papers, Short Notes & Review Papers* 1999, *38*, 5928.

158. K. G. Yager, C. J. Barrett, *Current Opinion in Solid State & Materials Science* 2001, *5*, 487.

159. P. Uznanski, M. Kryszewski, E. W. Thulstrup, *European Polymer Journal* 1991, *27*, 41.

160. J. J. de Lange, J. M. Robertson, I. Woodward, *Proceedings of the Royal Society (London)* 1939, *A171*, 398.

161. G. C. Hampson, J. M. Robertson, *Journal of the Chemical Society, Abstracts* 1941, 409.

162. C. J. Brown, *Acta Crystallographica* 1966, *21*, 146.

163. T. Naito, K. Horie, I. Mita, *Polymer* 1993, *34*, 4140.

164. R. G. Weiss, V. Ramamurthy, G. S. Hammond, *Accounts of Chemical Research* 1993, *25*, 530.

165. T. Hugel, N. B. Holland, A. Cattani, L. Moroder, M. Seitz, H. E. Gaub, *Science (Washington, DC)* 2002, *296*, 1103.

166. N. B. Holland, T. Hugel, G. Neuert, A. Cattani-Scholz, C. Renner, D. Oesterhelt, L. Moroder, M. Seitz, H. E. Gaub, *Macromolecules* 2003, *36*, 2015.

167. G. Neuert, T. Hugel, R. R. Netz, H. E. Gaub, *Macromolecules* 2005, *39*, 789.

168. M. Asakawa, P. R. Ashton, V. Balzani, C. L. Brown, A. Credi, O. A. Matthews, S. P. Newton et al., *Chemistry—A European Journal* 1999, *5*, 860.

169. V. Balzani, A. Credi, F. Marchioni, J. F. Stoddart, *Chemical Communications (Cambridge)* 2001, *18*, 1860.

170. S. Tsuchiya, *Journal of the American Chemical Society* 1999, *121*, 48.

171. S. Masiero, S. Lena, S. Pieraccini, G. P. Spada, *Angewandte Chemie International Edition* 2008, *47*, 3184.
172. M. Fujiwara, M. Akiyama, M. Hata, K. Shiokawa, R. Nomura, *ACS Nano* 2008, *2*, 1671.
173. C. Pakula, V. Zaporojtchenko, T. Strunskus, D. Zargarani, R. Herges, F. Faupel, *Nanotechnology* 2010, *21*, 465201-1–465201-6.
174. C. Raimondo, F. Reinders, U. Soydaner, M. Mayor, P. Samorì, *Chemical Communications* 2010, *46*, 1147.
175. A. Kimoto, K. Iwasaki, J. Abe, *Photochemical and Photobiological Sciences* 2010, *9*, 152.
176. M. Higuchi, N. Minoura, T. Kinoshita, *Colloid and Polymer Science* 1995, *273*, 1022.
177. L. M. Siewierski, W. J. Brittain, S. Petrash, M. D. Foster, *Langmuir* 1996, *12*, 5838.
178. B. Stiller, G. Knochenhauer, E. Markava, D. Gustina, I. Muzikante, P. Karageorgiev, L. Brehmer, *Materials Science and Engineering C* 1999, *8–9*, 385.
179. G. Moller, M. Harke, H. Motschmann, D. Prescher, *Langmuir* 1998, *14*, 4955.
180. C. L. Feng, Y. J. Zhang, J. Jin, Y. L. Song, L. Y. Xie, G. R. Qu, L. Jiang, D. B. Zhu, *Langmuir* 2001, *17*, 4593.
181. T. Chen, S. Xu, F. Zhang, D. G. Evans, X. Duan, *Chemical Engineering Science* 2009, *64*, 4350.
182. N. Delorme, J.-F. Bardeau, A. Bulou, F. Poncin-Epaillard, *Langmuir* 2005, *21*, 12278.
183. W. H. Jiang, G. J. Wang, Y. N. He, X. G. Wang, Y. L. An, Y. L. Song, L. Jiang, *Chemical Communications (Cambridge)* 2005, *28*, 3550.
184. A. Diguet, R.-M. Guillermic, N. Magome, A. Saint-Jalmes, Y. Chen, K. Yoshikawa, D. Baigl, *Angewandte Chemie, International Edition* 2009, *48*, 9281.
185. T. Fujita, N. Iyi, Z. Klapyta, *Materials Research Bulletin* 1998, *33*, 1693.
186. T. Fujita, N. Iyi, Z. Klapyta, *Materials Research Bulletin* 2001, *36*, 557.
187. K. G. Yager, O. M. Tanchak, C. Godbout, H. Fritzsche, C. J. Barrett, *Macromolecules* 2006, *39*, 9311.
188. C. D. Eisenbach, *Polymer* 1980, *21*, 1175.
189. K. G. Yager, C. J. Barrett, *Macromolecules* 2006, *39*, 9320.
190. K. G. Yager, O. M. Tanchak, C. J. Barrett, M. J. Watson, H. Fritzsche, *Review of Scientific Instruments* 2006, *77*, 045106-1–045106-6.
191. O. M. Tanchak, C. J. Barrett, *Macromolecules* 2005, *38*, 10566.
192. Y. Yu, M. Nakano, T. Ikeda, *Nature (London)* 2003, *425*, 145.
193. T. Ikeda, M. Nakano, Y. Yu, O. Tsutsumi, A. Kanazawa, *Advanced Materials (Weinheim, Germany)* 2003, *15*, 201.
194. Y. L. Yu, M. Nakano, T. Maeda, M. Kondo, T. Ikeda, *Molecular Crystals and Liquid Crystals* 2005, *436*, 1235.
195. H. F. Ji, Y. Feng, X. H. Xu, V. Purushotham, T. Thundat, G. M. Brown, *Chemical Communication* 2004, *22*, 2532.
196. Z. S. Mahimwalla, Y. Ngai, C. J. Barrett, Photomechanical effect of azobenzene thin polymer films measured with an AFM cantilever based sensor, *Proceedings of SPIE*, Brussels, Belgium, 2010, p. 7712.
197. T. J. White, N. V. Tabiryan, S. V. Serak, U. A. Hrozhyk, V. P. Tondiglia, H. Koerner, R. A. Vaia, T. J. Bunning, *Soft Matter* 2008, *4*, 1796.
198. T. J. White, S. V. Serak, N. V. Tabiryan, R. A. Vaia, T. J. Bunning, *Journal of Materials Chemistry* 2009, *19*, 1080.
199. S. Serak, N. Tabiryan, R. Vergara, T. J. White, R. A. Vaia, T. J. Bunning, *Soft Matter* 2010, *6*, 779.
200. W. Deng, M. H. Li, X. G. Wang, P. Keller, *Liquid Crystals* 2009, *36*, 1023.
201. U. Hrozhyk, S. Serak, N. Tabiryan, T. J. White, T. J. Bunning, *Optics Express* 2009, *17*, 716.
202. M. Kondo, T. Matsuda, R. Fukae, N. Kawatsuki, *Chemistry Letters* 2010, *39*, 234.
203. F. T. Cheng, Y. Y. Zhang, R. Y. Yin, Y. L. Yu, *Journal of Materials Chemistry* 2010, *20*, 4888.
204. S. Bai, Y. Zhao, *Macromolecules* 2001, *34*, 9032.
205. P. Rochon, E. Batalla, A. Natansohn, *Applied Physics Letters* 1995, *66*, 136.
206. D. Y. Kim, S. K. Tripathy, L. Li, J. Kumar, *Applied Physics Letters* 1995, *66*, 1166.
207. S. Yamaki, M. Nakagawa, S. Morino, K. Ichimura, *Applied Physics Letters* 2000, *76*, 2520.
208. S. Bian, L. Li, J. Kumar, D. Y. Kim, J. Williams, S. K. Tripathy, *Applied Physics Letters* 1998, *73*, 1817.
209. F. L. Labarthet, J. L. Bruneel, T. Buffeteau, C. Sourisseau, *Journal of Physical Chemistry B* 2004, *108*, 6949.
210. F. Lagugne-Labarthet, J. L. Bruneel, V. Rodriguez, C. Sourisseau, *Journal of Physical Chemistry B* 2004, *108*, 1267.
211. F. L. Labarthet, J. L. Bruneel, T. Buffeteau, C. Sourisseau, M. R. Huber, S. J. Zilker, T. Bieringer, *Physical Chemistry Chemical Physics* 2000, *2*, 5154.
212. O. Henneberg, T. Geue, U. Pietsch, B. Saphiannikova, B. Winter, *Applied Physics Letters* 2004, *84*, 1561.
213. U. Pietsch, P. Rochon, A. Natansohn, *Advanced Materials* 2000, *12*, 1129.

214. T. Geue, O. Henneberg, J. Grenzer, U. Pietsch, A. Natansohn, P. Rochon, K. Finkelstein, *Colloids and Surfaces A: Physicochemical and Engineering Aspects* 2002, *198–200*, 31.

215. T. M. Geue, M. G. Saphiannikova, O. Henneberg, U. Pietsch, P. L. Rochon, A. L. Natansohn, *Journal of Applied Physics* 2003, *93*, 3161.

216. U. Pietsch, *Physical Review B* 2002, *66*, 155430.

217. O. Watanabe, T. Ikawa, M. Hasegawa, M. Tsuchimori, Y. Kawata, C. Egami, O. Sugihara, N. Okamoto, *Molecular Crystals and Liquid Crystals* 2000, *345*, 629.

218. T. Ikawa, T. Mitsuoka, M. Hasegawa, M. Tsuchimori, O. Watanabe, Y. Kawata, C. Egami, O. Sugihara, N. Okamoto, *Journal of Physical Chemistry B* 2000, *104*, 9055.

219. C. D. Keum, T. Ikawa, M. Tsuchimori, O. Watanabe, *Macromolecules* 2003, *36*, 4916.

220. C. J. Barrett, A. L. Natansohn, P. L. Rochon, *Journal of Physical Chemistry* 1996, *100*, 8836.

221. X. Wang, S. Balasubramanian, J. Kumar, S. K. Tripathy, L. Li, *Chemistry of Materials* 1998, *10*, 1546.

222. J.-A. He, S. Bian, L. Li, J. Kumar, S. K. Tripathy, L. A. Samuelson, *Applied Physics Letters* 2000, *76*, 3233.

223. S.-H. Lee, S. Balasubramanian, D. Y. Kim, N. K. Viswanathan, S. Bian, J. Kumar, S. K. Tripathy, *Macromolecules* 2000, *33*, 6534.

224. S. Z. Yang, L. Li, A. L. Cholli, J. Kumar, S. K. Tripathy, *Journal of Macromolecular Science: Pure and Applied Chemistry* 2001, *38*, 1345.

225. S. Yang, M. M. Jacob, L. Li, K. Yang, A. L. Cholli, J. Kumar, S. K. Tripathy, *Polymer News* 2002, *27*, 368.

226. S. Z. Yang, L. Li, A. L. Cholli, J. Kumar, S. K. Tripathy, *Biomacromolecules* 2003, *4*, 366.

227. H. Nakano, T. Takahashi, T. Kadota, Y. Shirota, *Advanced Materials* 2002, *14*, 1157.

228. M.-J. Kim, E.-M. Seo, D. Vak, D.-Y. Kim, *Chemistry of Materials* 2003, *15*, 4021.

229. C. M. Chun, M. J. Kim, D. Vak, D. Y. Kim, *Journal of Materials Chemistry* 2003, *13*, 2904.

230. H. Ando, T. Takahashi, H. Nakano, Y. Shirota, *Chemistry Letters* 2003, *32*, 710.

231. F. Ciuchi, A. Mazzulla, G. Cipparrone, *Journal of the Optical Society of America B: Optical Physics* 2002, *19*, 2531.

232. I. Naydenova, T. Petrova, N. Tomova, V. Dragostinova, L. Nikolova, T. Todorov, *Pure and Applied Optics* 1998, *7*, 723.

233. E. Ishow, B. Lebon, Y. He, X. Wang, L. Bouteiller, L. Galmiche, K. Nakatani, *Chemistry of Materials* 2006, *18*, 1261.

234. N. C. R. Holme, L. Nikolova, S. Hvilsted, P. H. Rasmussen, R. H. Berg, P. S. Ramanujam, *Applied Physics Letters* 1999, *74*, 519.

235. M. Helgert, L. Wenke, S. Hvilsted, P. S. Ramanujam, *Applied Physics B: Lasers and Optics* 2001, *72*, 429.

236. T. Ubukata, T. Seki, K. Ichimura, *Advanced Materials* 2000, *12*, 1675.

237. T. Ubukata, T. Seki, K. Ichimura, *Colloids and Surfaces A: Physicochemical and Engineering Aspects* 2002, *198*, 113.

238. P. S. Ramanujam, N. C. R. Holme, S. Hvilsted, *Applied Physics Letters* 1996, *68*, 1329.

239. A. Archut, F. Vögtle, L. D. Cola, G. C. Azzellini, V. Balzani, P. S. Ramanujam, R. H. Berg, *Chemistry—A European Journal* 1998, *4*, 699.

240. T. Fukuda, H. Matsuda, T. Shiraga, T. Kimura, M. Kato, N. K. Viswanathan, J. Kumar, S. K. Tripathy, *Macromolecules* 2000, *33*, 4220.

241. L. Andruzzi, A. Altomare, F. Ciardelli, R. Solaro, S. Hvilsted, P. S. Ramanujam, *Macromolecules* 1999, *32*, 448.

242. V. Borger, O. Kuliskovska, K. G. Hubmann, J. Stumpe, M. Huber, H. Menzel, *Macromolecular Chemistry and Physics* 2005, *206*, 1488.

243. A. Natansohn, S. Xie, P. Rochon, *Macromolecules* 1992, *25*, 5531.

244. M. S. Ho, A. Natansohn, P. Rochon, *Macromolecules* 1995, *28*, 6124.

245. H. Nakano, T. Tanino, T. Takahashi, H. Ando, Y. Shirota, *Journal of Materials Chemistry* 2008, *18*, 242.

246. T. S. Lee, D.-Y. Kim, X. L. Jiang, L. Li, J. Kumar, S. Tripathy, *Journal of Polymer Science, Part A: Polymer Chemistry* 1998, *36*, 283.

247. J. P. Chen, F. L. Labarthet, A. Natansohn, P. Rochon, *Macromolecules* 1999, *32*, 8572.

248. Y. L. Wu, A. Natansohn, P. Rochon, *Macromolecules* 2001, *34*, 7822.

249. C. J. Barrett, P. L. Rochon, A. L. Natansohn, *Journal of Chemical Physics* 1998, *109*, 1505.

250. K. Sumaru, T. Yamanaka, T. Fukuda, H. Matsuda, *Applied Physics Letters* 1999, *75*, 1878.

251. T. Fukuda, K. Sumaru, T. Yamanaka, H. Matsuda, *Molecular Crystals and Liquid Crystals* 2000, *345*, 587.

252. D. Bublitz, B. Fleck, L. Wenke, *Applied Physics B: Lasers and Optics* 2001, *72*, 931.

253. D. Bublitz, M. Helgert, B. Fleck, L. Wenke, S. Hvilsted, P. S. Ramanujam, *Applied Physics B: Lasers and Optics* 2000, *70*, 863.
254. M. Saphiannikova, O. Henneberg, T. M. Gene, U. Pietsch, P. Rochon, *Journal of Physical Chemistry B* 2004, *108*, 15084.
255. M. Saphiannikova, T. M. Geue, O. Henneberg, K. Morawetz, U. Pietsch, *Journal of Chemical Physics* 2004, *120*, 4039.
256. G. Pawlik, A. C. Mitus, A. Miniewicz, F. Kajzar, *Journal of Chemical Physics* 2003, *119*, 6789.
257. A. C. Mitus, G. Pawlik, A. Miniewicz, F. Kajzar, *Molecular Crystals and Liquid Crystals* 2004, *416*, 113.
258. G. Pawlik, A. C. Mitus, A. Miniewicz, F. Kajzar, *Journal of Nonlinear Optical Physics & Materials* 2004, *13*, 481.
259. M. Matsumoto, D. Miyazaki, M. Tanaka, R. Azumi, E. Manda, Y. Kondo, N. Yoshino, H. Tachibana, *Journal of the American Chemical Society* 1998, *120*, 1479.
260. C. Hubert, C. Fiorini-Debuisschert, I. Maurin, J. M. Nunzi, P. Raimond, *Advanced Materials* 2002, *14*, 729.
261. C. Hubert, E. Malcor, I. Maurin, J. M. Nunzi, P. Raimond, C. Fiorini, *Applied Surface Science* 2002, *186*, 29.
262. N. Mechau, D. Neher, V. Borger, H. Menzel, K. Urayama, *Applied Physics Letters* 2002, *81*, 4715.
263. F. Lagugne-Labarthet, T. Buffeteau, C. Sourisseau, *Physical Chemistry Chemical Physics* 2002, *4*, 4020.
264. O. Henneberg, T. Geue, M. Saphiannikova, U. Pietsch, L. F. Chi, P. Rochon, A. L. Natansohn, *Applied Physics Letters* 2001, *79*, 2357.
265. F. Ciuchi, A. Mazzulla, G. Carbone, G. Cipparrone, *Macromolecules* 2003, *36*, 5689.
266. N. Zettsu, T. Ubukata, T. Seki, K. Ichimura, *Advanced Materials* 2001, *13*, 1693.
267. S. K. Tripathy, N. K. Viswanathan, S. Balasubramanian, J. Kumar, *Polymers for Advanced Technologies* 2000, *11*, 570.
268. P. Rochon, A. Natansohn, C. L. Callendar, L. Robitaille, *Applied Physics Letters* 1997, *71*, 1008.
269. R. J. Stockermans, P. L. Rochon, *Applied Optics* 1999, *38*, 3714.
270. J. Paterson, A. Natansohn, P. Rochon, C. L. Callendar, L. Robitaille, *Applied Physics Letters* 1996, *69*, 3318.
271. T. Nagata, T. Matsui, M. Ozaki, K. Yoshino, F. Kajzar, *Synthetic Metals* 2001, *119*, 607.
272. V. Dumarcher, L. Rocha, C. Denis, C. Fiorini, J.-M. Nunzi, F. Sobel, B. Sahraoui, D. Gindre, *Journal of Optics A: Pure and Applied Optics* 2000, *2*, 279.
273. L. Rocha, V. Dumarcher, C. Denis, P. Raimond, C. Fiorini, J. M. Nunzi, *Journal of Applied Physics* 2001, *89*, 3067.
274. C. Egami, Y. Kawata, Y. Aoshima, S. Alasfar, O. Sugihara, H. Fujimura, N. Okamoto, *Japanese Journal of Applied Physics Part 1: Regular Papers Short Notes & Review Papers* 2000, *39*, 1558.
275. P. S. Ramanujam, M. Pedersen, S. Hvilsted, *Applied Physics Letters* 1999, *74*, 3227.
276. Y. Gritsai et al., *Journal of Optics A: Pure and Applied Optics* 2008, *10*, 125304.
277. J. Neumann, K. S. Wieking, D. Kip, *Applied Optics* 1999, *38*, 5418.
278. X. T. Li, A. Natansohn, P. Rochon, *Applied Physics Letters* 1999, *74*, 3791.
279. M.-H. Kim, J.-D. Kim, T. Fukuda, H. Matsuda, *Liquid Crystals* 2000, *27*, 1633.
280. A. Parfenov, N. Tamaoki, S. Ohnishi, *Journal of Applied Physics* 2000, *87*, 2043.
281. A. Parfenov, N. Tamaoki, S. Ohnishi, *Molecular Crystals and Liquid Crystals* 2001, *359*, 487.
282. F. Kaneko, T. Kato, A. Baba, K. Shinbo, K. Kato, R. C. Advincula, *Colloids and Surfaces A: Physicochemical and Engineering Aspects* 2002, *198*, 805.
283. E. Ishow, A. Brosseau, G. Clavier, K. Nakatani, R. B. Pansu, J.-J. Vachon, P. Tauc, D. Chauvat, C. R. Mendonça, E. Piovesan, *Journal of American Chemical Society* 2007, *129*, 8970.
284. X. Chen, B. Liu, H. Zhang, S. Guan, J. Zhang, W. Zhang, Q. Chen, Z. Jiang, M. D. Guiver, *Langmuir* 2009, *25*, 10444.
285. Y. H. Ye, S. Badilescu, V. V. Truong, P. Rochon, A. Natansohn, *Applied Physics Letters* 2001, *79*, 872.
286. D. K. Yi, M. J. Kim, D. Y. Kim, *Langmuir* 2002, *18*, 2019.
287. D. K. Yi, E.-M. Seo, D.-Y. Kim, *Langmuir* 2002, *18*, 5321.
288. B. Liu, M. Wang, Y. He, X. Wang, *Langmuir* 2006, *22*, 7405.
289. G. Ye, X. Li, X. Wang, *Chemical Communications* 2010, *46*, 3872.
290. S. Noel, E. Batalla, P. Rochon, *Journal of Materials Research* 1996, *11*, 865.
291. M. Hasegawa, T. Ikawa, M. Tsuchimori, O. Watanabe, Y. Kawata, *Macromolecules* 2001, *34*, 7471.
292. M. Hasegawa, C.-D. Keum, O. Watanabe, *Advanced Materials* 2002, *14*, 1738.
293. T. Fukuda, K. Sumaru, T. Kimura, H. Matsuda, Y. Narita, T. Inoue, F. Sato, *Japanese Journal of Applied Physics Part 2: Letters* 2001, *40*, L900.

294. I. Naydenova, L. Nikolova, T. Todorov, N. C. R. Holme, P. S. Ramanujam, S. Hvilsted, *Journal of the Optical Society of America B: Optical Physics* 1998, *15*, 1257.

295. F. L. Labarthet, T. Buffeteau, C. Sourisseau, *Journal of Applied Physics* 2001, *90*, 3149.

296. M. Camacho-Lopez, H. Finkelmann, P. Palffy-Muhoray, M. Shelley, *Nature Materials* 2004, *3*, 307.

297. M. Chen, H. Huang, Y. Zhu, Z. Liu, X. Xing, F. Cheng, Y. Yu, *Applied Physics A: Materials Science and Processing* 2010, 100, 1, 39–43.

298. M. Yamada, M. Kondo, J. I. Mamiya, Y. Yu, M. Kinoshita, C. J. Barrett, T. Ikeda, *Angewandte Chemie, International Edition* 2008, *47*, 4986.

299. S. Tanaka, H. K. Kim, A. Sudo, H. Nishida, T. Endo, *Macromolecular Chemistry and Physics* 2008, *209*, 2071.

300. C. Zhang, X. Zhao, D. Chao, X. Lu, C. Chen, C. Wang, W. Zhang, *Journal of Applied Polymer Science* 2009, *113*, 1330.

301. H. K. Kim, W. Shin, T. J. Ahn, *IEEE Photonics Technology Letters* 2010, *22*, 1404.

302. K. T. Kim, N. I. Moon, H. K. Kim, *Sensors and Actuators, A: Physical* 2010, *160*, 19.

303. J. Liu, Y. He, X. Wang, *Polymer* 2010, *51*, 2879.

304. J. H. Liu, Y. H. Chiu, *Journal of Polymer Science, Part A: Polymer Chemistry* 2010, *48*, 1142.

305. J. Liu, Y. He, X. Wang, *Langmuir* 2008, *24*, 678.

306. J. Liu, Y. He, X. Wang, *Langmuir* 2009, *25*, 5974.

307. J. Küpfer, E. Nishikawa, H. Finkelmann, *Polymers for Advanced Technologies* 1994, *5*, 110.

308. H. Wermter, H. Finkelmann, *e-Polymers* 2001, 013.

309. P. G. D. Gennes, M. Hebert, R. Kant, *Macromolecular Symposia* 1997, *113*, 39.

310. T. Ikeda, J. Mamiya, Y. L. Yu, *Angewandte Chemie, International Edition* 2007, *46*, 506.

311. H. Finkelmann, E. Nishikawa, G. G. Pereira, M. Warner, *Physical Review Letters* 2001, *87*, 015501.

312. P. M. Hogan, A. R. Tajbakhsh, E. M. Terentjev, *Physical Review E* 2002, *65*, 041720.

313. J. Cviklinski, A. R. Tajbakhsh, E. M. Terentjev, *European Physical Journal E* 2002, *9*, 427.

314. H. Finkelmann, G. Rehage, Liquid crystal side chain polymers, in *Liquid Crystal Polymers II/III*, N. Platé, Ed., Springer, Berlin, Germany, 1984, p. 99.

315. Y. Zhang, J. Xu, F. Cheng, R. Yin, C. C. Yen, Y. Yu, *Journal of Materials Chemistry* 2010, *20*, 7123.

316. K. M. Lee, H. Koerner, R. A. Vaia, T. J. Bunning, T. J. White, *Macromolecules* 2010, *43*, 8185.

317. Y. Yu, M. Nakano, A. Shishido, T. Shiono, T. Ikeda, *Chemistry of Materials* 2004, *16*, 1637.

318. S. Tazuke, S. Kurihara, T. Ikeda, *Chemistry Letters* 1987, *16*, 911.

319. T. Ikeda, S. Horiuchi, D. B. Karanjit, S. Kurihara, S. Tazuke, *Chemistry Letters* 1988, *17*, 1679.

320. T. Ikeda, S. Horiuchi, D. B. Karanjit, S. Kurihara, S. Tazuke, *Macromolecules* 1990, *23*, 36.

321. T. Ikeda, S. Kurihara, D. B. Karanjit, S. Tazuke, *Macromolecules* 1990, *23*, 3938.

322. T. Sasaki, T. Ikeda, K. Ichimura, *Macromolecules* 1992, *25*, 3807.

323. O. Tsutsumi, Y. Demachi, A. Kanazawa, T. Shiono, T. Ikeda, Y. Nagase, *Journal of Physical Chemistry B* 1998, *102*, 2869.

324. T. Seki, Y. Morikawa, T. Uekusa, S. Nagano, *Molecular Crystals and Liquid Crystals* 2009, *510*, 21.

325. P. Camorani, L. Cristofolini, M. P. Fontana, L. Angiolini, L. Giorgini, F. Paris, *Molecular Crystals and Liquid Crystals* 2009, *500*, 1.

326. S. Kurihara, T. Ikeda, T. Sasaki, H.-B. Kim, S. Tazuke, *Journal of the Chemical Society, Chemical Communications* 1990, 24, 1751.

327. T. Ikeda, T. Sasaki, H. B. Kim, *The Journal of Physical Chemistry* 1991, *95*, 509.

328. M. H. Li, P. Keller, B. Li, X. Wang, M. Brunet, *Advanced Materials* 2003, *15*, 569.

329. Y. Yu, M. Nakano, T. Ikeda, *Pure and Applied Chemistry* 2004, *76*, 1435.

330. Y. L. Yu, T. Maeda, J. Mamiya, T. Ikeda, *Angewandte Chemie-International Edition* 2007, *46*, 881.

331. N. Tabiryan, S. Serak, X. M. Dai, T. Bunning, *Optics Express* 2005, *13*, 7442.

332. K. D. Harris, R. Cuypers, P. Scheibe, C. L. van Oosten, C. W. M. Bastiaansen, J. Lub, D. J. Broer, *Journal of Materials Chemistry* 2005, *15*, 5043.

333. M. Kondo, Y. Yu, T. Ikeda, *Angewandte Chemie, International Edition* 2006, *45*, 1378.

334. M. Kondo, M. Sugimoto, M. Yamada, Y. Naka, J. I. Mamiya, M. Kinoshita, A. Shishido, Y. Yu, T. Ikeda, *Journal of Materials Chemistry* 2010, *20*, 117.

335. T. Yoshino, M. Kondo, J. Mamiya, M. Kinoshita, Y. L. Yu, T. Ikeda, *Advanced Materials* 2010, *22*, 1361.

336. A. Kausar, H. Nagano, T. Ogata, T. Nonaka, S. Kurihara, *Angewandte Chemie, International Edition* 2009, *48*, 2144.

337. N. Hosono, T. Kajitani, T. Fukushima, K. Ito, S. Sasaki, M. Takata, T. Aida, *Science* 2010, *330*, 808.

338. K. Milam, G. O'Malley, N. Kim, D. Golovaty, T. Kyu, *Journal of Physical Chemistry B* 2010, *114*, 7791.

339. H. Koshima, N. Ojima, H. Uchimoto, *Journal of American Chemical Society* 2009, *131*, 6890.

340. K. Ichimura, *Chemical Communications* 2010, *46*, 3295.
341. K. Ichimura, *Chemical Communications* 2009, *12*, 1496.
342. H. Nakano, S. Seki, H. Kageyama, *Physical Chemistry Chemical Physics* 2010, *12*, 7772.
343. F. Toda, *Organic Solid State Reactions*, Kluwer Academic Publishers, Boston, MA, 2002.
344. K. Tanaka, *Solvent-Free Organic Synthesis*, Wiley-VCH Verlag GmbH & Co. KGaA, Weinheim, Germany, 2009.
345. G. Kaupp, *Angewandte Chemie, International Edition in English* 1992, *31*, 592.
346. G. Kaupp, *Angewandte Chemie, International Edition in English* 1992, *31*, 595.
347. M. Irie, S. Kobatake, M. Horichi, *Science* 2001, *291*, 1769.
348. H. Koshima, Y. Ide, N. Ojima, *Crystal Growth & Design* 2008, *8*, 2058.
349. S. Kobatake, S. Takami, H. Muto, T. Ishikawa, M. Irie, *Nature* 2007, *446*, 778.
350. M. Morimoto, M. Irie, *Journal of American Chemical Society* 2010, *132*, 14172.
351. K. Nakayama, L. Jiang, T. Iyoda, K. Hashimoto, A. Fujishima, *Japanese Journal of Applied Physics* 1997, *36*, 3898.
352. H. Nakano, T. Tanino, Y. Shirota, *Applied Physics Letters* 2005, *87*, 061910.
353. Y. Bar-Cohen, *Electroactive Polymer (EAP) Actuators as Artificial Muscles: Reality, Potential, and Challenges*, SPIE Press, Bellingham, WA, 2004.
354. T. Mirfakhrai, J. D. W. Madden, R. H. Baughman, *Materials Today* 2007, *10*, 30.
355. B. Kim et al., *Smart Materials and Structures* 2005, *14*, 1579.
356. E. W. H. Jager, O. Inganäs, I. Lundström, *Science* 2000, *288*, 2335.
357. M. Yamada, M. Kondo, R. Miyasato, Y. Naka, J. I. Mamiya, M. Kinoshita, A. Shishido, Y. Yu, C. J. Barrett, T. Ikeda, *Journal of Materials Chemistry* 2009, *19*, 60.
358. J. I. Mamiya, A. Yoshitake, M. Kondo, Y. Yu, T. Ikeda, *Journal of Materials Chemistry* 2008, *18*, 63.
359. M. Kondo, J. I. Mamiya, M. Kinoshita, T. Ikeda, Y. Yu, *Molecular Crystals and Liquid Crystals* 2007, *478*, 245.
360. M. Kondo, R. Miyasato, Y. Naka, J. Mamiya, M. Kinoshita, Y. Yu, C. J. Barrett, T. Ikeda, *Liquid Crystals* 2009, *36*, 1289.
361. R. Y. Yin, W. X. Xu, M. Kondo, C. C. Yen, J. Mamiya, T. Ikeda, Y. L. Yu, *Journal of Materials Chemistry* 2009, *19*, 3141.
362. Y. Naka, J. Mamiya, A. Shishido, M. Washio, T. Ikeda, *Journal of Materials Chemistry* 2011, *21*, 1681.
363. F. Cheng, R. Yin, Y. Zhang, C. C. Yen, Y. Yu, *Soft Matter* 2010, *6*, 3447.
364. F. T. Cheng, J. X. Xu, R. Y. Yin, Y. L. Yu, *Cailiao Kexue yu Gongyi/Material Science and Technology* 2008, *16*, 159.
365. M. Chen, X. Xing, Z. Liu, Y. Zhu, H. Liu, Y. Yu, F. Cheng, *Applied Physics A: Materials Science and Processing* 2010, *100*, 39.
366. H. Liu, Y. Zhu, Z. Liu, M. Chen, *Yadian Yu Shengguang/Piezoelectrics and Acoustooptics* 2010, *32*, 417.
367. C. L. van Oosten, C. W. M. Bastiaansen, D. J. Broer, *Nat Mater* 2009, *8*, 677.
368. P. Palffy-Muhoray, *Nature Materials* 2009, *8*, 614.

5 Photostrictive Microactuators

Kenji Uchino

CONTENTS

5.1 INTRODUCTION

The continuing thrust toward greater miniaturization and integration of microrobotics and microelectronics has resulted in significant work toward the development of piezoelectric actuators. One of the bottlenecks of the piezo-actuator is its necessity of the electric lead wire, which is too heavy for a miniaturized self-propelling robot less than 1 cm³. The important reason is a drastic reduction of the propelling friction force due to the increase in specific area; that is, surface area/ volume or weight ratio. "What if you, an expert on actuators, could produce a remote-controlled actuator that would bypass the electrical lead?" To many people, "remote control" equals control by radio waves, light waves, or sound. Light-controlled actuators require that light energy be transduced twice: first from light energy to electrical energy, and second from electrical energy to mechanical energy. These are "photovoltaic" and "piezoelectric" effects. A solar cell is a well-known photovoltaic device, but it does not generate sufficient voltage to drive a piezoelectric device; in other words, this combination fails due to the electric impedance mismatch. The key to success is to adopt a high-impedance photovoltaic effect (so-called anomalous or bulk photovoltaic effect in piezoelectrics), which is totally different from the p–n junction-based solar cell.

In the recent days, photostrictive actuators—which directly convert the photonic energy to mechanical motion—have drawn significant attention for their potential usage in microactuation and microsensing applications. Optical actuators are also anticipated to be used as the driving component in optically controlled electromagnetic-noise free systems. The photostrictive effect will also be used in fabricating a photophonic device, where light is transformed directly into sound from the mechanical vibration induced by intermittent illumination at a human-audible frequency.

The photostrictive effect has been studied mainly in ferroelectric polycrystalline materials for potential commercial applications. Lanthanum-modified lead zirconate titanate (PLZT) ceramic is one of the most promising photostrictive materials due to its relatively high piezoelectric coefficient and ease of fabrication. However, previous studies have shown that for commercial applications, improvements in photovoltaic efficiency and response speed of the PLZT ceramics are still essential. The improvement in photostrictive properties requires consideration of several parameters, such as material parameters, processing condition and microstructure, and sample configuration and performance testing conditions.

This chapter reviews theoretical background for the photostrictive effect first, then, enhanced performance through the composition modification, sample preparation technique (thickness and surface characteristics of the sample). Finally, its potential future applications are briefly described.

5.2 PHOTOVOLTAIC EFFECT

The photostriction phenomenon was discovered by Dr. P.S. Brody and the author independently almost at the same time in 1981.[1] In principle, photostrictive effect arises from a superposition of the "bulk" photovoltaic effect, i.e., generation of large voltage from the irradiation of light, and the converse-piezoelectric effect, i.e., expansion or contraction under the voltage applied.[2] The photostrictive phenomenon has been observed in certain ferroelectric/piezoelectric materials. By doping suitable ionic species, the photovoltaic effect is introduced in the material. The figure of merit (FOM) for photostriction magnitude is generally expressed as the product of photovoltage (electric field), E_{ph}, and the piezoelectric constant, d_{33}, while the FOM for response speed is determined by the photocurrent (current density), I_{ph}, as $d_{33}I_{ph}/C$ (C is the capacitance of the photostrictive device). Therefore, for application purposes, enhancement and/or optimization of photostrictive properties requires consideration of both the terms in the FOM; that is, photovoltaic voltage and current, as well as its piezoelectric d constant. Recently, PLZT ceramics have gained considerable attention due to their excellent photovoltaic properties, high d_{33}, and ease of fabrication. We will review the background of photovoltaic effect first in this section.

5.2.1 PRINCIPLE OF THE BULK PHOTOVOLTAIC EFFECT

5.2.1.1 "Bulk" Photovoltaic Effect

When a non-centrosymmetric piezoelectric material (with some dopants) is illuminated with uniform light having a wavelength corresponding to the absorption edge of the material, a steady photovoltage/photocurrent is generated.[3] Some may be suspicious about the distinction between the photovoltaic effect and pyroelectric effect (i.e., voltage/charge generation due to the temperature change). Figure 5.1 demonstrates the difference, where illumination responses of photovoltaic current are plotted under two different external resistances in 1.5 mol% MnO_2-doped $0.895PbTiO_3$-$0.105La(Zn_{2/3}Nb_{1/3})O_3$ ceramic.[4] Mercury lamp illumination on this ceramic sample slightly increased the sample temperature, leading to the initial voltage peak (up to 8 mV through 10 $M\Omega$ resistor) for a couple of tens of seconds. However, note that the output voltage is stabilized

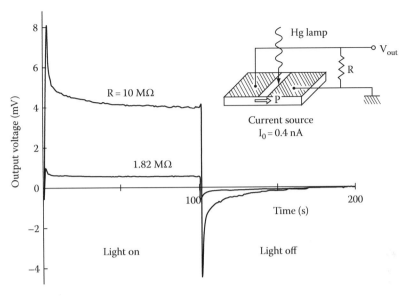

FIGURE 5.1 Illumination responses of photovoltaic current for 1.5 mol% MnO_2-doped 0.895 $PbTiO_3$-0.105 $La(Zn_{2/3}Nb_{1/3})O_3$ ceramic.

around 4 mV after the temperature stability was obtained. The magnitude of the steady current is independent of the externally connected resistance. When the illumination was shut off, the negative pyrocurrent was observed due to a slight temperature decrease again for a 10 s period. But, the output voltage became completely zero after the saturation, which verified that there was no junction (piezoelectric ceramic–metal electrode) effect. The reader can now clearly understand the difference between the photovoltaic and pyroelectric effects from this demonstration. Note that we can eliminate the pyroelectric effect when we use an IR blocking filter for cutting the longer wavelength light intensity (refer to Figure 5.2).

In some materials, the photovoltage generated is greater than the band-gap energy, and can be of the order of several kV/cm. This phenomenon, thus referred to as the "bulk" or "anomalous"

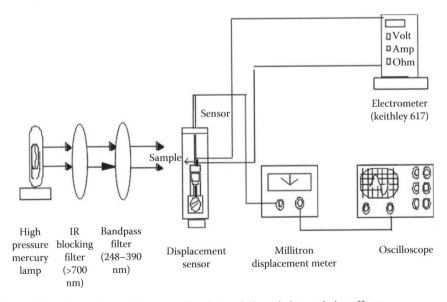

FIGURE 5.2 Experimental setup for measuring photovoltaic and photostrictive effects.

photovoltaic effect (APV), seems to be totally different from the corresponding phenomenon in the p–n junction of semiconductors (e.g., solar battery).[5,6] The APV effect is observed primarily in the direction of the spontaneous polarization (P_S) in the ferroelectric material, and the generated photovoltage is proportional to the sample length along the P_S direction.

The origin of photovoltaic effect is not yet clear, even though several models have been proposed on the mechanism of photovoltaic effect. The key features of the APV effect are summarized as follows:

1. This effect is observed in a uniform crystal or ceramic having non-centrosymmetry, and is entirely different in nature from the p–n junction effect observed in semiconductors.
2. A steady photo voltage/current is generated under uniform illumination.
3. The magnitude of the induced voltage is greater than the band-gap of the crystal.

Even our group previously proposed two models: current source model and voltage source model.

5.2.1.2 Experimental Setup

Prior to the detailed discussion, the measuring setup is described here. Refer to Figure 5.2. PLZT ceramic samples are cut into the standard sizes of 5×5 mm^2 and polished to 1 mm thickness. The samples are poled along the length (5 mm) under a field of 2 kV/mm at 120°C for 10 min. The ceramic preparation methods will be described in Sections 5.2.3 and 5.3.3.

The radiation from a high-pressure mercury lamp (Ushio Electric USH-500D) is passed through infrared-cut optical filters in order to minimize the thermal/pyroelectric effect. The light with the wavelength peak around 366 nm, where the maximum photovoltaic effect of PLZT is obtained, is then applied to the sample. A xenon lamp is alternatively used to measure the wavelength dependence of the photovoltaic effect. The light source is monochromated by a monochromator to 6 nm HWHM.

The photovoltaic voltage under illumination generally reaches several kV/cm, and the current is on the order of nA. The induced current is recorded as a function of the applied voltage over a range −100 to 100 V, by means of a high-input impedance electrometer (Keithley 617). The photovoltaic voltage and current are determined from the intercepts of the horizontal and the vertical axes, respectively. An example measurement is shown in Figure 5.3. The photovoltage (typically kV) is estimated by the linear-extrapolation method. Photostriction is directly measured by a differential transformer or an eddy current displacement senor.

5.2.1.3 Current Source Model

Taking into account the necessity of both doping and crystal asymmetry, we proposed a current source model, as illustrated in Figure 5.4, which is based on the electron energy band model for

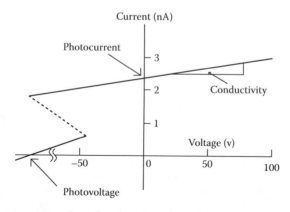

FIGURE 5.3 Photocurrent measured as a function of applied voltage under illumination.

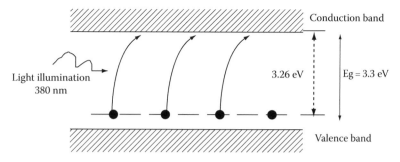

FIGURE 5.4 Energy band-gap model of excited electron transition from deep donor impurity level in PLZT.

(Pb, La)(Zr, Ti)O$_3$ (PLZT).[7,8] The energy band is basically generated by the hybridized orbit of p-orbit of oxygen and d-orbit of Ti/Zr. The donor impurity levels induced in accordance with La doping (or other dopants) are present slightly above the valence band. The transition from these levels with an asymmetric potential due to the crystallographic anisotropy may provide the "preferred" momentum to the electron. Electromotive force is generated when electrons excited by light move in a certain direction of the ferroelectric/piezoelectric crystal, which may arise along the spontaneous polarization direction. The asymmetric crystal exhibiting a photovoltaic response is also piezoelectric in principle, and therefore, a photostriction effect is expected as a coupling of the bulk photovoltaic voltage (E$_{ph}$) with the piezoelectric constant (d).

The photocurrent J$_{ph}$ varies in proportion to the illumination intensity I:

$$J_{ph} = \kappa\alpha I, \tag{5.1}$$

where
α denotes the absorption coefficient
κ is a Glass constant (named according to Glass's contribution to the APV effect)[9]

On the other hand, the photovoltage E$_{ph}$ shows saturation caused by a large photoconductive effect, represented by

$$E_{ph} = \kappa\alpha I / (\sigma_d + \beta I), \tag{5.2}$$

where
σ$_d$ is the dark conductivity
β is a constant relating to the photoconductivity

This model is validated:

1. The photovoltaic current is constant in Figure 5.1, regardless of the externally connected resistance.
2. The photocurrent J$_{ph}$ is strongly dependent on the wavelength under constant intensity of illumination, suggesting a sort of band-gap, as shown in Figure 5.5. A sharp peak is observed at 384 or 372 nm near the absorption edge for 0.895PT-0.105LZN or PLZT (3/52/48), respectively. The donor level seems to be rather deep, close to the valence band level.
3. The linear relationship of the photocurrent with light intensity (Equation 5.1) is experimentally verified in Figure 5.6, where photo-induced short-circuit current J$_{ph}$ (a) and the open-circuit electric field E$_{ph}$ (b) are plotted as a function of illumination intensity I for pure and MnO$_2$-doped 0.895PT-0.105LZN.[4]

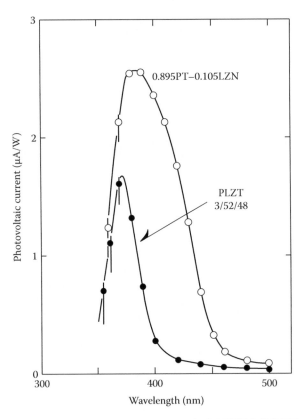

FIGURE 5.5   Wavelength dependence of photovoltaic current in 0.895PT-0.105LZN and PLZT (3/52/48).

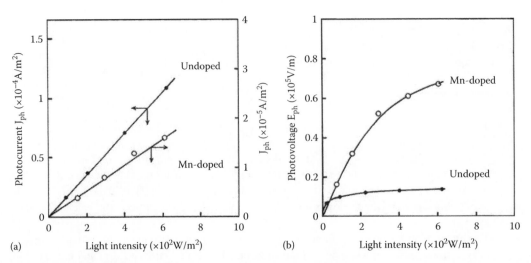

FIGURE 5.6   Short-circuit current J_{ph} (a) and open-circuit electric field E_{ph} (b) as a function of illumination intensity I for pure and MnO_2-doped 0.895PT-0.105LZN.

5.2.1.4 Voltage Source Model

In this model, the photovoltaic properties are attributed to the photocarriers and internal electric fields generated by near-UV illumination. The optical nonlinearity of the second order, which is popularly introduced in ferroelectrics, is proposed as the origin of photo-induced dc field generation.[10] The expression for the polarization of dielectrics, considering the nonlinear effect up to the second order is given by[11]

$$P = \varepsilon_0 \left(\chi_1 E_{op} + \chi_2 E_{op}^2 \right), \tag{5.3}$$

where

ε_0 is the permittivity of vacuum
χ_1 is the linear susceptibility
χ_2 is the nonlinear susceptibility of the second order
E_{op} is the electric field of the illumination beam at an optical frequency (THz)

In dielectrics, the value of the local electric field is different from the value of the external electric field. For simplicity, the local field in dielectrics has been approximated using the Lorentz relation for a ferroelectric material as Ref. [12]:

$$E_{local} = E + \frac{\gamma P}{3\varepsilon_0} \tag{5.4}$$

where

E is the external electric field
γ is the Lorentz factor

When an alternating electric field at an optical frequency is applied (i.e., light illumination), the average of the local electric field E_{local} is not zero, but can be calculated as

$$\overline{E_{local}} = \frac{1}{6} \gamma \chi_2 E_{op}^2 \tag{5.5}$$

It must be noted that Equation 5.5 has been derived for a coherent propagation of the light wave at a single frequency. However, the condition of coherent illumination may not be satisfied in our experimental conditions, where a mercury lamp is used as a light source. The nonlinear effect will be affected by the degree of coherence. Therefore, considering the depression of nonlinear effect due to the incoherency, the expression for the effective dc field induced by incoherent light source may be modified as

$$\overline{E_{local}} = c_1 \gamma \chi_2 \left(E_{op}^2 \right)^\beta \tag{5.6}$$

where

c_1 is a constant
β is a parameter expressing the depression effect

The value of parameter β is expected to lie between 0 and 1. Replacing the variable E_{op}^2 with the intensity (I_{op}) [Ref. 11], the following expression for the average induced (dc) field due to the incoherent light can be obtained:

$$E_{dc} = \overline{E_{local}} = c_2 \gamma \chi_2 \left(I_{op}\right)^\beta \tag{5.7}$$

where

c_2 is a constant

E_{dc} is the effective dc field for photo-induced carriers

Note that the induced field, E_{dc}, is proportional to the nonlinear susceptibility as well as the Lorentz factor, γ.

The photoconductivity can be obtained as a function of light intensity, I_{op},

$$\sigma_{op} = c_3 q\mu\sqrt{\frac{I_{op}}{R}} \tag{5.8}$$

where

q is the charge of the photocarrier

μ is the carrier mobility

R is the recombination rate of the carrier

c_3 is a constant

Since the photocurrent is provided by the product of the photoconductivity and the photo-induced dc field $\left(J_{ph} = \sigma_{op}E_{dc}\right)$, we finally obtain

$$J_{ph} = c_4 q\mu\gamma\chi_2\sqrt{\frac{1}{R}}\left(I_{op}\right)^{\beta+\frac{1}{2}} \tag{5.9}$$

where c_4 is another constant. Equations 5.8 and 5.9 provide a correlation for the photovoltaic response of ferroelectrics on the basis of optical nonlinearity.

The model validation and analysis are made by the light intensity dependence of photovoltaic properties. The experiments were made on PLZT 3/52/48 samples with 1 mm and 140 μm in thickness. Figure 5.7a shows the plot of photoconductivity (σ_{op}) as a function of light intensity (I_{op}). The exponent relating the photoconductivity and the light intensity was calculated to be 0.54. This is in good agreement with the value of 0.5 derived for the recombination process of the carriers (Equation 5.8). Note the difference from Equation 5.2, where we assume the photoconductivity directly in proportion to the intensity (Refer to Figure 5.6). Figure 5.7b shows the experimental results of the open-circuit photovoltage (E_{ph}) as a function of light intensity. The photovoltage was found to be proportional to the square root of the light intensity, leading to $\beta = 0.5$ (Equation 5.7). Figure 5.7c shows the results of short-circuit photocurrent (J_{ph}) as a function of I_{ph}. The parameter β based on Equation 5.9 was calculated to be 0.46, which is very close to the aforementioned β value. The depression in β value can be attributed to the incoherent illumination of the mercury lamp. Note again that J_{ph} is almost directly proportional to I_{ph}, in accordance with Equation 5.1 (Figure 5.6).

Investigation was further made in terms of the illumination coherency. Since a partial coherence of light can be achieved in a very small area, an increase in β value is expected in thinner photovoltaic samples. The photocurrent measured as a function of light intensity in a very thin (140 μm) PLZT sample (Figure 5.7d) resulted in the parameter β (Equation 5.9) to be 0.80, which is higher than the β value of 0.46 in the thicker sample (1 mm thickness). These results suggest that the parameter β increases with a decrease in the thickness of photovoltaic sample, due to higher coherency of illumination in thinner samples. This suggests that an enhancement in the photovoltaic properties may be achieved in a very thin sample or by using coherent illumination. As suggested already, we cannot conclude at present, which model fits better for the experiments, the current source or the voltage source.

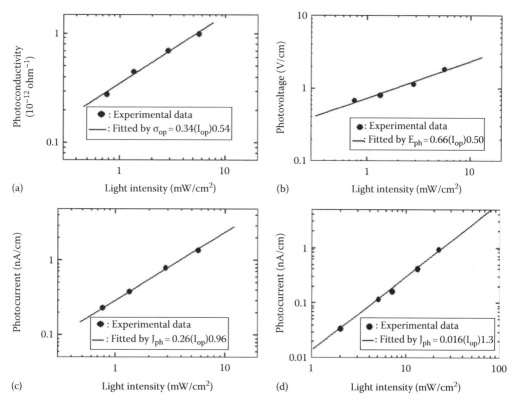

FIGURE 5.7 Dependence of (a) photoconductivity, (b) photovoltage, and (c) photocurrent on illumination intensity in a PLZT 3/52/48 sample with 1 mm in thickness; (d) the result for a sample with 140 μm in thickness.

5.2.2 EFFECT OF LIGHT POLARIZATION DIRECTION

Effect of the light polarization direction on the photovoltaic phenomenon also helps with understanding the mechanism. Figure 5.8 shows the measuring system of the dependence of photovoltaic effect on light polarization direction (a), and the photovoltaic voltage and current as a function of the rotation angle measured for the PLZT (3/52/48) polycrystalline sample (b). The rotation angle θ was taken from the vertical spontaneous polarization direction. Even in a polycrystalline sample, both the photovoltaic voltage and current provide the maximum at θ=0 and 180° and the minimum at θ=90°; this also indicates that the contributing electron orbit may be the p-d hybridized orbit mentioned previously (i.e., the perovskite Zr/Ti-O direction). This experiment is also important when the photostriction is employed to "photophones," where the sample is illuminated with the polarized light traveling through an optical fiber.

5.2.3 PLZT COMPOSITION RESEARCH

Since the FOM of the photostriction is evaluated by the product of the photovoltaic voltage and the piezoelectric constant, i.e., $d \cdot E_{ph}$, $Pb(Zr,Ti)O_3$ (PZT) based ceramics are focused primarily because of their excellent piezoelectric properties, i.e., high d values. Lanthanum-doped PZT (PLZT) is one of such materials with La^{3+} donor doping in the A-site, which is also famous as a transparent ceramic (good sinterability) applicable to electro-optic devices.

PLZT (x/y/z) samples were prepared in accordance with the following composition formula:

$$Pb_{1-x}La_x \left(Zr_y Ti_z\right)_{1-x/4} O_3 \left(y + z = 1\right)$$

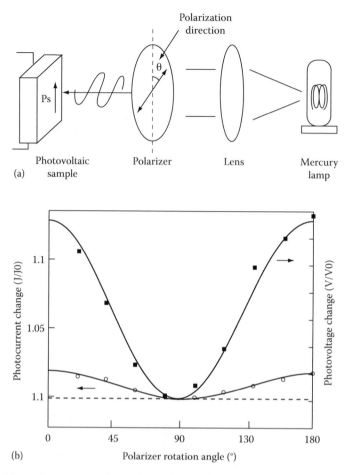

FIGURE 5.8 (a) Measuring system of the dependence of photovoltaic effect on light polarization direction. (b) Photovoltaic voltage and current as a function of the rotation angle.

As discussed in Figure 5.12 in the details, the piezoelectric d coefficient exhibits the maximum around the morphotropic phase boundary (MPB) between the tetragonal and rhombohedral phases, and our composition search was also focused around the MPB compositions. Figure 5.9 shows the photocurrent J_{ph} for various PLZT compositions with tetragonal and rhombohedral phases, plotted as a function of their remanent polarization P_r.

1. Significantly large photocurrent is observed for the tetragonal composition PLZT (3/52/48).[13] This is the major reason why many data in this paper were taken for this composition. The details will be discussed in Section 5.3.2.

2. The relation $J_{ph} \propto P_r$ first proposed by Brody[14,15] appears valid for the PLZT system. Further, it is worth noting that the P_r value capable of producing a certain magnitude of J_{ph} is generally larger in the rhombohedral symmetry group than in the tetragonal group. The average remanent polarization exhibiting the same magnitude of photocurrent differs by 1.7 times between the tetragonal and rhombohedral phases, which is nearly equal to $\sqrt{3}$, the inverse of the direction cosine of the (111) axis in the perovskite structure. This suggests the photo-induced electron excitation is related to the (001) axis-oriented orbit, i.e., the hybridized orbit of p-orbit of oxygen and d-orbit of Ti/Zr.[16]

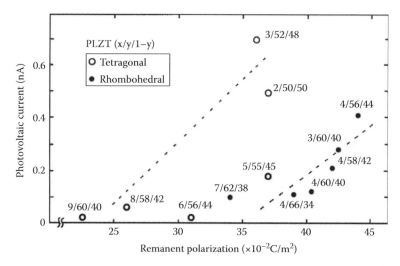

FIGURE 5.9   Interrelation of photovoltaic current with remanent polarization in PLZT family.

5.2.4 Dopant Research

Photovoltaic effect is caused by the dopant in a ferroelectric/piezoelectric crystal, as we discussed in Section 5.2.1. La^{3+} seems to be the primary dopant in $Pb(Zr,Ti)O_3$. Additional impurity doping on PLZT also affects the photovoltaic response significantly. Figure 5.10 shows the photovoltaic response for various dopants with the same concentration of 1 atomic% into the base PLZT (3/52/48) under an illumination intensity of 4 mW/cm² at 366 nm.[8] The dashed line in Figure 5.10 represents the constant power curve corresponding to the non-doped PLZT (3/52/48). Photovoltaic power is

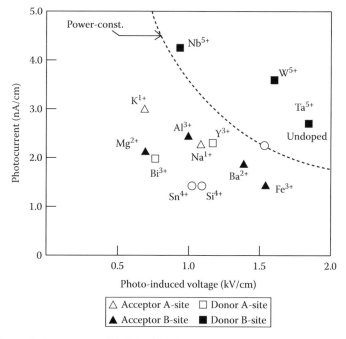

FIGURE 5.10   Photovoltaic response of PLZT (3/52/48) for various impurity dopants (illumination intensity: 4 mW/cm²).

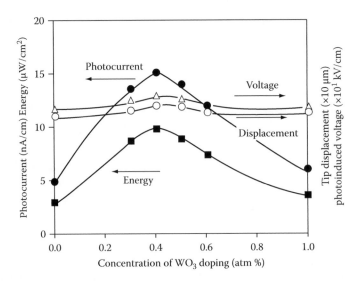

FIGURE 5.11 Photovoltaic current, voltage, power, and tip displacement of a bimorph specimen as a function of dopant concentration in WO$_3$ doped PLZT (3/52/48).

enhanced by donor doping onto the B-site (Nb^{5+}, Ta^{5+}, W^{6+}). On the contrary, impurity ions substituting at the A-site and/or acceptor ions substituting at the B-site, whose ionic valences are small (1–4), degrade the effect on the performance. Figure 5.11 shows the photovoltaic response plotted as a function of at.% of WO$_3$ doping concentration.[6] Note that the maximum power is obtained at 0.4 at.% of the dopant, due to a significant enhancement in the current density.

5.3 PHOTOSTRICTIVE EFFECT

5.3.1 FIGURES OF MERIT

The figures of merit for photostriction are derived here. The photostriction is induced as a function of time, t, as

$$x_{ph} = d_{33}E_{ph}\left(1 - \exp\left(\frac{-t}{RC}\right)\right),$$ (5.10)

where

x$_{ph}$ is the photo-induced strain
d$_{33}$ is the piezoelectric constant of the materials
E$_{ph}$ is the photovoltage
I$_{ph}$ is the photocurrent
t is the time
R and C are resistance and capacitance of the material, respectively

1. For t ≪ 1, we obtain

$$x_{ph} = d_{33}E_{ph}\left(\frac{t}{RC}\right).$$ (5.11)

Thus, the FOM for response speed should be provided by $d_{33}E_{ph}(1/RC)$.

Taking account of the relation $I_{ph} = E_{ph}/R$, this FOM is transformed to $d_{33}I_{ph}/C$. Or, it can be given by $d_{33}I_{ph}/\varepsilon$ (ε: permittivity).

2. On the other hand, for $t \gg 1$, the saturated strain is provided by

$$x_{ph} = d_{33}E_{ph}. \qquad (5.12)$$

Thus, the FOM for the magnitude of strain is defined by $d_{33}E_{ph}$.

In order to obtain high photo-induced strain, materials with high d_{33} and E_{ph} are needed. On the contrary, for high response speed such as photophonic applications, materials with high d_{33}, I_{ph} and low dielectric constant ε are required.

5.3.2 MATERIALS CONSIDERATIONS

We reconsider the optimum compositions in the PLZT system from the photostrictive actuator's viewpoint. Figure 5.12a through c show contour maps of photovoltaic voltage E_{ph}, photocurrent I_{ph}, and piezoelectric constant d_{33} on the PLZT (x/y/1−y) phase diagram, respectively.[17] There is the MPB between the rhombohedral and tetragonal phases around 52%–56% of Zr concentration y. As well known, the piezoelectric coefficient exhibits the maximum along the MPB. The photovoltaic effect is also excited around the MPB. However, precisely speaking, the photovoltage was found to be maximum at PLZT 5/54/46, while the maximum photocurrent was found at PLZT 4/48/52. In Figure 5.12a and b, the solid circles indicate the location of PLZT 3/52/48 which had been reported earlier to exhibit the maximum photovoltage and current. In the finer measurement, the maximum photovoltage and current have been found at different compositions of the PLZT system, both still being in the tetragonal phase. In conclusion, the FOM $d_{33}E_{ph}$ shows maximum for PLZT 5/54/46, while the maximum of the FOM $d_{33}I_{ph}/C$ is for PLZT 4/48/52. Refer to a similar composition study by Nonaka et al.[18]

5.3.3 CERAMIC PREPARATION METHOD EFFECT

5.3.3.1 Processing Method

Fabrication and processing methods have been reported to profoundly influence the photovoltaic properties and strain responses of PLZT ceramics.[16,19,20] This effect comes through the influence of processing methods on the microstructure, and other physical properties such as density, porosity, and chemical composition. Ceramic materials with high density, low porosity, better homogeneity and a good control of stoichiometry are desired for enhanced photovoltaic and photostrictive properties. Coprecipitation and sol–gel techniques are two of the chemical routes which have the inherent advantage in producing high density homogeneous ceramics with a greater control of stoichiometry. Therefore, processes to fabricate photostrictive ceramics via chemical routes with suitable non-oxide precursors are attractive. PLZT ceramics prepared by sol–gel and coprecipitated techniques exhibit better photovoltaic and photostrictive properties as compared to the oxide mixing process.[19,20] Ceramics prepared by solid state reaction have compositional variation and inhomogeneous distribution of impurities whereas the ceramics prepared by chemical synthesis exhibited high purity with good chemical homogeneity at the nanometer scale.

5.3.3.2 Grain Size Effect

Even when the composition is fixed, the photostriction still depends strongly on the sintering condition, or in particular, grain size.[16,21] Figure 5.13 shows the dependence of the photostrictive characteristics on the grain size. As is known well, the piezoelectric coefficient d_{33} gradually decreases with decreasing the grain size down to 1 μm range. On the contrary, photovoltage increases drastically with a decrease in grain size, and the photocurrent seems to exhibit the maximum at around 1 μm. Thus, the photostriction exhibits a drastic increase similar to the photovoltage change. The smaller grain sample is preferable, if it is sintered to a high density.

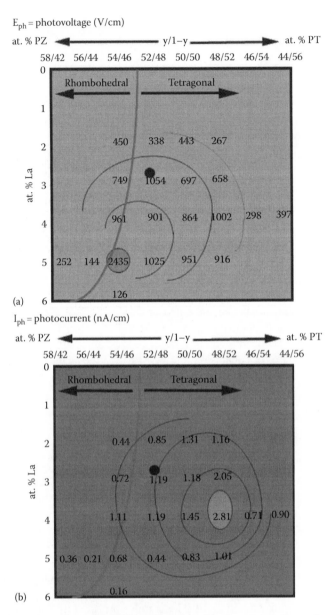

FIGURE 5.12   Contour maps of (a) photovoltaic voltage E_{ph}, (b) photocurrent I_{ph}, and

5.3.3.3   Surface/Geometry Dependence

Since the photostrictive effect is excited by the absorption of illumination in the surface layer of ceramics, it is apparent that the surface geometry of the photostrictive material will have a strong bearing on the generation of photocurrent and photovoltage. Using a sample thickness closer to the penetration depth will ensure that the entire film will be active and efficiently utilized. We also discussed on the light coherency for the "thin" sample in Figure 5.7. Therefore, investigation of photovoltaic response as a function of sample thickness is desired in determining the optimal thickness range with maximum photovoltaic effect. In addition, studying the effect of surface roughness will provide an insight on the absorption dependence of photostriction.

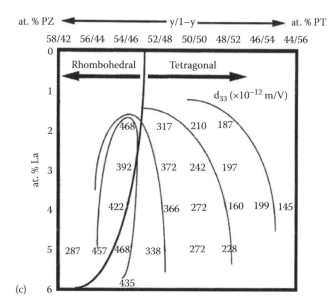

FIGURE 5.12 (continued) Contour maps of (c) piezoelectric constant d_{33} in the PLZT (x/y/1−y) system.

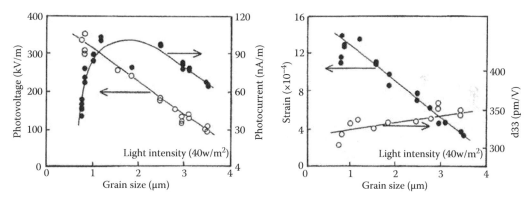

FIGURE 5.13 Grain size dependence of photostrictive characteristics in PLZT (3/52/48).

In order to determine the optimum sample thickness, dependence of photovoltaic effect on sample thickness of PLZT (3/52/48) ceramics doped with 0.5 at.% WO_3 was examined.[22] Photovoltaic response was found to increase with a decrease in sample thickness in PLZT ceramics (refer to Figure 5.14).

A model was proposed in Figure 5.15 to explain and quantify the observed influence of sample thickness on photovoltaic response,[22] where the absorption coefficient is assumed to be independent of light intensity and the photocurrent density is taken to be proportional to light intensity. The sample is assumed to comprise of thin slices along the thickness direction of the sample. Figure 5.14 shows the plot between the normalized photocurrent (i_m) and sample thickness calculated for the external resistance ($R_m = 200$ TΩ). The computed result shows good agreement with the experimental data (□ is for the measured photocurrent, and • for the computed results from the proposed model). With increasing in sample thickness, i_m increases, reaches a maxima, and subsequently decreases with the sample thickness. The decrease in i_m can mainly be attributed to the dark conductivity (σ_d). The optimum thickness (for the present set of samples) which yields maximum

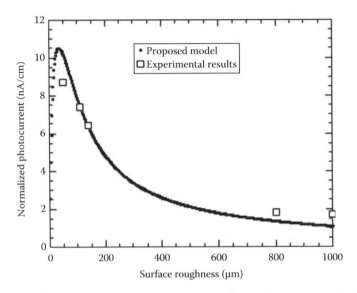

FIGURE 5.14 Comparison of measured and computed normalized photocurrent with photovoltaic coefficient (i_m/k) in 0.5 at.% WO_3 doped PLZT (3/52/48).

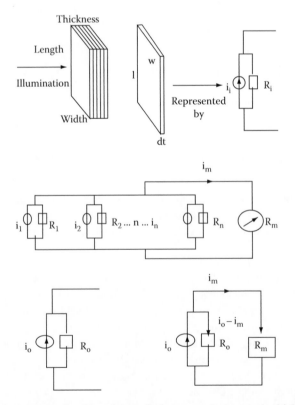

FIGURE 5.15 Model to compute the dependence of photocurrent on sample thickness. The sample was modeled as thin slices along the thickness direction and the corresponding circuit diagrams are also shown.

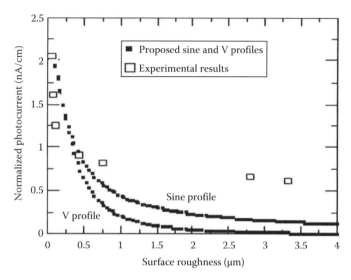

FIGURE 5.16 Variation of photocurrent with surface roughness in 0.5 at.% WO_3 doped PLZT. Comparison with the normalized computed photocurrent for the two surface profiles is also made.

photocurrent is found at 33 μm, which is close to the light (366 nm) penetration depth of the PLZT (absorption coefficient α of PLZT (3/52/48) = 0.0252 μm^{-1} at 366 nm; the inverse of α = 39 μm). The relatively low value of optimum thickness implies that the lower sample thickness will be expected to give better photovoltaic response.

The effect of surface roughness on photovoltaic and photostrictive properties was also examined in the PLZT sample, with different surface roughness obtained by polishing to different surface finishes. The surface roughness was measured by a profilometer (Tencor, Alpha-Step 200) and the average surface roughness was determined using the graphical center line method. The variation of photovoltaic current with surface roughness is plotted in Figure 5.16.[10] The photocurrent increases exponentially with decreasing surface roughness. This is due to the fact that with an increase in surface roughness, the penetration depth of the illumination decreases, while contributions from multiple reflections increase. A model based on the effect of multireflection has been proposed for two different shapes, a sine profile and a "V" profile roughness. In both these shapes, half of the up-down amplitude was taken as a roughness (r) and the cyclic distance period as a roughness pitch (g). The normalized photocurrents (i_m) computed for the aforementioned two surface profiles are plotted also in Figure 5.16 as a function of surface roughness. A distance pitch (wavelength) of roughness at 1 μm gave the best fit of the experimental results, which is close to the size of the grain of this PLZT sample.

In conclusion, the optimum profile of the photostrictive PLZT actuator is a film shape with the thickness around 30 μm and the surface roughness less than 0.2 μm.

5.4 PHOTOSTRICTIVE DEVICE APPLICATIONS

In this section, we introduce the possible applications of photostriction to photo-driven relay, a micro walking machine, a photophone, and the micro propelling robot, which are designed to function as a result of light irradiation, having neither lead wires nor electric circuits. Refer Ref. [23] for the details of applications of photostrictive devices.

5.4.1 Displacement Amplification Mechanism

Since the maximum strain level of the photostriction is only 0.01% (1 order of magnitude smaller than the electrically induced piezostriction, and this corresponds to 1 μm displacement from a

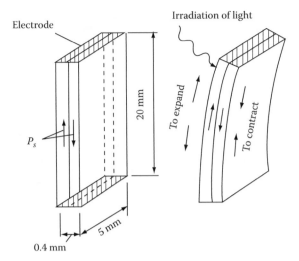

FIGURE 5.17 Structure of the photo-driven bimorph and its driving principle.

10 mm sample), we need to consider a sophisticated amplification mechanism of the displacement. We employed a bimorph structure, which is analogous to a bi-metal consisting of two metallic plates with different thermal expansion coefficients bonded together to generate a bending deformation according to a temperature change.

Two PLZT plates were pasted back to back, but were placed in opposite polarization, then connected on the edges electrically, as shown in Figure 5.17.[8] A purple light (366 nm) was shined to one side, which generated a photovoltaic voltage of 7 kV across the length (along the polarization direction). This caused the PLZT plate on that side to expand by nearly 0.01% of its length, while the plate on the other (unlit) side contracted due to the piezoelectric effect through the photovoltage. Since the two plates were bonded together, the whole device bent away from the light. Figure 5.18 demonstrates the tip deflection of the bimorph device made from WO_3 0.5 at.% doped PLZT under a dual beam control (illumination intensity: 10 mW/cm²). For this 20 mm long and 0.35 mm thick bi-plate, the displacement at the edge was ±150 μm, and the response speed was a couple of seconds.

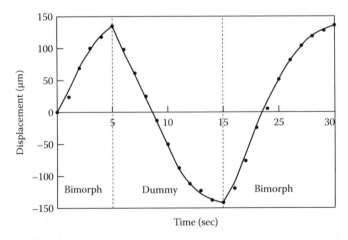

FIGURE 5.18 Tip deflection of the bimorph device made from WO_3 0.5 at.% doped PLZT under a dual beam control (illumination intensity: 10 mW/cm²).

5.4.2 PHOTO-DRIVEN RELAY

A photo-drive relay was constructed using a PLZT photostrictive bimorph as a driver which consists of two ceramic plates bonded together with their polarization directions opposing each other (Figure 5.19).[8] A dummy PLZT plate was positioned adjacent to the bimorph to cancel the photovoltaic voltage generated on the bimorph. Utilizing a dual beam method, switching was controlled by alternately irradiating the bimorph and the dummy. The time delay of the bimorph that ordinarily occurs in the off process due to a low dark conductivity could be avoided, making use of this dual beam method. Displacement of ±150 μm was transferred to a snap action switch, with which on/off switching was possible. The on/off response of the photo-driven relay was demonstrated with a typical delay time of 1–2 s.

5.4.3 MICRO WALKING MACHINE

A photo-driven micro walking machine was also developed using the photostrictive bimorphs.[24] It was simple in structure, having only two PLZT bimorph legs (5 mm×20 mm×0.35 mm) fixed to a plastic board, as shown in Figure 5.20. When the two legs were irradiated with purple light alternately, the device moved like an inchworm. The photostrictive bimorph as a whole was caused to bend by ±150 μm as if it averted the radiation of light. The inchworm built on a trial basis exhibited rather slow walking speed (several tens μm/min), since slip occurred between the contacting surface of its leg and the floor. The walking speed can be increased to approximately 1 mm/min by providing some contrivances such as the use of a foothold having microgrooves fitted to the steps of the legs.

5.4.4 "PHOTOPHONE"

The technology to transmit voice data (i.e., a phone call) at the speed of light through lasers and fiber optics has been advancing rapidly. However, the end of the line—interface speaker—limits the technology, since optical phone signals must be converted from light energy to mechanical sound via electrical energy at present. The photostriction may provide new photo-acoustic devices.

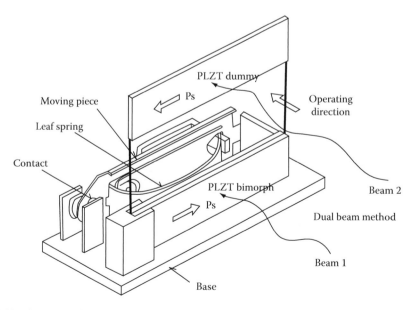

FIGURE 5.19 Structure of the photo-driven relay.

FIGURE 5.20 Photo-driven micro walking machine made of two photostrictive bimorphs. Alternating irradiation provides a walking motion.

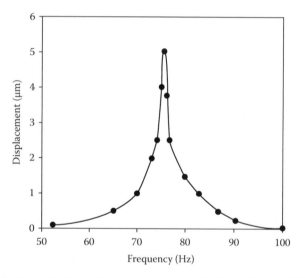

FIGURE 5.21 Tip deflection of the bimorph device made from WO_3 0.5 at.% doped PLZT under a dual beam control (illumination intensity: 10 mW/cm^2).

Photo-mechanical resonance of a PLZT ceramic bimorph has been successfully induced using chopped near-ultraviolet irradiation, having neither electric lead wires nor electric circuits.[25] A thin cover glass was attached on the photostrictive bimorph structure to decrease the resonance frequency so as to easily observe the photo-induced resonance. A dual beam method was used to irradiate the two sides of the bimorph alternately with an optical chopper; intermittently with a 180° phase difference. The mechanical resonance was then monitored by changing the chopper frequency. Figure 5.21 shows the tip displacement of the thin-plate-attached sample as a function of chopper frequency. Photo-induced mechanical resonance was successfully observed. The resonance frequency was about 75 Hz with the mechanical quality factor Q_m of about 30. The maximum tip displacement of this photostrictive sample was about 5 μm at the resonance point. Though the sound level is low, the experiment suggests the promise of photostrictive PLZT bimorphs as photo-acoustic components, or "photophones," for the next optical communication age.

5.4.5 MICRO PROPELLING ROBOT

A new application of highly efficient, photostrictive PLZT films on flexible substrates has been conceived for usage in the new class of small vehicles for future space missions.[26] Micro propelling robot can be designed into arch-shaped photo-actuating composite films (unimorph type) with a triangular top (Figure 5.22). In order to maximize the photostrictive properties of the sample, the

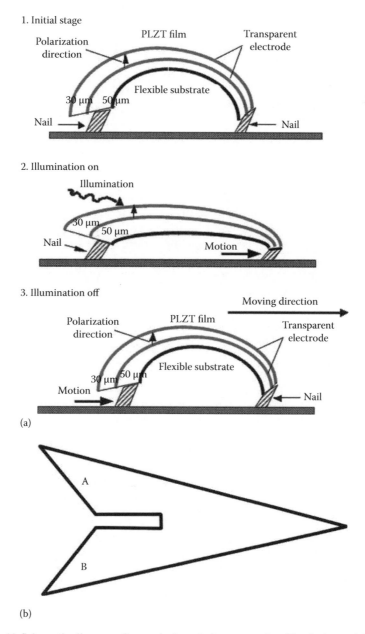

FIGURE 5.22 (a) Schematic diagram of an arch-shaped photo-actuating film device and (b) its triangular top shape.

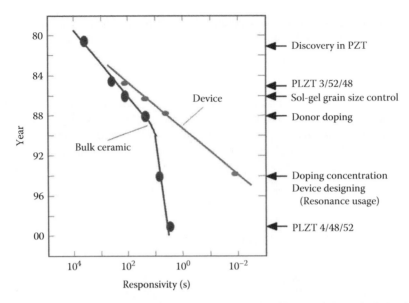

FIGURE 5.23 Response speed improvement of the photostrictive bulk ceramic and of the device in the sequence of year and the key technology development.

sample thickness was determined to 30 μm. This device is driven at their resonance mode under an intermittent illumination. Photo-actuating films may be fabricated from PLZT solutions and coated on one side of a suitable flexible substrate which will then be designed to have a curvature of 1 cm⁻¹. A slight difference in length/width between the right and left legs is designed in order to provide a slight difference between their resonance frequencies. This facilitates in controlling the device in both clockwise and counterclockwise rotations (i.e., right and left steering). A light chopper operating at a frequency close to resonance can be used to illuminate the device, in order to maximize the vibration of the bimorph which will then provide the capability to turn by applying different resonance frequencies of two legs.

5.5   CONCLUDING REMARKS

Photostrictive actuators can be driven only by the irradiation of light, so that they will be suitable for use in actuators, to which lead wires can hardly be connected because of their ultra-small size or of their employed conditions such as ultra-high vacuum or outer space. The photostrictive bimorphs will also be applicable to "photophones." Also note their remote control capability without being interfered by electromagnetic noise. Figure 5.23 summarizes response speed improvement of the photostrictive bulk ceramic and of the device in the sequence of year and the key technology development. Compared to the speed 1 h at the discovery age with PZT, 2 orders-of-magnitude improvement (up to 10 s) has been achieved in materials, and even photo-induced resonance was realized in the devices. The new principle actuators have considerable effects upon the future micro-mechatronics.

REFERENCES

1. Brody, P.S., *Ferroelectrics*, **50**, 27 (1983).
2. Uchino, K. and M. Aizawa, Photostrictive actuators using PLZT ceramics, *Jpn. J. Appl. Phys. Suppl.*, **24**, 139–141 (1985).
3. Fridkin, V.M., in *Photoferroelectrics*, eds., M. Cardona, P. Fulde, and H.-J. Queisser, Solid-State Sciences 9 (Springer-Verlag, New York), pp. 85–113 (1979).

4. Uchino, K., Y. Miyazawa, and S. Nomura, High-voltage photovoltaic effect in $PbTiO_3$-based ceramics, *Jpn. J. Appl. Phys.*, **21**(12), 1671–1674 (1982).
5. Uchino, K., New applications of photostriction, *Innovations Mater. Res.*, **1**(1), 11–22 (1996).
6. Chu, S.Y. and K. Uchino, Impurity doping effect on photostriction in PLZT ceramics, *J. Adv. Perform. Mater.*, **1**, 129–143 (1994).
7. Uchino, K., M. Aizawa, and S. Nomura, Photostrictive effect in (Pb,La)(Zr,Ti)O₃, *Ferroelectrics*, **64**, 199 (1985).
8. Tanimura, M. and K. Uchino, Effect of impurity doping on photo-strictive in ferroelectrics, *Sens. Mater.*, **1**, 47–56 (1988).
9. Glass, A.M., D. von der Linde, and T.J. Negran, *Appl. Phys. Lett.*, **25**, 233 (1974).
10. Poosanaas, P., K. Tonooka, and K. Uchino, Photostrictive actuators, *Mechatronics* **10**, 467–487 (2000).
11. Hecht, E., in *Optics*, eds., A. Zajac, 2nd edn. (Addison-Wesley Publishing, Boston, MA), pp. 44, 81–104, 610–616 (1987).
12. Kittel, C., in *Introduction to Solid States Physics*, 7th edn. (John Wiley & Sons, Inc., New York), p. 388 (1996).
13. Uchino, K., Y. Miyazawa, and S. Nomura, Photovoltaic effect in ferroelectric ceramics and its applications, *Jpn. J. Appl. Phys.*, **22**, 102 (1983).
14. Brody, P.S., *Solid State Commun.*, **12**, 673 (1973).
15. Brody, P.S., *J. Solid State Chem.*, **12**, 193 (1975).
16. Sada, T., M. Inoue, and K. Uchino, Photostriction in PLZT ceramics, *J. Ceram. Soc. Jpn. Int. Ed.*, **95**, 499–504 (1987).
17. Poosanaas, P. and K. Uchino, Photostrictive effect in lanthanum-modified lead zirconate titanate ceramics near the morphotropic phase boundary, *Mater. Chem. Phys.*, **61**, 31–41 (1999).
18. Nonaka, K., M. Akiyama, A. Takase, T. Baba, K. Yamamoto, and H. Ito, Nonstoichiometry effects and their additivity on anomalous photovoltaic efficiency in lead lanthanum zirconate titanate ceramics, *Jpn. J. Appl. Phys.*, **34**, 5380–5383 (1995).
19. Poosanaas, P., A. Dogan, A.V. Prasadarao, S. Komarneni, and K. Uchino, Effect of ceramic processing methods on photostrictive ceramics, *Adv. Perform. Mater.*, **6**, 57–69 (1999).
20. Poosanaas, P., A. Dogan, A.V. Prasadarao, S. Komarneni, and K. Uchino, Photostriction of sol-gel processed PLZT ceramics, *J. Electroceram.*, **1**, 105–111 (1997).
21. Sada, T., M. Inoue, and K. Uchino, Photostrictive effect in PLZT ceramics, *J. Ceram. Soc. Jpn.*, **5**, 545–550 (1987).
22. Poosanaas, P., A. Dogan, S. Thakoor, and K. Uchino, Influence of sample thickness on the performance of photostrictive ceramics, *J. Appl. Phys.*, **84** (3), 1508–1512 (1998).
23. Uchino, K., New applications of photostrictive ferroics, *Mater. Res. Innovations*, **1**, 163–168 (1997).
24. Uchino, K., Micro walking machine using piezoelectric actuators, *J. Rob. Mech.*, **124**, 44–47 (1989).
25. Chu, S.Y. and K. Uchino, *Proceedings of the 9th International Symposium on Applications of Ferroelectrics*, State College, PA, p. 743 (1995).
26. Thakoor, S., J.M. Morookian, and J.A. Cutts, The role of piezoceramics microactuation for advanced mobility, *Conf. Proc. 10th IEEE Int. Symp. Appl. Ferroelectr.*, **1**, 205–211 (1996).

6 Science and Applications of Photomechanical Actuation of Carbon Nanostructures

Balaji Panchapakesan

CONTENTS

6.1 INTRODUCTION

The direct conversion of different types of energy to mechanical energy is of prime importance in a wide variety of actuation applications such as robotics, artificial muscles, valves, optical displays, sensors, optical telecommunication, micro-electro-mechanical systems (MEMS), and micro-opto-mechanical systems (MOMS), spanning various engineering disciplines. Materials that have the capability of changing their physical dimensions in response to external stimuli such as heat, electrical field, magnetic field, and light are used as actuators. The best-known materials used today for actuators are piezoelectrics, electrostrictive materials, conducting polymers, and shape-memory alloys (SMA) (Kaneto et al. 1995; Smela et al. 1995; Kovacs 1998), which are primarily driven by electrical or thermal stimuli. Piezoelectric and electrostrictive materials are limited by the high driving voltages, low work density per cycle, low maximum allowable operational temperatures, and low strain output. Various polymer-based actuation materials exhibit excellent actuation performance with stroke, force, and efficiency similar to that of human muscles. They are also low cost and have a wide variety of choices of materials; however, either they suffer from Faradaic processes involving ionic diffusion, which present limitations on the actuation rate and cycle life (Baughman et al. 1999), or they require a high electrical field in operation. A further limitation for many conducting polymer-based actuators is the need for a liquid environment for their operation, presenting difficulties in dry and vacuum applications. SMA-based actuation has been widely used in many applications and can be configured to give high strain and stress responses; however, a cyclic deformation mechanism is needed for repeatable operation, which presents limitations for reversible actuation. It must be appreciated that all these actuation technologies have already been successfully employed in a wide variety of practical applications demonstrating that different actuation materials match specific application requirements while their drawbacks are not a critical issue in such applications. Most of the actuation technologies mentioned earlier, however, are not suitable for fabricating low-cost, high-performance miniature actuators in micro- or nanoscales for future nanotechnology, bio-nanotechnology, and biomedical applications where batch fabrication capability, scalability into nanoscales, ease of operation, and high performance are all critical issues. Therefore, actuators based on new material systems need to be developed to meet such future application requirements.

From the aspect of energy management, as the physical dimensions of an actuation system scale down, the performance of micro- or nanosystems will critically depend on the ways of power supply and energy transfer to the small systems. Traditional approaches of energy supply to most of the actuation systems rely on wired electricity; however, in many micro- and nanoscale applications, the energy transmission and storage may create problems due to their small size. This is particularly true when applications involve wireless devices and systems that need to be powered and controlled remotely. Examples of such applications vary from space applications to in situ real-time biomedical monitoring. One way of solving the energy management problem is creating the power locally in micro- or nanoscales in the small actuation systems, so the power transmission from the macroscopic world could be minimized or even eliminated. Pioneering work of piezoelectric nanogenerators based on a zinc oxide nanowire array, which could potentially convert mechanical energy into electricity at the nanoscale, demonstrated the possibility of such a concept (Wang and Song 2006). Technical difficulties of constructing highly efficient nanogenerators and integrating those generators with the functional nanoactuation systems, however, need to be resolved before a self-powered small actuation system becomes a reality. Another approach is to transfer energy to the small systems wirelessly where light is an excellent medium for energy transmission. This idea has already been employed in cancer therapeutics (Hirsch et al. 2003; Pitsillides et al. 2003; Panchapakesan et al. 2005; ShiKam et al. 2005) where infrared radiation has been demonstrated for remote energy transmission to heat up the nanostructured materials for cancer cell destruction. Similarly, constructing an optically powered actuation system will resolve the energy management problems in small actuators (Koerner et al. 2004; Ahir and Terentjev 2005; Lu and Panchapakesan 2005). Further, besides the capability of remote energy coupling into the small systems, the "optically driven" approach also offers distinctive advantages such as remote controllability, ease of system design and construction, electrical–mechanical decoupling, low electromagnetic noise, elimination of electrical circuits, better scalability, and capability of working in harsh environments.

Recent development of actuation materials has already taken advantage of the nanomaterials. Nanomaterial-based actuators may have novel actuation properties based on quantum effects, which are not present in the bulk type of materials. At the same time, nanomaterials may provide the opportunity to construct actuators in macro, micro, and nanoscales following both bottom-up and top-down approaches. Several types of nanomaterials being investigated for actuation applications include porous metallic nanoparticles (Baughman 2003; Weissmuller et al. 2003), nanowires (Husain et al. 2003; Lu and Panchapakesan 2006a; Lu et al. 2007b), and carbon nanotubes (Baughman et al. 1999, 2002; Spinks et al. 2002; Barisci et al. 2003). Electrochemical actuators capable of working at low driving voltages can be constructed based on these nanomaterial ensembles, which show better strain responses than piezoelectric and ferroelectric materials. Since these actuators work due to the electrochemical, double layer charging process on the high surface areas of the nanomaterials, they need electrolytes for operation.

Besides the actuators based on pure nanomaterial ensembles, recent developments of nanomaterial-based actuators attempt to look at the actuation behavior of nanocomposites from polymers, incorporating nanomaterials such as carbon nanotubes (Landi et al. 2002; Tahhan et al. 2003; Koerner et al. 2004). Important as they are, these studies have concentrated on accentuating the already existing features of the host matrix by adding nanotubes. The nanotubes, in essence, only serve to exaggerate actuation behavior of the host by either improving electromechanical responses or heating efficiency because of their inherent high thermal conductivity (Naciri et al. 2003; Koerner et al. 2004; Ahir and Terentjev 2005). While this approach proved effective, new development of actuation techniques based on the polymer–nanotube nanocomposite has departed from this traditional improvement scheme and taken the route of introducing new composite actuation behaviors to the polymers, which otherwise would not occur in these systems (Ahir et al. 2006). As a good example, an electrically driven mechanical response was discovered in liquid crystal elastomers embedded with multi-wall carbon nanotubes, while actuation does not show up in pure elastomers (Courty et al. 2003).

Meanwhile, down to the nanoscale, actuator systems have also been constructed based on single-carbon nanotubes and nanowires. Nanotube nanotweezers made by attaching two individual nanotubes onto a sharp tip were reported capable of nanomanipulation and electrical detection, which was actuated by electrostatic forces between the nanotubes (Kim and Lieber 1999). A low-friction, nanoscale, linear bearing from an individual multi-wall carbon nanotube was reported (Cumings and Zettl 2000). Static and dynamic mechanical deflections were electrically induced in cantilevered multi-wall carbon nanotubes in a transmission electron microscope (TEM) (Poncharal et al. 1999), while doubly clamped, suspended, single nanotube and nanowire behaved as high-frequency resonators (Husain et al. 2003; Li and Chou 2003; Li et al. 2004b; Sazonova et al. 2004). Nanotube-based, electrically driven, torsional actuators have also been reported (Williams et al. 2002; Fennimore et al. 2003). These studies showed the potential to construct nanoscale actuation systems based on individual nanostructures; however, they are difficult to construct and not compatible with batch fabrication techniques which presents limitations. This chapter details the development of carbon nanotube-based photomechanical actuators that are compatible with batch fabrication techniques. Such actuators have been developed for manipulation of small objects such as micro-spheres and rotation of small micromirrors. We believe the photomechanical actuation of carbon nanotubes is a rich area for development of new types of electronic, optical, and mechanical devices at the micrometer and nanometer length scales.

6.2 CARBON NANOTUBES

6.2.1 PHYSICAL PROPERTIES

The discovery of carbon nanotubes can be historically dated back to the early 1970s and 1980s when carbon fibers were grown through the thermal decomposition of hydrocarbons at high temperatures in the presence of transition metal catalyst nanoparticles (Oberlin et al. 1976a,b; Iijima 1980). The real burst of research regarding carbon nanotubes, however, did not come until the observation of multi-wall carbon nanotubes by Ijima in 1991 using high-resolution transmission electron microscopy (HRTEM) (Iijima 1991). Ijima's hallmark work heralded the entry of many scientists into the field of carbon nanotubes, who were stimulated at first by the remarkable 1-D quantum effects predicted for their electronic properties and subsequently by the promise that the remarkable structure and properties of carbon nanotubes might give rise to some unique applications (Dresselhaus et al. 2001). In the following 2 years, single-wall carbon nanotubes were also experimentally characterized by Ijima and Bethune (Ajayan and Lijima 1992; Dresselhaus et al. 1992; Bethune et al. 1993; Iijima and Ichihashi 1993). Single-wall carbon nanotubes can be thought of as rolling up a single graphene sheet composed of only one layer of carbon atoms to form a hollow cylinder along the tube axis, while multi-wall carbon nanotubes can be envisioned by rolling up more layers of graphene sheet to form concentric cylinders with each wrapped around the other layers. The graphene layers are coaxially arranged around the central axis of the nanotube, and the spacing between graphene layers remains at a constant of 0.339 nm (Charlier and Michenaud 1993). Single-wall and multi-wall carbon nanotubes are intimately related to each other in structure and in physical properties, as single-wall nanotubes become multi-wall nanotubes by adding more shells around the primary tubes, and each shell within the multi-wall nanotube can be viewed as a single-wall nanotube with different diameters. Single-wall tubes are generally smaller in diameter (0.3–10 nm); however, multi-wall tubes can be as big as tens to hundreds of nanometers in diameter. While the dimensions of carbon nanotubes are in nanometer scale in radial direction, the length of nanotubes along the tube axis can reach into the millimeter or centimeter scale, thus resulting in extremely high aspect ratios of 10^5–10^6. This offers the opportunity of studying the intriguing physical properties at nanometer length scales where quantum mechanical effects dominate in these 1-D systems.

Another important feature in the carbon nanotube atomic structure that is vital in deciding various physical properties is the arrangement of the carbon hexagons around the tube surfaces—in other words, the "helicity" of the honeycomb lattice of graphene layer with respect to the tube axis (Iijima 1991; Zhang et al. 1993). When a layer of graphene is rolled to form a nanotube, the edges of the atomic layer that carry un-bonded carbon atoms must join perfectly with each other in order to form a seamless cylinder. Thus, the pairing of the edges of graphene layers results in specific directions of rolling, and only from those directions can a tubular structure be formed. These specific directions can be characterized by the crystallographic orientations of the surface hexagons, which are noted by helicity indices (Ajayan and Ebbesen 1997).

As shown in Figure 6.1, carbon nanotubes are formed by rolling the graphene sheet in the direction of $\overrightarrow{OA}$, which is expressed by a chiral vector:

$$C_h = na_1 + ma_2 \tag{6.1}$$

"a_1" and "a_2" are two crystallographically equivalent lattice vectors on a 2-D graphene sheet and "m" and "n" are a pair of integer indices. A chiral angle θ is also defined in Figure 6.1, which is the angle between the chiral vector C_h and the "zigzag" direction ($\theta = 0$) parallel to the direction of unit vector of "a_1":

$$\theta = tg^{-1}\left[\frac{\sqrt{3}n}{(2m+n)}\right] \tag{6.2}$$

Three types of nanotube structures can be classified based on the chiral vector and chiral angle. "Zig zag" and "armchair" nanotubes are defined when the corresponding chiral angles are $\theta = 0$ and $\theta = 30°$, respectively (White et al. 1993). These nanotubes are considered achiral nanotubes because they have mirror plane symmetry in the structures. The other type of nanotubes is chiral nanotubes, corresponding to chiral angles in the range $0 < \theta < 30°$. If the helicity indices (n,m) are used to label these nanotubes, armchair nanotubes are those corresponding to n=m. When m=0,

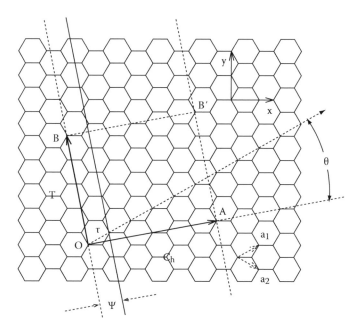

FIGURE 6.1   Illustration of how nanotubes are formed by rolling a sheet of graphite.

a zigzag nanotube is formed. 3-D models of the two types of achiral nanotubes are shown in Figure 6.2. Chiral nanotubes correspond to helicity indices "n" and "m" in other permutations. In practice, most carbon nanotubes do not form achiral structures with high symmetry. Instead, they like to form chiral nanotubes, whose helicities can be individually addressed experimentally using electron microscopes and electron diffraction techniques (Odom et al. 1998; Sfeir et al. 2006). For multi-wall nanotubes, constant inter-layer separation of continuous tubes puts an additional constraint on the tube structures, requiring each tube layer to have its own helicity. Multi-wall nanotubes, however, generally show a smaller number of helicities than the number of wall layers, indicating

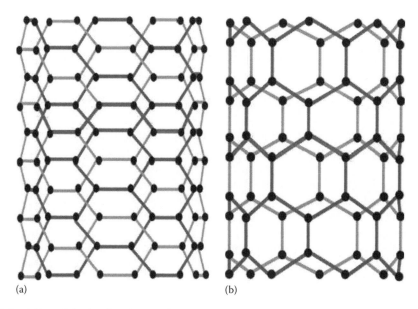

(a)   (b)

FIGURE 6.2   3-D models showing the atomic structure of (a) armchair and (b) zig-zag carbon nanotubes.

that some walls in multi-wall tubes share the same helicity (Ajayan and Ebbesen 1997). Because a wide variety of helicity index permutations exist, many types of nanotubes with different helicities can be formed in real nanotube samples, which further leads to a wide range of nanotube geometries and thus a wide range of properties of nanotubes. More or less, most current applications of carbon nanotubes, especially in micro- and macroscales, utilize the carbon nanotubes with properties averaged over the entire samples. Precise controlling of the helicities of carbon nanotubes within a narrow distribution through the nanotube growth is one of the critical issues to develop nanotubes with uniform properties (Blase et al. 1999; Mauricio 2005; Wiltshire et al. 2005).

Further, the diameters of carbon nanotubes are also related to their helicities and can be expressed as

$$d_t = \frac{\sqrt{3}a_{c-c}(m^2 + mn + n^2)^{1/2}}{\pi} = \frac{C_h}{\pi} \tag{6.3}$$

In Equation 6.3, C_h is the length of the chiral vector and a_{c-c} is the carbon bond length (1.42 Å). More detailed treatment of the translational and rotational vectors can be found in Saito et al. (1998). Table 6.1 summarizes the important structural parameters of carbon nanotubes, which can be used to further deduce the materials' properties of nanotubes (Saito et al. 1998; Dresselhaus et al. 2001). It would become more complicated if one considers a nanotube with changing helicity along the nanotube axis (Bretz et al. 1994), which happens if structural defects are present on the nanotube surface. Generally, single-wall nanotubes remain straight and relatively defect-free for most of the tube length before reaching the ends of the tubes where cones and polyhedral cap structures occur due to the presence of pentagonal defects (Ajayan and Ebbesen 1997). Further, smaller nanotubes (diameters <2 nm) tend to have less structural defects and are structurally perfect than bigger nanotubes (diameter >2 nm). As has been experimentally observed (Endo et al. 1995; Kiang et al. 1995), multi-wall nanotubes generally have more defect populations along the tube axis due to their larger tubular dimensions. The presence of structural defects in both single-wall and multi-wall tubes has profound impacts on their physical properties such as the electrical conductivity, the thermal conductivity, the mechanical strength, optical absorption, and emission properties (Dresselhaus et al. 2001). Through proper thermal treatments, most of the defects can be removed to leave relatively perfect carbon nanotubes (Martínez et al. 2003; Osvath et al. 2005).

6.2.2 ELECTRONIC PROPERTIES OF CARBON NANOTUBES

The nanometer dimensions of the carbon nanotubes that possess strong quantum confinement to the electronic structures of graphene sheet render nanotubes' unique electron transport properties. Theoretical work has shown that the electrical properties of carbon nanotubes depend sensitively on their helicities and geometrical arrangements (Mintmire et al. 1992; Saito et al. 1992; Blase et al. 1994; Chico et al. 1996b; Dresselhaus et al. 2001). When the nanotube diameter and helicity is different, which is the case for different helicity indices (n,m) permutations, the carbon nanotube can be either metallic or semiconducting with different energy bandgaps although their intimate relative-graphene is a zero bandgap semiconductor. This sensitive dependence of electronic properties on the geometrical structures can be understood based on a band-folding scheme that owns to the unique electronic structure of graphene sheet. When a layer of graphene sheet is rolled to form a tubular structure, the electronic structure along the circumferential direction of the tube must become quantized due to periodic boundary conditions. At the same time, the wave vector along the tube axis direction still remains continuous (Jorio et al. 2004). The quantization of wave vectors results in different electronic behaviors for different nanotubes (n,m). General rules can be stated as follows: (n,n) tubes which are so-called armchair nanotubes are metallic; (n,m) tubes where $n - m = 3j$ (j is a nonzero integer) are also metallic tubes; nanotubes with all other (n,m) permutations are semiconductors with appreciable large bandgaps (Dresselhaus et al. 2001). Statistically,

TABLE 6.1

Structure Parameters of Carbon Nanotubes

Symbol	Name	Formula	Value
a	Length of unit vector	$a = \sqrt{3}a_{C-C} = 2.49$ Å,	$a_{C-C} = 1.44$ Å
$\hat{a}_1, \hat{a}_2$	Unit vectors	$\left(\dfrac{\sqrt{3}}{2}, \dfrac{1}{2}\right)a, \ \left(\dfrac{\sqrt{3}}{2}, -\dfrac{1}{2}\right)a$	x, y coordinate
$\hat{b}_1, \hat{b}_2$	Reciprocal lattice vectors	$\left(\dfrac{1}{\sqrt{3}}, 1\right)\dfrac{2\pi}{a}, \ \left(\dfrac{1}{\sqrt{3}}, -1\right)\dfrac{2\pi}{a}$	x, y coordinate
C_h	Chiral vector	$C_h = na_1 + ma_2 \equiv (n,m)$,	$(0 \leq |m| \leq n)$
L	Length of C_h	$L = |C_h| = a\sqrt{n^2 + m^2 + nm}$	
d_t	Diameter	$d_t = L/\pi$	
θ	Chiral angle	$\sin\theta = \dfrac{\sqrt{3}m}{2\sqrt{n^2 + m^2 + nm}}$	$0 \leq |\theta| \leq \dfrac{\pi}{6}$
		$\cos\theta = \dfrac{2n + m}{2\sqrt{n^2 + m^2 + nm}}$,	$\tan\theta = \dfrac{\sqrt{3}m}{2n + m}$
d	gcd (n,m)[a]		
d_R	gcd(2n+m,2m+n)[a]	$d_R = \begin{cases} d & \text{if} \quad (n-m) \text{ is not multiple of } 3d \\ 3d & \text{if} \quad (n-m) \text{ is multiple of } 3d \end{cases}$	
T	Translational vector	$\mathbf{T} = t_1 a_1 + t_2 a_2 \equiv (t_1, t_2)$ $t_1 = \dfrac{2m + n}{d_R}, \quad t_2 = -\dfrac{2n + m}{d_R}$	$\gcd(t_1, t_2) = 1$[b]
T	Length of **T**	$T = |\mathbf{T}| = \dfrac{\sqrt{3}L}{d_R}$	
N	Number of hexagons in the nanotube unit cell	$N = \dfrac{2(n^2 + m^2 + nm)}{d_R}$	
R	Symmetry vector	$\mathbf{R} = p\hat{a}_1 + q\hat{a}_2 \equiv (p, q)$ $t_1 q - t_2 p = 1$, $(0 < mp - nq \leq N)$	$\gcd(p, q) = 1$[b]
τ	Pitch of **R**	$\tau = \dfrac{(mp - nq)T}{N} = \dfrac{MT}{N}$	
ψ	Rotation angle of **R**	$\psi = \dfrac{2\pi}{N}$	In radians
M	Number of **T** in N**R**.	$N\mathbf{R} = C_h + M\mathbf{T}$	

[a] In this table n,m, t_1, t_2, p, q are integers and d, d_R N and M are integer functions of these integers.
[b] gcd(n,m) denotes the greatest common divisor of the two integers n and m.

there are 1/3 metallic nanotubes and 2/3 semiconducting nanotubes in a random as-produced single-wall nanotube samples. In many applications of carbon nanotubes, using only one type of nano-tubes is preferred over using the mixtures. For example, carbon nanotube-based field effect devices, which are vital for nanoelectronics and sensing applications, rely on the capability of conductance modulation within semiconducting tubes (Martel et al. 1998; Tans et al. 1998; Heinze et al. 2002; Javey et al. 2003; Keren et al. 2003; Misewich et al. 2003), while metallic tubes are preferred for use in interconnections in nanoelectronics due to their high electronic conductance (Homma et al. 2002; Naeemi and Meindl 2005). Thus, selective enrichment of certain types of nanotubes in a nanotube product is a critical issue in many nanotube-related applications. It was reported that semiconduct-ing nanotubes can be preferentially produced at the ratio of >85% through plasma-enhanced chemi-cal vapor deposition (PECVD) processes (Li et al. 2004a). Other strategies involve separation of two types of nanotubes (Krupke et al. 2003; Zheng et al. 2003a,b) and selective removal of certain types of nanotubes from the ensembles (Hassanien et al. 2005; Zhang et al. 2006). An important property of semiconducting nanotubes is that the bandgap energy, which is essential in determin-ing the electronic and optical properties of carbon nanotubes, is inversely proportional to the tube diameters (Odom et al. 1998). The energy bandgap can be numerically expressed as

$$E_g \sim \frac{0.9}{d_t} \text{ (eV)} \tag{6.4}$$

where d_t is the diameter of the nanotube as shown in Equation 6.3 and Table 6.1. The bandgap tun-ability through diameters is an important merit of semiconducting nanotubes. The density of elec-tronic carriers as well as the absorption bandgap of nanotubes can be potentially tuned by the tube diameters to match certain electronic and optical applications.

As stated before, one of the most important applications of semiconducting nanotubes is to construct field effect devices. Many experimental studies have been conducted on nanotube field effect transistors (Martel et al. 1998; Tans et al. 1998; Antonov and Johnson 1999; Postma et al. 2001; Heinze et al. 2002; Javey et al. 2003; Keren et al. 2003; Li et al. 2004b). In a typical nanotube transistor, the modulation of gate voltage induces large channel conductivity modulation with on/off ratio approaching 10^6. Such field effect modulation would be sufficient for many applications in nanoelectronics. For example, prototype nanotube-based logic gates circuits have already been demonstrated (Soh et al. 1999; Rueckes et al. 2000; Bachtold et al. 2001; Collins et al. 2001; Derycke et al. 2001; Chen et al. 2006). Generally, as-produced carbon nanotubes behave like p-type semiconductors. P-type doping may be induced by metal electrodes due to the high work function of the metals (Dresselhaus et al. 2001). The processing environment may also result in chemical and gas absorption (especially oxygen) for acceptors doping (Collins et al. 2000; Dresselhaus et al. 2001). Additional doping processes such as alkaline metal doping can be utilized to change the nanotubes into N-type semiconductors (Jing et al. 2000).

The unique electronic transport properties of carbon nanotubes also bring high electrical conducting capabilities of carbon nanotubes, exceeding the best of any metal of the same size, with charge carrier mobility reaching 100,000 cm^2/Vs (Durkop et al. 2004) and current carrying capacity reaching 10^9 A/cm^2 (Wei et al. 2001). Ballistic electron transport was observed in defect-free multi-wall and single-wall carbon nanotubes (Frank et al. 1998; Heer 2004). Superconductivity of carbon nanotubes was also reported (Terrones 2003). Another important aspect of the electrons' transport of carbon nanotubes is the effects of defect-related structures on carbon nanotubes. A good example is that the introduction of pentagon–heptagon pair defects into the hexagonal network of a single nanotube can change the helicity of the tube and fundamentally alter its electronic structure (Dunlap 1994; Ebbesen and Takada 1995; Lambin et al. 1995; Charlier et al. 1996; Chico et al. 1996a; Saito et al. 1996). Based on such an idea, the defective carbon nanotube can be configured into functional nanodevices to behave as nanoscale metal–semiconductor Schottky barriers, semiconductor heterojunctions, or metal–metal junctions with novel properties (Dresselhaus et al. 2001).

Furthermore, extrinsic defects introduced onto nanotubes such as gas molecules and chemical groups attached to the surface through covalent or non-covalent bonding will have significant impacts on the electron transport of carbon nanotubes, thus opening a completely new area of electronic sensing applications based on carbon nanotubes (Collins et al. 2000; Kong et al. 2000, 2001; Chen et al. 2003; Lin et al. 2004; Punit Kohli 2004; Joseph 2005).

6.2.3   Thermal Properties of Carbon Nanotubes

Carbon nanotubes are nanostructures relating to diamond and graphite, which are well known for their high thermal conductivities. The stiff sp^3 bonds in diamond structures result in high phonon speed and, consequently, high thermal conductivities in the material. In carbon nanotubes, carbon atoms are held together by even stronger sp^2 bonds, so nanotubes are expected to have even higher thermal conductivities than their other carbon relatives. The rigidity of these nanotubes combined with the virtual absence of atomic defects or coupling to soft phonon modes of the embedding medium should make isolated nanotubes extraordinarily efficient thermal conductors (Berber et al. 2000). The high thermal conductivity of carbon nanotubes was first proved by theoretical calculations (Berber et al. 2000; Osman and Srivastava 2001). It was shown that the thermal conductivity of an isolated (10, 10) nanotube has an extraordinarily high value of ~6600 W/m-K at room temperature. From bulk carbon nanotube samples, Hone et al. (Hone et al. 1999b) estimated the thermal conductivity for a single tube to be 1800–6000 W/m-K at room temperature. On the other hand, experimental studies have also been performed to measure the thermal conductivity. Kim et al. (2001) measured the thermal conductivity of an individual carbon nanotube by attaching single nanotubes or small bundles to the microfabricated suspended devices. Thermal conductivity of more than 3000 W/m-K for a single nanotube at room temperature was reported from such experiment. Recently, Fujii et al. (2005) measured the thermal conductivity of a single carbon nanotube using a suspended sample of attached T-type nanosensor and reported values exceeding 2000 W/m-K for a single tube of 9.8 nm in diameter. These values were in the range of theoretical calculations and experimentally proved the existence of high thermal conductivity in carbon nanotubes. Due to their high thermal conductivities, carbon nanotubes or nanotube-based nanocomposites may be promising candidates for thermal management in many applications such as integrated circuits, optoelectronic devices, and MEMS structures. More details on the thermal properties of carbon nanotubes can be found in Dresselhaus et al. (2001).

6.2.4   Mechanical Properties of Carbon Nanotubes

As the smallest carbon fibers, carbon nanotubes have been discovered to have extraordinary mechanical properties. The sp^2 covalent bonds holding the carbon nanotube and graphite structures are one of the strongest bonds in nature. The as-produced σ bonds due to sp^2 hybridization lead to the formation of honeycomb carbon lattices. The remaining unhybridized bonds form π bonds that are further delocalized to form delocalized π-electron arrangements from which unique material properties of carbon species originate. Due to the existence of these bonds in nanotubes, σ bonding force and π bonding force exist between carbon atoms in the tubes. An additional van der Waals force would also exist in carbon nanotubes. These three forces are different in magnitude but are all important in deciding the mechanical properties of nanotubes (Ahir and Terentjev 2006). The strength of the carbon fibers would increase with graphitization along the fiber axis, and carbon nanotubes should presumably have the best mechanical properties in the carbon fibers species, showing high Young's moduli and high tensile strength (Krishnan et al. 1998). Theoretical calculations have predicted high Young's moduli for single-wall nanotubes to be 0.5–5.5 TPa, much higher than that of high-strength steel (~200 GPa) (Robertson et al. 1992; Yakobson et al. 1996). The first experimental investigation of Young's modulus for multi-wall tubes was conducted by measuring thermal vibrations of the tubes using TEM, yielding Young's modulus of 1.8 ± 0.9 TPa

(Treacy et al. 1996). By using a similar approach, Krishnan et al. (1998) measured the Young's modulus of single-wall nanotubes, resulting in an average value of 1.25–0.35/+0.45 TPa. Atomic force microscope (AFM) has also been employed to measure the Young's modulus of the carbon nanotubes (Wong et al. 1997). This was realized by bending an anchored carbon nanotube with an AFM tip while simultaneously recording the force experienced by the tube as a function of the displacement from its equilibrium position. The resultant Young's modulus was 1.28 ± 0.5 TPa. The values of Young's moduli measured from different ways were all in the range of theoretical prediction, proving the existence of high elastic modulus in carbon nanotubes.

The tensile strength of carbon nanotubes has been measured (Yu et al. 2000a,b). An individual multi-wall nanotube was mounted between two AFM tips, one on rigid cantilever and the other on soft cantilever. By recording the tensile loading, both the deflection of the soft cantilever from which the force was applied on the nanotube and the length change of the nanotube were simultaneously obtained. The nanotubes broke in the outermost layer ("sword-in-sheath" failure), and the tensile strength of this layer ranged from 11 to 63 GPa. The measured strain at failure was as high as 12%. For comparison, the tensile strength of high-strength steel is only 1–2 GPa (Yu et al. 2000a,b).

Another remarkable property of carbon nanotubes is that they exhibit extreme structural flexibility (Iijima et al. 1996; Qian et al. 2002; Sazonova et al. 2004) and can be repeatedly bent through large angles and strains without structural failure (Falvo et al. 1997). Single-wall nanotubes can be reversibly buckled under a load applied perpendicularly to the tube axis. They are essentially flexible structures because of their hollow cores. When large and mostly reversible bending is applied to the tubes, the σ-bonded skeleton remains unbroken when stress is applied perpendicularly to the long axis of the tube (Endo et al. 1993; Ruoff and Lorents 1995; Yakobson et al. 1996; Ahir and Terentjev 2006). Electron microscopic investigations of multi-wall nanotubes have also shown buckling/rippling distortions along the inner arc of bent nanotubes, suggesting a possible strain relaxation mechanism (Ruoff and Lorents 1995; Poncharal et al. 1999). It was also reported that freestanding films of vertically aligned multi-wall nanotubes exhibit super compressible foam-like behavior. Under compression, the nanotubes collectively form zigzag buckles that can fully unfold to their original length upon load release (Cao et al. 2005). It is the ability of nanotubes to adopt and switch between various buckled morphologies that makes them capable of accommodating and sustaining large local strains while maintaining structural integrity (Yakobson et al. 1996; Lourie et al. 1998). Such flexibilities of nanotube structures would have important implications in applications involving large nanotube dimensional change such as in various nanotube actuator applications (Baughman et al. 1999; Koerner et al. 2004; Ahir and Terentjev 2005; Lu and Panchapakesan 2005).

6.2.5 Optical Properties of Carbon Nanotubes

The optical properties of carbon nanotubes are closely related to their electronic structures, which are dependent on the nanotube's diameter and helicity. As stated before, the structure of a carbon nanotube can be specified by the chiral vector (Equation 6.1). The electronic transport properties are determined by the helicity that $n - m = 3j$ (j is a nonzero integer) defines metallic carbon nanotubes, while others are semiconducting ones, affecting the corresponding optical responses in carbon nanotubes. The 1-D density of states (DOS) is given by the energy dispersion of carbon nanotubes, which can be obtained by the zone folding of 2-D energy dispersion relations of graphite due to the tubular nature of the nanotubes. The periodic boundary conditions applied to carbon nanotubes quantize the wave vector along the circumferential direction of the tube into N discrete values, where N is the number of hexagons of the graphite honeycomb lattice within the nanotube unit cell. At the same time, wave vectors along the tube axis direction remain continuous. Thus, the energy dispersion relations of carbon nanotubes can only adopt certain 1-D parallel line traces in the 2-D energy dispersion relations of graphite whose positions near the K points in the Brillouin zone are determined by the chiral indices of the nanotubes. When the energy dispersion relations as a function of wave vector κ in the nanotube become flat in the Brillouin zone, the DOS becomes large due to an inverse dependence of DOS on the first derivatives of energy dispersion in nanotubes. The 1-D van Hove singularities (vHs) in the DOS, which are proportional to $(E^2 - E_0^2)^{-1/2}$ at both the energy minima and maxima $(\pm E_0)$ of the dispersion relations for carbon nanotubes, are important for determining many physical properties of carbon nanotubes such as optical absorption (Ajiki and Ando 1994; Kazaoui et al. 1999) and resonant Raman spectroscopy (Kasuya et al. 1997; Rao et al. 1997; Pimenta et al. 1998; Dresselhaus et al. 2001). Calculations from the energy dispersion relations revealed the energy differences (bandgap) of metallic and semiconducting nanotubes between the highest valence band singularity and the lowest conduction band singularity in the 1-D electronic DOS, which are expressed as $E_{11}^M(dt)$ and $E_{11}^S(d_t)$, respectively:

$$E_{11}^M\left(d_t\right) = 6a_{c-c} \frac{\gamma_0}{d_t} \quad \text{and} \quad E_{11}^S(d_t) = \frac{2a_{c-c}\gamma_0}{d_t} \tag{6.5}$$

d_t is the diameter of the nanotube and a_{c-c} is the carbon lattice constant. The double subscripts in the expressions denote the corresponding orders of the valence π and conduction π^* energy bands, which are symmetrically located with respect to the Fermi energy. E_{nn}, meaning the energy bandgap between the nth valence band and nth conduction band, could lead to optical transitions between these bands. For carbon nanotube samples mixed with metallic and semiconducting nanotubes with similar diameters, optical transitions determined by the energy gaps from both types of nanotubes may follow the order: $E_{11}^S(d_t)$, $2E_{11}^S(d_t)$, $E_{11}^M(d_t)$, $4E_{11}^S(d_t)$ ... , which starts from the lowest energy levels. In Figure 6.3, typical energy gaps for both metallic and semiconducting nanotubes are plotted as a function of tube diameters that cover all the nanotubes of different helicities at similar diameters. This plot shows a close relationship between the carbon nanotube electronic and optical transitions and the nanotube structures and provides plenty of information for both optical characterization and structural sorting of carbon nanotubes (Ichida et al. 1999; Kataura et al. 1999; Bachilo et al. 2002; Sfeir et al. 2006).

An important, so-called trigonal, warping effect due to the distortion of energy contour from circular patterns away from the K points induces splitting of van Hove singularities in metallic nanotubes (Saito et al. 2000), thus affecting electronic and optical transportation properties of the nanotubes. More information about the fundamental electronic and optical structures of carbon

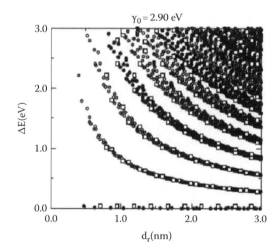

FIGURE 6.3 The energy separations $E_{nn}(d_t)$ for all (n,m) values as a function of the nanotube diameter between $0.7 < d_t < 3.0$ nm. The open and solid circles denote the energy gaps of semiconducting and metallic nanotubes, respectively. Squares denote the $E_{nn}(d_t)$ values for zigzag nanotubes which determine the width of each $E_{nn}(d_t)$ curve.

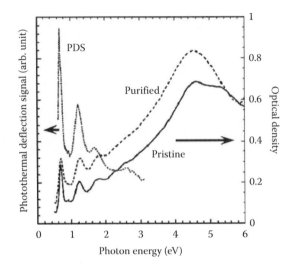

FIGURE 6.4 Optical absorption spectra of single-wall nanotubes synthesized by the electric arc method using NiY catalyst. Broken line and dotted line are the optical density and PDS signals for purified carbon nanotubes, and solid line is the optical density for pristine nanotubes.

nanotubes could be found in some references (Lin and Shung 1994; Saito et al. 1998; Dresselhaus et al. 2001; Ichida et al. 2004; Reich et al. 2004).

Many experimental studies have been conducted to characterize the optical absorption properties of carbon nanotubes (Charlier and Michenaud 1993; Charlier et al. 1995; Kataura et al. 1999; Kazaoui et al. 1999; Alvarez et al. 2000). A typical carbon nanotube optical absorption spectrum is shown in Figure 6.4 (Kataura et al. 1999). Clearly, three absorption peaks can be identified with transition energies located at 0.68, 1.2, and 1.7eV, which correspond to the first and second optical transitions in semiconducting nanotubes and subsequently, the first optical transition in metallic nanotubes, as discussed previously. The three peaks are superposed on the broadband absorption background due to π plasmon. To confirm that these optical absorption peaks are the intrinsic nature of carbon nanotubes, absorption spectra of as-produced and purified carbon nanotubes were compared, showing almost the same absorption structure as each other. Photothermal deflection spectrum (PDS) was also used to characterize the optical transitions. The PDS signal is not affected by light-scattering effects, and carbon black was used as a black body reference. As a result, the PDS spectrum reflects the differences in electronic structures and optical properties between carbon nanotubes and amorphous carbon. The positions of the peaks in the PDS spectrum coincide with that of the absorption spectrum, indicating these peaks are not due to light-scattering losses. As shown in Figure 6.3, when the diameter distributions of carbon nanotubes are different, optical transition peaks should also shift accordingly although the spectrum patterns remain the same. Experimental work has proved this assumption (Kataura et al. 1999). Furthermore, absorption properties of carbon nanotubes can also be modified by doping processes (Kazaoui et al. 1999). It was reported that the intensity of the absorption peaks decreased, especially for lower energy absorption peaks, with the increase of dopant concentration. At a high level of doping, new doping-induced absorption peaks appeared, which may be due to conduction to conduction inter-subband transitions for N-type nanotubes and valence to valence inter-subband transitions for P-type nanotubes (Kazaoui et al. 1999).

Newly discovered optical properties of carbon nanotubes such as strong optothermal effects (Ajayan et al. 2002; Smits et al. 2003; Lien et al. 2006) and optoelastic effects (Zhang and Iijima 1999), although not fully understood yet, raise intensive research interests in discovering the under-lying physical mechanisms and present important implications on many potential optical-related applications such as the optical actuators as will be discussed in this dissertation. The strong optothermal effects were discovered through a simple experimental work of shining a camera flash toward fluffy carbon nanotube samples in the air. It was discovered that carbon nanotubes experienced significant oxidation in the presence of oxygen or experienced structural reconstruction in inert environments (Ajayan et al. 2002); however, for oxidation of nanotubes to occur, the temperature must rise above 600°C, while at least 1500°C is needed for nanotube structure reformation, showing the strong localized thermal effects of carbon nanotubes under light exposure (Ajayan et al. 1993; Tsang et al. 1993; Nikolaev et al. 1997). Efficient light absorption and high thermal confinement of nanotube structures were employed to explain the optothermal effects. This optothermal effect from nanotubes has already shown applications such as carbon nanotube segmentation (Lien et al. 2006) and thermal agents for cancer therapeutics (Panchapakesan et al. 2005; ShiKam et al. 2005).

The elastic response of carbon nanotubes under light was first discovered from the as-grown, loosely bundled single-wall carbon nanotube webs, and networks (Zhang and Iijima 1999). When the nanotube ensembles were under halogen light or laser exposure, they showed distinct movements such as stretching, bending, and repulsion. When the light illumination was removed, the nanotube filaments and networks were restored to their original position and morphology, showing fully elastic responses. Although the motions of nanotube ensembles were not characterized in detail to expose the underlying mechanism fully, they unambiguously showed the unique light-induced mechanical responses in carbon nanotubes. Determination of the underlying physical mechanisms is complicated because they relate to the optical, thermal, electrostatic, and elastic effects of carbon nanotubes in micro- and nanoscales. The photon pressure effect can be excluded because the movement direction is not always identical with that of the light. The thermal expansion can also be ruled out because it is hard to account for the large movement amplitude. Instead, based on the similarity between light-induced motion and electrical field induced-actuation, it was argued that the elastic behavior might be due to the electrostatic interaction of the nanotube bundles as a result of a photovoltaic or light-induced thermoelectric effect physically related to the modification of the electronic structure during the bundle formation.

The existence of the high thermoelectric power effect in single-wall carbon nanotubes has been reported, which was explained by the break of electron–hole symmetry due to the intertube interaction, the charge transfer from metallic tubes to the semiconductor ones, and interbundle or intertube barriers (Hone et al. 1998; Kaiser et al. 1998; Tian et al. 1998; Bradley et al. 2000; Romero et al. 2002). A similar situation should also happen in nanotube filaments and networks under light exposure, where a network of numerous photovoltaic cells or thermoelectric cells exists to support the necessary energy for mechanical motion. The as-produced carbon nanotubes are highly curved in ensembles, where the inherent deformation may lead to an increase of local DOS near the Fermi level and a shift of valence band edge due to increased σ-π hybridization in the deformed region (Rochefort et al. 1998). Charge carrier separation between individual nanotubes or between nanotube bundles could also be induced, which could lead to localized charge accumulation due to the high-resistance barriers at the intertube and interbundle contacts. Thus, the electrostatic effects induced by the local electrical field would cause a mechanical response of the nanotube networks. On the other hand, theoretical work on photomechanical deformations of carbon nanotubes has predicted that polarons generated by light illumination could also induce lattice strains as well as nanotube contortions (Verissimo-Alves et al. 2001; Piegari et al. 2002). Based on these initial results on light-induced elastic response in carbon nanotubes, the carbon nanotube photomechanical actuation and their potential applications will be explored in detail in this dissertation. Besides the aforementioned properties of carbon nanotubes relating to photon–nanotube interactions, many other optical properties of carbon nanotubes have also been revealed in recent years. Specifically, nonlinear optical properties (Xuchun et al. 1999; Lauret et al. 2004), optical limiting behavior (Chen et al. 1999), photoluminescence (O'Connell et al. 2002), electroluminescence (Misewich et al. 2003; Chen et al. 2005), photon-induced molecular desorption (Robert et al. 2001), and radioactive properties (Wadhawan et al. 2003; Wang et al. 2004b) of carbon nanotubes have been studied. Many prototype devices and possible applications such as ultrafast optical switching (Hipplera et al. 2004), nanotube antennas (Wang et al. 2004b), large area transparent electrodes (Wu et al. 2004), photo detectors (Lehman et al. 2005), and solar cells (Landi et al. 2005; Pasquier et al. 2005) have also been demonstrated, which are good indications of the promising potentials of carbon nanotubes in optics, optoelectronics, and optical-related applications.

6.3 PHOTOMECHANICAL ACTUATORS

Actuators perform useful work in exhibiting certain functions to the environment in response to external or internal stimuli. The amount of work they can perform, the functions they fulfill, and the energy expenditures they require to do the desired work depend drastically on the method of actuation and of energy input. Several methods in actuation applications can be classified: electrical, magnetic, thermal/phase, mechanical/acoustic, chemical/biological, and optical (Tabib-Azar 1998). Actuators based on the optical transduction principles may operate under optical stimuli such as optical intensity, wavelength, phase, and polarization changes. Optical actuation can be divided into direct and indirect optical methods. Direct optical actuation uses light to interact with the active parts of the actuator and cause actuation, while indirect optical methods take advantage of the secondary effects of light such as heating, photon-generated electricity, or photoconductivity of photo-responsive materials. Comparing these two types of optical actuation, direct optical processes can be much faster than indirect ones. Furthermore, they can take place at much lower light power level, and they permit greater simplicity, versatility, and parallelism of the controller (Stuchlik et al. 2001). Due to the optical nature of photomechanical actuation technologies, they offer many advantages over traditional actuation technologies. Besides the capability of remote energy transfer and coupling into the small systems, optical actuation also enables remote controllability, easy system design and construction, electrical–mechanical decoupling, free of electromagnetic noise, elimination of electrical circuits, better scaling capability, and ability to work in harsh environments. Such merits of optical actuation could potentially lead to many applications in the actuation and sensing fields (Thakoor et al. 1998; Poosanaas et al. 2000). Potential applications of optical actuators include an alternative mechanism for converting solar energy directly into mechanical motion for planetary exploration, direct corrective control in adaptive optics/ interferometer, optical micro- and nanopositioning and control, solar tracking actuator/shutter for self alignment of the spacecraft to the sun for optimal power generation using solar sails, optically controlled valves for space applications, optically controlled microrobots, nanorobots, MOMS, and photophones. Potential sensing applications of optical actuators include a variety of tunable sensors for incident radiation (UV, visible) based on the detection of incident radiation intensity and indirect micro- and nanochemical and biological sensors based on photo detection when the device is loaded with foreign materials such as chemicals, particles, and cells (Thakoor et al. 1998). In developing these applications, photomechanical actuation can be incorporated with advanced optical components and systems such as fiber optics, waveguides, and integrated optoelectronic and optical systems to produce integrated "smart" systems, where many actuators are multiplexed in time or in frequency and located over large areas for parallel operations (Venkatesh and Novak 1987; Culshaw 1996; Stuchlik et al. 2001).

Despite the advantages of photomechanical actuation and the promising potentials of this novel actuation technique, only few material systems exhibit optical-mechanical energy conversion and actuation. The available material systems and techniques for photomechanical actuation include (1) photostrictive materials that work based on photovoltaic effects and inverse piezoelectric effects (Thakoor et al. 1998; Poosanaas et al. 2000; Takagi et al. 2004); (2) chalcogenide glasses due to mechanical polarization effects (Krecmer et al. 1997; Stuchlik et al. 2001); (3) nematic liquid-crystal elastomers due to polarized photon-induced reorientation of liquid-crystal components (Finkelmann et al. 2001; Hogan et al. 2002; Yu et al. 2003; Tabiryan et al. 2005); (4) material actuation due to optothermal effects (Jones and McKenzie 1993; McKenzie et al. 1995; Sarkisov et al. 2004; Zanardi Ocampo et al. 2004); (5) actuation due to photo-generated charges which induce stress on the surface or in bulk of the semiconductors (Suski et al. 1990; Datskos et al. 1998); (6) microscale actuation due to radiation pressure (Sulfridge et al. 2002, 2004); and (7) optical triggering of other actuation mechanisms (Koerner et al. 2004). Each of these optical actuation systems will be discussed in the following sections.

6.3.1　Photostrictive Actuators

Photostriction is the phenomenon in which strain is induced by incident light, directly causing physical motion of the sample (Poosanaas et al. 2000). The mechanism of the photostriction effect can be understood as a superposition of photovoltaic and inverse piezoelectric effects. It occurs in noncentrosymmetric materials such as ferroelectric single crystals or polarized ferroelectric ceramics. When the material is illuminated with photons corresponding to the absorption edge of the material, a steady photovoltage is generated (Fridkin 1980). In some materials, the generated photovoltage can be greater than the bandgap energy, which is approximately several kV/cm (10^5 V/m), caused by a so-called anomalous photovoltaic (APV) effect. This APV effect is different from the photovoltaic effects in normal p–n junctions in semiconductors in that it is observed only in the direction of the spontaneous polarization in the ferroelectric materials (Chu and Uchino 1994; Uchino 1997; Poosanaas et al. 2000). The generated photovoltage is proportional to the sample length in the P_s direction, and it disappears in the paraelectric phases. Due to the generation of this photovoltage across the sample, mechanical deformation is also induced because of the simultaneous existence of an inverse piezoelectric effect. The figure of merit of photostriction is generally expressed as the product of photovoltage E_{ph} and the piezoelectric constant d_{33}. Both of these parameters need to be optimized in order to improve the photostrictive effect in a material for practical applications. The photostrictive effects have been studied mainly in ferroelectric polycrystalline materials. Lanthanum-modified lead zirconate titanate (PLZT) ceramic is one of the most promising photostrictive materials due to its relatively high piezoelectric coefficient and ease of fabrication. The photon-induced strain in photostrictive materials shares similar amplitude with common piezoelectric materials, which are smaller than 0.2%. Example applications of photostrictive materials such as photo-driven relay (Sada et al. 1987) and a photo-driven micro-walking machine (Uchino 1989) have been demonstrated.

6.3.2　Polarized Photomechanical Actuators

Upon illumination of polarized light, chalcogenide glass was found to exhibit a photo-induced mechanical deformation. This anisotropic mechanical effect induced by polarized light was first reported by Krecmer (Krecmer et al. 1997). Under polarized irradiation, it was shown that a thin amorphous film of $As_{50}Se_{50}$ deposited on a clamped AFM cantilever exhibited reversible nanocontraction parallel to the direction of the electric field of the light and nanodilatation along the axis orthogonal to the electric field of the light. Further, a direct correlation of this optomechanical effect with the reversible photon-induced optical dichroism was observed. Measurements on a cantilever of 200 μm long and 0.6 μm thick, which bear an amorphous $As_{50}Se_{50}$ film of 250 nm thick, revealed reversible bending either up or down with displacement about ±1 μm under polarized light illumination. The optomechanical effect in chalcogenide glasses is in principle linked to the well-known photo-induced anisotropy (PA), from which polarized light can cause preferential absorption and reflection of inducing light in a previously isotropic chalcogenide bulk or film sample (Zhdanov et al. 1979; Krecmer et al. 1997). The structural origins of PA and associated photon-induced mechanical polarization effects are not clear. Krecmer et al. (Krecmer et al. 1997) suggested that the microscopic origin of the effect arose from some anisotropic structural elements that can be aligned by linearly polarized light although these structural elements still remain to be identified. The absorption of polarized light can occur when the electrical field of the polarized light is parallel to the main axis of the triangle consisting of a triplet of As-Se-As atoms. After excitation of a lone-pair electron (LP), an electron–hole pair is created, which no longer has the spatial symmetry of the LP orbital, and because of the change of interatomic potential, this leads to a displacement swing of the chalcogen atom. The cooperative swing of many atoms in one direction is then supposed to cause both the mechanical and optical anisotropy (Krecmer et al. 1997). In Krecmer's work, upon light illumination polarized either parallel or perpendicular to the main axis of the cantilever deposited with chalcogenide thin film, it bent in opposite directions to two "extreme" positions. Further experiments revealed that it is possible to drive the cantilevers to intermediate positions between the two "extremes" simply by changing the angle of polarized incident light relative to the cantilever. This means that a step-like actuation scheme could be obtained between the "extremes" by step-like tuning the polarization angles. Such a controllability of the cantilever positions will enable precise position control in nanoscale by simply controlling the rotation angles of the light sources, which could be a useful approach in optical nanomanipulation applications. Optical actuation of chalcogenide glass requires polarized light. Its strain response is small and slow (Stuchlik et al. 2001).

6.3.3 Liquid-Crystal-Based Photomechanical Actuators

Nematic elastomers have the remarkable property of being able to change their shape by up to 400% in a relatively narrow temperature interval straddling their nematic isotropic (NI) transition temperature (Finkelmann et al. 2001). This strong effect is a reversible elongation or contraction along the director, n, on entering or leaving the nematic states. Simultaneous measurements of the nematic order parameter with length change confirmed that the molecular shape is coupled with the macroscopic sample shape and that the molecular shape depends dramatically on the state of nematic order (Warner et al. 1988). It would then be supposed that if the nematic order could be suppressed or restored in some way, there would be equally dramatic accompanying mechanical responses happening in the materials. It was found that such drastic mechanical responses could be optically introduced into nematic elastomers, resulting in optomechanical actuation of the material. Polymer networks containing azobenzene liquid-crystalline (LC) moieties have the optomechanical effects due to this principle (Finkelmann et al. 2001; Hogan et al. 2002; Ikeda et al. 2003; Yu et al. 2003, 2004; Camacho-Lopez et al. 2004). Optical illumination can mechanically deform the azobenzene polymer films because of LC ordering due to photoisomerization of azo molecules, which are incorporated into the chemical structure of LCs or are present in the LC network as dopants. Two distinct processes happen in LC due to *trans–cis* photoisomerization of azobenzene chromophores: reorientation normal to the light polarization (Yaroshchuk et al. 2001; Kempe et al. 2003) and a decrease in the LC ordering followed by phase transition in the polymer network (Tsutsumi et al. 1998; Lee et al. 2000; Tabiryan et al. 2004). Both these two processes can contribute to the optomechanical properties of azo LC networks with their relative contribution and strength determined by the radiation wavelength. UV radiation corresponds to maximum absorption of azo molecules. Consequently, the orientational effect of *trans–cis* isomerization dominates, leading to nematic-isotropic phase transition. Radiation at longer wavelengths in visible range is at the edges of the absorption band of *trans*-isomers and is substantially absorbed by *cis*-isomers as well, resulting in reverse *cis–trans* isomerization. Many experimental studies have been conducted on evaluating the optomechanical response of the nematic elastomers. Finkelmann et al. (2001) found that the fractional contraction of elastomers achieved a large value of 22% at a temperature of 313 K, which is all the mechanical response that the elastomer would experience on heating from 313 K to the isotropic state. Yu et al. (2003) demonstrated that the large bending of a single film of liquid-crystal network containing an azobenzene chromophore can be induced by UV light. Tabiryan et al. (2005) reported a reversible bi-directional bending of the azo LC polymer by switching the polarization of the light beam between orthogonal directions where not only the magnitude but also the sign of photo-induced deformation can be controlled by the polarization state of the light beam. At the nanoscale, a single molecule level optomechanical cycle has been realized by optically lengthening or contracting individual polymers through switching the azobenzenes between their *trans* and *cis* configurations (Hugel et al. 2002). Photomechanical actuation of nematic elastomers is slow, non-elastic, and requires polarized light at specific wavelength ranges.

6.3.4 PHOTOMECHANICAL ACTUATORS BASED ON OPTOTHERMAL TRANSITIONS

Optothermal effect prevails in most of the material systems under illumination. When light illuminates a medium without 100% reflectivity, part of the photonic energy is absorbed by the medium and converted into thermal energy by multi-phonon or charge relaxation processes. The accumulated thermal energy in turn increases the temperature of the material, which causes mechanical deformation due to thermal expansion. This mechanism can be utilized to construct optical actuation systems. This technique is essentially the direct optical equivalent of electrothermal actuation, which could share similar designs with the later except for the source of heating (Jacobs-Cook 1996). In fact, optothermal effect in materials is almost inevitable in most of the optical devices, including the optical actuators, based on various operation principles. As a result, optomechanical actuation may involve combined actuation mechanisms with optothermal effect as a partial contribution, so the question is not whether an optical actuator has optothermal contribution, but how much this contribution could be. On the other hand, optomechanical systems can be solely constructed on the optothermal actuation principle. Various ways of implementations of optothermal actuation have been proposed, including bimorph structures, expansion of solids, liquids, or gases, and also phase changes between states of matter (Jones and McKenzie 1993; Jacobs-Cook 1996). Because of the thermal nature of the optothermal actuation, it is slow and inefficient, and needs relatively higher optical power to work. One promising approach of improving the actuation performance is to minimize the actuator structures, which will consequently reduce the thermal load and the required optical energy due to reduced mass and thermal capacity, thus increasing the actuation speed due to faster thermal balance within smaller structures; therefore, a combination of optothermal actuation with the current microfabrication techniques will greatly improve the performance of the actuator devices. An opto-pneumatic converter, which is useful in fluid and pressure regulation applications, has been demonstrated by using micromachining techniques (Jones and McKenzie 1993; McKenzie et al. 1995; Jacobs-Cook 1996). Other micro-mechanisms employing optothermal actuation such as optically excited micro-resonators (Jacobs-Cook 1996) and optical micromirrors (Zanardi Ocampo et al. 2004) have also been demonstrated. Recently, an optical actuator using optothermal actuation of polymer film polyvinylidene fluoride (PVDF) was reported (Sarkisov et al. 2004, 2006). This actuator essentially works based on the bimorph principle, which comes from the uneven distribution of the thermal coefficient of linear expansion across the film due to the film storage in a form of a roll. Possible applications of this optical PVDF actuator such as in a photonic switch and optical mechanical clock have been conceptually demonstrated (Sarkisov et al. 2006).

6.3.5 CHARGE-INDUCED PHOTOMECHANICAL ACTUATORS

In a depleted surface layer of polar semiconductors, piezoelectric coupling of electrical and mechanical properties can result in an important photomechanical response produced by light-induced electronic transitions (Gatos and Lagowski 1973; Lagowski et al. 1973; Lagowski and Gatos 1974; Suski et al. 1990). This effect occurs when sub-bandgap optical illumination stimulates the depopulation and population of surface states, while keeping the overall number of bulk-free carriers unchanged. Under such illumination, the barrier height of the depleted layer is modified, which in turn results in a variation of the surface stress. The photomechanical effect is consistent with the surface piezoelectric effect where the external stress applied to polar semi-conductors leads to a modification of the surface barrier height and causes pronounced changes in contact potential difference (Lagowski et al. 1974). In a micro-cantilever structure, if the modulation of barrier height by light illumination matches the natural frequency of the structure, a resonant vibration of the cantilever could be obtained (Lagowski and Gatos 1974). Suski et al. (1990) demonstrated that when a micro Si/SiO$_2$/ZnO cantilever of 10 mm × 1.5 mm × 50 μm in size (ZnO film ~5 μm thick) was activated in an argon laser beam (λ ~ 520 nm), it resonated at ~350 Hz with maximum deflection of 160 nm under 130 μW light power. Photomechanical effect was also reported in bulk semiconductor materials under light illumination with photon energies larger than the material bandgap. When the photons are absorbed in a semiconductor, free electrons are excited from valance band into conduction band and leave holes in the lattice, which creates local mechanical strain in the material. Following this principle, Datskos et al. (1998) studied the photon-induced stress in a microsilicon cantilever of 100 μm × 20 μm × 0.5 μm in size. Upon absorption 9 nW infrared light at 780 nm, the cantilever deflected ~1 nm. Furthermore, it was determined that photomechanical actuation due to photon-induced stress is of opposite direction and about four times larger than that from optothermal actuation. Photomechanical actuation due to charge modulation can be used in actuating the microstructures, however, only in nanoscale actuation ranges.

6.3.6 PHOTOMECHANICAL ACTUATORS BASED ON RADIATION PRESSURE

In the past, photomechanical actuators based on radiation pressure have been developed. This approach is based on the momentum transfer between light and actuation structures. When light interacts with a media, whether it is absorbed or reflected, it also exerts radiation pressure on the media. In macroscopic samples, radiation pressure is minimal and normally neglected; however, under some extreme conditions, a significant effect could result due to radiation pressure. For example, radiation pressure helps to counteract the crushing gravitational force at the cores of stars (Schwarzschild 1958). It also affects satellites and space probes with asymmetric cross-sections in that it creates a torque around the center of mass of these objects, causing them to rotate (Parvez 1994). Further, in micro- and nanoscales where objects bear small mass and small dimensions, the gentle radiative force may have significant effect on these objects, especially when the optical intensity is high. MEMS structures could be actuated based on such possibility. The momentum of a photon is given by

$$\vec{p} = \frac{\hbar\omega}{c} = \frac{h\nu}{c}, \quad \hbar = \frac{h}{2\pi} \tag{6.6}$$

where
  h is Planck's constant
  ν, ω, and c are the frequency, angular frequency, and the speed of light, respectively

When large numbers of photons are absorbed by a media, from Newton's second law, the force imparted is given by

$$F = \frac{dp}{dt} = \frac{(dE/dt)}{c} = \frac{W}{c} = \frac{IA}{c} \tag{6.7}$$

where
  W is the power
  E is the energy of light
  I and A are the average intensity of light and illumination area, respectively

In the case of reflection, the photons are reflected off the body, and their momentums are reversed, doubling the radiative force (Sulfridge et al. 2002). The magnitude of the radiation force is directly proportional to the optical power from the calculations presented earlier. Thus, in microscale applications, the radiation force could be kept constant with the scaling of the microsystems, providing a way to give better optical collimation to focus the optical energy into a smaller area; however, due to the diffraction limit of light and limitations of optical setups in collimation, a physical limitation is given for the size of optical focus below which further focus is barely possible. As a result, in smaller structures such as devices in nanoscale, the optical power and the radiation forces scale as l^2 with the scaling of the system dimensions. Koehler first proposed that electromagnetic momentum could be used as actuation mechanisms in microstructures in a theoretical aspect (Koehler 1997). The experimental demonstration of such actuation principle has lately been reported in microcantilever structures (Dragoman and Dragoman 1999; Yang et al. 2000). Further, it was reported that radiation pressure could be used to switch the states of a bistable MEMS beam, causing it to toggle as much as 23 μm (Sulfridge et al. 2002, 2004). Optical actuation by radiation force may suffer from severe optothermal effects due to the high optical intensity involved, thereby affecting the stability of the actuator over long periods of time.

6.3.7 PHOTON-INDUCED ACTUATION OF SHAPE-MEMORY POLYMERS

A newly developed method of photomechanical actuation takes advantage of optical triggering of actuation in a shape-memory polymer, which is discovered in multi-wall nanotube-thermoplastic elastomer (Morthane) nanocomposites (Koerner et al. 2004). Morthane exhibits a low glass-transition temperature ($T_g = -45°C$) and a near-ambient melting temperature of soft-segment crystallites ($T_{m,s} = 48°C$). Severe deformation in the rubbery state at room temperature ($T_g < RT < T_{m,s}$) induces crystallization of the flexible segments and creates physical cross-links in addition to the ubiquitous hard-segment crystallites, which prevent the polymer from strain recovery on removal of the applied stress. If providing some ways of melting the strain-induced soft-segment crystallites, the polymer could be released from the deformed state and return to the stress-free conformation. It was shown that uniform dispersion of 1–5 vol.% of carbon nanotubes in morthane resulted in a nanocomposite, which can be heated by infrared radiation due to non-radiative decay of infrared photons absorbed by the nanotubes in the composite. The internal temperature rise of the nanocomposite under illumination is enough to melt the strain-induced polymer crystallites that act as physical cross-links to hold the deformed shape and, consequently, trigger the release of stored mechanical energy to recover the original shape. Through remote actuation by infrared light, the nanocomposite can store and subsequently release up to 50% more recovery stress than the pristine polymer. The actuation is essentially based on the optothermal effect in a material (carbon nanotube), which is inherently a slow process. Further, the optical stimulus is used solely for triggering stored mechanical energy, so it is better to be viewed as a control signal rather than as an energy conversion mechanism. Thus, for this type of actuator to work continuously, another source of energy such as cyclic mechanical stretching must be supplied to this uni-directional actuator.

Several optical actuation techniques were discussed based on different systems. The corresponding underlying mechanisms were also explained accordingly; however, it may be possible for a certain actuation system, multiple actuation mechanisms, to coexist due to the complicated nature of material responses under light illumination. While it is possible to determine one mechanism with the largest contribution, the optical responsive behavior could be the competing results between several optically induced effects. Each optical actuation method differs from the others in terms of actuation performances such as efficiency, strokes, speed, optical sources, reversibility, and stability. Most of the available optical actuation schemes, however, cannot satisfy the requirements for constructing high-performance and low-cost optical actuators. The desires for high-performance optical actuators, especially in micro- and nanoscales, warrant further research on performance improvements on individual techniques and warrant the search of new optical responsive materials for new actuation schemes.

6.3.8 PHOTOMECHANICAL ACTUATION OF CARBON NANOTUBES

Light-induced elastic responses from single-wall nanotube bundles and fibrous networks were first reported by Zhang and Iijima (1999). It was explained as localized electrostatic effects due to uneven distributions of photo-generated charges. The discovered irregular elastic response was random rather than a configurable experimental set up. It was unknown whether such light-induced movements could be converted into actuation technologies of practical import. At the same time, theoretical work on photomechanical deformation of carbon nanotubes have predicted that lattice strain and conformational distortions of nanotubes can be optically induced due to polaron (electron–hole pair) generations under illumination (Verissimo-Alves et al. 2001; Piegari et al. 2002). As important as they are in both physical and engineering aspects, little experimental work has been conducted on the photomechanical responses of carbon nanotubes following the initial discoveries. Due to the high anisotropic nature of the nanotubes, the orientation and alignment of carbon nanotubes is also an important factor in determining their photomechanical properties. Both unaligned (films) and partially aligned nanotube samples (fibers) were studied to reveal the effects of nanotube alignment on optical actuation. Pure nanotube samples studied include (1) unaligned SWCNT film, (2) unaligned MWCNT film, (3) partially aligned SWCNT fiber, and (4) partially aligned SWCNT fiber. Such characterization revealed unique photomechanical responses from pure carbon nanotubes depending on the alignment. In addition, studies on the relaxation of pure carbon nanotubes revealed the non-thermal behaviors of multi-wall nanotubes and polymer-like behaviors of single-wall nanotubes, which further expand the science of photomechanical actuation of carbon nanotubes.

6.3.8.1 Photomechanical Actuation of Pristine Carbon Nanotubes

To study the photomechanical actuation of pure carbon nanotubes, both SWCNT and MWCNT were purchased from Nanolab Inc. SWCNT diameter is 1–1.5 nm, length > 10 μm and purity > 90%. MWCNT diameter is 15–45 nm, length is 5–20 μm, and purity > 95%. Unaligned carbon nanotube ensembles were prepared in the form of nanotube films. The fabrication of both SWNT and MWNT films follows the same procedure. Nanotubes were uniformly dispersed in isopropyl alcohol by ultrasonic agitation. Standard vacuum filtrating of the nanotube suspensions was used to fabricate the unaligned films on the filters, which were rinsed with isopropyl alcohol and DI water and then dried at 80°C for 2 h to remove the remaining organic residues in the film. After drying, the nanotubes were peeled from the filters as freestanding nanotube films. The film thickness (~20 μm) was readily controlled by the concentration and the amount of nanotube suspension in filtration. The films produced in this way have randomly oriented nanotubes and small tube bundles throughout the films. To produce partially aligned fibers for both SWCNT and MWCNT, CNT suspensions were diluted to ~5 μg/mL to better separate the nanotube bundles by long ultrasonic agitation. A vacuum filtration setup was modified to accommodate an airbrushing head for injecting CNT suspensions to the funnel. A mixed cellulose ester (MCE) filter was used for the ease of filter removal in post-processing. When the CNT suspensions were introduced to the filter surface by airbrushing in a fixed direction, nanotubes tended to align in this flow direction, while the solution was pumped off from the filter by the vacuum, leaving a mass amount of partially aligned CNTs on the filter surface. After ~300 nm of CNT films were deposited on the filter, the nanotube bearing MCE filter was removed from the system, dried, and cut into long strips in the flow direction, which was followed by filter dissolving in multi-baths of acetone and isopropyl alcohol. Then the resulting CNT strips floating in the solution were dragged out of the solution while rotating in a fixed direction. The strong surface tension of the solution served to collapse the wide CNT strips to form partially aligned CNT fibers. Both partially aligned SWCNT and MWCNT fibers were produced by this procedure with a fiber length up to 10 cm and a diameter of tens of microns.

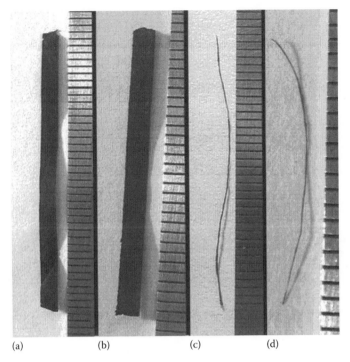

(a)   (b)   (c)   (d)

FIGURE 6.5   Optical images of pure carbon nanotube samples. (a) Unaligned SWCNT film, (b) unaligned MWCNT film, (c) partially aligned SWCNT fiber, and (d) partially aligned MWCNT fiber. The spacing between grid lines is 1 mm.

The optical images of each sample fabricated according to the aforementioned process are shown in Figure 6.5. Nanotube films produced directly from standard vacuum filtration compose randomly oriented nanotubes and nanotube bundles without any specific orientation or ordering, which are ideal for studying photomechanical actuation in unaligned systems. As shown in Figure 6.5 for SWCNT film and (b) for MWCNT film, these unaligned nanotube films have dimensions of ~3 mm in width, ~30 mm in length, and ~20 μm in thickness. SEM images in the lower inserts of Figures 6.6 and 6.7 showed the typical random orientation of nanotubes and small tube bundles for SWCNT and MWCNT films, respectively. Such entanglements of carbon nanotubes are typical in nanotube samples, and they are important in determining both the elastic responses and photomechanical actuation of carbon nanotube ensembles. The nanotube films were used in the characterization without further optimization, which deliberately retain the original physical properties of carbon nanotubes. To test the effects of nanotube alignment on the overall photomechanical response, we also need CNT ensembles with orientational orders in specific directions.

Appropriate sample candidates are the CNT fibers produced to have partial alignment along the fiber axis. Because of the high aspect ratio and highly anisotropic nature of the CNTs, ensembles composed of highly aligned CNTs have remarkable anisotropic macroscopic electromechanical and optical properties compared to their randomly aligned counterparts. Figures 6.6 and 6.7 present the SEM images of SWCNT and MWCNT fibers, respectively. The upper inserts in these two figures show the typical microscopic views of the SWCNT and MWCNT fibers, respectively, indicating the partial alignment of nanotubes and small nanotube bundles along the fiber axis. We estimated that more than 60% of the nanotubes were aligned in ±20° along the fiber axis from extensive SEM investigations. To characterize the photomechanical stress, an experimental setup composed of a precise weight scale (dynamometer) with resolution of ~100 μg and a micrometer was used to test the samples (Ahir et al. 2006). The setup is schematically shown in Figure 6.8, which enables

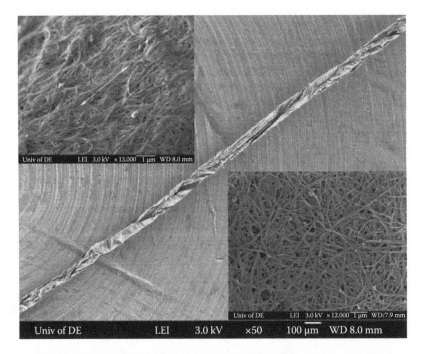

FIGURE 6.6 SEM image of a SWCNT fiber with the partial alignment shown in the upper insert. For comparison, the lower insert shows the random alignment (unaligned) of nanotubes in a typical unaligned SWCNT film. The arrow in the image indicates the direction of the alignment.

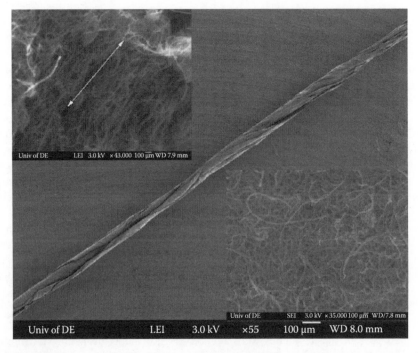

FIGURE 6.7 SEM image of a MWCNT fiber with the partial alignment shown in the upper insert. For comparison, the lower insert shows the random alignment (unaligned) of nanotubes in a typical unaligned MWCNT film. The arrow in the image indicates the direction of the alignment.

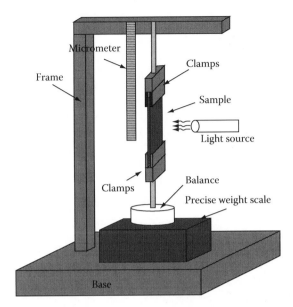

FIGURE 6.8 Schematic of experimental setup for characterizing the photomechanical actuation. It is composed of a precise weight scale for measuring the stress and a micrometer for measuring the prestrain of the sample. The sample is connected between the upper frames and the fixed-mass balance on the weight scale with two clamps.

both the precise stress (calculation is needed to average the force reading over the cross-section area of the sample) and prestrain (sample length change) measurements. The sample is connected between the upper frames and a fixed-mass balance on the precise weight scale with two clamps as shown in the schematic.

Prestraining of the samples is applied by tuning the position (height) of the upper frame. The amount of length change of the sample was recorded by the micrometer, and the stress encountered by the sample was measured by the weight scale. All the data were monitored in real time, transferred to a computer, and processed using Labview software. In this experimental setup, the physical dimensions of samples were fixed during each individual measurement (enabling iso-strain measurements). The advantages of iso-strain configuration are that the stresses experienced by the sample can be directly read out from the dynamometer, while the complex issues of long-time stress relaxation after deforming the sample can also be avoided because of the constant sample length during testing (Ahir et al. 2006). During the testing of all the four pure CNT samples, very little prestrain of ~0.5% was intentionally applied to the samples to keep the samples straight between the clamps.

Typical photomechanical responses of the pure carbon nanotube ensembles are presented in Figures 6.9 through 6.12. The experimental data were corrected to account for the natural relaxation of the nanotube networks during the time of testing. Testing both single and multi-wall nanotube films (for unaligned systems) revealed large photomechanical actuation in response to light, as shown in Figures 6.9 and 6.10, respectively. In Figure 6.9, with the light cycled between "on" and "off" states as indicated by the dashed line, SWCNT film showed fast reversible increase of photomechanical stress upon light illumination, which corresponds to contraction of the film under light exposure. After the light source was switched off, the SWCNT film expanded back to the original stress levels.

For MWCNT films, the photomechanical response under illumination cycles is shown in Figure 6.10, which followed a more complicated response pattern. Upon light illumination, MWCNT film showed an initial fast contraction (stress increase) in the form of a small sharp stress peak, followed

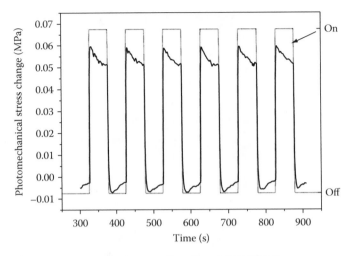

FIGURE 6.9 Typical photomechanical responses of unaligned SWCNT films.

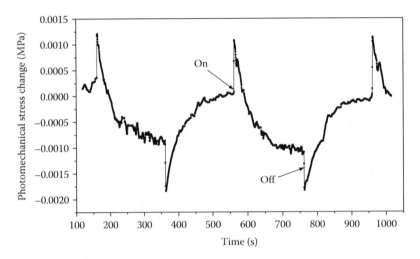

FIGURE 6.10 Typical photomechanical responses of unaligned MWCNT films.

immediately by a slow and large expansion (stress decrease), which acquired saturation gradually. When light was switched off, MWCNT film showed an initial fast expansion (stress decrease) in the form of a small, sharp stress valley (with nearly equal amplitude as the aforementioned sharp peak), followed immediately by a slow large contraction (stress decrease), which took ~3 min to relax back to the original stress value. It seemed as if the fast stress response happened in the first stage and soon was overwhelmed by another larger slow response in the second stage. This complicated stress pattern was highly repeatable and was no artifact of the experimental setup. It is hypothesized that the fast contraction is the result of better-aligned nanotubes within the bulk unaligned nanotube networks, which is taken over by the slow photomechanical response from the rest of the unaligned nanotube networks. In other words, this complicated stress pattern is the competing result of different parts of the nanotube ensembles in terms of nanotube alignment due to the difference in response speeds. Later discussions about the actuation speed in Section 6.6 support this hypothesis. From the aforementioned observations, it is clear that the photomechanical response of pure MWCNT film was quite different from that of SWCNT films in both stress patterns and actuation speed. To show the effects of nanotube alignment on their macroscopic photomechanical responses, both partially aligned SWCNT and MWCNT fibers were tested and presented in Figures 6.11 and 6.12, respectively.

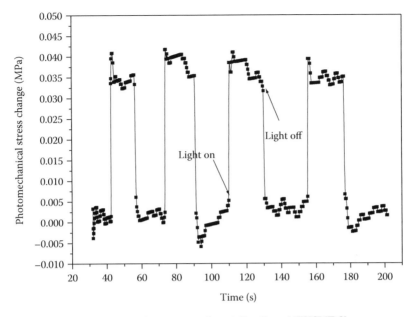

FIGURE 6.11 Typical photomechanical responses of partially aligned SWCNT fibers.

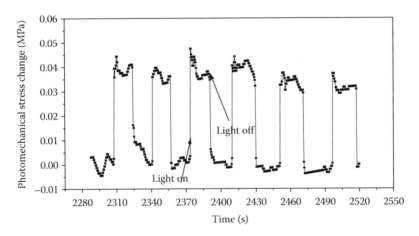

FIGURE 6.12 Typical photomechanical responses of partially aligned MWCNT fibers.

Showing similar response patterns, both MWCNT (aligned) and SWCNT (aligned) exhibited fast contractions during light "on" periods and experienced similarly fast expansions to relax to their original stress levels during "off" periods. All these results indicate that photomechanical actuation is an intrinsic property of carbon nanotubes.

6.3.8.2 Photomechanical Actuation of Nanotube–Polymer Composites

Actuators constructed using carbon nanotubes dispersed in a polymer matrix have been reported for several polymer systems. The carbon nanotubes in the polymer host acting as nanofillers serve to improve the actuation properties of the polymer host (Landi et al. 2002; Tahhan et al. 2003; Koerner et al. 2004). Doping nafion polymer with single-wall nanotubes resulted in composite actuators exhibiting superior performance when compared to metal-doped nafion films (Landi et al. 2002) due to the electrical conductivity of nanotubes. An addition of an electroactive polymer such as polyaniline (PAn) onto a CNT mat substantially enhanced the electrochemical actuation strain of the original CNT mat (Tahhan et al. 2003). As introduced in Section 6.2, mixing multi-wall nanotubes with a shape memory polymer (Morthane) produced a uni-directional optical actuator due to optical triggering of the shape memory effect in the nanocomposites (Koerner et al. 2004). Important as they are, these studies have concentrated on accentuating the already present features of the host matrix by adding nanotubes. The nanotubes, in essence, only serve to exaggerate actuator behavior of the host by either improving electromechanical responses or heating efficiency because of their inherent high thermal conductivity (Naciri et al. 2003; Koerner et al. 2004; Ahir and Terentjev 2005). While such approach was proved to be effective, new development of actuation techniques based on the polymer–nanotube nanocomposite has departed from this traditional "improvement" scheme and taken the route of introducing new composite actuation behaviors to the benign polymers, which otherwise would not occur in these systems (Ahir et al. 2006). As a good example, electrically driven mechanical response was discovered in liquid-crystal elastomers embedded with multi-wall carbon nanotubes, while pure elastomers do not show such effect (Courty et al. 2003).

The construction of nanotube polymer actuator was done using several steps. First, SWCNT film was first produced by a vacuum filtration technique. The SWCNT film is composed of highly entangled SWCNT bundles and network. Acrylic elastomer was purchased from 3 M, sold as 137DM-2. The material is available as a precast adhesive tape. A thin acrylic elastomer film was derived from the adhesive tape. Direct physical bonding of acrylic film with nanotube films of similar size finish the fabrication of the optical actuator, which could be tailored in various shapes depending on the application requirements. The resulting structure was used to study the light-induced elastic responses. As a first demonstration of actuation, a cantilever structure was fabricated by attaching the actuator to a 100 μm thick PVC film. Figure 6.13 shows the schematic arrangement of the entire setup with the cantilever structure anchored on a base that bends in a direction normal to the cantilever surface. The structure of the actuator is shown in the inset. A halogen lamp of tunable intensity was used as the light source and was incident normal to the surface of the cantilever. The light intensity was recorded using a Newport 1815-C intensity meter. A digital camera was used to characterize the displacement of the structure. The displacement of the cantilever under optical illumination is shown in Figure 6.14. An intensity of ~60 mW/cm² was used to actuate the cantilever

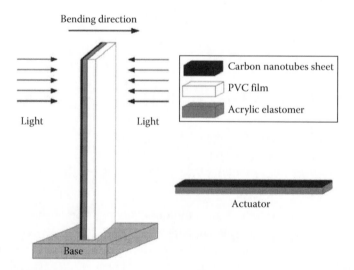

FIGURE 6.13 Setup for demonstrating photomechanical actuation of SWCNT/acrylic multilayer structure. A cantilever was vertically anchored on a base. The cantilever composed of an actuator (shown in the right lower part) and a 100 μm thick PVC film.

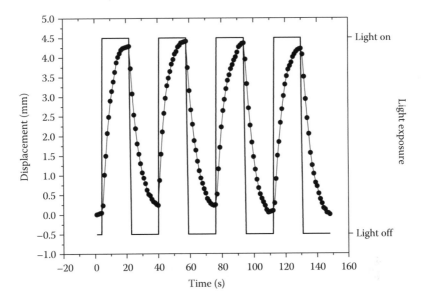

FIGURE 6.14   Displacement responses of the cantilever when light was switched "on" and "off."

for four cycles. During the period of light exposure, the displacement of the cantilever measured was to the side of the PVC indicating an increase in the length of the actuator. Once the light source was turned off, the actuator retracted back to its original length and the cantilever to its initial position, suggesting elastic deformations of the actuator upon illumination. The actuation was quite repeatable from cycle to cycle with nearly the same displacement amplitude showing the elastic nature of such actuation. A maximum displacement of 4.3 mm was achieved for a cantilever length of 30 mm. Such large reversible elastic responses from carbon nanotube/polymer systems are striking as they could potentially outperform many other optical actuation techniques, which only provide either low actuation stroke or lack of reversibility.

Experiments from the setup in Figure 6.13 cannot provide direct quantitative information on actuation strain of the multilayer actuator structure. In order to characterize the strain experienced by the actuator under optical illumination, an experiment was set up as shown in Figure 6.15, in which the actuator was doubly clamped between a vertical anchor and the PVC film. The PVC film was 100 μm thick and was fixed vertically to the base. The stress on the actuator (30 mm × 2 mm) due to light incident normal to its surface bent the PVC film. The displacement of the PVC film was recorded by a digital camera and was used to characterize the strain of the actuator. To balance the stress in the actuator, two carbon nanotube strips were attached to both sides of the polymer film. In the experiment, a thin reflective fiber of 15 μm in diameter was also attached to the top of the PVC film acting as a flag for easy monitoring of the actuation. Figure 6.16 shows six cycles of strain response of the actuator under different light intensities. The strain cycles were quite repeatable with nearly the same strain amplitude for any given intensity. It can be seen that the strain values are positive, suggesting that the actuator expands during optical exposure and returns to its original strain-free position when the source is turned off. At this point, if we check the alignment of carbon nanotubes in the actuator structure, nanotubes must be randomly aligned in the ensembles, as there is no apparent mechanism to induce biased alignment during nanotube film formation through vacuum filtration. Typical tube alignments of nanotube film used here are shown in the SEM image in Figure 6.6 where the random nature of tube orientation is clearly presented. Although we cannot relate the observed expansive actuation with the random orientations of nanotubes in the multilayered actuator at this time, as will be shown later in this chapter, nanotube alignment plays a significant role in deciding photomechanical responses of carbon nanotubes.

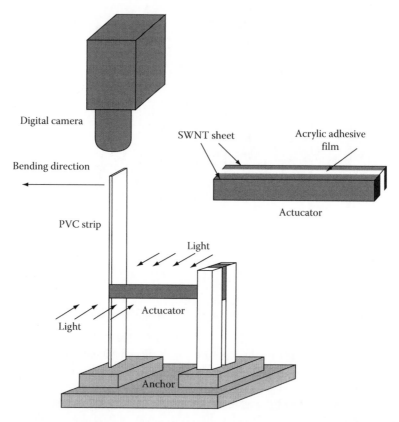

FIGURE 6.15 Experimental setup for strain characterization. The actuator was attached between a vertical anchor and a piece of PVC film, 100 μm in thickness. The stress from the actuator bent the PVC film and the displacement was recorded by a digital camera. To balance the stress in the actuator, two carbon nanotube strips were attached to both sides of the polymer film.

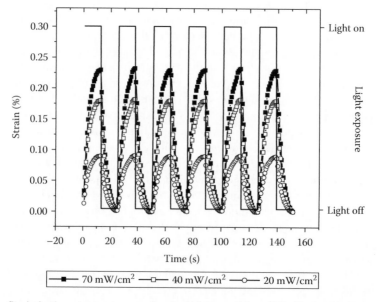

FIGURE 6.16 Strain in the actuator measured under light intensities of 70, 40, and 20 mW/cm².

Acrylic elastomers have previously been used as dielectric electroactive polymers as they produce higher strain and have a higher elastic energy density than other dielectric elastomers (Pelrine et al. 2000; Kofod et al. 2003; Ma and Cross 2004). When nanotube films are bonded with acrylic elastomers, a new type of optical actuator results, which operates solely based on the energy conversion from optical to mechanical energy; however, further experiments have revealed that an acrylic elastomer is not the determining factor in the optical actuation. Pure acrylic elastomers showed no apparent actuation under similar testing conditions in contrast to the larger photomechanical response in nanotube-acrylic multilayer actuators. The presence of carbon nanotubes in the actuator causes the actuation in otherwise benign systems; therefore, a more reasonable conclusion that could be drawn is that photomechanical actuation arises due to the intrinsic photo response of carbon nanotubes or due to the interplay between nanotubes and polymers. Experimental results from pure nanotube ensembles and nanotube–polymer nanocomposites indicate that both of the mechanisms are important in determining the overall photomechanical responses. Besides the acrylic elastomer, other types of benign polymers such as silicone rubber (polydimethylsiloxane: PDMS) and SU-8 epoxy resin have been used to construct multilayered actuators incorporating with carbon nanotubes. While these pure polymers show no apparent actuation in response to light, similar optical actuation arises from those structures having carbon nanotubes in them, suggesting that the photomechanical responses in these systems are "universal" responses of such systems and exist independent of certain polymers. Figure 6.16 shows the strain in the nanotube-acrylic multilayer actuator under light intensities of 70, 40, and 20 mW/cm^2. It is apparent that the higher the intensity of light incident on the sample, the greater the amplitude of strain measured. Figure 6.17 shows the strain versus incident light intensity ranging from 0 to 120 mW/cm^2. A monotone increase of strain with the increase of light intensity within the testing range is clearly shown in this figure. A strain value of ~0.3% was measured using light intensity of 120 mW/cm^2. Although electrically driven actuators have been shown to experience higher strains, these results show that similar strains could be obtained from optical actuation. Furthermore, this strain value also depends on the polymers in the actuator. Much higher strain value could be obtained when PDMS silicone rubber is used to construct the actuator, which will be discussed later. In order to study the robustness of the actuator and its capability of actuation

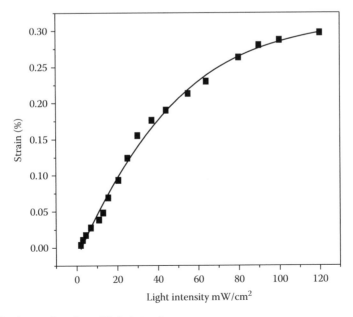

FIGURE 6.17   Strain as a function of light intensity.

in wet environments, the setup shown in Figure 6.15 was placed in a beaker containing deionized water. Upon exposure to a light of intensity 80 mW/cm^2, a strain of 0.06% was measured which is about 25% of the value measured in dry room temperature ambient. Although the strain value is small, this result shows that optical actuators capable of working in liquid environments could potentially be constructed based on carbon nanotubes through proper design.

In the experiments mentioned earlier, a halogen lamp was used as the light source, which covers a broad spectrum of wavelengths ranging from visible to near-infrared light. A study of the wavelengths' dependence on the optical actuation response would provide a better insight into the actuation behavior of the actuator. Eight different semiconductor lasers of various wavelengths (635, 690, 784, 808, 904, 980, 1310, and 1550 nm) were used as the optical source in the actuation. The average light intensity illuminating on the actuator surface was tuned to range from 0 to 65 mW/cm^2, depending on the maximum output power of the lasers. Figure 6.18 shows the strain characterization of the actuator under different laser sources. The strain responses again followed a trend of monotone increase with the increase of light intensity, similar to the case with the halogen lamp source. The lines indicated in the plots are linear fittings of the experimental data points. The curves can be observed following a fairly linear dependence within the testing ranges. Figure 6.18b is a magnification of Figure 6.18a in the intensity range of 3–28 mW/cm^2 to show clearly the difference in strain at lower intensities. To make a meaningful comparison of the photomechanical actuation under different wavelengths, photon-induced strains were compared when light intensities were deliberately kept constant under different wavelengths. Figure 6.19 shows the magnitude of optical responses as a function of photonic energies when light intensity was kept constant at 15 mW/cm^2. It is shown that as the photon energy is increased from 0.8 to 1.94 eV, the photo-induced strain increases from 0.0192% to 0.0365%, almost by a factor of two at this laser intensity level. Such wavelength dependence of photomechanical response gives us the opportunity of boosting actuation performance by simply employing optical sources with shorter wavelengths. Furthermore, the observed photomechanical actuation within a broad wavelength range covering both visible and near-infrared spectrums is beneficial, enabling a wide choice of optical sources for driving the optical actuators. Such a merit could lead to the construction of wavelength selective optical systems with the incorporation of optical filters.

6.4   MECHANISMS OF PHOTOMECHANICAL ACTUATION

Upon photon absorption in a material, temperature change is almost unavoidable. Carbon nanotubes are good light absorbers and show strong optothermal effects (Ajayan et al. 2002). As an important thermal property of carbon nanotubes, both theoretical and experimental work revealed that single-wall nanotubes have extremely high thermal conductivity along the tube axis (Hone et al. 1999a,b; Berber et al. 2000; Che et al. 2000; Osman and Srivastava 2001; Fujii et al. 2005). According to previous reports, the room temperature thermal conductivity of an isolated single-wall nanotube is 6600 W/m-K, much larger than that of a pure diamond (Berber et al. 2000). In the actuator, the global temperature rise due to photon absorption will cause thermal expansion/contraction of the structure and thus actuation, so it is true that thermal expansion/contraction due to global heating exists in nanotube optical actuators. The question, however, is how much is the contribution of global heating toward the overall photomechanical response. The global temperature rise in a typical actuator structure from nanotube–polymer systems has been measured under long-time light illumination (intensity in the range for optical actuation). With thermocouples placed at the surface or embedded in the center of the sample, maximum temperature change of up to 20°C was recorded, depending on the nanotube concentration and light intensity (white light of 5–50 mW/cm^2). An infrared thermometer was also used for measuring the surface temperatures of the samples with similar results obtained as that from the thermocouples.

Figures 6.20 and 6.21 show the temperature rise of a typical photomechanical actuator structure during the light transition intervals. The temperature traces are shown by the right axis in the

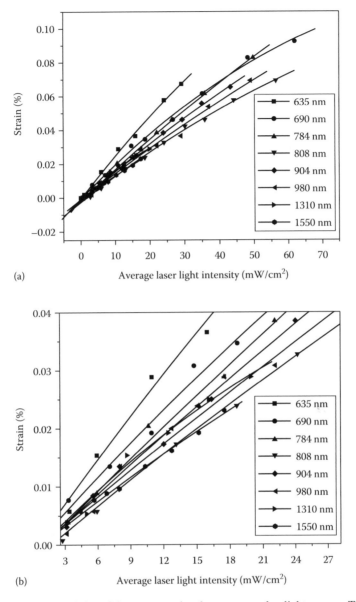

FIGURE 6.18 Strain characteristics of the actuator when lasers are used as light sources. The wavelength of lasers used range from 635 to 1550 nm with an average intensity varying from 0 to 65 mw/cm². (a) The strain responses corresponding to different laser sources. The lines are linear fittings of the experimental data points. (b) Magnified image at lower intensities ranging from 3 to 28 mW/cm².

figures. To compare the speed of temperature change, normalized actuation strokes were also plotted in the two figures labeled by the left axis. During both light "off-on" transition (Figure 6.20) and "on-off" transition (Figure 6.21), it is clear that photomechanical actuation followed a much faster rate of change than that of temperature change. The difference between the actuation speed and the speed of the global thermal effect suggests that sample heating is not the actuation mechanism or at least not the dominating mechanism. In addition, the global temperature change of ~20°C is also too small to be used to explain the extraordinary photomechanical responses observed in nanotube–polymer systems, which can be shown in a purely thermal actuation experiment. Actuator samples were studied for the mechanical response purely due to global thermal heating. In the

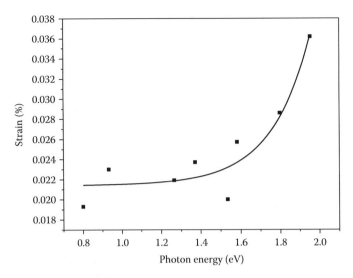

FIGURE 6.19 Strain response of actuator as a function of photon energies under a constant laser intensity of 15 mW/cm².

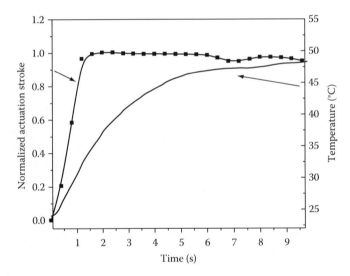

FIGURE 6.20 Global temperature change of a typical photomechanical actuator structure during the light "off-on" transition. Normalized photomechanical actuation stroke was also plotted to compare the speed.

experimental setup shown in Figure 6.6, the light source was removed from the sample. Instead, a small resistive heater was placed near the samples that were clamped between the micrometer and the dynamometer. The heater was programmed to cause cycled temperature changes on the sample at an amplitude of ~20°C.

Figure 6.22 compared the pure thermal actuation of single-wall nanotube–polymer multilayer structures with the photomechanical actuation cycles from the same actuator structure. The heating reproduced a slower actuation of the sample, however, by a much smaller amount than the optical actuation, which is a typical response in different actuator structures. In different sample constructions from nanotube–polymer systems, the contributions of global heating effects vary from ~12% in multi-wall nanotube–PDMS nanocomposites to ~23% in single-wall nanotube–polymer multilayer samples. By using a large area hotplate surface as the heating source to cause uniform

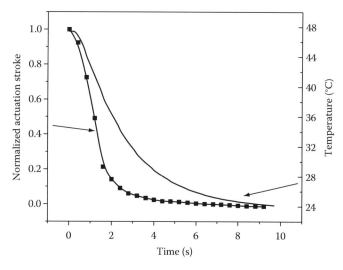

FIGURE 6.21 Global temperature change of a typical photomechanical actuator structure during the light "on-off" transition. Normalized photomechanical actuation stroke was also plotted to compare the speed.

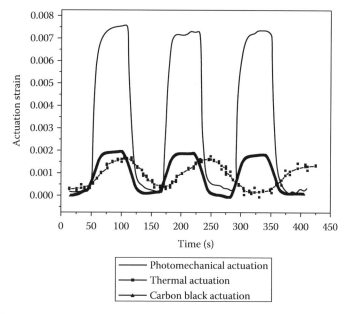

FIGURE 6.22 Comparison of photomechanical actuation in single-wall nanotube-PDMS multilayer actuators (solid line) with the thermal actuation of the same structure (■ line) and the photomechanical actuation of a carbon black-PDMS multilayer actuator (▲ line).

temperature rise across the sample, the single-wall nanotube–polymer multilayer structure was also studied for thermal actuation in a large temperature range from room temperature (∼22°C) to 128°C. Figure 6.23 shows the thermal actuation stroke of a single-wall nanotube-acrylic multilayer actuator as a linear function of the sample temperature. In order to obtain ∼0.3% thermal strain, the actuator structure needs to be heated to more than 100°C by the thermal source; however, the same amount of strain was obtained optically at ∼120 mW/cm² light intensity with the sample temperature rising up to 48°C. Such differences indicate the effectiveness of optical actuation and eliminate the global heating effect as the main contributing actuation mechanism. Moreover,

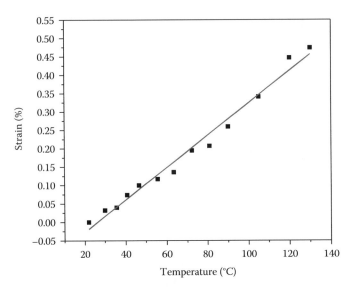

FIGURE 6.23 Thermal actuation in single-wall nanotube-acrylic multilayer actuators as a function of sample temperature.

considering the thermal energy exchange between the heating sources and the samples, there is still significant amount of energy transferred by infrared emission at long wavelengths from the heater. Thus, the thermal contribution derived from this experiment somehow still encompasses infrared optical actuation. It is hard to evaluate the exact amplitude of the thermally induced strains in the actuator; however, the percentage numbers from ~12% to 23% can be viewed as the upper bound of the contribution of the global heating effect in the samples toward the overall photomechanical actuation. Therefore, it is concluded that heating alone cannot be viewed as the leading actuation mechanism. It is worth mentioning here that independent studies on photomechanical actuation of multi-wall nanotube–polymer nanocomposites from other research groups have reported that thermal actuation only accounts for several percent of the optical actuation, much smaller than that from our investigation (Ahir and Terentjev 2005; Ahir et al. 2006). Further evaluation of global heating effects involves constructing an optical actuator using carbon black as the light absorber. When carbon black was used to make multilayered samples, optical stimulation yielded reversible actuation from the samples. Figure 6.24 also compared the photomechanical actuation in single-wall nanotube–PDMS multilayer actuators with that from a carbon black–PDMS multilayer actuator. It is clear that the photomechanical actuation strokes from carbon black samples were much smaller than that from carbon nanotube counterparts. This is better seen from Figure 6.24, which compared their photomechanical actuations under different prestrain conditions. Although the carbon black actuators also showed prestrain-dependent photomechanical patterns similar to that from carbon nanotube samples, the smaller actuation amplitude up to ~25% from nanotube samples are consistent with the result from the previous pure thermal actuation experiments (assuming that only pure optothermal effects happen in carbon black samples). It was argued that the limited optical actuation from carbon black–polymer systems was due to the trace amount of carbon nanotubes in the carbon black materials (Ahir et al. 2006). Even if this is not the case, optical actuation from nanotube actuators is still much larger than that from carbon black–polymer actuators. Such results indicated the big difference in optical responses between nanotubes and amorphous carbon nanostructures. Therefore, in the worst case, optical actuation due to global heating effects in nanotube/polymer systems would account for at most 25% of the observed photomechanical actuation.

In pure carbon nanotube ensembles, thermal expansion/contraction of carbon nanotubes also is not the dominate mechanisms in optical actuation. Anharmonicity of interatomic potentials causes

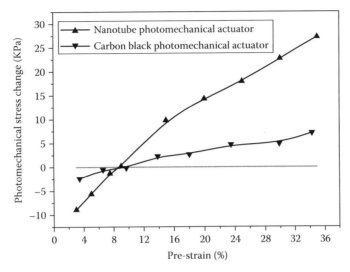

FIGURE 6.24 Comparison of photomechanical actuation in single-wall nanotube-PDMS multilayer actuators (▲ line) with that of a carbon black-PDMS multilayer actuator (▼ line).

thermal expansion/contraction in most material systems. Expansion or contraction is dependent on the balance between phonon modes and respective Gruneisen parameters (Schelling and Keblinski 2003). As one of the key properties of carbon nanotubes, thermal expansion determines many physical, optical, and mechanical behaviors of nanotubes (Li and Chou 2005); however, due to the challenge in nanoscale experiments and modeling, studies of thermal expansion of nanotubes are limited, and many of the studies gave contradictory results as to whether nanotubes experience thermal expansion or thermal contraction (Yosida 2000; Maniwa et al. 2001a,b; Jiang et al. 2004; Tang et al. 2004; Young-Kyun et al. 2004; Kwon et al. 2005; Li and Chou 2005; Mounet and Marzari 2005). The values of the coefficient of thermal expansion (CTE) also varied in a big range. Axial thermal contraction of an individual (10, 10) nanotube at room temperature has been shown in simulations (Young-Kyun et al. 2004). The x-ray scattering experiments by Maniwa (Maniwa et al. 2001a) registered an average lattice constant expansion of $\sim 0.75 \times 10^{-5}$/K. For the bulk nanotube systems, using these values, solely lattice potential effects cannot explain the photomechanical effect unless the extremely high local heating is assumed on photon absorption; however, the mean temperature change across the sample was only 20°C. Considering that the carbon nanotubes randomly orientated in the ensembles, the magnitude of optical actuation (contraction) is estimated to be at least an order of magnitude greater than predicted by thermal expansion ideas.

6.4.1 ELECTROSTATIC EFFECTS

Elastic deformation of carbon nanotubes due to electrostatic effect has been previously reported (Baughman et al. 1999; Kim and Lieber 1999; Zhang and Iijima 1999), where the electrostatic field can be introduced either electrically or optically. Carbon nanotube nanotweezers can be simply realized by electrostatically bending one nanotube against the other (Kim and Lieber 1999). The elastic responses of the carbon nanotube bundles and fibrous networks under visible light illumination are of much interest for us (Zhang and Iijima 1999). Single-wall nanotube bundles showed distinct movement such as stretching, bending, and repulsion when they were exposed to visible light. These elastic effects upon light exposure were well explained by electrostatic interaction induced by localized charge separation and imbalance as discussed previously. In carbon nanotube optical actuators, electrostatic interaction between the nanotubes may cause the nanotube conformational distortions and consequently actuation. When light is incident on carbon nanotubes,

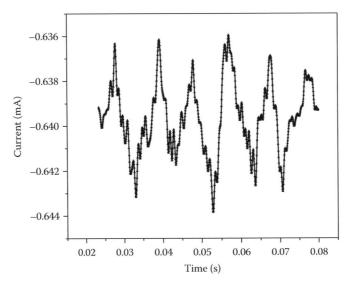

FIGURE 6.25   Photon-induced current flow in nanotube films.

it increases the conductivity of the nanotubes (Fujiwara et al. 2001; Freitag et al. 2003; Levitsky and Euler 2003). This photo-conducting nature of nanotubes arises due to charge carrier (electron–hole pairs) generation in the presence of light illumination. The photo carriers would diffuse in the nanotube ensembles to cause observable effects in conductivity. Experimentally, we have evaluated the photoconductivity of nanotube films during the optical cycles, which were shown in Figure 6.25. In this particular case, the current flowing through nanotube films experienced increases in amplitude with a response speed of several milliseconds under light pulses. One may be tempted to ask two questions here: (1) as the carrier generation is extremely fast, why is the photocurrent response slow, and (2) is the current change coming from the resistance change. The first question, which is quite important to reveal the nature of photoconductivity of carbon nanotube ensembles, is not well understood yet. Measurement of individual nanotubes has shown a fast photo-carrier generation in the scale of nanoseconds (Wang et al. 2004a; Sheng et al. 2005); however, in nanotube ensembles, the measurements were slow (Fujiwara et al. 2001; Freitag et al. 2003; Levitsky and Euler 2003). It may be understood that the charge relaxation throughout the nanotube networks is a slow process, and the relaxation of carriers depends on the slow thermal processes. The trapping of charge carriers due to defects and barriers in the nanotube ensembles may also be a partial reason. Further experiments are definitely needed to clear this matter.

For the second question, the photoconductivity is indeed due to charge separation. Recent studies indicate that a majority of photon-generated charge carriers are in the form of excitons rather than the free carriers because of the 1D nature of the nanotubes (Chen et al. 2005; Wang et al. 2005; Itkis et al. 2006). Various measurements on photoconductivity of single nanotube and nanotube ensembles, however, show the existence of a photocurrent, which is beyond the range of thermally induced pure resistance change. Especially, charge separation was observed at the vicinity of metal/nanotube contacts (Freitag et al. 2003; Moonsub and Giles 2003; Mohite et al. 2005; Lu and Panchapakesan 2006). This indicates that photon-generated excitons may also experience charge separations at certain locations in nanotube ensembles, where local field strength is large enough to free the bound carriers. We have conducted studies on charge separation in macro nanotube ensembles, in which we used point laser sources to excite selectively small portions of nanotubes of a long nanotube strip between the positive and negative electrodes. Electrical characterization revealed strong dependence between the photocurrent and the positions of the laser spot. In extreme cases, the current can even be reversed with respect to the applied voltage, suggesting that strong charge separations do occur between metal/nanotube barriers. The details about this study are beyond the scope of this dissertation. Interested readers are referred to Lu and Panchapakesan (2006). An implication of such strong charge separation is that a similar situation is true in photomechanical actuators from carbon nanotubes. Qualitatively, an electrostatic effect may be a significant contribution to the observed photomechanical actuation.

6.4.2 POLARONIC EFFECTS

Polaron generations due to light illumination may also be an important contribution to the photomechanical actuation (Verissimo-Alves et al. 2001; Piegari et al. 2002). The interplay between mechanical distortions and electronic structures render carbon nanotubes' unusual electronic and mechanical properties. Mechanical distortions can be generally classified as externally applied or spontaneous. Externally applied distortions such as twisting, bending, and axial compression of carbon nanotubes greatly affect their electronic structures (Heyd et al. 1997; Park et al. 1999; Mazzoni and Chacham 2000a,b). On the other hand, spontaneous distortions are usually related to strong electron–phonon interactions. In carbon nanotubes, an extra electron or hole would cause spontaneous distortion in the form of a combined radial (breathing-mode-like) and axial distortion (Verissimo-Alves et al. 2001). This perturbation causes the band edge energies to vary linearly and the elastic energy to increase quadratically with the distortion parameters. To show the magnitude of mechanical distortions due to polarons in carbon nanotubes, a hole polaron in a (7, 0) carbon nanotube will cause 0.02 Å change in nanotube length. Therefore, a modest number of 500 polarons would cause a sizable 10 Å variation in the nanotube length, large enough to be observed, for instance, in AFM or STM experiments where carbon nanotubes are used as probes. Electron–hole pairs can be generated in carbon nanotubes optically. Depending on the sign of axial distortions (C_z) for electrons and holes, polarons may bring strong or weak axial distortions in a carbon nanotube as an elastic response to light. For example, electron–hole pairs are generated by light in (11, 0) and (7, 0) nanotubes. From the signs of C_z in these nanotubes, an electron polaron causes an axial expansion in both (11, 0) and (7, 0) tubes, while a hole polaron causes an expansion in the (11, 0) and only a contraction in the (7, 0) tubes. Therefore, for the (11, 0) nanotubes, the axial mechanical effects of the electron and hole will add up, and the carbon nanotube will have a strong elastic response to light; however, mechanical distortions from the electron and hole will partly cancel each other in the (7, 0) nanotube, leading to weaker photomechanical effects in such tubes (Verissimo-Alves et al. 2001). Nanotubes with different helicities coexist in real carbon nanotube samples. The overall responses of the carbon nanotube ensembles would be the averaging result of polaron effects in individual nanotubes across the whole sample. This direct optical-mechanical coupling may cause lattice strain and conformational distortions of carbon nanotubes, thus contributing to the observed photomechanical actuation.

6.4.3 LOCALIZED THERMAL EFFECTS

Recent investigations on the dynamic behaviors of multi-wall nanotube–polymer nanocomposite have identified an alternative mechanism for the large microscopic contraction of individual nanotubes: the local thermal confinement of carbon nanotubes (Ahir and Terentjev 2006). We found that it could be used to account for the optical actuation in all the nanotube/polymer systems studied in this dissertation, including multilayers and nanocomposites for both single and multi-wall nanotubes. The effect of local thermal confinement is introduced next. The local extreme thermal confinement should be differentiated from the global thermal effect discussed earlier, which happens only in microscopic or nanoscales and is a physically discontinuous event in contrast to the uniform global heating. As mentioned previously, the high thermal confinement in the discontinuous nanotube networks would cause severe local effect, which can trigger the oxidation of carbon nanotubes and even nanotube structural reconstruction (Ajayan et al. 2002).

The dynamic responses of photomechanical actuation in nanotube/polymer systems can be well explained by the sharp local thermal effect, which is due to the local thermal confinement. Another possible consequence of the local thermal confinement is the charge generation and charge migration due to the thermoelectric power effect of nanotubes (Hone et al. 1998; Tian et al. 1998; Grigorian et al. 1999; Bradley et al. 2000; Romero et al. 2002; Savage et al. 2003). Large thermoelectric power effects in nanotubes may potentially result in effective charge generation and charge separation within the nanotube networks, especially when the local temperature is high. The nanotube ensembles can be envisioned as numerous small thermoelectric power cells powered by light. Consequently, strong electrostatic distortions may happen in nanotubes, which may be a partial contribution to the observed photomechanical responses.

6.5 APPLICATIONS OF PHOTOMECHANICAL ACTUATION OF CARBON NANOTUBES

The patterning of nanotube ensembles is essential in constructing multifunctional micro- and nanosystems based on carbon nanotubes. Broadly speaking, such multifunctional small systems not only include the carbon nanotube optical actuators as have been discussed in this dissertation but also include various nanotube-based electronic and optoelectronic devices such as chemical sensors (Valentini et al. 2003; Oakley et al. 2005), strain sensors (Prasad et al. 2004), photodetectors (Lehman et al. 2005), and thin-film transistors (Takenobu et al. 2006), transparent electrodes (Wu et al. 2004; Pasquier et al. 2005), and field emission devices (Groning et al. 2000). Whether carbon nanotubes would have success in these applications depends critically on whether nanotube ensembles could be patterned precisely in micro- or nanoscales with batch fabrication capability. While nanotubes have been processed into macroscopic sheets and films with controllable patterns, small devices based on nanotube ensembles employing batch fabrication techniques have been elusive. The lack of precise control of nanotube patterns in micro- and nanoscales has become one of the major obstacles for fulfilling these applications. The requirement of high-quality and high-resolution patterning of nanotube ensembles into various features at desired locations on specified substrates calls for advances in nanotube patterning techniques.

Small patterns of carbon nanotube ensembles can be produced by a "bottom-up" approach where carbon nanotubes are selectively grown on pre-patterned catalyst blocks (Ren et al. 1998; Fan et al. 1999; Bower et al. 2000; Sohn et al. 2001) through chemical vapor deposition (CVD) techniques. Generally, highly aligned "out-of-plane" nanotube ensembles would form within the nanotube patterns (Dai 2002). In addition, small "in-plane" carbon nanotube networks with well-defined orientations can also be grown through CVD processes. While catalyst nanoparticles provide the anchor for nanotube growth, the growth direction and nanotube alignment can be controlled by gas flow or by the electrical field (Zhang et al. 2001; Dai 2002; Huang et al. 2003). The high-temperature process involved in carbon nanotube growth restricts the choice of substrates to high-temperature materials, so glass or polymer substrates cannot be used without an extra pattern transfer process. The as-grown carbon nanotubes may have a low structural perfection and material purity with contaminants such as catalyst nanoparticles in the ensembles. The high-temperature CVD process is incompatible with CMOS/MEMS processing. These issues possess limitations to the nanotube pre-growth patterning technique.

Other approaches of fabricating patterned carbon nanotube ensembles follow the post-growth patterning route in which the purified nanotubes are reorganized to form desired patterns. Several techniques such as electrophoresis deposition (Choi et al. 2001; Gao et al. 2001), screen printing processes (Choi et al. 1999), self-assembly (Oh et al. 2003), and chemically anchored deposition (Jung et al. 2005) have been utilized to produce microscale nanotube patterns. The electrophoretic deposition could employ both DC and AC electrical fields to attract carbon nanotubes from solutions to desired electrically conducting locations of the substrates. The screen printing process involves squeezing the paste of well-dispersed carbon nanotubes with organic binders through metal mesh onto the substrate. In nanotube self-assembly and chemically anchored deposition, carbon nanotubes and the substrates are chemically modified so that nanotubes could be selectively deposited on substrates to form patterns. While these post-growth methods could be used to produce nanotube patterns with some success, they either only offer low feature resolutions or suffer from limitations of substrate restriction, requirement of substrate modification, use of electrical field gradients, or a tedious thickness control process. In this chapter, a new subtractive post-growth patterning technique is developed which offers simple and well-controllable patterning of carbon nanotubes with high resolutions. This miniaturization technique of carbon nanotube patterns would enable the construction of carbon nanotube-based micro-opto-mechanical systems (CNT-MOMS).

6.5.1   Nanotube Micro-Opto-Mechanical Actuators

To scale the carbon nanotube-based optical actuators down into micro- or nanodimensions for making the small multifunctional optically actuated systems, regular micro-size carbon nanotube patterns must be first fabricated for integration with the small systems. If polymer materials are involved in the actuation system, they also need to have the processability for easy fabrication of small (micro or nano) actuator structures. The miniaturization of nanotube optical actuators requires low temperature and high-resolution patterning of carbon nanotubes. A new patterning technique of carbon nanotubes has been developed, in which uniform thin nanotube films of desired thickness were formed by vacuum filtration. Then it was transferred to a substrate, followed by photolithography to define features. O_2 plasma etching was subsequently employed to remove the exposed carbon nanotubes to form patterns selectively. This method offers several advantages compared to other patterning techniques: (1) uniformity and reproducibility of carbon nanotube films (CNF) (Wu et al. 2004) within the patterns, (2) low-processing temperatures compatible with polymeric substrates, (3) high-feature resolutions even smaller than nanotube length due to the ability of plasma to etch the nanotubes precisely, (4) sharp pattern edges, and (5) compatible with MEMS fabrication technologies. The patterning process starts from carbon nanotube solutions. Single-wall nanotube suspension was vacuum-filtered through an MCE filter to produce thin nanotube films. A simple procedure described by Wu et al. was employed to transfer nanotube film onto a substrate (Wu et al. 2004), as shown in Figure 6.26 sequence (a) to (c). Briefly, the wet nanotube films still on top of the MCE filter was located on a silicon substrate by compressive loading. The nanotube films stuck to the substrate with enough adhesion strength for further processing after drying and subsequent annealing on a 75°C hotplate for 20 min. Then the MCE filter was dissolved in multiple baths of acetone, leaving clean, uniform, and wrinkleless carbon nanotube film on the substrate after drying. Figure 6.27 shows a uniform semitransparent single-wall nanotube film of ~3 cm × 3 cm in area and ~230 nm in thickness transferred to a silicon wafer. The thickness of the film was well controlled by the amount of carbon nanotube solution of known concentration during vacuum filtration. Several CNF of thickness ~40, 130, 230, 460, and 780 nm were fabricated with high film uniformity offered by the vacuum filtration process (Wu et al. 2004). As the film thickness was smaller than 230 nm, the nanotube film showed a high degree of transparency visible to naked eyes, as shown by the transparency in Figure 6.27.

Standard photolithographic process was then employed to define CNF patterns on the substrate. Several commercial photoresists of both positive and negative tones, including AZ5214E, NR7-1500, AZ4620, and SU8, have been tested, and all formed excellent features on top of the CNF. This indicates that randomly oriented nanotubes packed into thin films hardly affect the lithographic process. Through careful electron microscopic inspections, the photolithography process does not increase organic residuals in the CNF, indicating that lithography is a clean process for CNF patterning. The excellent compatibility of CNF with lithographic patterning offered the opportunity of defining precise and high-resolution features onto CNF through lithography with virtually only one requirement being the photo resin thickness. Since O_2 plasma attack photo resins, the etchmask from photo resins needs to be thick enough to sustain continuous O_2 plasma etching. For

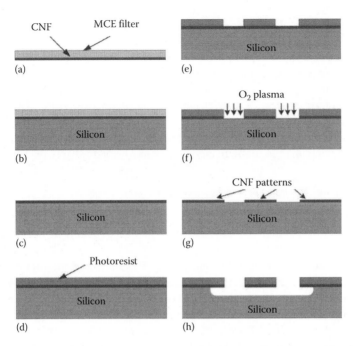

FIGURE 6.26 Sequence of CNF patterning. (a) CNF on MCE filter after vacuum filtration, (b) CNF with MCE filter being transferred onto silicon substrate, (c) MCE filter dissolved, (d) spin coating photoresist, (e) photo lithography, (f) O_2 plasma etching of CNF, (g) masking photoresist removed, (h) in case of CNF/SU8 active structure, XeF_2 etching was used to release the structure.

FIGURE 6.27 Rectangular semitransparent single-wall nanotube film of ~230 nm in thickness transferred onto glass substrate.

nanotube films with thickness smaller than 460 nm, ~1.5 μm, photoresist (AZ5214E) was used as the etch-mask. Commercial thick film photoresists such as AZ4620 was also used to pattern thick etch-masks up to tens of microns for etching thicker nanotube films. Sequence (d) to (e) in Figure 6.26 illustrated the lithography process.

O_2 plasma has been widely used to remove carbon-based organic materials such as photo resins from substrate surfaces. It forms volatile CO, CO_2, and H_2O which can be pumped out from the system during plasma etching (Madou 2002); however, O_2 plasma etching of carbon nanotubes to define patterns has not been reported until now. By employing a similar idea, we used O_2 plasma in an ICP system to etch the carbon nanotubes in order to form nanotube film patterns. At ICP power~200 W, bias power~100 W, and O_2 flow rate~50 sccm, an etch rate of CNF at ~4 nm/s was achieved, showing the fast etching of carbon nanotubes in strong O_2 plasma. Carbon nanotubes are well known for their chemical resistance and high stability in harsh environments; however, structural defects are often present in real carbon nanotube samples, especially after purification and ultrasonic dispersion steps (Liu et al. 1998).

At the same time, defects could also form on carbon nanotubes under bombardments of high-energy species in plasma ambient (Hassanien et al. 2005). It is possible that the oxygen plasma react with the carbon nanotubes starting from these defects, consuming the whole nanotube gradually. After etching, mild acetone rinsing served to dissolve the etch-mask to leave clean nanotube patterns. The etching process and subsequent etch-mask removal are schematically shown in Figure 6.26f through g. Well-defined carbon nanotube strip lines of ~4 μm in width and 130 nm thick were fabricated with all the unwanted nanotubes removed, as shown in Figure 6.28. The large area clear patterns showed the effectiveness of carbon nanotube patterning through O_2 plasma etching. In Figure 6.29, carbon nanotube lines as small as ~1.5 μm were also routinely produced on 130 nm thick nanotube films, with well-defined shapes and sharp feature edges. Moreover, we believe that higher resolution patterns with feature sizes even smaller than nanotube lengths are still possible to achieve because of the ability of O_2 plasma to "cut" exposed carbon nanotubes to leave sharp pattern edges, as witnessed in the insert of Figure 6.30. Electron beam lithography can reduce the size of nanotube patterns, potentially achieving a feature size in the sub-100 nm regime for nanoscale nanotube devices. The reason for such excellent pattern transfer may also be due to the lack of stresses in the CNF after processing, which is important in most surface microfabrication technologies.

One can use this carbon nanotube patterning technique to integrate carbon nanotubes into micro- or nanosystems, where carbon nanotube ensembles can be organized in a very configurable and controllable way. Due to the extraordinary mechanical properties of carbon nanotubes, micro- and nanoscale regular nanotube patterns will not only act as functional materials in the small systems but also could be employed as multifunctional structural materials. The introduction of new multifunctional materials will enable the construction of micro- and nanosystems with novel properties and behaviors beyond the reach of traditional small systems. On the other hand, while there

UDel   LEI   3.0 kV   ×600   10 μm   WD 7.4 mm

FIGURE 6.28 SEM image of large area (150 μm × 150 μm) patterns of carbon nanotube lines ~4 μm width fabricated by oxygen plasma etching.

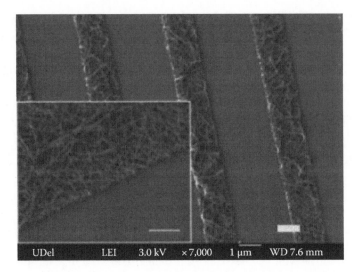

FIGURE 6.29 Clear patterns of ~1.5 µm carbon nanotube lines with ~2 µm spacing. Insert: sharp pattern edge formed by nanotube cutting in O_2 plasma.

is growing interest in the photomechanical properties of carbon nanotubes showing the feasibility of constructing small remotely powered and controlled wireless systems based on the nanomaterials, practical micro- and nanodevices that utilize this concept have not been reported. The lack of proper technical approaches for integrating such nanostructured materials into the micro- and nanosystems was the major obstacle. Through the nanotube integration process discussed earlier, carbon nanotube ensembles can be properly introduced into MEMS, and new CNT-MOMS can be demonstrated utilizing the novel optical actuation of carbon nanotubes.

As discussed in previous chapters, polymer materials may be needed to construct the carbon nanotube-based optical actuators. Then the polymer must also be compatible with the microfabrication process and can be used in batch fabricating small actuator structures. SU8 photo epoxy has been widely used in fabricating micro-mechanical structures in recent years because of its excellent mechanical properties, high glass-transition temperature, excellent biocompatibility, and excellent lithographic applications (Nguyen et al. 2004; Chronis and Lee 2005). Here, SU8 epoxy is found suitable to be combined with carbon nanotubes to construct microoptical systems. To illustrate the easy processing and compatibility of carbon nanotube and SU8 patterning in constructing nanotube–polymer multilayer optical actuators, an optically active CNF/SU8 multilayer cantilever structure was produced. This structure was made by a second standard lithography to pattern the SU8 onto the carbon nanotube patterns (SU8 patterned after sequence (g) in Figure 6.26).

Alternatively, one can use SU8 directly as the etch-mask in the carbon nanotube patterning (SU8 patterned at sequence Figure 6.26d through f). The second approach saves the second lithography; however, it only produces exactly the same patterns for carbon nanotubes and SU8 epoxy. After SU8 patterning, a simple CNT/SU8 bilayer cantilever actuator was produced. To finish the actuator fabrication, the structure needs to be set free from the substrate, which was done by isotropic silicon etching in a homemade pulse mode XeF_2 dry etching system (Madou 2002; Chronis and Lee 2005), as illustrated in Figure 6.26 sequence (h) (silicon used as substrate). Arrays of cantilevers are shown in the insert of Figure 6.30a. The magnified image of ~30 µm (width) × 300 µm (length) × 7 µm (thickness) cantilevers after releasing is also shown in Figure 6.30a. Figure 6.30b shows the cross-sectional area of the cantilever with the SU8 and carbon nanotube layers clearly observed. This indicates that high-quality nanotube layers can be formed from plasma etching and can be introduced into micro-devices to exhibit multiple functionalities of both structural and functional materials. Such nanotube/SU8 microcantilevers that were produced by the MEMS-compatible

(a)

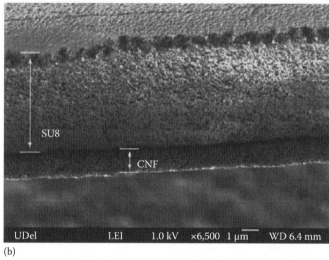

(b)

FIGURE 6.30   (a) SEM image of released CNF/SU8 actuators. Insert: SEM image of a $3 \times 3 \times 3$ actuator arrays. (b) SEM image of the squared region in (a) showing the bilayer cross-section of the actuator.

process can be optically actuated to retain the same photomechanical properties as the macrosize samples, showing the effectiveness of transferring carbon nanotube-based optical actuation into the microsales by the miniaturization technique presented earlier. With the ability of scaling the nanotube-based optical actuators, both in-plane (X-Y plane) actuation and out of-plane actuation (Z direction) mechanisms can be realized by proper design of the actuation structures, which shows the feasibility of constructing not only surface-machined microstructures but also microsystems with 3-D mobility through this CNT-MOMS technique (Lu and Panchapakesan 2006; Lu et al. 2007a). Such smart systems will see various applications in adaptive optics, all-optical networks, nano- and bionanotechnologies.

6.5.2   Nanotube Micro-Opto-Mechanical Grippers

Most of the microgrippers are operated by electrostatic or electrothermal principles; however, grippers based on optical actuation principles in carbon nanotubes have not been reported until now.

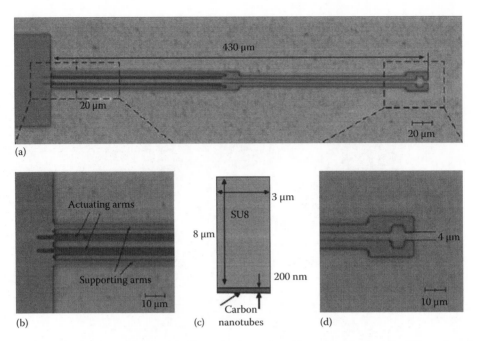

(a)

(b)   (c)   nanotubes   (d)

FIGURE 6.31 Design of the CNT MOMS gripper. (a) A "double arm" gripper with 430 μm length and 20 μm width. (b) The "supporting arm (2 μm in width)-actuating arm (3 μm in width)" structure. The region with deep gray color in the actuating arm indicates the CNT layer underneath the actuating arms. (c) Cross-sectional view of actuating arms showing the SU8 frame of 8 μm height and 3 μm width on top of a ~200 nm thick CNT films, and (d) the gripper tips with a 4 μm initial opening.

To design the CNT-MOMS gripper, a double arm structure consisting of "an actuating arm and a supporting arm" is employed to translate the photomechanical actuation in nanotube/polymer multilayer structures into the grippers operations. Figure 6.31 shows the gripper design, which is similar to the traditional "hot arm and cold arm" actuators (Nguyen et al. 2004; Chronis and Lee 2005). As shown in Figure 6.31a, the gripper bears the dimension of 430 μm in length and 20 μm in width. The long actuator beams serve to amplify the gripper openings. The double arm design is better seen in Figure 6.31b, where the inner two arms are actuating arms (3 μm in width) and the outer two arms are supporting arms (2 μm in width). The region with a deep gray color in the actuating arm indicates the nanotube layer underneath the actuating arms. In traditional "hot arm and cold arm" actuators, the hot arm expands more than the cold arm when there is a temperature difference between the two arms, so a length difference between the arms results, causing the bending and thus the actuation in the structure. In our gripper design, there is also a length difference developed between the actuating arm and the supporting arm during gripper operation; however, it is the photomechanical actuation of the single-wall carbon nanotube in the actuating arm instead of the resistive heating of the hot arm that results in the operation of the gripper. Simulation by ANSYS in Figure 6.32 shows the effectiveness of translating photomechanical actuation of the actuating beam into the gripper operation, while the out-of-plane bending of the gripper is negligibly small with optimized structure thickness.

In recent years, SU8 has been widely used in fabricating micro-mechanical structures because of its excellent material properties and extraordinary processability; however, the application of SU8 in actuators is quite limited due to the difficulties of implementing and integrating a reliable actuation mechanism into SU8 structures (Chronis and Lee 2005). Resistive heating for thermal actuation (Nguyen et al. 2004), and SMA (Roch et al. 2003) actuation mechanisms have been investigated with SU8 structures. Here by incorporating CNTs into SU8 structures,

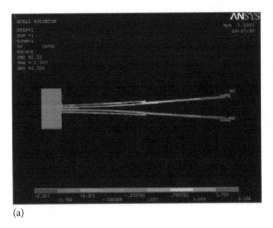 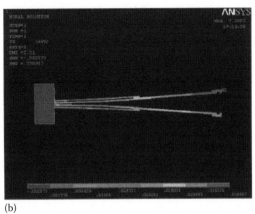

(a)   (b)

FIGURE 6.32 **(See color insert.)** ANSYS simulation of the gripper opening (a) and out-of-plane gripper beam bending (b) during actuation of the structure. By choosing the appropriate aspect ratio of the beams, the out-of-plane motion of the structure is negligible compared to the in-plane gripper opening.

the photomechanical actuation mechanism was effectively implemented to accomplish the SU8 actuation.

Figure 6.31c shows the schematic view of the cross-section area of an actuating arm, in which a thin layer of nanotubes is attached to the SU8 frame, the same as was discussed in the previous section. The slim actuating arms bear the height of ~8 μm and width of 3 μm, which gives an appropriate aspect ratio of the structure to minimize the out-of-plane bending of the gripper while avoiding beam warping or gripper "bite" problems in the released grippers. An optical image of the gripper tips in the shape of a holder is shown in Figure 6.31d. A 4 μm opening is designed in the gripper's "close" state, which can be tuned to match the size of objects for micro- and nanoscale manipulation. After fabrication, a blind cut of substrate was sometimes given to get a better view of the grippers during operation. By careful control of the fabrication process, the device yield can be as high as 95%, which is partially due to the small residue stress in the nanotube film after fabrication. The dry releasing process employed here also contributes to the high device yield, which minimizes the effects of the fabrication process and leaves intact device structures. Microgripper arrays were successfully fabricated from the process, in which all the grippers could operate simultaneously under illumination from a single light source and could be individually addressed by controlling the position of a small driving light spot. This is not possible with most MEMS-based tweezers operating on the electrostatic actuation principle if no complicated control circuits are involved. Figure 6.33a shows the SEM pictures of a gripper array after being released from the substrate. The wider actuating arm and narrower supporting arm are clearly seen in Figure 6.33b. Figure 6.33c depicts the cross-sectional view of the actuating arms, which shows the SU8 structure on top of the thin carbon nanotube layer. The gripper tips are shown in Figure 6.33d. Being far away from the actuating arms, the gripper tips are separated away from the light stimulus, ensuring minimal light-induced effects to the objects being manipulated during gripper operation.

Following the fabrication, the CNT-MOMS grippers were tested under light illumination. An 808 nm semiconductor laser focused into a ~5 mm × 1 mm spot was used as a light source. A CCD camera mounted on an optical microscope was used to measure the displacement of the gripper tips.

Successful operation of the microgrippers based on SU8 polymeric structures is guaranteed by the reliable photomechanical actuation from the nanotube/polymer double layer structure, which serves to couple the photonic energy directly into the microstructures, turns it into mechanical energy, and ensures enough mechanical deformation to occur in the rigid polymer structure. A typical optical image of the gripper tips in the "close" state is shown in the upper image of Figure 6.34a.

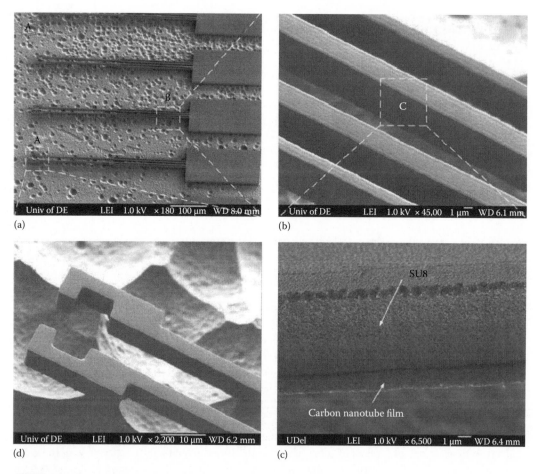

FIGURE 6.33 SEM images of CNT MOM grippers released from the substrate. (a) Gripper arrays, (b) magnified square B in (a) showing the actuating arms and the supporting arms, (c) magnified square C in (b) showing the cross-sectional view of the actuating arms, and (d) the magnified square A in (a) showing the structure of the gripper tips.

When infrared laser light of ~550 mW was directed to the actuating arms, the gripper tips open to ~20 μm, resulting in ~16.5 μm displacement, as shown in the lower image of Figure 6.34a. In our gripper, the photomechanical actuation of the actuating arm causes the elongation of this arm and a length difference between the actuating arm and the supporting arm, resulting in the gripper opening. By employing an "actuating arm and supporting arm" double arm structure, the photomechanical actuation was transferred into a horizontal "X-Y" plane as shown by the gripper example, which can be viewed as a motion mechanism compatible with surface micro-machined structures. Theoretically, many electrically driven MEMS structures that involve moveable structures such as micro-motors, micro-gears, micro-stages, and micromirrors can be implemented using this nanotube photomechanical actuation principle. The advantages of this actuation principle are the ease and simplicity of implementation, low power of operation, capability to work in liquid, air, and vacuum environments, capability to operate in a massively parallel fashion using one light source, and the elimination of complicated driving and control circuits compared to MEMS-based electrostatic actuators.

Figure 6.34b shows the gripper tip displacements as a function of light intensity. A nearly linear relationship is found, and there is no saturation of displacement in the whole testing range. When laser light is ~800 mW, a large displacement of ~24 μm is obtained from a gripper of 430 μm in length. This indicates that the performance of the MOMS grippers is comparable with their electrical-driven

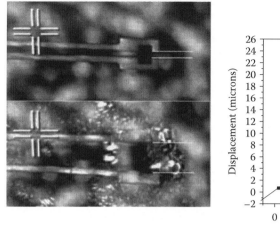

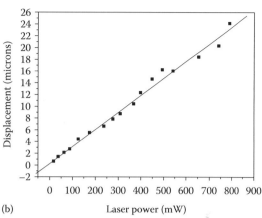

(a)   (b)

FIGURE 6.34 (a) Snapshots showing the closing and opening of the gripper tips under laser illumination. (b) Displacement of the CNT-MOMS gripper tips as a function of the laser illumination. The straight line is the linear fit of the experimental data.

counterparts although only driven by photomechanical actuation from the structures (Kim and Lieber 1999; Nguyen et al. 2004; Chronis and Lee 2005). The light intensity being used for the gripper operation is qualitatively high, ranging from 0.6–20 W/cm², taking into account the small laser spot size after focusing and collimation. Due to the extreme small area of carbon nanotube layers in the gripper (~600 μm²), however, the driving photonic energy is estimated to be as small as 240 μW per gripper during operation. This number is derived under the assumption that all the photonic energy illuminated on the actuating arm is absorbed. Low-power optical sources well focused into small spots would be enough for gripper operation, expanding the potential applications of this MOMS gripper.

Figure 6.35 shows the transient response of the gripper displacement during the light switching period. When the light is switched on, a fast increase of the displacement is witnessed, which follows a compressed exponential function:

$$y = 1 - \exp\left[-\left(\frac{t}{0.105}\right)^2\right] \tag{6.8}$$

When light is switched off, the relaxation of the gripper displacement follows a slower exponential function of time:

$$y = 0.023 + 0.96\exp\left(-\frac{t}{0.102}\right) \tag{6.9}$$

The time constant of the gripper operation is of interest over here. In both "light-on" and "light-off" transitions, it is about ~105 ms, progressively faster than that observed from macro-size samples in Section 6.3. The small sample size may be the dominant factor for the fast actuation in the MOM grippers. To further increase the switching speed of the MOM gripper, one can deliberately increase the alignment of the carbon nanotubes in the gripper structure. To test the reliability of the actuation mechanism in this MOMS gripper, the grippers were operated continuously under laser intensity of 350 mW. The laser pulses had 50% duty cycle and were operated at 5 Hz. Figure 6.36 shows the degradation of the MOMS gripper displacements that were normalized by the initial gripper displacement at the beginning of the testing. The gripper was continuously operated for more than

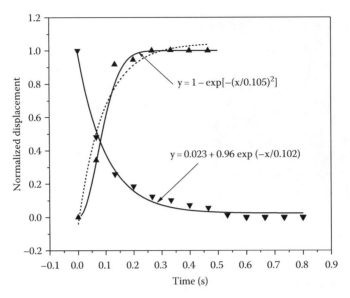

FIGURE 6.35 Transient response of normalized displacement (opening) of the gripper during light switching period. (▲) During light "on" period, the displacement data follow a compressed exponential function: $y = 1 - \exp\left[-\left(t/0.105\right)^2\right]$ as indicated by the solid line, while a simple exponential function cannot fit the data as shown by the dashed line. (▼) During light "off" period, the displacement data follow a simple exponential function: $y = 0.023 + 0.96\exp\left(-t/0.102\right)$.

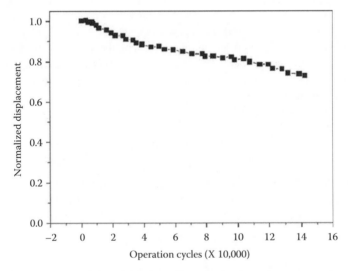

FIGURE 6.36 Stability of the CNT-MOM gripper during continuous operation under 350 mW laser stimuli at 5 Hz actuation rate.

140,000 cycles. The degradation of the gripper displacement was within 28% of the initial value, showing the limited stability of actuation in the nanotube/SU8 polymer structure. Multilayer structured actuators can be developed that can definitely improve the stability of the actuator.

The performance degradation of the gripper may be attributed to the relaxation of carbon nanotube networks as against thermal degradation, the degradation of the contacts between CNT and polymer at their interfaces, and mechanical fatigue of the SU8 polymer. The nanotube networks

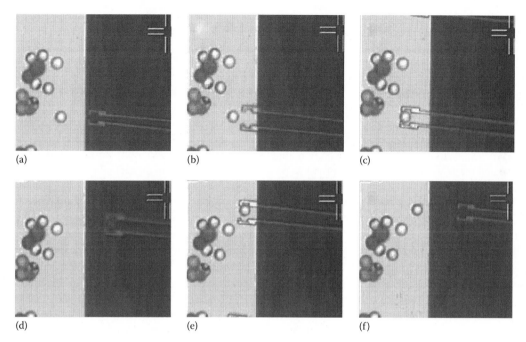

FIGURE 6.37 Manipulating a micro-polystyrene sphere using a MOMS gripper. (a) Closed gripper approaching the micro-sphere, (b) gripper opened under light illumination, (c) gripper closed to grasp the micro-sphere, (d) the micro-sphere lifted from the substrate and transferred to destination, (e) micro-sphere being released to the destination under light illumination on gripper, and (f) gripper removed and closed.

seem to relax their internal stresses slowly and infinitely through sliding at the tube junctions. Such network relaxation could deteriorate the actuation performance after long-time operation. The nanotube network relaxation may be minimized by a chemical functionalization of the nanotubes to induce strong interactions between tube and bundles to keep them in one place instead of sliding. By dynamic control of the light intensity used to drive the gripper, the degradation of the gripper displacements can be dynamically compensated to remain at a constant performance. Low-power light sources when employed to actuate the gripper can improve the stability even further. Apparently, more detailed investigations are needed to study the long-term stability of nanotube optical actuators in relation to the material parameters, the sample processing, and the operation conditions.

We successfully used the gripper to manipulate and position a polystyrene micro-sphere of ~16 μm diameter in air, as shown in Figure 6.37. Figure 6.37a through f shows the sequence of the gripper during manipulation of the micro-sphere. The released microgripper was attached to the arm of a mechanical manipulator, which first approaches the micro-sphere (Figure 6.37a). Upon approaching the micro-sphere, the gripper is then opened by a light stimulus as shown in Figure 6.37b. The gripper is then closed by switching off the light to grasp the objects (Figure 6.37c). Then the micro-sphere is lifted from the substrate and transported to the destination (Figure 6.37d), where the gripper is opened again by the light stimulus to release the object, as indicated in Figure 6.37e. Finally, it is closed and removed from the micro-sphere to finish the manipulation (Figure 6.37f). Although the manipulation shown here is 2-D, the gripper was released and fully free from the substrate unlike most MEMS-based electrostatic grippers that are anchored in plane. By improving the gripper design to accommodate the "Z" direction actuation mechanism, 3-D manipulations are also possible through CNT photomechanical actuation. While this prototype CNT-MOMS gripper is successful in manipulating the microscale objects, structural optimization may further improve the gripper performance and lower the power consumption. Such microoptically actuated grippers are ideal for micro- and nanomanipulation and sensing applications such as nanoparticle assembly and probing and live cell manipulation and sensing.

6.6 SUMMARY

The photomechanical actuation of carbon nanotubes is a rich area for research and device develop-ment at the micro- and nanometer length scales. The study of pure carbon nanotube systems includ-ing (1) unaligned SWCNT films, (2) partially aligned SWCNT fibers, (3) unaligned MWCNT films, and (4) partially aligned MWCNT fibers revealed extraordinary photomechanical responses from the carbon nanotubes. Both partially aligned SWCNT and MWCNT fibers and unaligned SWCNT film exhibited contraction in response to light illumination. Unaligned MWCNT film exhibited compli-cated stress patterns in response to light in the way that an initial fast contraction is followed by a slow sample expansion. Studies of the relaxation behaviors of both single and multi-wall nanotubes have revealed the non-thermal behaviors of the MWCNT network similar to a "sticky" granular network and the polymer-like behaviors from the SWCNT networks, which may help understand the photome-chanical responses in pure nanotube ensembles. Starting from a single-wall carbon nanotube-acrylic multilayer structure, several actuator configurations including carbon nanotube/polymer multilayers and nanotube–polymer nanocomposites for both single and multi-wall nanotubes have been studied. Important features of the "universal" photomechanical response were derived from the experimental characterizations: (1) It is unambiguously shown that all the samples follow similar photomechanical response patterns regardless of the type of nanotubes (single or multi-wall tubes) and regardless of the sample construction (multilayers or nanocomposites). (2) Under small prestrains, the nanotubes are randomly aligned. Photomechanical stress decreases upon illumination, equivalent to sample expan-sion. (3) Under large prestrains, the nanotubes are better aligned. Photomechanical stress increases upon illumination, translating into an equivalent sample contraction. (4) A transition point of pre-strain levels exists for all the samples, where samples show near-zero photomechanical response in response to light. Across this point, the photomechanical actuation changes signs from contrac-tion to expansion or vice versa. This point sits somewhere from 8.5% to 10% for different samples. (5) Photomechanical response increases with the increase of illumination levels, and nearly linear dependences were present for all the samples under almost all the testing prestrain levels. (6) The photomechanical properties of carbon nanotubes–polymer systems respond to light illumination in a broad spectrum covering both the visible and near-infrared regions and show strong dependence on the illumination wavelengths. (7) The photomechanical responses are independent of the types of the host polymer materials and exist in several polymers such as acrylic elastomer, silicone rubber, and SU8 epoxy resins. (8) During the illumination "off-on" transits, the transient dynamic photomechani-cal response follows a faster compressed exponential function than the classical simple exponential function in the illumination "on-off" transits. Comparison of actuation speeds revealed much faster actuation in pure nanotube ensembles compared to that of nanotube–polymer systems. Partially aligned nanotube fibers react to light much faster than that of unaligned nanotube films.

Several possible mechanisms of the photomechanical responses from carbon nanotubes can be identified. The global heating effect is unavoidable; however, it is not the leading contributing mechanism. Photo-induced carrier generation due to direct photon absorption or due to a thermo-electric power effect and subsequent uneven charge redistribution in carbon nanotubes will result in electrostatic nanotube distortions. Direct photomechanical deformations due to polaron generations may induce large lattice strain and most importantly conformational distortion of carbon nanotubes, which may also be a partial contributing mechanism. Local thermal confinement results in extreme thermal effect near nanotubes, which breaks the thermodynamic equilibrium state around nano-tubes. It may affect the way thermal energy contributes to the photomechanical actuation.

The photomechanical responses of carbon nanotubes can lead to the construction of multifunc-tional optically powered microsystems (CNT-MOMS). Multilayered actuator construction was employed because it offers the advantages of (1) easy sample fabrication, (2) intact material proper-ties, and (3) compatibility with the MEMS process. A new carbon nanotube patterning technique was developed from which well-defined high-resolution carbon nanotube patterns were achieved by a combination of nanotube film formation, transferring, photolithography patterning, and an O_2

plasma etching process. SU8 photo epoxy was incorporated with single-wall carbon nanotubes to construct micro nanotube–polymer optical actuators.

As the first example of the concept of CNT-MOMS, a CNT-MOM gripper has been demonstrated which employs the in-plane motion mechanism to drive the gripper. Tip opening of ~24 μm was obtained from a gripper of 430 μm in length under ~800 mW of laser stimulus, showing comparable performance to the electrically driven counterparts. Continuous operation of more than 140,000 cycles was acquired from the grippers. The optical power consumption of the gripper operation was estimated to be as small as ~240 μW. Manipulation of polystyrene micro-spheres in air was demonstrated using the CNT-MOMS grippers. Finally, we would like to conclude that photomechanical actuation of carbon nanotubes is not a fully explored area and therefore has potential for discovery and development based on optical actuation technologies.

Looking into the future, the area of graphene-based photomechanical actuators is yet to be explored. Recent reports show the change in stress of 50 MPa and strains over 100% on infrared-triggered graphene nanocomposites (Huang et al. 2009). Heat-treated graphene papers have recently shown to demonstrate superior hardness, 10 times that of synthetic graphite and twice that of carbon steel; besides, their yielding strength is significantly higher than that of carbon steel (Ranjbartoreh et al. 2011). Graphene paper shows extremely high modulus of elasticity during bending test; in the range of a few terapascal even higher than carbon nanotubes (~1TPA). Their extraordinary thermal conductivity of 3000 W/m/K would be highly useful for composite applications, heat sinks for micro- and nanoelectronics and finally as thermal actuators. The high strength and stiffness of graphene paper is ascribed to the interlocking-tile microstructure of individual graphene nanosheets in the paper unlike carbon nanotube papers that rely on intertube interactions and Van der Wals forces to keep the tubes together. The high stiffness of graphene paper 10–15 N/mm would be quite useful for advanced actuator applications. Recent measurements have shown that a continuously tunable band-gap of up to 250 meV can be generated in biased bilayer graphene, opening up pathway for possible graphene-based nanoelectronic and nanophotonic devices operating at room temperature (Louie and Park 2010). These properties would be highly useful for photomechanical actuation where the actuation can be controlled by tuning the band gap of the graphene layers. Mixing graphene with polymeric materials could also find useful applications as actuators and composites that can be electrically tunable, optically active, and mechanically robust for actuation applications. Patterning graphene papers should follow the same process as we developed for nanotube papers in the past that involves photolithography, oxygen DRIE, and finally XeF_2 etching for release from silicon. When implemented, these graphene-based actuators could be useful for both micro- and nanomanipulation as in nanogrippers, rotational actuators, and nano-cantilevers. Finally, hybrid materials such as liquid-crystal elastomers and solar-absorbing materials that are mixed with both nanotube and graphene could produce new types of actuators in the future that could just absorb sunlight and create useful mechanical work. Such light-absorbing and low-weight polymer composites with extraordinary strength could be highly attractive for future space missions and lengthy interplanetary missions. Therefore, the future of micro- and nano-optomechanical systems based on carbon and their composites is quite bright.

ACKNOWLEDGMENT

The author gratefully acknowledges the support of NSF CAREER AWARD ECCS: 0853066.

REFERENCES

Ahir, S. V., A. M. Squires et al. (2006). Infrared actuation in aligned polymer-nanotube composites. *Physical Review B* **73**(8): 085420.

Ahir, S. V. and E. M. Terentjev (2005). Photomechanical actuation in polymer-nanotube composites. *Nature Materials* **4**: 491–495.

Ahir, S. V. and E. M. Terentjev (2006a). Fast relaxation of carbon nanotubes in polymer composite actuators. *Physical Review Letters* **96**(13): 133902.

Ahir, S. V. and E. M. Terentjev (2006b). Polymers containing carbon nanotubes: Active composite materials. *Polymeric Nanostructures and Their Applications*, H. S. Nalwa, ed. American Scientific Publishers, Valencia, Spain, Vol. 1, p. 1.

Ajayan, P. M. and T. W. Ebbesen (1997). Nanometre-size tubes of carbon. *Reports on Progress in Physics* **60**(10): 1025.

Ajayan, P. M., T. W. Ebbesen et al. (1993). Opening carbon nanotubes with oxygen and implications for filling. *Nature* **362**(6420): 522.

Ajayan, P. M. and S. Lijima (1992). Smallest carbon nanotube. *Nature* **358**(6381): 23.

Ajayan, P. M., M. Terrones et al. (2002). Nanotubes in a flash—Ignition and reconstruction. *Science* **296**(5568): 705.

Ajiki, H. and T. Ando (1994). Aharonov-Bohm effect in carbon nanotubes. *Physica B: Condensed Matter* **201**: 349.

Alvarez, L., A. Righi et al. (2000). Resonant Raman study of the structure and electronic properties of single-wall carbon nanotubes. *Chemical Physics Letters* **316**(3–4): 186.

Antonov, R. D. and A. T. Johnson (1999). Subband population in a single-wall carbon nanotube diode. *Physical Review Letters* **83**(16): 3274.

Bachilo, S. M., M. S. Strano et al. (2002). Structure-assigned optical spectra of single-walled carbon nanotubes. *Science* **298**(5602): 2361–2366.

Bachtold, A., P. Hadley et al. (2001). Logic circuits with carbon nanotube transistors. *Science* **294**(5545): 1317–1320.

Barisci, J. N., G. M. Spinks et al. (2003). Increased actuation rate of electromechanical carbon nanotube actuators using potential pulses with resistance compensation. *Smart Materials and Structures* **12**: 549.

Baughman, R. H. (2003). Materials science: Muscles made from metal. *Science* **300**(5617): 268–269.

Baughman, R. H., C. Cui et al. (1999). Carbon nanotube actuators. *Science* **284**(5418): 1340–1344.

Baughman, R. H., A. A. Zakhidov et al. (2002). Carbon nanotubes—The route toward applications. *Science* **297**(5582): 787–792.

Berber, S., Y.-K. Kwon et al. (2000). Unusually high thermal conductivity of carbon nanotubes. *Physical Review Letters* **84**(20): 4613–4616.

Bethune, D. S., C. H. Klang et al. (1993). Cobalt-catalysed growth of carbon nanotubes with single-atomic-layer walls. *Nature* **363**(6430): 605.

Blase, X., L. X. Benedict et al. (1994). Hybridization effects and metallicity in small radius carbon nanotubes. *Physical Review Letters* **72**(12): 1878.

Blase, X., J. C. Charlier et al. (1999). Boron-mediated growth of long helicity-selected carbon nanotubes. *Physical Review Letters* **83**(24): 5078.

Bower, C., W. Zhu et al. (2000). Plasma-induced alignment of carbon nanotubes. *Applied Physics Letters* **77**(6): 830.

Bradley, K., S.-H. Jhi et al. (2000). Is the intrinsic thermoelectric power of carbon nanotubes positive? *Physical Review Letters* **85**(20): 4361–4364

Bretz, M., B. G. Demczyk et al. (1994). Structural imaging of a thick-walled carbon microtubule. *Journal of Crystal Growth* **141**(1–2): 304.

Camacho-Lopez, M., H. Finkelmann et al. (2004). Fast liquid-crystal elastomer swims into the dark. *Nature Materials* **3**: 307–310.

Cao, A., P. L. Dickrell et al. (2005). Super-compressible foamlike carbon nanotube films. *Science* **310**(5752): 1307–1310.

Charlier, J. C., T. W. Ebbesen et al. (1996). Structural and electronic properties of pentagon-heptagon pair defects in carbon nanotubes. *Physical Review B* **53**(16): 11108.

Charlier, J. C., X. Gonze et al. (1995). *Europhysics Letters* **29**: 43.

Charlier, J. C. and J. P. Michenaud (1993). Energetics of multilayered carbon tubules. *Physical Review Letters* **70**(12): 1858.

Che, J., Çagin, T., and Goddard, W. A. III, (2000). Thermal conductivity of carbon nanotube, *Nanotechnology* **11**(2), 65–70. DOI:10.1088/0957-4484/11/2/305.

Chen, Z., J. Appenzeller et al. (2006). An integrated logic circuit assembled on a single carbon nanotube. *Science* **311**(5768): 1735.

Chen, R. J., S. Bangsaruntip et al. (2003). Noncovalent functionalization of carbon nanotubes for highly specific electronic biosensors. *Proceedings of the National Academy of Sciences of the United States of America* **100**(9): 4984–4989.

Chen, J., V. Perebeinos et al. (2005). Bright infrared emission from electrically induced excitons in carbon nanotubes. *Science* **310**(5751): 1171.

Chen, P., X. Wu et al. (1999). Electronic structure and optical limiting behavior of carbon nanotubes. *Physical Review Letters* **82**(12): 2548–2551.

Chico, L., L. X. Benedict et al. (1996a). Quantum conductance of carbon nanotubes with defects. *Physical Review B* **54**(4): 2600.

Chico, L., V. H. Crespi et al. (1996b). Pure carbon nanoscale devices: Nanotube heterojunctions. *Physical Review Letters* **76**(6): 971.

Choi, W. B., D. S. Chung et al. (1999). Fully sealed, high-brightness carbon-nanotube field-emission display. *Applied Physics Letters* **75**(20): 3129–3131.

Choi, W. B., Y. W. Jin et al. (2001). Electrophoresis deposition of carbon nanotubes for triode-type field emission display. *Applied Physics Letters* **78**(11): 1547–1549.

Chronis, N. and L. P. Lee (2005). Electrothermally activated SU-8 microgripper for single cell manipulation in solution. *Journal of Microelectromechanical Systems* **14**(4): 857.

Chu, S. Y. and K. Uchino (1994). Impurity doping effect on photostriction in PLZT ceramics. *Journal of Advanced Performance Materials* **1**: 129–143.

Collins, P. G., M. S. Arnold et al. (2001). Engineering carbon nanotubes and nanotube circuits using electrical breakdown. *Science* **292**(5517): 706.

Collins, P. G., K. Bradley et al. (2000). Extreme oxygen sensitivity of electronic properties of carbon nanotubes. *Science* **287**(5459): 1801–1804.

Courty, S., J. Mine et al. (2003). Nematic elastomers with aligned carbon nanotubes: New electromechanical actuators. *Europhysics Letters* **64**(5): 654–660.

Culshaw, B. (1996). *Smart Structures and Materials*, Artech House Publishers, Norwood, MA.

Cumings, J. and A. Zettl (2000). Low-friction nanoscale linear bearing realized from multiwall carbon nanotubes. *Science* **289**(5479): 602–604.

Dai, H. (2002). Carbon nanotubes: Synthesis, integration, and properties. *Accounts of Chemical Research* **35**(12): 1035.

Datskos, P. G., S. Rajic et al. (1998). Photoinduced and thermal stress in silicon microcantilevers. *Applied Physics Letters* **73**(16): 2319.

Derycke, V., R. Martel et al. (2001). Carbon nanotube inter- and intramolecular logic gates. *Nano Letters* **1**(9): 453.

Dragoman, D. and M. Dragoman (1999). optical actuation of micromechanical tunneling structures with applications in spectrum analysis and optical computing. *Applied Optics* **38**(32): 6773–6778.

Dresselhaus, M. S., G. Dresselhaus et al. (1992). Carbon fibers based on C_{60} and their symmetry. *Physical Review B* **45**(11): 6234.

Dresselhaus, M. S., G. Dresselhaus et al. (2001). *Carbon Nanotubes Synthesis, Structure, Properties, and Applications*, Springer, Berlin, Germany.

Dunlap, B. I. (1994). Relating carbon tubules. *Physical Review B* **49**(8): 5643.

Durkop, T., B. M. Kim et al. (2004). Properties and applications of high-mobility semiconducting nanotubes. *Journal of Physics: Condensed Matter* **16**(18): R553.

Ebbesen, T. W. and T. Takada (1995). Topological and SP3 defect structures in nanotubes. *Carbon* **33**(7): 973.

Endo, M., K. Takeuchi et al. (1993). The production and structure of pyrolytic carbon nanotubes (PCNTs). *Journal of Physics and Chemistry of Solids* **54**(12): 1841.

Endo, M., K. Takeuchi et al. (1995). Pyrolytic carbon nanotubes from vapor-grown carbon fibers. *Carbon* **33**(7): 873.

Falvo, M. R., G. J. Clary et al. (1997). Bending and buckling of carbon nanotubes under large strain. *Nature* **389**(6651): 582.

Fan, S., M. G. Chapline et al. (1999). Self-oriented regular arrays of carbon nanotubes and their field emission properties. *Science* **283**(5401): 512–514.

Fennimore, A. M., T. D. Yuzvinsky et al. (2003). Rotational actuators based on carbon nanotubes. *Nature* **424**(6947): 408.

Finkelmann, H., E. Nishikawa et al. (2001). A new opto-mechanical effect in solids. *Physical Review Letters* **87**(1): 015501.

Frank, S., P. Poncharal et al. (1998). Carbon nanotube quantum resistors. *Science* **280**(5370): 1744.

Freitag, M., Y. Martin et al. (2003). Photoconductivity of single carbon nanotubes. *Nano Letters* **3**(8): 1067–1071.

Fridkin, W. M. (1980). *Photoferroelectrics*, Springer, Berlin, Germany.

Fujii, M., X. Zhang et al. (2005). Measuring the thermal conductivity of a single carbon nanotube. *Physical Review Letters* **95**(6): 065502.

Fujiwara, A., Y. Matsuoka et al. (2001). Photoconductivity in semiconducting single-walled carbon nanotubes. *Japanese Journal of Applied Physics* **40**(11B): L1229–L1231.

Gao, B., G. Z. Yue et al. (2001). Fabrication and electron field emission properties of carbon nanotube films by electrophoretic deposition. *Advanced Materials* **13**(23): 1770–1773.

Gatos, H. C. and J. Lagowski (1973). Surface photovoltage spectroscopy—A new approach to the study of high-gap semiconductor surfaces. *Journal of Vacuum Science and Technology* **10**(1): 130.

Grigorian, L., G. U. Sumanasekera et al. (1999). Giant thermopower in carbon nanotubes: A one-dimensional Kondo system. *Physical Review B* **60**(16): R11309–R11312.

Groning, O., O. M. Kuttel et al. (2000). Field emission properties of carbon nanotubes. *Journal of Vacuum Science and Technology B: Microelectronics and Nanometer Structures* **18**(2): 665–678.

Hassanien, A., M. Tokumoto et al. (2005). Selective etching of metallic single-wall carbon nanotubes with hydrogen plasma. *Nanotechnology* **16**(2): 278.

Heer, W. A. d. (2004). Nanotubes and the pursuit of applications. *MRS Bulletin* **29**(4): 281.

Heinze, S., J. Tersoff et al. (2002). Carbon nanotubes as schottky barrier transistors. *Physical Review Letters* **89**(10): 106801.

Heyd, R., A. Charlier et al. (1997). Uniaxial-stress effects on the electronic properties of carbon nanotubes. *Physical Review B* **55**(11): 6820.

Hipplera, H., A.-N. Unterreiner et al. (2004). Evidence of ultrafast optical switching behaviour in individual single-walled carbon nanotubes. *Physical Chemistry Chemical Physics* **6**(9): 2387–2390.

Hirsch, L. R., R. J. Stafford et al. (2003). Nanoshell-mediated near-infrared thermal therapy of tumors under magnetic resonance guidance. *Proceedings of the National Academy of Sciences of the United States of America* **100**(23): 13549–13554.

Hogan, P. M., A. R. Tajbakhsh et al. (2002). UV manipulation of order and macroscopic shape in nematic elastomers. *Physical Review E* **65**(4): 041720.

Homma, Y., T. Yamashita et al. (2002). Interconnection of nanostructures using carbon nanotubes. *Physica B: Condensed Matter* **323**(1–4): 122.

Hone, J., I. Ellwood et al. (1998). Thermoelectric power of single-walled carbon nanotubes. *Physical Review Letters* **80**(5): 1042–1045.

Hone, J., M. Whitney et al. (1999a). Thermal conductivity of single-walled carbon nanotubes. *Physical Review B* **59**(4): R2514–R2516.

Hone, J., A. Zettl et al. (1999b). Thermal conductivity of single-walled carbon nanotubes. *Synthetic Metals* **103**(1–3): 2498.

Huang, S., X. Cai et al. (2003). Growth of millimeter-long and horizontally aligned single-walled carbon nanotubes on flat substrates. *Journal of the American Chemical Society* **125**(19): 5636.

Huang, Y., J. J. Liang et al. (2009). Infrared-triggered actuators from graphene-based nanocomposites. *Journal of Physical Chemistry C* **113**(22): 9921–9927.

Hugel, T., N. B. Holland et al. (2002). Single-molecule optomechanical cycle. *Science* **296**(5570): 1103–1106.

Husain, A., J. Hone et al. (2003). Nanowire-based very-high-frequency electromechanical resonator. *Applied Physics Letters* **83**(6): 1240.

Ichida, M., S. Mizuno et al. (1999). Exciton effects of optical transitions in single-wall carbon nanotubes. *Journal of the Physical Society of Japan* **68**: 3131–3133.

Ichida, M., S. Mizuno et al. (2004). Anisotropic optical properties of mechanically aligned single-walled carbon nanotubes in polymer. *Applied Physics A* **78**: 1117–1120.

Iijima, S. (1980). Direct observation of the tetrahedral bonding in graphitized carbon black by high resolution electron microscopy. *Journal of Crystal Growth* **50**(3): 675.

Iijima, S. (1991). Helical microtubules of graphitic carbon. *Nature* **354**: 56–58.

Iijima, S., C. Brabec et al. (1996). Structural flexibility of carbon nanotubes. *The Journal of Chemical Physics* **104**(5): 2089.

Iijima, S. and T. Ichihashi (1993). Single-shell carbon nanotubes of 1-nm diameter. *Nature* **363**: 603–615.

Ikeda, T., M. Nakano et al. (2003). Anisotropic bending and unbending behavior of azobenzene liquid-crystalline gels by light exposure. *Advanced Materials* **15**(3): 201–205.

Itkis, M. E., F. Borondics et al. (2006). Bolometric infrared photoresponse of suspended single-walled carbon nanotube films. *Science* **312**(5772): 413.

Jacobs-Cook, A. J. (1996). MEMS versus MOMS from a systems point of view. *Journal of Micromechanics and Microengineering* **6**(1): 148.

Javey, A., J. Guo et al. (2003). Ballistic carbon nanotube field-effect transistors. *Nature* **424**(6949): 654.

Jiang, H., B. Liu et al. (2004). Thermal expansion of single wall carbon nanotubes. *Journal of Engineering Materials and Technology* **126**(3): 265.

Jing, K., Z. Chongwu et al. (2000). Alkaline metal-doped n-type semiconducting nanotubes as quantum dots. *Applied Physics Letters* **77**(24): 3977–3979.

Jones, B. E. and J. S. McKenzie (1993). A review of optical actuators and the impact of micromachining. *Sensors and Actuators A: Physical* **37–38**: 202.

Jorio, A., R. Saito et al. (2004). Carbon nanotube photophysics. *MRS Bulletin* **29**(4): 276.

Joseph, W. (2005). Carbon-nanotube based electrochemical biosensors: A review. *Electroanalysis* **17**(1): 7–14.

Jung, M. S., Y. K. Ko et al. (2005). Electrical and field-emission properties of chemically anchored single-walled carbon nanotube patterns. *Applied Physics Letters* **87**(1): 013114.

Kaiser, A. B., G. Dusburg et al. (1998). Heterogeneous model for conduction in carbon nanotubes. *Physical Review B* **57**(3): 1418.

Kaneto, K., M. Kaneko et al. (1995). Artificial muscle: Electromechanical actuators using polyaniline films. *Synthetic Metals* **71**(1–3): 2211.

Kasuya, A., Y. Sasaki et al. (1997). Evidence for size-dependent discrete dispersions in single-wall nanotubes. *Physical Review Letters* **78**(23): 4434.

Kataura, H., Y. Kumazawa et al. (1999). Optical properties of single-wall carbon nanotubes. *Synthetic Metals* **103**(1–3): 2555.

Kazaoui, S., N. Minami et al. (1999). Amphoteric doping of single-wall carbon-nanotube thin films as probed by optical absorption spectroscopy. *Physical Review B* **60**(19): 13339.

Kempe, C., M. Rutloh et al. (2003). Photo-orientation of azobenzene side chain polymers parallel or perpendicular to the polarization of red HeNe light. *Journal of Physics: Condensed Matter* **15**(11): S813.

Keren, K., R. S. Berman et al. (2003). DNA-templated carbon nanotube field-effect transistor. *Science* **302**(5649): 1380–1382.

Kiang, C.-H., W. A. Goddard et al. (1995). Carbon nanotubes with single-layer walls. *Carbon* **33**(7): 903.

Kim, P. and C. M. Lieber (1999). Nanotube nanotweezers. *Science* **286**(5447): 2148–2150.

Kim, P., L. Shi et al. (2001). Thermal transport measurements of individual multiwalled nanotubes. *Physical Review Letters* **87**(21): 215502.

Koehler, D. R. (1997). Optical actuation of micromechanical components. *Journal of the Optical Society of America B* **14**(9): 2197.

Koerner, H., G. Price et al. (2004). Remotely actuated polymer nanocomposites—Stress-recovery of carbon-nanotube-filled thermoplastic elastomers. *Nature Materials* **3**: 115–120.

Kofod, G., P. Sommer-Larsen et al. (2003). Actuation response of polyacrylate dielectric elastomers. *Journal of Intelligent Material Systems and Structures* **14**(12): 787–793.

Kong, J., M. G. Chapline et al. (2001). Functionalized carbon nanotubes for molecular hydrogen sensors. *Advanced Materials* **13**(18): 1384–1386.

Kong, J., N. R. Franklin et al. (2000). Nanotube molecular wires as chemical sensors. *Science* **287**(5453): 622.

Kovacs, G. T. (1998). *Micromachined Transducers Sourcebook*, McGraw-Hill, New York.

Krecmer, P., A. M. Moulin et al. (1997). Reversible nanocontraction and dilatation in a solid induced by polarized light. *Science* **277**(5333): 1799–1802.

Krishnan, A., E. Dujardin et al. (1998). Young's modulus of single-walled nanotubes. *Physical Review B* **58**(20): 14013–14019.

Krupke, R., F. Hennrich et al. (2003). Separation of metallic from semiconducting single-walled carbon nanotubes. *Science* **301**(5631): 344.

Kwon, Y.-K., S. Berber et al. (2005). Kwon, Berber, and Tomanek Reply. *Physical Review Letters* **94**(20): 209702.

Lagowski, J., I. Baltov et al. (1973). Surface photovoltage spectroscopy and surface piezoelectric effect in GaAs. *Surface Science* **40**(2): 216.

Lagowski, J. and H. C. Gatos (1974). Photomechanical vibration of thin crystals of polar semiconductors. *Surface Science* **45**(2): 353.

Lagowski, J., A. Morawski et al. (1974). Stress-induced amplification of the photovoltaic effect in non-centrosymmetric semiconductors: Cds. *Surface Science* **45**(1): 325.

Lambin, P., A. Fonseca et al. (1995). Structural and electronic properties of bent carbon nanotubes. *Chemical Physics Letters* **245**(1): 85.

Landi, B. J., R. P. Raffaelle et al. (2002). Single wall carbon nanotube-nafion composite actuators. *Nano Letters* **2**(11): 1329.

Landi, B. J., R. P. Raffaelle et al. (2005). Single-wall carbon nanotube-polymer solar cells. *Progress in Photovoltaics: Research and Applications* **13**(2): 165–172.

Lauret, J. S., C. Voisin et al. (2004). Third-order optical nonlinearities of carbon nanotubes in the femtosecond regime. *Applied Physics Letters* **85**(16): 3572–3574.

Lee, H. K., K. Doi et al. (2000). Light-scattering-mode optical switching and image storage in polymer/liquid crystal composite films by means of photochemical phase transition. *Polymer* **41**(5): 1757.

Lehman, J. H., C. Engtrakul et al. (2005). Single-wall carbon nanotube coating on a pyroelectric detector. *Applied Optics* **44**(4): 483–488.

Levitsky, I. A. and W. B. Euler (2003). Photoconductivity of single-wall carbon nanotubes under continuous-wave near-infrared illumination. *Applied Physics Letters* **83**(9): 1857–1859.

Li, C. and T.-W. Chou (2003). Single-walled carbon nanotubes as ultrahigh frequency nanomechanical resonators. *Physical Review B* **68**(7): 073405.

Li, C. and T.-W. Chou (2005). Axial and radial thermal expansions of single-walled carbon nanotubes. *Physical Review B* **71**(23): 235414.

Li, Y., D. Mann et al. (2004a). Preferential growth of semiconducting single-walled carbon. *Nano Letters* **4**(2): 317–321. DOI: 10.1021/nl035097c.

Li, S., Z. Yu et al. (2004b). Carbon nanotube transistor operation at 2.6 GHz. *Nano Letters* **4**(4): 753.

Lien, D.-H., H.-F. Kuo et al. (2006). Segmentation of single-walled carbon nanotubes by camera flash. *Applied Physics Letters* **88**(9): 093113.

Lin, M. F. and K. W. K. Shung (1994). Plasmons and optical properties of carbon nanotubes. *Physical Review B* **50**(23): 17744.

Lin, Y., S. Taylor et al. (2004). Advances toward bioapplications of carbon nanotubes. *Journal of Materials Chemistry* **14**: 527–541.

Liu, J., A. G. Rinzler et al. (1998). Fullerene pipes. *Science* **280**(5367): 1253–1256.

Louie, S. G. and C. H. Park (2010). Tunable excitons in biased bilayer graphene. *Nano Letters* **10**(2): 426–431.

Lourie, O., D. M. Cox et al. (1998). Buckling and collapse of embedded carbon nanotubes. *Physical Review Letters* **81**(8): 1638.

Lu, S., Y. Liu et al. (2007a). Nanotube micro-opto-mechanical systems. *Nanotechnology* **18**(6): 065501.

Lu, S. and B. Panchapakesan (2005). Optically driven nanotube actuators. *Nanotechnology* **16**(11): 2548.

Lu, S. and B. Panchapakesan (2006a). Hybrid platinum/single-wall carbon nanotube nanowire actuators: metallic artificial muscles. *Nanotechnology* **17**(3): 888.

Lu, S. and B. Panchapakesan (2006b). Nanotube micro-optomechanical actuators. *Applied Physics Letters* **88**(25): 253107.

Lu, S. and B. Panchapakesan (2006c). Photoconductivity in single wall carbon nanotube sheets. *Nanotechnology* **17**(8): 1843.

Lu, S., K. Sivakumar et al. (2007b). Sonochemical synthesis of platinum nanowires and their applications as electro-chemical actuators. *Journal of Nanoscience and Nanotechnology* **7**(5): 1–7.

Ma, W. and L. E. Cross (2004). An experimental investigation of electromechanical response in a dielectric acrylic elastomer. *Applied Physics A* **78**: 1201–1204.

Madou, M. J. (2002). *Fundamentals of Microfabrication: The Science of Miniaturization*, CRC Press, New York.

Maniwa, Y., R. Fujiwara et al. (2001a). Thermal expansion of single-walled carbon nanotube (SWNT) bundles: X-ray diffraction studies. *Physical Review B* **64**(24): 241402.

Maniwa, Y., R. Fujiwara et al. (2001b). Multiwalled carbon nanotubes grown in hydrogen atmosphere: An x-ray diffraction study. *Physical Review B* **64**(7): 073105.

Martel, R., T. Schmidt et al. (1998). Single- and multi-wall carbon nanotube field-effect transistors. *Applied Physics Letters* **73**(17): 2447–2449.

Martínez, M. T., M. A. Callejas et al. (2003). Modifications of single-wall carbon nanotubes upon oxidative purification treatments. *Nanotechnology* **14**(7): 691.

Mauricio, T. (2005). Controlling nanotube chirality and crystallinity by doping. *Small* **1**(11): 1032–1034.

Mazzoni, M. S. C. and H. Chacham (2000a). Atomic restructuring and localized electron states in a bent carbon nanotube: A first-principles study. *Physical Review B* **61**(11): 7312.

Mazzoni, M. S. C. and H. Chacham (2000b). Bandgap closure of a flattened semiconductor carbon nanotube: A first-principles study. *Applied Physics Letters* **76**(12): 1561.

McKenzie, J. S., C. Clark et al. (1995). Design and construction of a multiple-wafer integrated micromechanical optical actuator. *Sensors and Actuators A: Physical* **47**(1–3): 566.

Mintmire, J. W., B. I. Dunlap et al. (1992). Are fullerene tubules metallic? *Physical Review Letters* **68**(5): 631.

Misewich, J. A., R. Martel et al. (2003). Electrically induced optical emission from a carbon nanotube FET. *Science* **300**(5620): 783–786.

Mohite, A., S. Chakraborty et al. (2005). Displacement current detection of photoconduction in carbon nanotubes. *Applied Physics Letters* **86**(6): 061114.

Moonsub, S. and P. S. Giles (2003). Photoinduced conductivity changes in carbon nanotube transistors. *Applied Physics Letters* **83**(17): 3564–3566.

Mounet, N. and N. Marzari (2005). First-principles determination of the structural, vibrational and thermodynamic properties of diamond, graphite, and derivatives. *Physical Review B* **71**(20): 205214.

Naciri, J., A. Srinivasan et al. (2003). Nematic elastomer fiber actuator. *Macromolecules* **36**(22): 8499.

Naeemi, A. and J. D. Meindl (2005). Impact of electron-phonon scattering on the performance of carbon nanotube interconnects for GSI. *IEEE Electron Device Letters* **26**(7): 476–478.

Nguyen, N. T., S. S. Ho et al. (2004). A polymeric microgripper with integrated thermal actuators. *Journal of Micromechanics and Microengineering* **14**(7): 969.

Nikolaev, P., A. Thess et al. (1997). Diameter doubling of single-wall nanotubes. *Chemical Physics Letters* **266**(5–6): 422.

Oakley, J. S., H. T. Wang et al. (2005). Carbon nanotube films for room temperature hydrogen sensing. *Nanotechnology* **16**(10): 2218.

Oberlin, A., M. Endo et al. (1976a). Filamentous growth of carbon through benzene decomposition. *Journal of Crystal Growth* **32**(3): 335.

Oberlin, A., M. Endo et al. (1976b). High resolution electron microscope observations of graphitized carbon fibers. *Carbon* **14**(2): 133.

O'Connell, M. J., S. M. Bachilo et al. (2002). Band gap fluorescence from individual single-walled carbon nanotubes. *Science* **297**(5581): 593–596.

Odom, T. W., J.-L. Huang et al. (1998). Atomic structure and electronic properties of single-walled carbon nanotubes. *Nature* **391**(6662): 62.

Oh, S. J., Y. Cheng et al. (2003). Room-temperature fabrication of high-resolution carbon nanotube field-emission cathodes by self-assembly. *Applied Physics Letters* **82**(15): 2521–2523.

Osman, M. A. and D. Srivastava (2001). Temperature dependence of the thermal conductivity of single-wall carbon nanotubes. *Nanotechnology* **12**: 21.

Osvath, Z., G. Vertesy et al. (2005). Atomically resolved STM images of carbon nanotube defects produced by Ar[sup +] irradiation. *Physical Review B* **72**(4): 045429.

Panchapakesan, B., S. Lu et al. (2005). Single-wall carbon nanotube nanobomb agents for killing breast cancer cells. *NanoBiotechnology* **1**(2): 133–140.

Park, C.-J., Y.-H. Kim et al. (1999). Band-gap modification by radial deformation in carbon nanotubes. *Physical Review B* **60**(15): 10656.

Parvez, S. A. (1994). Solar pressure disturbance on GSTAR and SPACENET satellites. *Journal of Spacecraft and Rockets* **31**(3): 482–488.

Pasquier, A. D., H. E. Unalan et al. (2005). Conducting and transparent single-wall carbon nanotube electrodes for polymer-fullerene solar cells. *Applied Physics Letters* **87**(20): 203511.

Pelrine, R., R. Kornbluh et al. (2000). High-speed electrically actuated elastomers with strain greater than 100%. *Science* **287**(5454): 836–839.

Piegari, E., V. Cataudella et al. (2002). Comment on polarons in carbon nanotubes. *Physical Review Letters* **89**(4): 049701.

Pimenta, M. A., A. Marucci et al. (1998). Raman modes of metallic carbon nanotubes. *Physical Review B* **58**(24): R16016.

Pitsillides, C. M., E. K. Joe et al. (2003). Selective cell targeting with light-absorbing microparticles and nanoparticles. *Biophysical Journal* **84**(6): 4023–4032.

Poncharal, P., Z. L. Wang et al. (1999). Electrostatic deflections and electromechanical resonances of carbon nanotubes. *Science* **283**(5407): 1513.

Poosanaas, P., K. Tonooka et al. (2000). Photostrictive actuators. *Mechatronics* **10**(4–5): 467–487.

Postma, H. W. C., T. Teepen et al. (2001). Carbon nanotube single-electron transistors at room temperature. *Science* **293**(5527): 76–79.

Prasad, D., L. Zhiling et al. (2004). Nanotube film based on single-wall carbon nanotubes for strain sensing. *Nanotechnology* **15**(3): 379.

Punit, K., M. Wirtz. C. Martin. (2004). Nanotube membrane based biosensors. *Electroanalysis* **16**(1–2): 9–18.

Qian, D., G. J. Wagner et al. (2002). Mechanics of carbon nanotubes. *Applied Mechanics Reviews* **55**(6): 495.

Ranjbartoreh, A. R., B. Wang et al. (2011). Advanced mechanical properties of graphene paper. *Journal of Applied Physics* **109**(1): 014306.

Rao, A. M., E. Richter et al. (1997). Diameter-selective Raman scattering from vibrational modes in carbon nanotubes. *Science* **275**(5297): 187.

Reich, S., C. Thomsen et al. (2004). *Carbon Nanotubes: Basic Concepts and Physical Properties*, Wiley-VCH, Berlin, Germany.

Ren, Z. F., Z. P. Huang et al. (1998). Synthesis of large arrays of well-aligned carbon nanotubes on glass. *Science* **282**(5391): 1105–1107.

Robert, J. C., R. F. Nathan et al. (2001). Molecular photodesorption from single-walled carbon nanotubes. *Applied Physics Letters* **79**(14): 2258–2260.

Robertson, D. H., D. W. Brenner et al. (1992). Energetics of nanoscale graphitic tubules. *Physical Review B* **45**: 12592–12595.

Roch, I., B. Ph et al. (2003). Fabrication and characterization of an SU-8 gripper actuated by a shape memory alloy thin film. *Journal of Micromechanics and Microengineering* **13**(2): 330.

Rochefort, A., D. R. Salahub et al. (1998). The effect of structural distortions on the electronic structure of carbon nanotubes. *Chemical Physics Letters* **297**(1–2): 45.

Romero, H. E., G. U. Sumanasekera et al. (2002). Thermoelectric power of single-walled carbon nanotube films. *Physical Review B* **65**(20): 205410.

Rueckes, T., K. Kim et al. (2000). Carbon nanotube-based nonvolatile random access memory for molecular computing. *Science* **289**(5476): 94.

Ruoff, R. S. and D. C. Lorents (1995). Mechanical and thermal properties of carbon nanotubes. *Carbon* **33**(7): 925.

Sada, T., M. Inoue et al. (1987). Photostriction in PLZT ceramics. *Journal of the Ceramic Society of Japan International Edition* **95**: 499–504.

Saito, R., G. Dresselhaus et al. (1996). Tunneling conductance of connected carbon nanotubes. *Physical Review B* **53**(4): 2044.

Saito, R., G. Dresselhaus et al. (1998). *Physical Properties of Carbon Nanotubes*, Imperial College Press, London, U.K.

Saito, R., G. Dresselhaus et al. (2000). Trigonal warping effect of carbon nanotubes. *Physical Review B* **61**(4): 2981.

Saito, R., M. Fujita et al. (1992). Electronic structure of chiral graphene tubules. *Applied Physics Letters* **60**(18): 2204.

Sarkisov, S. S., M. J. Curley et al. (2004). Photomechanical effect in films of polyvinylidene fluoride. *Applied Physics Letters* **85**(14): 2747.

Sarkisov, S. S., M. J. Curley et al. (2006). Light-driven actuators based on polymer films. *Optical Engineering* **45**(3): 034302.

Savage, T., S. Bhattacharya et al. (2003). Photoinduced oxidation of carbon nanotubes. *Journal of Physics: Condensed Matter* **15**(35): 5915.

Sazonova, V., Y. Yaish et al. (2004). A tunable carbon nanotube electromechanical oscillator. *Nature* **431**(7006): 284.

Schelling, P. K. and P. Keblinski (2003). Thermal expansion of carbon structures. *Physical Review B* **68**(3): 035425.

Schwarzschild, M. (1958). *Structure and Evolution of the Stars*, Princeton University Press, Princeton, NJ.

Sfeir, M. Y., T. Beetz et al. (2006). Optical spectroscopy of individual single-walled carbon nanotubes of defined chiral structure. *Science* **312**(5773): 554.

Sheng, C. X., Z. V. Vardeny et al. (2005). Exciton dynamics in single-walled nanotubes: Transient photoinduced dichroism and polarized emission. *Physical Review B* **71**(12): 125427.

Shi Kam, N. W., M. O'Connell et al. (2005). Carbon nanotubes as multifunctional biological transporters and near-infrared agents for selective cancer cell destruction. *Proceedings of the National Academy of Sciences of the United States of America* **102**(33): 11600–11605.

Smela, E., O. Inganas et al. (1995). Controlled folding of micrometer-size structures. *Science* **268**(5218): 1735.

Smits, J., B. Wincheski et al. (2003). Response of Fe powder, purified and as-produced HiPco single-walled carbon nanotubes to flash exposure. *Materials Science and Engineering A* **358**(1–2): 384.

Soh, H. T., C. F. Quate et al. (1999). Integrated nanotube circuits: Controlled growth and ohmic contacting of single-walled carbon nanotubes. *Applied Physics Letters* **75**(5): 627.

Sohn, J. I., S. Lee et al. (2001). Patterned selective growth of carbon nanotubes and large field emission from vertically well-aligned carbon nanotube field emitter arrays. *Applied Physics Letters* **78**(7): 901–903.

Spinks, G. M., G. G. Wallace et al. (2002). Pneumatic carbon nanotube actuators. *Advanced Materials* **14**(23): 1728–1732.

Stuchlik, M., P. Krecmer et al. (2001). Opto-mechanical effect in chalcogenide glasses. *Journal of Optoelectronics and Advanced Materials* **3**(2): 361–366.

Sulfridge, M., T. Saif et al. (2002). Optical actuation of a bistable MEMS. *Journal of Microelectromechanical Systems* **11**(5): 574.

Sulfridge, M., T. Saif et al. (2004). Nonlinear dynamic study of a bistable MEMS: Model and experiment. *Journal of Microelectromechanical Systems* **13**(5): 725.

Suski, J., D. Largeau et al. (1990). Optically activated ZnO/Sio2/Si cantilever beams. *Sensors and Actuators A: Physical* **24**(3): 221.

Tabib-Azar, M. (1998). *Microactuators: Electrical, Magnetic, Thermal, Optical, Mechanical, Chemical and Smart Structures*, Springer, New York.

Tabiryan, N., U. Hrozhyk et al. (2004). Nonlinear refraction in photoinduced isotropic state of liquid crystalline azobenzenes. *Physical Review Letters* **93**(11): 113901.

Tabiryan, N., S. Serak et al. (2005). Polymer film with optically controlled form and actuation. *Optics Express* **13**(19): 7442–7448.

Tahhan, M., V.-T. Truong et al. (2003). Carbon nanotube and polyaniline composite actuators. *Smart Materials and Structures* **12**(4): 626.

Takagi, K., S. Kikuchi et al. (2004). Ferroelectric and photostrictive properties of fine-grained PLZT ceramics derived from mechanical alloying. *Journal of the American Ceramic Society* **87**(8): 1477–1482.

Takenobu, T., T. Takahashi et al. (2006). High-performance transparent flexible transistors using carbon nanotube films. *Applied Physics Letters* **88**(3): 033511.

Tang, Y., H. Cong et al. (2004). Thermal expansion of a composite of single-walled carbon nanotubes and nanocrystalline aluminum. *Carbon* **42**(15): 3260.

Tans, S. J., A. R. M. Verschueren et al. (1998). Room-temperature transistor based on a single carbon nanotube. *Nature* **393**(6680): 49.

Terrones, M. (2003). Science and technology of the twenty-first century: Synthesis, properties, and applications of carbon nanotubes. *Annual Review of Materials Research* **33**(1): 419.

Thakoor, S., P. Poosanaas et al. (1998). Optical microactuation in piezoceramics. *Smart Structures and Materials 1998: Smart Electronics and MEMS*, SPIE, San Diego, CA.

Tian, M., F. Li et al. (1998). Thermoelectric power behavior in carbon nanotubule bundles from 4.2 to 300 K. *Physical Review B* **58**(3): 1166–1168.

Treacy, M. M. J., T. W. Ebbesen et al. (1996). Exceptionally high Young's modulus observed for individual carbon nanotubes. *Nature* **381**(6584): 678.

Tsang, S. C., P. J. F. Harris et al. (1993). Thinning and opening of carbon nanotubes by oxidation using carbon dioxide. *Nature* **362**(6420): 520.

Tsutsumi, O., Y. Demachi et al. (1998). Photochemical phase-transition behavior of polymer liquid crystals induced by photochemical reaction of azobenzenes with strong donor-acceptor pairs. *Journal of Physical Chemistry B* **102**(16): 2869–2874.

Uchino, K. (1989). Micro walking machine using piezoelectric actuators. *Journal of Robotics and Mechatronics* **124**: 44–47.

Uchino, K. (1997). New applications of photostriction. *Materials Research Innovations* **1**(3): 63–168. DOI: 10.1007/s100190050036.

Valentini, L., I. Armentano et al. (2003). Sensors for sub-ppm NO[sub 2] gas detection based on carbon nanotube thin films. *Applied Physics Letters* **82**(6): 961–963.

Venkatesh, S. and S. Novak (1987). Micromechanical resonators in fiber-optic systems. *Optics Letters* **12**: 129.

Verissimo-Alves, M., R. B. Capaz et al. (2001). Polarons in carbon nanotubes. *Physical Review Letters* **86**(15): 3372.

Wadhawan, A., D. Garrett et al. (2003). Nanoparticle-assisted microwave absorption by single-wall carbon nanotubes. *Applied Physics Letters* **83**(13): 2683–2685.

Wang, F., G. Dukovic et al. (2004a). Time-resolved fluorescence of carbon nanotubes and its implication for radiative lifetimes. *Physical Review Letters* **92**(17): 177401.

Wang, F., G. Dukovic et al. (2005). The optical resonances in carbon nanotubes arise from excitons. *Science* **308**(5723): 838.

Wang, Y., K. Kempa et al. (2004b). Receiving and transmitting light-like radio waves: Antenna effect in arrays of aligned carbon nanotubes. *Applied Physics Letters* **85**(13): 2607–2609.

Wang, Z. L. and J. Song (2006). Piezoelectric nanogenerators based on zinc oxide nanowire arrays. *Science* **312**(5771): 242.

Warner, M., K. P. Gelling et al. (1988). Theory of nematic networks. *Journal of Chemical Physics* **88**(6): 4008.

Wei, B. Q., R. Vajtai et al. (2001). Reliability and current carrying capacity of carbon nanotubes. *Applied Physics Letters* **79**(8): 1172.

Weissmuller, J., R. N. Viswanath et al. (2003). Charge-induced reversible strain in a metal. *Science* **300**(5617): 312–315.

White, C. T., D. H. Robertson et al. (1993). Helical and rotational symmetries of nanoscale graphitic tubules. *Physical Review B* **47**(9): 5485.

Williams, P. A., S. J. Papadakis et al. (2002). Torsional response and stiffening of individual multiwalled carbon nanotubes. *Physical Review Letters* **89**(25): 255502.

Wiltshire, J. G., L.-J. Li et al. (2005). Chirality-dependent boron-mediated growth of nitrogen-doped single-walled carbon nanotubes. *Physical Review B* **72**(20): 205431.

Wong, E. W., P. E. Sheehan et al. (1997). Nanobeam mechanics: Elasticity, strength, and toughness of nanorods and nanotubes. *Science* **277**(5334): 1971–1975.

Wu, Z., Z. Chen et al. (2004). Transparent, conductive carbon nanotube films. *Science* **305**(5688): 1273–1276.

Xuchun, L., S. Jinhai et al. (1999). Third-order optical nonlinearity of the carbon nanotubes. *Applied Physics Letters* **74**(2): 164–166.

Yakobson, B. I., C. J. Brabec et al. (1996). Nanomechanics of carbon tubes: Instabilities beyond linear response. *Physical Review Letters* **76**: 2511.

Yang, J., T. Ono et al. (2000). Surface effects and high quality factors in ultrathin single-crystal silicon cantilevers. *Applied Physics Letters* **77**(23): 3860.

Yaroshchuk, O., A. D. Kiselev et al. (2001). Spatial reorientation of azobenzene side groups of a liquid crystalline polymer induced by linearly polarized light. *The European Physical Journal E - Soft Matter* **6**: 57–67.

Yosida, Y. (2000). High-temperature shrinkage of single-walled carbon nanotube bundles up to 1600 K. *Journal of Applied Physics* **87**(7): 3338.

Young-Kyun, K., B. Savas et al. (2004). Thermal contraction of carbon fullerenes and nanotubes. *Physical Review Letters* **92**(1): 015901.

Yu, M.-F., B. S. Files et al. (2000a). Tensile loading of ropes of single wall carbon nanotubes and their mechanical properties. *Physical Review Letters* **84**: 5552–5555.

Yu, M.-F., O. Lourie et al. (2000b). Strength and breaking mechanism of multiwalled carbon nanotubes under tensile load. *Science* **287**(5453): 637–640.

Yu, Y., M. Nakano et al. (2003). Photomechanics directed bending of a polymer film by light. *Nature* **425**(6954): 145.

Yu, Y., M. Nakano et al. (2004). Effect of cross-linking density on photoinduced bending behavior of oriented liquid-crystalline network films containing azobenzene. *Chemistry of Materials* **16**(9): 1637–1643.

Zanardi Ocampo, J. M., P. O. Vaccaro et al. (2004). Characterization of GaAs-based micro-origami mirrors by optical actuation. *Microelectronic Engineering* **73–74**: 429.

Zhang, Y., A. Chang et al. (2001). Electric-field-directed growth of aligned single-walled carbon nanotubes. *Applied Physics Letters* **79**(19): 3155.

Zhang, Y. and S. Iijima (1999). Elastic response of carbon nanotube bundles to visible light. *Physical Review Letters* **82**(17): 3472–3475.

Zhang, G., P. Qi et al. (2006). Selective etching of metallic carbon nanotubes by gas-phase reaction. *Science* **314**(5801): 974.

Zhang, X. F., X. B. Zhang et al. (1993). Carbon nano-tubes; their formation process and observation by electron microscopy. *Journal of Crystal Growth* **130**(3–4): 368.

Zhdanov, V. G., B. T. Kolomiets et al. (1979). Photoinduced optical anisotropy in chalcogenide vitreous semiconducting films. *Physica Status Solidi (a)* **52**(2): 621–626.

Zheng, M., A. Jagota et al. (2003a). DNA-assisted dispersion and separation of carbon nanotubes. *Nature Materials* **2**: 338–342.

Zheng, M., A. Jagota et al. (2003b). Structure-based carbon nanotube sorting by sequence-dependent DNA assembly. *Science* **302**(5650): 1545.

7 Light-Induced Phase Transition of Gels for Smart Functional Elements

Atsushi Suzuki

CONTENTS

7.1 INTRODUCTION

Polymer gels are known to exist in two distinct phases, swollen and collapsed, and to exhibit unique properties, particularly with respect to the volume phase transition and critical behavior [1–14]. The transition between two phases is defined as a reversible, discontinuous volume change in response to infinitesimal changes in the external conditions, that is, physical and chemical stimuli (Figure 7.1). It has been observed in various gels made of synthetic and natural polymers.

As for experimental studies of the phase transition gels, it is a fundamental technique to observe the transition by measuring the changes in volume [11–13]. This is because the observed volume should be reflected by the average network structure, and the macroscopic volume change should be determined by the interactions between the monomers; the phase transition can exaggerate the microscopic conformation changes at the molecular level. In other words, a small change in the external conditions can induce a macroscopic volume change. Variables that trigger the phase transition include the temperature, solvent composition, pH, ionic composition, electric field, light, and particular molecules [4,8–10,13–22].

Among the parameters to induce the volume phase transition in gels, visible light is readily available, inexpensive, safe, clean, and easily manipulated to change the gel volume in contrast to other parameters, such as the solvent composition and electric field. It can also be delivered in specific amounts with high accuracy. This function of light-sensitive gels is of technological importance in developing various applications in not only the engineering but also the biochemical and biomedical fields.

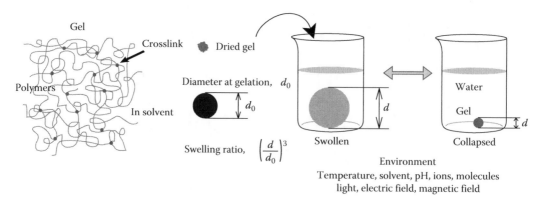

FIGURE 7.1 Network structure of gels. Swelling behavior of gels and volume phase transition in response to external conditions. The swelling ratio is defined here.

When a gel is illuminated by light, the light can be transmitted, absorbed, or scattered in general. If the system is designed so that the absorption is essential, the volume phase transition in gels can be induced; a thermoresponsive gel with a chromophore exhibits a local volume phase transition upon illumination with visible light. In 1990, a light-sensitive poly(*N*-isopropylacrylamide) gel (NIPA gel), impregnated slightly with the trisodium salt of copper chlorophyllin (chromophore), was first prepared to demonstrate the phase transition due to the local heating of the polymer network upon illumination with visible light [17] (Figure 7.2). In accordance with the equation of states of gels on the basis of the Flory–Huggins mean field theory [23], a phenomenological model was derived to predict the light-induced phase transition of gels. When the light-sensitive gel is illuminated by visible light, the chromophore absorbs the light energy, which subsequently dissipates as heat to the local environment, thereby inducing an increase in the local temperature. The incremental rise of the local temperature is proportional to the incident light intensity as well as the concentration of chromophores. This simple model successfully explained the light-sensitive phase transition of gels.

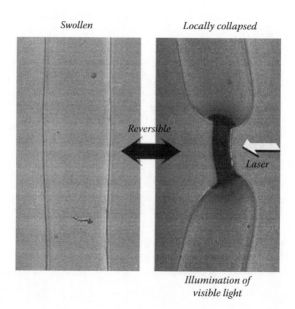

FIGURE 7.2 Light-induced phase transition of cylindrical gels with submillimeter diameter.

This chapter deals with the studies of the phase transition of gels in response to external stimuli and the mechanism of the induction of a phase transition due to local heating of the gel by illumination of visible light. The method to control the states of switch on (collapsed) and switch off (swollen) by light is presented. The states were observed in gels with a hysteresis during the phase transition. Volume change induced by visible light might be applicable to various optical devices, such as optical switches, display units, and optical actuators. Possible applications optically driven by light will be presented here on the basis of the principles.

7.2 VOLUME PHASE TRANSITION OF GELS

7.2.1 PHYSICAL APPROACH

An equilibrium swelling curve of gels was successfully described on the basis of the Flory–Huggins mean field theory [23]; the swelling curve is given by equaling the osmotic pressure of a gel to zero, and the equation of states of gels is described as follows [1,2]:

$$\tau \equiv 1 - \frac{\Delta F}{k_B T} = -\frac{\nu_e \nu_1}{\phi^2}\left\{(2f+1)\left(\frac{\phi}{\phi_0}\right)-2\left(\frac{\phi}{\phi_0}\right)^{1/3}\right\}+1+\frac{2}{\phi}+\frac{2\ln(1-\phi)}{\phi^2} \tag{7.1}$$

where

  T is the absolute temperature
  ΔF is the free-energy decrease associated with the formation of contact between polymer segments ($\Delta F=2k_B T\chi$, where χ is the polymer-solvent interaction parameter, called the Flory–Huggins χ parameter [23])
  k_B is the Boltzmann constant
  ν_1 is the volume of a water molecule
  ν_e is the total number of the effective polymer chains in the gel
  ϕ is the polymer network density
  ϕ_0 is the density at Θ temperature where the gel network has a random walk configuration
  f is the number of counterions per chain

This equation is the relationship between χ and ϕ if the materials parameters ν_1, νe, ϕ_0, and f are given. In other words, all factors which affect the equilibrium state of gels are simply included in χ. The swelling curves calculated using this formula satisfactorily describe the qualitative swelling properties of gels [10,20].

Among the phase transition gels, chemically cross-linked NIPA gel and its copolymer gels in water undergo a phase transition in response to very small temperature changes [11]: NIPA is thermoresponsive and has a lower critical solution temperature (LCST) with the cloud point at around 31°C [24], and weakly cross-linked NIPA gel can exhibit the volume phase transition at around 33.6°C in water only if the temperature is changed [11]. To calculate the swelling curve of this thermoresponsive gel, the following simple modification was used. If the temperature dependence of ΔF is assumed to be (ΔH–$T\Delta S$), where ΔS and ΔH are the entropy and enthalpy of polymer-polymer contact, respectively, the relation between T and ϕ is expressed as follows [12]:

$$\frac{1}{T} = \frac{\Delta S}{\Delta H} + \frac{k_B}{\Delta H}\left[\frac{\nu_e \nu_1}{\phi^2}\left\{(2f+1)\left(\frac{\phi}{\phi_0}\right)-2\left(\frac{\phi}{\phi_0}\right)^{1/3}\right\}-\frac{2}{\phi}-\frac{2\ln(1-\phi)}{\phi^2}\right] \tag{7.2}$$

For NIPA gels, the interaction parameters ΔS and ΔH are negative [12], and the gels are swollen at lower temperatures and collapsed at higher temperatures. The transition mechanism of NIPA gels has been attributed to the change in the balance of hydrophilic and hydrophobic interactions, and such multiple factors are represented by ΔF (i.e., χ) in this expression. Although the model is very simple and the internal and external conditions are not explicitly included in the equation, many swelling properties of gels have been quite successfully described by the equation of states of gels [23]. It is not necessary to consider the uniqueness of molecules or emphasize on particular molecules in the polymer network within this physical approach.

7.2.2 CHEMICAL APPROACH

From a chemical point of view, on the other hand, it is absolutely important to consider the unique character of every chemical element and relate it to an actual property of gels. This is because the gel phase transition is the result of a competitive balance between a repulsive force that acts to expand the polymer network and an attractive force that acts to shrink the network. The most effective repulsive force is the electrostatic interaction between polymer charges of the same kind, which can be imposed upon a gel by introducing ionization into the network. The osmotic pressure by counterions adds to the expanding pressure. The attractive interactions can be van der Waals, hydrophobic interaction, ionic interaction with opposite kinds, and hydrogen bonds. The phase transition driven mainly by each one of the fundamental biological interactions was well established [25], which allows us to draw a general picture of how polymers interact with each other through these interactions.

The fundamental interactions noted earlier have been believed to play an essential role in biology regarding the structures and specific functions of macromolecules and their assemblies. The magnitude, temperature dependence, and behavior in an aqueous environment are totally different among the fundamental forces, and it is those differences that allow the variety in the biological functions created by these forces. Understanding how these forces determine the phase behavior of polymers is of fundamental importance. A detailed examination of the gel phase behavior provides a deep insight into these problems [26]. The knowledge on the chemical fundamentals of the gel phase transition will play the role of guiding principles for a wide variety of technological applications of gels as functional elements.

7.2.3 REENTRANT VOLUME PHASE TRANSITIONS

In addition to a simple phase transition in response to changes in external stimuli, the reentrant volume phase transition was reported in acrylamide derivative copolymer gels and NIPA gels in dimethyl sulfoxide (DMSO)-water or methanol-water mixed solvents [11,27–30]. It is noteworthy that the reentrant volume phase transition was reflected by the cononsolvent behavior of polymers [31] since the swelling behavior can exaggerate the microscopic conformation changes at the molecular level. The driving force of cononsolvent behavior is the strong interaction between water and DMSO or methanol to form the stable hydrates, which excludes carbonyl groups of NIPA in the vicinity of the hydrate [32]; therefore, the gel shrank in the mixture. Very recently, it was suggested that the cosolvent behavior of polymers could realize the reentrant volume phase transition in the case of chemically cross-linked poly(vinyl alcohol) gel (PVA gel) [33].

Moreover, the intermediate phase and the re-swelling phase transition have been reported in a closed system on polyelectrolyte hydrogels [34–36]. For example, the swelling ratio and the phase transition behavior of strongly ionized NIPA gels co-polymerized with sodium acrylate were extensively investigated under the continuous exchange of solvent water. The size and the transition behavior were easily and precisely controlled by the number of water exchanges at room temperature. The re-swelling transition was observed on the heating process in closed systems, and the gel state could be reversibly controlled by the combination of a water exchange and a temperature change. These phenomena have been interpreted in terms of the exchange of counterions by protons and the formation and destruction of the dimeric form of a hydrogen bond between the carboxyl and carbonyl groups. The intermediate phase and the re-swelling phase transition are believed to be common properties of hydrogels when they have the carboxyl groups in their polymer networks to form hydrogen bonds.

7.3 LOCAL HEATING OF THERMORESPONSIVE POLYMER NETWORKS

The original copolymer gels consist of NIPA (main polymer constituent, Kohjin Co., Ltd.), trisodium salt of coppered chlorophyllin (chromophore, Aldrich Chem.), and N,N'-methylene-bis-acrylamide (BIS, cross-linker, Wako Pure Chem.) and were prepared by a free radical copolymerization in water under nitrogen atmosphere at ice temperature [17]. The pregel solution was brought into the thin capillaries with capillary action. After gelation was completed, the gels were taken out of the capillaries and immersed in a large amount of deionized, distilled water to wash away residual chemicals and unreacted monomers from the polymer networks. Pure NIPA gels made with standard concentrations undergo a discontinuous phase transition [11]. In contrast, the present system showed a continuous volume change in response to the temperature change. This continuous change may be reasonable since the incorporation of chromophore increases the hydrophobic interaction and requires a larger amount of BIS to form a network, both of which decrease the sharpness of the volume transition in spite of the increase of the ionization by the carboxyl groups in the chromophore.

The thin cylindrical gel with a 100 µm diameter was placed in water, and the temperature was precisely regulated. The light of a wavelength of 480 nm from an Argon ion laser was used as the light source. The radiation power of light at the gel was adjusted using a polarizer and ranged from 0 to 130 mW. The incident beam was focused on the gel with a lens to produce a focused beam and a focal depth. Two-minute illumination was sufficient for the gel to reach a swelling equilibrium. In order to avoid unnecessary bleaching of the chromophore, illuminations with microscope incandescent light and with laser light were used only when the measurements were carried out.

As shown in Figure 7.3a, there are two interesting and noteworthy features in the observation of the light-induced phase transition of the gel. First, the originally continuous transition without light became discontinuous upon illumination of light. Second, the transition temperature was lowered upon light illumination. These two features can be explained in terms of the equation of states of gels, which is given in Equation 7.2. For the sake of simplicity, let us rewrite Equation 7.2 as

$$T = T(\phi) \tag{7.3}$$

This function is monotonically increasing with network density ϕ and has a large Maxwell's loop for large ionization parameter f. When the gel is illuminated by light, the chromophore absorbs the light energy and the local temperature rises. The increment of the temperature at the immediate vicinity of polymer chains should be proportional to the intensity of the incident light as well as to the concentration of the chromophore and, therefore, to the polymer network density ϕ. It may thus be assumed that

$$T = T_0 + \alpha I \phi \tag{7.4}$$

where
  T_0 is the ambient temperature
  I is the intensity of light
  α is a constant

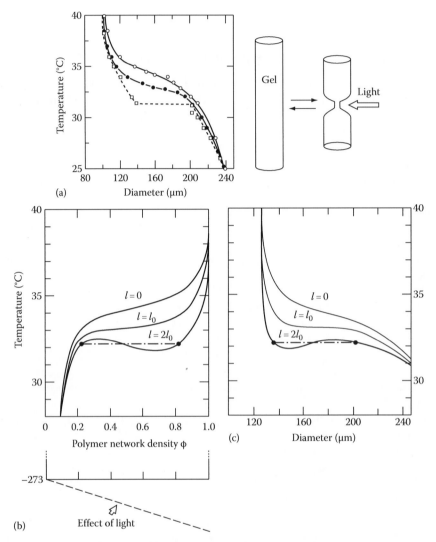

FIGURE 7.3 (a) Diameter of *N*-isopropylacrylamide-chlorophyllin copolymer gel as a function of the temperature. The open circles are without light illumination. The solid circles are under illumination of argon ion laser light (wavelength 480 nm) of intensity 60 mW, which was sharply focused on the gel with an approximate focal diameter of 20 μm. The triangles are with light of 120 mW. The solvent was NaOH solution with pH 11.9. (b) Theoretical swelling curves of a gel, polymer density versus temperature, for different light intensities are constructed using the equations shown in the text. The swelling curve on the top is without light, the second with light of a specific intensity, and the third with twice as much light as that in the second. The effect of light illumination is to add a linear line in proportion to the polymer density with a negative slope, as shown in the lower portion of the figure. The results agree excellently with those obtained in experimental observations. (c) Theoretical swelling curves of a gel, diameter versus temperature, are plotted for different light intensities constructed using the equations in the text. (Reproduced with permission from Macmillan Publishers Ltd., *Nature*, Suzuki, A. and Tanaka, T., Phase transition in polymer gels induced by visible light, 346, 345–347, 1990, copyright 1990.)

Combining Equations 7.3 and 7.4, we have the relation between the controlled temperature and the gel density ϕ:

$$T_0 = T(\phi) - \alpha I \phi \qquad (7.5)$$

Equation 7.5 shows that the light illumination enhances Maxwell's loop and, therefore, brings the gel state deeper into the unstable region, inducing the discontinuous transition. At the same time, it lowers the transition temperature. The theoretical swelling curves, which make use of the earlier equations for various illumination intensities, and the contribution by light ($+\alpha I \varphi$) to the swelling curves are shown schematically in Figure 7.3b.

The transformation of the swelling curves from the use of the polymer network density to that of the gel diameter is straightforward and is given by equation

$$d = \left(\frac{\phi_0}{\phi} \right)^{1/3} d_0 \qquad (7.6)$$

where d_0 is the gel diameter at Θ temperature. The swelling curves using the diameter are also plotted in Figure 7.3c.

The phase transition induced by visible light was semiqualitatively described using a simple phenomenological model based on the mean field theory [17]. All of these observations of light-sensitive gels are based on the local heating of polymer networks of gels: heat (photon absorbency) raises the local temperature of a thermoresponsive gel and thereby the osmotic pressure within the gel, which causes the phase transition. The phenomenological model was applicable to develop different types of gels with advanced flexibility and higher functional performance by selecting polymer, a solvent, and a light source. In fact, an infrared light-induced volume phase transition was reported on the NIPA gel [37,38]; the transmission of visible light could be controlled by an infrared light, which was also attributed to the local heating mechanism of the polymer networks. This is an example of the application of the principle of the phase transition induced by the local heating of polymer networks.

As for the relaxation time of the swelling and shrinking processes, the speed in the stimuli-induced phase transition of gels based on chemical reactions is slow, since these processes are governed by ionization, ion diffusion, and recombination with molecules. This could be a drawback of the system when used for technological applications. In contrast, in the systems presented here, the phase transition is induced by a local temperature increment, which is established as a balance of the absorption of light and heat dissipation. Reaching such a stationary state is much faster than the reactions involving ionization and ion migration. Since thermal diffusion is much faster than collective diffusion of the gel network, the swelling or shrinking process of the gel in response to light is solely governed by the motion of the polymer network. The time course of the change of diameter was plotted in a semilogarithmic scale. In accordance to the kinetic theory of collective diffusion of a gel, there is a fast decay followed by a single exponential change with time [4,13]. For 1 μm-diameter gels, the response time is expected to be approximately 10 ms, since the time is proportional to the square of the diameter.

7.4 POTENTIAL APPLICATIONS OF LIGHT-SENSITIVE HYDROGELS

7.4.1 OPTICAL SHUTTER

By using the mechanism, it is possible to design an optical shutter, which can change the transmitted light intensity in response to the strength of irradiated light intensity. For simplicity, NIPA-chlorophyllin copolymer gels were prepared in thin plate form as follows. A set of slides with thin

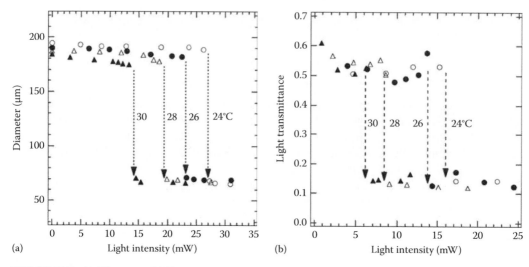

FIGURE 7.4 (a) Diameter of N-isopropylacrylamide-chlorophyllin copolymer gel as a function of illuminated light intensity, I_0, at each of the fixed temperatures, 24°C, 26°C, 28°C, and 30°C. (b) Transmittance ratio, I/I_0, through the thin film N-isopropylacrylamide-chlorophyllin copolymer gels is plotted as a function of the illuminated light intensity, I_0. Each arrow indicates the discontinuous decrease of d and I/I_0 when I_0 was increased.

spacers was inserted into the pregel solution, and the solution was brought into the thin spaces via capillary elevation. The gelation was carried out in the thin spaces at ice temperature overnight. After gelation was complete, one slide glass was removed from the thin plate gel that had adhered to the other glass using a bind-coupling chemical. The gels were then immersed in a large amount of deionized, distilled water to wash away residual chemicals and unreacted monomers from the polymer networks. Using a similar setup to that described in Section 7.3, the laser light was illuminated along the normal direction to the surface of the plate gels. As a result, the gel could change the transmitted light intensity in response to the strength of the irradiated light intensity, as shown in Figure 7.4.

According to the Lambert–Beer equation, the transmitted light intensity, I, in the present system is expressed by

$$I = I_0 \exp\left(-\alpha_0 \phi l\right) \tag{7.7}$$

where

α_0 is a constant, proportional to the molar absorption coefficient of the chromophore
l is the thickness of the gel

If $(\alpha_0 \phi l)$ is negligibly small, the sum of the absorption and scattering of light energy is proportional to $(I_0 \phi l)$

$$I_0 - I \cong \alpha_0 I_0 \phi l \tag{7.8}$$

It is noteworthy that, in the present experiment, the beam diameter is constant and one of the gel surfaces is adhered to the glass plate; therefore, the increase of ϕ by the illumination should be compensated for by a decrease in l if the network shrinks along the direction normal to the surface. However, the situation is much more complicated than this prediction since the gel is a three-dimensional polymer network, that is, a continuous medium. The experimental observations indicated that the increase of ϕ by the illumination of light is much larger than the decrease of l; therefore, the expression of the temperature increment in Equation 7.4 is applicable to this system.

To understand the implications of light-sensitive hydrogels, it was essential to establish the physical basis of the phase behavior under a mechanical constraint [39–41] as well as the dynamic properties or the kinetics of the structural change of gels. Generally, the relaxation time of gels under a mechanical constraint in the swelling and shrinking processes is larger than that of free gels [39–41]. However, the kinetics of the gel under a constraint in the present case, in which only one surface is mechanically constrained, depends on the thickness since the gel has a large free surface. Therefore, the response time of the shutter can be controlled by adjusting the thickness of the gel.

7.4.2 Optical Switching and Memory

In the previous samples, the gel underwent a sharp but continuous volume change at approximately body temperature. Upon illumination of weak light, the volume change was continuous, but the transition temperature was lowered. With strong light illumination, the gel showed a discontinuous volume transition at lower temperature. The effect of light was the largest near the transition temperature; therefore, it is desirable to use a phase transition gel, which exhibits a discontinuous volume change without light illumination.

NIPA or BIS molecules are not ionizable in water, and their contributions to the osmotic pressure of gels are not expected to be influenced by pH. A chlorophyllin molecule has three carboxyl groups, on the other hand, which are ionized in the pure water used in the earlier experiment (pH = ca. 5.8). Nevertheless, due to the large hydrophobic interaction in a chlorophyllin molecule, the volume change becomes continuous, as described in the previous section. To develop a gel that exhibits several degrees of swelling, sodium acrylate was incorporated in NIPA-chlorophyllin gels, and the swelling curves were determined for the gels in pure water as a function of pH. As a result, a hysteretic behavior and irreversibility of the polymer gels were found by pH change [42], which were qualitatively described on the basis of a simple Landau model [43]. Moreover, the volume phase transitions of the ionic NIPA-chlorophyllin gels were governed by the extent of hydrophobic interactions or the degree of hydrogen bonds induced between the polymers within the gel in response to varying either the temperature/visible light intensity or the pH, respectively. In other words, within the ionic gel, the pH of the system mainly affects the degree of hydrogen bonds, whereas the temperature and/or visible light intensity determines the extent of the hydrophobic interactions. Independently of the external condition used to induce a volume phase transition, the ionic gel exhibited a large hysteresis upon undergoing a reversible transition between the swollen and collapsed states (Figure 7.5a). Hysteresis was indicative of a first-order phase transition of the gel and the existence of two free-energy minima separated by a free-energy barrier.

On the basis of the findings of these unique changes in the swelling ratio in response to temperature and pH, a gel system, which was a covalently cross-linked network of NIPA, sodium acrylate, and a chromophore, was developed with a switch function that undergoes a local shrinking transition upon illumination with visible light that is stable even after the light source is removed (Figure 7.5b); a soft, optically transparent material using polymer gels was developed, which can be not only activated by visible light ("switched on") but also deactivated ("switched off") by altering the local environment using three different means: pH, temperature, and light [43]. In each system, between the transitions for swelling and shrinking, the gel can show either a swollen or a collapsed state, which can be selected according to the history of the variables. By making use of this phenomenon, the phase in which a gel exists has been successfully controlled with visible light. Without light illumination, the gel remains in a swollen state. Upon illumination that exceeds a specific threshold intensity, however, a volume transition is locally induced, thereby forming a material that is in a state of coexistence with both phases and stable for at least several hours after removal of the light source. It has been concluded that, at appropriate

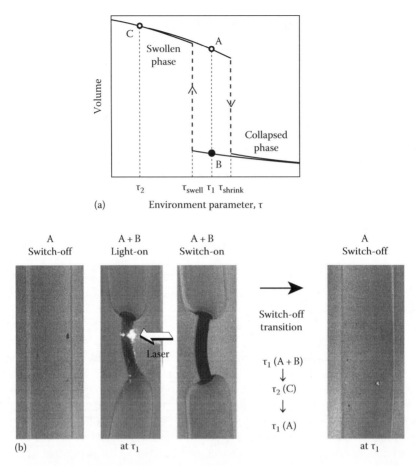

FIGURE 7.5 (a) Illustration of a swelling curve with a large hysteresis in response to the external parameter, τ (pH, temperature, and the light itself). (b) Photographic sequences showing how the gel can act analogous to a switch that is turned on by visible light and turned off by pH; the initial swollen state (first photograph) at pH = 7.4 between the swelling and shrinking transition pH can enter a coexistence state (switch on, second photograph) in response to light illumination from the right to left direction. The state is stable for several hours after the illumination is removed (third photograph). Swelling occurs at greater pH than pHswell (fourth photograph) and returns to the initial swollen state at pH = 7.4 (switch off).

locations within the volume phase diagram between the swelling and shrinking transitions, the locally collapsed state would be stable when the collapsed state remains in the local potential minimum; in other words, it would be stable enough not to enter the swollen state as a result of the external stress [43]. It would, therefore, be possible to control the phase using these phenomena by manipulation of visible light exposure.

It is noteworthy that the size of a possible switching structure is an important parameter. If the size of the collapsed gel is comparable to the beam diameter of light, the whole gel should shrink without showing the coexistence state. This is the real switch "on" and switch "off" of gels. One possible application based on this concept is the use of microscopic domains and hierarchical network structures (Figure 7.6), which have been observed experimentally on surfaces and in the internal structure of NIPA gel [44–46]. The mesoscopic structure of the network was constructed of highly cross-linked microgels connected by a loose network. The unique structure is attributed to the permanent inhomogeneity in the bulk NIPA gel, which can be introduced by the gelation process. The magnitude of the structures should change during the volume phase transitions [47,48]. Domains

Tapping mode AFM images

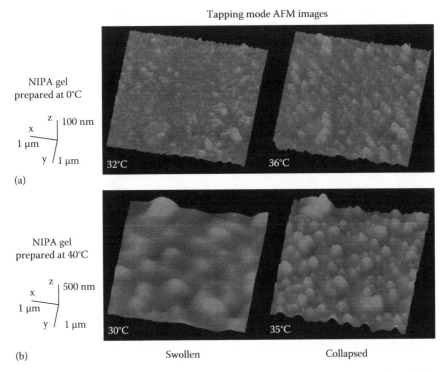

NIPA gel
prepared at 0°C

(a)

NIPA gel
prepared at 40°C

(b)   Swollen   Collapsed

FIGURE 7.6 AFM images (5×5 μm^2) in water of the 0°C and 40°C gels (prepared at below (0°C) and above (40°C) the cloud point of NIPA, respectively); (a) the 0°C gel at 32°C (swollen) and 36°C (collapsed) and (b) the 40°C gel at 30°C (swollen) and 35°C (collapsed). The vertical scales are greatly exaggerated to clearly display the spatial amplitude in the z-direction. Sponge-like domains are visible on a submicrometer scale. The domains indicated by the tops of the white tetragonals correspond to the same domains. (Reproduced with permission from Suzuki, A., Yamazaki, M., Kobiki, Y., and Suzuki, H., 1997. Surface domains and roughness of polymer gels observed by atomic force microscopy. *Macromolecules* 30: 2350–2354.)

of a submicrometer scale could be strongly affected by the nature of the gel network as well as the external conditions, including the condition of mechanical constraint. The reported results show that the microstructures of polymer gels in solvent, as well as the nanometer-scale structural changes, are associated with the gel phase transition. A submicrometer switch can be designed in a gel if the submicrometer domains can be controlled by the visible light. When the illuminated light is focused on a small portion of the gel network, a domain can shrink to a stable collapsed phase; this is similar to the submillimeter switch. In this way, an "on" or an "off" command could be stored or erased in a specific domain of the gel, which would serve as a "memory" or "display."

In order to realize this concept more effectively, the effects of guest NIPA microparticles on the phase transition of bulk gels of thermostable polyacrylamide (AAm gel) were examined by measuring the macroscopic swelling curves [48]. The guest particles were spheres of NIPA gel with a submicrometer diameter, which can be easily prepared by an emulsion-polymerized reaction in water [48]. The presence of guest particles does not strongly affect the swelling behavior in bulk gels when the concentration of incorporated particles does not exceed a threshold. These microparticles undergo a local volume phase transition in the host network only if the temperature is changed. Therefore, if the microparticles are incorporated by optically absorbing a chromophore, the phase transition could be induced by illumination of visible light. After elucidating the conditions to realize hysteresis and the effects of mechanical constraints on the swelling ratio of microparticles, this hybrid gel can have an optical memory function.

7.4.3 Optically Driven Actuators and Other Possible Applications

Light-sensitive hydrogels have been providing additional opportunities to design new actuators, transducers, controlled release systems, and selective pumps that can be triggered without disturbing the system. One example is a smart valve to control a water flow triggered by visible light. According to the literature [49–53], the friction between the polymer network of NIPA and water can be controlled by changing water temperature. The water flow in a gel corresponds with the prediction of the water flow in a capillary based on the Hagen–Poiseuille equation [51–56], assuming that the network consists of N microcapillaries per unit area ($N=n/S$, where n is the number of capillaries and S is the cross-sectional area) with inner diameter ξ. The flux of water, Q (volumetric flow rate; water volume per unit area per second), in a microcapillary can be expressed as

$$Q = N \times \left(\frac{\Delta p}{\alpha l} \right) \frac{\pi (\xi / 2)^4}{8\eta} \tag{7.9}$$

where
$\quad \Delta p$ is the applied pressure
$\quad l$ is the macroscopic length of the gel
$\quad \eta$ is the dynamic viscosity of the water

The length αl indicates the average distance of the total flow in the gel; therefore, $\alpha > 1$. Using the average flow velocity, u, Q can also be expressed as $N\pi(\xi/2)^2 u$; therefore, the following simple relation is obtained:

$$\Delta p = \frac{32\alpha\eta}{\xi^2} ul \equiv ful, \quad \text{where } f = \frac{32\alpha\eta}{\xi^2} \propto \frac{\eta}{\xi^2} \tag{7.10}$$

It is evident that u is proportional to Δp and inversely to f and l, while f depends on η, α, and ξ, but not on the gel size, S, and l. This equation was originally derived to model a steady laminar flow of an incompressible fluid through a cylindrical tube with a rigid wall. Although the real network structure is much more complicated and totally different from the capillary model, it is reasonable to consider the network of gels as a bundle of microcapillaries of inner diameter ξ (Figure 7.7a). The flow velocity of water can be controlled by changing water temperature when the gel is constrained mechanically in the flow path (Figure 7.7b). In this situation, the gel was mechanically constrained onto the inner surface of the microcapillary; therefore, large inhomogeneity should be introduced in the cross section of the network normal to the flow direction when the gel enters the collapsed state by the illumination of light. According to the earlier model, the inhomogeneity should enhance the water flow, since f is inversely proportional to ξ^2. In the present system, the inhomogeneity of a polymer network introduced at gelation as well as the total polymer density is responsible for determining the average pore size and, therefore, the absolute value of f. On the other hand, the effect of the mechanical constraint on the network structure can determine the change in f during the phase transition.

To examine the response time, the flow velocity was measured as a function of time in the respective isothermal process after a steplike temperature change. As a result, the flow velocity instantly changed and reached a steady state within several minutes, and the absolute velocity was reversibly changed [53]. The characteristic time, evidently, did not depend on the gel size but on the local conformation change. Although it is difficult to have a decisive picture to explain the origin of the shorter relaxation time, it could be related to the cooperative change of the length of the water path by the temperature-induced deformation of polymer networks.

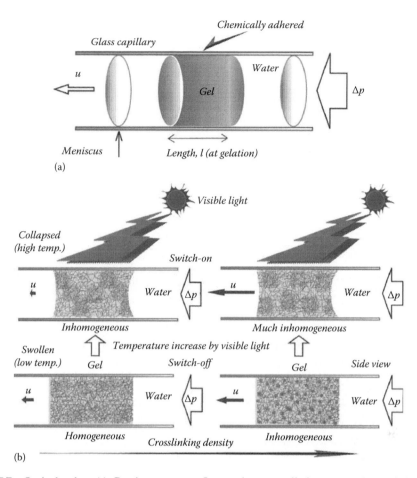

FIGURE 7.7 Optical valve. (a) Continuous water flow under an applied pressure, Δp, and (b) switching between the homogeneous and inhomogeneous networks triggered by visible light. The mechanical constraint that keeps the total network density constant in response to the temperature change and pore size increases or decreases when the gel is in the collapsed state through the phase transition.

Figure 7.7b is a schematic view of the microscopic inhomogeneity of gel networks depending on the BIS concentration and temperature [52,57]. In the present systems, the intrinsic inhomogeneity consists of dense (granular particle) and dilute (network) regions, and the water flow proceeds in the network regions. At a low molar fraction of BIS, the dense regions are isolated from each other, and the network density of the dilute region is still large. In the collapsed state, the dilute region separates into denser and more dilute parts, and the systems could form a percolated structure (a bicontinuous structure) of the denser parts along the flow direction; thus, the friction increases in the collapsed state. However, when the molar fraction of BIS increases above a certain threshold, the dilute parts in the network region will form a percolated network structure along the flow direction with a large water path, since the network density of the dilute region is too small to percolate into the denser parts in the collapsed state. In this case, the friction decreases in the collapsed state. It is noteworthy that the present switching on-off transition is completely reversible by repeated temperature jumps, and, in that sense, it is possible to realize a switching of the water flow triggered by an optical stimulus.

Another example is an optically controlled delivery system of target molecules, which is the most common application of hydrogels, such as a drug delivery system (DDS) [58]. In the case of hydrogels capable of a phase transition, the network structure controls the rate of molecular

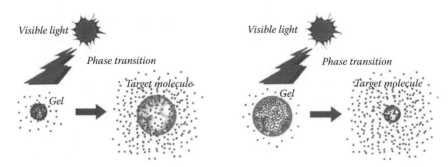

FIGURE 7.8 Optically driven controlled release of target molecules from gels consisting of an LCST (left) or UCST (right) polymer.

diffusion by releasing its target molecule triggered by specific external stimuli. The technique to realize a controlled release of molecules using the gel system can be proposed on the basis of the physical and chemical properties of gels. Optically absorbing chromophores can be embedded in a gel matrix with target molecules for the purpose of photothermally modulating the delivery system. For example, a thermoresponsive polymer exhibits an LCST or an upper critical solution temperature (UCST) that is desirable when slightly below a target temperature, for example, body temperature. When the temperature of the copolymer exceeds the LCST or UCST by illumination of light, the hydrogel shrinks or swells, causing a release of target molecules held within the hydrogel matrix (Figure 7.8). The possible applications presented here are based on the volume phase transition of gels, which is induced by a change in the osmotic pressure produced by local heating [23].

Recently, the development of light-sensitive materials and their applications have been proposed in the medical fields, such as biomaterials with absorbents other than the chromophore, which are designed to induce the deformations of gels by a photothermal effect [59–61]. The materials based on the photothermal effect can be used as smart elements that show optically "remote control" functions with only slight disturbance to the system. The practical applications of the photothermal effect are not necessarily based on the reversible volume phase transition of gels; for example, the irreversible change by erosion of gels has been applied to drug delivery and cartilage tissue engineering [62]. The elution [63] and erosion of gels in response to external stimuli are important not only for practical applications of smart functional gels but also for further understanding of the gel phase transition. This is because these phenomena are closely related with the definition of gels and the universality of the volume phase transition of gels.

7.5 SUMMARY

In this chapter, the fundamentals of the induction of phase transition due to local heating of the gel by illumination of visible light were described. The phase transition induced by light provided opportunity to carry out high-speed volume changes in gels with only slight disturbance to the system. In this regard, the increment of temperature in the immediate vicinity of polymer chains using light has been an ideal trigger for the induction of volume change. The method to control the states of switch on (collapsed) and switch off (swollen) by light is generic, and "on" and "off" commands could be stored or erased in a gel, which serves as a kind of switch or memory. Therefore, the light-induced phase transition of gels can be used in the development of photo-responsive switches and memory devices in specific fields. The potential application fields of visible light-responsive gels can spread not only in biomedical engineering but also in environment-conscious technologies. If the intermediate phase and the re-swelling phase transition can be used, the photothermal effect will have a wide application. If the principles of phases and phase transition of gels continue to be established, the results will enable us to develop a general method for the design and synthesis of gels as smart functional elements, such as those that exist in natural systems.

ACKNOWLEDGMENT

This work was supported in part by a grant-in-aid from the Ministry of Education, Science and Culture, Japan.

REFERENCES

1. Tanaka, T. 1978. Collapse of gels and the critical endpoint. *Phys. Rev. Lett.* 40: 820–823.
2. Tanaka, T. 1981. Gels. *Sci. Am.* 244: 124–136.
3. Tanaka, T. and Fillmore, D.J. 1979. Kinetics of swelling of gels. *J. Chem. Phys.* 59: 1214–1218.
4. Tanaka, T., Fillmore, D.J., Sun, S.-T., Nishio, I., Swislow, G., and Shah, A. 1980. Phase transitions in ionic gels. *Phys. Rev. Lett.* 45: 1636–1639.
5. Ilavsky, M., Plestil, J., and Dusek, K. 1980. The photoelastic behaviour and small-angle x-ray scattering of ionized gels of copolymers of 2-hydroxyethyl methacrylate with methacrylic acid. *Eur. Polym. J.* 16: 901–907.
6. Hrouz, J., Ilavsky, M., Ulbrich, K., and Kopecek, J. 1981. The photoelastic behaviour of dry and swollen networks of poly (*N,N*-diethylacrylamide) and of its copolymer with *N-tert*.butylacrylamide. *Eur. Polym. J.* 17: 361–366.
7. Ohmine I. and Tanaka, T. 1982. Salt effects on the phase transition of ionic gels. *J. Chem. Phys.* 11: 5725–5729.
8. Tanaka, T., Nishio, I., Sun, S.-T., and Ueno-Nishio, S. 1982. Collapse of gels in an electric field. *Science* 218: 467–469.
9. Ricka, J. and Tanaka, T. 1984. Swelling of ionic gels: Quantitative performance of the Donnan theory. *Macromolecules* 17: 2916–2921.
10. Tanaka, T., Sato, E., Hirokawa, Y., Hirotsu, S., and Peetermans, J. 1985. Critical kinetics of volume phase transition of gels. *Phys. Rev. Lett.* 55: 2455–2458.
11. Hirokawa, Y., Tanaka, T., and Matsuo, E-S. 1984. Volume phase transition in a nonionic gel. *J. Chem. Phys.* 81: 6379–6380.
12. Hirotsu, S., Hirokawa, Y., and Tanaka, T. 1987. Volume-phase transitions of ionized *N*-isopropylacrylamide gels. *J. Chem. Phys.* 87: 1392–1395.
13. Suzuki, A. 1993. Phase transition in gels of sub-millimeter size induced by interaction with stimuli. *Adv. Polym. Sci.* 110: 199–240.
14. Tanaka, T., Sun S-T., Hirokawa, Y., Katayama, S., Kucera, J., Hirose, Y., and Amiya, T. 1987. Mechanical instability of gels at the phase transition. *Nature* 325: 796–798.
15. Osada, Y. 1987. Conversion of chemical into mechanical energy by synthetic polymers (chemomechanical systems). *Adv. Polym. Sci.* 82: 1–46.
16. Hu, Z., Zhang X., and Li, Y. 1995. Synthesis and application of modulated polymer gels. *Science* 269: 525–527.
17. Suzuki, A. and Tanaka, T. 1990. Phase transition in polymer gels induced by visible light. *Nature* 346: 345–347.
18. Mamada, A., Tanaka, T., Kungwatchakun, D., and Irie, M. 1990. Photoinduced phase transition of gels. *Macromolecules* 23: 1517–1519.
19. Mafé, S., Manzanares, J.A., English, A.E., and Tanaka, T. 1997. Multiple phases in ionic copolymer gels. *Phys. Rev. Lett.* 79: 3086–3089.
20. Hirotsu, S. 1994. Static and time-dependent properties of polymer gels around the volume phase transition. *Phase Trans.* 47: 183–240.
21. Tanaka, K. 2000. Molecular recognition and minimization of frustration by imprinting in gels. PhD in Physics, Massachusetts Institute of Technology, Cambridge, MA.
22. English, A.E., Mafé, S., Manzanares, J.A., Yu, X., Grosberg, A.Y., and Tanaka, T. 1996. Equilibrium swelling properties of polyampholytic hydrogels. *J. Chem. Phys.* 104: 8713–8720.
23. Flory, P.J. 1953. *Principle of Polymer Chemistry*, Cornell University Press, Ithaca, NY.
24. Heskins, M. and Guillet, J.E. 1968. Solution properties of poly(*N*-isopropylacrylamide). *J. Macromol. Sci: Part A-Chem.* 2: 1441–1455.

25. Ilmain, F., Tanaka, T., and Kokufuta, E. 1991. Volume transition in a gel driven by hydrogen bonding. *Nature* 349: 400–401.

26. Tanaka T. and Annaka, M. 1993. Multiple phases of gels and biological implications. *Makromol. Chem. Macromol. Symp.* 70/71: 13–22.

27. Katayama, S., Hirokawa, Y., and Tanaka, T. 1984. Reentrant phase transition in acrylamide-derivative copolymer gels. *Macromolecules* 17: 2641–2643.

28. Amiya, T. and Tanaka, T. 1987. Phase transitions in crosslinked gels of natural polymers. *Macromolecules* 20: 1162–1164.

29. Hirotsu, S. 1988. Critical points of the volume phase transition in *N*-isopropylacrylamide gels. *J. Chem. Phys.* 88: 427–431.

30. Tanaka, F., Koga, T., and Winnik, F.M. 2009. Competitive hydrogen bonds and cononsolvency of poly (*N*-isopropylacrylamide)s in mixed solvents of water/methanol. *Prog. Colloid. Polym. Sci.* 136: 1–7.

31. Young, T.H. and Chuang, W.Y. 2002. Thermodynamic analysis on the cononsolvency of poly (vinyl alcohol) in water–DMSO mixtures through the ternary interaction parameter. *J. Membr. Sci.* 210: 349–359.

32. Young, T.H., Cheng, L.P., Hsieh, C.C., and Chen, L.W. 1998. Phase behavior of EVAL polymers in water-2-propanol cosolvent. *Macromolecules* 31: 1229–1235.

33. Kudo, S., Otsuka, E., and Suzuki, A. 2010. Swelling behavior of chemically crosslinked PVA gels in mixed solvents. *J. Polym. Sci. Part B: Polym. Phys.* 48: 1978–1986.

34. Hirashima, Y. and Suzuki, A. 2004. Roles of hydrogen bonding on the volume phase transition of ionized poly(*N*-isopropylacrylamide) gels. *J. Phys. Soc. Jpn.* 73: 404–411.

35. Hirashima, Y., Sato, H., and Suzuki, A. 2005. ATR-FTIR spectroscopic study on hydrogen bonding of poly(*N*-isopropylacrylamide-*co*-sodium acrylate) gel. *Macromolecules* 38: 9280–9286.

36. Sato, H., Hirashima, Y., and Suzuki, A. 2007. Reswelling transition of poly(sodium acrylate) gels due to destruction of hydrogen bonds observed by ATR FTIR spectroscopy. *J. Appl. Polym. Sci.* 105: 3809–3816.

37. Zhang, X., Li, Y., Hu, Z., and Littler, C.L. 1995. Bending of *N*-isopropylacrylamide gel under the influence of infrared light. *J. Chem. Phys.* 102: 551–555.

38. Hu, Z., Li, Y., Zhang, X., and Littler, C.L. 1995. CO_2 laser-controlled transmission of visible light in *N*-isopropylacrylamide gel. *Polym. Gels Networks* 3: 267–279.

39. Suzuki, A., Sanda, K., and Omori, Y. 1997. Phase transition in strongly stretched polymer gels. *J. Chem. Phys.* 107: 5179–5185.

40. Suzuki, A. and Hara, T. 2001. Kinetics of one-dimensional swelling and shrinking of polymer gels under mechanical constraint. *J. Chem. Phys.* 114: 5012–5015.

41. Suzuki, A., Wu, X.R., Kuroda, M., Ishiyama, E., and Kanama, D. 2003. Swelling properties of thin-plate hydrogels under mechanical constraint. *Jpn. J. Appl. Phys.* 42: 564–569.

42. Suzuki, A. and Suzuki, H. 1995. Hysteretic behavior and irreversibility of polymer gels by pH change. *J. Chem. Phys.* 103: 4706–4710.

43. Suzuki, A., Ishii, T., and Maruyama, Y. 1996. Optical switching in polymer gels. *J. Appl. Phys.* 80: 131–136.

44. Suzuki, A., Yamazaki, M., and Kobiki, Y. 1996. Direct observation of polymer gel surfaces by atomic force microscopy. *J. Chem. Phys.* 104: 1751–1757.

45. Suzuki, A., Yamazaki, M., Kobiki, Y., and Suzuki, H. 1997. Surface domains and roughness of polymer gels observed by atomic force microscopy. *Macromolecules* 30: 2350–2354.

46. Hirokawa, Y., Jinnai, H., Nishikawa, Y., Okamoto, T., and Hashimoto, T. 1999. Direct observation of internal structures in poly(*N*-isopropylacrylamide) chemical gels. *Macromolecules* 32: 7093–7099.

47. Suzuki, A., Kobiki, Y., and Yamazaki, M. 2003. Effects of network inhomogeneity in poly(*N*-isopropylacrylamide) gel on its surface structure. *Jpn. J. Appl. Phys.* 42: 2810–2817.

48. Suzuki, A., Okabe, N., and Suzuki, H., 2005. Effects of guest microparticles on the swelling behavior of polyacrylamide gels. *J. Polym. Sci. Part B: Polym. Phys.* 43: 1696–1704.

49. Tokita, M. and Tanaka, T. 1991. Reversible decrease of gel-solvent friction. *Science* 253: 1121–1123.

50. Tokita, M. and Tanaka, T. 1991. Friction coefficient of polymer networks of gels. *J. Chem. Phys.* 95: 4613–4619.

51. Yoshikawa, M., Ishii, R., Matsui, J., Suzuki, A., and Tokita, M. 2005. A simple technique to measure the friction coefficient between polymer network of hydrogel and water. *Jpn. J. Appl. Phys.* 44: 8196–8200.

52. Suzuki, A. and Yoshikawa, M. 2006. Water flow in poly(*N*-isopropylacrylamide) gels. *J. Chem. Phys.* 125: 174901 (6 pages).

53. Kondo, G., Oda, T., and Suzuki, A. 2008. Water flow through a stimuli-responsive hydrogel under mechanical constraint. *AIP Conf. Proc.* 982: 458–463.

54. Hecht, A.M. and Geissler, E. 1980. Gel deswelling under reverse osmosis. *J. Chem. Phys.* 73: 4077–4080.

55. Hecht, A.M. and Geissler, E. 1980. Pressure-induced deswelling of gels. *Polymer* 21: 1358–1359.

56. Geissler, E. and Hecht, A.M. 1982. Gel deswelling under reverse osmosis. II. *J. Chem. Phys.* 77: 1548–1553.

57. Doi, Y. and Tokita, M. 2005. Real space structure of opaque gel. *Langmuir* 21: 5285–5289.

58. Okano, T., Bae, Y.H., Jacobs, H., and Kim, S.W. 1990. Thermally on-off switching polymers for drug permeation and release. *J. Control. Release* 11: 255–265.

59. Fujigaya, T., Morimoto, T., Niidome, Y., and Nakashima, N. 2008. NIR laser-driven reversible volume phase transition of single-walled carbon nanotube/poly(*N*-isopropylacrylamide) composite gels. *Adv. Mater.* 20: 3610–3614.

60. Sershen, S.R., Westcott, S.L., Halas, N.J., and West, J.L. 2000. Temperature-sensitive polymer-nanoshell composites for photothermally modulated drug delivery. *J. Biomed. Mater. Res.* 51: 293–298.

61. Katz, J.S. and Burdick, J.A. 2010. Light-responsive biomaterials: Development and applications. *Macromol. Biosci.* 10: 339–348.

62. Anseth, K.S., Metters, A.T., Bryant, S.J., Martens, P.J., Elisseeff, J.H., and Bowman, C.N. 2002. In situ forming degradable networks and their application in tissue engineering and drug delivery. *J. Control. Release* 78: 199–209.

63. Otsuka, E., Sasaki, S., Koizumi, K., Hirashima, Y., and Suzuki, A. 2010. Elution of polymers from physically cross-linked poly(vinyl alcohol) gels. *Soft Matter* 6: 6155–6159.

Part III

Harnessing Light and Optical Forces

The ability to manipulate and interact with particles at small (micro- and nano-) scales has been essential to many recent advances in science and technology. Optical particle manipulation represent one of the most influential technologies to date. In the field of optical manipulation, optical tweezers (OTs) and optoelectronic tweezers (OETs) have emerged as the dominant methods. Optical tweezing traps particles and objects through the optical gradient forces resulting from a tightly focused laser source. OETs, on the other hand, exploit the principle of light-induced dielectrophoresis (LIDEP). In this technique, the optical field is not used directly to manipulate the particles; instead, it interacts with a photoconductive material to create virtual electrodes, which form electric field gradient landscapes that trap the particles based on the DEP principle.

Ming C. Wu and his colleagues at the University of California, Berkeley, and the Bioelectronics Research Centre at the University of Glasgow provide a detailed look into these emerging OT and OET technologies in Chapter 8. OTs are capable of trapping micro- and nanoscale particles in three dimensions and have been a critical tool in studying detailed biological and chemical mechanisms. To create a stable trap to hold the particle, it is necessary to use quite high optical power intensities. Unfortunately, these high intensities limit the OT's effectiveness and ability to perform large-scale optical manipulation tasks. Furthermore, the high intensities can damage biological materials. In contrast, OETs do not directly manipulate the particles; rather, they interact with a photoconductive material to create virtual electrodes that form electric field gradient landscapes that trap the particles based on the dielectrophoresis (DEP) principle. The LIDEP force requires optical power intensities approximately five orders of magnitude smaller than OT and is capable of manipulating multiple particles simultaneously, thereby significantly improving performance. The authors compare the two techniques and discuss the weaknesses and advantages of each method for various applications.

In Chapter 9, Halina Rubinsztein-Dunlop and her colleagues at the University of Queensland look deeper into these mechanisms not only for manipulating particles but driving functional micromotors and micromachines. Specifically, the authors look into the two-photon photopolymerization method as a powerful and versatile tool to fabricate arbitrarily shaped 3D microrotors and describe the use of form birefringence in microfabricated objects. Professor Shoji Maruo discusses in detail the design and multiphoton microfabrication of a variety of optically driven microfluidic devices such as micropumps and microvalves in Chapter 10. Since multiphoton microfabrication is capable of producing three-dimensional (3D) micro/nanostructures with transparent photopolymers, several types of optically driven micropumps, including a lobed pump and viscous pumps, were developed. In addition, it was demonstrated that movable micromachines were replicated by a membrane-assisted transfer molding technique. The 3D replication technique will open a way to mass produce all-optically controlled lab-on-a-chip devices.

Johtara Yamamoto and Toshiaki Iwai discuss in Chapter 11 how holographic optical tweezers (HOT) are a promising solution to improving the precision in manipulating and controlling living cells, industrial particles, droplets, and aerosols. A roadblock to exploiting this concept had been the time required to create appropriate holograms. Through time-division multiplexing and the phase-shifting of holograms, the authors demonstrate that this process can be significantly speeded up enabling highly controllable OTs to be possible.

8 Optical and Optoelectronic Tweezers

Arash Jamshidi, Steven L. Neale, and Ming C. Wu

CONTENTS

8.1 INTRODUCTION

The ability to manipulate and interact with particles at small (micro and nano) scales has been essential to many recent advances in science and technology. Several techniques such as mechanical manipulators,[1–3] electrophoresis,[4,5] dielectrophoresis[6] (DEP), electroosmosis,[7,8] microfluidics,[9–11] and magnetic[12–14] manipulation have been created in the past few decades to address this challenge. Optical manipulation of particles offers an attractive choice due to its inherent flexible and noninvasive nature. In the field of optical manipulation, only two technologies have emerged as most influential. The first technique, optical tweezers (OTs), which was invented by Ashkin et al.,[15] traps particles through the optical gradient forces[16–19] resulting from a tightly focused laser source. OT is capable of trapping micro- and nanoscale particles in three dimensions and has been a critical tool in studying detailed biological and chemical mechanisms. However, to stably trap the particles of interest, OT requires very high optical power intensities, which limits its effectiveness in performing high-throughput and large-scale optical manipulation functions and it can potentially damage the trapped objects,[16,20,21] especially biological materials. The second optical manipulation technique, called optoelectronic tweezers[22] (OET), works based on the principle of light-induced dielectrophoresis (LIDEP) force. In this technique, the optical field is not used directly to manipulate the particles; instead, it interacts with a photoconductive material to create virtual electrodes which will form electric field gradient landscapes that will trap the particles based on the DEP principle. Therefore, OET requires optical power intensities approximately 5 orders of magnitude smaller than OT and is capable of massively parallel manipulation of particles over larger areas. As a result, it is possible to achieve large-scale and high-throughput optical manipulation of particles using OET.

DEP is the force that a neutral but polarizable particle experiences in a nonuniform electric field. The force on one side of the induced dipole is different than the force on the opposite pole due to the nonuniformity in the field creating a net force. This is identical to the case for an OT trap where the particle is much smaller than the wavelength of light used to create the trap. Where DEP and OT differ is in the frequency (and hence wavelength) of the electric field applied. For OT, visible or near-infrared light is usually used with an associated electric field in the range of 10^{14} Hz, whereas the DEP used in OET uses an applied AC voltage typically in the range of 10^3–10^5 Hz, many orders of magnitude lower. This means that for manipulating microparticles, DEP is always in the limit where the particles are much smaller than the wavelength of the electric field. However, when the particle increases in size from nanoparticles to microparticles, it quickly becomes similar in size to the wavelength of the electric field associated to OT. When this happens, the particle can perturb the field causing the trapping much more strongly and the assumptions usually used to calculate the force, such as the assumption that the electric field does not vary greatly over the size of the particle, are obviously no longer valid. When the DEP force is controlled by an OET device, the experiment can appear to be exactly the same as an OT experiment with the particle being attracted to the light beam, and in the limit where the particle is smaller than the light used, the physics of the two trapping modalities is also similar; however, the great difference in the frequency of the field used to produce the force provides many differences in the forces produced which will be examined later.

The OT technology has been studied in depth in the past few decades, and there are several excellent review articles[18,19,23,24] on the detailed implementation and theory of this technique. Here, we start by briefly reviewing the OT technique followed by a more in-depth discussion of OET. Finally, we will focus on comparing the two techniques and will discuss the weaknesses and advantages of each method for various applications.

8.2   OPTICAL TWEEZERS

OTs, first proposed by Ashkin[15] in 1970, are a powerful optical manipulation technique that have been used for trapping of cells,[17] beads, nanoparticles,[16,25] and characterization of biomolecules.[24,26,27] In this method, the optical field of a highly focused laser light is used to trap particles. In the case of particles with sizes much smaller than the wavelength of the trapping laser (the Rayleigh limit), the particles are treated as electrical dipoles that are interacting with the optical trap's electric field gradient. For particles much larger than the wavelength of the trapping laser, (the Mie limit), the particles change the momentum of the photons by refracting the laser light, therefore experiencing an equal and opposite force. The radiation force resulting from bombardment of the particles with photons opposes the trapping of particles in both cases. Therefore, to create a stable trap, the trapping light source needs to be tightly focused, using a lens with high numerical aperture, to overcome the radiation pressure. Figure 8.1 depicts the forces acting on a trapped particle using OTs.

OTs usually rely on the optical gradient force which is given by[24]

$$F_g = -\frac{n_m^2 r^3}{2}\left(\frac{n_p^2 - n_m^2}{n_p^2 + 2n_m^2}\right)\nabla E^2 \tag{8.1}$$

where
  n_m and n_p are the refractive indices of the medium and the particle, respectively
  r is the radius of the particles
  E is the electric field applied

OTs' trapping capability is limited by the high optical power density required to create a stable trap. In addition, the high required numerical aperture (NA) for focusing the laser source limits the effective working area of OTs to about 100×100 μm using a spatial light modulator, commonly referred to as holographic optical tweezers (HOT).[28,29]

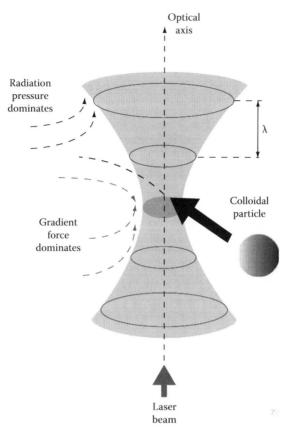

FIGURE 8.1 OT trapping mechanism: a tightly focused laser source is used to trap different types of particles. To create a stable trap, the gradient forces need to overcome the radiation pressure. (Reproduced from Grier, D.G., *Nature*, 424, 810, 2003.)

8.3 OPTOELECTRONIC TWEEZERS

OETs[22] are a powerful optical manipulation technique for manipulation, sorting, organization, and patterning of micro- and nanoparticles. Figure 8.2 shows the OET device structure. In this technique, 2D patterns of low-intensity light are projected onto a plane of photoconductive material sandwiched between transparent parallel-plate electrodes. The light excites carriers in the photoconductive layer (hydrogenated amorphous silicon), reducing the local impedance to create an inhomogeneous electric field across the liquid layer. In the presence of the nonuniform electric field, a polarization is induced in particles between the parallel plates. The polarized particles are either attracted to or repelled from the field gradient (according to the DEP principle) produced by the projected light pattern, yielding a powerful optoelectronic method of particle manipulation based on dynamic, light-actuated virtual electrodes.

The optical power intensity required for trapping in the OET device is reduced by approximately 5 orders of magnitude relative to OTs since the optical field is not directly used to trap the particles; instead, the optical field is used to create virtual electrodes in the photoconductive layer which locally transfer the AC voltage to the liquid media. Since the virtual electrodes are defined optically, it is possible to perform real-time and flexible trapping of particles using a spatial light modulator. In addition, the reduced optical power intensity required for trapping and relaxed optical requirements enable a large working area. These capabilities of OET have been demonstrated through massively parallel manipulation of 15,000 particles over a large area, 1.3 mm × 1.0 mm, shown in Figure 8.3.[22]

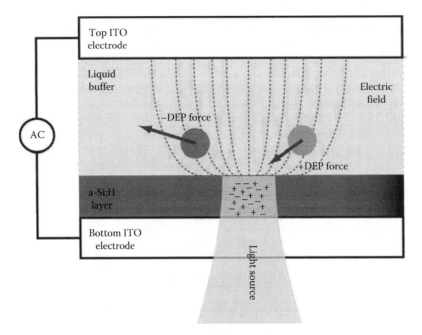

FIGURE 8.2 OET device structure. The OET device consists of a top transparent ITO electrode and a bottom ITO electrode. There is a layer of photoconductive material (hydrogenated amorphous silicon) on top of the bottom electrode. An AC voltage is applied between the two electrodes. The liquid medium containing the particles of interest is sandwiched between the top and bottom layers. The interaction of the light source with the photoconductive layer reduces the impedance locally, transferring the voltage to the liquid layer in the area that the light is present. This nonuniform electric field traps the particles by light-induced dielectrophoresis principle. (From Jamshidi, A. et al., Optofluidics and optoelectronic tweezers, *Proceedings of SPIE* 69930A-69930A-10, 2008. Copyright SPIE.)

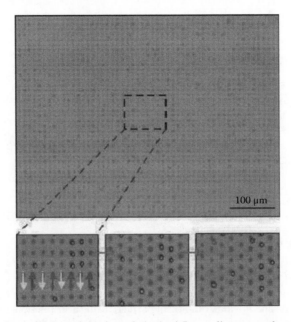

FIGURE 8.3 Massively parallel manipulation of single 4.5 μm diameter polystyrene particles over 1.3 mm × 1.0 mm area using 15,000 traps created by a digital micromirror device. The inset shows the transport of particles in the direction depicted by the arrows. (From Chiou, P.Y. et al., *Nature*, 436, 370, 2005.)

Since the DEP force is dependent on the properties of the particles in relation to the surrounding media, OET is capable of distinguishing between particles with differing complex permittivities such as dead and live cells[22] and semiconducting and metallic materials.[31] This ability is particularly important in cell separation and sample purification.

In addition to the conventional OET device, various photoconductive materials and device configurations have been used to realize the LIDEP force. For example, hydrogenated amorphous silicon,[22] silicon phototransistor,[32] CdS,[33] metallic plasmonic nanoparticles,[34,35] and polymers[36] have been used as photoconductive materials. OET device configurations such as phototransistor OET[32] (phOET) for manipulation of cells in high conductivity cell culture media, lateral-field OET (LOET) and planar lateral-field OET (PLOET)[37–40] for manipulation of particles in a lateral fashion using interdigitated photoconductive electrodes, floating electrode OET (FLOET)[41] for manipulation of aqueous droplets in oil, double-sided OET[42] for reduction of the nonspecific stiction of particles, and OET integrated with electrowetting-on-dielectric[43] devices for enabling manipulation of particles inside droplets have been invented. Moreover, dynamic actuation of these devices has been accomplished using devices such as scanning lasers,[44] digital micromirror devices,[22] and LCD flat panel displays.[45,46]

Various modes of operation such as light-induced DEP,[22] light-actuated AC electroosmosis,[47] and electrothermal heating[48] have been observed and studied in the OET devices. Manipulation of microparticles such as polystyrene beads,[22,37,45,46,49,50] semiconductor microdisks,[51] red blood cells,[46,52] *Escherichia coli* bacteria,[44] white blood cells,[22,37] Jurkat cells,[37] HeLa cells,[32,37] yeast cells,[49] mouse embryos,[53] sperm cells,[54] and neuron cells[55] and nanoparticles such as semiconducting and metallic nanowires,[31,39,40] carbon nanotubes,[56] metallic spherical nanoparticles,[57] and DNA[58,59] have been demonstrated using OET. Other functionalities such as dynamic single-cell electroporation[60] and cell lysis,[61] optically induced flow cytometry,[62] and large-scale, dynamic patterning of nanoparticles[63] have been demonstrated through integration of OET with microfluidics or using various operational regimes in the OET device.

OET device operation can be modeled as a simple lumped circuit element model shown in Figure 8.4. In the absence of the light pattern (Figure 8.4a), the photoconductive layer (a-Si:H) impedance, Z_{PC}, is higher than the liquid layer impedance, Z_L; therefore, the majority of the AC voltage is dropped across the photoconductive layer. However, once the light pattern is projected on the device (Figure 8.4b), the photoconductive layer impedance is locally reduced, forming a "virtual electrode" and transferring the majority of the AC voltage to the liquid layer. Due to a-Si:H's small ambipolar diffusion length of 115 nm,[64] the actuated area is confined to the illumination region causing the photoconductive layer impedance to be reduced only where the light source is present. As a result, OET device resolution is fundamentally limited by the light source diffraction limit given by

$$\text{Diffraction limit} = 1.22 \frac{\lambda}{2NA} \qquad (8.2)$$

where
  λ is the wavelength of illumination
  the NA is the numerical aperture of the objective lens

For a typical illumination wavelength of 630 nm and $NA = 0.6$, a diffraction limited spot of approximately 630 nm can be achieved. Moreover, the transfer of the AC voltage to the liquid layer only in the illuminated area creates a nonuniform electric field in the liquid layer. It is the interaction of this nonuniform field with the particles, liquid media, and the virtual electrodes which creates the various electrokinetic effects observed in the OET device.

The main electrokinetic effect used in the OET device is the DEP force. Here, we will briefly describe the DEP theoretical principles. DEP is a technique that uses the interaction of a nonuniform

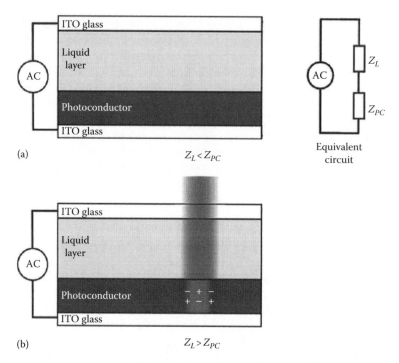

FIGURE 8.4 OET device operation principle. (a) In the dark state, the impedance of the photoconductive layer is higher than the impedance of the liquid layer and most of the voltage is dropped across the photo-conductive layer. (b) Once the light source is applied, it generates electron-hole pairs in the photoconductive layer, reducing its impedance below that of the liquid layer and transferring the voltage to the liquid layer in the illuminated area.

electric field with the induced dipoles on the particles to attract or repel the particles from areas of highest electric field intensity gradient. In the presence of a nonuniform electric field (E), a dipole moment (p) is induced in the particles with unequal charges on two ends. Therefore, the dipole feels a net force toward or away from areas of highest field intensity gradients depending on the AC bias frequency and properties of the particles and the liquid solution. By taking the difference between the forces experienced by the charges at the two ends of the dipole, we can approximate the DEP force as[6]

$$F = (p \cdot \nabla)E \tag{8.3}$$

To calculate the DEP force expression on various objects, the particles are assigned an effective dipole moment which is the moment of a point dipole that creates an identical electrostatic potential when immersed in the imposed electric field. By comparing the electrostatic potential of the particle of interest to the electrostatic potential of a point dipole, and by ignoring the higher order terms in the Taylor expansion of the electric field, the following formula is derived for the DEP force on a spherical particle:[6]

$$\langle F_{DEP} \rangle = 2\pi r^3 \varepsilon_m Re\left\{K^*(\omega)\right\} \nabla E^2 \tag{8.4}$$

$$K^*(\omega) = \frac{\varepsilon_p^* - \varepsilon_m^*}{\varepsilon_p^* + 2\varepsilon_m^*}, \varepsilon_m^* = \varepsilon_m - j\frac{\sigma_m}{\omega}, \varepsilon_p^* = \varepsilon_p - j\frac{\sigma_p}{\omega} \tag{8.5}$$

where

    r is the radius of the particle

    ε_m and ε_p are the permittivities of the medium and the particle, respectively

    σ_m and σ_p are the conductivities of the media and the particle, respectively

    ω is the frequency of the AC potential

    $Re\{K^*\}$ is the real part of the Clausius–Mossotti (CM) factor (K^*)

    ∇E^2 is the gradient of the electric field intensity

The CM factor is a function of the permittivity and conductivity of the particle and the medium and, in the case of spherical particles, has a value between −0.5 and 1. For particles that are more polarizable than the surrounding medium, the CM factor is positive and the particles experience a positive DEP force. However, the particles less polarizable than the surrounding medium experience a negative DEP force and are repelled from regions of highest electric field intensity gradient. The other important parameter that determines the magnitude of the DEP force is the gradient of the field intensity. In the OET device, ∇E^2 has its maximum close to the OET surface and falls off rapidly as we move further away from the surface. Figure 8.5 shows the normalized field intensity gradient as a function of distance from the OET surface at the center of the illuminated area. It is evident that the ∇E^2 magnitude drops by 1–2 orders of magnitude 10 μm away from the OET surface and it falls by roughly 3 orders of magnitude 50 μm away from the OET surface. Therefore, particles feel the maximum trapping force near the OET surface.

Many of the nanostructures of interest such as nanowires. carbon nanotubes, and biomaterials such as bacteria, viruses, and DNA have anisotropic geometries and can typically be modeled as cylindrical objects. Therefore, it is important to analyze the DEP force expression for particles with anisotropic geometries. Cylindrical objects with radius r and length l can be modeled as elongated ellipsoids with a, b, and c dimensions, where $b = c = r$ and $a = l/2$. The effective dipole moment along the length of the ellipsoid is given by[6]

$$(p_{eff})_l = \frac{2\pi r^2 l}{3} \varepsilon_m \left[\frac{\varepsilon_p^* - \varepsilon_m^*}{\varepsilon_m^* + (\varepsilon_p^* - \varepsilon_m^*)L_l} \right] E_l \qquad (8.6)$$

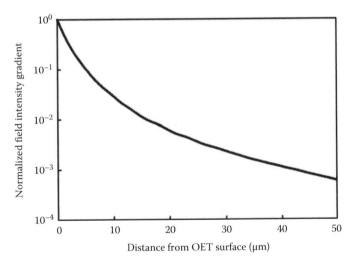

FIGURE 8.5 Normalized gradient of field intensity (∇E^2) as a function of distance from the OET surface in the center of the illuminated area.

where

$\varepsilon^* = \varepsilon - j\sigma/\omega$

Subscripts m and p denote complex permittivity of medium and particle, respectively

L_l is the depolarization factor along the long axis of the ellipsoid

Similar expressions can be derived for polarization along the short axis of the ellipsoid. The depolarization factor is defined by an elliptical integral, and for an elongated ellipsoid, it can be approximated to be[6]

$$L_l = \frac{lr^2}{4} \int_0^\infty \frac{ds}{(s+(l/2)^2)^{3/2}(s+r^2)} \approx \frac{2r^2(\ln(1+f/1-f)-2f)}{l^2 f^3} \tag{8.7}$$

where

$$f = \sqrt{\left(1 - \ln\left(\frac{2r}{l}\right)\right)^2}$$

For ellipsoids with high aspect ratio ($l \gg r$), we have $L_l \ll L_r$ which results in much stronger polarization along the long axis of the elongated ellipsoids. Once the expression of the ellipsoid's effective dipole moment is known, the DEP force can be approximated as

$$F_{DEP} = \left[\frac{2\pi r^2 l}{3}\varepsilon_m\left[\frac{\varepsilon_p^* - \varepsilon_m^*}{\varepsilon_m^* + (\varepsilon_p^* - \varepsilon_m^*)L_l}\right]E_l.\nabla\right]E_l = \left(\frac{\pi r^2 l}{3}\right)\varepsilon_m K\nabla(E_l^2) \tag{8.8}$$

where

$$K = \frac{\varepsilon_p^* - \varepsilon_m^*}{\varepsilon_m^* + (\varepsilon_p^* - \varepsilon_m^*)L_l}$$

which is the CM factor for ellipsoidal particles. Therefore, the time-averaged DEP force is given by

$$\langle F_{DEP} \rangle = \left(\frac{\pi r^2 l}{6}\right)\varepsilon_m Re\{K\}\nabla(E_l^2) \tag{8.9}$$

For high aspect ratio ellipsoids, the depolarization factor along the long axis of the particle is much smaller than depolarization factor for spherical particles (1/3). Therefore, the real part of the CM factor can be much larger than one for elongated ellipsoids ($Re\{K_{Ellipsoid}\} \gg 1$), as opposed to the CM factor of spherical particles, which has a value between −0.5 to 1 ($-0.5 \le Re\{K_{Spherical}\} \le 1$). Therefore, the DEP force for high aspect ratio ellipsoids is considerably enhanced compared to spherical particles with similar dimensions. Figure 8.6 shows the value of the CM factor as a function of the aspect ratio of the particle, assuming a frequency of 100 kHz, liquid relative permittivity $\varepsilon_m = 80$ and conductivity $\sigma_m = 10$ mS/m, and particle relative permittivity $\varepsilon_p = 12$ and conductivity $\sigma_p = 10$ S/m. For an aspect ratio equal to two, such as spherical particles, the CM factor is 1 (due to high polarizability of the particle assumed here). However, as the aspect ratio of the particle increases, corresponding to particles with ellipsoidal and cylindrical geometries, the value of the CM factor increases by 2–3 orders of magnitude, for aspect ratios over 100.

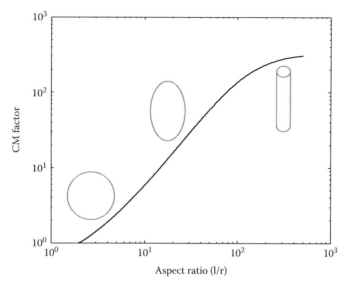

FIGURE 8.6 The value of the Clausius–Mossotti factor as a function of the aspect ratio of the particle.

This enhancement in the CM factor value has important implications in DEP manipulation of particles with anisotropic geometries such as nanowires and carbon nanotubes.

Finally, when objects move in a liquid medium, a net hydrodynamic force opposes the motion, and this force is typically referred to as the drag force. The approximate drag forces for a spherical particle and a cylindrical object moving perpendicular to its long axis are given, respectively, by[6]

$$F_{drag} = 6\pi r \eta v_{drag} \tag{8.10}$$

$$F_{drag} = \frac{8\pi \eta l v_{drag}}{(2\ln(2l/r) - 1)} \tag{8.11}$$

where η is the viscosity of the surrounding medium. When objects are transported using the DEP force in a liquid medium, we can calculate the velocity of the particles due to the DEP force by equating the DEP force to the drag force:

$$F_{DEP} = F_{drag} \tag{8.12}$$

Using this formula and the equations for the DEP force on spherical and elongated ellipsoidal particles, respectively, we can calculate the velocity of these particles due to the DEP force in the OET device.

8.4 DISCUSSION

In this section, we will compare the relative advantages and disadvantages of OT and OET for manipulating particles. First, we will construct a qualitative measure of how much force each tweezing modality can produce on a specific particle, and then we will look at the other discriminating factors including the sizes of particles that can be moved, 2D versus 3D control, limits on the liquid media, and the effect of trapping on the particles that can be used and how the materials of the particles influence the trapping.

8.4.1   Trap Stiffness Measurements

One way to quantify the difference in the forces that can be placed onto particles is to measure the stiffness of the trap. The stiffness measures the force a particle will experience as a function of its distance from the center of the trap with a unit of Nm⁻¹. For both OT and OET, if the trap is created with a Gaussian light pattern, then the force is roughly proportional to the distance from the center of the trap so the stiffness is constant at least for small displacements (see Figure 8.7). It would also be possible to use the maximum force that a trap can create on a particle as a measure of how strong the trap is; however, this would favor large diameter traps that allow the particle to be moved a large distance from the trap center when an external force (such as fluid drag) is placed on the particle. A smaller trap capable of the same maximum force would keep the particle close to the center of the trap and so provides better control over the particle. As OT requires a high light intensity gradient to produce trapping, a high NA objective is usually used which then produces a small diameter trap. As OET traps can be formed with a low NA objective, they are usually larger in diameter so that an OET trap with a similar stiffness to an OT trap will usually be able to produce a larger maximum force.

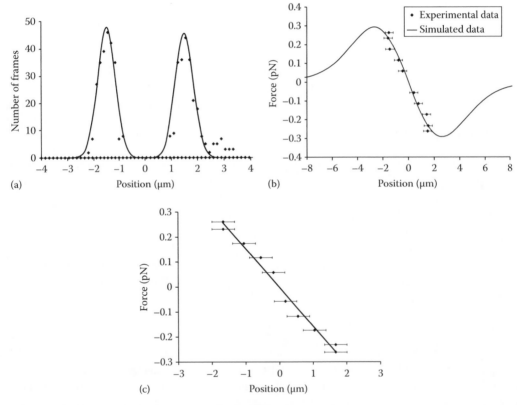

FIGURE 8.7   (a) When the OET device is moved with respect to the optical spot, the trapped particle experiences a dynamic equilibrium at a specific position given by the balance of the trapping force and the drag force exerted by the moving liquid. This position follows a Gaussian distribution around a point to the right of the trap center when the chamber is moved to the right and vice versa. The example shown was taken for a chamber velocity of 4.5 μm/s. (b) Each velocity corresponds to a force that is plotted against the center of the Gaussian position distribution (using the sigma value of the Gaussians as a measure of the error in position). This is compared with simulated data. The graph shown in (a) corresponds to the two extreme points (at ≈1.6 μm position) in (b). (c) Shows a similar curve plotted for an OT in the same optical setup but with an optical power of 20 mW (the OET trap used just 0.04 mW). (Adapted from Neale, S.L. et al., *Opt. Express*, 15, 12619, 2007. Copyright Optical Society of America.)

To quantitatively compare the two systems, a comparison has been carried out where OT and OET traps have been created with the same optics. This creates traps of similar diameter so that similar stiffness traps will have similar maximum forces making the comparison easier. To find the force produced by each trap, the stage holding the particle was moved at a constant velocity with respect to the trap so that a constant drag force was placed onto the particle. This caused the particle to move a certain distance from the center of the trap so that by measuring this distance for various stage velocities and calculating the drag force from Stokes drag, a force versus position profile can be created.[50] A measure of the position of the particle can be seen in Figure 8.7a and a force profile for the OET trap in Figure 8.7b and one for the OT trap in Figure 8.7c. The stiffness of the OT trap is proportional to the light power used, so here 20 mW was used for the optical trap giving a similar stiffness trap to the OET trap with just 0.04 mW of optical power. When taking into account the slight difference in stiffness between the two traps (OET 1.49e–7 N/m, OT 1.57e–7 N/m), this shows that the OET trap is 500 times stiffer per mW of optical power used. Stiffness per mW is the best metric to use when the amount of light available is the limiting factor on what the experiment can achieve. This is true for the case where we wish to trap multiple particles with the number of particle it is possible to trap with OT being limited by the laser power available, thus the main advantage of OET is the ability to perform massively parallel manipulation of microparticles.[22]

Although true for this experiment, the figure of 500 times stiffer per mW is not applicable to all comparisons between OT and OET for several reasons. Firstly, this experiment was performed with 2 μm diameter silica colloid which has a relatively large refractive index difference with the water it is suspended in, making it a good target for OT. If we were to repeat the experiment with biological cells where the refractive index is only just higher than the liquid, the OT force would be greatly reduced whereas the OET force would depend on the relevant CM factor that can be tuned by changing the applied AC frequency and medium conductivity (it is much easier to change the medium conductivity to change the OET force than to change its refractive index to change the OT force, see Section 8.4.4). Another factor that can be changed is the NA of the objective. As the NA is increased, the OT force, which is proportional to the gradient of the light intensity (see Equation 8.1), will increase. The OET forces would also increase with a higher NA but only to a point where the a-Si is saturated with charge carriers and the conductivity will no longer increase. In practice, this means that most OT traps created are stiffer per mW than the one used for comparison here although they are smaller in diameter.

The traps considered earlier were both created with a Gaussian light profile; if different light profiles are used, the trap profiles will also change. For OT different laser beam profiles can be created using either a diffractive optical element or a spatial light modulator. Due to the reduced requirement for high light intensities, OET traps can be created using a commercial data projector as both light source and to pattern the light. If the data projector is used to project a circle of light with a constant light intensity across it, the force profile is very different to the case for the Gaussian profile described earlier. Instead of having a single constant trap stiffness, a trap that obeys Hookes law where the force is proportional to the distance, the particle only experiences a force at the edges of the light pattern where there is an optical gradient that results in an electrical gradient providing the DEP force.

Figure 8.8 shows the trap profiles created by circular light patterns with different diameters for HeLa cells. The larger circles of light do not produce any force on the particle until it reaches the edge of the light pattern where they feel a force of a certain stiffness. If the pattern is made smaller, this distance decreases until a 12 μm diameter light pattern produces a trap with a roughly constant spring constant. This shows that we can produce this ideal trapping condition which provides good control over the particles position if the light pattern used is similar to the size of the particle being trapped. Using simulations to provide the force at all distances from the trap center which can be seen to agree well with the experimentally measured forces close to the trap center in Figure 8.8a, the potential energy experienced by the trapped particle can be found, this is shown in Figure 8.8b.

The simplest way to create an optical trap is just to focus a Gaussian laser beam through an objective lens creating a single point trap like the one that has been used here for comparison

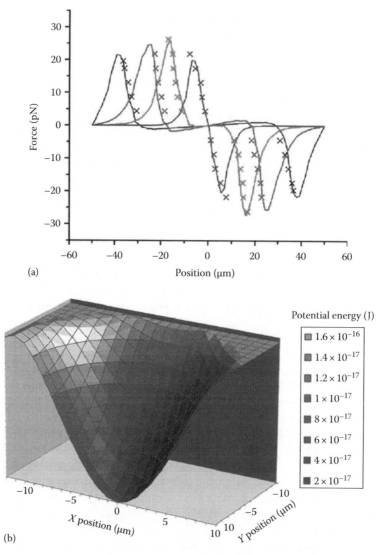

FIGURE 8.8   (See color insert.) (a) Shows simulated and experimentally measured forces created by four different-sized OET traps (trap diameters of 12 μm in blue to 73 μm in black) on a HeLa cell. (b) The potential energy well for a cell in the smallest trap is shown. (From Neale, S.L. et al., *Opt. Express*, 17, 5231, 2009. Copyright 2009 Optical Society of America.)

purposes. There are other forms of OTs that provide higher forces; one example of this is the optical stretcher which relies on two counter-propagating laser beams from a pair of optical fibers. This allows trapping to be produced with the optical scattering force rather than the gradient force and can deliver forces up to nN compared to the pN that are routinely found in standard OT and OET experiments. This device has been specifically developed to measure the Young's modulus of cells, and rather than the OT and OET traps discussed earlier it can't be used to arbitrarily position cells but instead can only trap them in a single position.[66]

8.4.2   Two- or Three-Dimensional Control

An OT trap created by focusing a laser beam through a high NA objective will focus the light much more tightly in the plane perpendicular to the light beam's direction of propagation (usually

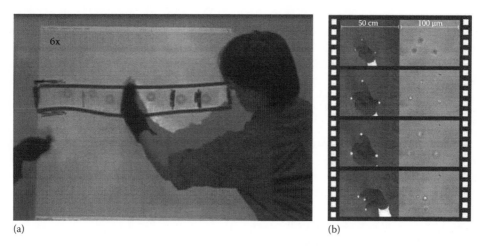

(a) (b)

FIGURE 8.9 (a) Shows an OET experiment where the view from the microscope is projected onto a wall and a camera is used to pick up black objects which are converted into a bright light pattern before being focused onto the OET device providing an intuitive user interface but only 2D control of the particles. (From Ohta, A.T. et al., *Proc. SPIE*, 6326, 632617, 2006.) (b) Shows an OT experiment where the traps are controlled by white dots placed on a gloved hand; this provides full 3D control. (From Whyte, G. et al., *Opt. Express*, 14, 12497, 2006. Copyright Optical Society of America.)

referred to as the x, y plane) than in the axis parallel to it (the z direction). As the light is focused in the z direction, a 3D trap is created so a particle can be moved in any direction; however, as the light is focused less strongly in the z direction, the trap stiffness in this direction is less. Most experiments involve moving particles in the x, y plane so this is rarely a limitation. In comparison, OET traps only create large electric field gradients close to the a-Si surface and so it is only possible to move particles around on this surface providing only 2D control. It is possible to repel particles from a surface using negative DEP and hold them there, and this has been demonstrated with metal electrode-based DEP experiments;[67] however, the level of control is not as refined as with OT as different particles can't be held at different heights. Figure 8.9 demonstrates this point with Figure 8.9a showing a intuitive user interface created for the OET trapping device. Here the OET device is set up under a microscope as usual however the view from the camera is projected onto a wall. A second camera is then placed so that it covers this image and can pick up anything placed onto the image. The image from this second camera is then inverted so that anything that is dark becomes bright and relayed back onto the OET device. This results in any dark object placed onto the projected image of the device becoming high electric filed regions on the device and hence being able to move the particles in the device allowing control of the particles by pushing at their projected images with black gloves.

Figure 8.9b shows a similarly intuitive control mechanism for an OT experiment.[69] Here white dots are placed onto a black glove which is moved around under a camera. The camera picks up the position and the apparent size of the white dots (giving the distance from the camera) and these 3D coordinates are then used to calculate the hologram required to produce this pattern of optical traps.

8.4.3 Medium Choice

Both OET and OT manipulations depend greatly on the medium being used to trap the particles in. For OT to create a trap, the medium must first be transparent to the wavelength of light being used and not contain too many particles as they will scatter the laser light (some research has been performed to mitigate the scattering of light by measuring the scattering and compensating for it[70]). Apart from this restriction, a wide range of media can be used, allowing the particles being manipulated to be suspended in anything from deionized water to highly conductive phosphate buffered saline (PBS) solutions which can be used to help keep biological cells closer to physiological conditions.

The OET device relies on switching the electric field from being across the photoconductive layer of the device when it is dark to being across the liquid layer where the device is illuminated. This puts a constraint on the conductivity of the liquid used to suspend the particles being manipulated. If the conductivity is too low, the majority of the field is always across the liquid, and if the conductivity is too high it is always across the photoconductor. The conventional OET device uses a one-micron-thick layer of a-Si as the photoconductor which results in the optimum liquid conductivity is around 10 mS/m. An iso-osmotic buffer solution can be created using sugars that has this low conductivity; however, the lack of ions means that cells can only be kept in this solution for short periods of time before it is unhealthy for them. To create an OET device that can be used at high conductivities, a phototransistor-based device has been created.[32] Here, ions have been embedded into a silicon substrate to create a phototransistor where the photoconductivity can be designed to match the conductivity of the liquid we wish to use.

As OT can also be performed in any transparent medium that has a refractive index difference with the particle being trapped, it can also be performed in air.[71] This is not possible in OET as the air is too insulating to form part of the electrical circuit.

8.4.4 PARTICLE COMPOSITION

The material composition of the particle being manipulated has a significant influence on the trapping force in both OT and OET traps. With OT, particles with a higher refractive index than the surrounding media experience a "positive" force, that is, toward the area of high intensity, whereas lower refractive index particles experience a "negative" force away from this region. This has been used to sort particles in a 3D optical lattice;[72] however, most biological particles have a refractive index that is slightly higher than the aqueous solution they are suspended in so they feel similar positive forces and hence in practice the challenge is to sort one species of particle that feel a certain force from another that feel a stronger or weaker positive force. As the size of the force depends on almost every physical aspect of the particle such as size, shape, composition, and granularity, the objects that are very physically different are easy to sort between, such as erythrocytes and leukocytes, whereas objects that are very similar in physical form are difficult to distinguish. OET can also be used to sort between different species of particles, and here the force is proportional to the relative polarizability of the particle and the liquid at the frequency of the applied AC as given by the CM factor (see Equation 8.5). As the complex permittivity of the particle will be frequency dependent, this allows us to tune the force experienced by the particle by changing the AC frequency. This is a very powerful technique if a condition can be found so that the CM factor is one sign for the species we wish to sort and the other sign for all the other particles in the solution. Similarly to OT, this condition is more likely to exist where there is a large physical difference between the different species of particle; however, unlike OT we can also have a large effect on the trapping conditions by changing the liquid's conductivity. This provides us with two parameters that can be tuned to try and find the optimal condition where we can get good separation between the two species. One example of where this has been demonstrated is with the separation of a single-celled blood-borne parasite called a trypanosome from the red blood cells that make up the vast majority of cells in whole blood. The diagnosis of infection by this parasite (the disease often known as African sleeping sickness) is made difficult by the variant surface glycoprotein (SVG) coating that continually changes, making it impossible to tag by the usual biological methods. Current separation methods use difference in the charge on the parasite and the other cells in the blood, but this technology also relies on centrifuging the sample and can't be miniaturized easily into a point of care device. As cheap and simple to administer drugs are available, if it is diagnosed early, a sensitive and selective point of care device is the missing tool needed to eradicate this disease.

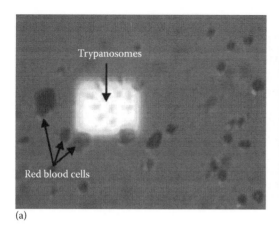 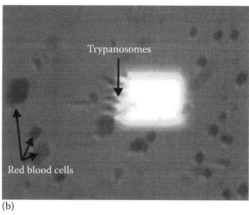

(a) (b)

FIGURE 8.10 (a) A sample of trypanosomes and red blood cells is placed into the OET device; the trypanosomes are attracted to the high field light region by positive DEP and the red blood cells are repelled. (b) When the light spot is moved, the trypanosomes follow it while the red blood cells are left behind. (From Clemens, K. et al., Manipulating blood borne parasites with optoelectronic tweezers (OET), Institute of Physics (IOP) Photon10, Southampton, UK, pp. 23–26, (August 2010).)

Figure 8.10 shows the selective manipulation of trypanosomes (these are *Trypanosoma cyclops*, a nonhuman infectious relative of the *Trypanosoma brucei* gambiense, which causes African sleeping sickness) from red blood cells.[73] Here a medium conductivity and AC frequency are used that result in a positive DEP force on the trypanosomes and a negative force on the red blood cells. This would allow the concentration of trypanosomes that could then be detected optically by observing their movement as in the current diagnostic method but would not require any lab-based procedures and could be miniaturized into a point of care device.

8.4.5 Effect of Trapping on the Particles

One of the drawbacks of the need for a high-intensity laser source in OT for stable trapping is the possibility of damaging the particle under manipulation. Several studies have quantified the effect of optical trapping on biological samples.[20,21,74–77] Moreover, prior experiments with a single-beam infrared laser trap has showed that high laser powers led to intense heating and scattering forces that prevented the optical trapping of silver nanowires.[16] OTs have been used to trap metallic spherical nanoparticles of different sizes;[25,78] however, the high optical power intensities required for stable trapping ($\sim 10^7$ W/cm^2) result in excessive heating in metallic nanoparticles ($\Delta T > 55°C$),[79] hampering the application of OT-trapped metallic nanoparticles in biological environments.

In the OET device, the main effect on the particles is the temperature increase during the particle manipulation process. The temperature increase during the OET manipulation process is due to two effects: (1) the absorption of trapping source illumination in the particles. This is the main factor contributing to high temperature increase in metallic nanoparticles trapped with OT. However, this effect is almost negligible in the case of OET since the trapping sources are orders of magnitude less in optical intensity relative to OT. (2) The source of temperature increase in OET is due to the joule heating of the liquid solution and the photoconductive layer. For typical OET operating conditions, the temperature increase[80] is approximately 2°C–3°C which is within acceptable range for biological experiments. Moreover, culturing of mammalian cells for 4–5 days while the cells were being trapped continuously in an open-access phOET device has been recently demonstrated.[81] OET is also capable of manipulation of metallic nanostructures such as silver nanowires,[31] carbon nanotubes,[56] and gold spherical nanoparticles[57] without damaging the structures or excessive heating. Further studies in this area to characterize the trapping effects can certainly benefit both techniques.

8.5   CONCLUSION

In conclusion, we have discussed the two main optical manipulation techniques for manipulation of micro- and nanoscale particles, namely, OT and OET in this chapter. Each technique has its own unique strengths and weaknesses as we reviewed in this chapter. However, these optical techniques are important tools for dynamic, flexible, and noninvasive manipulation of micro- and nanoparticles which play a critical role in the advancement of science and technology.

REFERENCES

1. Chronis, N. and Lee, L. Electrothermally activated SU-8 microgripper for single cell manipulation in solution. *Journal of Microelectromechanical Systems* **14**, 857–863 (2005).
2. Keller, C. and Howe, R. Hexsil tweezers for teleoperated micro-assembly. *Micro Electro Mechanical Systems, 1997. MEMS '97, Proceedings, IEEE, Tenth Annual International Workshop*, Nagoya, Japan, pp. 72–77 (1997). DOI:10.1109/MEMSYS.1997.581771.
3. Kim, C., Pisano, A., and Muller, R. Silicon-processed overhanging microgripper. *Journal of Microelectromechanical Systems* **1**, 31–36 (1992).
4. Barry, C.R., Gu, J., and Jacobs, H.O. Charging process and Coulomb-force-directed printing of nanoparticles with sub-100-nm lateral resolution. *Nano Letters* **5**(10), 2078–2084 (2005).
5. Cabrera, C.R. and Yager, P. Continuous concentration of bacteria in a microfluidic flow cell using electrokinetic techniques. *Electrophoresis* **22**, 355–362 (2001).
6. Jones, T.B. *Electromechanics of Particles*. Cambridge University Press, Cambridge, U.K. (1995).
7. Fu, A.Y., Spence, C., Scherer, A., Arnold, F.H., and Quake, S.R. A microfabricated fluorescence-activated cell sorter. *Nature Biotechnology* **17**, 1109–1111 (1999).
8. Loucaides, N.G., Ramos, A., and Georghiou, G.E. Trapping and manipulation of nanoparticles by using jointly dielectrophoresis and AC electroosmosis. *Journal of Physics: Conference Series* **100**, 052015 (2008).
9. Carlo, D.D., Wu, L.Y., and Lee, L.P. Dynamic single cell culture array. *Lab on a Chip* **6**, 1445 (2006).
10. Wheeler, A.R. et al. Microfluidic device for single-cell analysis. *Analytical Chemistry* **75**, 3581–3586 (2003).
11. Fu, A.Y., Chou, H., Spence, C., Arnold, F.H., and Quake, S.R. An integrated microfabricated cell sorter. *Analytical Chemistry* **74**, 2451–2457 (2002).
12. Lee, H., Purdon, A.M., Chu, V., and Westervelt, R.M. Controlled assembly of magnetic nanoparticles from magnetotactic bacteria using microelectromagnets arrays. *Nano Letters* **4**, 995–998 (2004).
13. Bentley, A.K., Trethewey, J.S., Ellis, A.B., and Crone, W.C. Magnetic manipulation of copper–tin nanowires capped with nickel ends. *Nano Letters* **4**, 487–490 (2004).
14. Tanase, M. et al. Magnetic alignment of fluorescent nanowires. *Nano Letters* **1**, 155–158 (2001).
15. Ashkin, A. Acceleration and trapping of particles by radiation pressure. *Physical Review Letters* **24**, 156 (1970).
16. Pauzauskie, P.J. et al. Optical trapping and integration of semiconductor nanowire assemblies in water. *Nature Materials* **5**, 97–101 (2006).
17. Ashkin, A., Dziedzic, J.M., and Yamane, T. Optical trapping and manipulation of single cells using infrared laser beams. *Nature* **330**, 769–771 (1987).
18. Dholakia, K. and Reece, P. Optical micromanipulation takes hold. *Nano Today* **1**, 18–27 (2006).
19. Grier, D.G. A revolution in optical manipulation. *Nature* **424**, 810–816 (2003).
20. Mohanty, S.K., Rapp, A., Monajembashi, S., Gupta, P.K., and Greulich, K.O. Comet assay measurements of DNA damage in cells by laser microbeams and trapping beams with wavelengths spanning a range of 308 nm to 1064 nm. *Radiation Research* **157**, 378–385 (2002).
21. Neuman, K., Chadd, E., Liou, G., Bergman, K., and Block, S. Characterization of photodamage to *Escherichia coli* in optical traps. *Biophysical Journal* **77**, 2856–2863 (1999).
22. Chiou, P.Y., Ohta, A.T., and Wu, M.C. Massively parallel manipulation of single cells and microparticles using optical images. *Nature* **436**, 370–372 (2005).
23. Neuman, K.C. and Block, S.M. Optical trapping. *Review of Scientific Instruments* **75**, 2787 (2004).
24. Svoboda, K. and Block, S.M. Biological applications of optical forces. *Annual Review of Biophysics and Biomolecular Structure* **23**, 247–285 (1994).
25. Svoboda, K. and Block, S.M. Optical trapping of metallic Rayleigh particles. *Optics Letters* **19**, 930–932 (1994).

26. Block, S.M., Goldstein, L.S.B., and Schnapp, B.J. Bead movement by single kinesin molecules studied with optical tweezers. *Nature* **348**, 348–352 (1990).

27. Wang, M., Yin, H., Landick, R., Gelles, J., and Block, S. Stretching DNA with optical tweezers. *Biophysical Journal* **72**, 1335–1346 (1997).

28. Dufresne, E.R., Spalding, G.C., Dearing, M.T., Sheets, S.A., and Grier, D.G. Computer-generated holographic optical tweezer arrays. *Review of Scientific Instruments* **72**, 1810 (2001).

29. Dufresne, E.R. and Grier, D.G. Optical tweezer arrays and optical substrates created with diffractive optics. *Review of Scientific Instruments* **69**, 1974 (1998).

30. Jamshidi, A. et al. Optofluidics and optoelectronic tweezers. *Proceedings of SPIE,* **6993**, 69930A.1–69930A.10 (2008). DOI:10.1117/12.787019.

31. Jamshidi, A. et al. Dynamic manipulation and separation of individual semiconducting and metallic nanowires. *Nature Photonics* **2**, 86–89 (2008).

32. Hsu, H. et al. Phototransistor-based optoelectronic tweezers for dynamic cell manipulation in cell culture media. *Lab on a Chip* **10**, 165 (2010).

33. Higuchi, Y. et al. Manipulation system for nano/micro components integration via transportation and self-assembly. *IEEE 21st International Conference on Micro Electro Mechanical Systems, 2008, MEMS 2008*, Tuscon, AZ, pp. 836–839 (2008). DOI:10.1109/MEMSYS.2008.4443786.

34. Miao, X., Wilson, B.K., Pun, S.H. and Lin, L.Y. Optical manipulation of micron/submicron sized particles and biomolecules through plasmonics. *Optics Express* **16**, 13517–13525 (2008).

35. Miao, X. and Lin, L. Trapping and manipulation of biological particles through a plasmonic platform. *IEEE Journal of Selected Topics in Quantum Electronics* **13**, 1655–1662 (2007).

36. Wang, W., Lin, Y., Wen, T., Guo, T. and Lee, G. Selective manipulation of microparticles using polymer-based optically induced dielectrophoretic devices. *Applied Physics Letters* **96**, 113302 (2010).

37. Ohta, A. et al. Optically controlled cell discrimination and trapping using optoelectronic tweezers. *IEEE Journal of Selected Topics in Quantum Electronics* **13**, 235–243 (2007).

38. Ohta, A., Neale, S., Hsu, H.-Y., Valley, J. and Wu, M. Parallel assembly of nanowires using lateral-field optoelectronic tweezers. *IEEE/LEOS International Conference on Optical MEMs and Nanophotonics 2008*, Freiburg, Germany, pp. 7–8 (2008). DOI:10.1109/OMEMS.2008.4607801.

39. Ohta, A. et al. Trapping and transport of silicon nanowires using lateral-field optoelectronic tweezers. *Conference on Lasers and Electro-Optics, 2007, CLEO 2007*, Baltimore, MD, pp. 1–2 (2007). DOI:10.1109/CLEO.2007.4452753

40. Neale, S. et al. Optofluidic assembly of red/blue/green semiconductor nanowires. *Conference on Lasers and Electro-Optics, 2009 and 2009 Conference on Quantum electronics and Laser Science Conference, CLEO/QELS 2009*, Baltimore, MD, pp. 1–2 (2009).

41. Park, S. et al. Floating electrode optoelectronic tweezers: Light-driven dielectrophoretic droplet manipulation in electrically insulating oil medium. *Applied Physics Letters* **92**, 151101 (2008).

42. Hwang, H. et al. Reduction of nonspecific surface-particle interactions in optoelectronic tweezers. *Applied Physics Letters* **92**, 24108 (2008).

43. Shah, G.J., Ohta, A.T., Chiou, E.P., Wu, M.C., and Kim, C. EWOD-driven droplet microfluidic device integrated with optoelectronic tweezers as an automated platform for cellular isolation and analysis. *Lab on a Chip* **9**, 1732 (2009).

44. Chiou, P.Y., Wong, W., Liao, J., and Wu, M. Cell addressing and trapping using novel optoelectronic tweezers. *17th IEEE International Conference on. Micro Electro Mechanical Systems 2004 (MEMS)*, Maastricht, the Netherlands, pp. 21–24 (2004). DOI:10.1109/MEMS.2004.1290512.

45. Choi, W., Kim, S., Jang, J. and Park, J. Lab-on-a-display: A new microparticle manipulation platform using a liquid crystal display (LCD). *Microfluid Nanofluid* **3**, 217–225 (2006).

46. Hwang, H. et al. Interactive manipulation of blood cells using a lens-integrated liquid crystal display based optoelectronic tweezers system. *Electrophoresis* **29**, 1203–1212 (2008).

47. Chiou, P.-Y., Ohta, A., Jamshidi, A., Hsu, H.-Y., and Wu, M. Light-actuated AC electroosmosis for nanoparticle manipulation. *Journal of Microelectromechanical Systems* **17**, 525–531 (2008).

48. Valley, J., Jamshidi, A., Ohta, A., Hsu, H.-Y., and Wu, M. Operational regimes and physics present in optoelectronic tweezers. *Journal of Microelectromechanical Systems* **17**, 342–350 (2008).

49. Lu, Y., Huang, Y., Yeh, J.A., and Lee, C. Controllability of non-contact cell manipulation by image dielectrophoresis (iDEP). *Optical and Quantum Electronics* **37**, 1385–1395 (2006).

50. Neale, S.L., Mazilu, M., Wilson, J.I.B., Dholakia, K., and Krauss, T.F. The resolution of optical traps created by light induced dielectrophoresis (LIDEP). *Optics Express* **15**, 12619–12626 (2007).

51. Tien, M., Ohta, A.T., Yu, K., Neale, S.L., and Wu, M.C. Heterogeneous integration of InGaAsP microdisk laser on a silicon platform using optofluidic assembly. *Applied Physics A* **95**, 967–972 (2009).

52. Ohta, A. et al. Dynamic cell and microparticle control via optoelectronic tweezers. *Journal of Microelectromechanical Systems* **16**, 491–499 (2007).

53. Valley, J.K. et al. Preimplantation mouse embryo selection guided by light-induced dielectrophoresis. *PLoS ONE* **5**, e10160 (2010).

54. Ohta, A.T. et al. Motile and non-motile sperm diagnostic manipulation using optoelectronic tweezers. *Lab on a Chip* **10**, 3213 (2010).

55. Hsu, H. et al. Sorting of differentiated neurons using phototransistor-based optoelectronic tweezers for cell replacement therapy of neurodegenerative diseases. *International Solid-State Sensors, Actuators and Microsystems Conference, 2009, TRANSDUCERS 2009,* Denver, CO, pp. 1598–1601 (2009). DOI:10.1109/SENSOR.2009.5285764.

56. Pauzauskie, P.J., Jamshidi, A., Valley, J.K., Satcher, J.H., and Wu, M.C. Parallel trapping of multiwalled carbon nanotubes with optoelectronic tweezers. *Applied Physics Letters* **95**, 113104 (2009).

57. Jamshidi, A. et al. Metallic nanoparticle manipulation using optoelectronic tweezers. *IEEE 22nd International Conference on Micro Electro Mechanical Systems, 2009, MEMS 2009,* Sorrento, Italy, pp. 579–582 (2009). DOI:10.1109/MEMSYS.2009.4805448.

58. Lin, Y., Chang, C., and Lee, G. Manipulation of single DNA molecules by using optically projected images. *Optics Express* **17**, 15318–15329 (2009).

59. Hoeb, M., Rädler, J.O., Klein, S., Stutzmann, M., and Brandt, M.S. Light-induced dielectrophoretic manipulation of DNA. *Biophysical Journal* **93**, 1032–1038 (2007).

60. Valley, J.K. et al. Parallel single-cell light-induced electroporation and dielectrophoretic manipulation. *Lab on a Chip* **9**, 1714 (2009).

61. Lin, Y. and Lee, G. An optically induced cell lysis device using dielectrophoresis. *Applied Physics Letters* **94**, 033901 (2009).

62. Lin, Y. and Lee, G. Optically induced flow cytometry for continuous microparticle counting and sorting. *Biosensors and Bioelectronics* **24**, 572–578 (2008).

63. Jamshidi, A. et al. NanoPen: Dynamic, low-power, and light-actuated patterning of nanoparticles. *Nano Letters* **9**, 2921–2925 (2009).

64. Schwarz, R., Wang, F., and Reissner, M. Fermi level dependence of the ambipolar diffusion length in amorphous silicon thin film transistors. *Applied Physics Letters* **63**, 1083 (1993).

65. Neale, S.L. et al. Trap profiles of projector based optoelectronic tweezers (OET) with HeLa cells. *Optics Express* **17**, 5231–5239 (2009).

66. Guck, J. Optical deformability as an inherent cell marker for testing malignant transformation and metastatic competence. *Biophysical Journal* **88**, 3689–3698 (2005).

67. Mike Arnold, W. and Franich, N.R. Cell isolation and growth in electric-field defined micro-wells. *Current Applied Physics* **6**, 371–374 (2006).

68. Ohta, A.T., Chiou, P., and Wu, M.C. Optically controlled manipulation of live cells using optoelectronic tweezers. *Proceedings of SPIE* **6326**, 632617–632617-11 (2006). DOI:10.1117/12.679710.

69. Whyte, G. et al. An optical trapped microhand for manipulating micron-sized objects. *Optics Express* **14**, 12497–12502 (2006).

70. Cizmar, T., Mazilu, M., and Dholakia, K. in situ wavefront correction and its application to micromanipulation. *Nature Photonics* **4**, 388–394 (2010).

71. Burnham, D.R. and McGloin, D. Holographic optical trapping of aerosol droplets. *Optics Express* **14**, 4175–4181 (2006).

72. MacDonald, M.P., Spalding, G.C., and Dholakia, K. Microfluidic sorting in an optical lattice. *Nature* **426**, 421–424 (2003).

73. Clemens, K., Neale, S., Barrett, M., and Cooper, J. Manipulating blood borne parasites with optoelectronic tweezers (OET). *Institute of Physics (IOP) Photon10*, Southampton, UK, pp. 23–26, (August 2010).

74. Liang, H. et al. Wavelength dependence of cell cloning efficiency after optical trapping. *Biophysical Journal* **70**, 1529–1533 (1996).

75. Liu, Y., Sonek, G.J., Berns, M.W., and Tromberg, B.J. Physiological monitoring of optically trapped cells: Assessing the effects of confinement by 1064-nm laser tweezers using microfluorometry. *Biophysics Journal* **71**, 2158–2167 (1996).

76. Leitz, G. Stress response in *Caenorhabditis elegans* caused by optical tweezers: Wavelength, power, and time dependence. *Biophysical Journal* **82**, 2224–2231 (2002).

77. Wang, M.M. et al. Microfluidic sorting of mammalian cells by optical force switching. *Nature Biotechnology* **23**, 83–87 (2005).

78. Hansen, P.M., Bhatia, V.K., Harrit, N., and Oddershede, L. Expanding the optical trapping range of gold nanoparticles. *Nano Letters* **5**, 1937–1942 (2005).

79. Seol, Y., Carpenter, A.E., and Perkins, T.T. Gold nanoparticles: Enhanced optical trapping and sensitivity coupled with significant heating. *Optics Letters* **31**, 2429–2431 (2006).

80. Pauzauskie, P.J. et al. Quantifying heat transfer in DMD-based optoelectronic tweezers with infrared thermography. *Proceedings of SPIE,* **7596**, 759609-759609-8 (2010). DOI:10.1117/12.846247

81. Hsu, H.Y. et al. Open-access phototransistor-based optoelectronic tweezers for long-term single-cell heterogeneity study. *24th International MEMS Conference*, Cancun, NM, pp. 63–66, 2011.

9 Design of Optically Driven Microrotors

*Halina Rubinsztein-Dunlop, Theodor Asavei, Alexander
B. Stilgoe, Vincent L. Y. Loke, Robert Vogel, Timo
A. Nieminen, and Norman R. Heckenberg*

CONTENTS

9.1   INTRODUCTION

9.1.1   MICROMACHINES

Richard Feynman anticipated the idea of micromachines in his 1959 talk, "There's plenty of room at the bottom" (Feynman 1960). Feynman suggested as a possible micromachine a micromechanical "surgeon," which could be swallowed and operate inside a faulty blood vessel. At the end of his talk

Feynman offered two prizes, one for building an operating electric motor only the size of a 1/64 in.[3] and the other for writing a page with letters 1/25,000 smaller than in normal text. Not long after, in November 1960, the first micromachine (the 1/64 in.[3] motor) was built by James McLellan, an electrical engineer. The second prize was won in 1985 by Tom Newman, a Stanford graduate student, who wrote the first page of "A Tale of Two Cities" in polymethyl methacrylate (PMMA) resist on a silicon nitride membrane by means of electron beam lithography (Newman et al. 1987).

At the end of the 1980s and the beginning of the 1990s, microelectromechanical systems (MEMS) came into play. MEMS, commonly known as micromachines, are electromechanical machines that range in size from a micrometer to a millimeter. They are built using techniques based on integrated circuit (IC) fabrication methods. The first reported MEMS were silicon electrical micromotors with a diameter of 100 μm (Fan et al. 1989, Mehregany et al. 1990). Nowadays, MEMS are used as sensors and actuators in various applications. Some of the common applications include accelerometers that trigger airbags in cars, inkjet printers, and optical switches for data communications. Although not yet very common, MEMS are finding their ways into medical applications such as blood pressure sensors (Benzel et al. 2004, Goh and Krishnan 1999, Ishiyama et al. 2002).

However, MEMS are not the only available micromachines. Another type of micromachine is the optically fabricated type, based on two-photon photopolymerization (2PP) of UV curing resins. They emerged as a consequence of the rapid prototyping techniques that became available in the 1980s, especially the stereolithography technique (Deitz 1990). Since then there have been several groups around the world that have demonstrated the production of very high–resolution 3D structures.

9.1.2 Drive Mechanisms for Micromachines

The most common drive mechanism encountered in surface-micromachined silicon-based elements is the electrical drive. This is achieved by means of electrostatic forces that convert electrical energy to mechanical energy. The basic elements of the electrically driven micromachine are the stator and the rotor. By applying a three-phase voltage difference between the stator and the rotor, one can achieve continuous motion of the rotor. However, micromachines can also be driven by thermal energy through thermal expansion (Oliver et al. 2003), by magnetic fields (Nishimura et al. 2003), by chemical energy (Hiratsuka et al. 2006), or by a tightly focused laser beam (optical drive). Optical drive is based on the fact that microscopic particles (25 nm–10 μm) can be stably trapped in a tightly focused laser beam due to gradient forces. The advantage of optical drive is that no contact is required and so it is particularly well suited to biological applications.

This type of drive is particularly suited to optically fabricated micromachines because they can be made out of transparent polymers. However, microstructures built by surface micromachining can be optically driven too. The earliest reported optically driven micro-objects were fabricated by means of reactive ion-beam etching of a 10 μm thick silicon dioxide (SiO_2) layer (Higurashi et al. 1994), resulting in 10–25 μm diameter microrotors which could be rotated in the optical trap. Our group also reported rotation in optical tweezers of surface-micromachined SiO_2 micromachine elements (Friese et al. 2001). These structures were produced using electron beam lithography and double lift off. If the micromachines are to be used in biological applications, they need to be made out of biocompatible materials and have a size that can be easily tailored to the studied systems, in which case optically fabricated and driven micromachines are most suitable. The first optically fabricated and optically driven micromachines were produced in 2001 (Galajda and Ormos 2001). They produced chiral micro-objects (helices, sprinklers, propellers) that could be spun in an optical trap.

9.1.3 Phase Singular Fields and Orbital Angular Momentum

From an experimental point of view, modal patterns in lasers in which the output beam takes the form of a circularly symmetric ring with a phase singularity seen as a dark spot of zero intensity, also called "donut" modes, can be observed in stable-resonator lasers in which the mode-controlling

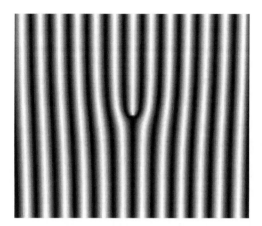

FIGURE 9.1 Interference pattern between a plane wave and an LG_{02} mode used to create an off-axis computer-generated hologram.

aperture is made slightly too large (Siegman 1986). Phase singularities in laser fields can also be identified by interferometry (White et al. 1991). The phase singularity is defined as the point where the closed path phase gradient integral around it is an integer multiple of 2π (Coulett et al. 1989):

$$\oint \nabla \phi \cdot dl = 2\pi m \tag{9.1}$$

where the integer m is the topological charge of the singularity.

These modes, also called Laguerre–Gauss (LG) modes, are of importance because they are free-space modes that propagate with constant shape and carry well-defined orbital angular momentum (OAM) (Allen et al. 1992). The amplitude of LG modes has an azimuthal angular dependence in the form $\exp(i\ell\phi)$ where ℓ is the azimuthal mode index or the topological charge. The OAM is associated with the helical phase structure of LG modes.

LG modes can be experimentally generated and implemented in an optical tweezers setup by using off-axis computer-generated holograms (He et al. 1995b, Heckenberg et al. 1992). These are created by simulating an interference pattern between a slightly inclined plane wave and the desired LG mode. Once the pattern is formed (Figure 9.1), the image is transferred to a 35 mm photographic film and then it is transferred to a holographic plate with thick emulsion using contact printing. By bleaching the emulsion layer on the plate, a phase hologram is made with regions of different optical thickness, which is determined by the exposure time during printing.

When illuminated by a Gaussian beam, the hologram will generate the desired LG mode in the first diffracted order. Computer-generated holograms have been used to produce beams carrying OAM, which is transferred to optically trapped absorbing microscopic particles (Friese et al. 1996, He et al. 1995a). Later in this chapter, work is described where they are used to transfer OAM to non-absorbing micromachines with discrete rotational symmetry.

9.1.4 Symmetry and Optically Driven Micromachines

The shape of the designed micromachines plays an important role in the interaction with the driving beam, and the most important aspect is the symmetry of the microstructure (Nieminen et al. 2004). If a scatterer has pth-order discrete rotational symmetry, an incident LG beam with azimuthally periodic wavefronts and with $m_0\hbar$ angular momentum per photon about the symmetry axis of the particle will be scattered into modes with $m_i\hbar$ angular momentum per photon, where $m_i = m_0 + ip$ and i is an integer. This result comes as a consequence of Floquet's theorem and can be understood from the following analogy. The azimuthal dependence of the fields that are incident on a rotationally symmetric object is analogous to the case of a planar wave incident on a planar grating. If the component of the wavevector of the incident wave along the grating is k_x and the component of the reciprocal lattice vector of the grating is q_x, the scattered waves have a discrete spectrum with wavevectors given by $k_{kj} = k_x + jq_x$, where j is the order of scattering. Thus, in the case of symmetric microstructures, a discrete azimuthal spectrum with azimuthal mode indices m_i is obtained.

Throughout this chapter, the micromachine elements discussed are driven by means of optical tweezers in which the beam is either a Gaussian beam or an LG beam of order ℓ. These beams have a well-defined OAM per photon, and we can consider the beam to be composed of two circularly polarized components with $(\ell \pm 1)\hbar$ total angular momentum per photon for each of the two components. Since a rotationally symmetric optical system will not change the total angular momentum per photon (Nieminen et al. 2008, Waterman 1971), this will also be the total angular momentum of the highly focused trapping beam at the focus. A non-paraxial beam can be represented by a superposition of vector spherical wavefunctions (VSWFs) having an azimuthal phase variation of $\exp(im\phi)$, where m is the azimuthal mode index such that the angular momentum component in the beam direction of propagation is equal to $m\hbar$ per photon. Thus, the non-paraxial VSWF representation of the beam will consist of modes with $m = l \pm 1$. Considering a pth-order rotationally symmetric microstructure, the driving incident modes will scatter into modes with $m = \ell \pm 1$, $\ell \pm 1 \pm p$, $\ell \pm 1 \pm 2p$, $\ell \pm 1 \pm 3p$, etc. In the case of $p = 2$, such as elongated or flattened particles, the left and right circularly polarized incident modes will scatter into the same set of modes, resulting in interference and affecting the polarization of the scattered light, which is responsible for the shape birefringence of such particles. If $p > 2$, the scattered modes corresponding to the two incident circular polarizations are distinct, and hence the incident polarization will only be affected weakly, meaning that the torque will be primarily due to OAM contribution. Also, if the incident beam is linearly or elliptically polarized, the total torque due to the combination of left and right circular modes will be the sum of the torques due to each circular polarization.

We can assume that the scattering to the lowest orders ($i = 0, \pm 1$) is usually the strongest and hence most light is likely to be scattered to the azimuthal orders with $m = m_0$, $m_0 \pm p$. If the object is achiral, meaning that the object and its mirror image are identical, the scattering is independent of the handedness of the angular momentum, that is to say the coupling of m_1–m_2 will be the same as from $-m_1$ to $-m_2$, and therefore an achiral rotor with $p > 2$ will experience no torque in a linearly polarized Gaussian beam but will be ideal for rotation in LG beams with equal torques in both directions. On the other hand, a rotor that is chiral, meaning it is not mirror symmetric, will rotate in a Gaussian beam and also in an LG beam, but in the latter case at different speeds depending on the handedness of the beam.

It is worth mentioning here that the micromachines produced by 2PP are weakly reflecting in optical tweezers, and hence the majority of the scattered light is the transmitted light. The coupling to the different scattered modes will depend on the phase variations in the transmitted light and therefore the optically driven microrotors can be viewed as microscopic holograms.

As seen in Figure 9.2, a simple possible design of microrotors is based on the on-axis holograms used for conversion of Gaussian beams into LG beams (Heckenberg et al. 1992). For the chiral structure the scattering to positive and negative orders will differ, and a torque will result from the generation of OAM by the structure and hence it could be rotated by a Gaussian beam. In this case, the direction of rotation is dependent on the chirality of the microstructure as seen by the incident Gaussian beam.

However, if the desired outcome is a microstructure that can be rotated with equal torque in both directions, the suitable shape is one that has mirror symmetry about a plane containing the axis of p-fold rotational symmetry, which is the achiral structure depicted in Figure 9.2. This type of structure is ideal for driving by a beam containing OAM. If the incident beam has $m_0 = 2$, there

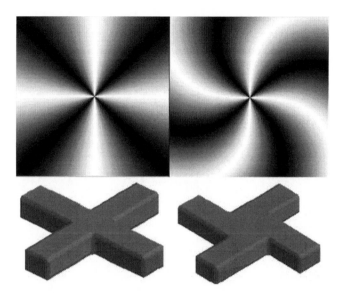

FIGURE 9.2 Upper row: on-axis holograms for the generation of $LG_{0,\pm4}$ Laguerre–Gauss beams. The achiral hologram (left) will produce equal amplitude LG_{04} and $LG_{0,-4}$ components in the transmitted beam when illuminated by a Gaussian beam and therefore no torque will be exerted. The chiral hologram (right) is the result of the interference between an LG_{04} beam and a diverging spherical wave or an $LG_{0,-4}$ beam and a converging spherical wave. In this case, the curvature of the wavefronts of an illuminating Gaussian beam will determine the direction of the torque. Lower row: planar structures that are binary approximations of the holograms in the upper row; achiral (left) and chiral (right).

will be scattering to VSWF modes with $m=-2$ and a torque will result with a direction that can be reversed by changing the handedness of the driving beam.

Based on the above principles, the practical realization of fourfold symmetry microrotors and their performance in optical tweezers will be discussed later. From a practical point of view, when trapped, the symmetry axis of the rotors should be parallel to the beam axis. This can be ensured by fabricating a central stalk, which will align with the beam axis. Also, from the "hologram picture" it is desirable that the thickness of the microstructure should be such that the phase difference between light that passes through the structure and light that misses it is half the wavelength of the trapping beam. In this case, the thickness of the rotor arms should be $\lambda_{med}/(2(n_r - n_w))$, where λ_{med} is the wavelength of the trapping beam in the trapping fluid, n_r is the refractive index of the microfabricated object, that is, of the photopolymerized resin ($n_r = 1.56$), and n_w is the refractive index of the fluid, in our case, water ($n_w = 1.33$). For a 1070 nm trapping beam the desired thickness should be approximately 2 μm. This can be readily achieved by 2PP technique.

9.1.5 EXPERIMENTAL SETUP

The microstructures are fabricated by photopolymerization of NOA resins (Norland Products, Inc.), which are based on a mixture of photoinitiator molecules and thiol-ene or acrylic monomers. The photopolymerization setup is based on an "in-house" built inverted microscope shown in Figure 9.3. For 2PP we use the infrared light ($\lambda = 780$ nm) produced by a femtosecond pulse Ti:Sapphire laser (Tsunami, Spectra Physics) pumped by a 532 nm solid-state laser (Millenia, Spectra Physics). The pulse length is 80 fs with 80 MHz repetition rate. The laser beam is attenuated to the power needed for polymerization by a variable neutral density filter wheel (ND filter in Figure 9.3) and then passes through a computer controlled shutter and is reflected into the objective lens by a dichroic mirror. The objective lens is an Olympus 100×oil immersion lens with high numerical aperture (N.A. = 1.3) to achieve high spatial resolution for polymerization.

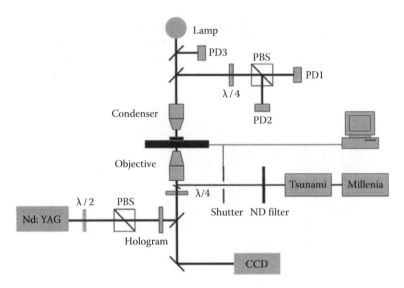

FIGURE 9.3 Schematic diagram of the photopolymerization setup for fabrication of microstructures combined with an optical tweezers setup for optical trapping of the microstructures and torque measurements (λ/2—half-wave plate; λ/4—quarter-wave plate; CCD—charged coupled device; PBS—polarizing beam splitter; ND—neutral density; PD—photodetector).

The sample is mounted on a computer-controlled high-precision piezo stage (model P-611.3S, Physik Instrumente) and is imaged onto a CCD with the same objective lens. The travel range of the piezo stage is 100 μm in all X, Y, and Z directions with a resolution of 2 nm.

The resin sample is sandwiched between two glass cover slips, which are separated by an adherent spacer (Parafilm, Structure Probe Inc., West Chester, PA) with a thickness of 127 μm. The 3D structures are fabricated by raster scanning the resin sample over the laser beam using the piezo stage. A scheme of the fabrication method is shown in Figure 9.4.

The 3D object is sliced into 2D layers (bitmap files) corresponding to the areas that need to be scanned. The program controlling the scanning stage reads the bitmap files; the resin is exposed (the shutter is opened) when the pixel in the bitmap is black (has value 0) and the shutter closes when the pixel is white (has value 1). The 3D structures are obtained by moving the sample in the Z direction after each XY scan.

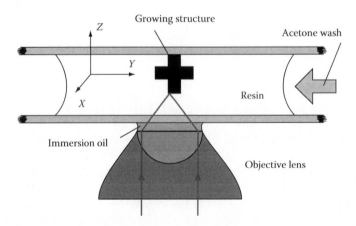

FIGURE 9.4 Schematics of the two-photon photopolymerization technique for fabrication of micron-sized objects out of UV-curable resins.

The bitmap resolution is set to 100×100 pixel^2, which corresponds to 10×10 μm^2 travel in X and Y directions so that each individual pixel is 100×100 nm^2 in size giving a lateral resolution of 100 nm. The offset in the Z direction is set to be 200 nm. The structures are grown upside down on the upper cover slip. This top down scanning method has the advantage that the laser beam does not pass through already exposed resin. After polymerization, the unexposed resin is washed off with acetone, leaving the 3D structure attached to the cover slip. The washing procedure is as follows: the top cover slip is carefully detached from the bottom one and the spacer, and then the cover slip is slightly tilted and afterwards acetone is pipetted onto the resin sample. Due to the small volume of the resin sample (around 0.1 mm^3), only three to four drops of acetone are enough to wash away the unpolymerized resin.

Optical trapping is performed in the same inverted microscope shown in Figure 9.3. The trapping laser is a 5 W, 1070 nm Nd:YAG fiber laser (IPG Photonics, Oxford, MA). It is focused on the sample by a $100 \times$ Olympus oil immersion objective lens with high numerical aperture (N.A. = 1.3). The output power is controlled by a half wave plate ($\lambda/2$) and a polarizing beam splitter cube (PBS). The sample is imaged onto a CCD camera by the same objective lens. Incident angular momentum of the trapping beam is controlled by using a quarter wave plate ($\lambda/4$) for spin angular momentum (SAM) and a computer-generated hologram creating LG modes with OAM.

The spin torque measurement is based on the fact that any coherent beam can be represented as a sum of two circularly polarized components with opposite handedness with a coefficient of circular polarization σ given by $\sigma_s = (P_L - P_R)/P$, where P_L and P_R are the powers of the left and right circularly polarized components, respectively, and P is the total power of the beam (Nieminen et al. 2001). Therefore, the spin torque τ_s can be written as $\tau_s = \Delta\sigma_s P/\omega$, with $\Delta\sigma_s$ being the change in the coefficient of circular polarization due to the spin angular momentum transfer, where P is the incident beam power and ω the beam angular frequency. Hence, by measuring $\Delta\sigma_s$ and knowing P and ω, the spin torque can be found. The change in the coefficient of circular polarization is measured by two photodetectors (PD1 and PD2), which are placed after a quarter wave plate and a polarizing beam splitter cube (see Figure 9.3). The outgoing beam is collimated by the condenser and then split into two orthogonal linearly polarized components by the quarter wave plate ($\lambda/4$) and the cube. The two linearly polarized components correspond to the left and right circularly polarized components of the outgoing beam. The two detectors measure the power of each beam, and hence the coefficient $\Delta\sigma_s$ can be found, which is directly related to the spin torque per photon. From the aforementioned equation $\tau_s = \Delta\sigma_s P/\omega$, writing $P = N\hbar\omega$ with N being the number of photons per unit time, one can find out that the spin torque per photon per second has the value of $\Delta\sigma_s \hbar$.

A small polarization-dependent fraction of the outgoing beam is also directed onto a third photodetector (PD3), which measures the intensity variation of the scattered beam and hence the rotation rate of the trapped particle. This is a much easier method for measuring the rotation rate in comparison to frame-by-frame video microscopy.

Similarly to the spin component of the torque, one can write the orbital torque τ_o as being $\tau_o = \Delta\sigma_o P/\omega$, with $\Delta\sigma_o$ being a coefficient related to the orbital torque per photon in the same way as $\Delta\sigma_s$ is related to the spin torque per photon. Thus, the orbital torque per photon on the object has the value of $\Delta\sigma_o \hbar$.

9.2 MICROROTORS DRIVEN BY SPIN ANGULAR MOMENTUM

9.2.1 Design and Microfabrication of Form Birefringent "Holey" Microspheres

For macroscopic objects, birefringence is a consequence of anisotropy at the molecular level and is an intrinsic property of the bulk material itself. However, birefringence can occur also in an optically isotropic material in certain conditions, for example, if there is an ordered arrangement of isotropic particles with sizes larger than molecular scale but comparable to the wavelength of light interacting with them. This type of birefringence, due to the shape of the object, is called shape or form birefringence (Born and Wolf 1999), and in this case, one can talk about an induced polarizability tensor rather than an intrinsic permittivity tensor. In this sense it is worth mentioning the behavior of elongated or flattened isotropic particles in an external electric field. They have different dielectric polarizabilities along their short and long axes and hence they behave as birefringent particles. A classical example is an ellipsoid in an external electric field for which the analytical solution of the polarizability tensor exists (Jones 1945, Stratton 1941).

One can see that the shape of the isotropic objects in an arrangement is crucial for the type of form birefringence in the following examples: an array of equally spaced thin parallel plates is negative uniaxial while an array of parallel cylindrical rods is positive uniaxial (Born and Wolf 1999). Based on these considerations, photolithographic fabrication of microgears with geometric anisotropy was reported (Neale et al. 2005). Due to their form birefringence, they could be rotated in a circularly polarized trapping beam. In our case, the motivation for fabricating form birefringence microstructures by means of 2PP came from previous work done on manipulation and rotation of biological molecules.

There has been an ongoing effort during the past decade to manipulate and probe individual biological molecules such as DNA or molecular motors (Oroszi et al. 2006, Ryu et al. 2000, Smith et al. 1992). In this type of study, forces and torques exerted on single molecules are measured in order to understand processes at the cellular level in which these molecules are involved. In order to perform these experiments one end of the molecule has to be attached to a fixed surface and the other to a micrometer-sized object that could be optically trapped. The most common particles used in optical trapping for these purposes are isotropic spheres, which are only suitable for force and displacement measurements through transfer of linear momentum. However, rotation of biomolecules plays a big role in cellular mechanisms like DNA supercoiling, transcription, or repair. In order to measure torques in an optical trap, one needs either birefringent or shape-birefringent particles.

Rotation measurements of biological molecules have been performed in our group using vaterite microspheres (Bishop et al. 2004). These microspheres are not very stable in the commonly used biological buffers unless some coating technique is used. Another type of birefringent micrometer-sized object was reported by Deufel et al. They used quartz microcylinders fabricated by standard photolithography and functionalized with biomolecules for DNA attachment in order to detect DNA supercoiling torque. A limitation of the cylindrical shape over the spherical shape in this type of measurements is the fact that for a cylinder the torque can only be measured optically while for a sphere it can be measured also through viscosity since the viscous drag torque for a sphere has a simple analytical solution. By measuring the rotation rate of a known radius sphere in a known fluid, one can easily calculate the torque. While it has been shown recently (Vogel et al. 2009) that by an appropriate silica and organosilica coating the stability in biological buffers can be considerably enhanced, the problem could be avoided entirely by using polymer materials, and it is interesting to determine whether objects could be fabricated using 2PP which exhibit sufficient birefringence to be useful in biological experiments.

We have carried out fabrication of polymer microspheres by means of 2PP, which are suitable for the same type of measurements as described earlier. Due to its symmetry a homogenous isotropic sphere does not experience a torque in an optical trap. However, one can break the symmetry by designing a sphere with an elongated cavity inside a "holey sphere," which can exhibit shape birefringence. One can then optimize the angular momentum transfer between the laser beam and the object, and hence the torque, by changing the shape of the cavity, which can be easily done with 2PP technique. The idea of designing "holey spheres" came about from previous work done in our laboratory on form birefringence (Bishop et al. 2003) in which rod-shaped glass microparticles were trapped and rotated in circularly polarized beams. However, a drawback of such elongated particles is the fact that they align along the beam axis if they are 3D trapped and the shape birefringence is not "seen" by the trapping beam unless they are only 2D trapped.

The 3D CAD object seen in Figure 9.5 is sliced into bitmaps, which are then input into the scanning program. Due to the versatility of the technique, we were able to vary the diameter of

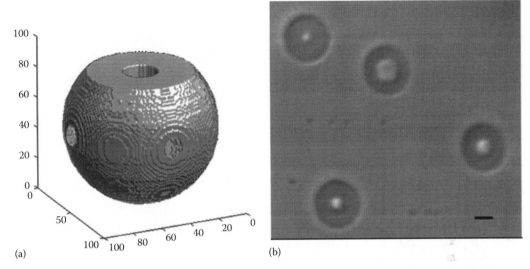

FIGURE 9.5   (a) CAD model of a "holey sphere" in pixel units. (b) Bright field microscope image of micro-fabricated hollow spheres in the unpolymerized resin. The spheres are 5 μm in diameter with different void diameters (1, 1.5, 2, and 2.5 μm). The scale bar is 2 μm.

the spheres as well as the diameter of the void. An optical microscope image of 5 μm diameter spheres with various void sizes in unpolymerized resin is shown in Figure 9.5. Due to the high travel range of the scanning stage, one could produce a large number of microstructures in one go, which is a big advantage in terms of fabrication efficiency. The fabrication time for each structure is about 7 min.

The structures were further characterized with the scanning electron microscope (SEM). Typical SEM images of the fabricated structures with different void diameters are shown in Figure 9.6. The layer-by-layer formation of the microstructure can be clearly seen from SEM images. The apparent slight ellipticity (approximately 1.1) of the holey spheres seen in SEM images is due to the inherent difference in the image pixel size in X and Y directions. The actual circularity was checked by a standard copper grid etalon.

9.2.2   Rotation in Optical Tweezers

After being detached from the cover slip (Figure 9.7) the objects were trapped, as expected, with the hole horizontal as that maximized the amount of high refractive index material in the most intense part of the beam.

The holey spheres rotate in a circularly polarized beam, reversing rotation direction when the handedness of the beam is reversed. Snapshots of a microsphere in circularly polarized light are shown in Figure 9.8. As expected, the rotation rate scales linearly with the laser power as seen in Figure 9.9. These measurements were performed on a 5 μm diameter microfabricated sphere with a 2.5 μm void diameter. Torque measurements were performed in circularly polarized light on 5 μm diameter microspheres. The measured torque efficiency of the microspheres as a function of the void diameter is shown in Figure 9.9.

We observe the maximum torque efficiency for a void diameter of around 2 μm, which presumably reflects an optimum trade-off between increasing anisotropy and decreasing total amount of material as the void diameter is increased. The maximum value of 4% for the spin torque efficiency can be compared to around 90% for a 5 μm diameter vaterite, 1% for the quartz cylinders (Deufel et al. 2007), and about 5% for 0.8 μm diameter form birefringent glass rods (Bishop et al. 2003). We see that the vaterites outperform in terms of torque efficiency the form birefringent microstructures.

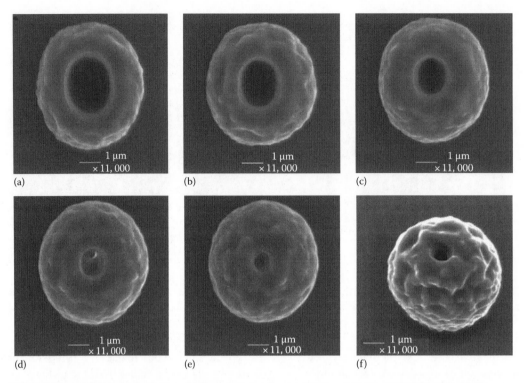

FIGURE 9.6 SEM images of microfabricated 5 μm diameter holey spheres with different void diameters (a) 3 μm, (b) 2.5 μm, (c) 2 μm, (d) 1.5 μm, and (e) 1 μm, and (f) an image of a tilted sample showing the spherical structure of the microstructure in picture (e).

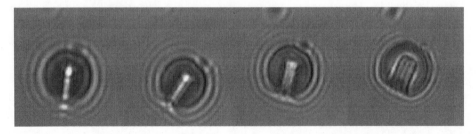

FIGURE 9.7 Bright field microscope images of a 5 μm diameter microfabricated sphere free floating in water at different z positions after being detached from the cover slip (the diameter of the cylindrical void is 2 μm). As can be seen from these snapshots, the overall sphericity of the object is altered at the starting point of the photopolymerization process.

However, we also notice that micro-objects made out of materials exhibiting weak birefringence have lower efficiencies than form birefringent microstructures. Another advantage of form birefringent objects is the possibility of size and shape control as well as the fact that they are a chemically safe tool in biological environments.

A closer look at the rotation of the holey sphere in circularly polarized light as seen in Figure 9.8 reveals the fact that it has an eccentric rotation. An analysis of this type of rotation was performed by using the center of mass of the rotating particle at each snapshot for a complete rotation cycle. By superimposing each center of mass on the same image, we could fit the trajectory to a circle with a radius of 1.1 μm as seen in Figure 9.10. The distance from the center of rotation to the closest edge of the void was found to be approximately 0.4 μm, which is also the value of the radius of the

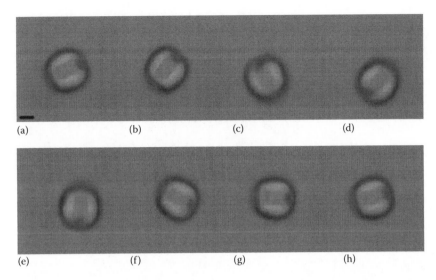

FIGURE 9.8   Snapshots of one full rotation cycle (a–h) of a 5 μm diameter "holey sphere" with a 2 μm diameter void rotating anticlockwise at 2 Hz in the focus of the circularly polarized trapping beam. The scale bar is 2 μm.

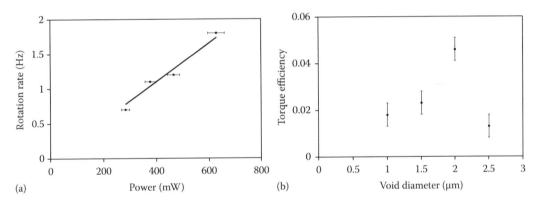

FIGURE 9.9   (a) Rotation rate as a function of the trapping beam power for a 5 μm diameter "holey sphere" with a 2.5 μm diameter core. (b) Torque efficiency of microspheres as a function of the void diameter.

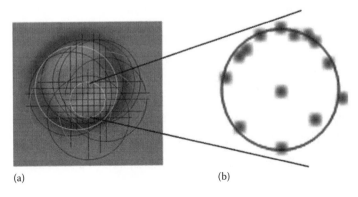

FIGURE 9.10   (a) Rotation dynamics of the "holey sphere." The center of mass of the microparticle is determined for each individual snapshot during one complete rotation and then they are all superimposed on the starting point microscope image. The center of mass trajectory could then be fitted to a circle (the smaller white circle in the middle). (b) The center of mass trajectory and the circular fit.

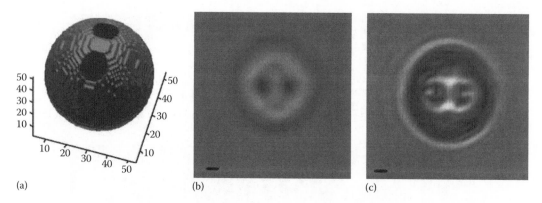

(a)   (b)   (c)

FIGURE 9.11 CAD model of the two-channel microsphere in pixel units (a) and bright field microscope images of a microfabricated sphere in unpolymerized resin (b) and after acetone wash (c). The scale bars are 1 μm.

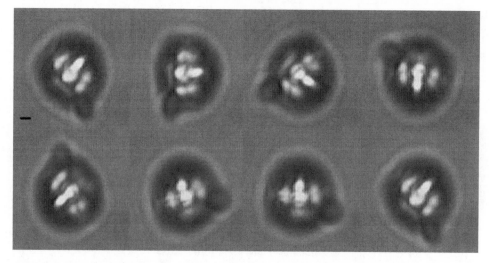

FIGURE 9.12 Snapshots of one full rotation cycle of a 5 μm diameter two-channel sphere with 1 μm diameter channels rotating clockwise at 0.1 Hz in the focus of the circularly polarized trapping beam. The scale bar is 1 μm.

trapping beam focal spot. Thus we can infer that the center of the trap is located in the center of the eccentric rotation.

To overcome the eccentric rotation, we have designed and fabricated an improved version of a holey sphere with two parallel voids one in each hemisphere such that there will be a trappable solid core of the sphere as shown in Figure 9.11. The diameter of the sphere is still 5 μm, with the two channels having diameters of 1 μm and the distance between the channel centers being 2 μm. As expected, this type of microsphere trapped centered on the beam axis and rotated around its center of mass in circularly polarized light as seen in Figure 9.12.

The spin torque efficiency was found to be around $0.01\hbar$ per photon. Taking into account that the solid core is slightly larger than the focal spot of the trapping beam, it is expected that the torque can be increased by decreasing the size of the core and also by varying the size of the channel diameter.

9.2.3 FUNCTIONALIZATION WITH BIOMOLECULES

To be useful in biological experiments the holey spheres need to be functionalized to have relevant biomolecules attached to them; so a number of tests were performed to show how this could be done.

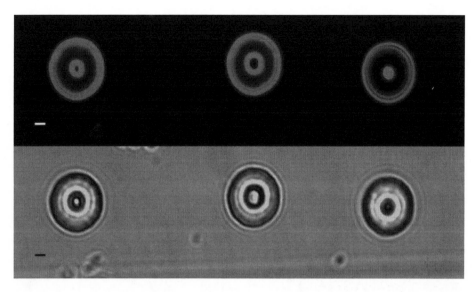

FIGURE 9.13 Upper row: laser scanning confocal fluorescence microscope image of functionalized holey spheres after being treated with fluorescent-labeled biotin. Lower row: bright field microscope images of the functionalized microspheres. The void diameters are 1, 1.5, and 2 μm and the sphere diameter is 5 μm. Scale bars are 1 μm.

Due to the thiol (−SH) functionalities present on the surface of the microfabricated holey spheres, the molecule of choice for coating the surface was maleimide-streptavidin (Sigma-Aldrich) as a result of the strong bond between thiols and maleimide. The procedure for streptavidin coating was as follows. The maleimide-streptavidin was dissolved in a phosphate buffered saline (PBS) solution with a concentration of 10^{-5} M. After that, the microspheres on the cover slip were incubated in the maleimide-streptavidin solution for 10 min and then the cover slip was washed thoroughly with PBS. To verify the attachment of the streptavidin to the microspheres, we used fluorescent-labeled biotin as it has a well-known strong bond to streptavidin (Vanzi et al. 2006).

Our choice of biotin was ATTO 550 labeled biotin (Sigma-Aldrich). ATTO 550 is a novel fluorescent label related to Rhodamine 6G and Rhodamine B tailored for use in life sciences. It has strong absorption with an excitation maximum at 550 nm, high fluorescence quantum yield with an emission maximum at 576 nm, and high thermal- and photostability (www.atto-tec.com).

The streptavidin-coated holey spheres were further incubated in the fluorescent-labeled biotin solution (~10^{-5} M) in PBS for another 10 min and then thoroughly washed with PBS. They were then imaged by a laser scanning confocal fluorescence microscope. Excitation was performed at 543 nm by a HeNe laser, and fluorescence was collected above 580 nm using a long-pass filter. A typical fluorescence image is shown in Figure 9.13, proving the functionalization of the holey spheres with streptavidin. The confocal image shows an equatorial slice and confirms the presence of the streptavidin coating on the outer surface and within the hole.

9.3 MICROROTORS DRIVEN BY ORBITAL ANGULAR MOMENTUM

9.3.1 OPTICALLY DRIVEN CHIRAL MICROROTOR

9.3.1.1 Design and Microfabrication

In order to obtain transfer of angular momentum between the trapping beam and the trapped object one needs to take into account the shape of the trapped object. As discussed earlier, the most important aspect of the shape of the object is its rotational symmetry, which can be tailored in order to

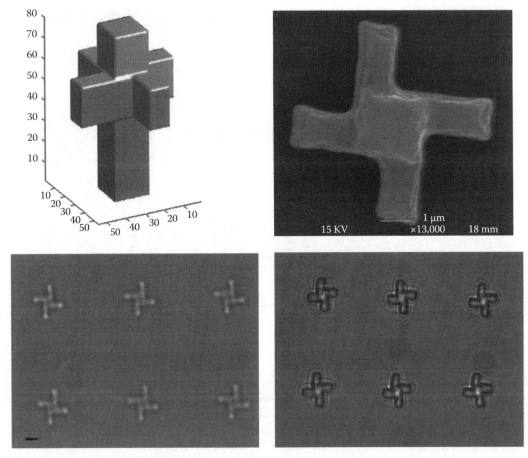

FIGURE 9.14 Upper row: CAD model in pixel units (left) and SEM image (right) of the "offset" cross. Lower row: bright field microscope images of microfabricated "offset" crosses in unpolymerized resin (left) and in water after rinsing with acetone (right). The black scale bar is 2 μm.

optimize the torque. So, depending on the angular momentum of the incident beam, one can find a suitable order of rotational symmetry for optimum torque efficiency. Furthermore, if chirality is introduced, torque can be achieved by using incident beams carrying no angular momentum.

We designed and microfabricated a fourfold rotational symmetry chiral object, namely an "offset" cross with a stalk. The reason for the stalk is the proper alignment of the object in the optical trap as was mentioned earlier.

Figure 9.14 shows the CAD model of the object. The size of the transverse square defining the stalk is chosen to be 18 × 18 pixels and the offset cross is 54 × 54 pixels wide with an arm thickness of 10 pixels. The whole object is composed of 41 layers, giving a physical length of 8 μm. The resin used was NOA63 and the parameters for successful photopolymerization were 25 mW of average laser power measured before entering the microscope and a scanning speed of 15 μm/s. Optical microscope images of the structures produced in unpolymerized resin and after rinsing with acetone are also shown in Figure 9.14. Due to the high travel range of the scanning stage, a number of microstructures can be produced in one go, which is a big advantage in terms of fabrication efficiency. The fabrication time for each structure is about 15 min.

A typical SEM image of the fabricated chiral structure is shown in Figure 9.14. The layer-by-layer fabrication can be clearly seen as well as the resemblance to the CAD model. It is interesting to note that the shape of the object after acetone washing is dependent on the amount of acetone used for rinsing the unpolymerized resin. In order to obtain a structure with smooth surfaces, only three to five drops of acetone need to be used; otherwise, the structure is corroded by the surplus amount of acetone.

9.3.1.2 Rotation in Optical Tweezers

After fabrication, the chiral microstructures were released from the cover slip by pushing them with a needle attached to a *XYZ* manually controlled translation stage. The freed microstructures float close to the cover slip since the density of the polymerized resin is slightly larger than that of water ($\rho_r = 1.3$ g/cm^3) and hence they can be readily used for optical trapping experiments. Proper alignment of the microstructures in the optical trap is of great importance if one wants to study the exchange of angular momentum from the beam to the trapped object. With these types of "offset" crosses, the desired outcome would be that the chiral structure would orient itself normal to the direction of propagation of the trapping beam.

By means of the 2PP technique, one could easily introduce a different long axis to the chiral microstructure such that it would have the desired orientation in the optical trap by adding a stalk. An "offset" cross with a stalk always has a stable equilibrium position in the optical trap, with the stalk parallel to the propagation direction and hence the chiral part is always perpendicular to the trapping beam. This is illustrated in Figure 9.15 where the microstructure is firstly trapped in an unstable equilibrium position and rapidly orients itself with the stalk parallel to the beam axis.

We note here the fact it is seldom that the microstructure is trapped in the unstable position shown in Figure 9.15a. In the vast majority of cases it is trapped directly in the stable position in Figure 9.15h.

Once trapped, the microstructure started to rotate in the linearly polarized beam. We distinguish two different configurations in which the microrotor could be trapped with the chiral structure normal to the beam axis. The most encountered situation is schematically depicted in Figure 9.16. In this case the "offset" cross rotates clockwise with a rotation frequency of 2 Hz at a laser power of 25 mW.

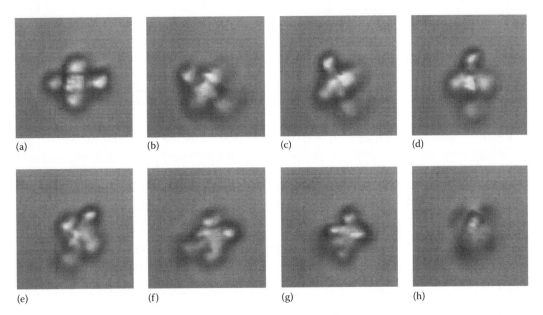

(a)   (b)   (c)   (d)

(e)   (f)   (g)   (h)

FIGURE 9.15 Bright field microscope images showing an "offset" cross trapped in an unstable equilibrium position (a) with the stalk normal to the beam axis, which rapidly changes its position in the trap (b–g) pivoting around the focus of the trapping beam until it reaches the stable equilibrium position (h). The whole process (a–h) takes about 2 s.

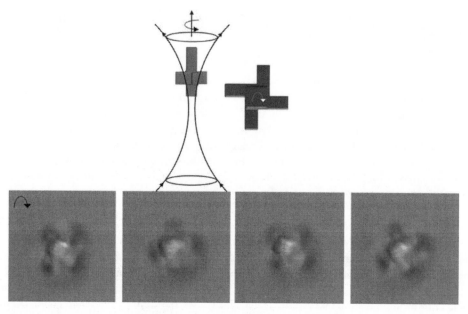

FIGURE 9.16 Upper row: schematics of a possible orientation of the microstructure in the optical trap and the corresponding clockwise (relative to the viewing direction from below) direction of rotation. Lower row: video microscopy snapshots showing one complete rotation of the chiral microstructure.

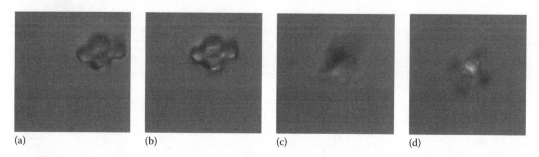

(a)   (b)   (c)   (d)

FIGURE 9.17 Bright field microscope images of the free-floating "offset" cross (a and b) of the microstructure heading toward the trap (c) and of the stably trapped microstructure (d).

Depending on the orientation of the free floating microstructure with respect to the focus of the trapping beam as seen in Figure 9.17, it could be trapped in a second possible stable configuration in which the chirality of the microstructure as seen by the incoming trapping beam is inverted.

In the second possible orientation as depicted in Figure 9.18, the microstructure rotates anticlockwise in the same linearly polarized trapping beam as seen by the CCD camera.

9.3.1.3 Discussion

The fact that the "offset" cross has opposite directions of rotation in a linearly polarized Gaussian beam depending on the chirality of the microstructure can, on one hand, be justified qualitatively by purely ray optics means (Higurashi et al. 1994). In their picture, the forces due to the change in linear momentum of light as the rays are refracted by the micro-object act on the side faces of the microstructure, leading to clockwise or anticlockwise rotation depending on its chirality. They used 10–25 µm wide chiral objects made by reactive ion-beam etching of a 10 µm thick SiO_2 layer. While they could not be 3D trapped due to the inherent planarity of the objects, when pushed against the cover slip, the same type of rotation was observed as with the "offset" crosses.

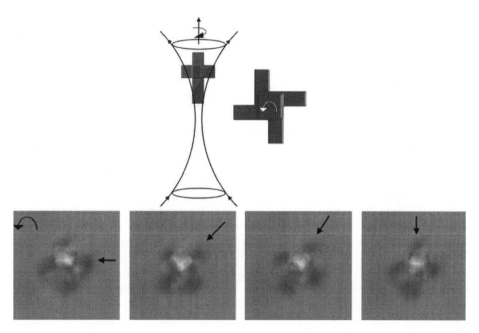

FIGURE 9.18 **(See color insert.)** Upper row: schematics of the second possible orientation of the microstructure in the optical trap and the corresponding anticlockwise direction of rotation. Lower row: video microscopy snapshots showing a quarter of a rotation of the chiral microstructure.

These microrotors are quite efficient in terms of the rotation rate achieved. A comparison with purely birefringent rotating micro-objects is worthwhile. As discussed, vaterite microspheres can achieve high spin angular momentum transfer from a circularly polarized trapping beam and therefore rotate at high rates. For example, a 5 μm diameter vaterite particle spins with a frequency of 35 Hz with 150 mW of laser power. The same rotation frequency can be achieved by the "offset" crosses with just three times the power needed for comparable size vaterite particles.

Another interesting fact about these chiral microrotors is the reversing of rotation direction depending on whether they are below or above the focus of the trapping beam. As first reported by Galajda and Ormos in 2002, if the rotor is held by some means in the converging part of the beam focus, rather than the diverging stable trapping area, it can be observed rotating in the opposite direction. Galajda and Ormos explained the effect in the following way: light propagating in or out (depending on the position of the microrotor with respect to the focus) in the radial direction is reflected by the structure giving rise to torques with opposite directions. This effect can also be explained by means of the "hologram picture," by viewing the microrotors as microscopic holograms. A chiral pattern such as this is like the interference pattern between a converging or diverging Gaussian beam and LG_{04} or $LG_{0,-4}$ modes, and the mode produced depends on the convergence or divergence of the incident beam. Thus, the curvature of the wavefronts of the trapping Gaussian beam will determine the direction of the exerted torque on the chiral microstructure.

9.3.2 MEASURING OPTICAL ANGULAR MOMENTUM TRANSFER TO OPTICALLY DRIVEN MICROROTORS

Here we describe the transfer of optical angular momentum from a trapping beam carrying OAM (LG_{02}) to a microfabricated object with fourfold rotational symmetry. The total optical torque exerted on the micron-sized object can be measured experimentally by trapping it in a beam with three different polarizations.

The experimental results are found to be in good agreement with the hydrodynamic simulations of counteracting torque generated by the fluid flow around the trapped object. These results clearly show that by only measuring the rotation frequency of the trapped object, one obtains quantitative measure of the optical angular momentum transfer to arbitrary shaped objects in optical tweezers.

9.3.2.1 Design and Microfabrication

The microrotor design is shown in Figure 9.19. The rationale for the shape was as follows: as before, there is a central stalk to ensure that the structure is stably trapped with the long axis along the axis of the beam. The four arms scatter the light with efficient exchange between modes with +2 and −2 angular indices.

The size of the squares defining the stalk is chosen to be 18×18 pixels and the cross is 54×54 pixels. The whole object is composed of 41 layers. The parameters for successful resin polymerization are 25 mW of average infrared laser power entering the microscope, 80 fs for the pulse duration, and 8 W of pumping laser power. These parameters were determined in a series of trial and error experiments. The scanning speed of the piezo stage is 15 μm/s, yielding a production time of 15 min per structure. After being produced, the structures were characterized with SEM. The cover slip with the produced structures attached is mounted on an adhesive carbon pad and then coated with a 10 nm platinum layer. A typical SEM image of the fabricated structure is shown in Figure 9.20, where the layer-by-layer formation of the microstructure can be clearly seen, as well as the firm attachment to the cover slip. We used a field emission SEM (JEOL JSM-6300F) operated at 5 kV for these measurements.

9.3.2.2 Rotation and Performance in Optical Tweezers

After production, the structures were used for rotation in optical tweezers. The goal of these experiments is to measure the torque exerted on the microstructures by the trapping beam through transfer of optical angular momentum. In our experiments, we used a hologram that generates LG_{02} modes in the first order when the incident beam is the TEM_{00} Gaussian beam of the laser. These modes have an OAM of $2\hbar$ per photon. The hologram was mounted on a *XYZ* translation stage for fine adjustments. The output mode is imaged onto a CCD camera while adjusting the hologram. Figure 9.21 shows CCD camera photographs of an LG_{02} mode and incident Gaussian beam, respectively.

In order to trap the structure, we first have to detach it from the cover slip. The microstructure can be easily detached by touching it with the tip of a needle mounted on a translation stage. The freed

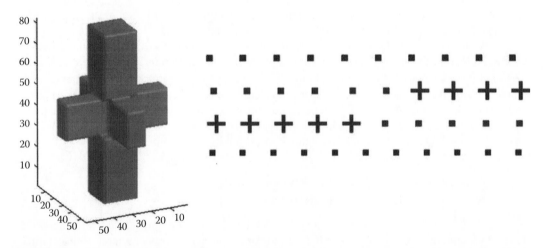

FIGURE 9.19 Three-dimensional computer-aided design (CAD) model of the microfabricated object on the left and the sequence of 2D bitmaps defining the 3D object on the right.

FIGURE 9.20   SEM image of the microfabricated structure attached to the cover slip.

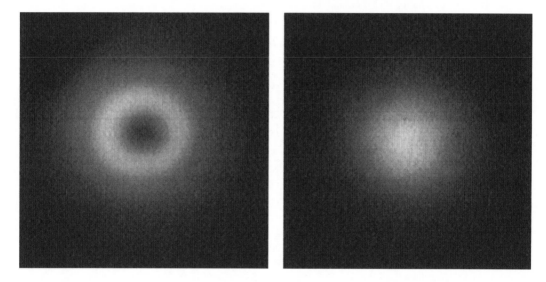

FIGURE 9.21   CCD camera photographs of the LG_{02} mode on the right and the TEM_{00} Gaussian beam on the left.

structure then floats close to the cover slip surface. Thus, we can observe the shape of the formed structure and the resemblance with the CAD model by means of bright field microscopy as shown in Figure 9.22.

The structure is easily trapped and starts to rotate in the focus of the beam carrying angular momentum. The rotation rate is in the order of 1 Hz for a trapping beam power of 20 mW at the sample. All the trapping experiments were performed in demineralized water. In Figure 9.23 are shown video microscopy frames of a rotating microstructure.

9.3.2.3   Discussion

Qualitatively, the rotor performs as expected in the sense that the structure rotates when trapped in a beam carrying OAM, does not rotate in a Gaussian beam, and the torque efficiency of 0.2 is much higher than that typical for shape-birefringent objects (e.g., 0.02–0.05) (Bishop et al. 2003).

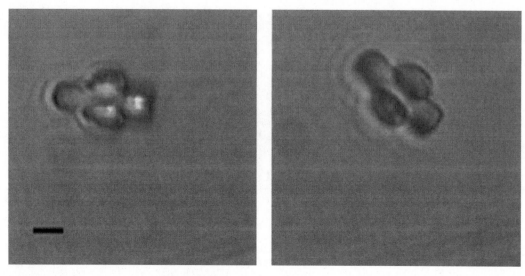

FIGURE 9.22 Bright field microscope images of the microstructure floating freely in solution after being detached from the cover slip. Scale bar is 2 μm.

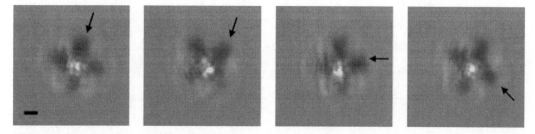

FIGURE 9.23 Video microscopy frames of a microstructure rotating at 1 Hz showing a quarter of a rotation period. The frames are 80 ms apart. Scale bar is 1 μm.

Quantitatively, we can compare the measured optical torque (Asavei et al. 2009) and the computationally predicted optical torque, as well as the viscous drag torque found from the measured rotation rate and the computationally determined viscous drag coefficient.

The optical force and torque acting on the microstructure due to the interaction with the electromagnetic field of the focused laser beam were calculated using computational methods. A variation of the discrete dipole approximation (DDA) was used, modified to generate the T-matrix (Mishchenko et al. 2000, Waterman 1971) and making use of optimizations based on the discrete rotational symmetry (Loke et al. 2009). The resulting T-matrix can be used in the Optical Tweezers Toolbox (Nieminen et al. 2007) to calculate the optical forces and torques.

An important element in the calculation, which is difficult to measure precisely, is the structure of the incident illumination (Figure 9.21). While the beam carries $2\hbar$ OAM per photon, the radial variation of the intensity is not exactly that of a pure LG_{02} mode (Heckenberg et al. 1992). From our measurements of the intensity profile of the beam, and the width of the back aperture of the objective, which is slightly underfilled, we estimate that the beam is approximately equivalent to a pure LG_{02} mode focused by an optimally filled objective of numerical aperture 1.26. The resulting forces and torques are shown in Figure 9.24. The equilibrium trapping positions are where the axial force goes to zero, and are different for each polarization. However, the torque efficiency predicted at the equilibrium positions turns out to be 0.2 for all three polarizations. While the torque versus

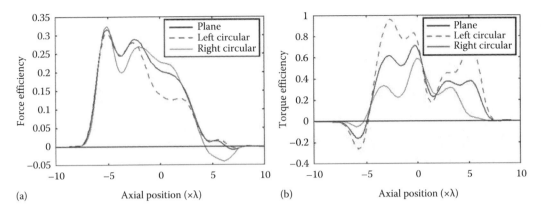

FIGURE 9.24 (a) Axial optical force acting on the rotor as a function of position along the beam axis. The predicted equilibrium points are where the curves cross the zero force efficiency line. (b) Optical torque acting on the rotor as a function of position along the beam axis. The torque efficiency at the equilibrium positions is 0.2 for all three polarizations.

axial position is different for each polarization, the different axial equilibrium positions result in very similar torques for the three polarizations, in this particular case. This agrees well with the optically measured torque.

The fluid flow field around the rotating microstructure and hence the drag torque can be calculated from the Stokes equation, which has an analytical solution for only a limited number of objects (sphere, spheroid, infinite cylinder). Thus, in order to compare the experimental results of the optical torque with the counteracting drag torque, we need to perform hydrodynamic simulations of the fluid flow generated by the rotating object of arbitrary shape.

We used the program FlexPDE (PDE Solutions Inc., Antioch, CA), which is a partial differential equation solver based on finite element numerical analysis. The program constructs a tetrahedral finite element mesh over the geometry specified by the user and then solves the differential equation numerically refining the mesh and the solution until the user-defined error bound is achieved.

Thus, the flow field can be simulated and the drag torque τ_d, which is the shear torque exerted by the fluid, can be found by integrating the shear stress tensor multiplied by the distance from the center of rotation, over the rotating surface of the microstructure.

The geometry chosen for our system was such that the microstructure was embedded in a cylinder with a diameter and height of 12 µm.

The accuracy of the method was checked by simulating the drag torque experienced by a rotating sphere in a similar geometry, for which the analytical solution is known. The difference between the simulation and the analytical solution was less than 0.5%. Typical graphical outputs of the program are shown in Figure 9.25 where the flow field around the rotating object is simulated as well as the shear stress field (the shear force acting per unit area), which is used to calculate the torque.

Calculations of the drag torque were carried out for the measured rotation frequencies with the corresponding error margins. When plotting the simulated drag torque as a function of frequency, we obtained, as expected, a linear dependence with a varying slope corresponding to the experimental error in the frequency (see Figure 9.26).

As we can see in Figure 9.27, the experimental values for the total optical torque agree with the predicted values of the drag torque within 10% error.

The torques determined from all three methods—optical measurement (orbital torque efficiency of 0.2 and spin torque efficiencies of −0.02 (right circular), 0 (plane), and 0.03 (left circular), or an orbital torque of 4.8 ± 0.7 pN µm, computational electrodynamics (total torque efficiency of 0.2), and computational fluid dynamics (torque of 5.4 pN µm at 2.75 Hz, which was the rotation speed with the plane-polarized light)—were in close agreement.

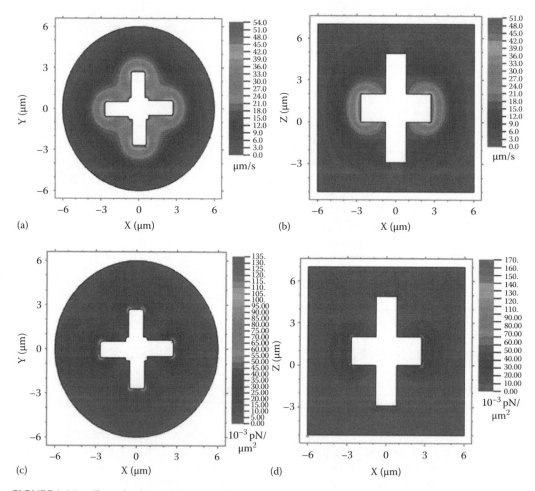

FIGURE 9.25 (See color insert.) Typical graphical outputs of the FlexPDE program. Cross sections through *XY* (a) and *XZ* (b) planes, respectively, of the simulated flow field around the rotating particle. Similarly, the cross sections through the same planes for the shear stress field are depicted in (c) and (d), respectively.

One potential error in the electrodynamic calculation of the optical torque is the incident beam. The calculation as performed used a beam that approximated the actual beam as well as could be determined from measurement of the beam, but the results are sensitive to the details of the beam. This can be seen from the rapid variation of torque efficiency with axial position in the vicinity of the predicted axial equilibrium position shown in Figure 9.24.

Another source of error is heating due to absorption by the structure, which will result in changes in the viscosity of the surrounding fluid, reducing the drag torque, and will also result in convective flow in the fluid, which will displace the rotor upwards through viscous drag. The close agreement between the observed viscous drag (through the observed optical torque) and the computationally determined drag coefficient (which assumed a uniform temperature) suggests that any heating was minor.

This is expected, since measurement of the absorption coefficient of the photopolymerized resin shows low absorption in the infrared region. Nonetheless, there will be some heating present, and the effect can become important at high powers. The fact that the experimental data fit is close to the lower limit of the predicted values is likely to be due to heating of the structure and surrounding fluid by the trapping beam, which would decrease the viscosity, and hence the torque required for rotation at a given frequency.

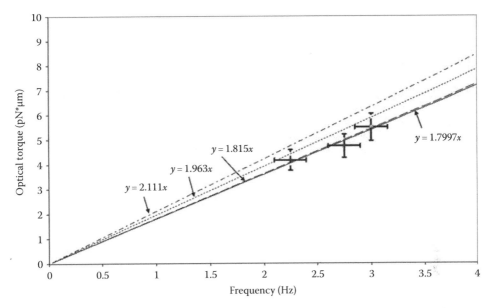

FIGURE 9.26   The total optical torque plotted as a function of the rotation frequency and the linear fit of the data (solid line). In the same graph is shown the simulated drag torque as a function of frequency (• • •) as well as the simulations corresponding to the experimental error in the frequency (- • - and - - -).

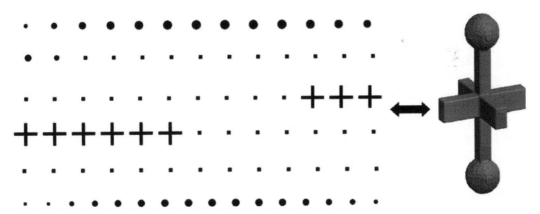

FIGURE 9.27   Three-dimensional model of the paddle-wheel and 2D bitmap slices of the object for microfabrication.

The orbital torque obtained with the fourfold symmetry microrotor corresponds to transfer of 10% of the OAM carried by the light. However, this is still less than the torque that can be achieved using strongly birefringent materials such as calcite or vaterite (Bishop et al. 2004), even though it is much higher than that obtained using shape birefringence or weakly birefringent materials (Deufel et al. 2007).

From Figure 9.24, we can see that torque efficiencies about three times higher can be obtained if the rotor is in the optimum axial position. If the rotor is mounted on a substrate, for example, by an axle (Kelemen et al. 2006) or if it is pushed towards the focus by a counter propagating independent beam, this can be achieved.

There are also fundamental physical limits to the available torque efficiency: a sufficiently large moment arm is still required, even if the azimuthal force acting on part of the rotor is produced by a beam of light. Thus, larger particles combined with larger focal spots can in principle allow higher torques (Courtial and Padgett 2000).

9.4 OPTICALLY DRIVEN MICROSCOPIC PADDLE-WHEEL

9.4.1 MOTIVATION

As an optically trapped micro-object spins in a fluid, there is a consequent flow. Since a free-floating optically driven microrotor can be moved to a desired position, it can allow the controlled application of a directed flow in a particular location. Here we demonstrate the control and rotation of such a device, an optical paddle-wheel, using a multiple-beam trap. In contrast to the usual situation where rotation is around the beam axis, here we demonstrate rotation normal to this axis.

The paddle-wheel can be optically driven and moved to any position in the field of view of the microscope, which can be of interest for various biological applications where controlled application of a fluid flow is needed in a particular location and in particular direction. This is of particular interest in signal transduction studies in cells (Yu et al. 2008), especially when a cell is flat and spread out on a surface.

9.4.2 DESIGN AND MICROFABRICATION

The paddle-wheel was designed so that it could be trapped and rotated such that one could obtain a flow in a plane with its normal parallel to the rotation axis. Trapping the paddle-wheel horizontally can be achieved by means of dual optical tweezers if the paddle-wheel has a dumbbell shape. Once trapped, a third beam could push the paddle and hence rotate it.

Figure 9.27 shows the 3D computer-aided design of the object as well as the 2D bitmap input for the photopolymerization computer program.

The object consists of 81 layers each separated by 200 nm in the Z direction, resulting in a length of 16 μm. The two spheres at the ends were chosen to have a diameter of 3 μm. The arms of the paddle have a length of 4 μm, a height of 2 μm, and a width of 1 μm, and the stalks on each side of the paddle have a 1×1 μm^2 rectangular profile. A single microstructure is produced in 25 min with a threshold photopolymerization power of 18 mW.

After photopolymerization, the sample is washed with acetone in order to get rid of the unpolymerized resin. Figure 9.28 shows a microfabricated paddle-wheel immediately after its fabrication and after the acetone wash. The images confirm the programmed geometry. The fact that the ends are not perfectly spherical is due, on the one hand, to the starting point of the photopolymerization process (lower end) and, on the other hand, to the inherent shape of the voxels creating a slightly prolate structure (upper end). However, these changes from the initial design do not affect the actual experiment.

SEM images of typical paddle-wheels, both attached to the cover slip and fallen due to hydrodynamic forces during the acetone wash, are shown in Figure 9.29. A closer look at the surface of the object reveals a smooth surface with small variations less than 100 nm.

The paddle-wheel can be lifted from the cover slip by mechanical force, using the tip of a needle, and then used in trapping and rotation experiments in water.

9.4.3 PADDLE-WHEEL OPTICAL ROTATION

The experimental setup used for trapping and rotating the paddle-wheel is based on a fully steerable dual-trap optical tweezers system (Fällman and Axner 1997), which allows independent movement of the trapped object in X and Y directions and most importantly in the Z direction. A scheme of the experimental setup is shown in Figure 9.30. For dual trapping, the output from a CW Yb doped fiber laser ($\lambda = 1070$ nm, IPG Photonics) is split into two beams by means of a polarizing beam splitter cube (PBS) and then recombined through a second PBS and the two gimbal mounted mirrors (GMM) in order to enter the microscope objective.

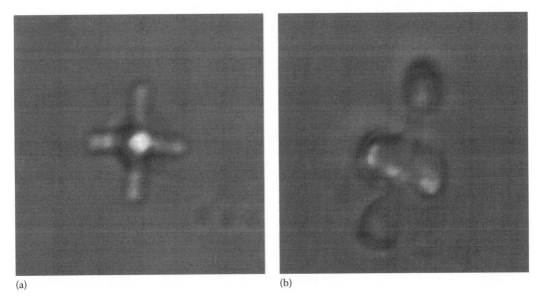

FIGURE 9.28 A bright field microscope image of a microfabricated paddle-wheel in unpolymerized resin (a) and after acetone wash (b). The scale bar is 1 µm.

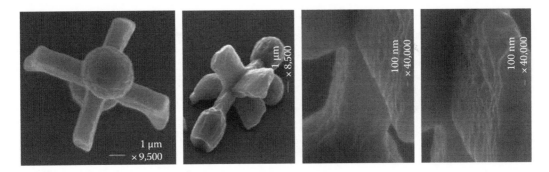

FIGURE 9.29 SEM images of microfabricated paddle-wheels.

By imaging the objective back aperture onto the center of each GMM by the lenses L_3 and L_4, the two traps can be steerable in the X and Y directions. Furthermore, if an afocal telescope is placed in each of the two arms (between PBS and GMM, see Figure 9.30), the two traps can also be moved in the Z direction.

Rotation of the paddle-wheel is achieved by a third beam, independent of the two traps. We used the IR output from the Ti:Sapphire laser ($\lambda = 780$ nm, Tsunami, Spectra Physics) pumped by a CW solid-state laser at 532 nm (Millenia, Spectra Physics) and a power of 8 W because it was readily available. For successful rotation of the paddle-wheel, the Ti:Sapphire laser has to be operated in CW mode. If operated in the pulsed mode, it was found that the femtosecond laser produced ablation of the paddles, hence destroying the microstructure. For convenience, the third beam was also steered in X and Y in the same manner as described earlier.

As mentioned earlier, the steering in the Z direction of the two traps is of importance due to the chromatic aberration of the microscope objective. We found that the difference between the focal point at 1070 nm and at 780 nm is 6 µm. In this configuration, even if we could trap the paddle-wheel, we could not achieve any rotation from the third beam due to the large difference in Z between the

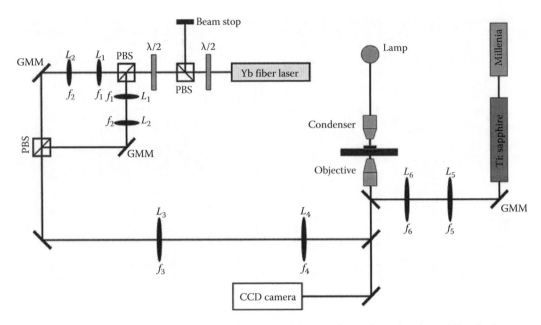

FIGURE 9.30   Schematics of the experimental setup used for trapping and rotating the paddle-wheel.

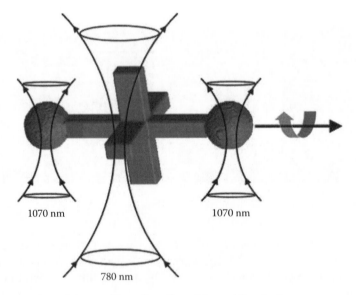

FIGURE 9.31   Schematics of the trapping and rotation of the paddle-wheel.

foci. Therefore, to rotate the paddle-wheel one needs to bring the foci of the three beams closer in Z direction. To do so, we trap 4.5 μm polystyrene beads in the three traps and we adjust the Z position of the dual traps with respect to the "pushing" beam until all three beads are in focus. In this configuration, the trapped paddle-wheel could be rotated by the "pushing" beam impinging on the edge of the paddles as shown in the schematic of the experiment in Figure 9.31.

In Figure 9.32 are shown consecutive frames of the rotating paddle-wheel with a rotation rate of 2 Hz. The power of the "pushing" beam was 40 mW at the sample. Knowing the rotation rate we could perform hydrodynamic simulations of the fluid flow created by the spinning paddle-wheel and could calculate the total drag torque opposing the optical torque.

FIGURE 9.32 A sequence of bright field microscope images of the rotating paddle-wheel. The frames are 140 ms apart, giving a rotation rate of 2 Hz. In the first frame, the position and direction of propagation of the "pushing beam" are indicated by the marker on the paddle. The scale bar is 3 μm.

The geometry chosen for our system was such that the microstructure was embedded in a cylinder with a diameter of 16 μm and a height of 20 μm. Typical graphical outputs of the program are shown in Figure 9.33 where the flow field around the rotating object is simulated as well as the shear stress field (the shear force acting per unit area), which is used to calculate the torque. The calculated drag torque for the paddle-wheel rotating at 2 Hz was $\tau_d = 10$ pN μm.

9.4.4 DISCUSSION

Since the paddle-wheel is rotating with uniform angular velocity, it means that the drag torque and the optical torque due to the beam should be equal in magnitude. Due to the geometry of our system we can state that the optical torque is in fact a mechanical torque due to the force created by the reflection and absorption of the "pushing" beam at the water–resin interface impinging on the arm of the paddle-wheel. Therefore, we can write

$$t_o = 2n_w P\left(\frac{2R}{c} + \frac{A}{c}\right)d \tag{9.2}$$

where
n_w is the refractive index of water
P is the power of the beam impinging on the paddle
R and A are the reflectance and absorbance at the resin–water interface
d is the paddle arm

We can neglect the absorption contribution to the torque due to the low absorption coefficient of the resin and we can approximate the reflectance R as follows:

$$R = \left(\frac{n_r - n_w}{n_r + n_w}\right)^2 \tag{9.3}$$

with n_r being the refractive index of the polymerized resin ($n_r = 1.56$).

We obtain for the optical torque in this approximation $\tau_o = 9$ pN μm, which is in good agreement with the simulated drag torque.

However, we should note that this approximation assumed that the whole beam was at normal incidence on the paddle. The actual "pushing" beam is highly focused (NA = 1.25), and hence in order to accurately describe the optical force, accurate simulations of the real beam would need to be performed.

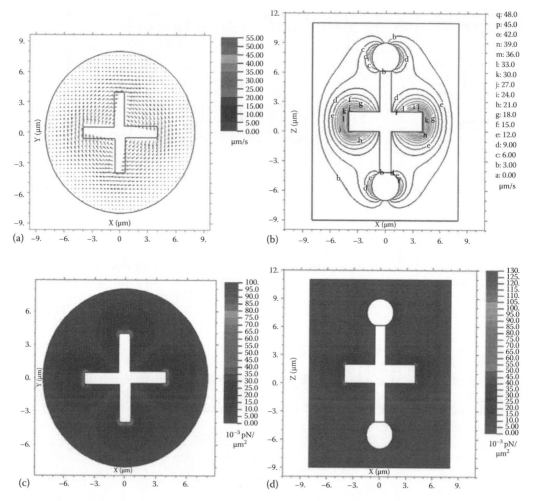

FIGURE 9.33 **(See color insert.)** Typical graphical outputs of the FlexPDE program. Cross sections through *XY* (a) and *XZ* (b) planes, respectively, of the simulated flow field around the rotating particle. Similarly, the cross sections through the same planes for the shear stress field are depicted in (c) and (d), respectively.

9.5 SUMMARY AND CONCLUSION

In this chapter, we have shown that 2PP method is a powerful and versatile tool to fabricate arbitrary shaped 3D microrotors. We described the use of form birefringence in microfabricated objects. We have demonstrated the transfer of spin angular momentum in the "holey spheres," which are microfabricated spheres containing cylindrical voids that induce birefringence. We have also shown that these microstructures can be functionalized with biomolecules, making them great candidates for biological experiments in optical tweezers. Based on the results on "holey spheres," namely the fact that they have an eccentric rotation in the optical trap, we have designed and fabricated the next generation of such form birefringent objects, the two-channel spheres, and we showed that they trap on-center.

We have also discussed the transfer of spin and OAM from the driving beam to the micromachines and thus the ability of rotational control, and we have seen the role played by the rotational and mirror symmetry of the microstructure in an efficient angular momentum transfer and hence the torque exerted by the beam on the driven microrotor. We have shown that a chiral rotationally symmetric object is best driven by a Gaussian beam while an achiral rotationally symmetric object has an optimal performance in beams carrying OAM. For the latter case we used a computer-generated hologram to produce an LG trapping beam carrying OAM (LG_{02}). We experimentally measured the transfer of OAM from the LG_{02} beam to a fourfold rotationally symmetric achiral microrotor. Finally, we have presented a different type of microstructure, an optical paddle-wheel, with potential applications in cell studies. The drive of the paddle-wheel is different than the other examples described in that no optical angular momentum about the beam axis is involved in the rotation. In this case the torque was generated by reflection on the paddles of a laser beam in a multiple-beam trap. In contrast to the usual situation where rotation is around the beam axis, we demonstrated rotation normal to this axis.

REFERENCES

Allen, L., Beijersbergen, M.W., Spreeuw, R.J.C. et al. 1992. Orbital angular momentum of light and the transformation of Laguerre-Gaussian laser modes. *Physical Review A* 45: 8185–8189.

Asavei, T., Loke, V.L.Y., Barbieri, M. et al. 2009. Optical angular momentum transfer to microrotors fabricated by two-photon photopolymerization. *New Journal of Physics* 11: 093021.

Benzel, E., Ferrara, L., Roy, S., and Fleishman, A. 2004. Micromachines in spine surgery. *Spine* 29: 601–606.

Bishop, A.I., Nieminen, T.A., Heckenberg, N.R. et al. 2003. Optical application and measurement of torque on microparticles of isotropic nonabsorbing material. *Physical Review A* 68: 033802.

Bishop, A.I., Nieminen, T.A., Heckenberg, N.R. et al. 2004. Optical microrheology using rotating laser-trapped particles. *Physical Review Letters* 92: 198104.

Born, M. and Wolf, E. 1999. *Principles of Optics*. Cambridge, U.K.: University Press.

Coullet, P., Gil, L., and Rocca, F. 1989. Optical vortices. *Optics Communications* 73: 403–408.

Courtial, J. and Padgett, M.J. 2000. Limit to the orbital angular momentum per unit energy in a light beam that can be focussed onto a small particle. *Optics Communications* 173: 269–274.

Deitz, D. 1990. Stereolithography automates prototyping. *Mechanical Engineering* 112(2): 34–39.

Deufel, C., Forth, S., Simmons, C.R. et al. 2007. Nanofabricated quartz cylinders for angular trapping: DNA supercoiling torque detection. *Nature Methods* 4: 223–225.

Fällman, E., and Axner, O. 1997. Design for fully steerable dual-trap optical tweezers. *Applied Optics* 36: 2107–2113.

Fan, L.S., Tai, Y.C., and Müller, R.S. 1989. IC-processed electrostatic micromotors. *Sensors and Actuators* 20: 41–47.

Feynman, R.P. 1960. There's plenty of room at the bottom. *California Institute of Technology Journal of Engineering and Science* 4: 23–36.

Friese, M.E.J., Enger, J., Rubinsztein-Dunlop, H. et al. 1996. Optical angular momentum transfer to trapped absorbing particles. *Physical Review A* 54: 1593–1596.

Friese, M.E.J., Rubinsztein-Dunlop, H., Gold, J. et al. 2001. Optically driven micromachine elements. *Applied Physics Letters* 78: 547–549.

Galajda, P. and Ormos, P. 2001. Complex micromachines produced and driven by light. *Applied Physics Letters* 78: 249–251.

Galajda, P. and Ormos, P. 2002. Rotors produced and driven in laser tweezers with reversed direction of rotation. *Applied Physics Letters* 80: 4653–4655.

Goh, P. and Krishnan, S.M. 1999. Micromachines in endoscopy. *Bailliere's Clinical Gastroenterology* 13: 49–58.

He, H., Friese, M.E.J., Heckenberg, N.R. et al. 1995a. Direct observation of transfer of angular momentum to absorptive particles from a laser beam with a phase singularity. *Physical Review Letters* 75: 826–829.

He, H., Heckenberg, N.R., and Rubinsztein-Dunlop, H. 1995b. Optical particle trapping with higher-order doughnut beams produced using high efficiency computer generated holograms. *Journal of Modern Optics* 42: 217–223.

Heckenberg, N.R., McDuff, R., Smith, C.P. et al. 1992. Generation of optical phase singularities by computer generated holograms. *Optics Letters* 17: 221–223.

Higurashi, E., Ukita, H., Tanaka, H. et al. 1994. Optically induced rotation of anisotropic micro-objects fabricated by surface micromachining. *Applied Physics Letters* 64: 2209–2210.

Hiratsuka, Y., Miyata, M., Tada, T. et al. 2006. A microrotary motor powered by bacteria. *Proceedings of the National Academy of Sciences of the United States of America* 103: 13618–13623.

Ishiyama, K., Sendoh, M., and Arai, K.I. 2002. Magnetic micromachines for medical applications. *Journal of Magnetism and Magnetic Materials* 242: 41–46.

Jones, R.C. 1945. A generalization of the dielectric ellipsoid problem. *Physical Review* 68: 93–96.

Kelemen, L., Valkai, S., and Ormos, P. 2006. Integrated optical motor. *Applied Optics* 45: 2777–2780.

Loke, V.L.Y., Nieminen, T.A., Heckenberg, N.R. et al. 2009. T-matrix calculation via discrete dipole approximation, point matching and exploiting symmetry. *Journal of Quantitative Spectroscopy and Radiative Transfer* 110: 1460–1471.

Mehregany, M., Bart, S.F., Tavrow, L.S. et al. 1990. Principles in design and microfabrication of variable-capacitance side-drive motors. *Journal of Vacuum Science and Technology A* 8: 3614–3624.

Mishchenko, M.I., Hovenier, J.W., and Travis, L.D. (Eds.) 2000. *Light Scattering by Nonspherical Particles: Theory, Measurements, and Applications*. New York: Academic Press.

Neale, S.L., MacDonald, M.P., Dholakia, K. et al. 2005. All-optical control of microfluidic components using form birefringence. *Nature Materials* 4: 530–533.

Newman, T.H., Williams, K.E., and Pease R.F.W. 1987. High resolution patterning system with a single bore objective lens. *Journal of Vacuum Science and technology B* 5: 88–91.

Nieminen, T.A., Heckenberg, N.R., and Rubinsztein Dunlop, H. 2001. Optical measurement of microscopic torques. *Journal of Modern Optics* 48: 405–413.

Nieminen, T.A., Loke, V.L.Y., Stilgoe, A.B. et al. 2007. Optical tweezers computational toolbox. *Journal of Optics A* 9: S196–S203.

Nieminen, T.A., Parkin, S.J., Heckenberg, N.R. et al. 2004. Optical torque and symmetry. *Proceedings of SPIE* 5514: 254–263.

Nieminen T.A., Stilgoe, A.B., Heckenberg, N.R. et al. 2008. Angular momentum of strongly focused laser beams. *Journal of Optics A* 10: 115005.

Nishimura, K., Uchida, H., Inoue, M. et al. 2003. Magnetic micromachines prepared by ferrite plating technique. *Journal of Applied Physics* 93: 6712–6714.

Oliver, A.D., Vigil, S.R., and Gianchandani, Y.B. 2003. Photothermal surface-micromachined actuators. *IEEE Transactions on Electron Devices* 50: 1156–1157.

Oroszi, L., Galajda, P., Kirei, H. et al. 2006. Direct measurement of torque in an optical trap and its application to double-strand DNA. *Physical Review Letters* 97: 058301.

Ryu, W.S., Berry, R.M., and Berg, H.C. 2000. Torque generating units of the flagellar motor of E-coli have a high duty ratio. *Nature* 403: 444–447.

Siegman, A.E. 1986. *Lasers*. Oxford, U.K.: Oxford University Press.

Smith, S.B., Finzi, L., and Bustamante, C. 1992. Direct mechanical measurements of the elasticity of single DNA molecules by using magnetic beads. *Science* 258: 1122–1126.

Stratton, J.A. 1941. *Electromagnetic Theory*. New York: McGraw-Hill.

Vanzi, F., Broggio, C., Sacconi, L. et al. 2006. Lac repressor hinge flexibility and DNA looping: single molecule kinetics by tethered particle motion. *Nucleic Acids Research* 34: 409–3420.

Vogel, R., Persson, M., Feng, C. et al. 2009. Synthesis and surface modification of birefringent vaterite microspheres. *Langmuir* 25: 11672–11679.

Waterman, P.C. 1971. Symmetry, unitarity, and geometry in electromagnetic scattering. *Physical Review D* 3: 825–839.

White, A.G., Smith, C.P., Heckenberg, N.R. et al. 1991. Interferometric measurements of phase singularities in the output of a visible laser. *Journal of Modern Optics* 38: 2531–2541.

Yu, L.F., Mohanty, S., Liu, G.J. et al. 2008. Quantitative phase evaluation of dynamic changes on cell membrane during laser microsurgery. *Journal of Biomedical Optics* 13: 050508.

10 Optically Driven Microfluidic Devices Produced by Multiphoton Microfabrication

Shoji Maruo

CONTENTS

10.1 INTRODUCTION

Radiation pressure from a tightly focused laser beam has been widely used as optical tweezers to confine, position, and transport microparticles in liquid, after Ashkin's group demonstrated this technique in 1987 (Ashkin et al. 1987). Optical tweezers provide unique features such as remote manipulation of micro/nano particles in liquid, noninvasive manipulation of biological samples, precise manipulation in sealed environment, and extremely small torque of the order of 10^{-12} nm. For these reasons, optical tweezers and its related techniques have been widely applied to studies on biological cells and DNA molecules, microchemistry with microdroplets and microbeads, and atom cooling. The history and previous works on optical tweezers were introduced in some review reports (Ashkin 2000, Grier 2003, Molloy and Padgett 2002, Neuman and Block 2004).

In most applications of optical manipulation techniques, target samples such as microparticles, droplets, and cells are merely trapped or transferred to the desired position. By contrast, sophisticated three-dimensional (3D) microstructures produced by microfabrication techniques such as lithography and multiphoton microfabrication (Galajda and Ormos 2001, Higurashi et al. 1994, 1997, Neale et al. 2005, Terray et al. 2002) can be also maneuvered according to the desired motions. The optically driven micromachines offer high-precision, remotely controlled microdevices such as microstirrer (Galajda and Ormos 2001, 2002, Higurashi et al. 1994, 1997, Neale et al. 2005) micropumps (Leach et al. 2006,

Maruo 2007, Maruo and Inoue 2006, Maruo and Inoue 2007, Maruo et al. 2009c, Terray et al. 2002), microvalves (Maruo 2007, Terray et al. 2002), and manipulators (Maruo and Hiratsuka 2007, Maruo 2007, Maruo et al. 2003a,b) for lab-on-a-chip applications. In this chapter, optically driven micromachines produced by a two-photon microfabrication technique are introduced in detail.

10.2 TWO-PHOTON MICROFABRICATION FOR PRODUCTION OF 3D MICROMACHINES

10.2.1 Fabrication Principle and Apparatus

The optically driven micromachines were fabricated by a two-photon microfabrication technique (Kawata et al. 2001, Maruo 2008, Maruo et al. 1997, Maruo and Fourkas 2008). As shown in Figure 10.1, in the two-photon process, a photopolymer absorbs two near infrared photons simultaneously in a single quantum event whose energy corresponds to the ultraviolet (UV) region. The rate of two-photon absorption is proportional to the square of the intensity of light, so that near infrared light is strongly absorbed only at the focal point within the photopolymer. The quadratic dependence of two-photon absorption assists to confine the solidification to a submicron volume. This virtue of the two-photon process enables us to create a 3D microstructure by scanning a focus inside a photopolymer.

Two-photon microfabrication was first demonstrated in 1997 (Maruo et al. 1997). It has been used to make various types of 3D microstructure such as microsprings (Kawata et al. 2001, Sun et al. 2001, Maruo 2008) and photonic crystals (Cumpston et al. 1999, Seet et al. 2006). However, almost all the microstructures were fixed on a base. Some types of free micro rotators have been demonstrated (Galajda and Ormos 2001, 2002), but they were simple, wire-frame components. By contrast, in recent years, we reported the fabrication of more sophisticated movable micromachines such as a pair of micromanipulators (Maruo 2007, Maruo et al. 2003b), microgears (Maruo et al. 2003a), and a micropump (Maruo 2007, Maruo and Inoue 2006, Maruo et al. 2009c).

Figure 10.2 shows our current fabrication system of the two-photon microfabrication. In this fabrication system, a mode-locked Ti: sapphire laser (Mira-900F, Coherent Inc., wavelength: 752 nm; repetition, 76 MHz; pulse width, 200 fs) is used to generate the two-photon absorption. The beam from the laser is introduced into the galvano-scanner system (M2 scanners, GSI Lumonics Inc.) to deflect its direction in two dimensions, and then it is focused with an objective lens set at a numerical aperture of 1.35. The beam scans laterally in the photopolymer while the stage (Mark-204, Sigma KK, Japan) that supports the photopolymer is scanned vertically, thereby moving the point of focus in three dimensions. By controlling both the galvano-scanner system and stage with 3D computer-aided design (CAD) data, a 3D microstructure is fabricated. The system has attained a peak resolution of 140 nm, thus exceeding the diffraction limit of light. After the 3D fabrication process, the unsolidified photopolymer is washed out with a rinse (EE-4210, Olympus Optical

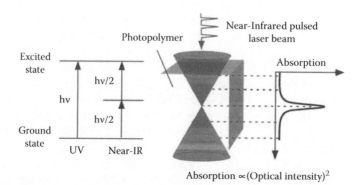

FIGURE 10.1   Principle of two-photon photopolymerization excited by a focused laser beam.

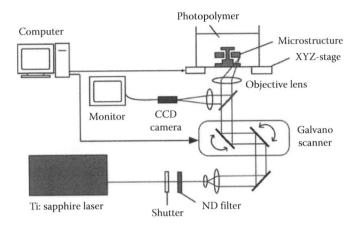

FIGURE 10.2 Fabrication system of two-photon microfabrication.

Co., Ltd.), leaving only the resultant microstructure. The photopolymer used was an epoxy resin for general UV stereolithography (SCR-701, Japan Synthetic Rubber Co., Ltd.).

10.2.2 ASSEMBLY-FREE, SINGLE-STEP FABRICATION PROCESS OF MOVABLE MICROPARTS

In conventional laser microfabrication, although simple raster scanning of a laser beam has been normally used to make a microobject fixed to a base, it is not suitable for the assembly-free fabrication of freestanding microparts because of shrinkage of the photopolymer during polymerization. The problem is overcome by optimizing the scanning pattern of a laser beam to reduce the deformation of the solidified object. For example, circularly scanning of a laser beam is utilized to make a rotation-symmetric movable microstructure such as a microgear (Maruo et al. 2003b).

The process of fabricating a microgear is illustrated in Figure 10.3 (Maruo et al. 2003b). An attached shaft and stopper are first fabricated by scanning a laser beam circularly at each layer, while the stage is lowered step-by-step with a constant interval along a central axis. Next, the stage is moved to the position to make a movable gear wheel. The circular part of a gear wheel is fabricated by scanning the laser beam circularly with increasing the radius. Finally, the teeth of the gear wheel are fabricated by circularly scanning the laser beam with increasing the radius, while the scanning speed is changed periodically. In this case, the nonlinear response of two-photon absorption makes possible to solidify the photopolymer only in the region where the scanning speed is low.

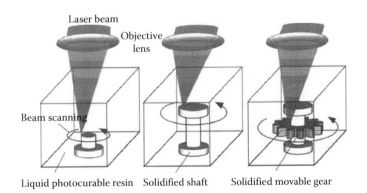

FIGURE 10.3 Single-step fabrication process of a movable microstructure.

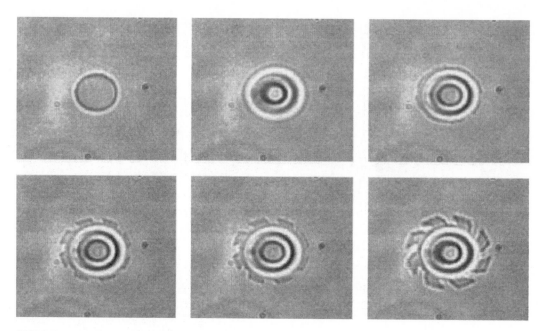

FIGURE 10.4 Sequential optical image of fabrication process of a microgear fabricated in photopolymer.

Figure 10.4 shows sequential optical microscopic images of the process of fabricating a micro-turbine (diameter: 14 μm). These optical microscopic images were observed in real time with a charge coupled device (CCD) camera. This result demonstrates that circularly scanning of a laser beam is useful for making such rotation-symmetric structure. Actually, through this approach, a rotation-symmetric microstructure whose inner diameter is smaller than about 20 μm can be fabricated without undesirable deformation. By using this single-step fabrication method, several types of movable micromachines including a microturbine and microtweezers have been fabricated as shown in Figure 10.5. This method enables to produce sophisticated polymeric movable micromachines driven by light, because the photopolymer is transparent for visible and near infrared light.

10.3 DRIVING METHOD OF OPTICALLY DRIVEN MICROMACHINES

One of the most promising applications of polymeric movable micromachines produced by two-photon microfabrication is optically driven micromachine for lab-on-a-chip application. The driving force of the optically driven micromachines results from momentum of photon. If the direction of light propagation is changed due to the reflection and refraction, a momentum change of photon is generated. According to Newton's third law, photon pressure is exerted on the interface between two media as a reaction to the momentum change. In electromagnetic theory of light, the photon force is well-known as optical radiation pressure. Since the net radiation pressure from a tightly focused laser beam applied to a microparticle is directed to the focus, the microparticle can be optically trapped in liquid. This technique named optical trapping has been widely used as optical tweezers to confine, position, and transport microparticles in liquid (Ashkin 2000, Ashkin et al. 1987, Grier 2003, Molly and Padgett 2002, Neuman and Block 2004).

By using optical trapping, polymeric movable microstructures that are produced by two-photon microfabrication can also be driven and controlled remotely. Figure 10.6 illustrates the driving method of a microgear. As shown in Figure 10.6b, when the laser beam is focused on the center

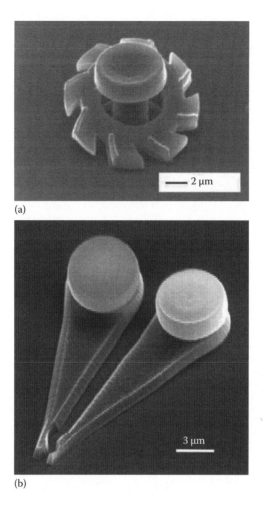

FIGURE 10.5   SEM images of movable microstructures. (a) Microturbine and (b) microtweezers.

of a gear tooth, the radiation pressure exerted on the gear tooth is balanced both in lateral and in depth. As a result, the tooth is stably trapped at the focus. By contrast, as shown in Figure 10.6c, if the focus is slightly moved to the side of the gear tooth, the net radiation pressure is directed to the focus. As a result, the microgear is attracted and moved to the focus. Therefore the circular scanning of a single laser beam makes possible to rotate the microgear. In the same way, various motions of micromachines can be generated according to the desired trajectory.

For the simultaneous driving of multiple micromachines using a single laser beam, there are several methods: holographic optical tweezers (Leach et al. 2004, Seet et al. 2006, Rodrigo et al. 2005), continuous laser scanning method (Sasaki et al. 1991), and time-shared laser scanning method (Arai 2004, Maruo 2008). In particular, the time-shared laser scanning method is a simple, feasible way to control multiple micromachines with a low-cost apparatus using galvano scanners or piezoelectric mirrors. In most cases, we use galvano scanners to scan the laser beam; we can generate any trajectory of the laser beam with a vector scanning method. The trajectory of vector scanning consists of a series of discrete foci separated with a constant distance. In our method, on–off control of the laser power is not employed for the simplification of the optical scanning system. The drawback owing to the simplification is easily canceled by the optimization of the laser scanning conditions such as trajectories, division distances, and waiting times of each focus.

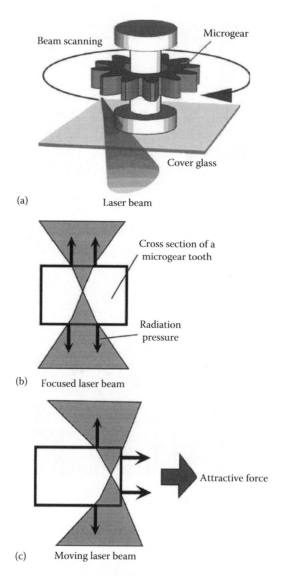

FIGURE 10.6 Optical driving of a microgear. (a) A microgear driven by scanning a laser beam. (b) Optical trap of a blade of a microgear at the focus. (c) Movement of a blade by scanning the focused laser beam.

10.4 OPTICALLY DRIVEN MICROPUMPS PRODUCED BY TWO-PHOTON MICROFABRICATION

10.4.1 LOBED MICROPUMP

In the development of functional lab-on-a-chip devices, built-in micropumps based on pressure-driven flow are necessary for the further extension of application fields. Although various kinds of pressure-driven micropumps have been developed using a range of micromachining techniques (Laser and Santiago 2004, Unger et al. 2000), most of the previously developed micropumps are diaphragm types, in which piezoelectric actuators or pneumatic actuation are utilized to deform an elastic membrane that drives fluid transport. To obtain adequate displacement of the membrane, the diameter of the membrane must range from 100 to 500 μm. As a result, the overall dimensions of the micropumps are much larger than the microchannels. This prevents further miniaturization and integration of microfluidic components. In addition, the use of built-in high-precision microactuators makes lab-on-a-chip devices expensive. For these reasons, most current disposable microfluidic chips for chemical synthesis and cell analysis still utilize external pumps such as syringe types rather than built-in micropumps.

As a promising method of addressing the aforementioned issues, we have developed an optically driven lobed micropump by using two-photon microfabrication (Maruo and Inoue 2006). The micropump consists of two lobed rotors that are incorporated into a microchannel. The lobed rotors are individually confined to a microchannel by their own shaft, thus preventing the rotors from moving unless continuously irradiated with light. This allows the micropump to be easily integrated into a mass-produced biochips made from plastics and glass.

Figure 10.7a shows a schematic diagram of the optically driven lobed micropump. This micropump is driven by means of radiation pressure generated by focusing a laser beam. When the laser beam is focused on the side of the rotor, the net radiation pressure applied to the rotor points toward the focus, allowing the rotor to be controlled by scanning the laser beam along two circular trajectories alternately. Figure 10.7b shows the time-shared laser scanning method to control the two rotors simultaneously. The black points shown in Figure 10.7b are trapping points for each status. The laser beam is divided and scanned along two circular trajectories in the opposite direction. In this case, on–off control of the laser beam is not performed when the laser beam alternates between the two trajectories. While the rotors are tightly engaged and rotated, fluid is trapped in the spaces between the rotors and the cage, and carried around in them as shown in Figure 10.7b.

Figure 10.8 shows a scanning electron microscope (SEM) image of a prototype of the lobed micropump. The length of the major axis of each rotor is 9 μm. The thickness of the rotor is 2.5 μm.

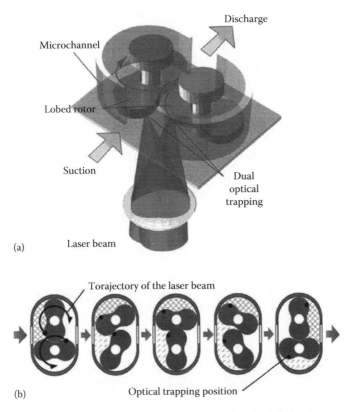

FIGURE 10.7 Optically driven lobed micropump. (a) Schematic of a lobed micropump driven by time-shared optical trapping. (b) Fluid transport by optically rotating two rotors.

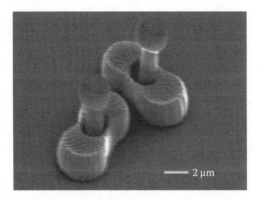

FIGURE 10.8   SEM image of a lobed micropump produced by two-photon microfabrication.

The lobed rotors were incorporated into a microchannel whose inlet and outlet are in 5 μm wide and 7 μm high. The microchannel was also fabricated by using two-photon microfabrication.

We demonstrated that fluid was transported with the optically driven micropump. Figure 10.9a and b shows sequential images of the micropump being driven. A tracer particle was also fabricated in the microchannel using two-photon microfabrication. The tracer particle was optically trapped to prevent it from moving around while unsolidified photopolymer was being washed out. After the washing process, the tracer particle was released, after which the rotors were interlocked using the time-shared laser scanning technique. As a result, the tracer particle was successfully brought into the microchamber by the pumped flow. When the rotation direction was reversed, the tracer particle was pushed to the outlet. Figure 10.10 shows the dependence of the velocity of the tracer particle on the rotational speed of the rotors. It is clear that the velocity of the tracer particle is proportional to

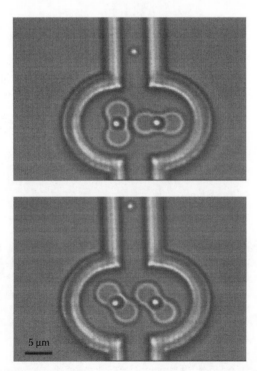

FIGURE 10.9   Sequential images taken while driving the optically driven micropump.

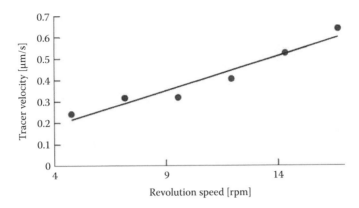

FIGURE 10.10   Dependence of the velocity of the tracer particle on the rotational speed of the rotors.

the rotational speed of the rotors. The flow rate was estimated at less than 1 pL/min. Ultralow flow rates such as this give our mechanism potential for use in micro/nano fluidic devices.

10.4.2   VISCOUS MICROPUMP USING A SINGLE ROTATING DISK

To produce high-performance lobed micropumps, high-precision microfabrication technique is required to engage two microrotors to prevent leakage between movable rotors and chamber. On the other hand, since viscous force is dominant in microfluidics, viscous force around a rotating disk is useful for making a viscous micropump. Figure 10.11 shows a schematic of the single-disk micropump driven by the time-shared optical trapping technique (Maruo 2007). A disk microrotor, which has three columns as targets for the optical trap, is confined to a U-shaped microchannel. A three-point optical trap using a single laser beam enables the disk microrotor to be rotated without a shaft. The laser beam, focused on the right and left columns, is scanned circularly, while the central laser beam is focused on the center column. Since the optically trapped center column acts as the axis of rotation, the disk microrotor is stably rotated without a shaft. The rotation of the disk microrotor induces both pressure gradient and flow along the curved channel. Although the pressure gradient is opposite to the desired flow direction, the viscous flow surrounding the rotor surmounts the resistance of the backpressure owing to the large viscous drag around the side of the rotor. In this micropump, the rotating disk microrotor provides continuous flow without pulsation unlike conventional diaphragm micropumps with check valves. Steady continuous flow such as this is useful for continuous-flow chemical processes and the gentle sorting of biological samples.

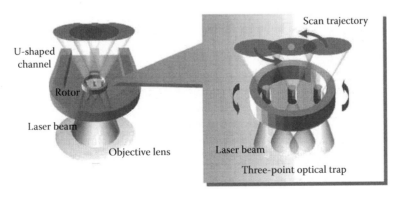

FIGURE 10.11   Schematic of a viscous micropump using a single rotating disk.

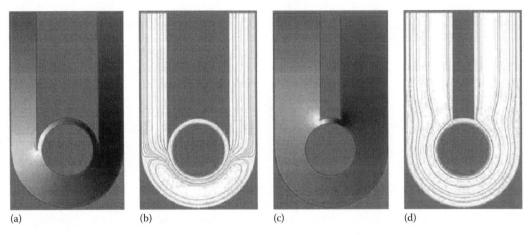

(a)   (b)   (c)   (d)

FIGURE 10.12 Pressure fields and streamlines of single-disk micropumps with different channel widths. (a) Pressure field channel width of 5 μm. (b) Pressure field channel width of 9 μm. (c) Streamlines channel width of 5 μm. (d) Streamlines channel width of 9 μm.

To optimize the design of the micropump, the flow field inside the U-shaped microchannel was analyzed using the two-dimensional (2D) finite element method (FEM) based on the Navier–Stokes equation. A commercially available multiphysics simulator (COMSOL Version 3.3) was used for FEM analysis. In the simulation, the diameter of the disk microrotor was 10 μm. The clearance between the rotor and the U-shaped microchannel was 6 μm in the lateral direction. The clearance between the rotor and the top of the center partition was 1 μm. The pressure at the inlet and outlet of the linear channel was 0 Pa. To analyze the flow generated by rotating the disk microrotor, velocity was applied to the fluid around the disk rotor. The density and viscosity of the fluid were 0.96×10^3 kg/m^3 and 1.92×10^{-3} Pa·s. These values agree with the parameters of glycol ether ester that was used in the following experiments.

Figure 10.12 shows examples of pressure fields and streamlines of micropumps with different channel widths. These simulations were performed using 2D FEM analysis. The flow velocity of 3 μm/s was assumed at the circumference of the rotor. As shown in Figure 10.12a and b, when the channel width is small, a large backflow is generated around the microrotor owing to the pressure gradient against the flow direction. On the other hand, when the channel width is wide, the pressure gradient is shallow, as shown in Figure 10.12c. In this case, the viscous force applied by the rotating rotor can be transferred efficiently from the inlet to the outlet through the U-shaped microchannel. Fluid can therefore be transported in the forward direction owing to sufficient viscous force against the backpressure caused by the rotation of the rotor, as shown in Figure 10.12d. At the linear regions of the U-shaped microchannels, a steady laminar flow is generated within 5 μm from the end of the semicircular channel. The pressure gradient is in the order of 10 mPa. This continuous laminar flow without pulsation is useful for continuous-flow chemical analysis and synthesis and gentle sorting of biological samples such as cells and proteins.

The dependence of the maximum flow velocity of the laminar flow on the channel width was examined. The results indicated that a microchannel of 9 μm width is suitable for a micropump with a 10 μm diameter microrotor. In this case, the diameter of the microrotor is smaller than the width of the microchannel, so the microrotor is confined to the U-shaped microchannel. This feature is advantageous for practical use, since the microrotor is enclosed inside the microchannel without irradiation of a laser beam.

Figure 10.13a shows a SEM image of a prototype of the viscous micropump with a disk of 10 μm in diameter. Both the microrotor with three columns and the U-shaped microchannel were fabricated using two-photon microfabrication. We demonstrated that the single-disk micropump could be used for pumping fluid and transportation of micro-objects. Figure 10.13b shows a sequence of

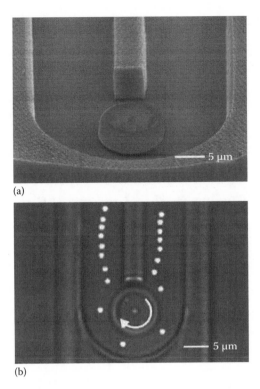

(a)

(b)

FIGURE 10.13 Prototype of the single-disk viscous micropump. (a) SEM image of a disk microrotor and a U-shaped microchannel without upper cover. (b) Particle transportation by rotating the single-disk microrotor at 27 rpm. The white points show the positions of the microparticle taken at 2 s intervals.

images taken at 2 s intervals showing the tracer particle being pumped through the channel with the disk microrotor being rotated at a speed of 27 rpm. The tracer particle was entrained with the high flow velocity. The trajectory of the tracer particle movement was similar to the streamlines obtained by the FEM simulation shown in Figure 10.12d. The flow velocity around the microrotor is higher than that of the linear microchannel since the viscous drag around the microrotor is dominant.

Figure 10.14 shows the dependence of the flow velocity on the rotational speed of the microrotor. The flow velocity is proportional to the rotational speed of the microrotor. The measured flow velocity of the tracer particle agreed with the average flow velocity obtained by 3D FEM analysis using COMSOL. In the 3D simulations, the thickness of the disk microrotor was 3 μm. The clearance between the microrotor and the microchannel was 2 μm in depth. Other parameters were the same as in our previous 2D simulations. These parameters agree with the experimental conditions. Therefore it was confirmed that the viscous micropump was precisely fabricated by two-photon microfabrication and driven by optical trapping stably.

10.4.3 Viscous Micropump Using a Twin Spiral Microrotor

The first prototype of the viscous micropump using a single-disk microrotor needs scanning of a laser beam to rotate the disk-like microrotor as shown in Figure 10.11. Recently, on the other hand, a spinning microrotor with twin spiral blades has been utilized as a microrotor for viscous micropump. Figure 10.15 shows an optically driven viscous micropump using a spinning microrotor with twin spiral blades (Maruo et al. 2009c). The outer cylinder attached to the twin spiral microrotor plays an important role in that it generates a continuous unidirectional flow without flow separation. The twin spiral microrotor is confined to a U-shaped microchannel to generate a laminar flow.

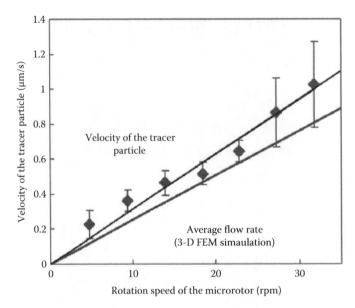

FIGURE 10.14 Dependence of the velocity of tracer particle movement and average flow velocity obtained by 3D FEM simulations on the rotational speed of the disk microrotor.

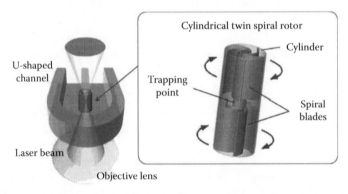

FIGURE 10.15 Schematic of an optically driven viscous micropump using a spinning microrotor with twin spiral blades.

When the twin spiral microrotor is optically trapped by focusing a laser beam at the corner of the U-shaped microchannel, it is stably rotated at a high speed of several hundred micrometers per minute while the microrotor is trapped in three dimensions at the focus. As a result, the microrotor rotation provides a continuous unidirectional flow without pulsation like the micropump using a rotating single disk.

The optically induced rotation of a microrotor with a single spiral blade was first reported by Galajda and Ormos (2002). In their experiments, they demonstrated that the rotation direction of the spiral microrotor is reversed when the trapped position of the microrotor is changed from below to beyond the focus along the optical axis. We employ the phenomenon of the reversed rotation of the single spiral microrotor to create a novel microrotor that has higher rotational speed than a single spiral microrotor. As shown in Figure 10.15, in our twin spiral microrotor, two spiral blades with inversed directions are connected along the optical axis inside an outer cylinder. When a laser beam is focused on the center of the twin spiral microrotor, optical torque induced by the optical radiation pressure exerted on the spiral blades is generated in the same direction, because the spiral

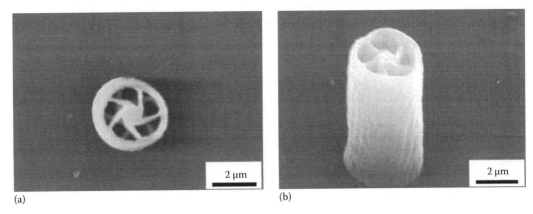

(a)   (b)

FIGURE 10.16 SEM images of a twin spiral microrotor with an outer cylinder. The diameter and height of the microrotor were 4 and 9 μm.

microrotors are mounted in opposite directions. Therefore, the net optical torque generated by the twin spiral microrotor can exceed that of the single spiral microrotor.

To demonstrate the validity of our twin spiral microrotor, we examined the driving performance of single and twin spiral microrotors experimentally. We fabricated three types of microrotors: single and twin spiral microrotors, and a twin spiral microrotor with an outer cylinder. Figure 10.16 shows SEM images of the twin spiral microrotor with an outer cylinder. The diameter and height of the microrotor were 4 and 9 μm, respectively. The opposite facing spiral blades were fabricated by two-photon microfabrication. When extracting the microrotor from unsolidified photopolymer, we used a supercritical CO_2 drying process to reduce harmful deformation caused by the surface tension of the rinse (Maruo et al. 2009b).

To drive the three types of microrotors, we also used the optical system of two-photon microfabrication, while operating the Ti:sapphire laser in the CW mode. The surrounding liquid was glycol ether ester, which we used for washing out the liquid photopolymer. Figure 10.17 shows the microrotor rotation speed dependence on laser power. The rotation speed of each microrotor was proportional to the intensity of the induced laser beam. We found that the rotation speed of the twin spiral microrotor was three-and-a-half times that of the single spiral microrotor. The maximum speed reached 560 rpm at a laser power of 500 mW. Although the rotation speed of the twin spiral microrotor with an outer cylinder was slower than that of the twin spiral microrotor alone, it

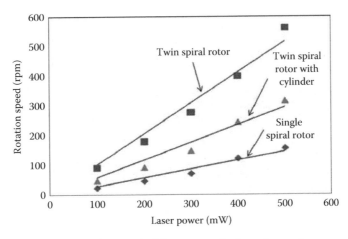

FIGURE 10.17 Dependence of rotation speed of three types of microrotors on laser power.

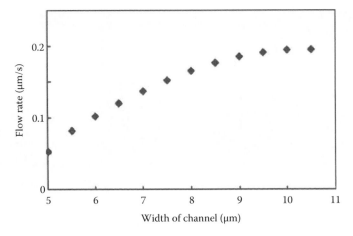

FIGURE 10.18 Dependence of the maximum flow velocity on the channel width.

exceeded 300 rpm. These results indicate that twin spiral blades are useful for increasing the rotation speed of a microrotor.

The pumping performance of the viscous micropump with the twin spiral microrotor depends on several parameters including the width of the U-shaped microchannel, and the diameter, and position of the microrotor like a viscous micropump using a single rotating disk. We examined the pressure field and streamlines of the micropump by changing the width of the microchannel. Figure 10.18 shows the dependence of the maximum flow velocity of the laminar flow on the channel width. In this simulation, we found that the backflow disappeared when the channel width was larger than 6 μm. Fluid can therefore be transported in the forward direction because there is sufficient viscous force against the backpressure caused by the rotation of the microrotor. A steady laminar flow is generated in the linear regions of the U-shaped microchannels. The results indicated that the flow rate was saturated at a microchannel width of 10 μm with a 4 μm diameter microrotor.

A viscous micropump containing a twin spiral microrotor (diameter: 4 μm) was fabricated by two-photon microfabrication. Figure 10.19 demonstrates that the viscous micropump with a twin spiral microrotor can be used for pumping fluid and the transportation of microobjects. The sequence of images was taken at 2 s intervals showing a tracer particle being pumped through the channel when the microrotor was rotating at a speed of 300 rpm. The trajectory of the tracer particle movement was similar to the streamlines obtained by the FEM simulation.

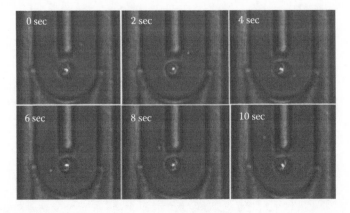

FIGURE 10.19 Optical drive of the viscous micropump with a twin spiral microrotor.

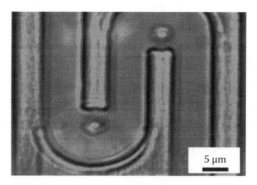

FIGURE 10.20   Tandem micropump driven by dual optical trapping using the spatial light modulation technique.

10.4.4   TANDEM MICROPUMP USING HOLOGRAPHIC OPTICAL TRAPPING

By connecting multiple viscous micropumps with a crooked microchannel, we can construct a tandem micropump using multiple twin spiral microrotors. As an example, we fabricated a tandem micropump using two microrotors (Maruo et al. 2009c). Since the twin spiral microrotors are rotated at a high speed of over several hundred rpm, multiple focal spots are needed to rotate the two microrotors simultaneously. Therefore, we constructed an optical micromanipulation system using a spatial modulation technique. The optical system consists mainly of a green laser (Verdi-5, Coherent Inc., wavelength: 532 nm, maximum laser power: 5W), and a liquid crystal spatial light modulator (PPM-X8267, Hamamatsu Photonics). The laser beam is diffracted at the spatial light modulator and forms the desired multiple spots at the Fourier plane of a lens (focal length: 300 mm). The multiple laser spots are then reduced with a lens (focal length: 100 mm) and an objective lens with a numerical aperture of 1.45. Finally, the multiple laser spots are focused on the twin spiral rotors. In our experiments, we inputted the phase distribution of the multiple Fresnel lens into the spatial light modulator to generate multiple focuses at the focal plane. The simultaneous focusing at multiple points makes it possible to drive multiple microrotors at high speed. Figure 10.20 shows an optical microscope image of the tandem micropump driven by dual optical trapping using the spatial light modulation technique. Both microrotors were stably rotated at each corner of the microchannel. The tandem micropump is useful for the long-distance transportation of microobjects such as living cells and microparticles.

10.4.5   PROTOTYPE OF ALL OPTICALLY CONTROLLED LAB-ON-A-CHIP

Use of optically driven micromachines such as micromanipulators and micropumps makes possible to construct all optically controlled lab-on-a-chip. Figure 10.21 shows the concept of all optically controlled lab-on-a-chip proposed by Maruo (2008). In this chip, micromachines such as microtweezers, microstirrers, and micropumps are cooperatively controlled by scanning a single laser beam. Target biological samples are transported into a chamber by pressure-driven flow, and examined with the microtweezers. To analyze the biological sample, chemical reagents are supplied to it using micropumps and microvalves.

The all optically controlled lab-on-a-chip offers unique advantages as follows: The combination of various types of micromachines can provide custom-made lab-on-a-chip for versatile applications. Since the optically controlled lab-on-a-chip does not need expensive and sophisticated microactuators, it is suitable for disposable use. Two-photon microfabrication can provide 3D microstructures by a direct laser scanning inside photopolymer, so both micromachines and a microchannel are easily fabricated together. For these reason, disposable, low-cost lab-on-a-chip made from photopolymer can be provided. In addition, the micromachines can also be built in a microchannel made from

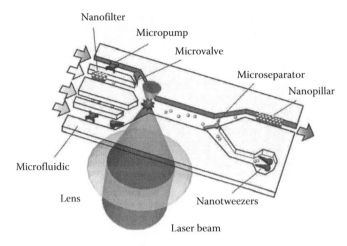

FIGURE 10.21 Schematic of all optically controlled lab-on-a-chip.

glass and other plastics. The integration of optically controlled micromachines into mass-produced microchannels can provide low-cost functional lab-on-a-chip devices.

10.4.6 EVANESCENT-WAVE DRIVEN MICROPUMP

Most of the previously developed optically driven micropumps were operated by an optical trapping technique with a focusing laser beam. For this reason, they require a laser scanning system and focusing optics to generate a tightly focused laser spot. In addition, since the area of the laser scanning is limited to less than about 100 μm² owing to the field of view of a high-numerical aperture objective lens, the number of built-in micromachines driven at the same time is also limited. This restricts to provide practical lab-on-a-chip devices containing optically driven micromachines.

To overcome these problems, we propose optically driven microrotors driven by evanescent waves as another type of optically driven micromachines (Murakami et al. 2011). It has already been demonstrated that evanescent waves generated at the surface of a high-refractive-index prism or a waveguide can propel microparticles by exerting optical radiation pressure on the microparticles (Kawata and Sugiura 1992, Kawata and Tani 1996). Recently, the propulsion of microparticles using waveguides has been applied to the sorting of microparticles (Grujic et al. 2005, Yang and Erickson 2010) and the transportation of nanomaterials, such as nanoparticles (Ng et al. 2002) and semiconductor nanowires (Néel et al. 2009). However, the rotation of microrotors driven by evanescent waves has not been demonstrated yet, as far as we know.

Figure 10.22 shows the basic concept of an evanescent-wave driven microrotor using the total internal reflection. In this case, the evanescent wave is generated at the surface of the prism by illuminating a laser beam from the prism side at an angle greater than the critical angle. When the

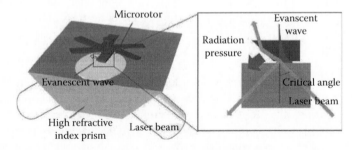

FIGURE 10.22 Schematic of an evanescent-wave driven microrotor using the total internal reflection.

evanescent wave is scattered by the blade of the microrotor, the blade is affected by optical radiation pressure and propelled. As a result, the microrotor restricted to a shaft was rotated. After the blade is moved away from the evanescent field, the next blade reaches the evanescent field and is affected by optical radiation pressure because of the rotation of the microrotor. Therefore, the microrotor can be rotated continuously.

The evanescent-wave driven micromachines have some advantages. For example, since evanescent waves are localized at the boundary between a high-refractive-index substrate and sample solutions, evanescent-wave driven micromachines have the potential to be driven even in colloidal or opaque solutions, in which a laser beam cannot be focused owing to scattering or absorption. In addition, the optical driving system of evanescent-wave driven micromachines using a high-refractive-index prism is compact compared with a conventional optical driving system, because it does not require a laser scanning system. Furthermore, if we employ the evanescent waves generated at the surface of an optical waveguide, micromachines, such as microrotors, can be integrated into the optical waveguide. The use of the optical waveguide allows us to drive multiple micromachines without the limitation of the field of view of a focusing optics unlike in the case of conventional optical trapping. The combination of an optical waveguide and a microchannel can provide a novel type of lab-on-a-chip devices integrated into an optical waveguide.

We analyzed the dependence of radiation pressure on the cross-sectional shapes of the blades of microrotors. In our simulation, we used multiphysics simulation software (COMSOL Version 3.5a) based on the 2D finite element analysis of electromagnetic field.

Figure 10.23a shows an analytical model of a blade. In our simulation, we assume that the refractive indices of the high-refractive-index glass substrate, blade, and surrounding liquid are

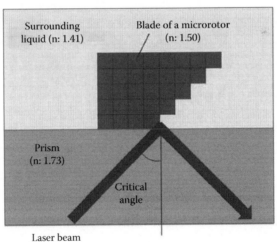

(a)   Laser beam
       (λ: 1064 nm, TE wave)

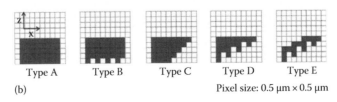

FIGURE 10.23 Analytical model of radiation pressure exerted on blades of microrotors. (a) Schematic of analytical model. (b) Five types of cross sections of blades. The unit cell of the blades is 500 nm^2.

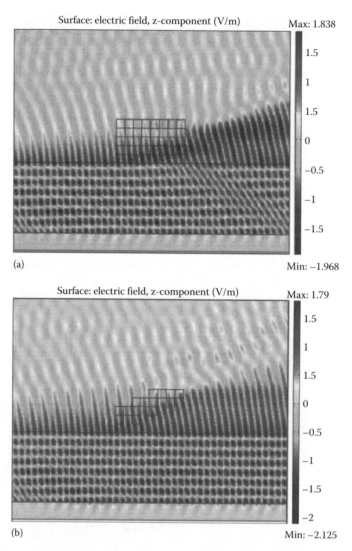

FIGURE 10.24 Electrical field distributions of types A and E blades. (a) Type A. (b) Type B. (Copyright (2011) The Japan Society of Applied Physics.)

1.73, 1.5, and 1.41, respectively. These values are the same as those of our experimental materials. The incident laser beam has a wavelength of 1064 nm and exhibits transverse electric wave polarization. The transverse electric wave has the electric field element that is vertical to the incidence plane. Since it was demonstrated that the driving force of a transverse electric wave is larger than that of a transverse magnetic wave (Kawata and Sugiura 1992), we used a transverse electric wave in our simulation. To generate an evanescent field at the boundary between the substrate and the liquid, the laser beam is illuminated at an incident angle of 55.8° greater than the critical angle between the substrate and the liquid. To avoid the reflection of the propagating waves at the circumference of the analytical area, a perfectly matched layer (PML) is applied to the simulations using COMSOL software.

Figure 10.23b shows five types of cross sections of blades. We analyzed not only the rectangular blade (type A) but also blades with a grating-like shape and a taper (types B, C, D, E). Figure 10.24a and b shows the electrical field distributions of types A and E. The unit cell of the blades is square, 500 nm on each side (500 nm²). The size of the unit cell depends on the depth resolution of our

TABLE 10.1

Calculated Relative Values of the Horizontal (X-Direction) and Vertical (Z-Direction) Components of Radiation Pressure Exerted on Blades

	Type A	Type B	Type C	Type D	Type E
F_X	2.36	4.24	3.15	8.48	12.5
F_Z	26.8	13.4	17.2	2.44	3.57

two-photon microfabrication system. When we fabricate a microrotor, the thickness of each cross-sectional layer is 500 nm. Therefore, the unit cell of the blades is 500 nm^2.

Using the electromagnetic field at the circumference of the blades, the optical radiation pressure exerted on each blade can be calculated using a Maxwell stress tensor. The calculated relative values of the horizontal (x-direction) and vertical (z-direction) components of radiation pressure are summarized in Table 10.1. In the case of the rectangular blade (type A), since the evanescent wave is refracted or scattered at the front and backside of the blade, the positive propulsive force generated at the front side is constrained by the negative force generated at the backside. As a result, the net propulsive force is relatively small. In addition, since the vertical force affected by the top part of the blade is larger than those in the case of other blades, the blade easily floats during rotation. This causes the fluctuation of the rotation of the microrotor under evanescent-wave illumination. In the case of type B, the grating-like structure fabricated at the bottom part of the rectangular blade enhances the scattering of the evanescent wave. As a result, the net propulsive force of type B is greater than that of type A. In addition, the vertical force of type B is reduced to half that of type A. The taper structure shown in type C can also make the propulsive force larger than that of type A. These results indicate that both the grating-like structure and the taper structure are useful for the improvement of not only the rotational speed but also the stability of rotation. Therefore, as shown in types D and E, we combine the grating-like structure and the taper structure to increase further the propulsive force. As a result, we succeeded in increasing the propulsive force of types D and E compared with that of type A. In particular, type E has a propulsive force that is about five times larger than that of type A. The vertical force of type E is also reduced to about 13% of that of type A. From these simulation results, it is expected that the type E microrotor will have the fastest rotation among these microrotors.

We fabricated a microrotor on a high-refractive-index glass substrate using a two-photon microfabrication system. Since a high-refractive-index substrate is not suitable for generating a tight focus through a high-numerical-aperture objective lens owing to spherical aberration, we introduced an inverted setup, in which a high-refractive-index substrate is set opposite to the objective lens. The microrotor is fabricated from the bottom part of the shaft on the upper high-refractive index substrate in the inverted setup, and each layer of the microrotor is sequentially fabricated using the sliced data of a 3D CAD model. In this case, since the previously fabricated structure does not disturb the intensity distribution of the focus, the microrotor can be fabricated precisely under optimal exposure conditions with an ideal laser focus.

Figure 10.25 shows SEM images of a prototype of type E. The diameter and thickness of the microrotor are 60 and 2 μm, respectively. To drive the microrotors, a YAG laser (wavelength: 1064 nm, output laser power: 1.0 W) is used. The microrotor is located on the surface of a high-refractive-index substrate attached on a prism. The prism and substrate are fabricated from high-refractive-index glass (SF 10). The solvent is glycol ester acetate. The incident angle of the laser beam is adjusted at an angle (55.7°) over the critical angle between the prism and the solvent by tilting the mirror in front of the focusing lens. Then, the laser beam is focused by a lens (focal length: 50 mm). The rotation of the microrotor under evanescent-wave illumination is observed using a CCD camera.

FIGURE 10.25   SEM images of a prototype of type E microrotor. (Copyright (2011) The Japan Society of Applied Physics.)

  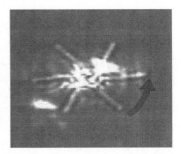

FIGURE 10.26   Sequential optical images of a rotating microrotor of type E at 2 s intervals. (Copyright (2011) The Japan Society of Applied Physics.)

We examined the rotation speed of microrotors of types A and E. Figure 10.26 shows sequential optical images of a rotating microrotor of type E at 2 s intervals. In our experiments, the type A microrotor was rotated at 1 rpm, and the type E microrotor was rotated at 2.5 rpm. As a result, it was demonstrated that the type E microrotor could be rotated faster than the type A microrotor. However, the increase in rotation speed is smaller than the analytical result. One of the reasons for this smaller increase is the lack of the fabrication accuracy of the complex blades. In addition, although the radiation pressure exerted on the cross-sectional shape of a single blade was calculated and compared in our simulation, the microrotors had six blades actually. Therefore, the actual rotation speed of the microrotors depends on the net radiation pressure exerted on the six blades. This causes the difference between the experimental and simulation results of the radiation pressure.

We also examined the effect of the number of blades on the rotation speed. To realize a continuous rotation of a microrotor, it is preferable to increase the number of blades. In our experiments, however, we found that the rotation speed of an eight-blade microrotor was lower than that of a six-blade microrotor. One reason for this is that the net drag force affecting the microrotor increases owing to the increase in the number of blades.

Finally, a micropump using an evanescent-wave driven microrotor (type E) was developed as shown in Figure 10.27. The diameter of the rotor was 60 μm. The width of the microchannel was 35 μm. The evanescent-wave driven micropump will be useful for the large-scale optically controlled lab-on-a-chip devices using optical circuits.

10.5   REPLICATION OF MOVABLE MICROMACHINES TOWARD MASS PRODUCTION OF OPTICALLY DRIVEN MICROFLUIDIC DEVICES

Two-photon microfabrication can offer novel microdevices such as high-precision micropumps (Maruo 2007, Maruo and Inoue 2006, Maruo et al. 2009c) and microtweezers (Maruo 2007,

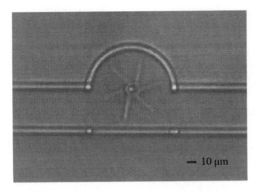

FIGURE 10.27 Prototype of a micropump using an evanescent-wave driven microrotor of type E.

Maruo et al. 2003b). However, since two-photon microfabrication is based on direct writing of a femtosecond pulsed laser beam, the throughput is lower than that of conventional lithographic approach. To overcome the intrinsic drawback, several types of parallel processing, including microlens array method (Kato et al. 2005) and spatial light modulation method (Hayasaki et al. 2005), were developed in multi-photon microfabrication. These techniques are useful for mass production of 3D microdevices and large-scale devices including 3D micro/nano structures.

By contrast, replication technique with poly (dimethylsiloxane) (PDMS) mold, which is produced from a 3D master structure, was also proposed by LaFratta et al. (2004, 2006) for improving the throughput of multi-photon microfabrication. In their method called membrane-assisted microtransfer molding (MA-µTM), not only high-aspect ratio microstructures but also sophisticated 3D structures with closed loops can be reproduced by using a 3D master structure with partition membranes.

We intended to reproduce 3D movable microstructures on the order of micrometers by using MA-µTM with thinner membranes whose thickness is less than 300 nm (Maruo et al. 2009a). In our method, we reproduce a movable microstructure restricted to a shaft by using a cylindrical membrane attached to the movable micropart according to the replication procedure shown in Figure 10.28. In this process, self-adhesion property of PDMS mold allows to inject photopolymer into the mold of the movable microring separated from the substrate. Figure 10.29a and b shows a master model of a micro ring model with a cylindrical membrane and a replica of movable micro ring (inner diameter of the ring: 8 µm, thickness of the ring: 2 µm, diameter of the shaft: 4 µm). The thickness and height of the cylindrical membrane, which separates the inside and outside of the movable ring, is important parameters to reproduce the freely movable ring by separating the PDMS mold without

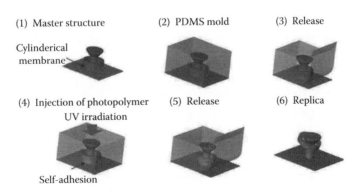

FIGURE 10.28 Replication process of movable microparts using a membrane-assisted transfer molding technique. (Copyright (2011) The Japan Society of Applied Physics.)

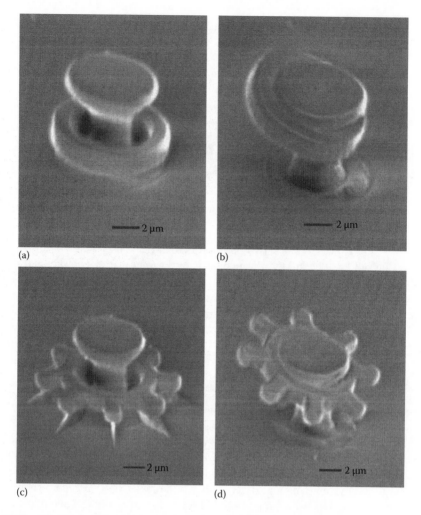

(a)   (b)

(c)   (d)

FIGURE 10.29   Replication of movable microparts. (a and b) Master model and replica of a microring. (c and d) Master model and replica of a microgear. (Copyright (2011) The Japan Society of Applied Physics.)

extreme tensile force. In case of the replication of the micro ring, we found that the optimal height and thickness of the cylindrical membrane was 2.5 μm and 300 nm. The optimal cylindrical membrane offers both good sealing of the membrane and easy release to reproduce the movable ring.

Microrotor models were also replicated by this method. Figure 10.29c and d shows SEM images of a master structure and a replicated microrotor with eight blades. In the master structure, the additional plane membranes attached under the blades make it easy to release the PDMS mold around the blades. By using this approach, we succeeded in replicating not only a rotor model but also microtweezers model. Figure 10.30a and b shows SEM images of the master structure and a replica of microtweezers. The replicated microtweezers could be driven by optical trapping in a liquid. Figure 10.31 demonstrates the optical driving of replicated microtweezers in water containing surfactant at 0.5 wt%. Although the replicated movable arms were adhered to the shaft after replication process, the movable micropart could be released by a glass needle and driven by light in a liquid.

The current yield rate of the replication process of movable micromechanism depends on several factors such as complexity of the master model and layout and shape of the partition membranes. For example, the averaged yield rate of the replication of movable microrings attained to 42% within eight times replication. The durability of the mold is also an important parameter to obtain both

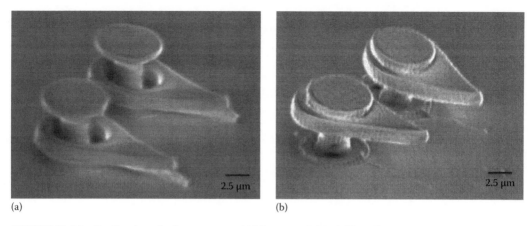

(a)   (b)

FIGURE 10.30   Replication of microtweezers. (a) Master model and (b) replica.

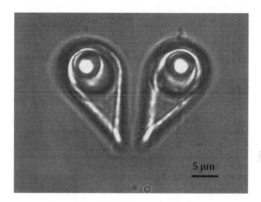

FIGURE 10.31   Optical driving of replicated microtweezers.

high yield rate and a number of replication. Therefore, not only modification of both mechanical characteristic and self-adhesion ability of the PDMS mold, but also improvement of the antiadhesion property between master polymer structure and PDMS mold are required for the further improvement of replication process of movable micromechanism. In the near future, this replication method of movable microstructures makes it possible to mass-produce optically driven microfluidic devices.

10.6   CONCLUSION AND FUTURE PROSPECT

We have developed several types of optically driven micromachines, including microtweezers and micropumps, using two-photon microfabrication. As one of the most promising applications of the optically driven micromachines, we proposed a concept of all-optically controlled lab-on-a-chip. In the lab-on-a-chip, since built-in micromachines are remotely controlled by light, the chip is suitable for disposable usage. The disposable lab-on-a-chip devices suit medical diagnosis and analysis from the standpoint of prevention of biohazard. Although conventional optical microscopes and lasers are still used for the operation of the current optically driven microfluidic devices, evanescent-wave driven microfluidic devices will make possible to construct a compact optically controlled lab-on-a-chip without laser scanning system. In addition, on-chip lasers or optical fiber systems are ideal for practical use in medical diagnosis and palm-top analysis systems. Some pioneer works have been reported in this context. For example, optical fibers have been utilized to manipulate microparticles inside a microchannel (Domachuk et al. 2005, Jensen-McMullin et al. 2005). Semiconductor lasers are also integrated into a microchannel for the sequential manipulation of microparticles (Cran-McGreehin et al. 2006). In the near future, all-optically controlled biochips that are fully integrated not only with light sources but also with detectors could be created by advanced micromachining techniques. The fully integrated biochips driven by light are expected to open a path toward the future biotechnology and medical care and diagnosis.

REFERENCES

Arai, F., Yoshikawa, K., Sakami, T. et al. 2004. Synchronized laser micromanipulation of multiple targets along each trajectory by single laser. *Appl. Phys. Lett.* 85: 4301–4303.

Ashkin, A. 2000. History of optical trapping and manipulation of small-neutral particle, atoms, and molecules. *IEEE J. Sel. Top. Quantum Electron.* 6: 841–856.

Ashkin, A., Dziedzic, J. M., and Yamane, T. 1987. Optical trapping and manipulation of single cells using infrared laser beams. *Nature* 330: 769–771.

Cran-McGreehin, S. J., Dholakia, K., and Krauss, T. F. 2006. Monolithic integration of microfluidic channels and semiconductor lasers. *Opt. Express* 14: 7723–7729.

Cumpston, B. H., Ananthavel, S. P., Barlow, S., Dyer, D. L. et al. 1999. Two-photon polymerization initiators for three-dimensional optical data storage and microfabrication. *Nature* 398: 51–54.

Domachuk, P., Cronin-Golomb, M., Eggleton, B. J. et al. 2005. Application of optical trapping to beam manipulation in optofluidics. *Opt. Express* 13: 7265–7275.

Galajda, P. and Ormos, P. 2001. Complex micromachines produced and driven by light. *Appl. Phys. Lett.* 78: 249–251.

Galajda, P. and Ormos, P. 2002. Rotors produced and driven in laser tweezers with reversed direction of rotation. *Appl. Phys. Lett.* 80: 4653–4655.

Grier, D. G. 2003. A revolution in optical manipulation. *Nature* 424: 810–816.

Grujic, K., Hellesø, O. G., Hole, J. P. et al. 2005. Sorting of polystyrene microspheres using a Y-branched optical waveguide. *Opt. Express* 13: 1–7.

Hayasaki, Y., Sugimoto, T., Takita, A. et al. 2005. Variable holographic femtosecond laser processing by use of a spatial light modulator. *Appl. Phys. Lett.* 87: 031101.

Higurashi, E., Ohguchi, O., Tamamura, T. et al. 1997. Optically induced rotation of dissymmetrically shaped fluorinated polyimide microobjects in optical traps. *J. Appl. Phys.* 82: 2773–2779.

Higurashi, E., Ukita, H., Tanaka, H. et al. 1994. Optically induced rotation of anisotropic micro-objects fabricated by surface micromachining. *Appl. Phys. Lett.* 64: 2209–2210.

Jensen-McMullin, C., Lee, H. P., and Lyons, E. R. 2005. Demonstration of trapping, motion control, sensing and fluorescence detection of polystyrene beads in a multi-fiber optical trap. *Opt. Express* 13: 2634–2642.

Kato, J., Takeyasu, N., Adachi, Y. et al. 2005. Multiple-spot parallel processing for laser micronanofabrication. *Appl. Phys. Lett.* 86: 044102.

Kawata, S. and Sugiura, T. 1992. Movement of micrometer-sized particles in the evanescent field of a laser beam. *Opt. Lett.* 17: 772–774.

Kawata, S., Sun, H.-B., Tanaka, T. et al. 2001. Finer features for functional microdevices. *Nature* 412: 697–698.

Kawata, S. and Tani, T. 1996. Optically driven Mie particles in an evanescent field along a channeled waveguide. *Opt. Lett.* 21: 1768–1770.

LaFratta, C. N., Baldacchini, T., Farrer, R. A. et al. 2004. Replication of two-photon-polymerized structures with extremely high aspect ratios and large overhangs. *J. Phys. Chem. B* 108: 11256–11258.

LaFratta, C. N., Li, L., and Fourkas, J. T. 2006. Soft-lithographic replication of 3D microstructures with closed loops. *Proc. Natl. Acad. Sci. USA* 103: 8589–8594.

Laser, D. J. and Santiago, J. G. 2004. A review of micropumps. *J. Micromech. Microeng.* 14: R35–R64.

Leach, J., Mushfique, H., Leonardo, R. et al. 2006. An optically driven pump for microfluidics. *Lab Chip* 6: 735–739.

Leach, J., Sinclair, G., Jordan, P. et al. 2004. 3D manipulation of particles into crystal structures using holographic optical tweezers. *Opt. Express* 12: 220–226.

Maruo, S. 2007. Manipulation of microobjects by optical tweezers. In *Microfluidic Technologies for Miniaturized Analysis Systems, MEMS Reference Shelf*, eds. Hardt, S. and Schonfeld, F., pp. 275–314, New York: Springer.

Maruo, S. 2008. Optically driven micromachines for biochip application. In *Nano- and Micromaterials Series: Advances in Materials Research*, eds. Ohno, K., Tanaka, M., Takeda, J. and Kawazoe, Y., Vol. 9, pp. 291–309, New York: Springer-Verlag.

Maruo, S. and Fourkas, J. T. 2008. Recent progress in multiphoton microfabrication. *Lasers Photonics Rev.* 2: 100–111.

Maruo, S., Hasegawa, T., and Yoshimura, N. 2009a. Replication of three-dimensional rotary micromechanism by membrane-assisted transfer molding. *Jpn. J. Appl. Phys.* 48: 06FH05.

Maruo, S., Hasegawa, T., and Yoshimura, N. 2009b. Single-anchor support and supercritical CO2 drying enable high-precision microfabrication of three-dimensional structures. *Opt. Express* 17: 20945–20951.

Maruo, S. and Hiratsuka, Y. 2007. Optically driven micromanipulators with rotating arms. *J. Rob. Mechatron.* 19: 565–568.

Maruo, S., Ikuta, K., and Korogi, H. 2003a. Submicron manipulation tools driven by light in a liquid. *Appl. Phys. Lett.* 82: 133–135.

Maruo, S., Ikuta, K., and Korogi, H. 2003b. Force-controllable, optically driven micromachines fabricated by single-step two-photon microfabrication. *J. Microelectromech. Syst.* 12: 533–539.

Maruo, S. and Inoue, H. 2006. Optically driven micropump produced by three-dimensional two-photon microfabrication. *Appl. Phys. Lett.* 89: 144101.

Maruo, S. and Inoue, H. 2007. Optically driven viscous micropump using a rotating microdisk. *Appl. Phys. Lett.* 91: 084101.

Maruo, S., Nakamura, O., and Kawata, S. 1997. Three-dimensional microfabrication with two-photon absorbed photopolymerization. *Opt. Lett.* 22: 132–134.

Maruo, S., Takaura, A., and Saito, Y. 2009c. Optically driven micropump with a twin spiral microrotor. *Opt. Express* 17: 18525–18532.

Molloy, J. E. and Padgett, M. J. 2002. Lights, action: Optical tweezers. *Contemp. Phys.* 43: 241–258.

Murakami, S., Ikegame, M., Okamori, K. et al. 2011. Evanescent-wave-driven microrotors produced by two-photon microfabrication. *Jpn. J. Appl. Phys.* 50: 06GM16.

Neale, S. L., Macdonald, M. P., Dholakia, K. et al. 2005. All-optical control of microfluidic components using form birefringence. *Nat. Mat.* 4: 530–533.

Néel, D., Gétin, S., Ferret, P. et al. 2009. Optical transport of semiconductor nanowires on silicon nitride waveguides. *Appl. Phys. Lett.* 94: 253115.

Neuman, K. C. and Block, S. M. 2004. Optical trapping. *Rev. Sci. Instrum.* 75: 2787–2809.

Ng, L. N., Luff, B. J., Zervas, M. N. et al. 2002. Propulsion of gold nanoparticles on optical waveguides. *Opt. Commun.* 208: 117–124.

Rodrigo, P. J., Gammelgaard, L., Bøggild, P. et al. 2005. Actuation of microfabricated tools using multiple GPC-based counterpropagating-beam traps. *Opt. Express* 13: 6899–6904.

Sasaki, K., Koshioka, M., Misawa, H. et al. 1991. Pattern formation and flow control of fine particles by laser-scanning micromanipulation. *Opt. Lett.* 16: 1463–1465.

Seet, K. K., Mizeikis, V., Juodkazis, S. et al. 2006. Three-dimensional horizontal circular spiral photonic crystals with stop gaps below 1 μm. *Appl. Phys. Lett.* 88: 221101.

Sun, H.-B., Takada, K., and Kawata, S. 2001. Elastic force analysis of functional polymer submicron oscillators. *Appl. Phys. Lett.* 79: 3173–3175.

Terray, A., Oakey, J., and Marr, D. W. M. 2002a. Fabrication of linear colloidal structures for microfluidic applications. *Appl. Phys. Lett.* 81: 1555–1557.

Terray, A., Oakey, J., and Marr, D. W. M. 2002b. Microfluidic control using colloidal devices. *Science* 296: 1841–1844.

Unger, M. A., Chou, H.-P., Thorsen, T. et al. 2000. Monolithic microfabricated valves and pumps by multilayer soft lithography. *Science* 288: 113–116.

Yang, A. H. J. and Erickson, D. 2010. Optofluidic ring resonator switch for optical particle transport. *Lab Chip* 10: 769–774.

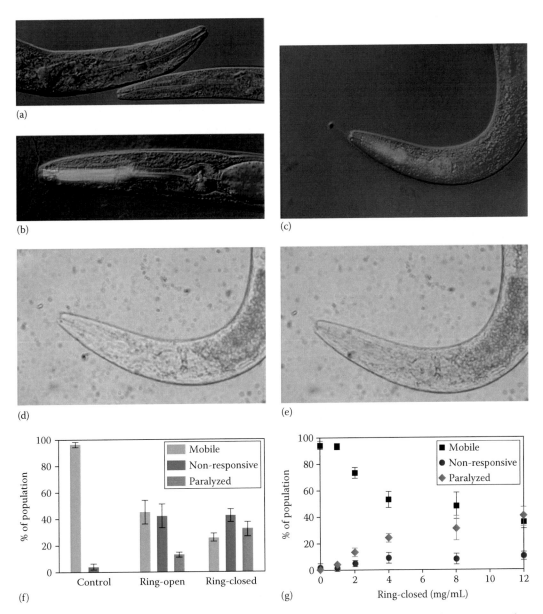

FIGURE 1.9 Photocontrolled molecular switch used to induce paralysis in *C. elegans* nematodes. (a) Fluorescence microscopy image of *C. elegans* incubated with only 10% dimethylsulfoxide. Images of the nematode fed the ring-closed isomer of the photoswitch at (b) 10 min and (c) 60 min. Microscopy images also show the photoswitch being converted between colorless ring-open isomer (d) to ring-closed blue isomer (e) when exposed to visible wavelengths greater than 490 nm and UV light (365 nm) for 20 and 2 min, respectively. The number of mobile, nonresponsive, and paralyzed nematodes for samples that have been treated with the ring-open or the ring-closed form of the photoswitch compared to controls after 60 min incubation is shown in (f). Finally, the samples exposed to varying amounts of ring-open isomer are presented in (g). (Reprinted with permission from Al-Atar, U., Fernandes, R., Johnsen, B., Baillie, D., and Branda, N.R., A photocontrolled molecular switch regulates paralysis in a living organism, *J. Am. Chem. Soc.*, 131, 15966–15967. Copyright 2009 American Chemical Society.)

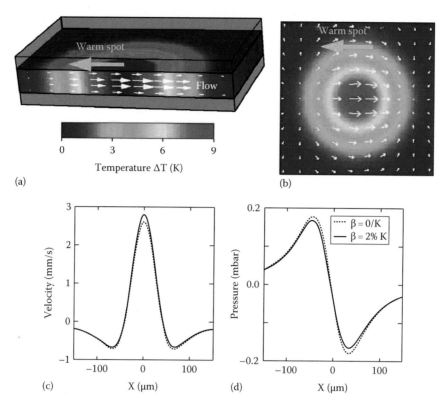

FIGURE 1.17 Finite element simulation of the fluid velocity for thermo viscous pumping around the laser spot in both (a) 3D and (b) 2D. The color-coded temperature spot is moving to the left producing fluid expansion in its front (left side) and contraction in its wake (right side). The velocity and pressure profiles created by the laser spot, along the x-direction, are shown in (c) and (d), respectively. The pressure profile is positive in front of the spot, as a result of the thermal expansion, and negative in its wake. (Reprinted with permission from Weinert, F.M. and Braun, D., Optically driven fluid flow along arbitrary microscale patterns using thermoviscous expansion, *J. Appl. Phys.*, 104, 104701–104701-10. Copyright 2008 American Institute of Physics.)

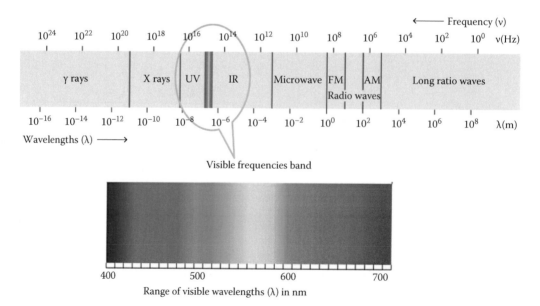

FIGURE 2.1 Electromagnetic Spectrum. (Adopted from: http://upload.wikimedia.org/wikipedia/commons/f/f1/EM_spectrum.svg)

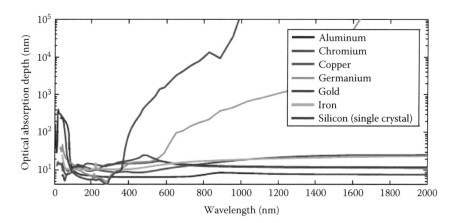

FIGURE 2.8 Optical absorption depth versus wavelength for different materials. (From Brown, M.S. and Arnold, C.B., Fundamentals of laser-material interaction and application to multiscale surface modification, in *Laser Precision Microfabrication*, Sugioka, K. et al. (Eds.). Springer-Verlag, Berlin, Germany, Chap. 4, 2010.)

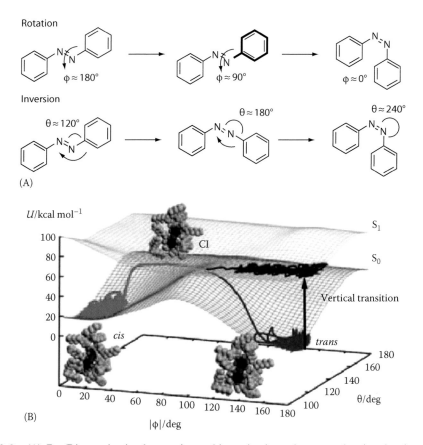

FIGURE 3.2 (A) $E \rightarrow Z$ isomerization by rotation and inversion in azobenzene, showing the changes in the C–N=N–C dihedral angle (φ) and the C–N=N angle (θ). (B) Molecular dynamics simulation in *n*-hexane (light gray molecules) of $S_0 \rightarrow S_1$ photoisomerization of *trans*-azobenzene to *cis*-azobenzene (dark gray, blue, and red molecule), proceeding through the conical intersection (CI) connecting the first excited singlet (S_1) and ground state singlet (S_0) potential energy surfaces; U is internal energy. (Reproduced from Tiberio, G. et al. *Chem. Phys. Chem.*, 11, 1018, 2010.)

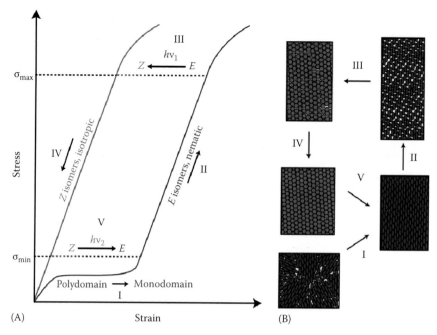

FIGURE 3.7 Model work cycle for a photoactuating polydomain LCE; the E isomers of the photoactive mesogens are rod-like, and the Z isomers are bent. (A) The work cycle depicted on a stress–strain plot. The descriptions of the steps (I–V) in the work cycle are described in the main text. (B) Illustrations of the material's microstructural changes in Steps I–V, depicting the shape distributions of polymer chains in the nematic (blue ovals) and isotropic (red circles) states.

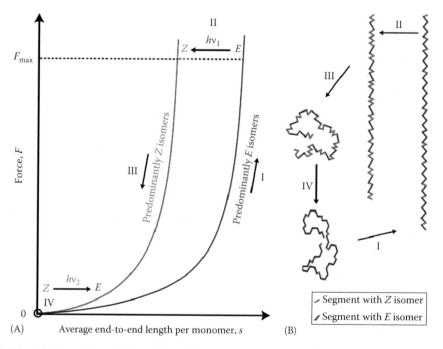

FIGURE 3.9 Model work cycle for a single molecule of a photoactuating polymer. (A) The work cycle depicted on a force–extension plot. The descriptions of the steps (I–IV) in the work cycle are described in the main text. (B) Illustrations of the polymer shape changes in Steps I–IV, indicating the E (blue) and Z (red) isomers in each state.

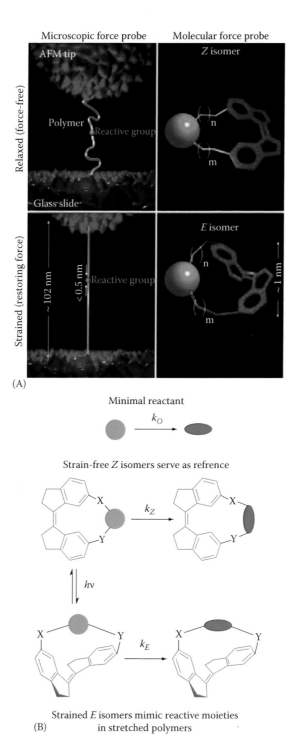

FIGURE 3.12 (A) Comparison of methods for measuring force-dependent kinetics of localized reactions. On the left, conventional single-molecule force spectroscopy requires the incorporation of the reactive moiety (blue sphere) into a long flexible polymer, attaching the polymer to a pair of microscopic force probes (here, the tip of the atomic force microscope cantilever and a glass slide on a piezoelectric stage) and stretching it by separating the probes. The size of the reactive moiety is typically less than the surface roughness of the probes or the magnitude of their thermal fluctuations, which significantly limits the accuracy of the measurements and the scope of reactions amenable to such studies. The right panels show a molecular force probe containing the same reactive moiety. (Reproduced from Yang, Q.Z. et al., *Nat. Nanotechnol.*, 4, 302, 2009.) (B) The general method for measuring force-dependent kinetics with molecular force probes. The strained *E* isomers are obtained by photoisomerization of strain-free *Z* analogs, which are synthesized using conventional chemistry. (Modified from Kucharski, T.J. et al., *J. Phys. Chem. Lett.*, 1, 2820, 2010.)

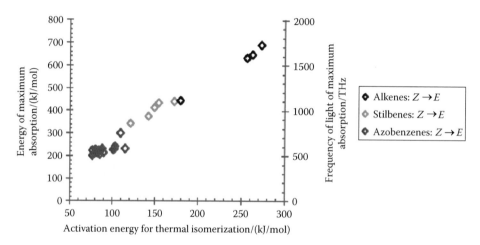

FIGURE 3.13 Apparent relationship between the activation energy for thermal $Z \rightarrow E$ isomerization and the energy of the maximum absorption band leading to photoisomerization of the Z isomer. Absorption energies for the Z isomers of azobenzenes was estimated as the absorbance energy in the E isomer + the absorbance energy of (Z)-azobenzene – the absorbance energy of (E)-azobenzene. Alkene data from Bouman (1985), Huh (1990), Walker (1986), Gary (1954), Molina (1999), and Kistiakowsky (1934); stilbene data from Görner (1995) and Ross (1971), and azobenzene data from Nishimura (1976).

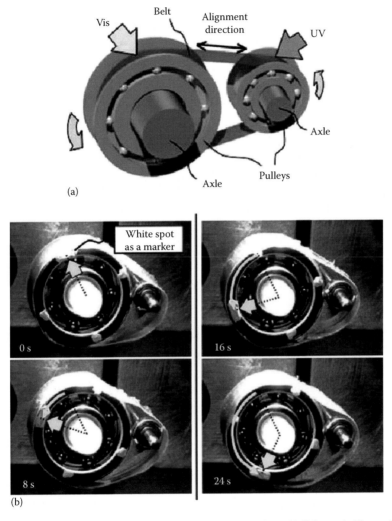

FIGURE 4.16 Light-driven plastic motor with the LCE laminated film. (a) Schematic illustration of a light-driven plastic motor system, showing the relationship between light irradiation positions and a rotation direction. (b) Series of photographs showing time profiles of the rotation of the light-driven plastic motor with the LCE laminated film induced by simultaneous irradiation with UV (366 nm, 240 mW cm^{-2}) and visible light (>500 nm, 120 mW cm^{-2}) at room temperature. Diameter of pulleys: 10 mm (left), 3 mm (right). Size of the belt: 36 mmH5.5 mm. Thickness of the layers of the belt: PE, 50 mm; LCE, 18 mm. (Reproduced from Yamada, M. et al., *Angew. Chem. Int. Ed.,* 47, 4986, 2008.)

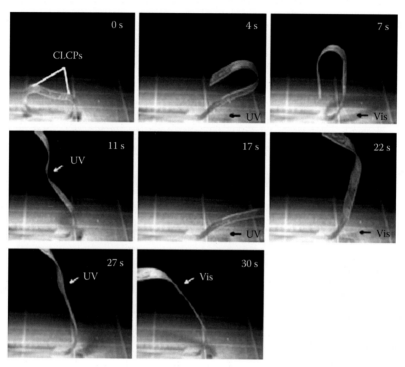

FIGURE 4.18 Series of photographs showing time profiles of the flexible robotic arm motion of the CLCP laminated film induced by irradiation with UV (366 nm, 240 mW cm^{-2}) and visible light (>540 nm, 120 mW cm^{-2}) at room temperature. Arrows indicate the direction of light irradiation. Spot size of the UV light irradiation is about 60 mm^2. Size of the film: 34 mm × 4 mm; the CLCP laminated parts: 8 mm × 3 mm and 5 mm × 3 mm. Thickness of the layers of the film: PE, 50 mm; CLCP layers, 16 mm. (Reproduced from Yamada, M. et al., *J. Mater. Chem.*, 19, 60, 2009.)

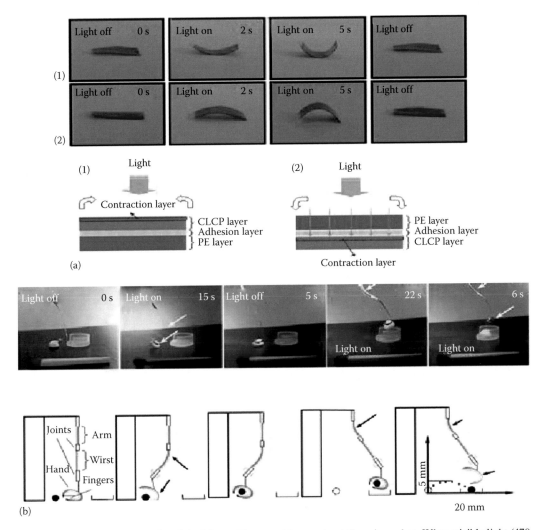

FIGURE 4.19 (a) Photographs of the bilayer films used to construct the microrobot. When visible light (470 nm, 30 mW cm^{-2}) is irradiated on the azobenzene-based cross-linked liquid crystalline polymer (CLCP) layer the films can either bend upward (1) or downward (2) depending upon the position of the CLCP layer. A schematic illustration of the bending is also included in the bottom with the number corresponding to the photographs. Size of the film: 7 mm×4 mm×30 mm. (b) (Top) Photographs showing the microrobot picking, lifting, moving, and placing the object to a nearby container by turning on and off the light (470 nm, 30 mW cm^{-2}). Length of the match in the pictures: 30 mm. Thickness of PE and CLCP films: 12 mm. Object weight: 10 mg. (Bottom) Schematic illustrations of the states of the microrobot during the process of manipulating the object. The insert coordinate indicates the moving distance of the object in vertical and horizontal directions. White and black arrows denote the parts irradiated with visible light. (Reproduced from Cheng, F. et al., *Soft Matter*, 6, 3447, 2010.)

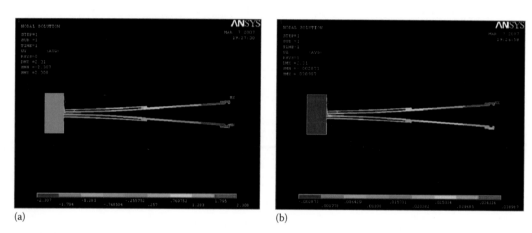

FIGURE 6.32 ANSYS simulation of the gripper opening (a) and out-of-plane gripper beam bending (b) during actuation of the structure. By choosing the appropriate aspect ratio of the beams, the out-of-plane motion of the structure is negligible compared to the in-plane gripper opening.

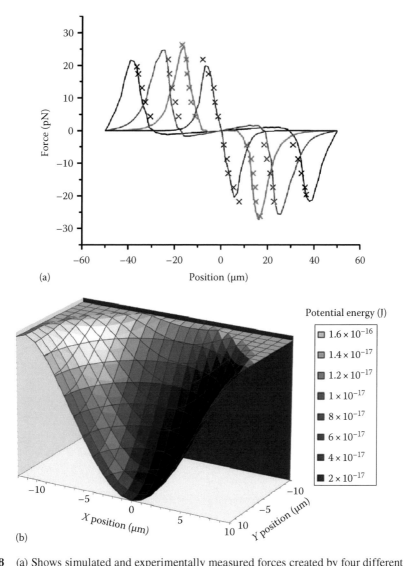

FIGURE 8.8 (a) Shows simulated and experimentally measured forces created by four different-sized OET traps (trap diameters of 12 μm in blue to 73 μm in black) on a HeLa cell. (b) The potential energy well for a cell in the smallest trap is shown. (From Neale, S.L. et al., *Opt. Express*, 17, 5231, 2009. Copyright 2009 Optical Society of America.)

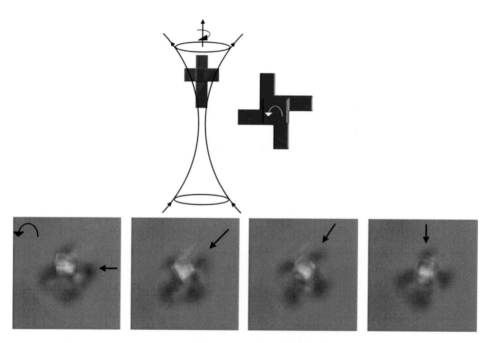

FIGURE 9.18 Upper row: schematics of the second possible orientation of the microstructure in the optical trap and the corresponding anticlockwise direction of rotation. Lower row: video microscopy snapshots showing a quarter of a rotation of the chiral microstructure.

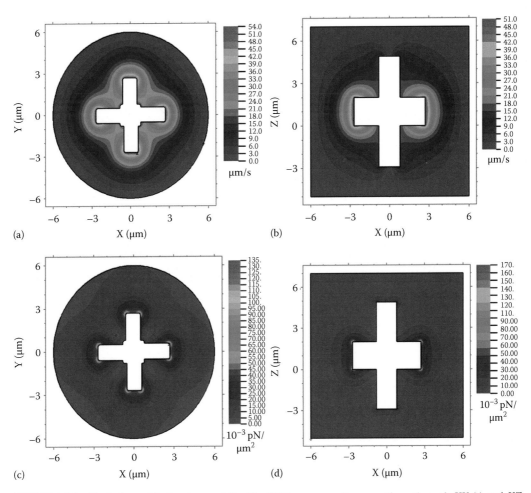

FIGURE 9.25 Typical graphical outputs of the FlexPDE program. Cross sections through *XY* (a) and *XZ* (b) planes, respectively, of the simulated flow field around the rotating particle. Similarly, the cross sections through the same planes for the shear stress field are depicted in (c) and (d), respectively.

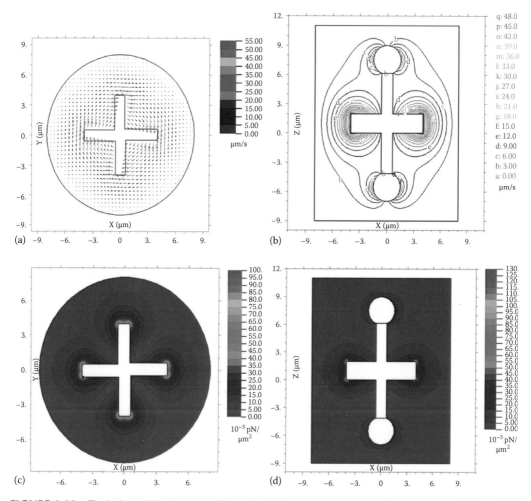

FIGURE 9.33 Typical graphical outputs of the FlexPDE program. Cross sections through *XY* (a) and *XZ* (b) planes, respectively, of the simulated flow field around the rotating particle. Similarly, the cross sections through the same planes for the shear stress field are depicted in (c) and (d), respectively.

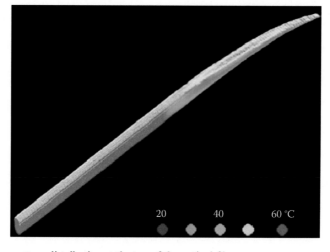

FIGURE 12.9 Temperature distribution at the top of the optical fiber.

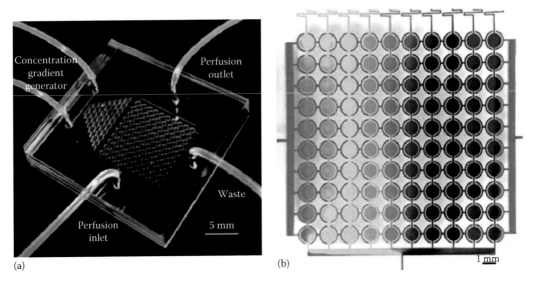

(a)   (b)

FIGURE 13.1 Microfluidic cell culture array for high-throughput cell-based assays. (a) Photograph of a 10×10 array of microchambers fabricated on a 2×2 cm device. The medium is introduced through the left port and exits at the right port (outlet). Cells and reagents are from the top and flow out through the bottom port. The device is capable of conducting cell-based assays with multiple concentrations of reagents. (b) Graphic showing the concentration gradient across 10 columns. A concentration gradient generator was connected to the 10 columns at the top of the device. Red dye was initially poured into the chambers. Blue and yellow dyes were then loaded from the two separate ports at the top of the gradient generator. (From Hung, P.J., Lee, P.J., Sabounchi, P., Lin, R., and Lee, L.P.: Continuous perfusion microfluidic cell culture array for high-throughput cell-based assays. *Biotech. Bioeng.* 2004. 89(1). 8pp. Copyright Wiley-VCH Verlag GmbH & Co. KGaA. Reprinted with permission.)

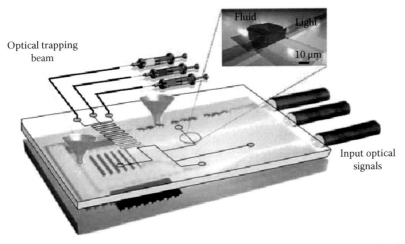

FIGURE 13.4 The basic design consists of a multifunctional optofluidic device. The device is comprised of a photonic chip (bottom part) coupled to optical fibers, a microfluidic circuit (on top of it) connected to fluid sources with different refractive indexes, and optical trapping beams to manipulate micro-mechanical components. (From Monat, C., Domachuk, P., Grillet, C., Collins, M., Eggleton, B.J., Cronin-Golomb, M., Mutzenich, S., Mahmud, T., Rosengarten, G., and Mitchell, A.: Optofluidics: A novel generation of reconfigurable and adaptive compact architectures. *Microfluid. Nanofluid.* 2008. 4. 81–95. Copyright Wiley-VCH Verlag GmbH & Co. KGaA. Reprinted with permission.)

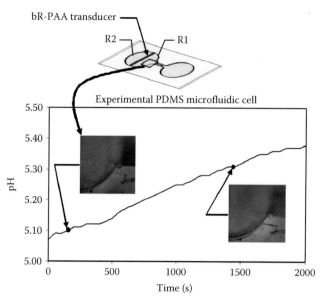

FIGURE 13.21 The change in pH of the target solution is sufficient to cause a phenolphthalein indicator dye to *change* color. The optically driven proton pumps of the bR-PAA pH gradient generator cause the solution in R1 to transform from a light greenish blue (pH 5.10) to a darkened, more intense blue (pH 5.31).

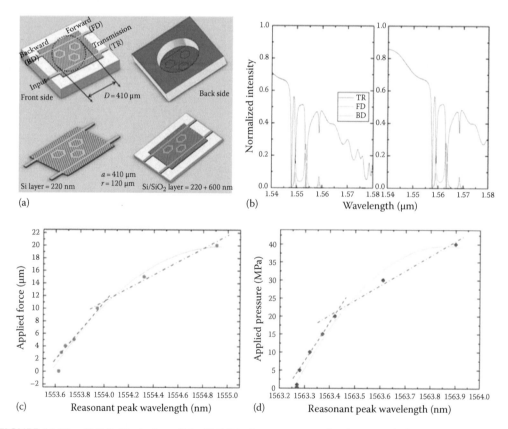

FIGURE 14.58 (a) 3-D illustration of the PhC Diaphragm sensor using hexagonal triple nano-ring (TNR) resonator. (b) Spectra of ports FD, BD, and TR for TNR channel drop filter for Si layer diaphragm and Si/SiO$_2$ layer diaphragm. (c) Resonant wavelength as function of applied force. (d) Resonant wavelength as function of applied pressure. (From Li, B. and Lee, C., *Sens. Actuat. A*, in press, doi:10.1016/j.sna.2011.02.028.)

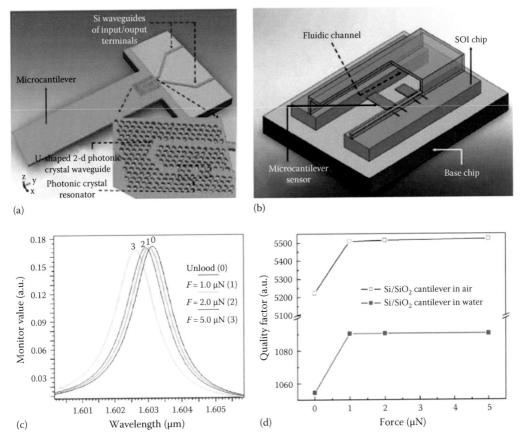

FIGURE 14.59 (a) 3-D illustration of a microcantilever sensor using PC resonator as a strain sensor located at the edge of SOI chip. Inset shows the 3-D illustration of the PC resonator. (b) Conceptual drawing of two microcantilevers packaged inside a fluidic channel. (c) Output resonant peaks of the Si/SiO$_2$ cantilever sensor cantilever sensor operated in water under different loading force. (d) Quality factor change of the resonant peaks of the Si/SiO$_2$ cantilever sensor operated in air/water under different loading force. (From Xiang, W. and Lee, C., *IEEE J. Sel. Top. Quantum Electron.*, 15(5), 202, 2009.)

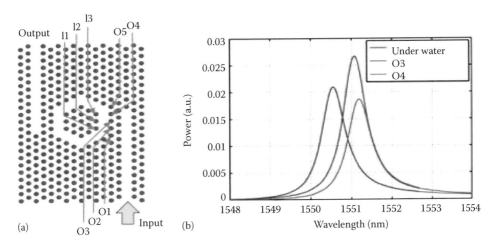

FIGURE 14.60 (a) Sketch of biosensing nano-ring resonator. (b) Spectra of output. (From Hsiao, F.-L. and Lee, C., *IEEE Sens. J.*, 10(7), 1185, 2010.)

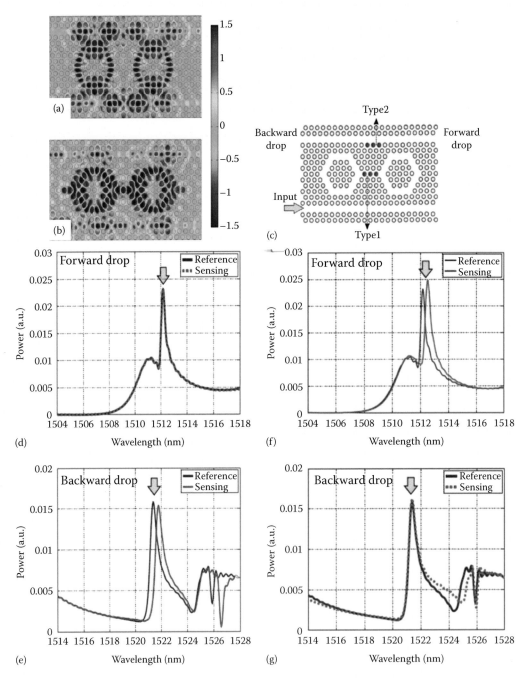

FIGURE 14.61 Steady resonant filed distributions of (a) forward and (b) backward drop. (c) Sketch of sensing-hole position for types 1 and 2. (d) Forward and (e) backward drop spectra of reference and sensing cases of type 1. (f) Forward and (g) backward drop spectra of reference and sensing cases of type 2. (From Hsiao, F.-L. and Lee, C., *SPIE J. Micro/Nanolith. MEMS, and MOEMS (JM3)*, 10(1), 013001, 2011.)

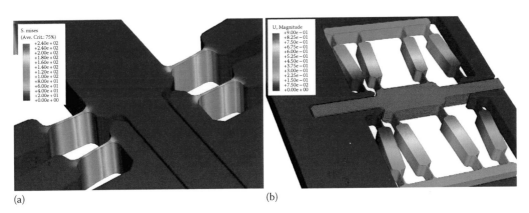

(a)   (b)

FIGURE 15.17 FEM analysis—stress distribution in four hinges (left) and displacement distribution of the entire structure (right).

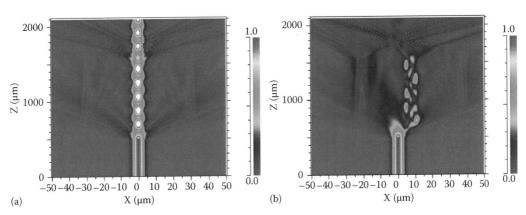

(a)   (b)

FIGURE 15.20 Wave propagation in the ILE for two configurations as the array is moved from the right to the left. The horizontal line indicates the free-space gaps.

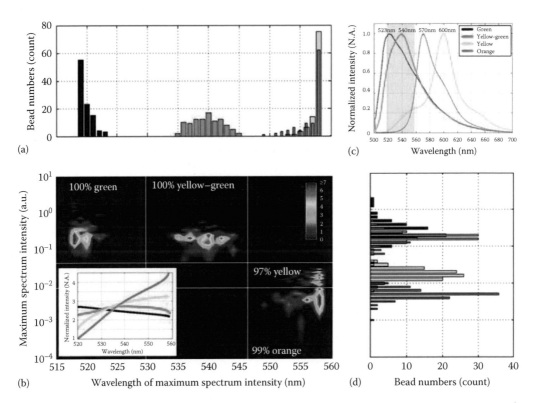

FIGURE 16.18 Four color discrimination of polystyrene microspheres in a 518–558 nm wavelength (λ) range (shadowed region in (C)) by MMFC. (a) A univariate histogram of the wavelength of maximum spectrum intensity. (b) A population density contour plot for the maximum spectrum intensity and the wavelength of maximum spectrum intensity. (c) Spectral data of the microspheres measured by a commercial spectrometer. (d) A univariate histogram of the maximum spectrum intensity. (Reprinted with permission from Huang, N.T. et al., *Anal. Chem.*, 82, 9506. Copyright 2010 American Chemical Society.)

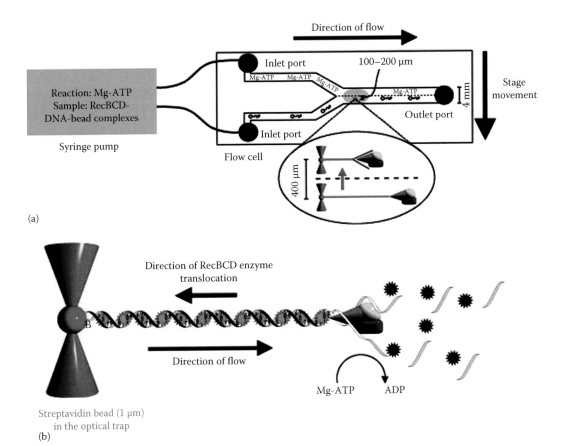

FIGURE 17.2 Visualization of DNA helicase action on individual DNA molecules. (a) Syringe pump and flow cell. The red arrow indicates movement of the trapped DNA–bead complex across the boundary between solutions. (b) A trapped and stretched fluorescent DNA molecule is shown. As RecBCD enzyme translocates, it both unwinds and degrades the DNA, simultaneously displacing dye molecules (black stars). (Reprinted by permission from Macmillan Publishers Ltd. *Nature*, Bianco, P.R., Brewer, L.R., Corzett, M. et al., Processive translocation and DNA unwinding by individual RecBCD enzyme molecules, 409, 374–378, Copyright 2001.)

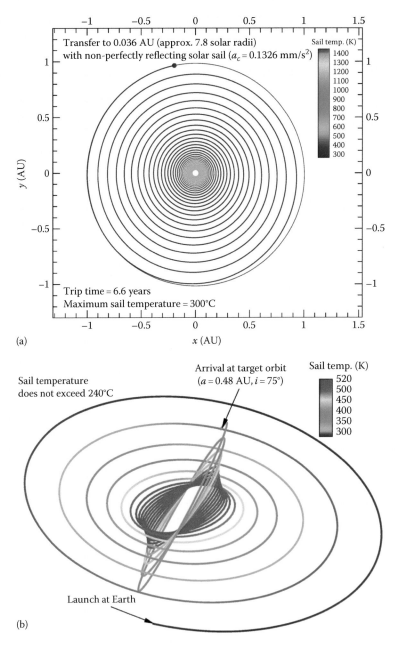

FIGURE 21.7 Trajectory to (a) a very close solar orbit and (b) to the SPI orbit.

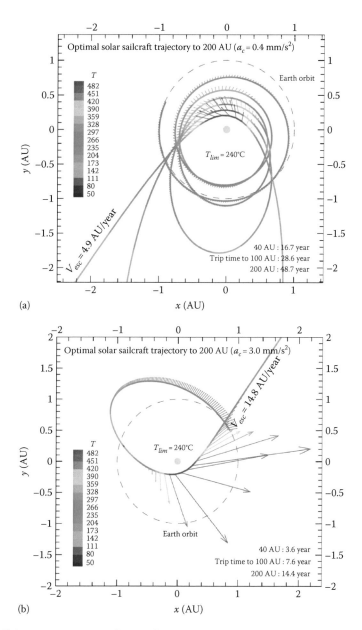

FIGURE 21.8 Solar system escape trajectory for a solar sail with (a) moderate performance and (b) high performance. The arrows indicate the SRP acceleration.

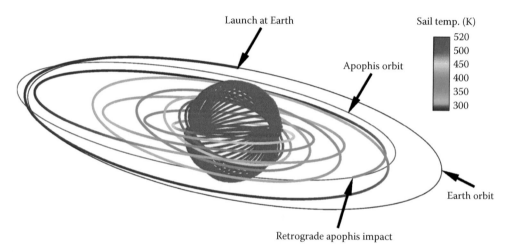

FIGURE 21.10 Trajectory to bring a solar sail onto a retrograde intercept orbit with asteroid Apophis.

11 On-Demand Holographic Optical Tweezers

Johtaro Yamamoto and Toshiaki Iwai

CONTENTS

11.1 INTRODUCTION

It is known that light exerts a pressure on an illuminated object. This light pressure was first predicted by Maxwell based on the Maxwell equation. Lebedev experimentally confirmed the existence of light pressure in 1901 [1]. After that, there were no significant studies on light pressure until 1970 when Ashkin demonstrated experimentally optical acceleration and trapping of dielectric particles with diameters in the range 0.59–2.68 µm [2]. A particle could be stably trapped within the focal spots of two counter-propagating laser beams. In 1978, it was experimentally demonstrated that nanoparticles can be trapped even by a single beam if the beam was tightly focused. Such the technique is known as *optical tweezers* [3].

Optical tweezers have been one of the most attractive research subjects since the experimental demonstration of laser trapping of a microparticle [4–6]. Optical tweezers are used in order to manipulate a wide variety of particles including dielectric particles [7], metal particles [8], aerosol particles [9–11], microbubbles [12], semiconducting nanowires [13,14], carbon nanotubes [15], and living cells [5,6,16]. All these particles were manipulated without any contact and they were not damaged.

Recently, research on optical tweezers has entered a new stage due to the commercialization of computer-addressable spatial light modulators (SLMs). By sending appropriately designed holograms to a SLM, it is possible to control the wavefront of a light beam and the resultant intensity pattern that forms in the focal plane of an objective lens. The SLM expands the functions of *holographic optical tweezers* (HOTs) by extending their applicability and increasing their flexibility.

New functional optical tweezers can be developed by designing the wavefront and using an SLM. This enables the multiple beam spots, the focal point, to be shifted along the optical axis [17] and special beam modes such as Bessel [18] and Laguerre–Gaussian [19] beams to be generated. Optical tweezers based on holograms are called HOT.

Such highly controllable HOT systems have been strongly in demand in cell biology. This is because cell biology is interested in investigating cell–cell signaling during reproduction, fusion, and differentiation of living cells, and to achieve this, it is critical to arrange cells precisely in a specified array. However, most optical tweezers systems which have been commercialized are based on galvano mirrors, and only one or two beam spots can be generated in the optical tweezers systems.

We had developed a highly controllable HOT that allows on-demand manipulation of microparticles to accommodate the demand, and we call the HOT system as the on-demand HOT. In the on-demand HOT system, a fixed beam-spot array and a moving beam spot are quasi-simultaneously generated in a sample plane. The carrier-beam spot carries particles to the fixed beam spots. As a result, microparticles are precisely arrayed in the fixed beam spots. In this chapter, we review on-demand HOT system; principles of optical trapping, numerical investigations on the conditions for stable particle trapping, construction of the HOT system, and experimental demonstration of its performance.

11.2 RADIATION PRESSURE BY A FOCUSED LASER BEAM

11.2.1 RADIATION PRESSURE BY RAYLEIGH SCATTERING

The optical electric field scattered by a sufficiently smaller particle compared with the wavelength is approximated by the Rayleigh scattering theory (RST). When the Rayleigh scattering is occurred, two kinds of forces are induced, a scattering force and a gradient force. The former is caused by the change of the incident light momentum due to the light scattering. The scattering force may be negligible when its particles are controlled only in the focal plane, because the scattering force is only in the direction of the optical axis. The latter is induced by the Lorentz force that is generated by effect of the electromagnetic field inside the beam spot to the dipoles in the particle. Figure 11.1 shows the schematic diagram of the relationship between the electric field of the focused Gaussian laser beam with the waist size of w_0 and a particle with the radius of a. When the focused Gaussian laser beam is incident to the particle located at the distance r away from the focal point, the transversal trapping force $F(r)$ is proportional to the gradient of the incident intensity $I(r)$ and given by [20]

$$F(r) = -\frac{2\pi n_2 a^3}{c}\left(\frac{m^2 - 1}{m^2 + 2}\right)\nabla I(r), \tag{11.1}$$

where

$$I(r) = \frac{2P}{\pi w_0{}^2}\exp\left(-\frac{2r^2}{w_0{}^2}\right). \tag{11.2}$$

In Equations 11.1 and 11.2, c is the velocity of light, n_2 the refractive index of the surrounding medium, m the relative refractive index of the particle against n_2, and P the incident laser power. Equation 11.1 together with Equation 11.2 holds under the Rayleigh scattering approximation of $a < \lambda/20$.

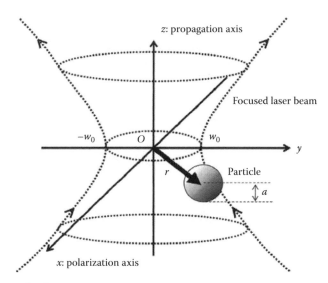

FIGURE 11.1 Schematic of geometry.

The description of the electromagnetic field for the extremely focused laser beam is complicated rather than Equation 11.2. For simplicity, the paraxial Gaussian beam description of Equation 11.2 held under the 0th-order approximation is used in this analysis. The measure of the approximation is defined by [21]

$$s = \frac{1}{kw_0} = \frac{\lambda_0}{2\pi n_2 w_0}, \tag{11.3}$$

where k and λ_0 denote the wave number in the surrounding medium and the wavelength in vacuum, respectively. It must be noted that the Gaussian beam description given by Equation 11.2 can provide an accurate result only when the s parameter is much smaller than unity. The 0th- to 5th-order Gaussian beam descriptions were evaluated by Barton and Alexander, and it is reported that the 0th-order Gaussian beam description contains the average errors of ~2.10% for $s=0.05$ and ~4.37% for $s=0.1$. Therefore, the average errors of several percent will be contained in the numerical results in the following. As indicated by Equation 11.1, the gradient force always attracts particles to the focal point. As a result, the particles are trapped in the focal spot of the laser beam.

We can qualitatively understand the gradient force by the RST, but the RST cannot be applied to particles larger than the wavelength including living cells. In order to investigate the radiation pressure induced on such the particles, we introduce more rigorous scattering theory in the following section.

11.2.2 RADIATION PRESSURE BY GENERALIZED LORENZ–MIE SCATTERING

The generalized Lorenz–Mie theory (GLMT) gives the electromagnetic field scattered by a spherical particle with an arbitrary size in an arbitrary electromagnetic field. Here, we use the coordinate system shown in Figure 11.1 again. Assuming that the origin is the focal position and the x-axis of the coordinate system is parallel to the linear polarization of the illuminating laser beam, $r=(x, y, z)$ denotes the position of the particle. The radiation pressure $F(r)$ obtained by the GLMT is given by

$$F(r) = \left(\frac{n_2}{c} \right) \frac{2P}{\pi w_0^2} \left[\hat{x} C_{pr,x}(r) + \hat{y} C_{pr,y}(r) + \hat{z} C_{pr,z}(r) \right]$$

$$= \hat{x} F_x(r) + \hat{y} F_y(r) + \hat{z} F_z(r)$$

(11.4)

where

(F_x, F_y, F_z) and $(C_{pr,x}, C_{pr,y}, C_{pr,z})$ are radiation pressures and corresponding cross sections in the directions defined by the unit vectors $\hat{x}$, $\hat{y}$, and $\hat{z}$ along, x, y, and z axes, respectively

c is the velocity of light

n_2 is the refractive index of a surrounding media

P is the incident laser power

The transversal radiation pressure F_x and F_y correspond to the gradient force in the RST. The detail of these cross sections is discussed in the Ref. [22].

In the GLMT, the TEM_{00}-mode Gaussian laser beam is the first-order approximation similar to the RST. The degree of approximation in GLMT is also expressed by a measure of the s parameter defined by Equation 11.3. It is shown that the accuracy of the GLMT reduces greatly for the tightly focused laser beam with the measures of $s > 0.15$ for on-axis scattering and $s > 0.10$ for off-axis scattering [23,24]. In the following section, the s parameter was estimated to be 0.079 for the conditions of the beam waist $w_0 = 0.81$ μm, the wavelength $\lambda = 0.532$ μm, the refractive index of the polystyrene particle $n_1 = 1.59$, and the refractive index of water as the surrounding medium $n_2 = 1.33$, so that the error for the first-order approximation can be neglected.

11.2.3 Comparison of Radiation Pressure Obtained by RST and GLMT

Figure 11.2a through d show the transverse radiation pressure F_x as a function of the position of the particle located on x-axis for the different four radii of 0.1, 0.5, 1.0, and 1.5 μm. In the figures, the solid and broken lines denote the calculation results by the GLMT model and the RST, respectively. The difference between them increases consistently with a particle radius. Therefore, errors inherent in the radiation pressures calculated by the RSM becomes more erroneous for a larger particle radius. The maximum radiation pressure calculated by the GLMT model is near the position of $x = w_0/2$ for the particle with smaller radius than the wavelength of the incident light and farther from the position of $x = w_0/2$ with larger radius than the wavelength. On the other hand, it is at the position of $x = w_0/2$ independently of the particle radius for the RSM. Therefore, the difference between calculation results by both the models becomes distinct for larger particle radius than the wavelength of the incident light.

Figure 11.3 shows the maximum radiation pressure as a function of the radius of a particle for the incident laser power of $P = 1.0$ mW and the beam waist of $w_0 = 0.81$ μm. The maximum radiation pressure calculated by the GLMT model is directly proportional to the particle radius to the same size as the wavelength of the incident light, continues to increase gradually, and turns down. On the other hand, the maximum radiation pressure calculated by the RSM is directly proportional to the particle radius and in good agreement with that of GLMT for smaller particle radius than the wavelength of the incident light. The radiation pressure obtained by the RSM has 10 times and 10^4 times in error larger than that by the GLMT for a radius of 2 and 10 μm, respectively. This comes from that the number of the mode of the electromagnetic field exerted on the surface of the particle increases with the enlargement of the particle surface.

The comparison between the analyses based on the GLMT and Rayleigh scattering models demonstrates that the GLMT model is applicable to the optical manipulation of particles from the Rayleigh scattering regime to the Mie scattering regime. A more detailed comparative study is discussed in the Ref. [25].

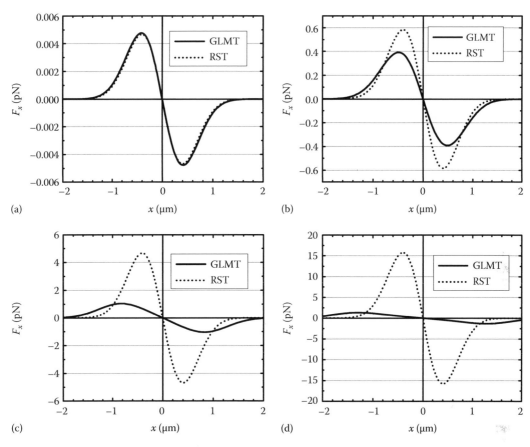

FIGURE 11.2 Transversal radiation pressure F_x as a function of the particle position on the x-axis for the four different radii of (a) 0.1, (b) 0.5, (c) 1.0, and (d) 1.5 μm, $w_0 = 0.81$ μm and $P = 1.0$ mW.

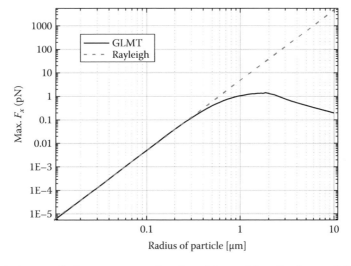

FIGURE 11.3 Maximum radiation pressure as a function of the particle radius for $w_0 = 0.81$ μm, $P = 1.0$ mW.

11.3 PRINCIPLE OF ON-DEMAND OPERATION

11.3.1. EXPERIMENTAL SETUP

Figure 11.4 shows a schematic diagram of an experimental setup. A trapping laser was a frequency-doubled Nd:YVO$_4$ laser source with the maximum output power of 2 W at the wavelength of 532 nm. The laser beam expanded and collimated by a collimator was incident on the active area of the SLM after being reflected by a nonpolarization beam splitter. The light phase was modulated by the SLM and was guided via the beam splitter to the back of a conventional microscope installing a dichroic beam splitter and the objective lens with the NA of 0.75 and the magnification of 40. The intensity pattern was produced in the focal plane of the objective lens, which coincides with the upper surface of a slide glass. The dilute suspension of polystyrene particles with the diameter of 3 μm was instilled on the slide glass. The optical manipulation of particles was observed microscopically by a CCD camera under illumination of a Halogen lamp, from which light at the wavelength of 532 nm was eliminated by a dichroic beam splitter.

The Holoeye LC-R 2500 SLM possesses the 19.5×14.6 mm active area and the 1024×768 (XVGA) pixels with the maximum frame rate of 75 frames per second. The phase pattern is fed into the SLM as a XVGA video signal with the 8 bit gray scale to generate the phase-only hologram. The Fourier transform hologram is generated by the SLM so that the carrier-beam spot and the beam-spot array are produced in the focal plane of the objective lens. Holograms fed into the SLM can be refreshed 75 frames per second. However, it is generally difficult to exploit its high refresh rate because most methods of hologram calculation take too long time compared with the refresh rate. In order to overcome this problem and realize the on-demand operation of particles, we introduce the linear-phase shifting of holograms and the time-division multiplexing method.

11.3.2 LINEAR-PHASE SHIFTING

The substantial time reduction for calculating the holograms corresponding to the displacement of the carrier-beam spot is absolutely essential to realize the on-demand optical tweezers. The carrier-beam spot which carries particles is displaced in real time by introducing the linear-phase-shift

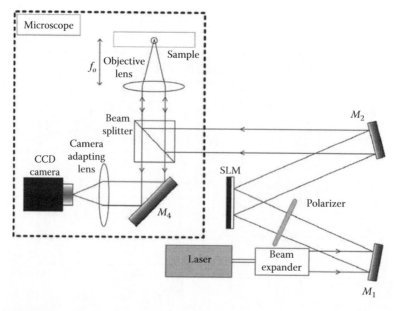

FIGURE 11.4 Schematic diagram of the optical setup.

method based on the frequency shift theorem in the Fourier transform theory [26]. Multiplying the hologram of the carrier-beam spot by the linear-phase-shift hologram and feeding the resultant hologram pattern to the SLM, the carrier-beam spot is displaced linearly in the focal plane of the objective lens as the Fourier transform plane. This procedure is mathematically expressed by

$$
\begin{aligned}
&F_G\left(u - u_0, v - v_0\right) \\
&= \iint \exp\left\{ j\left[\phi_{c0}(x,y) + \frac{1}{\lambda f_o}(u_0 x + v_0 y) \right] \right\} \exp\left[-\frac{j}{\lambda f_o}(ux + vy) \right] dxdy,
\end{aligned}
\tag{11.5}
$$

where λ and f_o represent the wavelength of the laser and the focal length of the objective lens, respectively. ϕ_{c0} and F_G are an original Fourier phase hologram of the carrier-beam spot and its reconstructed amplitude pattern, respectively. Two coordinate systems, (x,y) and (u,v), are defined in the SLM plane and the focal plane of the objective lens, respectively, whose origin is set on the optical axis. (u_0, v_0), the coordinates of the position assigned to reconstruct the intensity pattern, is reconstructed by the phase hologram ϕ_{c0}. As just described, the new phase pattern ϕ_c for the displaced spot pattern $F_G(u-u_0, v-v_0)$ is given merely by adding the phase function corresponding to the linear displacement to the original phase pattern ϕ_{c0} of the beam spot as follows:

$$
\phi_c(x,y) = \phi_{c0}(x,y) + \frac{u_0}{\lambda f_o} x + \frac{v_0}{\lambda f_o} y.
\tag{11.6}
$$

The linear-phase-shift method ensures the displacement of the carrier-beam spot by the on-demand operation because simple addition of the linear-phase-shift hologram to the original hologram reduces drastically the calculation time.

11.3.3 TIME-DIVISION MULTIPLEXING

The key mechanism of the HOTs system is how to multiplex the holograms to perform the complicated optical manipulation. In our system, the fixed beam-spot array to arrange the micro objects and the carrier-beam spot to transfer the objects need to be simultaneously generated. Recently, several hologram-multiplexing methods in the spatial domain have been proposed, for example, the random mask encoding method [27] and simple superimpose method [28,29]. The hologram multiplexing in the spatial domain degrades the quality of reconstructed intensity patterns though they are very simple and useful. From the recent technical development and commercialization of the SLM, the pixel density and the frame rate have been increasing drastically. Therefore, we adopted the time-division multiplexing of two holograms of the carrier-beam spot and the beam-spot array in the virtue of the fast frame rate. Two phase holograms are fed into the SLM alternately, which means the time-division multiplexing. This method provides much the better reconstructed intensity patterns compared with the spatial multiplexing methods because this method never processes the holograms to be multiplexed. The on-demand operation can be achieved with the high-quality holograms by combining the time-division multiplexing based on the fast frame rate of the SLM with the linear-phase-shift method mentioned in the following.

11.3.4 ON-DEMAND HOLOGRAPHIC OPTICAL TWEEZERS SYSTEM

Figure 11.5 shows the procedures of the on-demand HOTs system [30], by which the assigned micro object is transferred to the beam-spot array, as if it was dragged by a mouse cursor. The holograms of the carrier-beam spot and the beam-spot array are calculated as the originals by the Gerchberg–Saxton algorithm [31,32] in advance of experiments. Each is stored into a different buffer and fed to

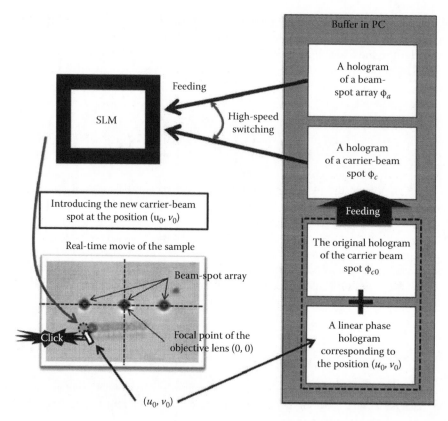

FIGURE 11.5 Schematic diagram of the proposed on-demand HOTs system. The real-time image of the sample is captured by the CCD camera. The holograms of the carrier and the beam-spot array are fed to the SLM, alternately, by means of the time-division multiplexing. By clicking the mouse button at an assigned position on the captured image, the carrier-beam spot is produced just at the position, and we can manipulate particles by the mouse cursor.

the SLM alternately. The microscopic image of micro objects suspending in the liquid is captured into the video memory in the computer and presented on the display. The mouse cursor is moved to the position at the distance less than the radius of the carrier-beam spot from the assigned micro object. When clicking the button on the mouse, the coordinates of the assigned position in the captured image is obtained. The new phase hologram of the carrier-beam spot is calculated by Equation 11.6, input to the buffer instead of the previous carrier-beam spot hologram, and as a result, the carrier-beam spot is generated at the assigned position. The micro object can be trapped again within the new carrier-beam spot. By performing such the operation in sequence toward the position of the beam-spot array, the micro object can be transferred from an arbitrary position to that of the beam-spot array as if it was dragged by the mouse cursor. Combination of time-division multiplexing in virtue of the high frame rate of the SLM with fast calculation of the phase shift for the new carrier-beam spot are essential in the on-demand operation of the proposed HOTs system to realize the reliable trapping and transfer of the micro object.

11.4 STABILITY OF OPTICAL TRAPPING

11.4.1 NUMERICAL ANALYSIS BY THE SMOLUCHOWSKI EQUATION

In the on-demand HOT system, a carrier-beam spot and a fixed beam-spot array are alternately generated; as a result, these beam spots pattern repeat "on" and "off." There was a problem that the trapped particle could free from the beam spot during the beam spots were not illuminated. We therefore investigated the stability of optical trapping in the time-divided beam spots [33,34].

In the case where a continuous beam spots are illuminated, Brownian particles will be trapped within the optical potential generated by the radiation pressure of Equation 11.1. In this case, the particle may still perform Brownian motion in the beam-spot area. The spatiotemporal distribution $N(r,t)$ of the position of such the constrained Brownian particle is governed by the Smoluchowski equation [35]:

$$\frac{\partial N(r,t)}{\partial t} = D \frac{\partial}{\partial r} \left[\frac{\partial N(r,t)}{\partial r} + \frac{1}{k_B T} \frac{\partial U(r,t)}{\partial r} N(r,t) \right], \tag{11.7}$$

where
  D is the diffusion coefficient of the particle
  k_B is the Boltzmann constant
  U is the optical potential field generated by the radiation pressure
  T is the absolute temperature

When there is no potential field, Equation 11.2 is equivalent to the conventional diffusion equation for a particle. In Equation 11.2, the diffusion coefficient is given by

$$D = \frac{k_B T}{6\pi \eta a} \tag{11.8}$$

where a and η are the radius of the particle and the viscosity coefficient of the surrounding medium, respectively. In the steady-state, $N(r)$ corresponds to the Boltzmann factor and is given by

$$N(r) \propto \exp\left[-\frac{U(r)}{k_B T} \right]. \tag{11.9}$$

In time-division multiplexing, the time-modulated optical potential is defined as

$$U(r,t) = \begin{cases} U_C(r) \ldots (\text{potential:on}) \\ 0 \ldots (\text{potential:off}) \end{cases} \tag{11.10}$$

where

$$U_C(r) = -\int F(r) dr. \tag{11.11}$$

In this case, the particle diffuses freely and is attracted toward the focal point of the incident laser beam when the potential is off and on, respectively. The dynamics of the trapped particle thus changes alternately between these two states within the potential.

The Smoluchowski equation given by Equation 11.7 was evaluated numerically by means of the explicit method that is one of the finite difference methods. Using a forward difference for the time derivative and second-order central differences for second-order space derivatives, the recurrence equation of Equation 11.7 can be obtained; thus the Smoluchowski equation can be numerically solved.

11.4.2 SIMULATION RESULTS

We show some results of the numerical simulations based on the GLMT and the Smoluchowski equation. In the following simulations, the position range from −1.5 to +1.5 μm is divided into 128 subintervals with its width 23.4 nm, and the time range from 0 to 2.5 s is divided into 25,600 subintervals with its period 24.4 μs.

Figure 11.6 shows the spatial distribution of the position for the constrained Brownian particle with the radius of $a = 0.40$ μm as a function of the displacement from the focal point under time-varying illumination with five different switching rates of 5, 15, 30, 60, and 0 Hz. The simulations were conducted for the incident laser power of 0.8 mW and the beam waist of 0.81 μm. Only the result for 0 Hz is obtained with the incident laser power of 0.4 mW. The central shape of the distribution around the origin for the time-varying potential coincides approximately with the continuous illumination (i.e., the switching rate of 0 Hz), and therefore, the particle is trapped stably when the illumination is on. On the other hand, the base broadens from the width for the continuous illumination with decrease of the switching rate. Such a broadening of the base for the lower switching rate means that the particle diffuses freely from the area of the beam spot during the off-illumination and, therefore, the particle trapping becomes more unstable with decrease of the switching rate. Increasing the switching rate furthermore, the distribution asymptotically approaches that for the continuous illumination with the half power. We experimentally validated such numerical results in Refs. [33] and [34].

Next, the settings to trap the particle stably are evaluated numerically by the Smoluchowski equation together under the GLMT. Figure 11.7 shows the probability that the particle is stably trapped inside the beam waist area as functions of particle radius and incident laser power with the beam waist of 0.81 μm at the switching rate of 37.5 Hz that corresponds to the maximum switching rate of the SLM (LC-R 2500 SLM, Holoeye) used in our on-demand HOT system. In Figure 11.7a and b, the probabilities were plotted for the particle radii in the ranges from 0.1 to 1.0 μm and from 1.0 to 10.0 μm and for the incident laser powers in the range from 50 to 1500 μW and from 50 to 500 μW, respectively. As seen from Figure 11.7a, the particle is trapped more stably within the beam waist by increasing the incident laser power due to increasing of the gradient force and the particle radius due to slowing down of the Brownian motion. If particle trapping is assumed to be stable when the probability is beyond 95%, the incident laser power is required to be larger than 1.0 mW for the particle

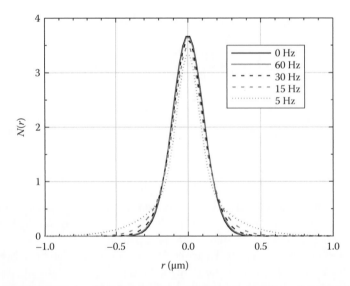

FIGURE 11.6 Spatial distribution $N(r)$ of the position of a constrained Brownian particle as a function of displacement r from the center for $a = 0.4$ μm under switching illuminations at five different rates of 5, 15, 30, 60, and 0 Hz with $w_0 = 0.81$ μm and $P = 0.8$ mW. The result of 0 Hz is obtained with $P = 0.4$ mW.

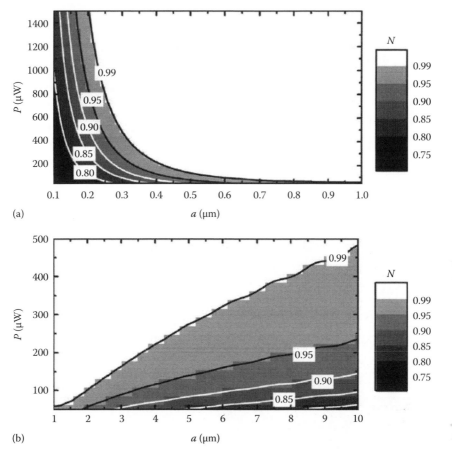

FIGURE 11.7 Probability N that the particle is stably trapped inside the beam waist area as functions of particle radius a and laser power P with the beam waist of 0.81 μm at the switching rate of 37.5 Hz. (a) The results for particles smaller than 1 μm. (b) The results for particles larger than 1 μm.

larger than 0.2 μm. As shown in Figure 11.7b, the incident power required to trap stably particles increases consistently with the particle radius because the maximum radiation pressure decreases in the Mie scattering regime as shown in Figure 11.3. However, the incident power required for stable trapping in this region is much lower than that in the radius from 0.10 to 1.0 μm because the Brownian motion becomes slower increasing the particle radius.

Figure 11.8 shows the probability that the particle is stably trapped inside the beam waist as functions of the particle radius and the switching rate for the incident power of 1.0 mW and the beam waist of 0.81 μm. In this case, the probabilities were calculated for the radii of particles in the range from 0.1 to 1.0 μm and the switching rate in the range from 2 to 40 Hz. By increasing the switching rate and the particle radius, trapping the particle within the beam waist becomes more stable. This comes from the facts that the speed of the Brownian motion becomes slower so that the particle can be illuminated in succession before it diffuses toward the outside of the beam spot. If it is assumed that the particle is stably trapped when the probability is beyond 95%, a switching rate is required to be faster than 10 Hz for the particle with the radius larger than 0.2 μm. On the other hand, trapping the particle smaller than 0.2 μm may be unstable at any switching rate because the probability becomes lower than 95%. To realize the stable trapping of such the particle, the incident laser power needs to be increased. On the other hand, the particle larger than 0.7 μm can be trapped stably at very low switching rate because the probability is always beyond 99.0%.

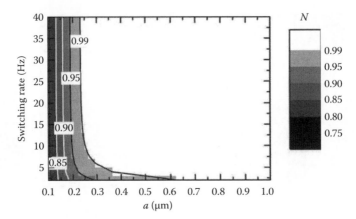

FIGURE 11.8 Probability N that the particle is stably trapped inside the beam waist as functions of the particle radius a and the switching rate for the incident power of 1.0 mW and the beam waist of 0.81 μm.

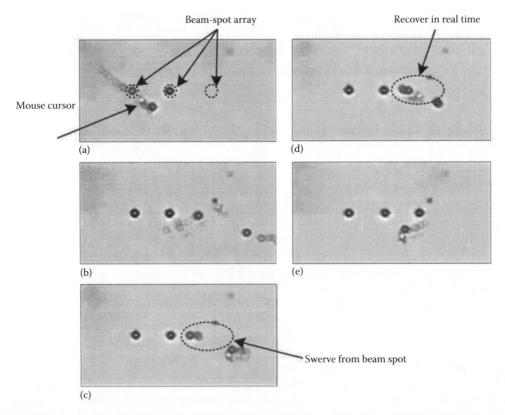

FIGURE 11.9 Particle arrangement in the designed beam-spot array by the on-demand operation. The carrier-beam spot produced at the position of the mouse cursor. (a) A particle was carried by the carrier-beam spot to the right fixed beam spot. (b) An array of three particles was completed. (c) The particle trapped in the right beam spot was swerved. (d) The swerved particle was recovered in real time. (f) The array of three particles was completed again.

11.5 EXPERIMENTAL DEMONSTRATION

At this stage, the selection, the transfer, and arrangement of the micro objects are performed by the on-demand operation of the proposed HOTs system. Figure 11.9 shows the optical arrangement of the polystyrene particles which are transferred one by one by the carrier-beam spot to the fixed three beam spots located straightly at equal distances. In the figure, the positions of the fixed beam spots are shown by the dashed circles and the carrier-beam spot is generated at the position that is assigned by the cursor. The trajectories of transferring the particle by the carrier-beam spot are shown in the figures. Notice especially the virtue of the on-demand operation shown in Figure 11.9c through e; the particle swerves unexpectedly from the right-hand spot (Figure 11.9c), is assigned by the cursor (Figure 11.9d), and can be flexibly recovered (Figure 11.9e). Figure 11.10 shows simultaneous

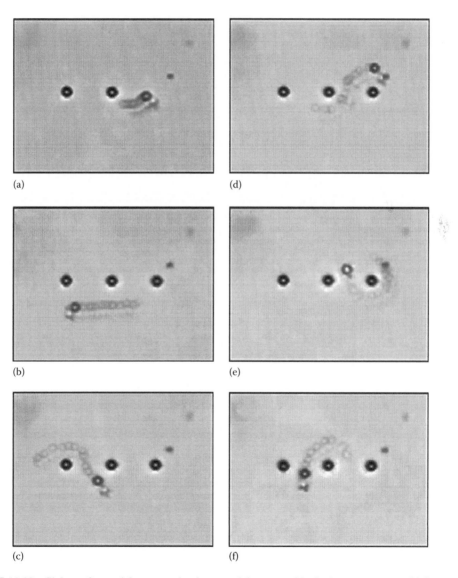

(a)   (d)

(b)   (e)

(c)   (f)

FIGURE 11.10 Slalom of a particle among the three particles arrayed in the beam-spot array (a) An array of three particles was constructed. (b) An additional particle was carried. (c)-(f) The particle was made to move around three fixed particles just like slalom..

operations of fixing and transferring the particles just like slalom by displacing the trapped particle in the space among the three particles arrayed in the previous figure. The trajectory of slaloming the particle is also shown in the figure.

11.6   CONCLUDING REMARKS

In this chapter, we have reviewed our on-demand HOT system that enables interactive operation in optical trapping. The system conditions for stable optical trapping of the particle were evaluated by numerical analysis based on the GLMT together with the Smoluchowski equation for diffusion of a particle constrained in an optical potential. The effectiveness of the new on-demand HOT system was confirmed experimentally, and its applicability and flexibility for optical trapping were demonstrated. We intend to expand our system to array living cells. In future, we will realize the investigation of communication system between living cells to establish a cell network.

REFERENCES

1. P.N. Lebedev, Untersuchungen über die Druckkräfte des Lichetes, *Phisik* **6** (1991), 433–458.
2. A. Ashkin, Acceleration and trapping of particles by radiation pressure, *Phys. Rev. Lett.* **24** (1970), 156–159.
3. A. Ashkin, J.M. Dziedzic, J.E. Bjorkholm, and S. Chu, Observation of a single-beam gradient force optical trap for dielectric particles, *Opt. Lett.* **11** (1986), 288–290.
4. A. Ashkin, Trapping of atoms by resonance radiation pressure, *Phys. Rev. Lett.* **40** (1978), 729–732.
5. A. Ashkin and J.M. Dziedzic, Optical trapping and manipulation of viruses and bacteria, *Science* **235** (1987), 1517–1520.
6. A. Ashkin, J.M. Dziedzic, and T. Yamane, Optical trapping and manipulation of single cells using infrared laser beams, *Nature* **330** (1987), 769–771.
7. H. Misawa, M. Koshioka, K. Sasaki, N. Kitamura, and H. Masuhara, Laser trapping, spectroscopy, and ablation of a single latex particle in water, *Chem. Lett. (Jpn)* **19** (1990), 1479–1482.
8. Q. Zhan, Trapping metallic Rayleigh particles with radial polarization, *Opt. Express* **12** (2004), 3377–3382.
9. A. Ashkin and J.M. Dziedzic, Observation of light scattering from nonspherical particles using optical levitation, *Appl. Opt.* **19** (1980), 660–668.
10. D.R. Burnham and D. McGloin, Holographic optical trapping of aerosol droplets, *Opt. Express* **14** (2006), 4175–4181.
11. M.D. Summers, D.R. Burnham, and D. McGloin, Trapping solid aerosols with optical tweezers: A comparison between gas and liquid phase optical traps, *Opt. Express* **16** (2008), 7739–7747.
12. S.-Y. Sung and Y.-G. Lee, Trapping of a micro-bubble by non-paraxial Gaussian beam: Computation using the FDTD method, *Opt. Express* **16** (2008), 3463–3473.
13. A.v.d. Horst, A.I. Campbell, L.K.v. Vugt, D.A. Vanmaekelbergh, M. Dogterom, and A.v. Blaaderen, Manipulating metal-oxide nanowires using counter-propagating optical line tweezers, *Opt. Express* **15** (2007), 11629–11639.
14. S.-W. Lee, G. Jo, T. Lee, and Y.-G. Lee, Controlled assembly of In_2O_3 nanowires on electronic circuits using scanning optical tweezers, *Opt. Express* **17** (2009), 17491–17501.
15. J. Plewa, E. Tanner, D. Mueth, and D. Grier, Processing carbon nanotubes with holographic optical tweezers, *Opt. Express* **12** (2004), 1978–1981.
16. T. Buican, M.J. Smith, H.A. Crissman, G.C. Salzman, C.C. Stewart, and J.C. Martin, Automated single-cell manipulation and sorting by light trapping, *Appl. Opt.* **26** (1987), 5311.
17. T. Haist, M. Schonleber, and H.J. Tiziani, Computer-generated holograms from 3D-objects written on twisted-nematic liquid crystal displays, *Opt. Commun.* **140** (1997), 299–308.
18. T. Cižmár, V. Kollárová, X. Tsampoula, F. Gunn-Moore, W. Sibbett, Z. Bouchal et al., Generation of multiple Bessel beams for a biophotonics workstation, *Opt. Express* **16** (2008), 14024–14035.
19. N.B. Simpson, L. Allen, and M.J. Padgett, Optical tweezers and optical spanners with Laguerre–Gaussian modes, *J. Mod. Opt.* **43** (1996), 2485–2491.
20. P.W. Smith, A. Ashkin, and W.J. Tomlinson, Four-wave mixing in an artificial Kerr medium, *Opt. Lett.* **6** (1981), 284–286.

21. J.P. Barton and D.R. Alexander, Fifth-order corrected electromagnetic field components for a fundamental Gaussian beam, *J. Appl. Phys.* **66** (1989), 2800–2802.

22. K.F. Ren, G. Gréhan, and G. Gouesbet, Radiation pressure forces exerted on a particle arbitrarily located in a Gaussian beam by using the generalized Lorenz-Mie theory, and associated resonance effects, *Opt. Commun.* **108** (1994), 343–354.

23. J.A. Lock and G. Gouesbet, Rigorous justification of the localized approximation to the beam-shape coefficients in generalized Lorenz-Mie theory. I. On-axis beams, *J. Opt. Soc. Am. A* **11** (1994), 2503–2515.

24. G. Gouesbet and J.A. Lock, Rigorous justification of the localized approximation to the beam-shape coefficients in generalized Lorenz-Mie theory. II. Off-axis beams, *J. Opt. Soc. Am. A* **11** (1994), 2516–2525.

25. Y. Harada and T. Asakura, Radiation forces on a dielectric sphere in the Rayleigh scattering regime, *Opt. Commun.* **124** (1996), 529–541.

26. J.W. Goodman, *Introduction to Fourier Optics*, New York: McGraw-Hill (1968).

27. M. Montes-Usategui, E. Pleguezuelos, J. Andilla, and E. Martín-Badosa, Fast generation of holographic optical tweezers by random mask encoding of Fourier components, *Opt. Express* **14** (2006), 2101–2107.

28. M. Reicherter, T. Haist, E.U. Wagemann, and H.J. Tiziani, Optical particle trapping with computer-generated holograms written on a liquid-crystal display, *Opt. Lett.* **24** (1999), 608–610.

29. J. Liesener, M. Reicherter, T. Haist, and H.J. Tiziani, Multi-functional optical tweezers using computer-generated holograms, *Opt. Commun.* **185** (2000), 77–82.

30. J. Yamamoto and T. Iwai, On-demand optical tweezers using computer-generated phase holograms, *Rev. Laser Eng.* **36** (2008), 1331–1334

31. R.W. Gerchberg and W.O. Saxton, A practical algorithm for the determination of the phase from image and diffraction plane pictures, *Optik* **35** (1972), 237–246.

32. B. Liu and N. C. Gallagher, Convergence of a spectrum shaping algorithm, *Appl. Opt.* **13** (1974), 2470–2471.

33. J. Yamamoto and T. Iwai, Spatial stability of particles trapped by time-division optical tweezers, *Int. J. Optomech.* **3** (2009), 253–263.

34. J. Yamamoto and T. Iwai, Stability analysis of particle trapping in time-division optical tweezers by the generalized Lorenz-Mie theory, *Jpn. J. Appl. Phys.* **49** (2010), 092701-1–6.

35. F. Schwabl, *Statistical Mechanics*. Berlin, Germany: Springer (2002).

Part IV

Optically Driven Systems

Optical actuators can either *directly* or *indirectly* transform light energy into the desired structural displacement. These transducers can be designed to operate under different properties of light such as intensity, wavelength, phase, and polarization changes. This part of the volume explores how light-driven technologies can be used to manipulate liquids, control optofluidic systems, and improve the performance of nano-electro-mechanical system (NEMS) and micro-electro-mechanical system (MEMS) devices. Yukitoshi Otani from Utsunomiya University discusses the principles and mechanisms of photothermal actuation in Chapter 12.

In Chapter 13, George K. Knopf and Khaled Al-Aribe provide an overview of light-driven micro- and nanoscale fluidic systems. The miniaturization and integration of various analytical laboratory processes on a single, often disposable, platform is a key design requirement for many lab-on-a-chip (LoC) and micro-total analysis systems (μTAS) used for medicine and environmental monitoring. To compensate for the physics of "very small scale" and improve the overall system performance, it is necessary for engineers to fabricate submicron structures along the transport channels and storage reservoirs for efficient energy storage, thermal management, heat dissipation, or other energy-related requirements. The integration of optical technologies with microfluidic components to create high-performance devices is the goal of micro-optofluidic technologies. The key advantages of "on-the-chip" integration with optical technologies are the relative ease in which the optical properties of the device can be used to manipulate the fluid or, alternatively, change the fluid properties to manipulate the optics. The authors introduce a variety of methods for manipulating and controlling microflows, including the direct optical manipulation of micro-objects in fluid streams, transporting liquid droplets using the principle of opto-electrowetting, controlling microflow through thermoviscous expansion, opto-pneumatic valves and pumps, and direct laser micromixing. An alternative approach to manipulating fluid flow patterns in microstructures using environmentally sensitive hydrogels is also discussed. Some applications require nanoscale systems such as the transfer of protons to control the pH of solutions. A simple optically driven proton pump based on light-sensitive bacteriorhodopsin is introduced.

Optical MEMS technology has been an enabling tool for numerous cutting-edge devices for optical communication and microdisplay applications. In Chapter 14, Chengkuo Lee and Kah How Koh of the National University of Singapore discuss the cornerstones for the success of optical MEMS technology, including actuator technology, optics design, and development of movable or tunable micromechanical elements. The combination of these elements enables optical MEMS to perform unique and sophisticated functions such as beam steering and attenuation in optical communication applications and raster scanning in image display applications. The development trends and state-of-the-art technologies of optical switches, variable optical attenuators (VOA), and scanning mirror are introduced and discussed in this chapter.

The unconventional laser–matter interaction resulting from the very high-peak powers associated with femtosecond laser pulses provides novel ways to tailor material properties. Applied to dielectrics, such as fused silica, these localized changes of physical properties can be advantageously used to create miniaturized devices that combine optical, fluidic, and mechanical functions. In Chapter 15, Yves Bellouard from Eindhoven University of Technology and his colleagues from Translume Inc. (United States) discuss the opportunities of using femtosecond laser processes to form microsystems. The first part briefly introduces micro- and nanosystems with an emphasis on limitations of current fabrication practices. The second part reviews the work done to create miniaturized devices with femtosecond lasers. The third part presents a generalized concept of integrated microdevices produced using femtosecond lasers.

Finally, Katsuo Kurabayashi and his colleagues from the University of Michigan and Academia Sinica (Taiwan) provide a detailed look into the hierarchical integration of soft polymer micro- and nanostructures in optical MEMS in Chapter 16. Historically, MEMS technology has been seen as an offspring of silicon-based integrated circuit (IC) technology, which primarily relies on "top-down" photolithography techniques. In contrast, the technological promises that polymers hold in future micro- and nanosystems have recently attracted much attention due to their cost effectiveness, manufacturability, various material properties, and compatibility with biological and chemical systems. A wide variety of polymer-based fabrication techniques, including those based on "bottom-up" self-assembly and soft printing approaches, meet the demand for forming nanometer-sized structures rapidly and economically, thus driving nanomanufacturing research and nanotechnology. Nanoscale polymer structures, such as nanopatterned polymeric films and self-organized polymeric materials, provide important building blocks in nanotechnology, nanoelectronics, and photonics. Some of these structures manifest material behavior radically different from that in micro/mesoscale. Integration of nanostructures into a microelectromechanical system across multiple dimensional scales ranging from a few nanometers to several micro/millimeters is expected to open up new unexplored MEMS research frontiers. Multiscale integration is the key to combine MEMS technology and emerging nanomanufacturing technology, making it possible to implement technological fruits offered by nanosciences and nanotechnology in sensor and actuator applications. The technology described in this chapter meets the need for manufacturing methods to integrate nanostructures within micro- and mesoscopic devices and systems in functionally scaled products, permitting hierarchical, continuous structuring across multiple scales.

12 Photothermal Actuation

Yukitoshi Otani

CONTENTS

12.1 INTRODUCTION

Light wave has several properties such as wavelength, frequency, phase, polarization, energy, etc. The most widely used in this field is referred to as the communicative function such as optical measurement and optical communication. On the other hand, the light wave does not have any significant use in the area of energy function apart from laser processing. Since the research field of the optomechatronics is likely to grow, we consider that it is important to develop neither the research nor the development in the field of energy application of light. Some trials are ongoing to develop optical actuators and manipulators that assume light to be the driving energy. They convert photonics energy to mechanical motion. This section introduces the features of optical actuation and manipulation.

In recent years, there have been many new attempts for actuating and/or driving to a small object at nanometer to micrometer scale There are many requirements with the size of from nano meters up to meters order (Cho 2003). Our requirements have reached various areas of local minimum and local maximum technology, ranging from robots and motorcars to spacecrafts. The actuators can enumerate electrical, engine, rocket, hydraulic, and molecular motor as an example. The most popular and flexible method is represented by driving with the electrical method, classified as AC motors, DC motors, linear motors, stepping motors, and ultrasonic motors. Stepping motors and ultrasonic motors are especially useful for precision engineering. They are applied for positioning cameras, in robots, and for manipulation. The requirement now is for nano-micro-technology, which uses equipments that are too small to be controlled by electric motors. A micro-electro-mechanical system (MEMS) has attractive features for the next generation of actuators and manipulators. In addition, a nano-electro-mechanical system (NEMS) is also expected one after another as generation's actuator.

There are some disadvantages to electric motors. Moreover, the wiring for electric motors is obstructive for the drive and control. An optical driving technology for optomechatronics is one of the new fields of actuation. In the future, we may even try to build an optical actuator moving at the speed of light, but this is impossible right now.

The features of optical actuators and manipulators are indicated as follows:

1. Energy supplied remotely through wireless means
2. No generation of magnetic noise
3. Simple construction
4. Not only energy supply but also information exchange

Optical actuators have been proposed by using various methods, such as optical radiation and photo-electromotive and photothermal effects. In this chapter, a survey of optical actuators and manipulators is provided.

Table 12.1 shows examples of various optical driving methods that are classified by applications such as optical trapping and optical tweezers. The first trial of optical movement was to use optical radiation power. Optical trapping and tweezers are well known for their applications. Uchino first proposed using the photo-electromotive power of an optical actuator. This actuator consisted of PLZT; therefore, its moving speed is as slow as several tenths of a micrometer per minute. We have already proposed several studies about the photothermal effect. In most of them, light converts to mechanical movement by changing into heat, electricity, chemical reaction, etc. We now discuss them in detail.

Figure 12.1 illustrates an optical radiation power proposed by Ashkin (1970). A typical example is shown where the laser beam is focused on to a sample. A small particle is trapped at the balanced point between trapped power and the gravity by an optically driven laser. A particle with a large dielectric constant is drawn to the strong electric field. Therefore, we can manipulate a particle with accurate control.

This evanescent wave can also work to move a small particle. Figure 12.2 shows an example of evanescent photon power manipulating a nanoparticle (Kawata and Sugiura 1992). The evanescent field wave is made on the surface of a prism in case the laser beam reflected at illuminating more than critical angle. When the small particle goes into the evanescent wave, it is cased to illuminate

---

TABLE 12.1
Various Optical Driving Methods

Method	Applications	Proposed by
Optical radiation	Trapping	Ashikin (1970)
Evanescent photon power	Trapping	Kawata and Sugiura (1992)
Photo-electromotive power	Actuator	Uchino (1989)
	Gripper	Hattori et al. (1993)
	Rotatory motor	Morikawa et al. (2003)
Photothermal	Actuator	Fukushima et al. (1998)
	Manipulator	Otani et al. (2001)
	Microbubble	Otani et al. (2003)
Photothermal magnetic force	Rotatory motor actuator	Takizawa (2002) Otani et al. (2004)
Photothermal shape-memory alloy (SMA)	Actuator	Inaba and Hane (1993)
Photo-marangoni	Actuator	Kotz et al. (2004)
Photochemical change (polymer gel)	Micropump	Juodlkazis et al. (2000)
	Actuator Manipulator	Tatsuma et al. (2007)
Photochemical change (Azobenzene)	Actuator	Yu et al. (2000)

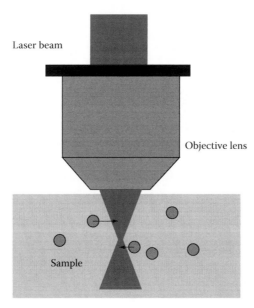

FIGURE 12.1 Laser manipulation by focusing objective lens.

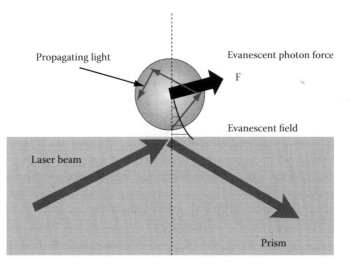

FIGURE 12.2 Evanescent photon power.

the scattered wave into the particle. The light propagates inside the particle and the evanescent photon force works in the direction of the arrow. Finally, the particle moves in the direction of the arrow.

Photo-electromotive power, called as photostrictive effect, is of interest for future actuators. It was the first trial of an optical actuator proposed by Uchino (Uchino 1985, Uchino and Aizawa 1989). Its material is usually lead zirconate titanate (PLZT) doped with WO_3. When ultraviolet light is exposed to the PLZT material, it undergoes mechanical strain, called photostrictive effect by photothermal effect. The photostrictive effect is a result of the photovoltaic effect, and it expands and contracts the material by piezoelectric effect. Uchino proposed a walking machine with two PLZT bimorph legs with cover glass as a foot connected to an acrylic holder, as shown in Figure 12.3. When the legs were illuminated by UV light alternately, it moved very slowly with the speed of several tenths of micrometers per minutes. Uchino's report demonstrated the possibility of optical actuators to many researchers.

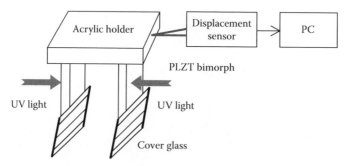

FIGURE 12.3 Walking machine with two PLZT biomorph legs with a cover glass as a foot connected to an acrylic holder.

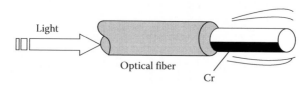

FIGURE 12.4 An optical fiber cantilever for pressure sensor.

After Uchino's optical actuator, the walking machine using photo-electromotive effect by PZT was investigated (Fukushima et al. 1998, Yoshizawa et al. 2000). The PZT whose electrode is removed makes electrostatic action to the stiction at parts such as feet in case the light illuminates it for photostrictive effect by photothermal effect. It can easily control stiction and stretch so that it moves like an inchworm. The stretching part is made of temperature-sensitive ferrite and magnet. The expansion and contraction can be made by temperature change.

Photothermal effect means that a light converts to a mechanical movement by changing into heat, electricity, and chemical. Thermal expansion by light is one of the methods to produce mechanical movement. An optical fiber cantilever for pressure sensor shown in Figure 12.4 was proposed by Hane (Inaba et al. 1995). It was removed and covered with laser and coated with Cr at half of the core part. After illuminating the light, the Cr-coated area was absorbed and it converted the heat. If we illuminate switched light, the end of the fiber is vibrated by photothermal effect. The end of the fiber can be vibrated by illuminated light. It means the optical fiber can be used both as a light guide and as an oscillator. However, in this case the amplitude of vibration is too small to move an object.

Photothermal magnetic force promotes an effect of the magnetic force by changing the magnetic susceptibility by the photothermal effect. A temperature-sensitive magnet is known to lose the magnetic force above the Curie temperature. We can easily move the magnet by controlling the temperature of the sample. This method was applied for a rotary motor and a linear motor (Takizawa 2002, Otani et al. 2004).

A shape-memory alloy (SMA) has a remarkable feature of temperature-dependant shape change. We can produce an increase in temperature by photothermal effect (Inaba and Hane 1993). This method is applied to an SMA active forceps for surgery using an endoscope (Nakamura and Shimizu 1999).

The Marangoni effect caused fluid flow inside the droplet. When photothermal made the temperature rise in the droplet, Kotz succeeded in moving the droplet by the Marangoni effect called photo-marangoni (Kotz et al. 2004).

Many reports of biology and neurology were published to study the reactions of optical driving. A bacteriorhodopsin for instance works the innervate nerve by different wavelength to control its motion and to communicate information with each other. Optical driving by photochemical reaction is a remarkable method to drive the motion from light energy directly. Two famous materials were

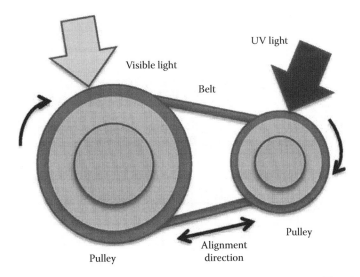

FIGURE 12.5   Light-driven plastic motor by Azobenzene-containing liquid-crystalline elastomers.

proposed for optical actuator (Juodlkazis et al. 2000). One is a polymer gel for photoelectrochemical actuator and the other is Azobenzene. A material made of polyacrylic acid (PAA) gel works to absorb and swell the water by illuminated UV light in the water. Moreover, it discharges and shrinks after illuminating the visible light. This cycle can make an optical actuator. In addition, an optical actuator was also proposed by photochemical reaction of optical induction oxidation-reduction reaction by silver nanoparticle–titanium oxide system (Tatsuma et al. 2007).

Ikeda proposed bending cross-linked photochromic polymer such as Azobenzene by illuminating UV light and returning to the initial position by visible light at a wavelength longer than 540 nm. The material can be shown the specification direction parallel to the polarized direction of the illuminated beam. It moves more than $90°$ in 10 s by 3.5 mW/cm (Yu et al. 2003). His group also proposed a light-driven plastic motor by Azobenzene-containing liquid-crystalline elastomers, as shown in Figure 12.5 (Yamada et al. 2008). They demonstrated optical rotation by light illumination of UV and visible alternately.

The optical radiation is really energy transfer from light power to mechanical power. However, there is a problem in converting energy efficiency. It is estimated at less than 10^{-5}% power presets. It means that the power of optical radiation is less than pico Newtons. In general, the requirement of efficiency for mechanical driving is more than 1%. Therefore, we have to use additional effect with light energy. The polymer materials have the possibility of applications in future optical actuators and manipulators. If the response time could become much faster, it might be used in many industrial areas.

12.2   PHOTOTHERMAL ACTUATOR COMPOSED OF OPTICAL FIBERS

An optical actuator by photothermal effect is one of the unique driving methods for small objects by using only optical energy supplied from a remote position. We also recognize it as a vibration drive. We have already proposed some optical actuators (Uchino 1989, Fukushima et al. 1998, Poosanaas et al. 2000). However, the size of the element is a little large. Therefore, a response time of several tenth seconds is required for one movement. In this section, a new type of a small optical actuator is shown that is driven by photothermal effect. We try to build up a novel optical actuator driven by only optical energy supplied from a remote position. Figure 12.6 shows how each cantilever be can moved by optical light. To control the direction of illuminated light, we can control the moving directions. Moreover, it is possible to develop the moving direction. We employ the methods to drive

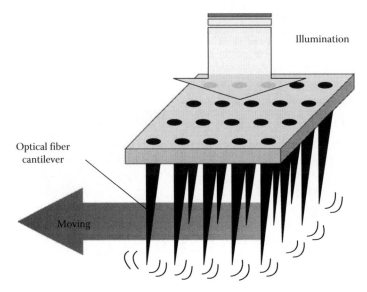

FIGURE 12.6 Ideal optical actuator with optical cantilever array.

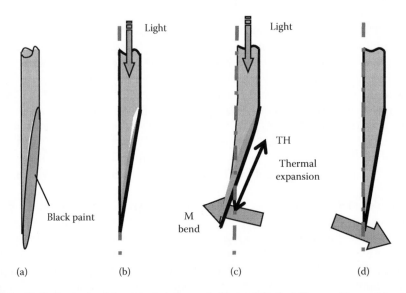

FIGURE 12.7 Optical actuator driven by photothermal effect. (a) Optical fiber cantilever, (b) convert illuminated light energy to heat, (c) bend by thermal expansion, and (d) back to initial position by natural radiation.

optical cantilevers by photothermal effect of an optical fiber and pyroelectric and piezo effects of polyvinylidine difluoride (PVDF) film (Mizutani and Otani 2008, Mizutani et al. 2008b).

Figure 12.7 shows a photothermal transducer using an optical fiber. The end of the optical fiber is cut for a bevel, and its surface is colored in black, as shown in Figure 12.7a. The light is illuminated through the optical fiber, and then it is absorbed to convert to heat, as shown in Figure 12.7b. The thermal expansion makes the surface of the blacked area expand to the arrowed direction marked "TH." Finally, the top of the optical fiber beds to the arrowed direction marked "M" in Figure 12.6c. After stopping the illumination, the top returns to the initial position, as in Figure 12.6d. The optical fiber also works like a flat spring because of its shape. This effect makes the end of the fiber a stretching vibration called "photothermal effect."

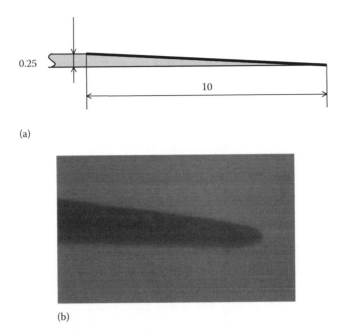

(a)

(b)

FIGURE 12.8   Optical fiber cantilever. (a) Size optical fiber cantilever and (b) picture at the point of end.

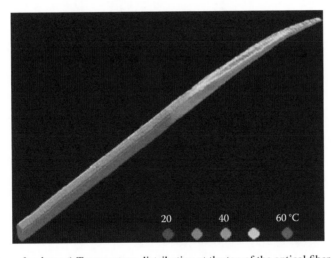

FIGURE 12.9   **(See color insert)** Temperature distribution at the top of the optical fiber.

Figure 12.8 shows an optical fiber cantilever. It is 10 mm in length and 0.25 mm in diameter, as shown in Figure 12.8a. Figure 12.8b shows a magnification picture at the point of end. After illuminating into the optical fiber by a laser diode with 35 mW, it deforms 70 μm of its displacement and 200 mN of its generative power. The temperature distribution at the top of the optical fiber just after illumination is simulated as shown in Figure 12.9. It is shown that the highest temperature is 60° in case of 35 mW of the laser power.

Figure 12.10 shows the construction of an optical actuator made of optical fiber. It is composed of three optical fibers for a leg joined to a body. Each of the fibers is cut and polished for a bevel as shown in Figure 12.10. This surface is colored in black so that it can absorb light to convert to heat. The optical fiber works also like a flat spring because of its shape. When the switching illumination is tuned to the resonant frequency, the optical actuator can be moved by repeating of the process as shown in Figure 12.10. Its balance is adjusted using a small weight on the body of the optical fiber. A rotation of the actuator can be controlled to illuminate to two rear legs as shown in Figure 12.11. The actuator moves to the straight line as in Figure 12.11a, rotates to the right as in Figure 12.11b, and to the left as in Figure 12.11c. After repeating these processes, it can freely move in two-dimensional space. The size of the optical actuator is $3 \times 3 \times 11$ mm, as shown in Figure 12.11. The illuminated light is modulated with 4 Hz of switching speed. Figure 12.12 is an experimental result at 0 s in (a) and 90 s in (b). It moves 2.3 mm with 25 μm/s of speed.

A new type of a small optical actuator that is driven by photothermal effect is proposed to expand to an optical robot for a micro optomechatronic machine.

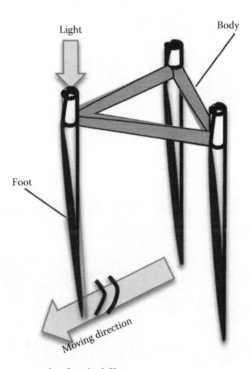

FIGURE 12.10   Optical actuator made of optical fiber.

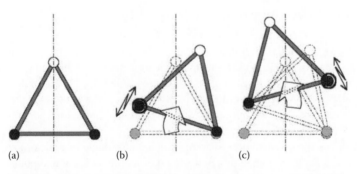

FIGURE 12.11   A rotation of the actuator controlled to illuminate two rear legs. (a) Movement straight in line, (b) rotation to the right, and (c) rotation to the left.

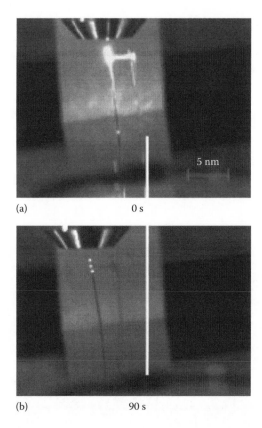

FIGURE 12.12 Experimental results of movement of optical actuator. (a) After 0 s (b) after 90 s.

12.3 MICROBUBBLE ACTUATOR USING PHOTOTHERMAL EFFECT

A microbubble by photothermal effect has an advantage that the process of appearance and disappearance is easy to control. It is also applied to move microscale objects and an optical switch by using microbubble (Otani et al. 2003). A similar work using Photo-marangoni has been proposed by Kotz. In the next section, the principle of microbubble generation is investigated and, then, three different methods of applications such as driving micro object, rotating micro object, and switching are proposed.

Figure 12.13 is a principle of microbubble generation by photothermal effect and its scanning process. The inside of a clear vessel whose bottom panel is coated with black is filled with limpid liquid shown in Figure 12.13a. After laser light is focused onto the bottom of the vessel by an objective lens (Figure 12.13b), a microbubble is generated in neighborhood of focal point by photothermal effect as shown in Figure 12.13c. The light beam is converted to heat on to the focused point. When the focal point moves to the right direction in Figure 12.13d, a temperature gradient happens in the liquid. Because a surface tension at the boundary of the bubble depends on the temperature, it becomes much weaker at high temperature. Therefore, the bubble behaves to gather at the center point to the focus point as shown in Figure 12.13d. The microbubble can be arbitrary controlled in the two-dimensional area by following the moving focus point. A laser diode with 810 nm of wavelength and 450 mW of its maximum power is used for the light source. The laser beam is focused by an objective lens with 0.24 of numerical aperture. A test cell is made of glass with 1 mm of thickness, and its space is 1 mm. The ethanol is enclosed inside of the test cell. A microscope is used to measure the size of the bubble. The maximum size of the bubble is 1000 μm and the

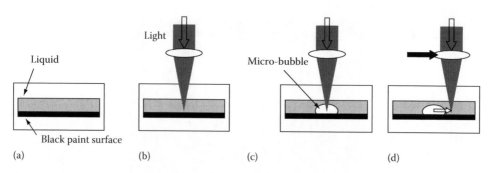

FIGURE 12.13 Principle of microbubble movement by photothermal effect. (a) Off, (b) light-on, (c) micro bubble, and (d) movement.

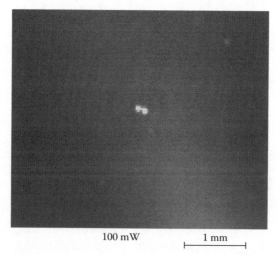

FIGURE 12.14 Generated microbubble.

minimum is 50 μm as shown in Figure 12.14. The behavior of the buildup time is almost the same as the temperature change. The microbubble can be archived at the speed of 7 mm/s. When the room temperature is below 26°C, the microbubble can easily disappear. The photothermal microbubble has reversibility; therefore, there are many possibilities for an optical actuator. Another application of the optical actuation has succeeded to move an aluminum block with a diameter of 500 μm by controlling a photothermal microbubble.

Figure 12.15 illustrates the experimental setup for an optical switch. A He–Ne laser at 633 nm is used as a signal beam, and a laser diode with 810 nm is adopted as a control beam. The cell is filled in black ink taken into the ethanol in the cell. The optical switch is achieved by generating a bubble. Figure 12.15a indicates that without the bubble it means it is switched off. When the microbubble is controlled to the signal beam, the signal beam travels through the cell. Figure 12.15b shows the signal beam transmitted through the cell. From this result, we know it is possible to make an optical switch using the photothermal microbubble.

The microbubble by photothermal effect is proposed as a novel type of optical actuator. The advantage of photothermal microbubble is that the process of appearance and disappearance is easy to control. The principle of microbubble generation is studied, and three different methods of applications such as driving a micro object, rotating a micro object, and switching are proposed.

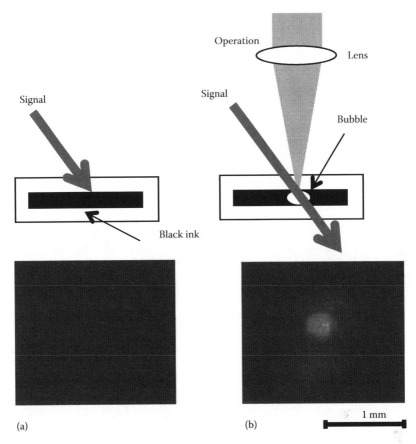

Operation

Lens

Signal

Signal

Bubble

Black ink

1 mm

(a)　　　　　　　　　　　　(b)

FIGURE 12.15 Optical switch by microbubble. (a) Switch on and (b) switch off.

12.4 TWO-DIMENSIONAL ACTUATOR USING TEMPERATURE-SENSITIVE FERRITE DRIVEN BY LIGHT BEAM

A temperature-sensitive ferrite (TSF) has a characteristic such as magnetizability depending on the temperature. Its characteristic is a ferromagnetism in room temperature, but its magnetizability decreases over 45°C at Curie temperature as shown paramagnets in Figure 12.16. The temperature-sensitive ferrite works to control magnetizing force by contact with a ferrite magnet changing the temperature. It is mostly used for a heat-sensitive switch. The optical-driven actuator is called as "photothermal magnetic force."

The possibility of a micro motor by a temperature-sensitive ferrite was reported in 1995 (Hashimoto et al. 1995). An optical actuator using a temperature-sensitive ferrite shown in Figure 12.17 was first proposed in 1998 (Fukushima et al. 1998). A stretching part as a body is made of temperature-sensitive ferrite, magnet, and springs. It is possible to control the expansion and contraction by changing its temperature illuminated by laser light. When the TSF is below the Curie temperature, the TSF and magnet are stacked together as shown in Figure 12.17a. When the temperature of TSF rises above the Curie temperature, the magnetic susceptibility of TSF disappears and the body expands as shown in Figure 12.17b. Figure 12.18 shows the moving process of a moving machine. It moves like as an inchworm. Each process from Figure 12.18a through d repeats the situation of expansion, contraction, and adsorption. The suction part as a foot is made of PZT whose electrode was removed. When the light is illuminated to the PZT, the photo-electromotive power caused by thermal stress makes electrostatic adsorption on a substrate. When the illumination stops,

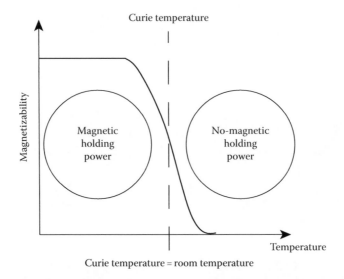

FIGURE 12.16 Characteristic of magnetizability along the temperature.

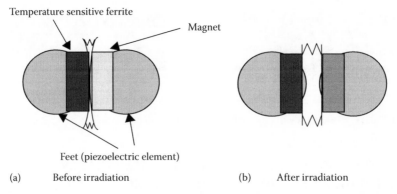

FIGURE 12.17 A photothermal actuator using temperature-sensitive ferrite. (a) Before Irradiation and (b) after irradiation.

the feet are free to move. Figure 12.19 is an experimental result of movement. Two conditions of slope angle with 0° and 5° are tested. The moving speed is very slow. Moreover, it can climb the slop because of adsorption. The walking machine using photo-electromotive by PZT can easily control adsorption and stretch so that it moves like an inchworm.

A two-dimensional optical actuator using temperature-sensitive ferrite (TSF) has some difficulty in manufacturing the ferrite to a free form because it is a brittleness material. Mizutani overcame this problem by using paste condition from fractured ferrite (Otani et al. 2004). Figure 12.20 is a construction of a 2D optical actuator made of TSF. The small TSFs are displayed to the matrix state on an acrylic substrate. The cross section of the optical actuator is shown in Figure 12.20a. A permanent magnet works as a moving object. At the first step, the laser is irradiated to two temperature-sensitive ferrites to the opposite side of the moving direction in Figure 12.20a. Their magnetizability is decreased by photothermal effect. As the balance of holding force between TSF and the object are collapsed, the object as a permanent magnet is moved to the arrowed direction in Figure 12.20b. When the illumination is shifted to the next set of TSFs, the object moves in the direction of the arrow further in Figure 12.20c. This movement of one cycle completes light

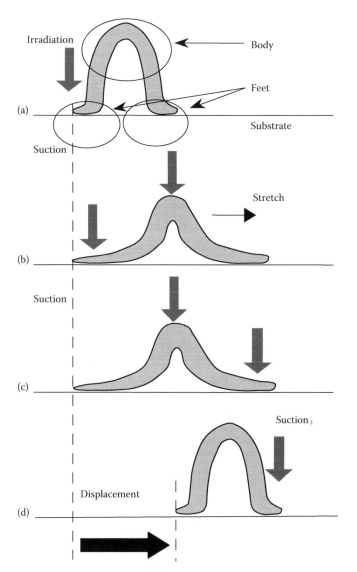

FIGURE 12.18 Movement of inch worm. (a) Initial position and suction at left feet after irradiation. (b) Expansion to right direction at body part after irradiation left feet and body. (c) Clamp at right feet after irradiation body and right feet. (d) Displacement by contraction of body part after stopped irradiation and continuous suction at right feet by irradiation.

conversion; therefore, it is possible to control the permanent magnet in the arbitrary position by changing a lighting position of the laser.

A method of two-dimensional movement of the object is important how to illuminate the light for photothermal effect as shown in Figure 12.21. Since a moving object is intended to move to the direction of arrow in Figure 12.21, the laser is irradiated to two TSFs just under or near to the moving object and the object moves the opposite side of direction simultaneously as shown in Figure 12.21c through e.

Figure 12.22 is a photograph of an experimental setup for 2D movement. A semiconductor laser with 450 mW of power and with 810 nm of wavelength is used as a light source, which is mounted on an xy electrical stage. A moving object is made of the neodymium permanent magnet with 200 mT of magnetic force and $2.4 \times 2.4 \times 1.8$ mm of its size. An acrylic plate is used with 0.8 mm of

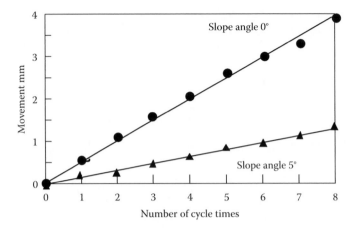

FIGURE 12.19 Experimental result of movement.

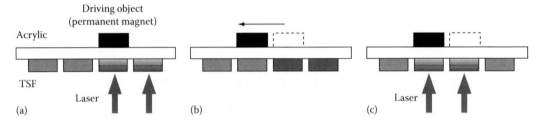

FIGURE 12.20 Principle of 2D movement. (a) Irradiation of laser onto two arrowed temperature-sensitive ferrites to the opposite side of the moving direction. (b) Movement of the object to arrowed direction after decreased magnetizability by photothermal effect of two TSFs and the unbalance of holding force between TSF and the object as a permanent magnet. (c) Further movement to direction of the arrow after shifting the illumination to the next set of TSFs.

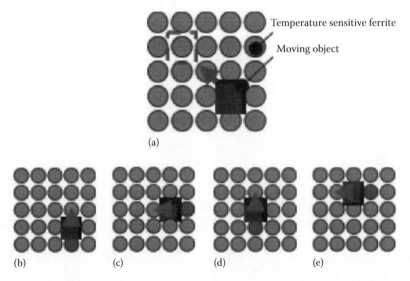

FIGURE 12.21 TSF optical actuator for 2D movement. (a) In case of moving object from initial point shown in solid squire to broken line along the arrow. (b) Irradiation to temperature sensitive ferrite (TSF) at lower side of moving object. (c) Irradiation to TSF at right side. (d) Irradiation to TSF at lower side. (e) Irradiation to TSF at left side.

thickness. The TSFs are made of MnCuZn ferrite and their Curie temperature is designed for 45°C. The holes with a size of 1.5 mm diameter on the plastic substrate are filled with the pasted TSF, which is made with fractured ferrite and the silicon grease for an arbitrary form and for adhering the laser powder condition in Figure 12.22. The TSFs are aligned as 10×8 pieces into matrix condition with 1.5 mm interval. A semiconductor laser mounted on the xy stage is irradiated to two TSFs alternately. The holding power between the neodymium magnet and the TSF is measured as 0.035 N at the room temperature and as 0.018 N at more than the Curie temperature. It is found that the holding power changes almost linearly from 35°C to 55°C of temperature.

Figure 12.23 is an experimental result of the 2D movement of an object driven by photothermal effect. It succeeds in moving almost linearly in 60 s to the distance of 20 mm. The moving speed of the object is estimated as 0.3 mm/s as shown in Figure 12.23.

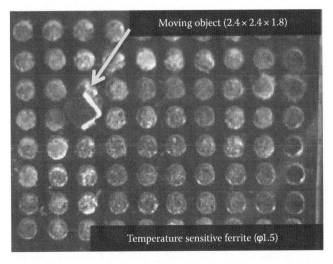

FIGURE 12.22 Photograph of an experimental setup for 2D movement.

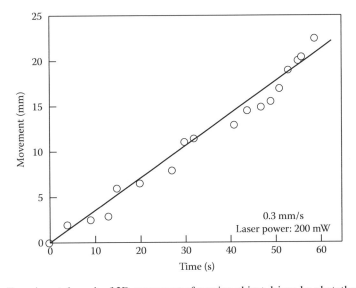

FIGURE 12.23 Experimental result of 2D movement of moving object driven by photothermal effect.

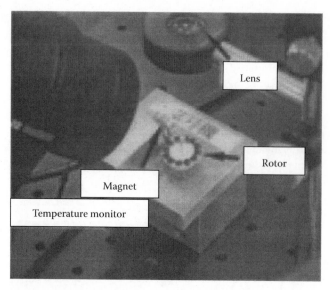

FIGURE 12.24   TSF motor.

The TSF motor is also proposed in Figure 12.24 (Takizawa 2002). The TSFs are aligned along the circle. Since the light beam is illuminated to the TSF, the rotor rotates in the direction of the unbalanced magnetic force. It works like the photothermal motor.

New types of 2D optical movement and rotary motor using temperature-sensitive ferrites by photothermal effect have succeeded to move in the field of two-dimensional area. It might apply to new industry in an optical robot for micro-optomechatronic machine and research areas such as a new field of biomedical sciences, as bio-mechanophotonics.

12.5   THREE-DIMENSIONAL OPTICAL CONTROL OF MAGNETIC LEVITATION BY TEMPERATURE-SENSITIVE FERRITE

In the previous section, noncontact movement of a small object was achieved in the two-dimensional field by the photothermal effect using temperature-sensitive ferrites. However, it moves very slowly with an angular gait. Because there are disadvantages of frictional force between floor and ferrite as an object, the best way to control the friction is to use magnetic levitation and also to control by light energy. Since the object is floated on a space by magnetic levitation using a diamagnetic material, there is no friction on a floor. In this article, we focus on building a novel optical actuator for moving small objects in the three-dimensional area. The magnetic Levitation is known to levitate a flog by producing extremely strong magnetic fields as demonstration because it is composed of primarily water and it acts as a weak diamagnet (Berry and Geim 1997). A 3D optical actuator using the magnetic levitation was proposed to utilize temperature-sensitive ferrites.

If a small permanent magnet is faced to a large magnet (Figure 12.25a), it is impossible to keep balance. That means it immediately sticks to the big magnet because it is difficult to keep the only one balanced point between the magnetic force and the gravity shown in Figure 12.25b. If the diamagnetic material is employed between the magnets in Figure 12.26a and b, it is possible to levitate the permanent magnet in the condition of a superconducting property such as the diamagnetic material shown in Figure 12.26.

The diamagnetic material works reversed polarity in Figure 12.26a. It is called "Meissner effect" (Beaugnon and Tournier 1991). In case the diamagnetic material is mounted between two faced permanent magnets in Figure 12.26b, it works as a damper. Therefore, the magnet is situated under keeping the balance between the suction and the gravity by diamagneticity in Figure 12.26c.

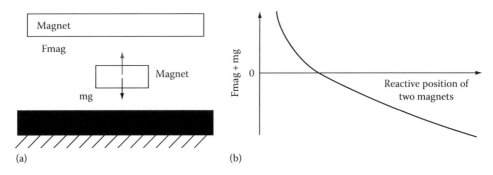

FIGURE 12.25 Equilibrium position between magnetic force and its gravity. (a) Reaction to two magnets and (b) relation to gravity and magnetic force.

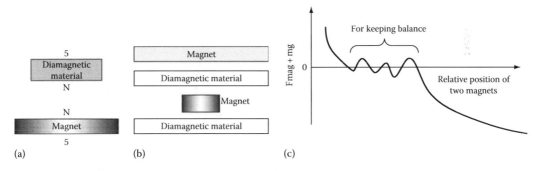

FIGURE 12.26 Magnetic levitation using diamagnetic material. (a) Relation to the magnet diamagnetic material, (b) set up to two magnets and diamagnetic materials, and (c) relation to gravity and magnetic force.

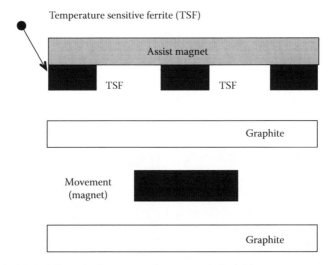

FIGURE 12.27 The idea of 3D optical actuator using magnetic levitation.

The idea of a 3D optical actuator using magnetic levitation is to utilize diamagnetic material and control the magnitude of the magnet by temperature-sensitive ferrite (TSF) between the two faced magnets shown in Figure 12.27 (Mizutani et al. 2008a). The diamagnetic material is made of graphite. Three TSF work to control the magnetizing force by contact with the ferrite magnet changing the temperature. The characteristic of temperature-sensitive ferrite has ferromagnetism in room temperature, and its magnetizability decreases over 45°C at Curie temperature. A moving object, a permanent magnet, is made of neodymium, which is set between the graphite sheets. When the ferrite magnet with the temperature-sensitive ferrite comes close to the object, the neodymium magnet is pulled to the ferrite by magnetic suction. A repulsive force happens between the neodymium and the ferrite magnet because the graphite is diamagnetic material. This means the magnetic suction comes up by closing the distance between the neodymium and the ferrite and works as a damper. The balanced point between the own weight, the gravity, and magnetic force can be adjusted by controlling the position of the ferrite magnet. The object can be levitated at well-balanced condition.

Figure 12.28 shows the principle of three-dimensional movement. When the temperature-sensitive ferrite is irradiated by a light source as shown in Figure 12.28a, the magnetic force between the neodymium and the ferrite is out of balance. The object in Figure 12.28a shows the object sets on an initial position. It moves along the vertical direction in Figure 12.28b in case irradiating the temperature-sensitive ferrite because the gravitational force of the object becomes stronger than the magnetic force. If the balance of magnetic power along the horizontal direction is broken by irradiation to the different direction in Figure 12.28c, the object moves along the horizontal direction.

Figure 12.29 shows an experimental result of 3D optical movement by magnetic levitation for vertical direction. The object is a neodymium magnet with 300 mT of magnetic force, 3 mm diameter, and 2 mm thickness. The thickness of graphite upon the object is 20 mm, and the magnetic

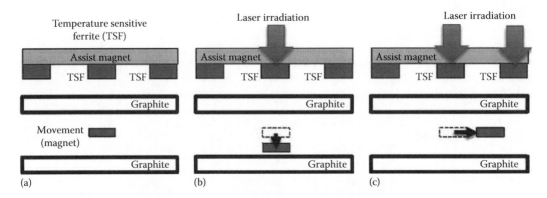

FIGURE 12.28 Two-dimensional movement of optical actuator using magnetic levitation. (a) Initial condition, (b) vertical movement, and (c) horizontal movement.

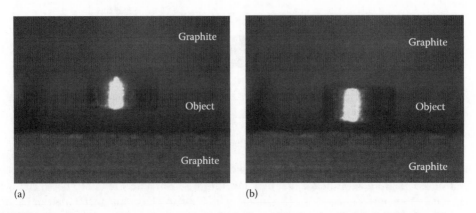

FIGURE 12.29 Experimental result of magnetic levitation. (a) Before irradiation and (b) after 60 s.

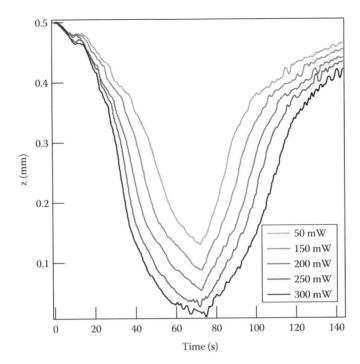

FIGURE 12.30 Experimental result of vertical movement.

force of the ferrite magnet is 90 mT as in Figure 12.29. Three TSFs are made of MnCuZn with 45°C at Curie temperature, and their size is $2 \times 2 \times 10$ mm. The distance between the ferrite magnet and the TSF is set to 10 mm. The temperature of TSFs is easily controlled by a laser diode with 810 nm wavelength. The object, the neodymium magnet, is levitated at the initial point by adjusting the vertical position of the ferrite magnet shown in Figure 12.29a. After illuminating the laser light to the middle of TSF during 60 s, the object is moved down from the initial point in Figure 12.29b. The measured result of vertical movement against time is shown in Figure 12.30. The laser exposure time was set in the first 90 s. The intensity variations were compared in the case of 50, 100, 150, and 200 mW. The movements of object depended on the laser power, and this effect was reversible.

Finally, one-dimensional movement of the horizontal direction was examined to illuminate to two TSFs simultaneously, as shown in Figure 12.31. Figure 12.31a shows the position before the irradiation. The balance of magnetic force was broken by illumination of TSFs, and the object moved to the x-direction smaller than z-direction as in Figure 12.31c. Figure 12.32 shows the measured result of horizontal movement against time after 70 s of illumination with the same intensities of the vertical movement. The maximum of its movement was 0.18 mm.

A new type of 3D optical movement was enplaned using temperature-sensitive ferrites by magnetic levitation. It succeeded in controlling the magnet in the 3D area by illuminating the temperature-sensitive ferrites. The graphite was employed for magnetic levitation using diamagnetic material. The optical driving technology would be applicable for new industries and research areas.

12.6 OPTICAL ACTUATOR BY PVDF

The optical actuator in Figure 12.10 is a modified form of the optical multileg actuator in Figure 12.6. The problem pointed out is that the legs move due to thermal expansion by the photothermal effect. The idea employed in the polyvinylidine difluoride (PVDF) film is a ferro-electric polymer that has the pyroelectric effect and the piezoelectric effect.

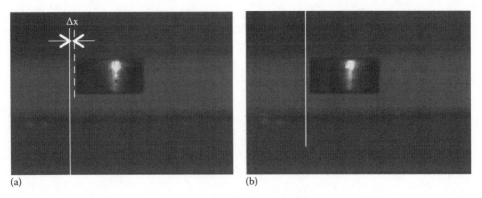

FIGURE 12.31 Experimental result of horizontal movement. (a) Before irradiation and (b) after 60 s.

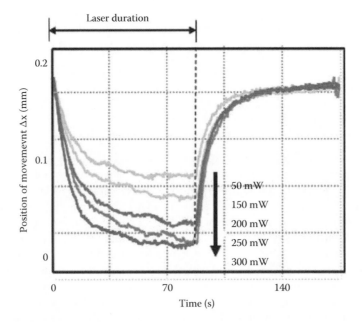

FIGURE 12.32 Horizontal movement against time.

There were many studies conducted on the ferroelectric polymer, PVDF. The ferroelectric polymer has two characteristic effects. One is the pyroelectric effect and other is the piezoelectric effect. The pyroelectric effect converts electric energy from absorbed light energy. The other effect is piezoelectric effect, which converts mechanical energy from electric energy. Therefore, it is possible to produce mechanical energy from photo energy. Moreover, thermal energy is not used for the moving energy, so the polymer can work an advantage of the response time. In addition, this material is easily produced. The possibility of PVDF polymer for an optical actuator has been reported by the University of Alabama group (Sarkisov et al. 2004).

Figure 12.33 illustrates the moving principle of an optical cantilever by photothermal effect of an optical fiber and pyroelectric and piezoelectric effects of PVDF film. A cross-sectional view of the PVDF cantilever is shown in Figure 12.33a. Both sides of the PVDF film are covered with silver coating as an electrode. One side of silver coating was removed. After light irradiation, it is polarized by pyroelectric effect toward the cross-sectional direction in Figure 12.33b. Then, electrons generated by the effect are dispersed on the surface of Ag coating in Figure 12.33c. The electric field toward the cross-sectional direction of PVDF film is inhomogeneous. Consequently, PVDF cantilever is bended toward the cross-sectional direction by the inverse piezoelectric effect in Figure 12.33d. The pyroelectric effect is faster than the photothermal effect. Therefore, the PVDF cantilever is expected to have high response time. To repeat this cycle, it is possible to build the photovoltaic repeatable cantilever.

Figure 12.34 shows an optically driven cantilever by PVDF whose thickness is 28 μm with 6 μm of Ag coating. The light source employs a He-Ne laser as a light with 10 mW of power. An acousto-optic modulator is used to modulate laser light. An edge of the PVDF cantilever was fixed on a wall.

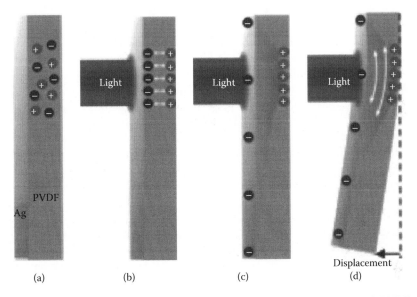

FIGURE 12.33 Moving principle of PVDF cantilever. (a) Cross-sectional view of PVDF cantilever, (b) polarization of pyroelectric effect, (c) dispersion of surface electrical charge, and (d) Bending by inverse piezoelectric effect.

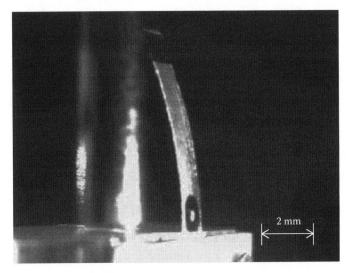

FIGURE 12.34 Optically driven cantilever by PVDF.

The cantilever can be vibrated by the switched light. The mechanism of the PVDF cantilever was measured in time response of its behaviors using a displacement sensor. The cantilever was irradiated onto a silver-coated surface by He–Ne laser with 0.8 ms of the interval and 5, 7, 9, and 11 mW of different powers. There were inflection points in the displacement curve at 0.1 ms followed by a slight rise. The first one was caused by pyroelectric and piezoelectric effect, and the second was caused by photothermal effect because of bimetal effect. There was no response of pyroelectric effect in the case of 5 mW. Therefore, the driving condition with 0.2 ms of the laser duration and 7 mW of laser power was used for the PVDF optical cantilever. One of the most important parameters for the optically driven actuator is the generative force of the PVDF cantilever. The force was measured by using a glass cantilever knowing its stress parameter. Its power was estimated as 107 mN/mW by comparing between its generative force and laser power.

There have been two types of the optical actuator using the PVDF cantilever. The idea of moving optical actuators was used for the stick-slip method, which was caused by different response time of rise and fall time. It was proposed to construct the PVDF cantilever as a leg. Figure 12.35 shows a side view of the constructed optical actuator using PVDF film as legs and on an acrylic plate. The size of the optical actuator is 5 mm in width and 10 mm in height. When a He–Ne laser as a light source with 10 mW and 1 Hz irradiates the optical actuator, the front leg repeats slipping and sticking. Its velocity was archived as 33.3 mm/s at 1 Hz and 76.7 mm/s at 2 Hz.

The other type of the optical actuator was proposed to construct the PVDF cantilever as a body. Figure 12.36 shows a multilegged actuator. Figure 12.36a shows that eight legs made of the optical fiber were mounted on the body. They were designed with a symmetrical appearance. When the laser light is illuminated at one third as long as the length of the body is at the opposite side of the wanted moving direction as shown in Figure 12.36b, it bends to the other side of the legs such as the vertical direction of the arrows by the pyroelectric effect and the leg at the point "P" was slipped to move the horizontal arrowed direction. After stopping the illumination shown in Figure 12.36c, the body returns quickly. Therefore, three legs except "P" move the horizontal direction by the stick-slip effect in Figure 12.36d. This movement of one cycle completes light conversion shown in Figure 12.37a through d. Figure 12.35 shows a photograph of the multilegged optical actuator. Its body size is 20 mm (length) × 5 mm (width) × 52 μm (thickness), and the size of its leg is 5 mm (length) and 500 μm (diameter). It was moved 20 μm/s with 1 Hz of illumination by laser diode with 810 nm of its power.

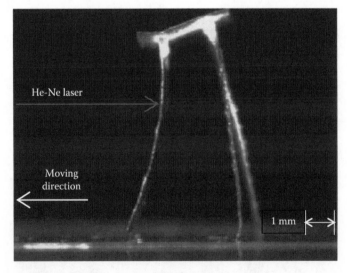

FIGURE 12.35 Optically driven actuator using PVDF cantilevers.

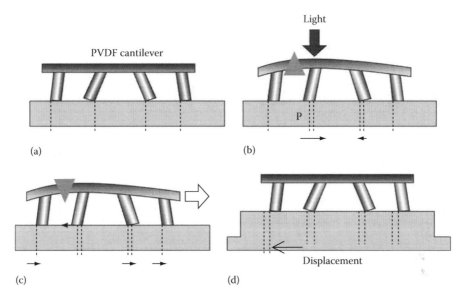

FIGURE 12.36 Principle of multilegged optical-driven actuator. (a) Initial position (side view), (b) light irradiation, (c) after irradiation, and (d) finish.

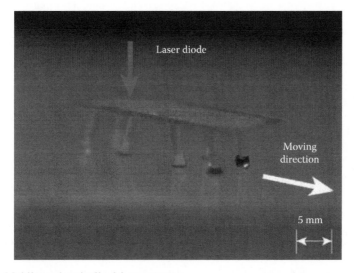

FIGURE 12.37 Multilegged optically driven actuator.

12.7 SUMMARY

In this session, the utilization of light energy for optomechatronic application, such as optical driving actuators in nano-micromachine, and possibility of optical movement by photothermal effect was reviewed. We have already developed the control of the magnetic levitation in a 3D area using temperature-sensitive ferrites. We can expect an optical driving technology for applications in new industries and research areas. We intend to expand to an optical robot for micro-optomechatronic machines. Moreover, these actuators are a powerful tool to handle biomedical samples.

REFERENCES

Ashkin, A. 1970, Acceleration and trapping of particles by radiation pressure, *Phys. Rev. Lett.* 24:156–159.

Beaugnon, E. and Tournier, R. 1991, Levitation of organic materials, *Nature* 349:470.

Berry, V. and Geim, A.K. 1997, Of flying frogs and levitrons, *Eur. J. Phys.* 18:307–312.

Cho, H. 2003, *Opto-Mechatronic Systems Handbook*, Boca Raton, FL: CRC Press.

Fukushima, K., Otani, Y., and Yoshizawa, T. 1998, An optical driving of a moving machine consisting of piezoelectric elements and temperature sensitive ferrite, *J. Jpn. Soc. Precis. Eng.* 64(10):1512–1516 (in Japanese).

Hashimoto, E., Uenishi, Y., Tanaka, H. et al. 1995, Development of a thermally controlled magnetization actuator (TCMA) for a micromachined motor, *Trans. Inst. Electron. Inf. Commun. Eng.* CJ78-C-2 6:350–356.

Hattori, S., Fukuda, T., Maruhashi, T. et al. 1993, Structure and motion characteristics of an optical microgripper, *Trans. Jpn. Soc. Mech. Eng.* 59(559):709–806 (in Japanese).

Inaba, S. and Hane, K. 1993, Miniature actuator driven photothermally using a shape-memory alloy, *Rev. Sci. Instrum.* 64(6):1633–1635.

Inaba, S., Kumazaki, H., and Hane, K. 1995, Photothermal vibration of fiber core for vibration type sensor, *Jpn. J. Appl. Phys.* 34:2018–2021.

Juodlkazis, S., Mukai, N., Wakaki, R. et al. 2000, Reversible phase transitions in polymer gels induced by radiation forces, *Nature* 408:178–181.

Kawata, S. and Sugiura, T. 1992, Movement of micrometer-sized particles in the evanescent field of a laser beam, *Opt. Lett.* 17:772–774.

Kotz, K.T., Noble, K.A., and Faris, G.W. 2004, Optical microfluidics, *Appl. Phys. Lett.* 85:2658–2661.

Mizutani, Y., Mizue Ebisawa, E., Otani, Y. et al. 2008a. Optically 3D controlled magnetic levitation using temperature sensitive ferrite, *Jpn. J. Appl. Phys.* 47:3461–3465.

Mizutani, Y. and Otani, Y. 2008. Optically driven actuators using Poly(vinylidene difluoride), *Opt. Rev.* 15(3):162–165.

Mizutani, Y., Otani, Y., and Umeda, N. 2008b. Optically controlled bimorph cantilever of poly(vinylidene difluoride), *Appl. Phys. Express* 1:041601-1–041601-2.

Morikawa, Y., Ichiki, M., and Nakada, T. 2003, Electrostatic optical motor using PLZT elements: Driving mechanism and fundamental experiment, *Trans. Jpn. Soc. Mech. Eng.* 69(684):2101–2106 (in Japanese).

Nakamura, Y. and Shimizu, K. 1999, Optical drive of SMA active forceps for minimally invasive surgery, *J. Rob. Soc. Jpn.* 17(3):439–448.

Otani, Y., Matsuba, Y., and Osaka, A. 2003, Micro bubble actuator using photo-thermal effect, *Proc. SPIE* 5264:150–153.

Otani, Y., Matsuba, Y., and Yoshizawa, T. 2001, Photothermal actuator composed of optical fibers, *Proc. SPIE* 4564:216–219.

Otani, Y., Mizutani, Y., Yoshizawa, T. et al. 2004, Two-dimensional actuator using temperature-sensitive ferrite driven by light beam, *Proc. SPIE* 5602:35–40.

Poosanaas, P., Tonooka, K., and Uchino, K. 2000, Photostrictive actuators, *Mechatronics* 10:467–487.

Sarkisov, S.S., Curley, M.J., Fields, A. et al. 2004, Photomechanical effect in films of polyvinylidene fluoride, *Appl. Phys. Lett.* 85(14):2747–2749.

Takizawa, S. 2002, Dynamic drive and control technique by the light. Photothermal magnetic motor, *O Plus E* 24(3):300–304 (in Japanese).

Tatsuma, T., Takada, K., and Miyazaki, T. 2007, Light-induced swelling and visible light-induced shrinking of a TiO2-containing redox gel, *Adv. Mater.* 19:1249–1251.

Uchino, K. 1989, Micro walking machine using piezoelectric actuators, *J. Rob. Mech.* 1:124.

Uchino, K. and Aizawa, M. 1985, Photostrictive actuator using PLZT ceramics: A: Applications and fundamentals, *Jpn. J. Appl. Phys.* 3(Suppl. 24):139–141.

Yamada, M., Kondo, M., Mamiya, J. et al. 2008, Photomobile polymer materials: Towards light-driven plastic motors, *Angew. Chem. Int.* 47(27):4986–4988.

Yoshizawa, T., Hayashi, D., and Otani, Y. 2000, Optical driving of a miniature walking machine composed of temperature-sensitive ferrite and shape memory alloy, *Proc. SPIE* 4190:212–219.

Yu, Y., Nakano, M., and Ikeda, T. 2003, Directed bending of a polymer film by light, *Nature* 425:145.

13 Light-Driven Micro- and Nanofluidic Systems

George K. Knopf and Khaled Al-Aribe

CONTENTS

13.1 INTRODUCTION

Microfluidics is the science and technology of systems that manipulate or process very small quantities of fluids (10^{-9}–10^{-18} L) using material structures and mechanical components with dimensions of tenths to hundreds of micrometers (Whitesides 2006). The miniaturization and integration of various analytical laboratory processes on a single, often disposable, platform (Shiu et al. 2010a) are a key design requirement for many lab-on-a-chip (LoC) and micrototal analysis systems (μTAS) used for medicine and environmental monitoring. Initially, microfluidics was developed to reduce sample consumption and increase efficiency in particle separation methods like electrophoresis. The reduction in physical size permitted very small volumes of liquid concentrations to be used for performing analysis and chemical synthesis.

The fluid transport channels, liquid reservoirs, and chemical processing chambers are typically a few hundred μm in size which enables the system to handle minute volumes of liquid in the nanoliter range. However, one important characteristic of most microflows is the absence of turbulence. Liquid flow at this dimensional scale is largely laminar and, therefore, molecular diffusion becomes the dominant mechanism for mixing fluids. In addition, the effects of liquid viscosity play a far greater role in the performance of these systems than what is observed in classical macrofluidics. To improve the mixing rate over shorter microchannels, researchers have developed various passive and active micromixer designs (Nguyen and Wereley 2002).

More recently, device miniaturization has greatly expanded the capabilities of bioassays, separation technologies, and chemical synthesis techniques (Dittrich and Manz 2006) without changing the nature of the molecular reactions. Today, microfluidic devices contain a variety of interconnected discrete components that introduce reagents and samples (fluids or powders) to the

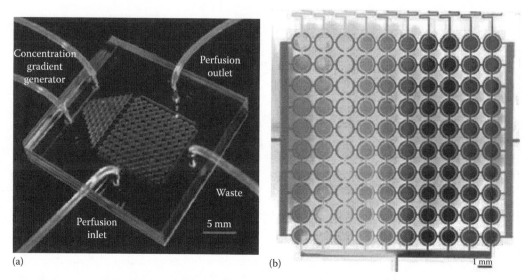

FIGURE 13.1 **(See color insert.)** Microfluidic cell culture array for high-throughput cell-based assays. (a) Photograph of a 10×10 array of microchambers fabricated on a 2×2 cm device. The medium is introduced through the left port and exits at the right port (outlet). Cells and reagents are from the top and flow out through the bottom port. The device is capable of conducting cell-based assays with multiple concentrations of reagents. (b) Graphic showing the concentration gradient across 10 columns. A concentration gradient generator was connected to the 10 columns at the top of the device. Red dye was initially poured into the chambers. Blue and yellow dyes were then loaded from the two separate ports at the top of the gradient generator. (From Hung, P.J., Lee, P.J., Sabounchi, P., Lin, R., and Lee, L.P.: Continuous perfusion microfluidic cell culture array for high-throughput cell-based assays. *Biotech. Bioeng.* 2004. 89(1). 8pp. Copyright Wiley-VCH Verlag GmbH & Co. KGaA. Reprinted with permission.)

channels and processing chambers, move one or more streams of fluid around on the network, combine and mix liquids to specification, detect substances in the stream, and perform complex chemical synthesis. Figure 13.1 is an example of a 2×2 cm microfluidic cell culture array for high-throughput cell-based assays (Hung et al. 2004). The goal is to maintain and continuously monitor cells while providing a stable microenvironment. The port located at the left provides continuous perfusion of the medium uniformly across the array, whereas the port at the right is the outlet for the medium. Reagents and cells were loaded from the top and flow out through the bottom port. Another practical example is the polycarbonate microfluidic system for detecting saliva-based bacteria developed by Chen et al. (2007; Figure 13.2). The $85 \times 35 \times 3$ mm device contains a variety of compartments and functional components for on-chip analysis including fluid modules for cell lysis, lateral flow strips for nucleic acid detection, and nucleic acid analysis.

To compensate for the physics of "very small scale" and improve the overall system performance, it is possible to engineer submicron structures along the transport channels and storage reservoirs for efficient energy storage, thermal management, heat dissipation, or other energy-related requirements. In addition, a number of different energy conversion devices such as microreactors, miniaturized heat engines, microbatteries, and microfuel cells can also be found on microfluidic systems (Pennathur et al. 2007). Microfuel cells and microreactors convert chemical and biochemical energy into electrical energy that provides portable power to the chip.

An important subdiscipline is *nanofluidics* which deals with the behavior and manipulation of fluids confined to structures in the nanometer scale. As the size of these fluidic nanostructures approach the Debye length and hydrodynamic radius of the liquid (Eijkel and van den Berg 2005), different physical constraints are imposed on the behavior of the fluid being transported and manipulated (Kuo et al. 2004). These constraints cause regions of the fluid to exhibit properties

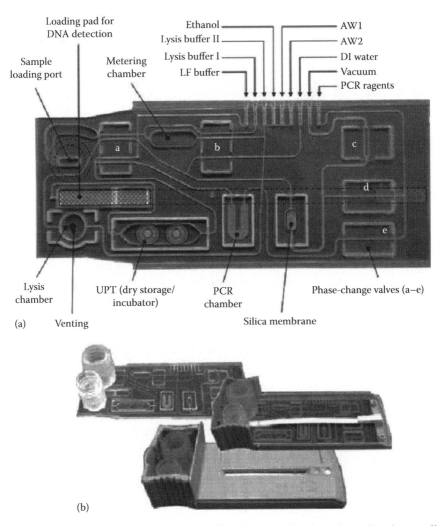

FIGURE 13.2 Microfluidic cassette for the Up*link*™ laser scanner/reader unit used to detect saliva-based bacteria. (a) The polycarbonate microfluidic device consists of various conduits, chambers, and valve sites for sample metering. (b) The insertion and mounting of the microfluidic "cassette" in the molded plastic cartridge. (From Chen, Z., Mauk, M.G., Wang, J., Abrams, W.R., Corstjens, P.L.A.M., Niedbala, R.S., Malamud, D., and Bau, H.H.: A microfluidic system for saliva-based detection of infectious diseases. *Ann. NY Acad. Sci.* 2007. 1098. 429–436. Copyright Wiley-VCH Verlag GmbH & Co. KGaA. Reprinted with permission.)

such as a vastly increased viscosity near the nanochannel wall. The large viscosity tends to alter the thermodynamic properties and even affect the chemical reactivity of the species at the fluid–solid interface.

The nanostructures can be created as single cylindrical channels, nanoslits, or nanochannel arrays (Figure 13.3) from materials such as silicon, polymers (PMMA, PDMS), and synthetic vesicles (Karlsson et al. 2004). Bulk and surface micromachining, chemical etching, photolithography, and replication techniques such as embossing, printing, casting, and injection molding have all been used to fabricate structures for nanofluidic devices (Mijatovic et al. 2005). Alternatively, bottom-up fabrication techniques such as self-assembled monolayers (SAM) can be used to create nanostructures (Mijatovic et al. 2005, Al-Aribe et al. 2011a,b). This technique utilizes biological materials to form a molecular monolayer on the substrate material. Nanochannels can also be formed by growing carbon nanotubes (CNT) and quantum wires. These bottom-up methods often

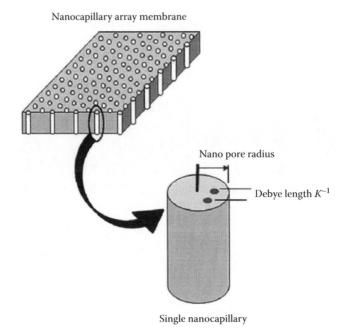

Nanocapillary array membrane

Nano pore radius

Debye length K^{-1}

Single nanocapillary

FIGURE 13.3 Simplified illustration of a nanocapillary array membrane used for nanofluidic applications. The membrane is comprised of numerous parallel nanocapillaries, each with a pore radius approximately equal to the Debye length (distance over which significant charge separation can occur).

provide well-defined shapes with characteristic lengths that are only a few nanometers. For these types of structures, the interconnection between the nanochannel structures and microfluidic systems become an important design issue.

Nanofluidic devices have been built for a variety of applications in chemistry, molecular biology, medicine, and environmental science. One of the main goals of these nanodevices is the separation and measurement of solutions containing nanoparticles for drug delivery, gene therapy, and nanoparticle toxicology on a μTAS (Stavis et al. 2009). An advantage of nanoscaled fluidic systems is the very small amount of sample or reagent used in the analysis. This reduction in sample size, even when compared to microsamples, significantly speeds up the time for processing. It also makes it possible to perform the analysis in an array which permits parallel processes to occur simultaneously and thereby significantly increase the throughput of the analysis. Another application for nanofluidics has been the development of nano-optics for tunable microlens arrays (Grilli et al. 2008).

Several optically driven methods for transporting, manipulating, and mixing small volumes of liquid on LoC devices and μTAS are presented in this chapter. Optofluidics is first introduced as the integration of optical and microfluidic technologies. This discussion is followed by examination of different methods for manipulating and controlling microflows. Many of the techniques exploit physical phenomena to induce optical gradient forces, manipulate surface tension of liquid droplets, alter thermoviscous behavior of flows, push liquid by driving bubbles through channels, take advantage of the ideal gas law, and use laser light to increase molecular diffusion. Light-sensitive hydrogel microstructures are also examined as a mechanism for controlling flow through microchannels. Sometimes the goal is not to transport the fluid but rather change a physical property of the liquid such as pH. In this regard, optically driven proton pumps that transport H^+ ions from adjacent reservoirs are introduced. Although this chapter represents only a brief introduction to light activated micro- and nanofluidic technologies, it does provide context for a rapidly evolving discipline.

13.2 OPTOFLUIDICS: FUSION OF MICROFLUIDICS AND OPTICS

Optofluidics involves the integration of optical technologies with microfluidic components to create high-performance LoC devices and µTAS (Psaltis et al. 2006, Erickson 2008, Monat et al. 2008). Some of the pioneers in optofluidics insist that this is a discipline unique and distinct from microfluidics (Erickson 2008), but for the purpose of this introductory chapter, it is viewed as an important subdiscipline. The primary advantage of "on-the-chip" integration with optical technologies is the relative ease in which the optical properties of the device can be used to manipulate the fluid or, alternatively, change the fluid properties to manipulate the optics. Unlike solids, the *optical properties* of the target liquid can be altered by mixing multiple streams of dissimilar fluids or replacing one fluid with another. As well, it is often possible to increase the rate of mixing two or more liquids by using a laser beam to alter the thermoviscous properties or increase molecular diffusion (Shiu et al. 2010b).

An optofluidic device is typically constructed from three interacting structural layers. The top layer contains the microfluidic controls, including microvalves and micropumps, while the middle layer is comprised of the microfluidic channels used to transport the fluid between reservoirs. The third, and final, layer is the optical structure used to guide the light to the target regions of the fluidic platform. It is this layer that contains the optical components such as photonic crystals, sensors, sources, and waveguides. The physical size of the overall fabricated optofluidic device can be merely tens of nanometers or hundreds of millimeters for complex microfluidic applications (Psaltis et al. 2006). Figure 13.4 is an illustration of an optofluidic device that combines optofluidic and microphotonic components (Monat et al. 2008).

The interface between two adjacent materials can be used to control the propagation of light. An example is the *total internal reflection* that occurs between the core and the cladding in an optical fiber. Similarly, the interface between a solid and liquid can also be used to perform specific functions in an optofluidic system. One example is to create a solid structure with voids that have dimensions larger than the wavelength of light. The transmission, reflection, and refraction of light through the void are then modified by filling the void cavity with a liquid.

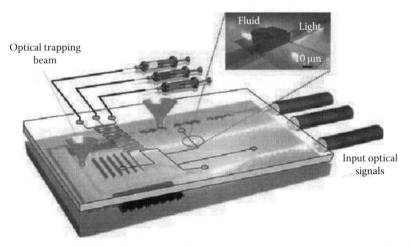

FIGURE 13.4 (See color insert.) The basic design consists of a multifunctional optofluidic device. The device is comprised of a photonic chip (bottom part) coupled to optical fibers, a microfluidic circuit (on top of it) connected to fluid sources with different refractive indexes, and optical trapping beams to manipulate micro-mechanical components. (From Monat, C., Domachuk, P., Grillet, C., Collins, M., Eggleton, B.J., Cronin-Golomb, M., Mutzenich, S., Mahmud, T., Rosengarten, G., and Mitchell, A.: Optofluidics: A novel generation of reconfigurable and adaptive compact architectures. *Microfluid. Nanofluid.* 2008. 4. 81–95. Copyright Wiley-VCH Verlag GmbH & Co. KGaA. Reprinted with permission.)

Evanescent waves, a phenomenon related to total internal reflection, has also found a variety of practical uses in optofluidics. These evanescent waves are produced in the lower index region when light undergoes *total internal reflection* at the interface. Although these evanescent modes cannot carry energy away from the interface, the light propagation in the region of higher refractive index can be strongly affected by the modes. This provides a nonintrusive mechanism for optofluidic control by introducing liquids with specific optical properties into the region surrounding the optical structure (Psaltis et al. 2006). A consequence is that the phase velocity of a dielectric waveguide depends on the refractive index of the cladding. By selecting a liquid with an appropriate refractive index to surround the dielectric waveguide, it is possible to adjust the light propagation delay through the waveguide. This concept is used for a variety of integrated optics applications including optical switching and modulation (Psaltis et al. 2006).

A second-class of optofluidic systems increase design flexibility by exploiting the interface properties between two adjacent fluids. Wolfe et al. (Wolfe et al. 2004) showed the versatility of this idea in developing liquid-core/liquid-cladding (L^2) optical waveguides. The L^2 waveguide consisted of a microfluidic channel where an optically dense fluid is allowed to flow inside of fluid with a lower refractive index. The method also can be used to create a tunable light source with the addition of dyes to the core liquid (Psaltis et al. 2006). In addition to providing a functioning waveguide, this type of fluidic system will have a profile independent of the underlying physical microfluidic system. In other words, the light traveling through a dynamic L^2 waveguide can be directed and steered by simply changing the relative flow profile and flow rate of the fluid. As well, it will be possible to obtain finer features in the composite fluid media such as a small jet stream (<10 μm) formed in a much larger microchannel (>100 μm). Finally, it is possible to create an optically smooth interface between the two liquids by choosing immiscible fluids or ensuring the fluids in the adjacent flows have low Reynolds numbers.

A third group of optofluidic systems exploit solid particles suspended in the fluid. These types of light-driven colloidal systems permit large changes to the liquid's optical properties. In addition, the small particle in the medium can be manipulated with ease and precision creating gradients in the optical properties or localized modifications. Particles significantly smaller than the wavelength have been used to change the refractive index or absorption coefficient of the liquid medium (Psaltis et al. 2006). Alternatively, particles that are larger than the wavelength were treated as a spherical focusing lens in the medium. In addition, quantum dots can be designed to fluoresce or scatter light.

13.3 MANIPULATING AND CONTROLLING MICROFLOWS

The building blocks of these integrated microfluidic device enable precision control and efficient transportation of liquids. Common components include micromixers, micropumps, separators, filters, reaction chambers, and waste disposal chambers. The reduction in physical size of these functional components have increased the speed of analysis, lowered operating costs due to the consumption of small quantities of reagents, and enabled the design of miniaturized portable devices. These integrated microfluidic systems must also be disposable and used only once to avoid the potential biological sample cross contamination.

One of the most important fluid control components on the microsystem platform is the micromixer. Efficient mixing of chemicals or biological substances, in order to create the desired reactions, is an essential step in preparing the sample for analysis. At a macroscale, the typical mechanism for mixing fluids creates turbulent flows at a high Reynolds (Re) number. In the microdomain, however, mixing liquids becomes more difficult because laminar flow dominates the process and molecular diffusion is the primary mixing mechanism (Hessel et al. 2005, Nguyen and Wu 2005). To improve the mixing rate over shorter length microchannels, researchers have developed various micromixer designs. These microfluidic mixers can be categorized into *passive* or *active mixing*, depending on whether an external source of energy is being used to assist mixing operation (Nguyen and Wereley 2002).

Passive mixers do not require external power sources and, typically, use specially designed microfluidic channel microstructures to improve the overall mixing rate (Kamholz et al. 1999, Ismagilov et al. 2000, Kamholz and Yagar 2002). The simplest passive micromixers designs are T- and Y-micromixers for combining the liquids in two adjacent streams. Unfortunately, a lengthy microchannel is required because the mixing mechanism is completely dependent on the molecular diffusion rate. To improve the mixing rate, Kamholz et al. (1999) use a parallel lamination micromixer to split the streams into a number of substreams. These substreams are then recombined to improve mixing over shorter channel lengths. The higher mixing rate of the split–join approach is due to the increased contact surfaces between fluids. Several works employed the similar approaches to improve mixing are also reported such as split–join (Branebjerg et al. 1996), split–split–join (Munson and Yager 2004), and multiple intersecting (He et al. 2001).

Instead of splitting both streams to increase contact surface between fluids, injection mixers that use a nozzle array to inject one stream into another are also reported to increase mixing rate (Voldman et al. 2000). Chaotic advection micromixers are also used to improve mixing via employing specially designed geometries to generate splitting, stretching, folding, and breaking the flows (Mizuno and Funakoshi 2002, Wang et al. 2002, Jen et al. 2003, Chen and Meiners 2004, Park et al. 2004). Wang et al. (2002) reported the simulation of micromixing where obstacles were placed in the channels to create the chaotic advection in turn to assist mixing. Three-dimensional (3D) complex microstructure micromixers were also introduced to generate a chaotic advection type of micromixing (Mizuno and Funakoshi 2002, Jen et al. 2003, Chen and Meiners 2004, Park et al. 2004), such as the connected out-of-plane L-shape microchannels (Chen and Meiners 2004), twisted microchannels (Park et al. 2004), and the Tesla micromixer design (Hong et al. 2004) as shown in Figure 13.5.

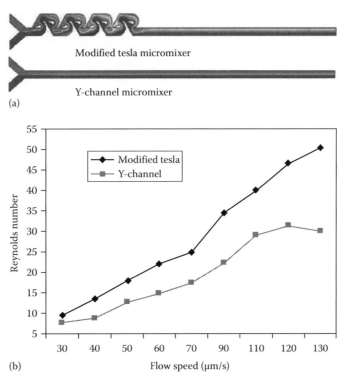

FIGURE 13.5 Comparison of a simple passive Y-channel and modified Tesla chaotic micromixer. (a) COMSOL™ simulation of the microflow through the two passive micromixers. (b) The change in flow speed of the two designs as a function of the Reynolds number (Re).

The advantages of passive mixers have been their relative operating simplicity, reliability, and robustness over the designed microflow range. However, to be efficient these mechanisms require lengthy microchannels or complex 3D microchannel structures. The additional material and fabrication steps for 3D structures can significantly increase the cost of any disposable LoC device that must be produced in large quantities.

In contrast, active micromixers use an external energy source to create disturbances in liquid streams to improve mixing rates (Liu et al. 2002a, Suzuki and Ho 2002, Tang et al. 2002, Tsai and Lin 2002, El Moctar et al. 2003, Fujii et al. 2003). Fujii et al. (2003) reported using an external micropump to generate pressure disturbance to improve mixing. Electrodydrodynamic and electrokinetic disturbance are also reported to improve mixing rates (Tang et al. 2002, El Moctar et al. 2003). Suzuki and Ho (2002) describe a micromixer integrated with electrical conductors that generate magnetic fields, which then move 1–10 μm diameter magnetic beads to create a disturbance in the flow stream. Liu et al. (2002a) reported using the acoustic streaming technique to induce air bubbles in streams where the bubbles disturb the flows to improve mixing. Tsai and Lin (2002) reported utilizing a microheater embedded in microchannels to thermally generate air bubbles that disturb the streams.

Active micromixers require peripheral hardware to introduce additional energy into the flow stream or to activate a moving mechanism. This approach results in higher mixing rates but may become unreliable because of increased opportunities for mechanism failure. Furthermore, the higher unit cost of fabricating disposable LoC devices with embedded active components has discouraged industry from commercializing microfluidic devices. Optical technology, however, provides a solution to improving the mixing rate by enhancing molecular diffusion (Shiu et al. 2010b), inducing thermoviscous expansion (Weinert and Braun 2009), or driving bubbles (Bezuglyi and Ivanova 2007).

Optical micro- and nano-actuators can either *directly* or *indirectly* transform light energy into the desired force or physical movement (Tabib-Azar 1998, Knopf 2006). The engineering benefits and performance capabilities of using an optically driven systems increases as the mechanisms are reduced in physical size. Furthermore, these optical transducers can be designed to operate under different properties of light such as intensity, wavelength, phase, and polarization characteristics.

13.3.1 DIRECT OPTICAL MANIPULATION OF MICRO-OBJECTS IN FLUIDS

Optical gradient forces that arise from the evanescent bonding between light waves have been used to manipulate objects and biological cells in microfluidic flows. The gradient optical force was initially described more than four decades ago by Ashkin (1970). Ashkin observed the forced acceleration of freely suspended particles by radiation pressure from continuous wave (CW) visible laser beam. Based on this observation, he was able to demonstrate that it was possible to lift a glass microsphere off a glass plate and levitate it above the surface (Ashkin 1970). This concept was used to develop an optical trap and "tweezer" that could manipulate molecules through a specially designed microscope (Dholakia and Reece 2006; Figure 13.6). In the late 1990s, Higurashi et al. (1997) demonstrated how carefully shaped fluorinated polyimide micro-objects with a 6–7.5 mm cross-sectional radius can be rotated using Ashkin's method of optical trapping and tweezing.

Optical driving provides a mechanism to remotely control microparts and biological samples that reside inside sealed compartments. The optically driven systems can employ elements and components such as micropumps, microvalves, microstirrers, and particle sorters (Maruo and Inoue 2007). Optically driven micropumps can be classified as either colloidal-based or micromachined components. In the first case, colloidal microparticles are cooperatively driven by time-shared laser micromanipulation techniques (Terray et al. 2002). The spinning of the dispersed microparticles in the colloidal liquid can be used to stir or pump the fluid. These colloidal microstirrers and micropumps are very simple; however, their performance is often limited by the shapes and properties of the microparticles. Micromachined pumps and stirrers have the potential for flexible design.

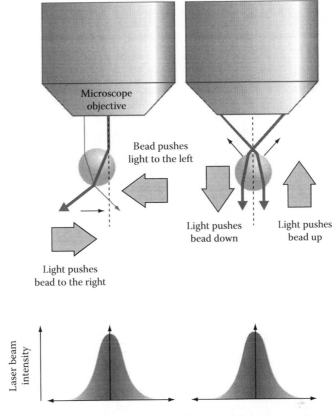

FIGURE 13.6 A geometric optical interpretation of the light-particle interaction that occurs in optical trapping. Optical forces are balanced when a transparent object with a high refractive index reaches the center of a tightly focused Gaussian beam. (Reprinted from *Nano Today*, 1(1), Dholakia, K. and Reece, P., Optical micromanipulation takes hold, 18–27, Copyright 2006, with permission from Elsevier.)

In terms of colloidal devices, single and multiple parallel-acting optical tweezers have been described in the literature for manipulating objects in microfluidic systems (Ozkan et al. 2003). These optical manipulating technologies have become critical in handling biological samples because cell membranes are highly deformable and excessive shear stresses induced during fluid flow can cause rupture of the membrane. Ozkan et al. (2003) describe a microfluidic separation (cytometry) system based on the gradient force optical switch as shown in Figure 13.7. The trapping force of an optical tweezer is used to move polystyrene microspheres or biological cells into the input port of a T-shaped fluidic channel. In this illustration, the authors identify each sample through selective fluorescence or by measuring the dielectric constant. Depending upon the type of sample, the micro-object is directed to one of the two branches of the T-channel. Cytometry or sorting is performed at the junction by capturing the sample in an optical trap and then moving it either left or right.

Figure 13.7b is a series of images showing the manipulation of the polystyrene microspheres using the optical gradient forces. The T-channel in this example is 40 μm wide and 20 μm deep. The test chip was fabricated in PDMS, and gold electrodes have been inserted in the reservoirs at each end of the channel to induce electroosmosis for fluid flow. The microfluidic device was placed in an optical tweezer system that used a 27 mW, 850 nm laser light source. The fluid flow at 10 μm/s was generated by applying a dc bias between the electrodes. In the images, the microspheres flow downstream from the top reservoir and are captured by the optical tweezer at trap A. The microsphere is then directed manually to the left of the right, and then released allowing it to follow the stream.

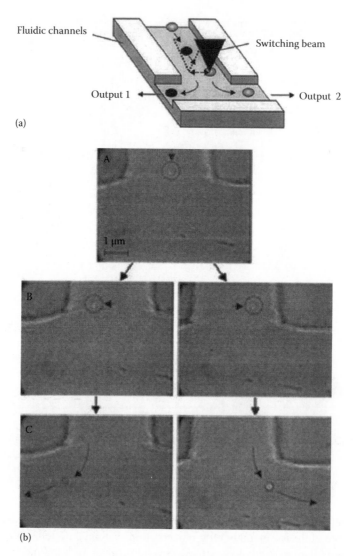

(a)

(b)

FIGURE 13.7 (a) Microfluidic separation (cytometry) system based on the gradient force optical switch. Objects are transferred into the input port and directed to one of the output ports by using the trapping force of an optical tweezer to drag and release the objects within the laminar flow stream. (b) Images demonstrating the gradient force optical switch. Microspheres flow into the downstream channel from a reservoir at the top. Spheres are first captured by the optical tweezers trap A. The microsphere is directed manually to either the left or right B, then released from the trap C, and allowed to follow the fluid stream toward left or right. The dotted circle indicates the position of the optical trap. (Reprinted with permission from Springer Science+Business Media: *Biomed. Microdevices*, Optical manipulation of objects and biological cells in microfluidic devices, 5(1), 61, Ozkan, M., Wang, M., Ozkan, C., Flynn, R., Birkbeck, A., and Esener, S., Copyright 2003.)

As an alternative, Maruo and Inoue (2006) introduced a micromachined optically driven micropump that used two interlocked spinning rotors incorporated in the microchannel. The micropump is driven by means of radiation pressure created by a tightly focused laser beam. Each individual rotor has two lobes (Figure 13.8) and is held within the microchannel by a shaft. This design prevents the rotor from moving unless it is irradiated by light. When light is focused on the side of the rotor, the net radiation pressure applied to the rotor points toward the focus, allowing the mechanism to be controlled by changing the trajectory of the scanning beam. Different techniques such as holographic optical tweezing (Padgett and Di Leonardo 2011) and time-divided laser scanning

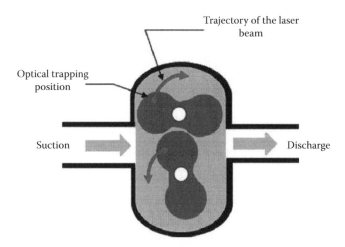

FIGURE 13.8 Simplified illustration of the Maruo's lobed micropump driven by the time-divided laser scanning of a single laser beam. Fluid is transported by optically rotating the two intermeshed rotors. (Adapted from Maruo, S. and Inoue, H., *Appl. Phys. Lett.* 89, 144101, 2006.)

methods (Arai et al. 2004) can be used to simultaneously control both rotors. The authors fabricated the optically driven lobed micropump using two 9 μm diameter rotors driven by a single laser beam. To control both lobed rotors simultaneously, a time-divided scanning technique was used. The velocity of tracer particles added to the fluid was observed and determined to be 0.2–0.7 μm/s, which was directly proportional to the rotation speed of the rotors.

Maruo and his colleagues at Yokahama University (Maruo and Inoue 2007, Maruo et al 2009) further developed this concept by introducing an optically driven micropump that uses a single-disk microrotor. Again, the micropump is driven by the time-shared optical trapping technique (Maruo and Inoue 2007), but the single 10 μm diameter disk rotor has three columns as targets for the optical trap. As well, the disk is confined to a U-shaped microchannel eliminating the need for a shaft. The laser beam is scanned circularly focusing on the right and left columns. The laser beam is simultaneously focused on the center column in an effort to hold the rotating disk without the need for a mechanical shaft. The rotation of the microrotor cause the fluid to flow around the disk but as produces a small pressure gradient in the opposite direction. However, the viscous flow surrounding the rotor overcomes the resistance of this backpressure because of the large viscous drag around the side of the microrotor (Maruo and Inoue 2007). The result is a continuous steady flow. Experiments by the authors have shown that the flow velocity is proportional to the rotational speed of the microrotor.

13.3.2 TRANSPORTING LIQUID DROPLETS USING OPTO-ELECTROWETTING

The ability to transport liquids in a droplet form along a substrate has a number of advantages over other methods because it requires no moving parts, is free of unwanted mixing caused by leakage, and eliminates the need for fabricating specialized micropumps, microstirrers, and microvalves (Chiou et al. 2003, Park et al. 2010). However, to realize this goal it will be necessary to control local surface tension. Unfortunately, the affect of surface tension on a fluid increases significantly as the droplet is reduced in size. Surface tension between the liquid–solid interface becomes the dominant force when attempting to manipulate and transport liquid droplets in a microenvironment. A number of different mechanisms have been studied to control surface tension between the liquid–solid interface including thermocapillary pumping of discrete drops (Sammarco and Burns 1999), irradiation of photoisomerizable monolayer covering surface (Ichimura et al. 2000), and direct electrical control of surface tension by electrowetting (Pollack et al. 2000).

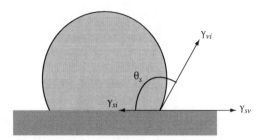

FIGURE 13.9 Wetting of surfaces is measured in terms of the contact angle, θ_s, between the droplet and the solid substrate surface.

Electrowetting has received significant attention in recent years because of its fast switching response and low power consumption. Wetting of surfaces is measured in terms of the contact angle of the droplet striking the surface. The contact angle is the angle between the droplet and the solid substrate surface as shown in Figure 13.9. A low contact angle indicates good wetting, and if the angle approaches 0°, a film is formed. A high contact angle indicates hydrophobic properties. In extreme cases, this angle approaches 180°, which means a droplet has formed that hardly wets the surface.

The surface tension acting on the droplet at the interface is in balance. The contact angle formed by the droplet and the surface is θ_s and called the equilibrium contact angle (Groenendijk 2008). This angle satisfies Young's formula:

$$\cos\theta_s = \frac{\gamma_{sv} - \gamma_{sl}}{\gamma_{vl}} \tag{13.1}$$

where
γ_{sv} is the surface tension between the solid surface and the vapor
γ_{sl} is the surface tension between the solid and liquid
γ_{vl} is the surface tension between the vapor and the liquid

It is important to note that Young's formula only holds true for smooth surfaces.

In the electrowetting process, the surface tension between the liquid–solid interface is altered by an external field which then reduces the contact angle. The general mechanism for electrowetting involves placing a droplet of polarized liquid on a substrate with an insulating laser between the liquid and electrode (Chiou et al. 2003). The surface tension at the solid–liquid interface is altered when an external voltage is applied. The voltage potential also changes the contact angle, $\theta_s(V_A)$, by

$$\cos[\theta_s(V_A)] = \cos[\theta_s(0)] + \frac{1}{2}\frac{\varepsilon}{d\gamma_{lv}}V_A{}^2 \tag{13.2}$$

where
V_A is the applied voltage
d is the thickness of the insulating layer
ε is the dielectric constant of the insulating layer
γ_{lv} is the interfacial tension between the liquid and vapor (Chiou et al. 2003)

For an *opto-electrowetting* (OEW) mechanism, the concept is modified by placing a photoconductive material below the electrode as shown in Figure 13.10. The electrical impedence of the liquid, insulator, and photoconductor are serially connected, and the contact angle change of the droplet is determined by the voltage drop across the insulating layer. The authors demonstrated that this

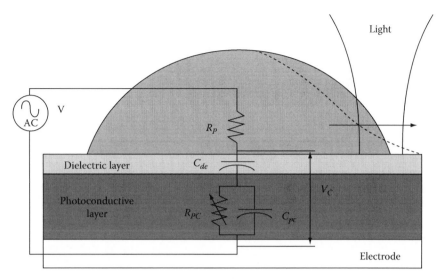

FIGURE 13.10 A typical OEW system consists of an electrode, a photoconductive layer, and a dielectric layer on which a droplet is positioned. An equivalent electric circuit of the system is superimposed on the schematic diagram. The liquid droplet is moved by locally varying the light intensity. (Reprinted from *Sens. Actuators A*, 141, Krogmann, F., Qu, H., Mönch, W., and Zappe, H., Push/pull actuation using opto-electrowetting, 499–505, Copyright 2008, with permission from Elsevier.)

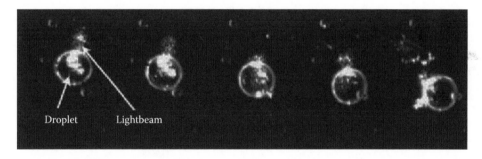

FIGURE 13.11 A series of images showing the movement of the liquid droplet when the light beam is applied to the top of the droplet forcing it downward. (Reprinted from *Sens. Actuators A*, 141, Krogmann, F., Qu, H., Mönch, W., and Zappe, H., Push/pull actuation using opto-electrowetting, 499–505, Copyright 2008, with permission from Elsevier.)

simple sandwich setup could move a liquid droplet. Figure 13.11 shows the pushing motion created at a frequency of 1.5 kHz and voltage of 75 V. The pushing velocity was observed in the range of several millimeters per second.

Chiou et al. (2003) further showed how this basic concept could be used to move a liquid droplet on an OEW surface along a light path (Figure 13.12). In this system, an ac voltage is applied between the top ITO electrode and the bottom Al grid. Since ac voltage is applied in this case, the VA is replaced by the root-mean-square (RMS) voltage. The voltage drop across the insulator and the electrode is then controlled by shining an optical beam on one edge of the liquid droplet. This illumination causes the contact angle to decrease and creates a pressure difference between the two ends of the droplet forcing the liquid droplet to follow the movement of the light beam. Figure 13.13 shows the movement of a droplet across a 1 × 1 cm OEW actuated by a 4 mW, 532 nm laser. The authors observed a droplet speed of 7 mm/s.

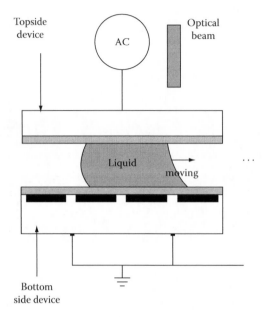

FIGURE 13.12   The basic principle used to actuate or move liquid droplets using the OEW device. (Reprinted from *Sens. Actuators A*, 104, Chiou, P.Y., Moon, H., Toshiyoshi, H., Kim, C.-J., and Wu, M.C., Light actuation of liquid by optoelectrowetting, 222–228, Copyright 2003, with permission from Elsevier.)

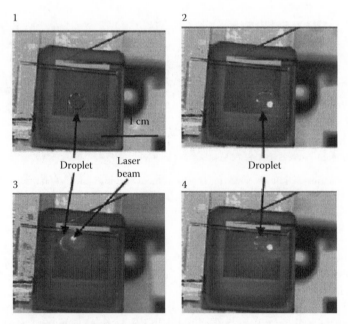

FIGURE 13.13   Images of a liquid droplet being transported across a 1 × 1 cm OEW area actuated by an optical beam. (Reprinted from *Sens. Actuators A*, 104, Chiou, P.Y., Moon, H., Toshiyoshi, H., Kim, C.-J., and Wu, M.C., Light actuation of liquid by optoelectrowetting, 222–228, Copyright 2003, with permission from Elsevier.)

13.3.3   CONTROLLING MICROFLOW USING THERMOVISCOUS EXPANSION

The precise control of fluid flow fields or the concentration of solutes is essential for investigating chemical, biological, and cellular processes (Weinert and Braun 2008, Weinert and Braun 2009, Weinert et al. 2009) in LoC technology. Unfortunately, transporting, mixing, and controlling fluid flow at small scales require complex chip designs that are often difficult to manufacture. An alternative approach is to implement a light pump that can transport fluid, or particles suspended in the liquid, without the need for microchannel walls to predefine the path or micromechanisms to create the pressure and forces necessary to move the liquid. In other words, light is used to locally drive fluids without requiring structural changes to the substrate, thereby eliminating some of the costly steps previously necessary for LoC microfabrication.

Weinert and Braun (2008) showed how small volumes of fluid can be moved by a laser scanning microscope. In this research work, selected portions of the fluid film were pumped along a predefined path of a moving warm spot that was created using an infrared laser. The repetitive motion of the laser beam remotely drove 2D fluid flows of water with a resolution of 2 μm. The experiments produced pump speeds of 150 μm/s with a maximal temperature increase in the local spot of 10 K. These experiments confirmed the notion that the fluid motion was the result of the dynamic thermal expansion in the gradient of liquid viscosity. The viscosity in the spot is reduced by the increase in local temperature, resulting in a broken symmetry between the thermal expansion and thermal expansion in the front and wake of the spot. Consequently, the fluid is observed to move opposite to the spot direction due to the asymmetric thermal expansion in the spot front and corresponding asymmetric thermal contraction in the wake. Simulations and experiments showed that the light-driven pumping could reach speeds up to 150 mm/s for moderate heating.

The authors also developed a theoretical model of the 2D laminar flow based on a thin-film approximation to predict the pump velocity (Weinert and Braun 2008). The pump velocity was approximated as

$$v_{\text{pump}} = -\frac{3\sqrt{\pi}}{4} f \alpha \beta b \Delta T^2 \tag{13.3}$$

where
  f is the repetition frequency of the laser
  b is the spot width
  α is the expansion coefficient
  β is the temperature dependence of the viscosity
  ΔT is the amplitude of the temperature spot

The temperature was obtained by imaging a temperature-sensitive fluorescent dye dissolved in the water. The shape of the moving warm spot could be measured by stroboscopic illumination.

The theoretical model was verified by experiments (Weinert and Braun 2009) in the low-frequency range of the laser ($f \ll 1$ kHz). Furthermore, the model predicts a linear response of the pump velocity to both the thermal expansion $\alpha \Delta T_o$ and the change in temperature-dependent viscosity $\beta \Delta T_o$ for a similar-shaped temperature spot. If the spot temperature is increased with a higher laser power, then the pump velocity increases proportionally to ΔT^2.

In addition to controlling the dynamic flow of fluid through microchannels, the ability to transport dissolved molecules is crucial in many microfluidic and LoC applications. Weinert and Braun (2008) used this basic principle to pump nanoparticles a distance over millimeters through agarose gel. The experiment involved preparing test device by sandwiching two droplets containing initially warm low melting temperature agarose gel between two slides. One droplet contained a fluorescent labeled biomolecule. The fluorescent molecules were then pumped into three pockets of different areas and mixed with the surrounding dark liquid with volume ratios of 4:1, 1:1, and 1:4.

The light pump based on the principle of thermoviscous expansion was also used to move fluids through an unstructured environment (Weinert and Braun 2009, Weinert et al. 2009). In this work, an infrared laser was used to melt liquid channels into a sheet of ice. Experiments demonstrated that the liquid water in the ice sheet would refreeze behind the laser spot. The repetition of the laser spot motion with high frequencies enabled the water to undergo a series of melting and freezing cycles, increasing the velocity of the pumping action. The solid ice boundaries would confine the liquid flow such that the movement is from the back of the molten spot to the front. The pumping action is performed with a repetition rate of $f = 650$ Hz and a chamber temperature of $T_o = -10°C$. With densities $\rho_{water} = 1000$ kg/m^3 and $\rho_{ice} = 917$ kg/m^3, the pump velocity is $\upsilon_{pump} = 9.5$ mm/s. Weinert and his colleagues (Weinert and Braun 2009, Weinert et al. 2009) were also able to measure the pump velocity at 11 mm/s. The length of the molten spot depends on the temperature of the ice sheet. At low ice temperatures, only a short molten spot is formed. At higher ice temperatures, the molten spot can reach lengths beyond 500 μm with the pump velocity exceeding 50 mm/s.

13.3.4 Micropumping Using Optically Driven Bubbles

Bezuglyi and Ivanova (2007) introduced a gas-bubble piston concept to pump fluids through a microchannel. The bubble movement was induced by converting the energy of optical radiation into the kinetic energy of capillary motion under the thermal action of light on the fluid near the bubble formation. As these bubbles move, they conform to the shape of the channel and do not leave "stagnant" amounts of liquid behind. Gas-bubble methods for pumping small amounts of fluid are an attractive approach for some applications because these bubbles can be readily produced by microheaters.

The method of optically driven microfluidic pumping has several important advantages over electrical microheaters. First, the air bubble is initially at the temperature of the fluid making it possible to pump at small temperature differences. Second, there are no observed fluctuations in the fluid flow during pumping. It is also not necessary to provide a high-power pulse and then switch off the heater in expectation of the bubble collapsing. Third, the use of optical radiation energy is efficient because the light radiation is released directly in the fluid volume and converted into thermal energy for $\sim 10^{-12}$ s (Bezuglyi and Ivanova 2007). Furthermore, the energy supply does not require conductors with high electrical conductivity of a given cross section. The problem is that the cross-section of the electrical conductors cannot be arbitrarily small as is the case of resistive methods. The diameter and wavelength of the laser beam can be controlled to provide an optimal solution for the desired applications. The performance can be further improved by producing smooth microchannel walls, increasing the thermal insulation properties of the channel material, applying light radiation with a wavelength that can be efficiently absorbed by the fluid, and selecting a fluid with a high-temperature surface tension coefficient.

13.3.5 Opto-Pneumatic Microactuators, Valves, and Pumps

Many indirect optical actuation methods that take advantage of the heat generated by the light source to create the desired force or pressure (Hale et al. 1988, Hale et al. 1990, Hockaday and Waters 1990, Jones and McKenzie 1993, Liu and Jones 1989, McKenzie and Clark 1992, McKenzie et al. 1995) have also been incorporated in microfluidic systems. Recall that when a gas is heated, it expands according to the ideal gas law:

$$P_g V_g = nRT \tag{13.4}$$

where
    P_g is the gas pressure
    V_g is the gas volume
    n is the number of moles
    R is the gas constant (0.0821 L atm/mol K)
    T is the absolute temperature

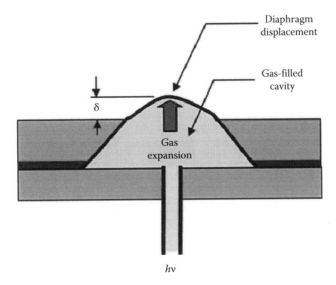

FIGURE 13.14 Light-driven diaphragm device based on optical heating of enclosed gas. The basic concept has been used for a variety of engineering applications. (Adapted from Hale, K.F. et al., *IEE Proc.*, 135(5), 348, 1988; Hale, K.F. et al., *Sens. Actuators.*, A21–A23, 207, 1990.)

In other words, when a simple gas confined to an enclosure is heated, it expands according to the ideal gas law which states that a change in fluid temperature produces a proportional change in pressure.

The optically actuated pump illustrated in Figure 13.14 exploits this simple principle to deflect a flexible diaphragm in order to perform mechanical work. The cavity is filled with a gas or oil that expands when heated from the light source. As the diaphragm expands under pressure, it produces the desired deflection δ (Tabib-Azar 1998). Mizoguchi et al. (1992) used this same simple concept to create a microcell that was of a predeflected 800 × 800 μm square membrane that was micromachined in 0.25 mm³ of silicon and filled with Freon 113. Freon 113 is a liquid with a boiling point of ~47.5°C. A carbon–wool absorber was placed inside the cell to help convert the 10 mW laser light into heat. The microcell exhibited a relatively large deflection, ~35 μm, when the cell's contents were heated and the Freon 113 has undergone a phase change from liquid to a gas. The fluid that is being transported by the pump is fed into a flow channel between the glass plate and deflecting membrane using very small harmonic movements. The micropump achieved a head pressure of ~30 mmag and flow rate of 30 nL per cycle. The small quantities of Freon in each cell allowed relatively low optical powers to be used to change the phase of the liquid to gas, giving the large membrane deflections needed to operate the pump.

13.3.6 LASER-DRIVEN MICROMIXING

Analytical microsystems often use molecular diffusion to mix small quantities of different liquids. However, this passive mixing process requires relatively long microchannels which impose design restrictions on the physical dimensions of the fluidic network. An active micromixer driven by a focused laser beam can be used to shorten the length of the mixing channels (Shiu et al. 2010b). The solution improves the mixing rate by using laser radiation to heat the disparate fluids being transported through the channels. The operating principle is based on the observation that the rate of molecular diffusion for nonreactive fluids increases with elevated temperatures. The advantage of using laser-assisted micromixing mechanism is that it requires only simple microchannel structures, without the need of complicated multiple split–join microchannels to improve mixing rate (Branebjerg et al. 1996, Munson and Yager 2004), or the specially designed 3D microchannel structures to create the chaotic advection type of passive mixing (Mizuno and Funakoshi, 2002, Jen et al 2003, Chen and Meiners 2004, Park et al. 2004, Wang et al. 2005).

Laminar flow predominates over turbulence in microscale liquid flow streams where the Re number is small, Re < 100. When Re is small, molecular diffusion becomes the primary mixing mechanism. Under this condition, the mixing action can be represented by Fick's law of molecular diffusion (Nguyen and Wereley 2002):

$$\tau = \frac{d^2}{2 \cdot D} \tag{13.5}$$

where
  τ is the time required to complete the diffusion (s)
  d is the channel width (μm)
  D is the diffusion coefficient (μm²/s)

The channel length required for mixing to be completed can be approximately described as

$$L = U \cdot \tau \tag{13.6}$$

where
  L is the microchannel length in μm
  τ is the diffusion time in seconds
  U is the average flow velocity in μm/s

The higher average flow rate U implies that a longer channel is required to complete the mixing. The diffusion coefficient D is defined as

$$D = \frac{k_B \cdot T}{6 \cdot \pi \cdot \mu \cdot R_A} \tag{13.7}$$

where
  k_B is the Boltzman constant (1.38×10^{-23} J/K)
  T is the system temperature (K)
  μ is the dynamic viscosity of the pure solvent
  R_A is the radius of the solute particle (Bird et al. 2002).

The diffusion coefficient D has a direct proportional relationship with the temperature T. As the temperature increases, the amount of energy available to the particles in the flow stream also increases causing them to move faster and, thereby, increasing the diffusion rate. A higher diffusion rate implies the mixing of the streams can be completed in a shorter period of time τ or over a shorter microchannel length.

Figure 13.15 shows a simple Y-channel micromixer designed with a larger chamber about 450 μm wide for the experiments (Shiu et al. 2010b). The beam from a 1 mW, 670 nm laser was focused on the microchannel using a 100 mm focal length objective lens. The laser-assisted mixing of the test fluids showed a 36.4% increase in the average diffusion coefficient value with 1–10 μL/min flow rates. The maximum percentage difference of diffusion distances had increased by ~7.85% over the non-laser-assisted conditions.

Although the laser wavelength used in these experiments did not have the highest absorption wavelength for water, the selected wavelength prevented the water from overheating and destroying biological samples at an elevated temperature. In fact, for some application, MEMS engineers can select a laser wavelength that is invisible to the targeted biological samples to avoid laser radiation

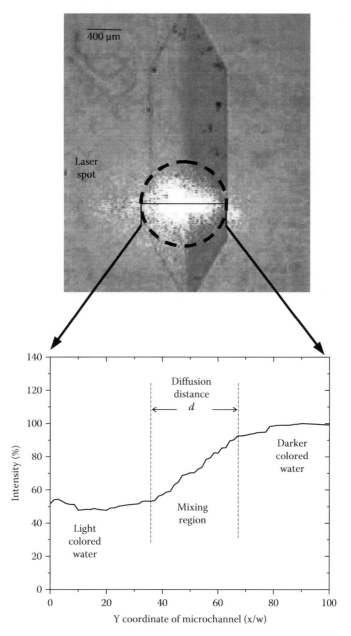

FIGURE 13.15 An optically assisted active micromixer where a focused laser beam delivers a small amount of energy that is absorbed by a microfluidic stream and alters the localized diffusion coefficient enabling increased mixing over a shorter channel distance.

damages, yet the selected wavelength is able to increase the buffer solution temperature appropriately to increase the mixing rate in order to shorten the analysis time.

13.4 LIGHT-DRIVEN POLYMER GEL MICROACTUATORS AND VALVES

An alternative approach to manipulating fluid flow patterns in microstrucures is to take advantage of materials that can undergo shape changes when exposed to very specific stimuli.

Functional microstructures constructed from environmentally sensitive polymers such as hydrogels have been proposed for a variety of applications in microfluidic systems and biomedical devices (Watanabe et al. 2002). Hydrogels undergo large conformational changes causing swelling/deswelling when exposed to a variety of environmental stimuli including changes in pH, temperature, and light.

Among the environmentally sensitive gels, pH-sensitive ionic hydrogels are one of the most commonly studied and often least intrusive for biomedical applications. One material that has studied extensively for microvalves and micropumps in fluidic systems is the pH-sensitive hydrogel containing hydroxyethyl methacrylate- acrylic acid (HEMA-AA). This environmentally sensitive hydrogel undergoes abrupt volumetric changes when the pH of the surrounding medium increases slightly above the phase transition point pKa. If the networked gel is immersed in an ionic aqueous solution, then the polymer chains will absorb water and the *association, dissociation,* and *binding* of the various ions to the chains will cause the hydrogel material to swell, producing an usable microforce. The expansion and contraction of the hydrogel, Figure 13.16, under environmental stimuli have been used to regulate the flow of liquids in a variety of microfluidic systems (Liu et al. 2002, Baldi et al. 2003). The advantages of hydrogel over other deformable microactuator shells are relatively simple fabrication, no external power requirements, no integrated electronics, significant displacements (up to 185 μm), and relatively large force generation (∼22 mN) (Al-Aribe and Knopf 2010).

A variety of LoC components such as microvalves and micropumps can be created using hydrogels as the actuating shell. To create these active components on microfluidic platforms, researchers have developed a number of different fabrication techniques including selectively photo-polymerizing the hydrogel around posts inside microchannels to adjust the liquid flow based on changes in the surrounding solution (Lui et al. 2002b). An alternative approach was proposed by Yu et al (2001) where two hydrogel bistrips were attached to a microchannel and changed their volume and shape based on local pH differences. More sophisticated microvalve designs have also been proposed

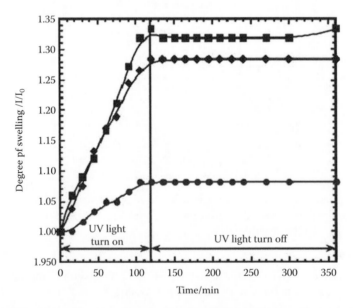

FIGURE 13.16 Degree of swelling versus time for AA/AAm/MBAAm hydrogel in water. AA content in gel: 1.97 mol% (●), 4.43 mol% (■), 9.09 mol% (◆). (From Watanabe, T., Akiyama, M., Totani, K., Kuebler, S.M., Stellacci, F., Wenseleers, W., Braun, K., Marder, S.R., and Perry, J.W.: Photoresponsive hydrogel microstructure fabricated by two-photon initiated polymerization. *Adv. Funct. Mater.* 2002. 12(9). 611–614. Copyright Wiley-VCH Verlag GmbH & Co. KGaA. Reprinted with permission.)

where the flow inside a microchannel is controlled by the change in concentration of environmental stimuli in the solution flowing through an adjacent channel (Baldi et al. 2003). The two channels are separated by a thin flexible silicone membrane that deflects in response to the change in volume of the hydrogel in one channel. The deforming membrane then regulates opening and closing of an inlet hole in the neighboring channel. All these techniques control the flow in response to external stimuli which change the characteristics of the solution inside the microchannel.

Although the absence of any external power source has been promoted as an advantage of a smart gel, the response times of direct chemical to energy conversion are often impractical often requiring several hours for the gel to fully respond to changes in solution pH. Furthermore, using the pH as a control mechanism limits the number of components that can be independently controlled and requires all the components to be compatible with the requisite changes in pH (Sershen et al. 2005). The downside is that many microfluidic applications are not compatible with a large range of pH values. In other applications it is beneficial to have independent external control.

Light has also been used to induce the phase transition in polymer gels (Suzuki and Tanaka 1990, Hirasa 1993, Watanabe et al. 2002, Sershan et al. 2005). For example, illumination by UV light initiates an ionization reaction in photosensitive gels creating internal osmotic pressure which induces swelling. This process is slow compared with the illumination with whiter light that causes the phase transformation by heating the gel. Very large volume changes occur in N-isopropylacrylamide/chlorophine copolymer gels as a function of optical power at 31.5°C (Suzuki and Tanaka 1990). This phenomenon can be used in microactuators to generate large displacements.

UV radiation has been found to induce significant swelling in hydrogels. Watanabe et al. (2002) showed that a hydrogel prepared from a comonomer solution containing acryloacetone, acrylamide, and N,N'-methylene bisacrylamide (AA/AAm/MBAAm) would exhibit significant change in shape if exposed to a 244 nm light source. Figure 13.17a shows the degree of swelling over time for the hydrogel in water. The degree of swelling is given as the ratio of length of the gel cylinder at a given time (L) to the initial cylinder length (l_0). When exposed to illumination at 244 nm, the degree of swelling increased with time but the dimension of the gel did not change under dark conditions after exposure. This observation confirmed that the swelling of the hydrogel was reversible. All gels

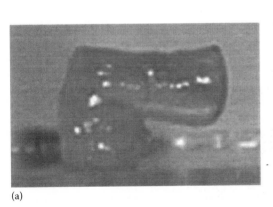

(a)   (b)

FIGURE 13.17 (a) An optical microscopy image of a microcantilever in water prior to UV irradiation. The microcantilever was fabricated on a fused silica substrate. (b) An optical microscopy image of a microcantilever in water after UV irradiation. The microcantilever was irradiated with UV light through the fused silica substrate. (From Watanabe, T., Akiyama, M., Totani, K., Kuebler, S.M., Stellacci, F., Wenseleers, W., Braun, K., Marder, S.R., and Perry, J.W.: Photoresponsive hydrogel microstructure fabricated by two-photon initiated polymerization. *Adv. Funct. Mater.* 2002. 12(9). 611–614. Copyright Wiley-VCH Verlag GmbH & Co. KGaA. Reprinted with permission.)

studied by the authors confirmed that the AA/AAm/MBAAm gels would experience significant swelling within 120 min, a relatively slow process. The equilibrium time, t_{eq}, of the hydrogel to undergo swelling, under diffusion-limited conditions (Watanabe et al. 2002), is

$$t_{eq} \propto \frac{L^2}{D} \tag{13.8}$$

where

L is the gel dimension
D is the solvent diffusion constant of the gel

In other words, the response time can be significantly shortened by reducing the size of the hydrogel. Watanabe et al. (2002) examined the behavior of the AA/AAm/MBAAm hydrogel by constructing a microcantilever in water, Figure 13.17b. A collimated UV light beam was directed to the underside of the beam and the microstructure deflected upward by ~45°. The 244 nm, 3 mW cm^{-2} light source was applied for 20 min.

An alternative approach is to have hydrogel microstructures that are responsive to the exposure of light at specific wavelengths (Sershen et al. 2005). To accomplish this, Sershan et al. (2005) introduced an opto-mechanically responsive nanocomposite hydrogel that has undergone pronounced and reversible changes in shape when exposed to different wavelengths of light (Figure 13.18). The materials are composites of a thermally responsive polymer (poly[N-isopro-pylacrylamide-co-acrylamide] with a 95:5 comonomer ratio) and particles that have distinct optical absorption profiles. Duff and Baiker (1993) used gold colloids for this purpose, while Oldenburg et al. (1998) used gold nanoshells with a 110 nm diameter silica core and 10 nm thick gold shell. The composites were formed by mixing the nanoparticles with the monomer solution thereby trapping the particles within the hydrogel matrix after polymerization. These nanocomposite materials responded to different wavelengths of light.

Sershan et al. (2005) further demonstrated the independent control of two valves along the T-junction as shown in Figure 13.19. One is constructed from gold colloid nanocomposite hydrogel and the other is gold nanoshell nanocomposite hydrogel. The channels are 100 μm wide. The entire device is illuminated with 532 nm, 1.6 W/cm^2 (green light). The gold colloid valve opened while the nanoshell valve remained closed. However, when the device is illuminated with 832 nm, 2.7 W/cm^2 (near infrared) light, then the opposite response was observed. In both cases the valves opened within 5 s.

13.5 OPTICALLY DRIVEN PROTON PUMPS

Monitoring and directly regulating the pH of target solutions are an important step in controlling a variety of biological and chemical processes on LoC and μTAS devices. The nano- and microscale control of pH has enabled new methods to be developed for on-chip protein identification, the transfer of large molecules (Suzurikawa et al. 2010), and ion exchange chromatography (Pabst and Carta 2008). Other LoC applications that depend on generated pH gradients are the fractionation, separation, and assembly of biologically based molecules such as the human salivary proteins and the collagen bundles (Cheng et al. 2008). From an engineering perspective, optically manipulating the pH of a solution in a microchannel, or reservoir, provides a mechanism for activating and controlling an ionic hydrogel microactuator, or the color of a phenolphthalein indicator dye for manually monitoring acidity changes in sample solutions (Al-Aribe and Knopf 2010, Al-Aribe et al. 2011a).

One potentially useful light-driven transducer that generates measureable pH gradients in very small quantities of ionic solutions is shown in Figure 13.20. The microscale planar transducer

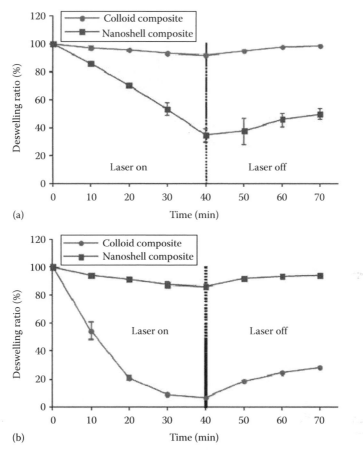

FIGURE 13.18 The collapse and reswelling of the gold colloid composite hydrogels (circles) and the gold nanoshell composite hydrogels (squares) during and after 40 min of irradiation at (a) 832 nm (2.7 W/cm²) and (b) 532 nm (1.6 W/cm²). Data are the mean deswelling ratios and the error bars are standard errors of the mean. (From Sershen, S.R., Mensing, G.A., Ng, M., Halas, N.J., Beebe, D.J., and West, J.L.: Independent optical control of microfluidic valves formed from optomechanically responsive nanocomposite hydrogels. *Adv. Mater.*, 2005. 17. 1366–1368. Copyright Wiley-VCH Verlag GmbH & Co. KGaA. Reprinted with permission.)

converts an optical signal into a flow of hydrogen ions from a reservoir to a target solution. The transducer or pH-gradient generator exploits the molecular proton pumps found in the purple membranes (PM) of wild type bacteriorhodopsin (bR) (Hampp 2000, Lanyi 2006, Wang et al. 2005, Al-Aribe et al. 2011b). The photo-electro-chemical transducer is an ultrathin layer (~13 nm) of oriented PM patches self-assembled on an Au-coated porous substrate. A biotin labeling and streptavidin molecular recognition technique are used to ensure that the extracellular side of all PM patches is attached to the porous substrate, enabling unidirectional and efficient transport of ions across the transducer surface. The photo-induced proton pumps generate a flow of ions that produce a measureable change in pH between the separated solutions. The self-assembly procedure is experimentally quantified based on the capacitance characteristics of the bR membranes.

An investigation by Al-Aribe et al. (2011b) demonstrated that the transducer was covered with the bR proton pumps at a mass density of 2.33 ng/cm². Experimental tests also showed that the proposed transducer could repeatedly generate pH gradients as high as 0.42 and absolute voltage differences as high as 25 mV when illuminated by an 18 mW, 568 nm light source. Furthermore, the ΔpH is observed to be nonlinear with respect to light intensity and exposure time. The ΔpH of

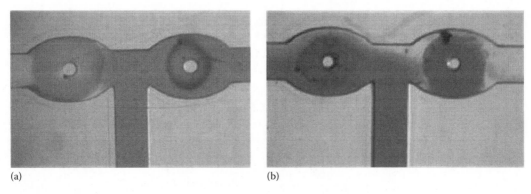

(a)   (b)

FIGURE 13.19 Two valves formed at a T-junction in the microfluidic device, one made of gold colloid nanocomposite hydrogel and the other of gold nanoshell nanocomposite hydrogel. The channels are 100 μm wide. (a) When the entire device is illuminated with green light (532 nm, 1.6 W/cm²) the gold colloid valve opened while the nanoshell valve remained closed. (b) However, when the device was illuminated with near infrared light (832 nm, 2.7 W/cm²), the opposite response was observed. In both cases the valves opened for 5 s. (From Sershen, S.R., Mensing, G.A., Ng, M., Halas, N.J., Beebe, D.J., and West, J.L.: Independent optical control of microfluidic valves formed from optomechanically responsive nanocomposite hydrogels. *Adv. Mater.*, 2005. 17. 1366–1368. Copyright Wiley-VCH Verlag GmbH & Co. KGaA. Reprinted with permission.)

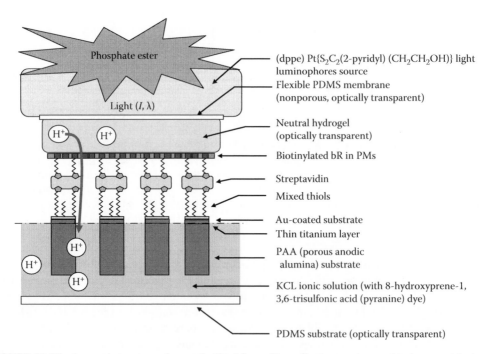

FIGURE 13.20 Layered structure of an optically driven pH-gradient generator used to transport hydrogen ions (H⁺) between two adjacent ionic solutions separated by a porous substrate.

the target solution is sufficient to cause a phenolphthalein indicator dye to change color or an ionic hydrogel microvalve to expand.

A simple demonstration of the bR-PAA transducer's ability to generate significant changes in target solution pH is shown in Figure 13.21. The ionic solution in $R1$ contains the 8-hydroxyprene-1,3,6-trisulfonic acid (pyranine) dye (Al-Aribe and Knopf 2010, Al-Aribe et al. 2011a). When an 18 mW,

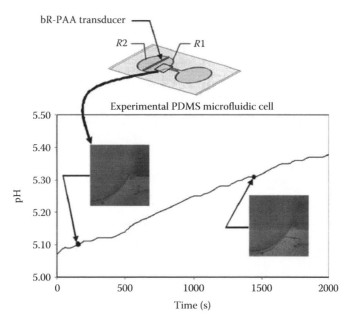

FIGURE 13.21 **(See color insert.)** The change in pH of the target solution is sufficient to cause a phenol-phthalein indicator dye to *change* color. The optically driven proton pumps of the bR-PAA pH gradient generator cause the solution in R1 to transform from a light greenish blue (pH 5.10) to a darkened, more intense blue (pH 5.31).

568 nm light source strikes the biofunctionalized surface of the transducer, the photon-activated bR proton pumps transport H⁺ ions from *R*1 to *R*2 and changed the color of the target electrolyte solution in *R*1 from a pale yellow (pH 5.10) to blue (5.31). The color change indicates a reduction in H⁺ ions and, thereby, a lowering of the solution's overall acidity.

13.6 SUMMARY

In recent years, the integration of optical technologies in fluidic systems has led to the development of a variety of LoC and microanalysis μTAS for chemical analysis, environmental monitoring, and medicine. Light has been used to directly control microflows in transport channels by manipulating micro-objects in the fluid stream, altering the wetting of surfaces, inducing thermoviscous expansion, and introducing gas bubbles that push fluids along the channel. Alternatively, indirect optical methods have also been used heat gases and liquids that undergo thermal expansion and produce an external pressure on thin mechanical films and flexible diaphragms. In a very simple sense, light can be used to heat a small region of a liquid to induce molecular diffusion for enhanced mixing of adjacent parallel streams. Light-sensitive polymers such as hydrogels can be fabricated into deformable microstructures that can be used to control flow through microchannels.

Sometimes the goal is not to transport the fluid, but rather change a physical property of the liquid such as pH. In this regard, optically driven proton pumps that transport H⁺ ions from adjacent reservoirs are created to regulate ion concentrations in the target solution. Although this chapter represents only a brief introduction to light activated micro- and nanofluidic technologies, it does provide context for a rapidly evolving discipline. The key to developing a successful micro- or nanofluidic system is the exploitation of material properties and the utilization of recent fabrication techniques that enable precision manufacturing at the micro- and nanoscale.

REFERENCES

Al-Aribe, K.M. and Knopf, G.K. 2010. Photoresponsive hydrogel microvalve activated by bacteriorhodopsin proton pumps. *Proc. SPIE* 7646: 11-1–11-12.

Al-Aribe, K.M., Knopf, G.K., and Bassi A.S. 2011a. Fabrication of an optically driven pH gradient generator based on self-assembled proton pumps. *Microfluid Nanofluid* 12: 325–335. (Available online: DOI 10.1007/s10404-011-0876-4)

Al-Aribe, K.M., Knopf, G.K., and Bassi A.S. 2011b. Photoelectric monolayers based on self -assembled and oriented purple membrane patches. *IEEE/ASME J. MEMS* 20(4): 800–810.

Arai, F., Yoshikawa, K., Sakami, T., and Fukuda, T. 2004. Synchronized laser micromanipulation Synchronized laser micromanipulation of multiple targets along each trajectory by single laser. *Appl. Phys. Lett.* 85(19): 4301–4303.

Ashkin, A. 1970. Acceleration and trapping of particles by radiation pressure. *Phys. Rev. Lett.* 24: 156–159.

Baldi, A., Gu, Y., Loftness, P.E., Siegel, R.A., and Ziaie, B. 2003. A hydrogel-actuated environmentally sensitive microvalve for active flow control. *J. Microelectromech. Syst.* 12(5): 613–621.

Bezuglyi, B.A. and Ivanova, N.A. 2007. Pumping of a fluid through a microchannel by means of a buble driven by a light beam. *Fluid Dyn.* 42(1): 91–96.

Bird, R.B., Stewart, W.E., and Lightfoot, N.E. 2002. *Transport Phenomen*, 2nd edn., John Wiley & Sons, Inc., New York.

Branebjerg, J., Graveson, P., Krog, J., and Nielsen, C. 1996. Fast mixing by lamination. In *Proceedings of 9th IEEE International Workshop Micro Electromechanical System (MEMS'96)*, San Diego, CA, pp. 441–446.

Chen, Z., Mauk, M.G., Wang, J., Abrams, W.R., Corstjens, P.L.A.M., Niedbala, R.S., Malamud, D., and Bau, H.H. 2007. A microfluidic system for saliva-based detection of infectious diseases. *Ann. NY Acad. Sci.* 1098: 429–436.

Chen, H. and Meiners, J.C. 2004. Topologic mixing on a microfluidic chip. *Appl. Phys. Lett.* 84: 2193–2195.

Cheng, X., Gurkan, U.A., Dehen, C.J., Tate, M.P., Hillhouse, H.W., Simpson, G.J., and Akkus, O. 2008. An electrochemical fabrication process for the assembly of anisotropically oriented collagen bundles. *Biomaterials* 29: 3278–3288.

Chiou, P.Y., Moon, H., Toshiyoshi, H., Kim, C.-J., and Wu, M.C. 2003. Light actuation of liquid by optoelectrowetting. *Sens. Actuators A* 104: 222–228.

Dholakia, K. and Reece, P. 2006. Optical micromanipulation takes hold. *Nano Today* 1(1): 18–27.

Dittrich, P.S. and Manz, A. 2006. Lab-on-a-chip: Microfluidics in drug discovery. *Nat. Rev.* 5: 210–218.

Duff, D.G. and Baiker, A. 1993. A new hydrosol of gold clusters. 1. Formation and particle size variation. *Langmuir* 9: 2301–2309.

Eijkel, J.C.T. and van den Berg, A. 2005. Nanofluidics: What is it and what can we expect from it. *Microfluid. Nanofluid.* 1: 249–267

El Moctar, A.O., Aubry, N., and Batton, J. 2003. Electro–hydrodynamic micro-fluidic mixer. *Lab Chip* 3: 273–280.

Erickson, D. 2008. Special issue on "optofuidics." *Microfluid. Nanofluid.* 4: 1–2.

Fujii, T., Sando, Y., Higashino, K., and Fujii,Y. 2003. A plug and play microfluidic device. *Lab Chip* 3: 193–197.

Grilli, S., Miccio, L., Vespini, V., Finizio, A., de Nicola, S., and Ferraro, P. 2008. Liquid micro-lens array activated by selective electrowetting on lithium niobate substrates. *Opt. Express* 16(11): 8084 (10pp).

Groenendijk, M.N.W. 2008. Fabrication of super hydrophobic surfaces by fs laser pulses. *Laser Tech. J.* 3: 44–47.

Hale, K.F., Clark, C., Duggan, R.F., and Jones, B.E. 1988. High-sensitivity optopneumatic converter. *IEE Proc.* 135(5): 348–352.

Hale, K.F., Clark,C., Duggan, R.F., and Jones, B.E. 1990. Incremental control of a valve actuator employing optopneumatic conversion. *Sens. Actuators* (A21–A23): 207–210.

Hampp N. 2000. Bacteriorhodopsin as a photochromic retinal protein for optical memories. *Chem. Rev.* 100: 1755–1776.

He, B., Burke, B.J., Zhang, X., Zhang, R., and Regnie, F.E. 2001. A picoliter–volume mixer for microfluidic analytical systems. *Anal. Chem.* 73: 1942–1947.

Hessel,V., Lowe, H., and Schonfeld, F. 2005. Micromixers—A review on passive and active mixing principles. *Chem. Eng. Sci.* 60: 2479–2501.

Higurashi E., Ohguchi O., Tamamura T., Ukita H., and Sawada R. 1997. Optically induced rotation of dissymmetrically shaped fluorinated polyimide micro-objects in optical traps. *J. Appl. Phys.* 82(6): 2773–2779.

Hirasa, O. 1993. Research trends of stimuli-responsive polymer hydrogels in Japan. *J. Intell. Mater. Syst. Struct.* 4: 538–542.

Hockaday, B.D. and Waters, J.P. 1990. Direct optical-to-mechanical actuation. *Appl. Opt.* 29(31): 4629–4632.

Hong, C.C, Choi, J.W., and Ahn, C.H. 2004. A novel in-plane microfluidic mixer with modified Tesla structures. *Lab Chip*, 4: 109–113.

Hung, P.J., Lee, P.J., Sabounchi, P., Lin, R., and Lee, L.P. 2005. Continuous perfusion microfluidic cell culture array for high-throughput cell-based assays. *Biotech. Bioeng.* 89: 1–8.

Ichimura, K., Oh, S.K., and Nakagawa, M. 2000. Light-driven motion of liquids on a photoresponsive surface. *Science* 288: 1624–1626.

Ismagilov, R.F., Stroock, A.D., Keins, P.J.A., Whiteside, G., and Stone, H.A. 2000. Experimental and theoretical scaling laws for transverse diffusive broadening in two-phase laminar flows in microchannel. *Appl. Phys. Lett.* 76: 2376–2378.

Jen, C.P., Wu, C.Y., Lin, Y.C., and Wu, C.Y. 2003. Design and simulation of the micromixer with chaotic advection in twisted microchannels. *Lab Chip* 3: 77–81.

Jones, B.E. and McKenzie, J.S. 1993. A review of optical actuators and the impact of micromachining. *Sens. Actuators A* (37–38): 203–207.

Kamholz, A.E., Weigl, B.H., Finlayson, B.A., and Yager, P. 1999. Quantitative analysis of molecular interactive in microfluidic channel: The T-sensor. *Anal. Chem.* 71: 5340–5347.

Kamholz, A.E. and Yager, P. 2002. Molecular diffusive scaling laws in pressure-driven microfluidic channels: Deviation from one-dimensional Einstein approximations. *Sens. Actuators B* 82: 117–121.

Karlsson, M., Davidson, M., Karlsson, R., Karlsson, A., Bergenholtz, J., Konkoli, Z., Jesorka, A., et al. 2004. Biometric nanoscale reactors and networks. *Annu. Rev. Phys. Chem.* 55: 613–649.

Knopf, G.K. 2006. Optical actuation and control. In *Handbook of Optoelectronics*, Dakin, J. and Brown, R.G.W. (Eds.), CRC Press, Boca Raton, FL, pp. 1453–1479.

Krogmann, F., Qu, H., Mönch, W., and Zappe, H. 2008. Push/pull actuation using opto-electrowetting. *Sens. Actuators A* 141: 499–505.

Kuo, T.C., Kim, H.K., Cannon, D.M., Shannon, M.A., Sweedler, J.V., and Bohn, P.W. 2004. Nanocapillary arrays effect mixing and reaction in multilayer fluidic structures. *Angew. Chem. Int. Ed.* 43: 1862–1865.

Lanyi, J.K. 2006. Proton transfers in the bacteriorhodopsin photocycle. *Biochim. Biophys. Acta* 1757: 1012–1018.

Liu, K. and Jones, B.E. 1989. Pressure sensors and actuators incorporating optical fiber links. *Sens. Actuators* 17: 501–507.

Liu, R.H., Yang, J., Pindera, M.Z., Athavale, M., and Grodzinski, P. 2002a. Bubble-induced acoustic micromixing. *Lab Chip* 2: 151–157.

Liu, R.H., Yu, Q., and Beebe, D., 2002b. Fabrication and characterization of hydrogel-based microvalves. *J. Microelectromech. Syst.* 11(1): 45–53.

Maruo, S. and Inoue, H. 2006. Optically driven micropump produced by three-dimensional two-photon microfabrication. *Appl. Phys. Lett.* 89: 144101 (3pp).

Maruo, S. and Inoue, H. 2007. Optically driven viscous micropump using a rotating microdisk. *Appl. Phys. Lett.* 91: 084101 (3pp).

Maruo, S., Takaura, A., and Inoue, H. 2009. Optically driven micropump with a twin spiral microrotor. *Opt. Express* 17(21): 18525–18532.

McKenzie, J.S. and Clark, C. 1992. Highly sensitive micromachined optical-to-fluid pressure converter for use in an optical actuation scheme. *J. Micromech. Microeng.* 2: 245–249.

McKenzie, J.S., Hale, K.F., and Jones, B.E. 1995. Optical actuators. In *Advances in Actuators*, Dorey, A.P. and Moore, J.H. (Eds.), IOP Publishing Ltd., Bristol, U.K., pp. 82–111.

Mijatovic, D., Eijkel, J.C.T., and van den Berg, A. 2005. Technologies for nanofluidic systems: Top-down vs bottom-up—A review. *Lab Chip* 5: 492–500.

Mizoguchi, H., Ando, M., Mizuno, T., Takagi, T., and Nakajima, N. 1992. Design and fabrication of light driven pump. In MEMS'92, Travemunde, Germany, Feb.4–7, 1992, pp. 31–36. DOI: 10.1109/MEMSYS. 187686

Mizuno, Y. and Funakoshi, M. 2002. Chaotic mixing due to a spatially periodic three-dimensional flow. *Fluid Dyn. Res.* 31: 129–149.

Monat, C., Domachuk, P., and Eggleton, B.J. 2007. Integrated optofluidics: A new river of light. *Nat. Photonics* 1: 106–114.

Monat, C., Domachuk, P., Grillet, C., Collins, M., Eggleton, B.J., Cronin-Golomb, M., Mutzenich, S., Mahmud, T., Rosengarten, G., and Mitchell, A. 2008. Optofluidics: A novel generation of reconfigurable and adaptive compact architectures. *Microfluid. Nanofluid.* 4: 81–95.

Munson, M.S. and Yager, P. 2004. Simple quantitative optical method for monitoring the extent of mixing applied to a novel microfluidic mixer. *Anal. Chim. Acta* 507: 63–71.

Nguyen, N.T. and Wereley, S.T. 2002. *Fundamentals and Applications of Microfluidics Fabrication Techniques for Microfluidics*. Artech House, Boston, MA.

Nguyen, N.T. and Wu, Z. 2005. Micromixers—A review. *J. Micromech. Microeng.* 15: R1–R16.

Oldenburg, S.J., Averitt, R.D., Westcott, S.L., and Halas, N.J. 1998. Nanoengineering of optical resonances. *Chem. Phys. Lett.* 288: 243–247.

Ozkan, M., Wang, M., Ozkan, C., Flynn, R., Birkbeck, A., and Esener, S. 2003. Optical manipulation of objects and biological cells in microfluidic devices. *Biomed. Microdevices* 5(1): 61–67.

Pabst, T.M. and Carta, G. 2008. Separation of protein charge variants with induced pH gradients using anion exchange chromatographic columns. *Biotechnol. Progr.* 24: 1096–1106.

Padgett, M. and Di Leonardo, R. 2011. Holographic optical tweezers and their relevance to lab on chip devices. *Lab Chip* 11: 1196–1205.

Park, S.J., Kim, J.K., Park, J., Chung, S., Chung, C., and Chang, J.K. 2004. Rapid three-dimensional passive rotation micromixer using the breakup process. *J. Micromech. Microeng.* 14: 6–14.

Park, S.Y., Teitell, M.A., and Chiou, E.P.Y. 2010. Single-sided continuous optowetting (SCOEW) for droplet manipulation with light patterns. *Lab Chip* 10: 1655–1661.

Pennathur, S., Eijkel, J.C.T., and van den Berg, A. 2007. Energy conversion in microsystems: Is there a role for micro/nanofluidics. *Lab Chip* 7: 1234–1237.

Pollack, M.G., Fair, R.B., and Shenderov, A.D. 2000. Electrowetting-based actuation of liquid droplets for microfluidic applications. *Appl. Phys Lett.* 77(11): 1725–1726.

Psaltis, D., Quake, S.R., and Yang, C. 2006. Developing optofluidic technology through the fusion of microfluidics and optics. *Nature* 442: 381–386.

Sammarco, T.S. and Burns, M.A. 1999. Thermocapillary pumping of discrete drops in icrofabricated analysis devices. *AIChE J.* 45(2): 350–366.

Sershen, S.R., Mensing, G.A., Ng, M., Halas, N.J., Beebe, D.J., and West, J.L. 2005. Independent optical control of microfluidic valves formed from optomechanically responsive nanocomposite hydrogels. *Adv. Mater.* 17: 1366–1368.

Shiu, P.P., Knopf, G.K., and Ostojic, M., 2010a. Fabrication of metallic molds for microfluidic devices via laser micromachining and microEDM. *Microsyst. Technol.* 16(3): 477–485.

Shiu, P.P., Knopf, G.K., and Ostojic, M. 2010b. Laser-assisted active microfluidic mixer. In *2010 International Symposium on OptomechatronicTechnologies (ISOT),* Toronto, Ontario, Canada, October 25–27, pp. 1–5 (DOI: 10.1109/ISOT.2010.5687346).

Stavis, S.M., Strychalski, E. A., and Gaitan, M. 2009. Nanofluidic structures with complex three-dimensional surfaces. *Nanotechnology* 20: 165302 (7pp).

Suzuki, H. and Ho, C.M. 2002. A magnetic force driven chaotic micro-mixer. In *Proceedings on MEMS'02, 15th IEEE International Workshop on Micro Electromechanical System,* Las Vegas, NV, pp. 40–43.

Suzuki, A. and Tanaka, T. 1990. Phase transition in polymer gels induced by visible light. *Nature* 346: 345–347.

Suzurikawa, J., Nakao, M., Kanzaki, R., and Takahashi, H. 2010. Microscale pH gradient generation by electrolysis on a light-addressable planar electrode. *Sens. Actuators B* 149: 205–211.

Tabib-Azar, M. 1998. *Microactuators: Electrical, Magnetic, Thermal, Optical, Mechanical, Chemical, and Smart Structures*. Kluwer Academic, Norwell, MA.

Tang, Z., Hong, S., Djukic, D., Modi, V., West, A.C., Yardley, J., and Osgood, R.M. 2002. Electrokinetic flow control for composition modulation in a microchannel. *J. Micromech. Microeng.* 12: 870–877.

Terray, A., Oakley, J., and Marr, D.W.M. 2002. Microfluidic control using colloidal devices. *Science* 296: 1841–1844.

Tsai, J.H. and Lin, L. 2002. Active microfluidic mixer and gas bubble filter driven by thermal bubble pump. *Sens. Actuators A* 97–98: 665–671.

Voldman, J., Gray, M.L., and Schmidt, M.A. 2000. An integrated liquid mixer/valve. *J. Microelectromech. Syst.* 9: 295–302.

Wang, H., Iovenitti, P., Harvey, E., and Masood, S. 2002. Optimizing layout of obstacles for enhanced mixing in microchannels. *Smart Mater. Struct.* 11: 662–667.

Wang, W.W., Knopf, G.K., and Bassi, A. 2005. Photoelectric properties of a detector based on dried bacteriorhodopsin film. *Biosens. Bioelectron.* 21: 1309–1319.

Watanabe, T., Akiyama, M., Totani, K., Kuebler, S.M., Stellacci, F., Wenseleers, W., Braun, K., Marder, S.R., and Perry, J.W. 2002. Photoresponsive hydrogel microstructure fabricated by two-photon initiated polymerization. *Adv. Funct. Mater.* 12(9): 611–614.

Weinert, F.M. and Braun, D. 2008. Optically driven fluid flow along arbitrary microscale patterns using thermo-viscous expansion. *J. Appl. Phys.* 104:104701–1047010 (DOI 10.1063/1.3026526).

Weinert, F. and Braun, D. 2009. *Light Driven Microfluidics- Pumping Water Optically by Thermoviscous Expansion*. Optomechatronic Technologies, Istanbul, Turkey, September 21–23, pp. 383–386.

Weinert, F.M., Wühr, M., and Braun, D. 2009. Light driven microflow in ice. *Appl. Phys. Lett.* 94:113901-1–113901-3.

Whitesides, G.M. 2006. The origins and future of microfluidics. *Nature* 442: 368–373.

Wolfe, D.B., Conroy, R.S., Gartecki, P., Mayers, B.T., Fischbach, M.A., Paul, K.E., Prentiss, M., and Whitesides, G.M. 2004. Dynamic control of liquid-core/liquid cladding optical waveguides. *Proc. Natl. Acad. Sci. USA* 101(34): 12434–12438.

Yu, Q., Bauer, J.M., Moore, J.S, and Beebe, D.J. 2001. Responsive biomimetic hydrogel valve for microfluidics. *Appl. Phys. Lett.* 78(17): 2589-2591.

14 Optical NEMS and MEMS

Chengkuo Lee and Kah How Koh

CONTENTS

14.1   INTRODUCTION

In the late 1990s and early 2000, significant progress in the optical microelectromechanical systems (MEMS) technology, alongside with the development of wavelength-division multiplexing (WDM) systems, has been made in the telecommunication industry. Enormous investments have been made on optical MEMS technology as it has been recognized to be an indispensable technology meant to fulfill the missing link that can help connect other existing technologies to form an all-optical communication network. Many crucial MEMS-based components such as variable optical attenuator (VOA), optical switch, tunable switch, and reconfigurable optical add/drop multiplexer for telecommunication applications have been demonstrated and commercialized. Comprehensive review articles have been made [1–5], while topical reviews on optical switches [6], VOA [7], and tunable lasers [8] have been reported.

On the other hand, optical MEMS have also been conceptualized for a large range of image display applications such as projection displays [9,10], biomedical imaging [11,12], and retinal scanning [13,14]. In recent years, MEMS display has formed a circle of growing interest, with development of hand-held picoprojectors based on scanning mirror technology becoming an intriguing killer application in consumer electronics, game consoles, and automobiles [15]. The most well-known commercial product is probably the digital micromirror device (DMD), which is the core of digital light processing (DLP) projection technology developed by Texas Instrument in 1996. In 1994, O. Solgaard et al. developed the grating light valve (GLV) technology, providing an alternative implementation in commercial projectors.

With the advancement of process and lithography capabilities in the semiconductor industry, transistors with dimensions in the tens of nanometers can now be fabricated. This enhanced fabrication competency of modern semiconductor foundries has also assisted optical MEMS devices to extend their dimension scalability to the nanometer regime. Optical nanoelectromechanical system (NEMS) devices such as nanophotonics biosensors and nanophotonics nanomechanical sensors have since been widely reported in the past 5 years. Such integration of silicon photonic crystals (PCs) with NEMS-based structures and function elements opened up a new paradigm of sensors and devices that achieve outstanding performance in terms of sensitivity, size, and novelty. Details of various forms of optical NEMS devices will be discussed later in the chapter. The rest of the chapter will be on the review of major micromechanisms for actuations, with representative optical M/NEMS devices being introduced thereafter.

14.2   ACTUATION MECHANISMS IN MEMS DEVICES

Recent developments in the rapidly emerging discipline of MEMS show special promise in sensors, actuators, and micro-optical systems. In fact, optics is an ideal application domain for the MEMS technology as photons have no mass and are much easier to be actuated than other microscale objects. In conjunction with proper designed mirrors, lenses, and gratings, various micro-optical systems driven by microactuators can provide many unique functions in light manipulations such as reflection, beam steering, filtering, focusing, collimating and diffracting, etc. Despite the force and displacement generated by the microactuators being small, however with displacement generated by microactuator on the order of a wavelength, these features can help make micro-optical systems to be a promising application area for MEMS technology. In the next few paragraphs, we will introduce concepts for the four major actuation micromechanisms.

14.2.1   Electrostatic Actuation

Let us consider a parallel plate capacitor with fixed gap distance, g, between two plates with overlap area, A. The energy stored in this capacitor when subjected to an applied dc bias, V, is given as

$$W = \frac{1}{2} CV^2 = \frac{\varepsilon_o \varepsilon_r A V^2}{2g} \qquad (14.1)$$

and the attractive force generated between two plates is

$$F = \frac{dW}{dg} = \frac{\varepsilon_o \varepsilon_r A V^2}{2g^2} \qquad (14.2)$$

where
  ε_o is the vacuum permittivity
  ε_r is the relative permittivity

In electrostatic actuation, a typical configuration consists of a movable electrode connected to suspended mechanical springs, while a fixed electrode is anchored onto a substrate. When a voltage is applied to the capacitive electrodes, the electrostatic attractive force actuates the movable electrode to the stationary electrode, causing the area of overlap and the capacitance between the two electrodes to increase. As a result, the spring suspending the movable electrode is deformed. Thus, the displacement of movable electrode, Δx, is determined by the force balance between the springs restoring force and the electrostatic force.

There are two major types of electrodes in the electrostatic actuators: parallel plate and interdigitated comb. In the case of lateral driven interdigitated comb drive, where the direction of finger movement is parallel to the direction of fingers, i.e., x-direction, the electrostatic forced generated is

$$F = \frac{1}{2} V^2 \frac{dC}{dx} = N \frac{\varepsilon_o \varepsilon_r t}{g} V^2 \qquad (14.3)$$

where
  N is the number of comb electrode finger
  t is the comb electrode thickness
  g is the gap distance between a pair of comb fingers
  V is the actuation voltage

In the lateral driven comb-drive actuation setup, the force is independent on the displacement, unlike the parallel plate actuator setup. In addition, the force is inversely proportional to the gap distance, hence making the force generated to be much smaller than that of parallel plate actuator. This can be compensated by having more fingers and applying a higher voltage.

In general, the fundamental trade-off in optical MEMS using electrostatic actuators is the choice between suitable process technology and actuation mechanism. Parallel plate actuator and comb actuator are the available designs to be used in bulk micromachined optical MEMS devices, while polysilicon-based comb actuator, as shown in Figure 14.1, is often used in surface micromachined structures [16]. Briefly speaking, parallel plate actuation can provide very large force ($\sim$100 μN) with small displacement ($\sim$5 μm), but the force is highly nonlinear and instable within the displacement range. Interdigitated comb actuation provides a moderate level of force ($\sim$10 μN) with large displacement ($\sim$30 μm).

14.2.2 Thermal Actuation

Thermal actuation makes use of the thermal expansion of materials to achieve mechanical actuation. The thermal expansion of a solid material is characterized by coefficient of thermal expansion (CTE),

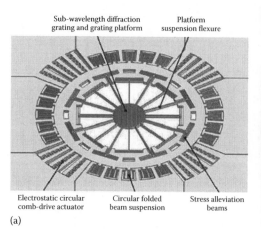

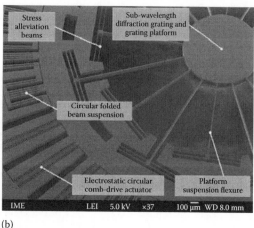

(a)   (b)

FIGURE 14.1 (a) Schematic of the micromachined 2-DOF electrical comb-driven circular resonator driven in-plane vibratory grating scanner. (b) SEM image showing the center part of the fabricated device. (From Du, Y. et al., *J. Microelectromech. Syst.*, 18(4), 892, 2009.)

α_T, and it has a unit of strain per change in temperature (K^{-1}). The CTE of a material is generally a function of temperature. With a small temperature change of ΔT, the introduced mechanical strain is defined as the product, $\alpha_T \cdot \Delta T$. One of the basic actuator structures for thermal actuation is a thermal bimorph, which consists of a cantilever with two or more layers [17]. The actuation relies on the difference in linear expansion coefficients of two materials, with one layer expanding by a different amount compared to the other. This results in thermal stress at the interface of these two layers, leading to bending of the cantilever. ΔT can be created by heating up the cantilever when a bias current flows through an embedded resistor in cantilever, i.e., Joule heating effect. Out-of-plane displacement is generated at the bended cantilever because of the difference in expansion volume in different layers. Given that parameters for materials 1 and 2 are denoted by subscripts "1" and "2," respectively, the radius of curvature for the bended bimorph cantilever composing of two material layers (material 1 has higher CTE than material 2) is

$$\frac{1}{R} = \frac{6 w_1 w_2 E_1 E_2 t_1 t_2 (t_1 + t_2)(\alpha_1 - \alpha_2)\Delta T}{(w_1 E_1 t_1^2)^2 + (w_2 E_2 t_2^2)^2 + 2 w_1 w_2 E_1 E_2 t_1 t_2 (2 t_1^2 + 3 t_1 t_2 + 2 t_2^2)} \tag{14.4}$$

where
   R is the radius of bending curvature
   w is the width of material
   E is the elastic modulus of material
   t is the thickness of the material
   α is the CTE of the material
   ΔT is the temperature change

Nevertheless, some applications demand in-plane displacement for actuation purposes. By using a single material, a U-shaped thermal actuator consisting of two arms of uneven widths, as shown in Figure 14.2, makes in-plane actuation possible [18,19]. The U-shaped thermal actuator can be made of metallic beam or conductive silicon beam. When an electrical current is applied from one anchor to the other, the arm with larger electrical resistance heats up more. This results in higher temperature and larger volume expansion, i.e., so called hot arm. The other arm is relatively cold and is referred as cold arm. The arm of larger electrical resistance is the one with longer arm or

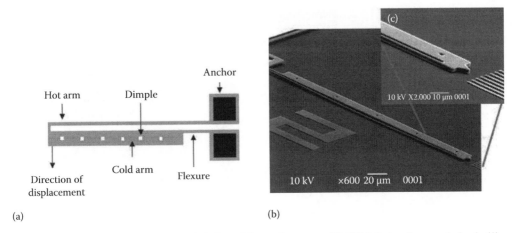

FIGURE 14.2 (a) Schematic drawing of U-shaped thermal actuator; (b) SEM photo of suspended polysilicon U-shaped thermal actuator, (c) close-up view of U-shaped thermal actuator. (From Riethmüller, W. and Benecke, W., *IEEE Trans. Electron Devices*, 35(6), 758, 1988; Comtois, J.H. and Bright, V.M., *Sens. Actuators A*, 58, 19, 1997.)

smaller cross-sectional area. The U-shaped thermal actuator will deflect laterally toward the cold arm side due to asymmetrical thermal expansion when the actuator is dc biased. Other designs for in-plane thermal actuators, such as V-beam [20,21] and H-beam thermal actuators [22], have also been reported.

The design of thermal actuators is a multiphysics problem [23] requiring thermal, structural, and electrical analysis. The heat transfer involved in a thermal analysis includes conduction and convection. Radiation heat transfer in thermal actuator can be ignored since it is generally not significant. The upper practical limit for temperature in the polysilicon and single-crystal-silicon-based thermal actuator is approximately 600°C and 800°C, above which material property changes such as localized plastic yielding and material grain growth become an issue. The ac operation of thermal actuators is generally limited to a frequency response less than 1000 Hz because of the time constants associated with heat transfer. For example, the thermally driven scanning mirrors demonstrate scanning frequency in the range of 100 ~ 600 Hz [24]. However, a complementary metal-oxide-semiconductor CMOS-MEMS-based scanning mirror operated at 2.6 KHz has been successfully developed because its thermal time response of the thermal actuator is less than 0.4 ms [25]. It shows a fact that thermal actuator operated at a few KHz is possible to achieve.

14.2.3 MAGNETIC ACTUATION

Lorentz force is generated when a current-carrying element is placed within a magnetic field and is given by

$$F_L = iL \times B \tag{14.5}$$

where L, i, and B refer to length of conductor, electric current flowing in conductor, and magnetic field, respectively. As indicated by the cross-product term, the Lorentz force occurs in a direction perpendicular to the current and magnetic field. Although Lorentz force actuation may be applied to MEMS devices in a number of ways, the prevailing approach is to have metal coils integrated on a mirror and actuated by an ac current at resonance when this mirror is placed near a permanent magnet [13,26,27]. Figure 14.3 shows a three-axis actuated micromirror developed by Cho et al., where actuation coils made of gold are electroplated on the mirror plate and cantilever actuators. Another approach is to integrate a permanent magnet (hard ferromagnet) or a permalloy layer

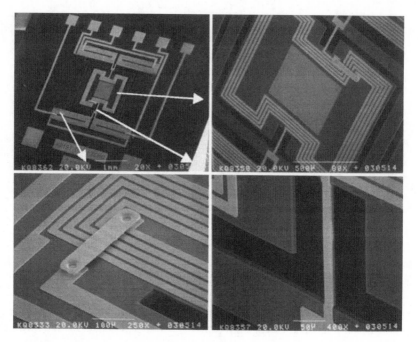

FIGURE 14.3 SEM pictures of the fabricated mirror structure, with electroplated gold coils integrated on the mirror plate and cantilever-type actuators. (From Cho, H.-J. and Yoon, E., *J. Micromech. Microeng.*, 19, 085007, 2009.)

(soft ferromagnet) on a movable mirror while the Lorentz force is introduced by the interaction between magnetic layer and surrounding ac magnetic field of an external solenoid [14,28]. The availability of permanent magnetic materials, which are compatible with MEMS processing, is limited and this brings necessary process development effort. Thus, it is common for the magnetic field to be generated externally, while the discrete and movable magnetic actuators often comprise metal coils.

14.2.4 PIEZOELECTRIC ACTUATION

Piezoelectric effect is understood as the linear electromechanical interaction between the mechanical and the electrical state in crystalline material. An applied dc voltage across the electrodes of a piezoelectric material will result in a net strain that is proportional to the magnitude of the electric field. The piezoelectricity is attributed to charge asymmetry within the primitive unit cell, resulting in the formation of a net electric dipole. Adding up these individual dipoles over the entire crystal gives a net polarization and an effective field within the material. Some of the popular piezoelectric materials include quartz, lithium niobate, aluminum nitride (AlN), zinc oxide (ZnO), and lead zirconate titanate (PZT), while the most well-known polymer-based piezoelectric material is polyvinylidene fluoride, which is a thermoplastic material.

The piezoelectric effect is often described in terms of piezoelectric charge coefficient, d_{ij}, which relates the static voltage or electric field in the i direction to displacement or applied force in the j direction. When we consider a PZT thin film actuator prepared on top of a Si cantilever, we define the axis 1 and axis 3 as longitudinal and normal direction, respectively, with respect to the cantilever. The piezoelectric charge coefficients are given as d_{33} for both voltage and force along the vertical axis (axis 3) and d_{31} for voltage along the vertical axis but force along the longitudinal axis (axis 1). The units of the piezoelectric charge coefficients are C/N or m/V, depending on whether the electrical parameter of interest is voltage or charge. The induced stress will bend the free end of the cantilever and introduce a displacement at cantilever end, δ, which is given by [29,30]:

$$\delta_{cantilever} = \frac{3AB}{K} L^2 V d_{31} \tag{14.6}$$

$$A = S_{si} S_{PZT} \ (S_{PZT} t_{Si} + S_{Si} t_{PZT}) \tag{14.7}$$

$$B = \frac{t_{Si}(t_{Si} + t_{PZT})}{S_{PZT} t_{Si} + S_{Si} t_{PZT}} \tag{14.8}$$

$$K = (S_{si})^2 (t_{PZT})^4 + 4 S_{si} S_{PZT} t_{Si} (t_{PZT})^3 + 6 S_{si} S_{PZT} (t_{Si})^2 (t_{PZT})^2 + 4 S_{si} S_{PZT} (t_{Si})^3 (t_{PZT}) + (S_{PZT})^2 (t_{Si})^4 \tag{14.9}$$

where
  L is the length of the cantilever
  V is the applied voltage
  S_{Si} and S_{PZT} are the compliances of the silicon cantilever (6.0×10^{-12} Pa^{-1}) and PZT actuator (1.43×10^{-11} Pa^{-1})
  t_{Si} and t_{PZT} are their respective thicknesses

The piezoelectric charge coefficient of d_{31} for various piezoelectric thin films PZT, ZnO, and AlN are reported as −110, 5, and 2–3 pC/N, respectively.

14.3 MEMS OPTICAL SWITCHES

Optical networks based on WDM systems have played a key role in increasing the capacity and flexibility of these networks. When the network architecture evolves from point-to-point WDM transmission systems into ring-type network, optical add/drop multiplexing (OADM) systems and optical cross-connect (OXC) systems are required to enable the networks to have more flexibility, thereby evolving the network into mesh-type architecture. Based on free-space optics, various OADM and OXC devices have been demonstrated by using MEMS technology. On the other hand, MEMS-based optical switches have various scale in sizes, from fundamental 1×2 and 2×2 switches to $N \times N$ switches, where N can be as large as one thousand. The basic 1×2 optical switch is often used for protection against equipment failure. A fundamental 2×2 switching element can be used as a stand-alone switch or within a multistage interconnection network architecture for constructing larger switch fabrics. A $1 \times N$ switch routes optical signals from one fiber to one of an array of N fibers. Finally, large $N \times N$ switches, often referred to as OXCs, are used to establish a desired connectivity pattern across many fibers. An OXC performs as an automated patch panel, whose connectivity can be changed without the need for a technician's visit to the equipment site of patch panel. Besides, the OXC can also be used to route individual WDM channels at a network node using opaque or transparent operating modes.

14.3.1 SMALL-SCALE OPTICAL SWITCHES

Optical add/drop switch, otherwise known as 2×2 crossbar optical switch, is a critical element for OADM device. The most common device configuration of the crossbar switch is that of a tiny mirror sliding in and out of the intersection point of light path. Thus, light beams either travel unimpeded to the fiber opposite them or get diverted into the fiber on the next channel. The early demonstrated gate switches and crossbar switches were realized by using polysilicon-related surface micromachining technology and bulk micromachining technology. The first crossbar switch was demonstrated by a surface micromachined and assembled polysilicon mirror [31]. Crossbar switches, based on bulk micromachining technology, have also been realized by using electromagnetic [32,33]

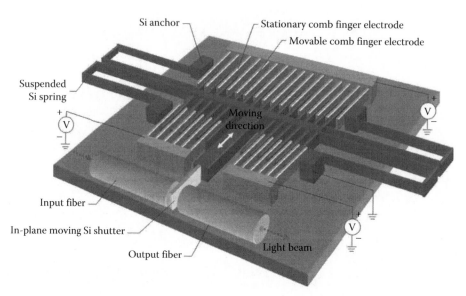

Si anchor — ⌐ Stationary comb finger electrode
⌐ Movable comb finger electrode

Suspended
Si spring

Moving
direction

Input fiber —

In-plane moving Si shutter —

Output fiber — Light beam

FIGURE 14.4 Schematic drawing of a surface micromachined silicon shutter connected with a lateral movable electrostatic comb actuator. This electrostatic actuated lateral movable shutter can perform both functions of gate switch and VOA. (From Juan, W.-H. and Pang, S.W., *IEEE J. Microelectromech. Syst.*, 7(2), 207, 1998.)

and electrostatic [34–36] actuation schemes as well. A crossbar switch derived from a silicon-on-insulator (SOI) substrate, combined with comb-drive actuators and deep reactive ion etching (DRIE)-made trenches for holding optical fibers, is shown in Figure 14.4. Such optical switch design comprises a high-aspect-ratio micromirror with vertical sidewall and an electrostatic comb-drive actuator for controlling the position of micromirror, i.e., the shutter [34–36]. Electrostatic force for moving the micromirror can be generated by applying voltage to the comb-drive actuators. The restoring force generated by the deformed spring will pull the actuated micromirror return to the initial position.

However, such crossbar switch designs require a continuous electrical bias to be applied on the MEMS actuator in order for the optical switch to stay in on-state. Therefore, a latching function to allow the mirror to maintain in a fixed position without power consumption is necessary. As such, bi-stable micromechanisms have been proposed to provide two relative positions that are both mechanically stable. Various latching mechanisms, such as buckled beam [37,38] and clamp gripper [39,40], were reported. In order to move the bi-stable structure from one stable position to the other position, an actuator with a large force output is preferred. Thermal (or denoted as electrothermal) actuator has been known as an alternative to provide higher force output under lower applied voltage than the electrostatic comb actuator, even though thermal actuator consumes more power than electrostatic one. Lee and Wu reported a bistable crossbar optical switch comprising two sets of movable V-beam actuators, a set of buckle beam springs connected to a suspended movable shutter beam with a reflective mirror shutter and a suspended movable translation link at ends of the suspended movable shutter beam [41]. Both ends of this set of buckle beam springs are anchored to the substrate, while the center of the buckle beam is connected to the suspended movable shutter beam (Figure 14.5a and b). Force generated by one of the two sets of V-beam thermal actuator, upon various values of the applied electrical load, acts against the restoration force from buckle beam springs. The buckle beam is deflected to a range where the force from the bended buckle beam spring is balanced by the force generated due to the actuated V-beam actuator under electrical load. The V-beam actuator can push or pull the suspended movable translation link to move the shutter beam when the buckled beam

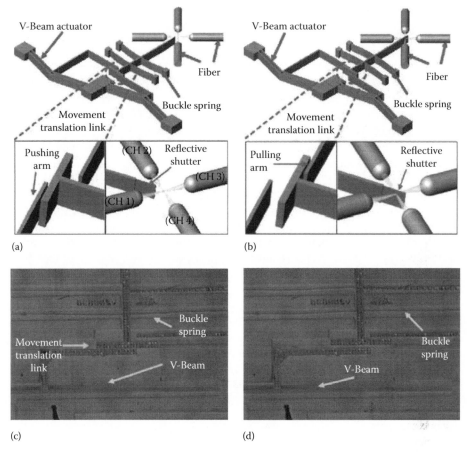

FIGURE 14.5 Schematic drawing of bi-stable optical crossbar switch driven by two sets of V-beam thermal actuators. (a) Transmission state, (b) switching state, (c) optical microscope photo of thermal actuator and movement translation link at transmission state, and (d) optical microscope photo of thermal actuator and movement translation link at switching state. (From Lee, C. and Wu, C.-Y., *J. Micromech. Microeng.*, 15, 11, 2005.)

spring is deflected into opposite direction due to the generated electrothermal force against the existing buckle beam spring force. Thereafter, the mirror and shutter beam will move from the initial position to another position of the bi-stable state (Figure 14.5c and d). On the other hand, the mirror and shutter beam will be moved by the suspended movable translation link back to the initial position of the bi-stable state, when another one of the two sets of V-beam actuator is actuated to pull or push the suspended movable translation link.

The next level of complexity involves using a two-dimensional (2-D) array of these mirrors to form a matrix switch, with rows of inputs and columns of outputs (or vice versa). Optical switches with 8×8 and 16×16 ports were demonstrated [42–44]. Mirror control for these 2-D switches is binary and thus straightforward, but the trade-off of this simplicity is optical loss. The substantial difference in length of optical paths through various switch configurations limit the scaling. Limits to the scaling also include the diameter of the mirrors and their maximum tilt angle. The mirrors are designed to be about 50% bigger than the optical beams to avoid excessive loss, and tilt is limited both by the method used to build the switch and the technique used to actuate the mirror. While the path length grows linearly with the number of ports, the optical loss also grows rapidly due to the Gaussian nature of light. Therefore, 2-D architectures are found to be impractical beyond 32 input and 32 output ports.

14.3.2 LARGE-SCALE OPTICAL SWITCHES

In contrast to the case of 2-D optical switches, i.e., all the light beams reside on the surface plane of MEMS substrate, and this feature leads to unacceptably high loss for large port counts, 3-D optical switches deploy an array of two-axis mirrors to steer the optical beams in 3-D free space. These require extremely fine analog control to align the optical beams because the beams must be accurately directed along two angles and then stop at precise intermediate positions, not just at fixed end-points. Key parameters of the two-axis mirror array include size, tilt angle, flatness, fill factor, and resonant frequency of the mirror. More importantly, the stability and repeatability of the actuated mirror under certain electric loads play a critical role in the complexity of the control schemes.

Early demonstrations of such switches relied mainly on surface micromachined two-axis mirrors [45,46]. The residual stress limits the mirror size to approximately 1 mm, and the different thermal expansion coefficients between the mirror and the metal coating also cause the mirror curvature to change with temperature. Bulk micromachined single-crystalline-silicon micromirrors are often used in high port-count 3-D optical switches that demand larger mirror size [47–49]. In early 2000s, research effort was focused on high port-count 3-D optical switches for OXC applications. For example, J. Kim et al. reported an OXC of 1100×1100 ports based on surface micromachined two-axis mirror array as the explosion of internet data transport demands in the boom of telecommunication industry [50]. R. Sawada et al. reported another unique design using electrostatic two-axis tilt mirrors as shown in Figure 14.6a through c in 2001 [51]. Two wafers are processed independently by silicon bulk micromachining first. By using Au/Sn-solder-based wafer bonding, a bonded structure of mirror and the underneath terraced electrode form a novel canted parallel electrode in Figure 14.6b and d. The 10 μm thick mirror is supported by folded torsion springs in two orthogonal axes and tilted two-dimensionally by electrostatic force (Figure 14.6c and e). The torsion spring has an aspect ratio of greater than 6, hence giving it a strong bending stiffness relative to torsion and strong support

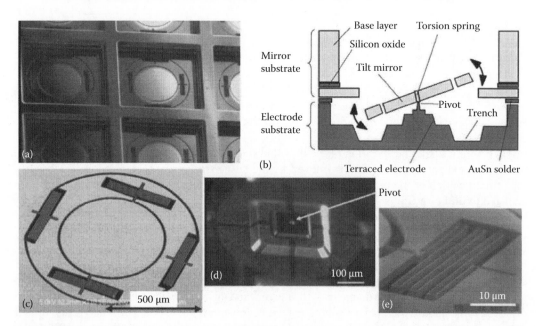

FIGURE 14.6 (a) Optical microscope photo of two-axis tilt mirrors with wafer bonded electrostatic actuator of terraced electrode, (b) schematic drawing of cross-sectional view, (c) SEM photo of a two-axis tilt mirror, (d) image of terraced electrode with a pivot, and (e) optical microscope photo of folded torsion spring of high aspect ratio. (From Sawada, R. et al., Single crystalline mirror actuated electrostatically by terraced electrodes with high-aspect ratio torsion spring, *Proceedings of the Optical MEMS 2001*, 2001, pp. 23–24.)

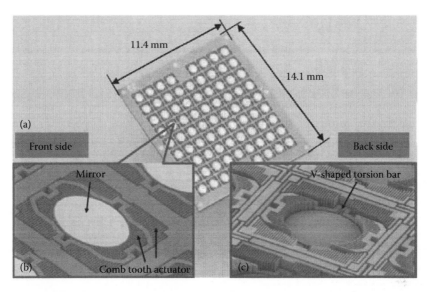

FIGURE 14.7 (a) Microfabricated two-axis mirror array of 79 channels, (b) SEM photo of top view of a two-axis mirror, and (c) SEM photo of backside view of a two-axis mirror. (From Mi, X. et al., *J. Opt. A: Pure Appl. Opt.*, 8, S341, 2006.)

for the mirror in order to achieve reliable switching operation. The mirror has a diameter of 600 μm and is integrated with the gimbal structures that provide freedom of tilt about two axes. The tilt angle of the mirror can be changed by controlling the applied voltage between the two substrates.

Another improved electrostatic actuation mechanism for mirrors is vertical comb actuator. The first micromirror with vertical comb actuator was reported in 2000 [52]. By leveraging the vertical comb actuators, tilt mirror devices can have a much larger electrostatic-force-induced torque such that one can use to reduce the operating voltage as well as increase the resonant frequency. A 3-D optical switch of 256×256 ports based on a 512-mirror array demonstrates a ±5° rotation of the two-axis stationary operation under a drive voltage of 160 V and a resonant frequency of 2 kHz in 2006 [53]. Figure 14.7 includes the photo of derived mirror array and scanning electron microscope (SEM) photos of both sides of a mirror with vertical comb. Mirror-based variations of parallel plate actuators and vertical comb actuators have also been reviewed [5]. Among these approaches, the angular vertical comb actuators can be easily integrated with mirror structure by using single-side lithography process on common wafers. Thus, the angular comb actuator gives a potential and improved solution of a new two-axis for 3-D optical switch applications.

14.4 MEMS VARIABLE OPTICAL ATTENUATORS

Among these optical communication applications, VOA and its array are crucial components for enabling the advanced optical network. MEMS VOA devices offer physical features like transparency (bit rate and protocol independent), tunability, scalability, low electrical operation power consumption, and small form factor. Currently, the dynamic gain equalizer is provided in conjunction with the wavelength-division multi/demultiplexers (MUX/DEMUXs) to perform the attenuation function and the function of reconfigurable and transparent add/drop at nodes. A multichannelled VOA device can be the channel-power equalizer in WDM cross-connect nodes and transmission networks. Thus, integrated multichannel VOAs with MUX/DEMUXs will be an alternative to fulfill this market. In view of the market requirements like small footprint and low power consumption, an array structure containing multiple MEMS attenuators in a single silicon chip is preferred for future DWDM applications. In this section, we conducted an extensive survey on the evolution of MEMS VOA technology in the past decade.

14.4.1 EARLY DEVELOPMENT WORKS

The first MEMS-based VOAs were demonstrated by two groups from Lucent Technology and the group led by Prof. N.F. Rooij at the University of Neuchâtel, Switzerland. In 1998, J.E. Ford and J.A. Walker developed a MEMS VOA using a surface-micromachined SiN suspended membrane with λ/4 optical thickness above a silicon nitride (SiN) substrate, with a fixed 3λ/4 spacing [54]. Voltage applied to electrodes on top of the membrane creates an electrostatic force and pulls the membrane close to the substrate, while membrane tension provides a linear restoring force. When the membrane gap is reduced to λ/2, the layer becomes an antireflection coating (ARC) with close to zero reflectivity, hence allowing intensity of reflected light to be controlled accordingly. Dynamic attenuation range of 25 dB was achieved under 50 V_{dc} load.

In contrast to the dielectric antireflection membrane, B. Barber et al. led another team at Lucent Technologies in 1998 to develop MEMS VOA using a surface-micromachined polysilicon microshutter arranged between ends of two fibers and aligned along the same axis [55,56]. This shutter is driven by an electrostatic parallel actuator, and attenuation is achieved by the percentage of light blocked by this shutter at various vertical positions. This surface-micromachined in-line type MEMS VOA can achieve attenuation as high as 50 dB under 25 V_{dc} and less than 1 dB insertion loss.

In addition to surface-micromachined polysilicon-based approach, DRIE technology is another major approach for making MEMS VOA structures from device layer of a silicon on insulator (SOI) wafer. The first demonstration was done by Prof. N. F. de Rooji's group at University of Neuchâtel, Switzerland in 1998 [57]. The fabrication also includes silicon trenches to accommodate optical fibers with photolithography process determined alignment accuracy regarding to microshutter. This feature makes the originally tedious assembly and alignment easier. This SOI-based VOA device comprises a movable comb finger electrode connected with microshutter via a suspended spring and a stationary comb finger electrode. The attenuation range is determined in terms of the in-plane position of Si microshutter, which is controlled via force balance between electrostatic force and spring force. A maximum attenuation of slightly over 50 dB was obtained at a dc voltage of 31 V.

14.4.2 SURFACE MICROMACHINED VOAs

In 2002, another surface-micromachined MEMS in-line type VOA using a pop-up microshutter based on electrostatic parallel plate actuation was reported by Prof. A. Q. Liu's group at NTU, Singapore [58–60]. The pop-up microshutter uses the same design as that by Lucent Technology. However, in this design, the shutter is fixed on a drawbridge plate and can be moved downward to substrate due to an applied dc bias. It demonstrates 45 dB attenuation under 8 V bias and 1.5 dB insertion loss. Compared with Refs [55,56], the driving voltage was reduced substantially through the use of this unique drawbridge structure.

In 2003, a group at Asia Pacific Microsystems, Inc. (APM, Inc.) developed a new movement translation micromechanism (MTM) to convert and amplify small in-plane displacement into large out-of-plane vertical displacement or large out-of-plane rotational angle [61,62]. As shown in Figure 14.8a and b, the in-plane displacement was provided by electrically controlled electrothermal actuator (ETA) array. Based on this MTM, only 3 dc volts was needed to generate 3.1 μm in-plane displacement and a rotational angle of 26.4°. This is equivalent to an out-of-plane vertical displacement of 92.7 μm for pop-up micromirror, which was subsequently derived. Using this MTM, the in-line-type VOA demonstrated 37 dB attenuation range under 3-V dc load, while return loss, polarization-dependent loss (PDL), and wavelength-dependent loss at attenuation of 3 dB are measured as 45, 0.05, and 0.28 dB, respectively. This work reveals the effort of reducing driving voltage for surface-micromachined VOA devices. It evidenced that a small planar displacement can be amplified into large 3-D displacement, i.e., large rotational angle or significant displacement of the micromirror.

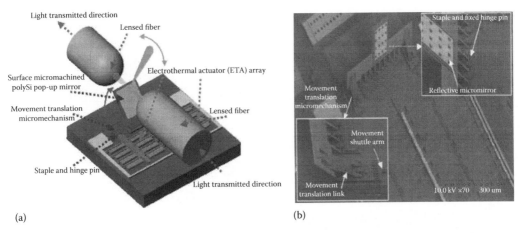

FIGURE 14.8 (a) Schematic drawings of a surface micromachined MEMS VOA comprising a pop-up micromirror and the input and output fibers. After wet-etching release process, the lens fibers are aligned to achieve the minimum insertion loss first. The attenuated light was reflected toward the out-of-plane direction, when a dc voltage is applied to the ETA array. (b) SEM photo of the MEMS VOA shows that in-plane displacement from the ETA array under dc voltage load is converted into out-of-plane rotation. Upper right inset shows close-up view of pop-up micromirror, staple, and fixed-hinge pin, while bottom left inset shows close-up view of lifted-up MTM structure. (From Lee, C. et al., *IEEE Photonics Technol. Lett.*, 16(4), 1044, 2004; Lee, C. and Lin, Y.-S., *IEEE Sens. J.*, 4(4), 503, 2004.)

14.4.3 DRIE-Derived Planar VOAs Using Electrostatic Actuators

14.4.3.1 Shutter and Single Reflection Mirror

DRIE-derived VOA from SOI substrates with fiber alignment trenches makes testing, alignment, and assembly works easier [57]. As shown in Figure 14.4, a DRIE-derived VOA developed by Prof. N. F. de Rooji's group really opens a window for new research activities. C.-H. Kim et al. reported a new electrostatic comb actuated VOA with off-axis misalignment-based light attenuation scheme, i.e., single reflection type, in 2002 [63]. These devices exhibited 2.5 dB insertion loss and 50 dB attenuation with respect to 14 μm displacement of comb actuator at 5 V. Meanwhile, C. Lee et al. at APM Inc. improved the design for electrostatic comb actuator and single reflection type VOA [64]. As illustrated in Figure 14.9a, the insertion loss is maintained initially at its minimum level, and incoming light signals are fully transmitted. Due to an applied electrical bias, the shutter approaches toward the light in transmission, causing a portion of incoming light to be blocked by the shutter. Figure 14.9b illustrates another fiber configuration setup, where the input fiber port and output fiber port are arranged in an orthogonally planar location. This causes the transmission light incident on the reflective mirror to be reflected toward the output port. As such, the reflected light path is changed according to the different mirror positions, which is in turn determined by the voltage applied to the comb-drive actuator. Therefore, the coupled intensity of reflected light to output port is dependent on the path of reflected light.

Briefly speaking, VOAs using single reflection demonstrate extremely well PDL and better return loss than shutter-based VOAs. With proper design of comb actuators, well-optimized DRIE process, and appropriate selection of lens fibers, reflective type VOA is superior except to a concern that package of reflective type VOA with 45° between input and output optical fiber ports is not a common layout configuration in application markets.

14.4.3.2 Dual Reflection Mirror

Due to the layout format concern mentioned in the previous paragraph, C. Lee et al. came up with a new retro-reflective type VOA in 2003 [65,66], while VOA of similar concept has been reported

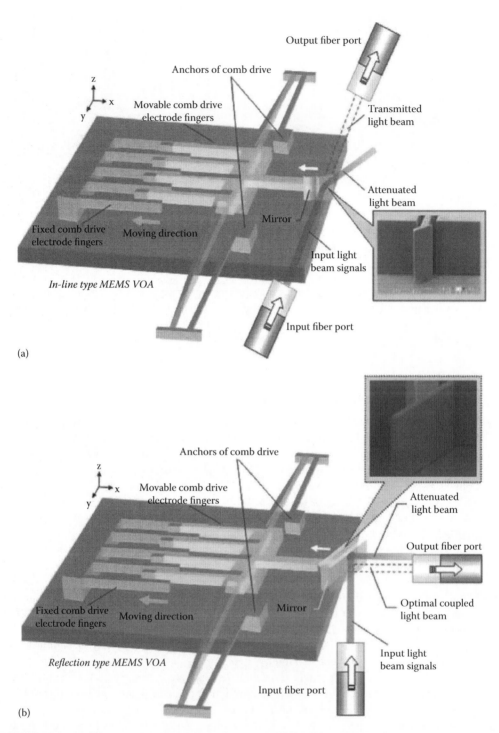

(a)

(b)

FIGURE 14.9 (a) Schematic drawing of the in-line type MEMS VOA, where the SEM photo of a microshutter with a mirror plane of a tilted angle is shown in inset. (b) Schematic drawing of the reflection type MEMS VOA, where the SEM photo of a reflective micromirror with a mirror plane of 45° tilted angle is shown in inset. (From Chen, C. et al., *IEEE Commun. Mag.*, 41(8), S16, 2003.)

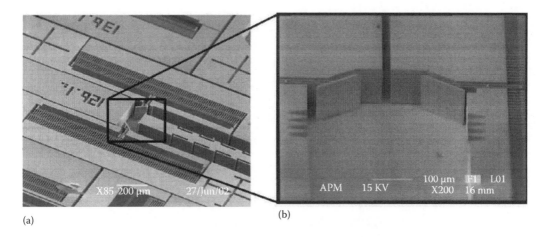

(a) (b)

FIGURE 14.10 SEM photographs of retro-reflective type MEMS VOA devices: (a) two reflective micromirrors and two coaxially arranged fiber trenches to form a retro-refractive type VOA and (b) close-up view of micromirrors made by using DRIE of SOI substrate. (From Lee, C. et al., Characterization of MOEMS VOA based on various planar light attenuation configurations, *Proceedings of the IEEE/LEOS International Conference on Optical MEMS*, 2004, pp. 98–99; Chen, C. et al., *IEEE Photonics Technol. Lett.*, 16(10), 2290, 2004.)

by T.-S. Lim et al. as well. T.-S. Lim et al. created a folded-mirror with 45° between two reflective mirrors, where this folded-mirror is connected with a set of comb actuators and is suspended via a silicon beam [67]. This VOA achieved 30 dB attenuation at 34 dc volts. In contrast to Lim's work, C. Lee's retro-reflective type VOAs comprise of two separately controlled reflective micromirrors that are allocated in the front of input and output fibers and assembled in a planar coaxial layout (Figure 14.10). The measured characteristics exhibit insertion loss of less than 0.9 dB, return loss of less than −50 dB, and wavelength dependent loss (WDL) of less than 0.35 and 0.57 dB at 20 and 30 dB attenuations, respectively. There are numerous advantages offered by the retro-reflective attenuation mechanism. First, a lower operation voltage is needed, since the intensity adjustment depends on the light path shift that is doubled after retro-reflection for the same actuator driving voltage in previous reflective type VOAs. Second, these two micromirrors are potentially capable of being feedback-controlled individuals; thus, attenuation curve could behave more linear with dedicated control design. Finally, users in optical communication industry are more familiar with the planar coaxial layout, rather than the 45° layout configuration.

14.4.3.3 Rotary Comb Actuator

T.-S. Lim et al. further explored the possibility of driving folded-mirror by using rotary comb actuator for VOA application. This VOA comprised a folded-mirror connected with rotary comb via a suspended beam. It reported attenuation of 45 dB at a rotation angle of 2.4° under 21 dc volts and response time of less than 5 ms [68]. J. A. Yeh et al. also reported a reflective-type VOA [69]. This new VOA device attenuates the optical power using a planar rotational tilted mirror driven by novel rotary comb-drive actuators. As shown in Figure 14.11a and b, the micromachined rotary comb-drive actuator is connected to the tilted mirror where the entire structure is suspended by four orthogonally cross-linked meander springs anchored at four corners. Four stripes of comb electrode pairs are packed in each quarter of the rotary comb-drive actuator. Meander springs are deployed to reduce the rotational spring constant K_Φ and to maintain the sufficiently large ratio of the radial spring constant K_r to the K_Φ. Hence, stable rotations at low driving voltages can be achieved along with high radial robustness in the device plane. The device reported 50 dB attenuation at a rotation angle of 2.5° under 4.1 dc volts. The response time from 0 to 40 dB attenuation and backward switching time were measured as 3 and 0.5 ms, respectively. The measured insertion loss was 0.95 dB, and the PDL

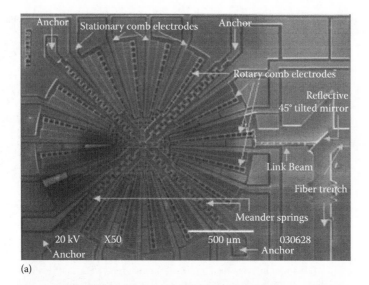

(a)

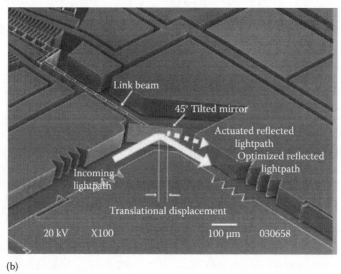

(b)

FIGURE 14.11 (a) SEM photograph of MEMS VOA device using rotary comb drive actuator. The VOA device has four anchors at the ends of two orthogonal located meander springs. (b) SEM photograph of closed-up view of the rotational 45° tilted mirror driven by a beam linked via a rotary comb drive actuator. (From Yeh, J.A. et al., *IEEE Photonics Technol. Lett.*, 18(10), 1170, 2006.)

was 0.3 dB at 20 dB attenuation with regard to a wavelength of 1550 nm. The WDL was measured to be 0.19, 0.25, 0.61, and 0.87 dB for attenuation at 0, 3, 10, and 20 dB, respectively.

14.4.3.4 Functional Shutter and Elliptical Mirror

In 2003, a wedge-shaped silicon optical leaker, i.e., a revised shutter, is proposed as a new refractive-type VOA by Y. Y. Kim of Korea Aerospace Research Institute, S. S. Yun et al., of the Gwang-Ju Institute of Science and Technology, Gwangju, Korea, and Y. G. Lee et al. of Samsung Electro-Mechanics Company, Suwon, Korea [70–72]. With proper design of wedge shape, this type of shutter allows multiple optical internal reflection to occur near the fiber core of input and output fibers. Thus, only a small portion of incoming light leaks out with a tilted propagation angle. It achieved return loss of less than −39 dB and wide attenuation range of 43 dB for optical fibers of 8° facet end, while the PDL was also measured to be less than 0.08, 0.43, 1.23, and 2.56 dB for the attenuation at 0.6, 10, 20, and 30 dB, respectively.

Without using 45° reflection mirror, Prof. A. Q. Liu's group at NTU, Singapore, proposed a novel elliptical mirror driven by an axial movable-comb actuator, where the input and output fibers were arranged and aligned in an orthogonal layout with respect to the elliptical mirror [73]. In this unique design, the input and output fibers were allocated at the two focal centers of the reflective elliptical mirror. The VOA enjoys low insertion loss while using the common single-mode fibers as the elliptical mirror can focus the light from one center to the other. As the mirror gets shifted in the direction of the axial direction of one fiber, the input beam is rapidly defocused, producing wide attenuation range without requiring large mirror displacement. The dynamic range of 44 dB was achieved at 10.7 dc bias, while the response time was measured as 0.22 ms, respectively.

The issue of nonlinearity of attenuation always surface when electrostatic actuation is involved. Two major nonlinear factors contribute to this issue. First of all, the moving distance of microshutter is proportional to square of driving voltage of electrostatic comb actuator. Second is that the collimated optical beam shows Gaussian distribution. J.-H. Lee et al. at the Gwang-Ju Institute of Science and Technology, Gwangju, Korea, have demonstrated an in-line shutter-type VOA with linear attenuation curve versus by using comb actuator with curved comb finger shape [74]. On the other hand, Prof. A. Q. Liu's group at NTU, Singapore, demonstrated another way to achieve the attenuation versus driving voltage with good linearity by using two curved shutters connected with a pair of individually controlled comb actuators [75]. It shows very good linearity within 25-dB attenuation regarding to 8 dc volts of driving voltage. Recently, B. Glushko et al. reported a two-shutter-type VOA with advantage in reduced driving voltage and compensation of nonlinearity [76].

14.4.4 DRIE-DERIVED PLANAR VOAs USING ELECTROTHERMAL ACTUATORS

All the devices demonstrated so far relied on electrostatic actuation mechanisms. Since there is no current flowing between the electrodes of electrostatic actuators, it implies no power consumption within actuator part during operation of electrostatic actuators. However, ETAs have been known for their large displacement and high force output. These characteristics make ETAs become an alternative to be the actuator for VOA device. The size and weight of MEMS elements are relatively small, and only a small amount of energy is required to operate the MEMS-based VOA. In other words, the power consumption of electrothermally driven MEMS VOA is expected to be low as well. Besides, the nature of ETA structure renders electrothermally driven MEMS VOAs with a smaller footprint and lighter weight than their counterpart based on electrostatic actuators. J. C. Chiou and W. T. Lin deployed two pairs of U-shaped ETAs linked together to push two separated microshutters on opposite sides [77]. Although these U-shaped actuators have been symmetrically located at four corners of the device, the robustness of this VOA device is still constrained by the mechanical weakness in the flexure beam of U-shaped actuator itself. The driving voltage for 45 dB attenuation was as low as 4.5 dc volts. This is due to the merit contributed by two shutters approach.

In another electrothermal-actuated VOA design, C. Lee reported a H-shaped silicon beam structure comprising two V-beam ETAs on both sides and a pair of reflective mirrors for VOA applications, where the pair of mirrors were arranged at the center of a linked beam between two V-beam ETAs [78]. When dc bias was applied onto the anchors of V-beam, the volume expansion due to joule heating will lead beams of both sides of H-shaped beam deformed toward the arched direction, i.e., denoted as moving direction in Figure 14.12. The position of retro-reflective mirrors depends on the applied bias voltage. If the same bias were applied to both V-beams, the generated displacement of both sides will be the same. Optical attenuation happens when the retro-reflective mirrors moved from its initial location, where they have already been optimized in terms of minimum insertion loss. The maximum dynamic range of attenuation was 50 dB under 9 dc volts. The PDL was measured as 0.15 dB at 20-dB attenuation.

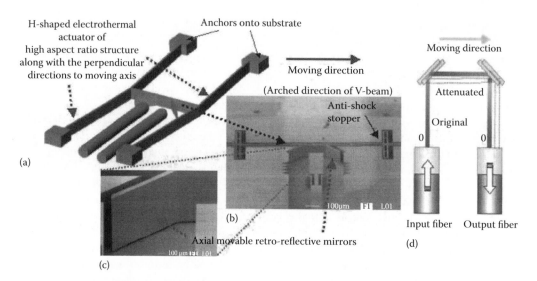

FIGURE 14.12 VOA device comprises retro-reflective mirrors driven by a linked electrothermal H-beam actuators pair, where four anchors of V-beam actuators pair are arranged symmetrically allocated on four corners of device occupied square area to form a robust VOA structure. (a) Schematic drawing, (b) SEM photo of retro-reflective mirrors and link beam, (c) SEM photo of closed-up view of mirror, and (d) the retro-reflective light path attenuation scheme. (From Lee, C., *IEEE Photonics Technol. Lett.*, 18(6), 773, 2006.)

14.4.5 THREE-DIMENSIONAL VARIABLE OPTICAL ATTENUATORS

Combining optics and a tilt mirror or an array of mirrors to be assembled in a 3-D configuration is also a key approach of making MEMS VOA devices. B. M. Andersen et al. in 2000 reported a surface micromachined tilted mirror using 3-D attenuation scheme [79], while N. A. Riza et al. reported a 3-D VOA by using DMD mirror array of Texas Instruments in 1999 [80]. However, these early demonstrations only show moderate performance. In principle, we may prepare large single-crystalline-silicon tilt mirror by using the state-of-the-art DRIE micromachining technologies. When tilt mirrors are deployed for 3-D VOA in conjunction with large micro-optics, like dual core collimator, etc., the resulted VOA device can gain in excellent data of return loss, PDL, and WDL under reasonable driving voltage, i.e., small rotational angle. H. Toshiyoshi et al. have demonstrated a 3-D VOA using electrostatic parallel plate actuator in which the derived performance is very good in this sense, i.e., only 4.5 dc volts is applied to achieve 0.3° rotational angle and 40-dB attenuation range [81–83].

Recently, C. Lee et al. developed a 3-D VOA device based on a pair of PZT actuators as shown in Figure 14.13 [84]. In the operation of the PZT 3-D VOA, various dc biases were applied to one or two of actuators. As a result, the reflected light deviated from the optimized light path corresponding to minimum insertion loss. The insertion loss increases with increased dc bias because the coupled reflected light intensity toward output fiber is reduced, i.e., attenuation operation. Only 1 V_{dc} is needed to achieve 42 dB dynamic range of attenuation, while the 50 dB attenuation is obtained at 1.2 dc V. The optical deflection angle is measured as 0.18° at 1 dc V. Referring to most of the commercial applications, 40 dB dynamic range is enough. It means the developed PZT 3-D VOA only requires operation voltage of 1 dc V.

14.4.6 VOAS USING VARIOUS MECHANISMS

Another type of 3-D VOAs relies on grating structure, i.e., the diffractive type of mechanism. By modifying the original design of GLV diffractive technology, Lightconnect has proposed a revised design using circularly symmetric membrane structure [85,86]. This novel diffractive

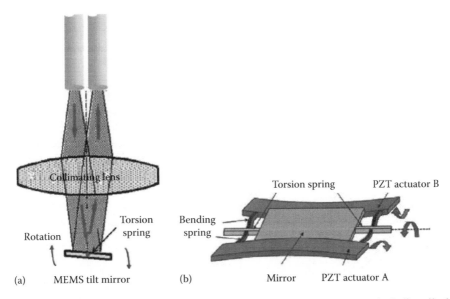

(a)   MEMS tilt mirror   (b)   Mirror   PZT actuator A

FIGURE 14.13 (a) Schematic drawings of optical path in a 3-D VOA arrangement including tilted mirror and optics and (b) schematic drawing of tilt mirror for 3-D VOA. The mirror rotates against the torsion spring due to static displacements introduced by the PZT actuator beams. (From Lee, C. et al., *IEEE J. Sel. Top. Quantum Electron.*, 15(5), 1529, 2009.)

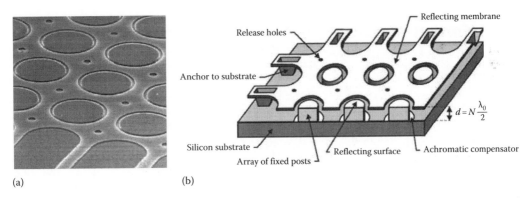

(a)   (b)

FIGURE 14.14 (a) SEM photo of a diffractive MEMS VOA. (b) Schematic drawing of the diffractive MEMS VOA. The architecture incorporates achromatic compensation and cylindrical symmetry to ensure low dependence on polarization. (From Godil, A., *Laser Focus World*, 38(5), 181, 2002; U.S. Patents 6,169,624, January 2, 2001, and 6,501,600, December 31, 2002.)

MEMS VOA is known as the first Telcordia-qualified MEMS VOA device. When the reflective membrane and reflective postsurface are coplanar, incident light is reflected back into the aperture without attenuation (Figure 14.14). When the reflective membrane is pulled down using electrostatic actuation by one quarter of a wavelength ($\lambda/4$) relative to its adjacent reflective postsurface, the incident energy diffracts into higher orders that are directed outside the aperture, and the incident beam is completely attenuated. When the separation gap is less than $\lambda/4$, the incident beam is partially attenuated because some energy is shifted into the higher diffracted orders. Another unique feature made in this diffractive MEMS VOA is the design of reduction of PDL. Without using the ribbon-like microstructures in GLV device, lightconnect's diffractive MEMS VOA deploys reference circular post to replace the reference ribbons and suspended reflective membrane to be the movable ribbons. This diffractive MEMS VOA has a dynamic range (attenuation range) of 30 dB, a wavelength dependence of attenuation of 0.25 dB, and a PDL of 0.2 dB. The total insertion loss, which included losses from fiber coupling, is 0.7 dB. The response time of the device is outstanding fast, i.e., 40 μs, in which it is attributed to the small mass and short actuation distance (e.g., $\lambda_0/4 \cong 400$ nm). The actuation voltage is less than 8 V.

On the other hand, to avoid the requirement of on-hold voltage or power, vernier-type latching mechanism for in-line shutter-type VOAs has been reported by R. R. A. Syms et al. [87,88]. Moving fibers mechanisms can cause the two aligned fibers to move to positions away from the optimized coupling position, hence achieving attenuation. Such mechanisms can be used as VOAs as well because of analogue control capability. Approaches using polymer waveguide or MEMS of polymer membrane structure have also been investigated, with potential advantages of small footprint or low cost [89–92]. Optofluidic technology has also been applied to VOA device as well [93]. From the preliminary results, rather good VOA characteristics including 38 dB dynamic range, 0.479 dB insertion loss and the PDL of less than 0.4 dB have been reported.

14.5 MEMS MICROSCANNER

With a number of advantage, including small size, light weight, and fast speed compared to conventional bulky scanners, MEMS-based optical microscanners have been drawing attention for a wide range of image display applications such as projection displays, biomedical imaging, and retinal scanning. MEMS and optics make a perfect match as MEMS devices have dimensions and actuation distances comparable to the wavelength of light. Besides using MEMS mirror, which are reflective-type devices, diffractive-type in the form of grating and steering-type in the form of lens have also been reported for scanning purposes.

14.5.1 COMMERCIAL PRODUCTS

14.5.1.1 Digital Micromirror Device

The most well-known commercial product is probably the digital micromirror device (DMD), which is based on the DLP technology developed by Texas Instrument in 1996 [94,95]. The DMD consists of a semiconductor-based array of fast, effective digital light switches that precisely control a light source using a binary pulse-width modulation technique. Unlike most MEMS devices, the DMD is fully integrated and monolithically fabricated on a mature static random access memory (SRAM) CMOS address circuitry. The DMD is currently in volume production for resolution formats such as super video graphics adapter (SVGA; 848×600), extended graphics adapter (XGA; 1024×768), super extended graphics adapter (SXGA; 1280×1024), and 1080 p high definition (HD; 1920×1080). As such, the various resolution display formats offered by DMD allow it to be used in very compact and light-weight applications such as portable projector or picoprojector, as well as in very high brightness fixed installation applications such as digital cinema, boardroom projectors, and HD television.

As shown in Figure 14.15, each DMD consists of hinge-mounted electrostatically actuated bistable micromirrors suspended on a chip. Each micromirror corresponds to an image pixel, and the pixel brightness can be controlled by switching between two tilt states. In the on state, light from the projector bulb or laser is reflected into the lens, making the pixel appear bright on the screen. In the off state, the light is directed elsewhere, making the pixel appear dark. The yoke, suspended by thin torsion hinges attached to posts, supports the whole mechanical structure. Two pair of electrodes control the position of the mirror by electrostatic attraction. Each pair has one electrode on each side of the hinge, with one positioned to act on the yoke and the other acting directly on the mirror. To move the mirror, bias charges will be applied to the electrodes, causing the aluminum mirror to be attracted to the biased electrode. First-generation DMD device with pixel pitch of 17 μm, 0.7 μm gap, and ±10° rotation has given way to 10.8 μm pitch, 0.7 μm gap, and ±12° rotation in their current latest 1080 p resolution product. Greater rotation can accommodate higher numerical aperture, while small pixel pitch shrinks the chip area, offering cost benefit to microdisplay and optical systems.

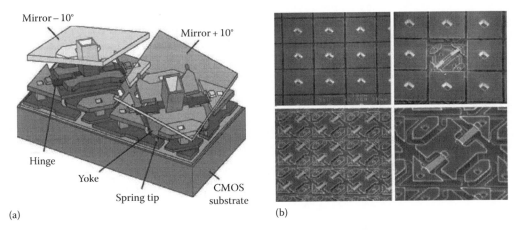

(a)   (b)

FIGURE 14.15 (a) Schematic illustration of the DMD, consisting of micromirrors, springs, hinges, yoke, and CMOS substrate. (b) SEM photos of the DMD. (From Sampsell, J.B., An overview of Texas instrument digital micromirror device (DMD) and its application to projection displays, *Proceedings of the International Symposium on Society of Information Display*, May 1993, Vol. 24, pp. 1012–1015; Hornbeck, L.J., *IEDM Tech. Dig.*, 381, 1993.)

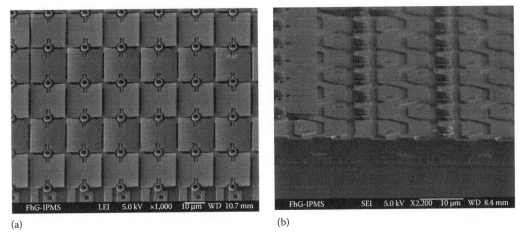

(a)   (b)

FIGURE 14.16 (a) Close-up SEM photo of the one-megapixel micromirror array, where the micromirrors were individually deflected to different tilting angles by underlying high-voltage CMOS electronics. (b) SEM photo of the electrode setup of the micromirror array. (From Zimmer, F. et al., *IEEE J. Microelectromech. Syst.*, 20(3), 564, 2011.)

Design similar to the DMD has also been reported by F. Zimmer et al., where high resolution spatial light modulator chip with 1 million tilting micromirrors made of monocrystalline silicon on analog high-voltage CMOS driving electronics is demonstrated and illustrated in Figure 14.16 [96]. Very large heterogeneous integration was used to fabricate this micromirror arrays (MMA), allowing virtually any solid-state material to be integrated with CMOS electronics.

14.5.1.2 Grating Light Valve

In 1994, O. Solgaard et al. from Stanford University developed the grating light valve (GLV) or deformable grating device, providing alternative MEMS-based technology for implementation in commercial projectors [97–99]. This technology was owned by Silicon Light Machines, which

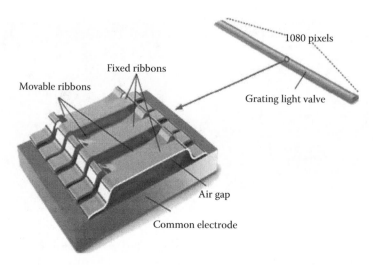

FIGURE 14.17   Schematic illustration of the GLV, where the color of each pixel is created by groups of movable and fixed ribbons. A group is set to diffracting state when alternate movable ribbons are deflected. (From Solgaard, O. et al., *Opt. Lett.*, 17(9), 688, 1992; Bloom, D., The grating light valve: Revolutionizing display technology, *Proceedings of the SPIE Projection Displays Symposium V*, Vol. 3634, pp. 132–138, 1999; Perry, T.S., *IEEE Spectr.* 41, 38, 2004.)

was acquired by Cypress Semiconductor in recent times. Alongside with DMD, GLV has also been incorporated in commercial products, with Sony Corp, Tokyo, having offered the public a tantalizing demonstration of its GLV-based 2005 in. ultra-wide HD screen of the Laser Dream Theatre at the central pavilion of Expo 2005 in Japan.

The key idea behind GLV technology is the use of movable ribbons to modulate the phase of light so that it can be regarded as a MEMS tunable phase grating (Figure 14.17). Each pixel consists of three movable and three fixed ribbon strips. Each pair of movable and fixed ribbons is responsible for the intensity of red, green, or blue color. The ribbons are made of silicon nitride with aluminum thin film coated on the surface of the ribbons to improve its reflectivity. The mechanism of the GLV employs diffraction and reflection mode to produce bright and dark states, respectively. When the six ribbons are not biased, each ribbon is stretched flat and lies in the same plane. They form a mirror that reflects light straight back to its source. The first order diffracted light from the GLV is nearly zero, causing no color to reach the screen. When the applied voltage is turned on, the movable ribbons are attracted down to the substrate by electrostatic attraction, causing a portion of the light aimed at the ribbon to diffract. Increasing the voltage, pulls the ribbon further down, causing differing amount of light to diffract and reflect. In the fully on state, the ribbons are deflected by a quarter of wavelength. As such, the color of a pixel on the screen is determined by the amount of red, blue, and green light being diffracted and incident collectively on the pixel as first-order light.

14.5.2   Microscanners Using Electrostatic Actuation

In addition to those electrostatic actuation mechanism demonstrated by both DLP and GLV technologies, there are other MEMS scanners that made use of electrostatic interdigitated vertical or rotary comb drives. In 2005, W. Piyawattanametha et al. from University of California, LA, designed a 2-D optical scanner with electrostatic angular vertical comb (AVC) actuators [100]. The scanner is realized by combining a foundry-based surface-micromachining process known as multiuser MEMS processes with DRIE postfabrication process. As shown in Figure 14.18a, there are two set of AVC actuators for each scanning axis, with the 1 mm diameter mirror in the center. All the movable comb actuators were manually assembled to a predefined angle (10°) and locked

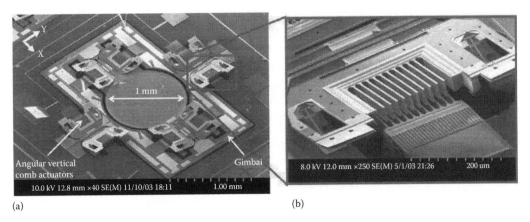

FIGURE 14.18 SEM of (a) 2-D angular vertical comb scanner and (b) close-up view of a movable comb. (From Piyawattanametha, W. et al., *IEEE J. Microelectromech. Syst.*, 14(6), 1329, 2005.)

in place by polysilicon latches (Figure 14.18b). The scanner achieved mechanical scanning ranges of ±6.2° (at 55 V_{dc}) and ±4.1° (at 50 V_{dc}) for the inner and outer gimbals, respectively. The resonant frequencies are 315 and 144 Hz for the inner and outer axes, respectively.

Y. Du et al. from National University of Singapore developed a novel micromachined electro-static double-layered vibratory grating scanner for high-speed high-resolution laser scanning applications [101]. The vibratory grating scanner achieved its scanning effect by coupling the rotation of the diffraction grating about the z-axis with the incident laser beam (Figure 14.19a). This causes the diffracted laser beam, except the zeroth order beam, to scan accordingly. However, the wavelength, incident angle of the laser beam, and pitch of the diffraction grating need to obey certain scanning conditions to achieve a bow-free scanning trajectory. In this work, the grating platform and driving actuator are located in different layers. This design is slightly different from their previous work in Ref. [16], where the grating platform and driving actuator are located in a single layer. This new configuration allows the device to have a larger grating platform and at the same time, a lower suspension flexure stiffness so that a bigger optical scan angle can be obtained. This allows the device to achieve a higher optical resolution, which is defined as the product of the optical scan angle and diameter of diffraction grating. The grating platform was attached to the connection pillar using epoxide resin (Figure 14.19b).

14.5.3 MICROSCANNERS USING ELECTROTHERMAL ACTUATION

One of the most common techniques for thermal actuated microscanner is the use of thermal bimoprh actuators. K. Jia et al. made modification to the simple bimorph actuator design and came up with a actuator structure termed as folded dual S-shaped (FDS) bimorph [102]. With a single bimorph, there exists a lateral shift and a tangential tilt angle at the tip of the bimorph. With the FDS design, both the lateral shift and tilt angle will be compensated. Figure 14.20a shows the SEM image of the device, with the mirror plate elevated at a height above the plane of the device. This is due to the residue stress remaining in the thermal actuator after the release process. Figure 14.20b shows the side view of the FDS bimorph actuator. When a current is passed through the aluminum layer of the FDS actuator, the actuator gets heated up by joule heating. This introduces a pure vertical downward displacement, actuating the mirror plate during the process. When same voltage is simultaneously applied to all four pairs of actuators, the mirror plate moves vertically without any lateral shift, i.e., piston motion. When different voltages are applied to the actuators, the mirror plate performs tip-tilt scanning. As a result, various kinds of Lissajous patterns can be generated by varying the ac signal frequency ratio and phase difference between the two orthogonal actuator

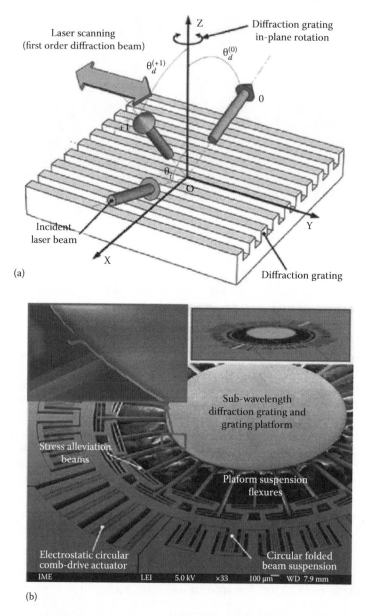

(a)

(b)

FIGURE 14.19 (a) Working principle of the MEMS vibratory grating scanner. (b) Whole view and part of the assembled electrostatic double-layered vibratory grating scanner. (From Du, Y. et al., *IEEE J. Microelectromech. Syst.*, 19(5), 1186, 2010.)

pairs (Figure 14.21). The micromirror shows an optical scan range of ±30° about both x and y axes and displaces 480 μm in the z-axis, all at dc voltages that are less than 8 V. Resonant frequencies of the piston and rotation motion are 336 Hz and 448 Hz, respectively.

In addition to single mirror device, K. Jia et al. also investigated a high fill factor 4×4 MMA device utilizing the same FDS bimorph actuator design [103]. The MMA device has an 88% area fill factor at normal incidence, 1.5 mm×1.5 mm subaperture size, and single-crystal-silicon-supported mirror plates fabricated based on a single SOI wafer, without additional bonding or transfer processes. SEM images of the fabricated 4×4 MMA device are shown in Figure 14.22. Each subaperture is actuated by four FDS actuator pairs with an initial downward displacement. Figure 14.23 shows

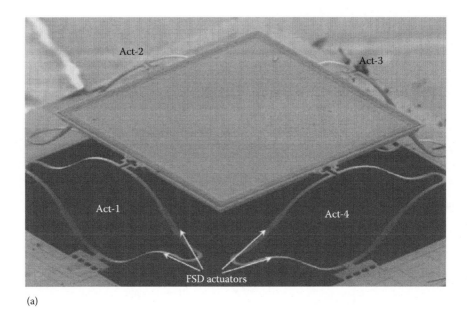

(a)

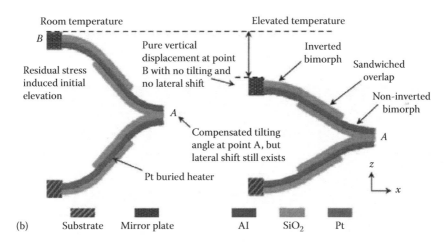

FIGURE 14.20   (a) SEM of fabricated micromirror device and (b) side view of the FDS bimorph actuator before and after actuation. (From Jia, K. et al., *IEEE J. Microelectromech. Syst.*, 18(5), 1004, 2009.)

the optical images of the surface-mounted MMA device wire bonded on to a ceramic package. Various modes of actuation, i.e., piston-tip-tilt capabilities are also demonstrated in Figure 14.23. This MMA device developed is particularly suitable for optical phased array applications as the array of subapertures can imitate a large combined optical aperture. Figure 14.24 demonstrates the motion capabilities of the mirror pixels, forming different letters by controlling the actuators to reflect the light spot of each subaperture.

Besides the out-of-plane actuation of mirror plate by bimorph actuator, in-plane actuation of the mirror plate is also possible. This is exemplified by the work done by Y. Eun and J. Kim in Yonsei University, Korea, where both one- and two-degree-of-freedom (DOF) torsional micromirror driven by in-plane thermal actuators were designed and fabricated [104]. Figure 14.25a shows the schematic layout of the 1-DOF micromirror with electrical connection. When the thermal actuators heat-up, a lateral force is applied along the section AB, causing the lateral motion to be converted to torsional motion of the mirror. This 1-DOF in-plane actuation concept was extended to a 2-DOF design, with

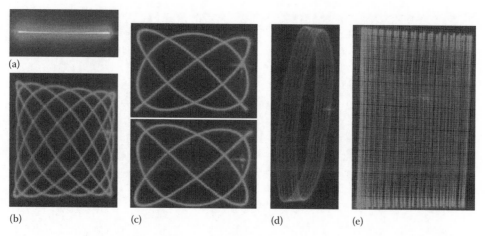

FIGURE 14.21 Various Lissajous patterns generated by varying ac signal frequency ratio and phase difference between the two orthogonal actuator pairs. (a) One-dimensional scan, (b) Lissjous patterns with frequency ratio of 5:6, (c) Lissajous pattern with frequency ratio of 5:6 and a phase difference of 30°, (d) band pattern with one axis at resonance, and (e) Raster scan with 40 lines (frequencies for two axes: 480 Hz and 12 Hz). (From Jia, K. et al., *IEEE J. Microelectromech. Syst.*, 18(5), 1004, 2009.)

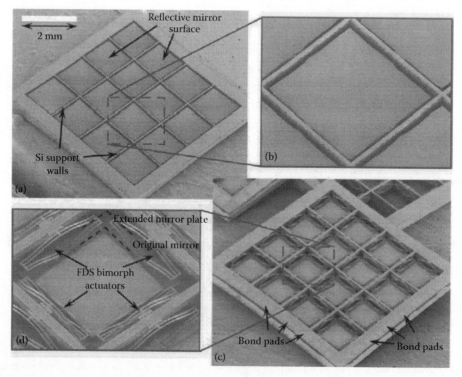

FIGURE 14.22 SEM images of the 4×4 micromirror array device. (a) Top view of the device and (b) Close-up view of one subaperture. (c) Bottom side of the device, showing actuators and bond pads. (d) Close-up view of the FDS bimorph actuators of one subaperture. (From Jia, K. et al., *IEEE J. Microelectromech. Syst.*, 20(3), 573, 2011.)

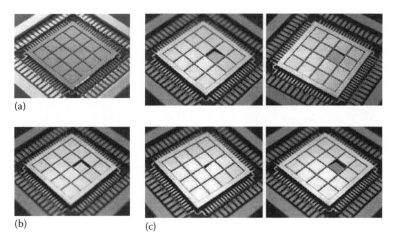

FIGURE 14.23 Optical images of a packaged surface mounted 4×4 micromirror array device. (a) Without actuation, (b) piston actuation of a single aperture, and (c) tip-tilt of a single subaperture in four directions. (From Jia, K. et al., *IEEE J. Microelectromech. Syst.*, 20(3), 573, 2011.)

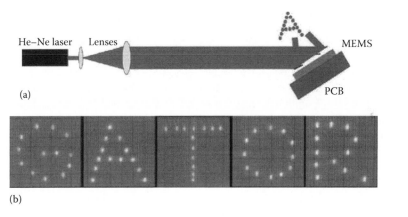

FIGURE 14.24 Demonstration of the motion capabilities of the mirror pixels. (a) Schematic of the experimental setup. (b) Letters achieved using the reflected light spots from the subapertures. (From Jia, K. et al., *IEEE J. Microelectromech. Syst.*, 20(3), 573, 2011.)

four in-plane thermal actuators pre-bent torsion bars, and an elastic square frame (Figure 14.25b). The static optical tilt angle of the 1-DOF micromirror is 6.5° under an applied dc voltage of 13 V. The 2-DOF micromirror can be actuated up to angles of 5.4° and 5.2° in each direction of rotation at 11 V_{dc}. The response time for the mirror angle to reach 98% of the final value is 0.0024 s.

14.5.4 MICROSCANNERS USING ELECTROMAGNETIC ACTUATION

Besides electrostatic actuated MEMS commercial microscanner being made available in the market, electromagnetic actuated ones have also been commercialized. An example is Microvision's Nomad head-worn display developed by A. D. Yalcinkaya et al. at Koç University and Microvision Inc. for retinal scanning display and imaging applications [13]. Figure 14.26a and b show the layout of the scanner, where a suspended outer frame with a multiturn spiral coil and an inner mirror (1.5 mm diameter) attached to the outer frame through fast scan flexures is designed and fabricated. The operation of the biaxial scanner depends heavily on super-imposing the drive torques of the slow and fast scan directions. A proposed superimposed torque is applied at 45° relative to the two

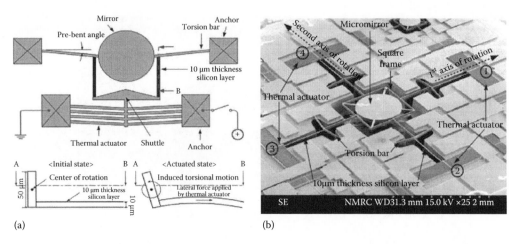

FIGURE 14.25 (a) At the top shows the schematic layout of the 1-DOF micromirror actuator with electrical connection. At the bottom shows the cross-sectional view of the torsion bar along the line A-B before and after actuation, showing how the lateral motion of the thermal actuator is converted to torsion motion. (b) SEM image of the fabricated 2-DOF micromirror. The device consists of four in-plane thermal actuators, four pre-bent torsion bars, and an elastic square frame. (From Eun, Y. and Kim, J., *J. Micromech. Microeng.*, 19(4), 045009, 2009.)

orthogonal scan axes, making use of the Lorentz's force excitation created by an external magnetic field. Full optical scan angles of 65° and 53° are achieved for slow (60 Hz sawtooth) and fast (21.3 kHz sinusoidal) scan directions, respectively. This 2-D magnetic actuation technique delivers sufficient torque to allow non-resonant operation in the slow scan axis while at the same time allowing 1 atm pressure operation even at fast-scan axis frequencies large enough to support SXGA (1280 × 1024) resolution scanned beam displays.

In 2007, another electromagnetic actuated 2-D scanner design was reported by the same research group in Istanbul [105]. S. O. Isikman et al. developed a scanner system as shown in Figure 14.27, where the device consists of a narrow cantilever beam supporting a rectangular mirror and an electromagnet for actuation. Polymer is used as the structural material because of its simple fabrication steps and mechanical properties. Magnetic permalloy is electroplated on one side of the mirror, while a silicon die with evaporated aluminum is attached as a reflecting mirror surface on the other side. Two modes of operation are possible with this design: bending (56.5 Hz) and torsional mode (340 Hz). Simultaneous excitation of the bending and torsion modes yields a 2-D raster-like resonant motion when the structure is energized with proper signals.

14.5.5 MICROSCANNERS USING PIEZOELECTRIC ACTUATION

In piezoelectric MEMS mirror, a piezoelectric material such as PZT film is often deposited between two metal electrode layers. When a voltage bias is applied at the electrodes, an electric field is generated between the two electrodes. As a result, stress is induced in the piezoelectric film, causing it to bend. A new mechanical design of piezoelectric unimorph actuator that generates large static deflection by accumulating angular displacement in a cascaded piezoelectric cantilever formed in a meandering shape [106]. The SEM image of the developed scanner, with a size of 4 mm × 6 mm, is shown in Figure 14.28a. The piezoelectric PZT actuators are folded into eight- and six-bars for the outer and inner actuator, respectively. The PZT film is deposited by arc-discharged reactive ion plating, with Pt metal electrodes sandwiching the PZT film. Figure 14.28b illustrates the mechanism of the novel PZT actuator that accumulates angular deflection in the cascaded cantilever. The PZT actuator is folded into a meandering form, in which the PZT film and metal electrodes are partitioned into two alternating groups, Piezo-A and Piezo-B. Voltages of opposite polarities are applied to the two actuator groups such that one of them deflects upward while the other bends downward.

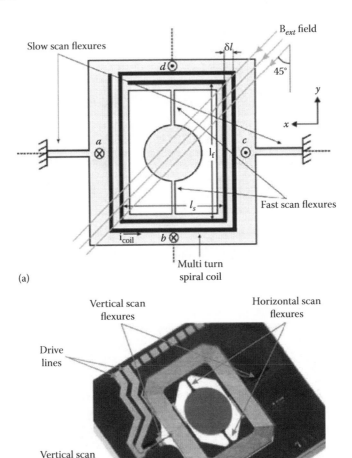

FIGURE 14.26 (a) Schematic of the biaxial microscanner. Only two turns of the actuation coils are shown for simplicity. Direction of the forces at points a and b are into the paper plane, at points c and d are out of the paper. (b) Die photo of the biaxial MEMS scanner. (From Yalcinkaya, A.D. et al., *J. Microelectromech. Syst.*, 15, 786, 2006.)

This angular motion is thus accumulated in the cascaded beam to give a large scan angle at the tip of the actuator where a mirror is attached. The scanner delivered a relatively large static mechanical angle of ±8.6° at an applied voltage of 20 V_{dc} at nonresonant operation.

K. H. Koh et al. from National University of Singapore reported a silicon micromirror driven by 1 × 10 PZT arrayed actuators, electrically connected in series, has been demonstrated for 2-D scanning mirror applications [107]. The schematic drawing of the PZT MEMS mirror is illustrated in Fig 14.29a. Ten patterned PZT thin film actuators are arranged in parallel along one of the sides of the micromirror. Figure 14.29b shows a photo of the packaged mirror, with the inset picture showing a gold-plated mirror surface to improve mirror reflectivity. The actuation concept for 2-D scanning is similar to that in Ref. [105], i.e., the mirror can operate in two modes: bending (34 Hz) and torsional (198 Hz). Bending mode occurs when the PZT actuators are electrically excited at 34 Hz, causing the mirror to rotate as shown in Figure 14.29a, while torsional mode occurs when the PZT actuators are electrically excited at 198 Hz, causing the mirror to twist orthogonally when compared to the rotational motion. Hence, 2-D Lissajous scan pattern, as shown in Figure 14.30, can be obtained when half of the 1 × 10 PZT arrayed actuators are biased at 34 Hz while the other half at 198 Hz such that both bending and torsional mode can be elicited simultaneously.

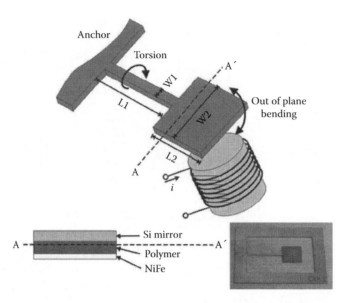

FIGURE 14.27 Schematic (top) and photo (lower right) of the electrocoil-actuated scanner. The system consists of a narrow cantilever beam supporting a rectangular mirror and uses an electromagnet for actuation. (From Isikman, S.O. et al., *IEEE J. Sel. Top. Quantum Electron.*, 12(2), 283, 2007.)

In another novel 2-D actuation mechanism proposed by K. H. Koh et al., a similar design of 1 × 10 PZT cantilever array, with separate electrical connection to each of the cantilevers, is explored and characterized [108]. With the separated electrical connections, the mirror can now achieve magnified torsional scanning based on the 180° phase difference in the applied bias to the ten piezoelectric cantilevers besides depending on the resonance phenomenon. A schematic diagram of the PZT MEMS scanning mirror is shown in Figure 14.31a, with the inset showing torsional mode where a set of five cantilevers 1–5 bend in one direction, while the other set of cantilevers 6–10 bends in the opposite direction. Figure 14.31b is a magnified photo showing the packaged device with a gold-coated mirror surface. In bending mode operation of the PZT scanning mirror, the sinusoidal ac driving voltage at 30 Hz frequency is applied simultaneously to all the top electrodes of the cantilevers, while the bottom electrodes of all the cantilevers are grounded. In torsional mode at 89 Hz actuation frequency, each cantilever will have different ac bias amplitude, with the largest and smallest cantilever displacements being introduced at the mirror edges and center, respectively. More importantly, the generated displacements for the two sets of cantilevers 1–5 and 6–10 are toward opposite directions, resulting in torsional deflection of the mirror plate. In mixed mode operation, bending and torsional actuation occur simultaneously to produce a 2-D Lissajous scanning pattern as shown in Figure 14.32. Two sets of cantilevers, 4–7 and 1–3, 8–10, are operating in bending and torsional actuations, respectively, during mixed-mode operation.

Besides 1 × 10 PZT cantilever array design, a novel dynamic excitation of an S-shaped piezoelectric actuator integrated with silicon micromirror is conceptualized for 2-D optical scanning illustration by having two superimposed ac voltages to bias the actuator [109]. Single actuator arranged in meandering style, compared to straight cantilever design, offers several advantages such as smaller footprint and larger displacement due to its lower mechanical stiffness. The proposed S-shaped PZT actuator integrated with a MEMS 2-D scanning mirror is shown in Figure 14.33a. Figure 14.33b illustrates the close-up photo of the device assembled onto a dual in-line package, with a spacer chip in between the device and the package. Bending and torsional modes occur when ac electrical signals with resonant frequencies of 27 and 70 Hz are used to excite the device, respectively. In mixed mode operation, two ac electrical signals of 27 and 70 Hz are applied simultaneously to the

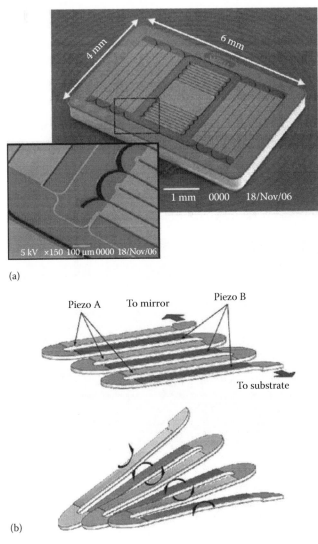

(a)

(b)

FIGURE 14.28 (a) SEM microphoto of the developed 2-D PZT scanner, where the inner frame and the mirror are driven by the meandering PZT actuators. (b) Principle of the PZT meandering actuator for large deflection. Upward and downward pair of deflection is accumulated in the cascaded beams. (From Tani, M. et al., A two-axis piezoelectric tilting micromirror with a newly developed PZT-meandering actuator, *IEEE 20th International Conference on Micro Electro Mechanical Systems*, January 21–25, 2007, pp. 699–702.)

S-shaped actuator using a summing amplifier, hence enabling the mirror to achieve 2-D scanning capability. An example of the Lissajous scan pattern is illustrated in Figure 14.34.

14.5.6 HIGH-SPEED MICROSCANNERS

In high-speed MEMS scanners for high-performance microdisplay, horizontal scanner frequency in the range of 10–20 kHz and frame rate of 60 Hz are often required [110]. Many such high-resolution MEMS scanners reported in literature are based on electromagnetic, electrostatic, and piezoelectric actuation, with electrothermal actuation highly unsuitable due to its slow thermal response [13,106,111–113]. H. M. Chu et al. demonstrated a microscanner that can be actuated in two orthogonal axes using slanted electrostatic comb-drive and silicon conductive V-shaped torsion hinges

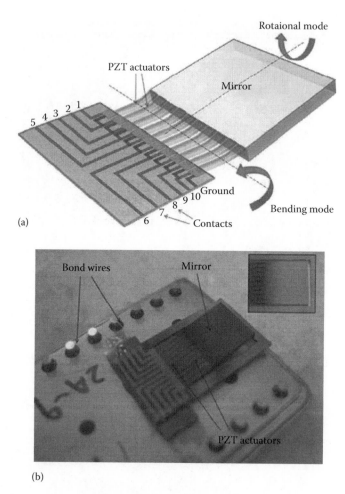

(a)

(b)

FIGURE 14.29 (a) Schematic drawing of MEMS scanning mirror where the mirror twists or bends due to ac voltage applied to the PZT actuators (b) A photo of the packaged MEMS micromirror device with a mirror area of 5 mm×5 mm. Inset picture shows a gold plated mirror surface of the device after gold sputtering. (From Koh, K.H. et al., *Sens. Actuators A*, 162, 336, 2010.)

FIGURE 14.30 Lissajous scan pattern obtained when half of the actuators are biased at bending mode, while the other half of the actuators are biased at torsional mode. (From Koh, K.H. et al., *Sens. Actuators A*, 162, 336, 2010.)

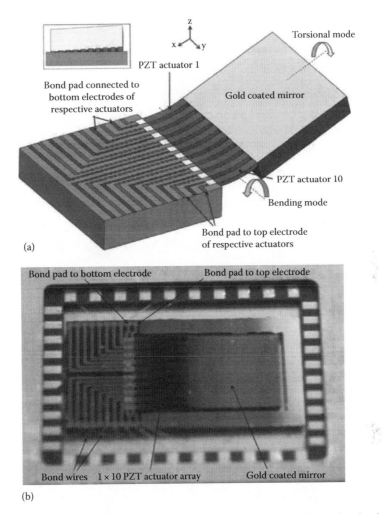

(a)

(b)

FIGURE 14.31 (a) Schematic diagram of a PZT 2-D scanning mirror where it twists or bends when different ac voltage combinations are applied separately to the 10 cantilevers. The inset shows torsional mode where a set of five cantilevers 1–5 bend in one direction, while the other set of cantilevers 6–10 bends in the opposite direction. (b) A magnified photo showing the packaged device with a gold-coated mirror surface. The bond pads are connected to the package via gold bond wires. (From Koh, K.H. et al., *J. Micromech. Microeng.*, 21(7), 075001, 2011.)

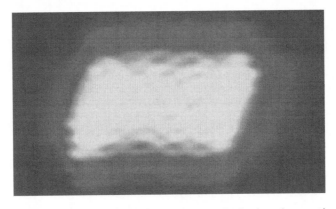

FIGURE 14.32 Lissajous scan pattern obtained in resonance of mixed mode operation. (From Koh, K.H. et al., *J. Micromech. Microeng.*, 21(7), 075001, 2011.)

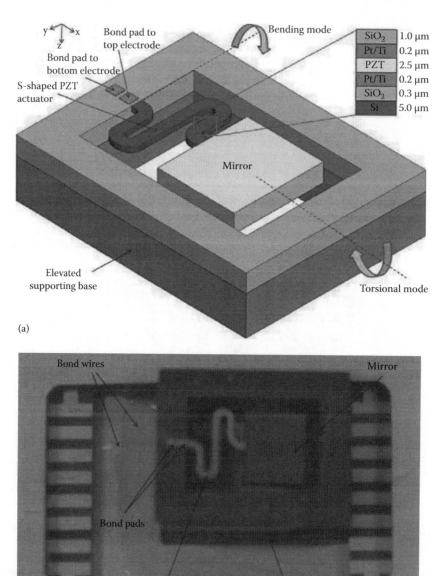

FIGURE 14.33 (a) Schematic diagram of 2-D scanning mirror actuated by S-shaped PZT actuator. Bending and torsional mode occur when the device is excited at their resonant frequencies, respectively. (b) Close-up photo showing the packaged MEMS mirror on a dual in-line package. (From Koh, K.H. et al., *Opt. Express*, 19(15), 13812, 2011.)

fabricated from a SOI wafer [112,113]. The design of the electrostatic vertical comb-drive actuated 2-D scanner is illustrated in Figure 14.35a. The resonant frequencies of the inner mirror and gimbal frame are 40 kHz and 162 Hz, respectively. The gimbal frame have a large mass and soft torsion hinge design so as to increase the resonant frequency ratio between the horizontal to vertical axes. The optical scanning angles for the inner mirror and the gimbal frame are 11.5° and 14° at operating voltages of 12 and 10 V in 1 Pa vacuum, respectively. Scanning image in 1 Pa vacuum at 75 and 6 V for the inner and outer gimbal frame, respectively, is demonstrated in Figure 14.35b.

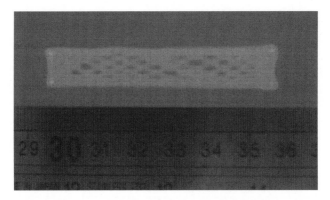

FIGURE 14.34 Example of Lissajous scan pattern obtained when ac signals of 27 and 70 Hz are applied simultaneously to the S-shaped actuator. (From Koh, K.H. et al., *Opt. Express*, 19(15), 13812, 2011.)

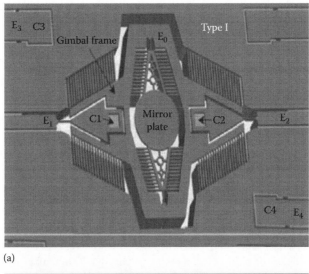

(a)

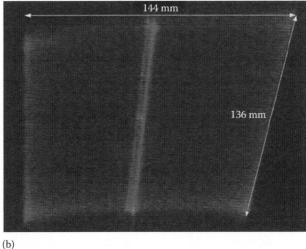

(b)

FIGURE 14.35 (a) Design of electrostatic vertical comb-drive actuated 2-D scanner. (b) Scanning image in 1 Pa vacuum at 75 V for inner mirror and 6 V for the gimbal frame, respectively. (From Chu, H.M. and Hane, K., *Sens. Actuators A*, 165, 422, 2011.)

14.6   MEMS MICROSPECTROMETER

Spectral analysis of an optical spectrum has been a well-established technique in all areas of science, ranging from emission and absorption spectra of a particular atom to the wavelength-dependent adsorption of an unknown chemical. Infrared (IR) absorption spectroscopy has also been well established in gas analysis, fire, and flame detection and in many other applications. A typical spectrometer consists of a broadband light source and spectral measurement channel with fixed narrow band-pass filters. These filters are selected to match the characteristic adsorption bands that are of interest in each specific application. Microspectrometers fabricated using MEMS technologies are versatile microsensors: small, lightweight, and fabricated using CMOS-compatible processes, thus featuring the possibility of realizing an intelligent optoelectronic system by an on-chip integration of optics and microelectronics. There are two main approaches to achieve spectroscopy using MEMS devices: interferometer-based spectrometers, such as Michelson, Mach–Zehnder, and Fabry–Perot (FP), or grating-based spectrometers. There have been extensive reviews done by R. F. Wolffenbuttel on various MEMS-based spectrometers [114,115] in 2006. In this section, we will only focus on MEMS spectrometers based on FP, grating, and Michelson architecture.

14.6.1   FABRY PEROT INTERFEROMETER

A classical FP interferometer is the key element of a MEMS-based tunable IR filter, which is built from an optical resonator consisting of two coplanar reflectors with a separation distance, d, and a material with a refractive index, n, in between them. By varying the separation distance d, the filter can be spectrally tuned. Two key figures are often used as the figures of merits for FP devices: the free spectral range (FSR) and the finesse (F). The FSR is defined as the spectral distance between two consecutive interference peaks, while F is the FSR divided by the full width half maximum (FWHM) of the interference peaks. In other words, FSR indicates the spectral tenability of the FP device, while F specifies its spectral resolution. Various actuation mechanisms have been proposed to drive the movable dielectric stacks of Bragg reflective mirror: electrothermal [116], piezoelectric [117], electromagnetic [118,119], and electrostatic [120–129].

In 1999, R. F. Wolffenbuttel et al. from Delft University of Technology, the Netherlands, developed a bulk-micromachined tunable FP filter for the visible spectral range [120]. As shown in Figure 14.36a and b, the FP filter is formed by two parallel 40 nm silver mirrors supported by a 300 nm low tensile stress silicon nitride membrane with a square aperture (2×2 mm^2) and an initial cavity gap of 1.2 µm. One of the mirrors is fixed, while the other is under tension on a movable Si frame, which is electrostatically deflected. Several distributed electrodes are used to control the cavity spacing and mirror parallelism. The photograph of the fabricated packaged FP filter is shown in Figure 14.36c. The FP filter has a finesse exceeding 30 and a FWHM smaller than 3 nm using 50 nm silver-coated mirrors.

Another research group from University of Western Australia designed and characterized a microspectrometer based on a voltage-tunable FP filter monolithically integrated on a HgCdTe IR photoconductor [121]. The device microphoto is illustrated in Figure 14.37a, with the cross-sectional view of it depicted in Figure 14.37b. The optical filter utilizes a FP structure that consists of a pair of vertically distributed Bragg mirrors (Ge and SiO) of size 100×100 µm^2, silicon nitride membrane, and a designed static air gap of 1.4 µm so as to achieve single-mode tuning from $\lambda = 2.5$ µm to 1.6 µm. Wavelength tuning is demonstrated from 1.7 to 2.35 µm (650 nm) with less than 9 V. Bandwidths of less than 55 nm ± 5 nm and switching times of 60 µs ± 10 µs have been achieved. In 2009, the same group demonstrated the use of strain stiffening in fixed–fixed electrostatic beam actuators to extend the wavelength tuning range of their FP device from 1.615 to 2.425 µm, which is the largest reported for such filter to date of publication [122]. The schematic diagram of the improved FP design and cross-section of the device is shown in Figure 14.38a and b, respectively. To address the curvature issues in the movable mirror, the authors did a series of low-power oxygen

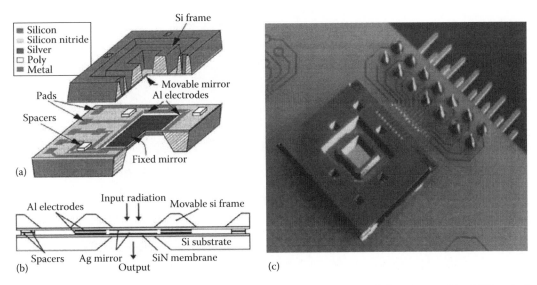

FIGURE 14.36 (a) Proposed FP microinterferometer. (b) Cross section of the micromachined FP optical filter. (c) Photograph of the fabricated FP. (From Correia, J.H. et al., *Sens. Actuators A*, 76, 191, 1999.)

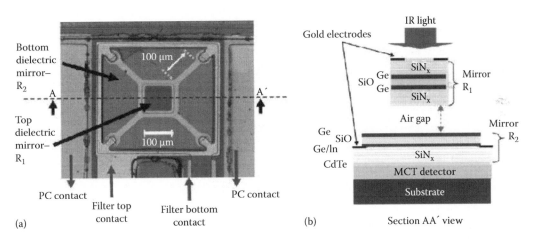

FIGURE 14.37 (a) Optical microscope photos of the MEMS spectrometer. (b) Cross-sectional view of AA′ showing the dielectric Bragg mirrors, gold electrodes, and the photoconducting IR detector under the suspended membrane. (From Keating, A.J. et al., *IEEE Photonics Technol. Lett.*, 18(9), 1079, 2004.)

plasma steps to alter the stress gradient and reduce the mirror curvature. The measured FWHM of the device is 52 nm at $\lambda = 2$ μm, which corresponds to a finesse of approximately 16, which is higher than the constrained finesse of many other MEMS FP filters with large tuning range.

In another research work from a German team, they reported results of a tunable pyroelectric detector with an integrated FP filter in the mid-wave IR range [123–126]. Figure 14.39a illustrates the design principle of the detector with tunable filter. To achieve the tuning of the resonator air cavity, an electrostatic actuation using a parallel plate design has been chosen. An optical aperture of 2×2 mm^2 is used, with the Bragg mirrors being made of alternative layers of silicon dioxide ($n = 1.38$ at $\lambda = 4$ μm) and polycrystalline silicon ($n = 3.33$ at $\lambda = 4$ μm), while silicon nitride was tested for use as single layer ARC. Figure 14.39b shows the schematic and actual picture of the packaged tunable pyroelectric detector with integrated filter, while Figure 14.39c depicts SEM images of the filter and the parallel spring suspensions, which serve to actuate the Bragg mirror ideally in a vertical

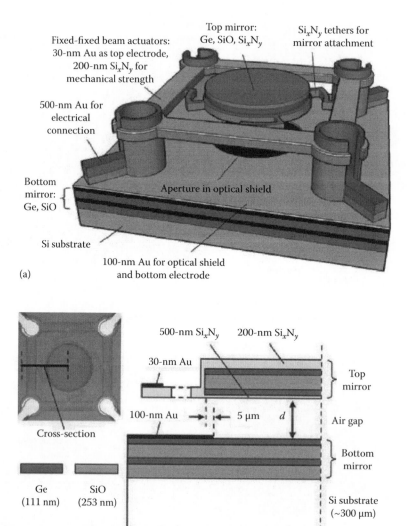

FIGURE 14.38 (a) Schematic illustration of the improved design that uses electrostatic fixed-fixed beam actuators to drive a suspended mirror. (b) Optical micrograph and cross section of filter, with the mirrors fabricated from quarter-wave dielectric stacks of Ge and SiO. (From Milne, J.S. et al., *IEEE J. Microelectromech. Syst.*, 18(4), 905, 2009.)

direction. An effective finesse of 45.6 is derived after taking into account of the warping and surface roughness of the Bragg mirrors. The German team followed up this design with a new improved tunable filter design in the IR (8–11 μm) range [127]. Instead of using only movable reflector to tune the gap of the resonant air cavity, they proposed a new design that has two movable reflectors driven by electrostatic forces instead (Figure 14.40a and b). This enhanced design reduces the influence of gravitation and vibration as both of these forces will act simultaneously on both reflector carriers, hence causing little disturbance to the optical cavity. In addition, the stiffness of the suspensions can be significantly lowered compared to a design with one movable reflector, which leads to a much lower control voltage. A FWHM bandwidth of less than 200 nm is obtained in the range from 8 to 10.2 μm. The maximum tuning range is achieved with voltages below 63 V, with simulation demonstrating the possibility to decrease that below 30 V. The influence of gravitation is determined with ±4 nm in comparison to their initial design with only one movable reflector, which shows ±28 nm.

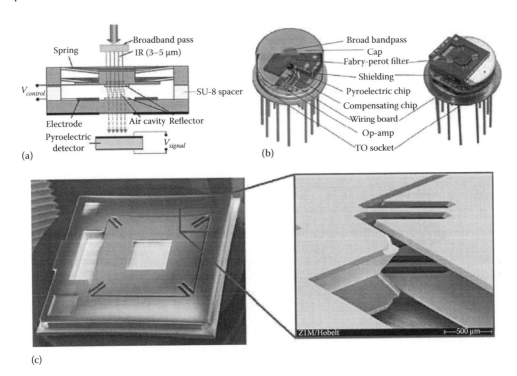

FIGURE 14.39 (a) Schematic cross-sectional diagram of the device, illustrating the design principle of the detector with tunable filter. (b) Schematic drawing (left) and picture of a packaged sample (right) with a tunable pyroelectric detector with integrated filter. (c) SEM images of the tunable filter and a zoom-in view of the parallel spring suspension. (From Neumann, N. et al., Tunable infrared detector with integrated micromachined Fabry-Perot filter, *Conference on MOEMS and Miniaturized Systems VII*, San Jose, CA, SPIE vol. 6466, 646606, 2007; Neumann, N. et al., *J. Micro/Nanolith. MEMS MOEMS*, 7(2), 021004, 2008; Neumann, N. et al., Novel MWIR microspectrometer based on a tunable detector, *Conference on MOEMS and Miniaturized Systems VIII*, San Jose, CA, SPIE vol. 7208, 72080D, 2009; Ebermann, M. et al., Recent advances in expanding the spectral range of MEMS Fabry-Perot filter, *Conference on MOEMS and Miniaturized Systems IX*, San Jose, CA, SPIE vol. 7594, 75940V, 2010.)

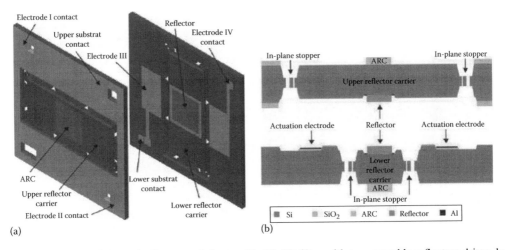

FIGURE 14.40 (a) Schematic diagram of the tunable IR FP filter with two movable reflectors driven by electrostatic forces. (b) Cross-sectional schematic diagram of the two movable reflectors and the resonant air cavity. (From Meinig, M. et al., Tunable mid-infrared filter based on Fabry–Pérot interferometer with two movable reflectors, *MOEMS and Miniaturized Systems X*, San Jose, CA, SPIE vol. 7930, 79300K, 2011.)

H. Toshiyoshi et al. from University of Tokyo also developed a wavelength tunable fiber-optic filter using a hybrid-assembly technique of glass reflector cubes (1 mm³) onto a bulk-micromachined electrostatic translation stage [128,129]. The schematic diagram and SEM images of the hybrid-assembled tunable filter is illustrated in Figure 14.41a and b, respectively. Figure 14.41c illustrates the closing up of the gap when the movable stage is not actuated, while Figure 14.41d shows the

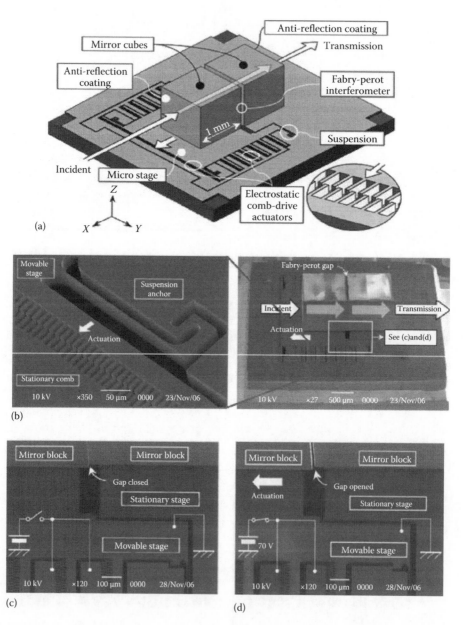

FIGURE 14.41 (a) Schematic diagram of the hybrid assembled wavelength tunable filter. SEM images of the tunable filter (b) after reflector cube assembly and close-up view of the suspension and lateral comb-drive actuators (c) during electrostatic operation when gap is closed without operation bias (d) during electrostatic operation when gap is opened with 70 V_{dc}. (From Yamanoi, T. et al., A hybrid-assembled MEMS Fabry-Perot wavelength tunable filter, *The 14th International Conference on Solid-State Sensors, Actuators and Microsystems (Transducers'07)*, Lyon, France, June 10–14, 2007; Yamanoi, T. et al., *Sens. Actuators A*, 145–146, 116, 2008.)

opening up of the gap when the movable stage is electrostatic actuated. More than 0.6 μm change of the FP interferometer gap can be electrostatically controlled with a tuning voltage upwards of 4 V with the aid of a dc bias voltage of 64 V. A wide tuning range from 1479 to 1590 nm was achieved with a bandwidth of 1.3–1.2 nm and insertion loss of 1.59–1.76 dB.

14.6.2 GRATING-BASED AND LAMELLAR GRATING INTERFEROMETER

A grating is commonly used for the dispersion of the spectrum of incident optical radiation into different wavelength components. Radiation incident upon the grating yields an interference pattern with an angular distribution determined by the spectral components. By measuring the distributed spectra, the incident light can be analyzed. A grating-based microspectrometer is developed by R. F. Wolffenbuttel et al. in 2001 [130]. The IR spectrometer, shown schematically in Figure 14.42a, consists of two independently processed wafers, which are bonded in the final step. The grating composed of 30 or 60 slits with the grating constant ranging from 4 to 20 μm. Thermal detector strips of 400 μm long and 100 μm wide are used, with one array of detectors consisting between 6 and 16 of them. Figure 14.42b shows the top view of the fabricated device, with the upper half showing the array of thermal detectors. The silicon bulk is used as optical path so as to take advantage of the transparency of silicon in the IR region, while the grating is realized using aluminum due to its excellent reflectance and adsorption properties in the IR range.

In recent years, Fourier transform spectrometers (FTS), which is a Michelson interferometer with a movable reference mirror, have been widely reported in literature. Faster detector arrays, efficient Fourier transfer algorithms, and the increasing computing power of microprocessors enabled FTS to become a well-known and widely used measurement technique. A lamellar grating interferometer (LGI) is a special type of FTS that utilizes a movable diffraction grating and operates in the zeroth order. In contrast with a Michelson interferometer that splits wave amplitude at the beam splitter, a LGI divides the wavefront to form the interference pattern. Such lamellar-grating-based interferometers offer several outstanding advantages such as the absence of beam splitters, extended wavelength range, robustness, and high efficiency. For example, C. Ataman et al. developed a novel LGI based on out-of-plane vertical resonant comb actuators with the comb fingers serving as both actuation and dynamic diffraction grating [131]. This greatly reduces the die size and eliminates the use of a beam splitter and other optics. An illustration of the designed grating structure and a photograph of the actual device are shown in Figure 14.43a and b, respectively. The comb sets are symmetrically placed on a rigid backbone, which enhances the robustness of the structure and helps

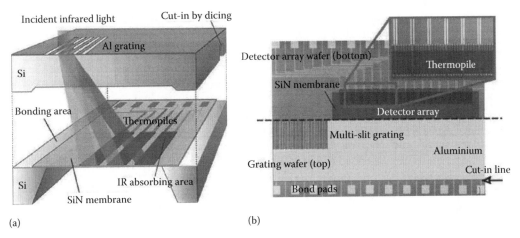

(a)   (b)

FIGURE 14.42 (a) Schematic view of the grating-based microspectrometer for IR. (b) Microfabricated IR microspectrometer in silicon. (From Kong, S.H. et al., *Sens. Actuators A*, 92, 88, 2001.)

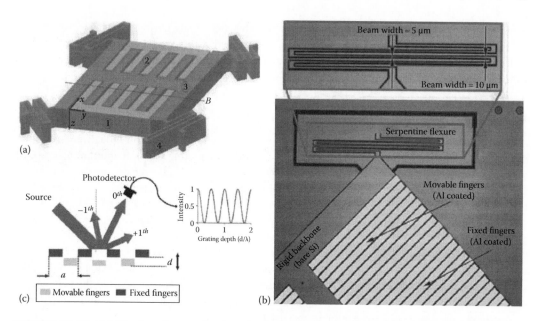

FIGURE 14.43 (a) The lamellar grating interferometer device structure: (1) fixed fingers, (2) movable fingers, (3) rigid backbone, and (4) folded flexures. (b) Microscopic images of the fabricated device and a close-up view of the serpentine flexure with nonuniform beam cross section. (c) Schematic of the comb-drive rectangular diffraction grating and modulation of the zeroth intensity as d changes. (From Ataman, C. et al., *J. Micromech. Microeng.*, 16(12), 2517, 2006.)

to keep the device surface flat while resonating. Experimental results show the device being capable of producing ±52 μm vertical deflection at 28 V square-wave excitation without requiring vacuum packaging. This yields a theoretical spectral resolution of 0.4 and 3.6 nm in the visible and IR wavelengths, respectively. Light collection efficiency is measured to be 15%. FTS results show a FWHM of 12.2 nm and a peak wavelength of 639 nm, while the reference measurement result is 638 nm.

G. Zhou et al. from National University of Singapore reported another lamellar-grating-based FTS, in which an electromagnetic actuator is used to move the mobile facets of the lamellar grating bi-directionally [132]. As shown in Figure 14.44a, the proposed FTS has a platform with four folded-beam suspensions arranged at each corner to accommodate the permanent magnet and movable grating facets. The lamellar grating consists of one set of light-reflecting fingers that is connected to the platform and driven up and down, while the other set is fabricated onto the substrate constituting the fixed part of the lamellar grating. All the finger surfaces are coated with gold to enhance reflectivity. Figure 14.44b shows a microscopic image of a micron-sized permanent magnet already assembled manually on the platform using ultraviolet epoxy, with magnetization perpendicular to the device surface. The image of the reflected light beams in Figure 14.44c shows light being propagated in two different directions with a small angle. This is due to the residual stresses in the structure material, thus preventing interference from occurring. To overcome this problem, a focusing lens is inserted between the collimator and the beam splitter to make the two reflected light beams diverge and overlap with each other to produce an interference pattern in the detection plane as shown in Figure 14.44d and e. Experimental results presented show that deflection of ±65 μm can be achieved by applying positive and negative current of 129 mA to the external actuation coil. The final Fourier transform spectrum show a spectral resolution of 3.8 nm at wavelength of 632.8 and 3.44 nm at 532 nm.

G. Zhou et al. also reported another novel SOI micromachined stationary LGI for FTS applications [133]. Unlike the previous spectrometer design developed in Refs. [131,132], which records interferograms as a function of time, the proposed stationary FT spectrometers record interferograms as a spatial distribution of light intensity. The device utilizes a set of tilted

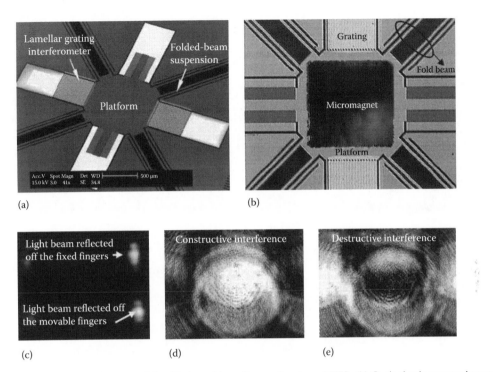

FIGURE 14.44 (a) SEM image of the fabricated lamellar-grating-based FTS. (b) Optical microscope image of the assembled micron-sized permanent magnet onto the platform using UV epoxy to realize the final device. Images of the reflected light beams captured by a detector. (c) Light reflected from grating with collimated incident light. (d, e) Interference patterns at two different optical path differences. (From Yu, H. et al., *J. Micromech. Microeng.*, 18(5), 055016, 2008.)

interdigitated light reflecting beams to produce an interferogram with spatially varying optical path difference recorded by a line of photodetector array. The SEM image of the device is shown in Figure 14.45a, with the inset showing a close-up view of the stationary lamellar grating, tilted platform, and platform locking plate. The lamellar grating is formed by tilting interdigitated silicon beams through assembly process illustrated in Figure 14.45b. The prototype FTS demonstrates a FWHM spectral resolution of 2 and 2.6 nm at wavelength of 532 and 632.8 nm, respectively.

14.7 OPTICAL NEMS: TUNABLE NANOPHOTONICS

There has been increasing interest in using PCs as a platform for compact photonics integrated circuits and optical communication because of their ability to confine and localize optical energy. Control of photonic micro/nanocavity resonance in conjunction with large spectral shifting is critical in achieving reconfigurable photonic devices. While various dynamic tuning approaches such as electro-optic, magneto-optic, liquid-crystal, and microfluidics have been vastly reported, the small tuning range and high power consumption pose as major limitations. To extend the application of these nanophotonic structures while overcoming these limitations, MEMS structures can be combined with nanophotonic devices to control the submicron distance between a PC resonator and PC waveguide (PCWG) driven by a MEMS actuator.

14.7.1 PC OPTICAL SWITCH

In 2006, M. Wu et al. demonstrated a MEMS-actuated 1-D PC switch, which consists of a PC slider connected to an electrostatic comb-drive actuator and inserted between two waveguides

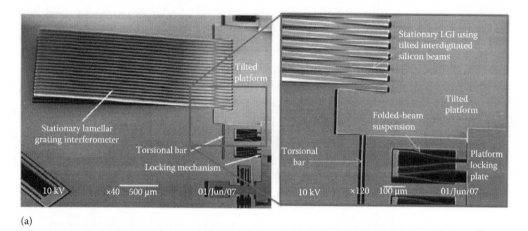

(a)

(b)

FIGURE 14.45 (a) SEM image of the stationary lamellar grating interferometer based on SOI micromachining technology. Inset shows a close-up view of the details for stationary lamellar grating interferometer, tilted platform, and platform-locking plate. (b) Schematic illustration of the assembly process for the stationary lamellar grating interferometer. (From Chau, F.S. et al., *J. Micromech. Microeng.*, 18(2), 025023, 2008.)

(Figure 14.46a) [134]. As shown in Figure 14.46b, the reflection state is made up of two Si quarter-wave slabs separated by a quarter-wave air gap, while the transmission state is made of a silicon piece, which mimics two periods PCs filled with silicon in the central gap as a defect. As such, when the reflection state is moved in between the waveguides, it forms a forbidden photonic bandgap and reflects the input light. On the other hand, when the transmission state is positioned between the waveguides, it creates a narrow passband within the photonic bandgap. Signals at the center wavelength, which is determined by the size of the defect, can thus be switched dynamically between two states by the MEMS actuator, achieving ON-OFF switching function. Figure 14.47 shows SEM images of the fabricated MEMS PC switch with a close-up view of the PC slider. Experimental

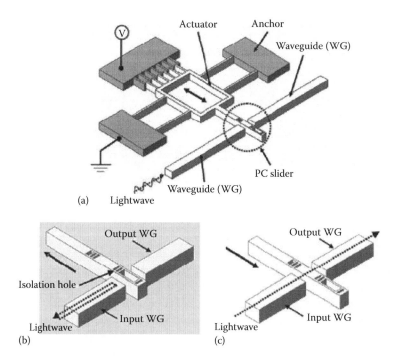

FIGURE 14.46 (a) Schematic of the MEMS actuated PC switch. The lateral electrostatic comb-drive actuators connect with PC slider, which is inserted between two waveguides. Two states are defined on tip of PC slider: (b) reflection state and (c) transmission state. (From Lee, M.-C.M. et al., *IEEE Photonics Technol. Lett.*, 18(2), 358, 2006.)

results show an extinction ratio of 11 dB at 1.56 μm wavelength and a 0.5 ms time constant of the step response. This work is followed up a year later, with a report on the first single-crystalline-silicon microtoroidal resonator with MEMS tunable optical coupler from the same research group [135]. The schematic of the tunable microtoroidal resonator (39 μm diameter resonator, 200 μm toroidal radius) is shown in Figure 14.48a, where two suspended waveguides are vertically coupled to a microtoroidal resonator. The initial spacing between the microtoroid and the waveguides is 1 μm so that there is negligible coupling at zero bias. As illustrated in Figure 14.48b, with increasing voltage bias between the waveguide and the fixed electrodes, the suspended waveguide is pulled down toward the microtoroid, increasing the optical coupling exponentially. The SEM image of the fabricated device is shown in Figure 14.48c, with the waveguides having width and thickness of 0.69 and 0.25 μm thick, respectively. The unloaded quality factor is as high as 110,000, with the loaded Q continuously tunable from 110,000 to 5,400. The extinction ratio of the transmittance is 22.4 dB, potentially making this device to be a building block of resonator-based reconfigurable photonic integrated circuits.

K. Hane et al. from Tohoku University, Japan, also demonstrated various works using MEMS technology to control the coupling efficiency of optical resonator [136–138]. They first proposed a wavelength-selective add-drop switch using a silicon microring resonator combined with an electrostatic microactuator [136]. The schematic diagram of the proposed device is illustrated in Figure 14.49a and b. The coupling between the microring and the buslines can be tuned by the electrostatic lateral comb-drive actuators. Input light is selectively transmitted to the drop port when the microactuator makes the microring approach to the buslines under the condition that the input wavelength is equal to one of the resonant wavelengths of the microrings. On the other instance, when the actuator is not biased, the microring remains far apart from the buslines, hence decreasing the coupling efficiency. As a result, the input light passes to the through port without interacting

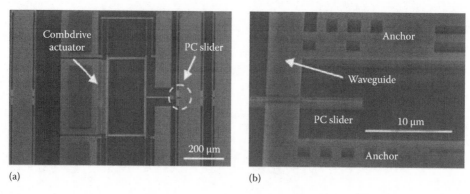

(a)   (b)

FIGURE 14.47   (a) SEM image of the fabricated MEMS PC switch and (b) close-up view of the PC slider, with two anchors supporting the suspended waveguides. (From Lee, M.-C.M. et al., *IEEE Photonics Technol. Lett.*, 18(2), 358, 2006.)

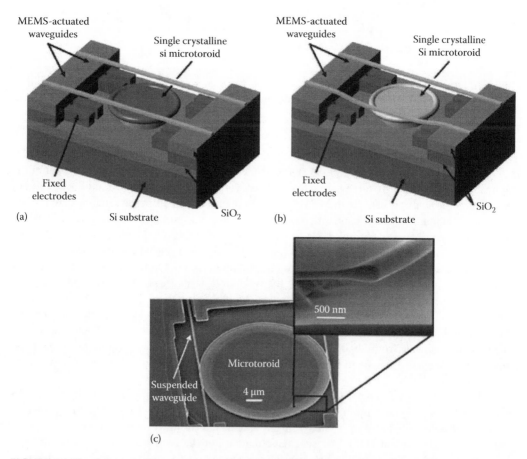

FIGURE 14.48   Schematic illustration of the microtoroidal resonator with integrated MEMS coupler at (a) zero bias, in which there is negligible coupling occurring between the resonator and the waveguide coupler due to large initial spacing, and (b) bias condition, where the lower waveguide is pulled downward to increase coupling, while the upper waveguide remains straight and uncoupled. (c) SEM images of the microtoroidal resonator and the suspended waveguide. Inset shows the cross-sectional view of the microtoroid. (From Yao, J. et al., *IEEE J. Sel. Top. Quantum Electron.*, 13(2), 202, 2007.)

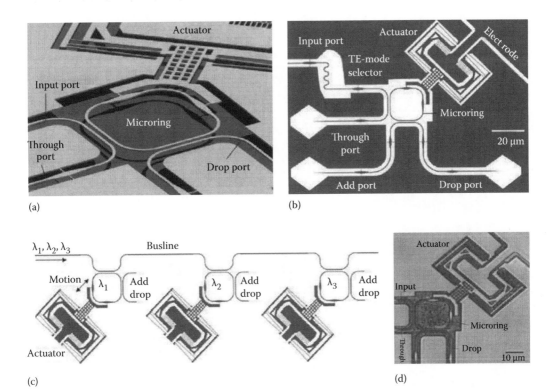

FIGURE 14.49 Schematic illustration of the (a) microring resonator wavelength-selective add-drop switch using a electrostatic-biased microactuator, (b) actual arrangement of the device components, (c) an application of the proposed device for wavelength division multiplex system, and (d) optical microphoto of the fabricated device. (From Takahashi, K. et al., *Opt. Express*, 16, 19, 2008.)

with the microring resonator. This two operation states allow the proposed device to work as a wavelength-selective add-drop switch (Figure 14.49c). When the applied voltage increases from 0 to 28.2 V, transmittance from the input port to drop port and through port decrease by 32.9 and 7.83 dB, respectively. The FWHM bandwidth of the dropped light was 0.5 nm, which corresponds to a Q-value of 3150.

In addition to using microring structure to tune the optical coupling efficiency, K. Hane et al. also demonstrated using ultra-small electrostatic comb actuators to tune nanomechanically the PC nanocavity [137]. The proposed device, as shown in Figure 14.50a through c, consists of PCWG and a movable PC nanocavity. The PCWG consists of an edge waveguide of a PC slab, in which the propagating wave is confined on the left side by a PC, as shown in Figure 14.50d, while on the right side, it is confined by a low refractive index ambient air. The nanocavity is a PC defect located in a movable PC slab, with the air gap between the PCWG and the nanocavity being adjustable through the translation of the movable PC slab. The light wave propagating in the PCWG can interact with the nanocavity under resonant condition, thus causing the propagated light to be dropped from the PCWG to the nanocavity if the coupling between them is strong. As shown in Figure 14.50d, little light is coupled to the nanocavity at a gap of 500 nm. By decreasing the gap, the light wave coupled to the nanocavity increases at gaps of 250 and 50 nm. Experimental results show a controlled drop efficiency of 12.5 dB with a gap change of 600 nm. Under the critical coupling condition using a $\lambda = 1.5724\ \mu m$, the absolute value of drop efficiency obtained from the measured spot intensities was −6.1dB, which corresponds to 24.56% of the input waveguide intensity.

Another similar in-plane nanomechanics approach to achieve reconfigurable PC device is proposed by G. Zhou et al. [139,140]. In their research, they developed three different perturbing tip

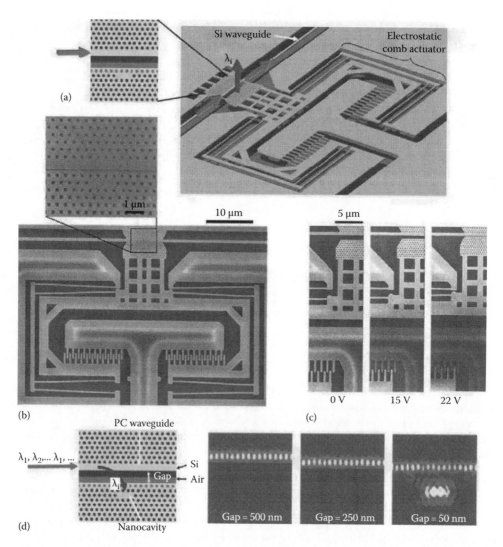

FIGURE 14.50 (a) Schematic illustration of the nanomechanical channel drop PC switch. (b) SEM of the proposed device fabricated by sacrificial etching of 2 μm thick SiO₂ layer under the movable structure. (c) SEM of the biased device obtained when 0, 15 and 22 V were applied to the actuator. (d) A schematic close-up view illustrating the PC waveguide, air gap, and nanocavity and the calculated electrical fields at the resonant wavelength of the nanocavity for different air gaps. (From Kanamori, Y. et al., *Appl. Phys. Lett.*, 95, 17, 2009.)

designs, in which the tips are driven by submicron MEMS systems to achieve reversible and low-loss resonance control on a 1-D PC nanocavity. SEM image of the proposed retractable nanomechanical actuating mechanism with integrated low loss nanowire waveguides is illustrated in Figure 14.51a, while Figure 14.51b and d show SEM images of coupled nanocavity nano-tip, rectangular nano-tip, and meniscus-like perturbing nano-tip, respectively. The retractable sub-micron MEMS push–pull structure shown in Figure 14.51a features two sets of electrostatic comb-drive actuators, with each set for an opposite direction of displacement, i.e., push-pull. This sub-micron MEMS structures can ensure mechanical stability and robustness of nanomechanical perturbing members when they are in the optical near-field region. Figure 14.51e shows the nanocavity nano-tip sticking to the 1-D PC resonator during near-field operating regime. This is due to the Van der Waals surface forces being much larger than the spring restoring force. Figure 14.51f shows the retracted mechanical structure actuated by the second set of pull electrostatic comb-drive actuators. The coupled nanocavities

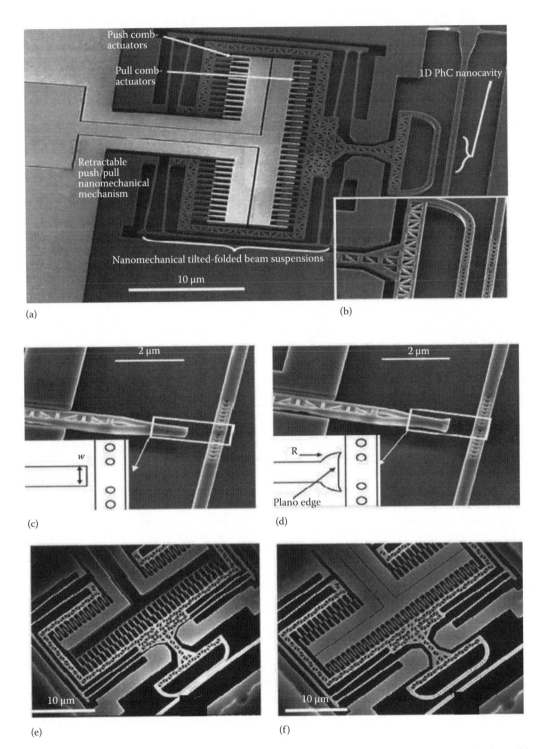

FIGURE 14.51 SEM images of the (a) proposed retractable nanomechanical actuating mechanism with integrated low loss nanowire waveguides, (b) coupled-nanocavity nanotip, (c) rectangular nanotip, (d) meniscus-like perturbing nanotip, (e) stiction incurred nanomechanical structure, and (f) retracted mechanical structure. (From Chew, X. et al., *Opt. Express*, 18(21), 22232, 2010; Chew, X. et al., *Opt. Lett.*, 35(15), 2517, 2010.)

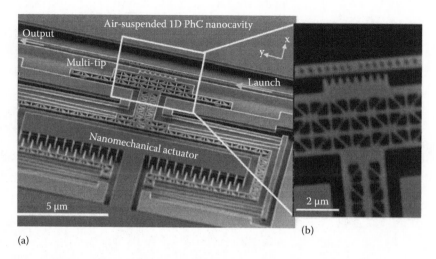

FIGURE 14.52 SEM images of the (a) MEMS driven multitip nanoprobe in the vicinity of an air-suspended 1-D PC nanocavity and (b) nanoprobe displaced to a gap of 100 nm. (From Chew, X. et al., *J. Nanophotonics*, 5(1), 059503, 2011.)

approach demonstrates a relatively large resonance shift of 18 nm with minimal degradation in Q, making it useful for tunable photonic devices such as photonic switches which require large tuning range. The in-plane rectangular nano-tip perturbation approach demonstrates small tuning properties and incurs vast broadening of Q and decreasing transmitted power. The meniscus-like perturbing tip resulted in 2.25 nm resonance shifts with minimal transmitted power loss and deterioration of Q. These results demonstrated that 1-D PC nanocavities are highly sensitive to geometries and size of perturbing tip. Reducing the width of the tip effectively reduces the scattering losses. However, there is a trade-off, in which less resonance perturbation is achieved.

To enhance the resonance tuning of PC nanocavities, G. Zhou et al. integrated a periodic multitip nanoprobe that consists of a total of eight nanotips that are designed to perturb into the 1-D PC resonance mode [141]. Figure 14.52a illustrates the fabricated MEMS-driven multitip nanoprobe in the vicinity of an air-suspended 1-D PC nanocavity, while Figure 14.52b shows a close-up view of the probe at an offset gap of 100 nm. Experimental results for the periodic multitip device having a periodicity of 470 nm show a resonance at 1608 nm with an experimental loaded Q of 1200 perturbation. When the perturbing multitip is actuated at 14.2 V, which corresponds to a 160 nm offset gap, the resonance is tuned by about 0.92 nm. However, the Q degrades to 580 due to energy leakage. Further testing has shown a resonance tuning up to 5.4 nm with minimal Q and transmission degradation.

14.7.2 OPTICAL-DRIVEN NANO-OPTO-MECHANICAL SYSTEMS

With the recent development of nanosize DRIE and micromachining technologies, nano-opto-mechanical systems (NOMS) present potential for various highly sensitive sensor applications. Most of the conventional actuators reported are driven by electrothermal, electrostatic, electromagnetic, and piezoelectric means, which occupy large size and consume high power with slow start-up speed. However, with the discovery of attractive and repulsive optical forces, optical-driven actuation with the development of nano-sensor and nano-actuator allows the application of optical force in NOMS.

A. Q. Liu et al. demonstrated various NOMS devices that utilize optical force for switching and sensing applications [142–144]. In one of their works, they reported a nano-opto-mechanical actuator that is driven by optical radiation force [142]. As shown in Figure 14.53a and c, the actuator consists of two waveguides, two identical ring resonators, and an actuator with Bragg reflector.

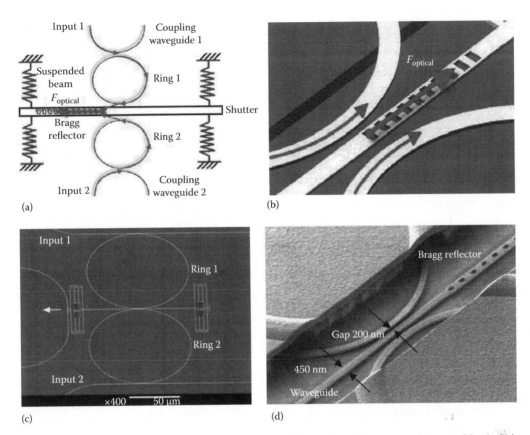

FIGURE 14.53 Schematic of the (a) NOMS actuator, which consists of two waveguides, two identical ring resonators, and four suspended folded beams and (b) Bragg reflector. (c) SEM images of the NOMS actuator and (d) Bragg reflector. (From Zhao, X. et al., A nano-opto-mechanical actuator driven by optical radiation force, *The 16th International Conference on Solid-State Sensors, Actuators and Microsystems (Transducers'11)*, Beijing, China, June 5–9, 2011, pp. 1468–1471, http://nocweba.ntu.edu.sg/laq_mems/conference/Transducers/AMEMSDIGITALMIRRORFORTUNABLELASERWAVELENGTHSELECTION.pdf.)

The input light is coupled through the waveguides to the central actuator with the Bragg reflector via the two ring resonators. The free-standing actuator is suspended by four folded beams and can be pushed forward via the radiation force introduced by the ring resonators. The Bragg reflector reflects the light coupled from the ring resonators and thus enlarges the momentum exchange between the photons and the actuator. The optical force actuator shows high resolution (2.501 nm/mW) and linear displacement to the input optical power, which leads to its potential applications on precise distance control, small mass measurement, and contact-free actuator. A. Q. Liu et al. also proposed a pressure sensor array based on optical force using NOMS [143]. Figure 14.54a shows the schematic diagram of the pressure sensor, which consists of a deformable diaphragm and a waveguide array. Part of the waveguide is suspended and formed a tunable gap with the diaphragm. By applying pressure on the diaphragm, the gap is changed and optical forces are generated subsequently to deform the waveguides. This causes a change of effective refractive index of the waveguide and the output power. Experimental results have shown that the device is capable of measuring up to 32 dB optical output modulation when the pressure is changed from 0 to 250 kPa, which is equivalent to a sensitivity of 0.128 dB/kPa.

Besides NOMS devices, which make use of optical force for sensing applications, A. Q. Liu et al. also revealed a NOM switch using electromagnetically induced transparency (EIT)-like effects of coupled-ring resonator and realized it by using optical force to drive a silicon waveguide [144].

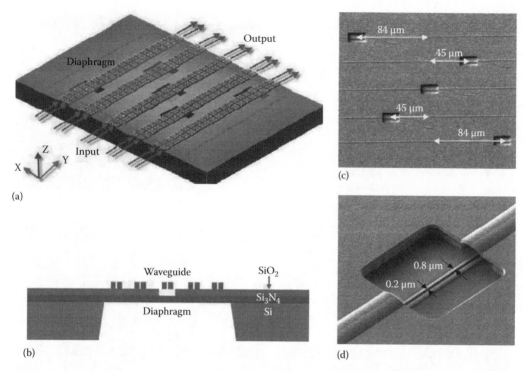

(a)

(b)

(c)

(d)

FIGURE 14.54 (a) Schematic diagram of the optical pressure sensor, consisting of a deformable diaphragm and a waveguide array. (b) Cross section of the pressure sensor. (c) SEM image of the pressure sensor array. (d) SEM image showing the top view of the coupled waveguides with a hard mask layer on the top. (From Zhao, X. et al., Pressure sensor using nano-opto-mechanical systems (NOMS), *The 16th International Conference on Solid-State Sensors, Actuators and Microsystems (Transducers'11)*, Beijing, China, June 5–9, 2011, pp. 1030–1033, http://nocweba.ntu.edu.sg/laq_mems/conference/Transducers/AMEMSDIGITALMIRRORFORTUNABLELASERWAVELENGTHSELECTION.pdf.)

Figure 14.55a and b show the schematic illustration and SEM image of the NOM switch, which consists of a free-standing spiderweb-like ring resonator (ring 1) and a fixed ring resonator (ring 2), respectively. Both rings are side-coupled to a bus waveguide. Ring 1 is supported by four spokes connected to an inner disk, which reduces the mechanical stiffness of the ring cavity. When light wave travels in the free-standing ring resonator, it couples evanescently with the oxide dielectric substrate, and attractive optical force between them is generated. This causes the spoke to bend toward the substrate with the gap between the ring 1 and dielectric substrate decreases. The effective index changes as a result and causes the EIT-like transmission peak to be switched off. Experimental results have shown that as the injection pump power increases from 0 to 6 mW, the normalized power of transmission peak at 1553.3 nm decreases from 1 to 0.02, and the position of the transmission peak has a 7.8 GHz red shift. This developed NOM switch has been proposed for use in all-optical filter, phase shifters, logic gates, and memory.

14.7.3 NANOPHOTONIC, NANOMECHANICAL, AND BIOMOLECULE SENSOR

The integration of silicon PCs with NEMS-based structures and function elements offer a new category of nanomechanical sensors. By measuring the wavelength shift of the output resonant peak, optical nanomechanical sensors can now be used to measure physical parameters such as force, stress, strain, and displacement. For example, C. Lee et al. from National University of Singapore designed and modeled a PCWG filter by introducing microcavities within the line defect

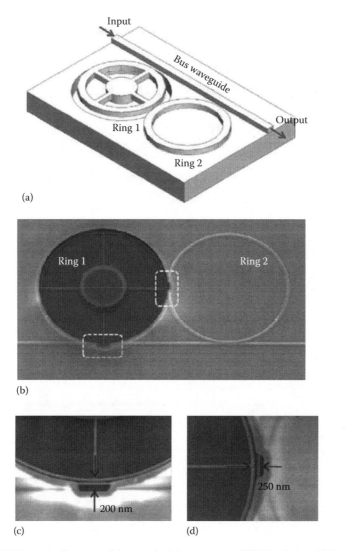

FIGURE 14.55 (a) Schematic diagram of the coupled ring resonator NOM switch, which consists of a freestanding spiderweb-like ring resonator and a fixed ring resonator. SEM image of the (b) SOI coupled ring resonator, (c) close-up view of ring-waveguide coupling region, and (d) ring-ring coupling region. (From Ren, M. et al., Design and experiments of a nano-opto-mechanical switch using EIT-like effects of coupled-ring resonator, *The 16th International Conference on Solid-State Sensors, Actuators and Microsystems (Transducers'11)*, Beijing, China, June 5–9, 2011, pp. 1436–1439, http://nocweba.ntu.edu.sg/laq_mems/conference/Transducers/AMEMSDIGITALMIRRORFORTUNABLELASERWAVELENGTHSELECTION.pdf.)

so as to form the resonant bandgap structure for PC. A suspended silicon bridge structure integrated with the PCWG filter structure is proposed and shown in Figure 14.56a and b. A shift of the output resonant wavelength is observed for suspended PCWG beam structure under particular force loading. In other words, the induced strain modifies the shape of air holes and the spacing among them. Such an effect leads to shift of resonant wavelength. Under optical detection limitation of 0.1 nm for resonant wavelength shift, the sensing capability of this nanomechanical sensor is derived as 20–25 nm of vertical deformation at the center, and the smallest strain is 0.005% for defect length. In addition to PCWG structure, C. Lee et al. also demonstrated a novel nanomechanical sensor using silicon cantilever embedded with a 2-D PC microcavity resonator [146]. As shown

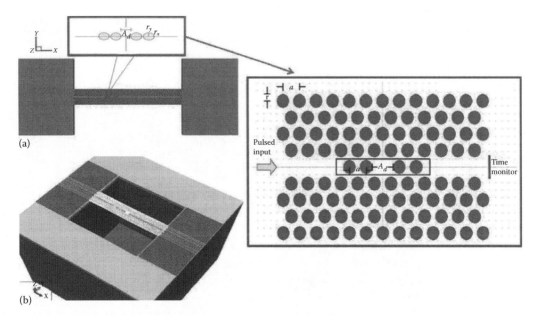

FIGURE 14.56 (a) Top view of the bridge structure integrated with proposed PCWG microresonator, with length = 20 μm, width = 5 μm, thickness of silicon bridge = 200 nm. The inset shows structure of the reflectors. (b) Top view of the entire device. (From Lee, C. et al., *J. Lightwave Technol.*, 26, 7, 2008.)

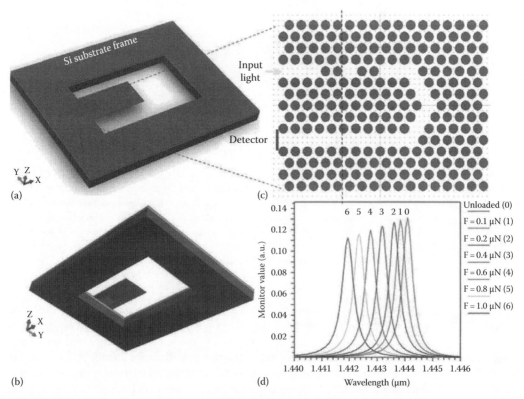

FIGURE 14.57 Tilted (a) top view drawing and (b) back-side view drawing of the PC cantilever. (c) Schematic drawing of PC microcavity resonator on U-shaped Si waveguide. (d) The resonant wavelength peaks of a 50 μm cantilever under various force loads. (From Lee, C. et al., *Appl. Phys. Lett.*, 93, 11, 2008.)

in Figure 14.57a through c, the cantilever is embedded with a PC microcavity resonator at the junction edge of cantilever and substrate. The PC structure contains an array of air holes in the silicon cantilever layer of 220 nm with a hexagonal lattice constant of $a = 500$ nm and the radius of all holes of $r = 180$ nm, where a silicon waveguide is formed by removing one row of air holes in a U-shaped layout. Figure 14.57d shows the simulated resonant wavelength peaks for 50 μm long cantilever under various applied force from 0.1 to 1 μN. The resonant wavelength peak of the unloaded case is simulated as 1444.097 nm. When the force load increases, the resonant wavelength moves to the shorter wavelength region, and the observed quality factor of the resonant wavelength peak decreases. The result also shows a linear relationship between the resonant wavelength and the force loads.

Besides implementing cantilever-type PC strain sensor, a diaphragm-type triple nano-ring (TNR) PC force or pressure sensor has also been investigated by C. Lee et al. using similar analysis methods. As shown in Figure 14.58a, the TNR PC resonator is formed by three triangular-arranged hexagonal nano-rings with size of 2.87 μm for each ring. The nano-rings, which are located in the middle of the diaphragm have two designs, i.e., either pure single layer of Si with thicknesses of 220 nm or bilayer of Si and SiO_2 with thicknesses of 220 and 600 nm. The TNR resonators give strong resonant peaks at the forward drop (FD) port, as shown in Figure 14.58b. The spectra are shown in similar fashion but with different location of the resonant peak wavelength for Si layer and Si/SiO_2

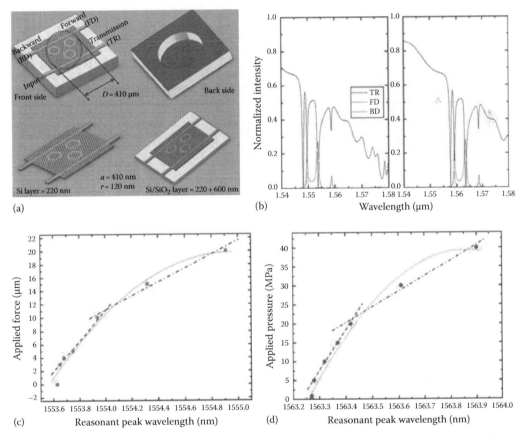

FIGURE 14.58 (See color insert.)   (a) 3-D illustration of the PhC Diaphragm sensor using hexagonal triple nano-ring (TNR) resonator. (b) Spectra of ports FD, BD, and TR for TNR channel drop filter for Si layer diaphragm and Si/SiO_2 layer diaphragm. (c) Resonant wavelength as function of applied force. (d) Resonant wavelength as function of applied pressure. (From Li, B. and Lee, C., *Sens. Actuat. A*, in press, doi:10.1016/j.sna.2011.02.028.)

layer cases in Figure 14.58b. The MEMS structures used for this TNR resonator are Si and Si/SiO$_2$ diaphragm for microforce sensing and pressure sensing, respectively. Figure 14.58c and d illustrate the resonance wavelength as functions of applied force and pressure, respectively. From the result, it can be seen that the Si diaphragm device shows wide sensing range and gives minimum detectable force of 0.847 μN, while the Si/SiO$_2$ diaphragm-based pressure sensor gives minimum detectable pressure of 4.17 MPa. Present designs reveal interesting device configurations for pressure and force sensing with ultra-compact footprint, as the sensing area can go to as small as 10 μm^2. With further integration of fiber optics, the investigated device concepts can find particular applications in guided-wire-based implanted biomedical sensors.

In the development of novel chemical and biological sensors, MEMS/NEMS have received a great deal of attention because of its superior detection capability for various chemical analytes. Cantilever beam with dimension in micro/nanometer scale is a well-adopted sensor structure due to its high sensitivity to cantilever deflection. When molecular adsorption is confined to the end of the microcantilever, it undergoes deflection due to a differential stress caused by the forces involved in the adsorption process. A Si-based cantilever sensor with PC resonator as readout for chemical sensing and analysis has been developed. Based on this concept, C. Lee and W. Xiang designed a cantilever-type PC resonator sensor with four holes defect [148]. Figure 14.59a shows a microcantilever

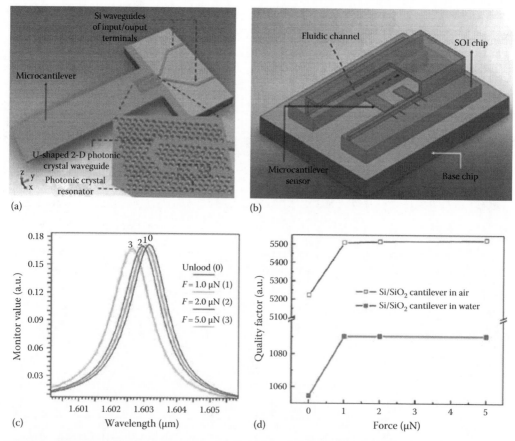

FIGURE 14.59 (See color insert.) (a) 3-D illustration of a microcantilever sensor using PC resonator as a strain sensor located at the edge of SOI chip. Inset shows the 3-D illustration of the PC resonator. (b) Conceptual drawing of two microcantilevers packaged inside a fluidic channel. (c) Output resonant peaks of the Si/SiO$_2$ cantilever sensor cantilever sensor operated in water under different loading force. (d) Quality factor change of the resonant peaks of the Si/SiO$_2$ cantilever sensor operated in air/water under different loading force. (From Xiang, W. and Lee, C., *IEEE J. Sel. Top. Quantum Electron.*, 15(5), 202, 2009.)

strain sensor with an integrated PC resonator at the edge of SOI chip. Figure 14.59b shows a schematic drawing of two microcantilevers of different dimensions being packaged in a microfluidic channel. Solution containing molecules flow through this channel such that the targeted molecules are bound on to cantilever surface due to the biomolecule selective binding, e.g., antigens to antibodies. The induced cantilever deflection is characterized in terms of the resonant wavelength shift due to deformation of PC resonator. Each one of the plural microcantilevers is designed for detection of a specific molecule, while multiple unknown molecules could be tested in the same fluidic channels according to different coated counterpart probes on these cantilevers. Figure 14.59c shows the simulated resonant peaks of the Si/SiO_2 cantilever sensor operation in water under different force loads at the cantilever end from 0 to 5 μN. From this result, it can be interpolated that the minimum detectable z-displacement and strain for Si/SiO_2 cantilever are 0.6 μm and 0.0098% when operated in water. Figure 14.59d shows the variation of quality factor of resonant peaks of the Si/SiO_2 cantilever sensor operated in air/water as the function of the loading force. As shown in Figure 14.59d, the loading force has no significant influence on the quality factor.

Besides Si-based cantilever sensor structure, the authors also studied the bio-sensing capability of a nano-ring resonator structure [149]. In contrast to micro-ring resonator, 2-D PC-based nano-ring resonators provide a smaller sensor area and very well optical confinement due to ultra low bending loss. The bio-sensing mechanism, in this case, is based on the change in effective refractive index of resonator when the analytes bind on it. This phenomenon will cause a red shift in resonant wavelength. Direct-write dip-pen-nano-lithography is used to perform immobilization of protein and DNA in nanometer scale holes, hence allowing biomolecules to be applied on a hole in the nanocavity of the PC resonators and use it as a sensing element. Figure 14.60a sketches the nano-ring resonator configuration for the biosensor investigated. The green and blue lines in Figure 14.60b show the spectra with biomolecule trapped in holes O_3 and O_4, respectively. The refractive index of sensing hole is regards as 1.45. The resonant wavelength is shifted to longer wavelength range, i.e., red shift. In this design, the minimum detectable biomolecule weight in a sensing hole for a nano-ring resonator of two-hole coupling distance is derived as 0.23 fg. This result demonstrates promising applications that demand detection of biomolecules down to the level of single copy of DNA. In addition to single nano-ring resonators, dual nano-ring resonator has also been studied by C. Lee et al. as the cascading of PC resonators can be adopted to control the propagated direction of dropped light [150]. The steady resonant field distributions of forward and backward drop are shown in Figure 14.61a and b, respectively. Biomolecule, e.g., DNAs or proteins, are chosen to be trapped inside the highest coupling sensing holes, as shown in Figure 14.61c. The resultant wavelength shifts are shown in Figure 14.61d through g.

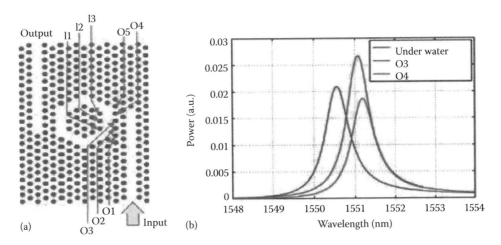

FIGURE 14.60 **(See color insert.)** (a) Sketch of biosensing nano-ring resonator. (b) Spectra of output. (From Hsiao, F.-L. and Lee, C., *IEEE Sens. J.*, 10(7), 1185, 2010.)

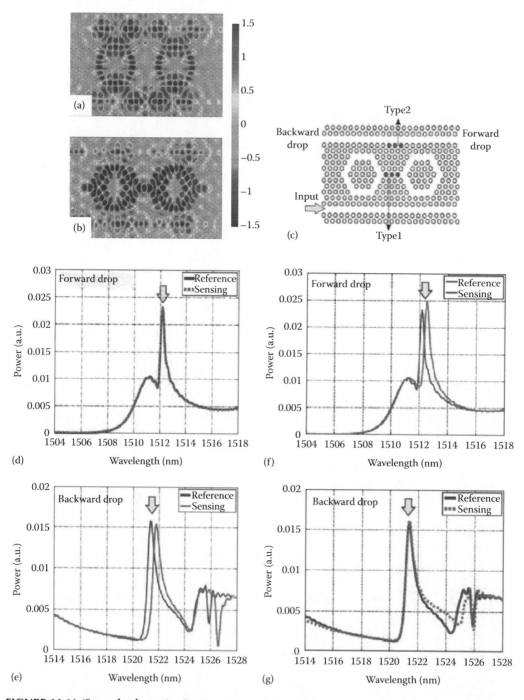

FIGURE 14.61 (See color insert.)   Steady resonant filed distributions of (a) forward and (b) backward drop. (c) Sketch of sensing-hole position for types 1 and 2. (d) Forward and (e) backward drop spectra of reference and sensing cases of type 1. (f) Forward and (g) backward drop spectra of reference and sensing cases of type 2. (From Hsiao, F.-L. and Lee, C., *SPIE J. Micro/Nanolith. MEMS, and MOEMS (JM3)*, 10(1), 013001, 2011.)

Type 1 and Type 2 holes resulted in forward and backward drop wavelength shift, respectively. The quality factors for these two resonant modes are 2100 and 1855, respectively. Compared to the single nano-ring biosensor architecture investigated in Ref. [149], this dual nano-ring resonator has a novel sensing mechanism as it is capable of sensing two different biomolecules simultaneously. For example, the concentration of two different target DNAs in the same solution can be detected at the same time by simultaneously monitoring the resonant wavelength shift at forward and backward drop ports. This application will be useful for study of certain biochemical interaction processes. As such, this dual nano-ring design provides more versatility compared to devices that can detect only one type of molecule. In view of this, multi-channel detection may be made possible by cascading more nano-rings with various independent resonant modes, wavelengths, and drop channels.

REFERENCES

1. J. A. Walker, The future of MEMS in telecommunications networks, *J. Micromech. Microeng.*, 10, 3, R1–R7 (2000).
2. C. R. Giles and M. Spector, The wavelength add/drop multiplexer for lightwave communication networks, *Bell. Labs Tech. J.*, 4, 1, 207–229 (1999).
3. A. Neukermans and R. Ramaswami, MEMS technology for optical networking, *IEEE Commun. Mag.*, 39, 1, 62–69 (2001).
4. L. Y. Lin and E. L. Goldstein, Opportunities and challenges for MEMS in lightwave communications, *IEEE J. Sel. Top. Quantum Electron.*, 8, 1, 163–172 (2002).
5. M. C. Wu, O. Solgaard, and J. E. Ford, Optical MEMS for Lightwave Communication, *J. Lightwave Technol.*, 24, 12, 4433–4454 (2006).
6. T.-W. Yeow, K. L. E. Law, and A. Goldenberg, MEMS optical switches, *IEEE Commun. Mag.*, 39, 11, 158–163 (2001).
7. C. Lee and J. A. Yeh, Development and evolution of MOEMS technology in variable optical attenuators, *J. Micro/Nanolith., MEMS, MOEMS*, 7, 2, 021003 (2008).
8. A. Q. Liu and X. M. Zhang, A review of MEMS external-cavity tunable lasers, *J. Micromech. Microeng.*, 17, 1, R1–R13 (2007).
9. L. Gizmo, The latest video projectors can fit inside tiny cameras or cellphones yet still produce big pictures, *IEEE Spectr.*, 47, 5, 41–45 (2010).
10. D. Graham-Rowe, Projectors get personal, *Nat. Photonics*, 1, 667–679 (2007).
11. J. Sun, S. Guo, L. Wu, L. Liu, S.-W Choe, Brian S. Sorg, and H. Xie, 3D in vivo optical coherence tomography based on a low-voltage, large-scan-range 2D MEMS mirror, *Opt. Express*, 18, 12065–12075 (2010).
12. J. Singh, J. H. S. Teo, Y. Xu, C. S. Premachandran, N. Chen, R. Kotlanka, M. Olivio, and C. J. R. Sheppard, A two axes scanning SOI MEMS micromirror for endoscopic bioimaging, *J. Micromech. Microeng.*, 18, 025001 (2008).
13. A. D. Yalcinkaya, H. Urey, D. Brown, T. Montague, and R. Sprague, Two-axis electromagnetic microscanner for high resolution displays, *J. Microelectromech. Syst.*, 15, 786–794 (2006).
14. S. D. Yalcinkaya, H. Urey, and S. Holmstrom, NiFe plated biaxial MEMS scanner for 2-D imaging, *IEEE Photon. Technol. Lett.*, 19, 330–332 (2007).
15. C.-D. Liao and J.-C. Tsai, The evolution of MEMS display, *IEEE Trans. Ind. Electron.*, 56, 1057–1065 (2009).
16. Y. Du, G. Zhou, K. L. Cheo, Q. Zhang, H. Feng, and F. S. Chau, A 2-DOF circular-resonator-driven in-plane vibratory grating laser scanner, *J. Microelectromech. Syst.*, 18, 4, 892–904 (2009).
17. W. Riethmüller and W. Benecke, Thermally excited silicon microactuators, *IEEE Trans. Electron. Devices*, 35, 6, 758–763 (1988).
18. J. H. Comtois and V. M. Bright, Application for surface micromachined polysilicon thermal actuators and arrays, *Sens. Actuators A*, 58, 19–25 (1997).
19. C. S. Pan and W. Hsu, An electro-thermally and laterally driven polysilicon microactuator, *J. Micromech. Microeng.*, 7, 7–13 (1997).
20. L. Que, J.-S. Park, and Y. B. Gianchandani, Bent-beam electrothermal actuators—Part I: Single beam and cascaded devices, *J. Microelectromech. Syst.*, 10, 2, 247–254 (2001).
21. J.-S. Park, L. L. Chu, A. D. Oliver, and Y. B. Gianchandani, Bent-beam electrothermal actuators—Part II: Linear and rotary microengines, *J. Microelectromech. Syst.*, 10, 2, 255–262 (2001).

22. C. Lee, Novel H-beam electrothermal actuators with capability of generating bi-directional static displacement, *Microsyst. Technol.*, 12, 717–722 (2006).

23. C. D. Lott, T. W. McLain, J. N. Harb, and L. L. Howell, Modeling the thermal behaviour of a surface-micro-machined linear-displacement thermomechanical microactuator, *Sens. Actuators A*, 101, 239–250 (2002).

24. S. Schweizer, S. Calmes, M. Laudon, and P. H. Renaud, Thermally actuated optical microscanner with large angle and low consumption, *Sens. Actuators A*, 76, 470–477 (1999).

25. A. Jain, H. Qu, S. Todd, and H. Xie, A thermal bimorph micromirror with large bi-directional and vertical actuation, *Sens. Actuators A*, 122, 9–15 (2005).

26. H. Miyajima, N. Asaoka, T. Isokawa, M. Ogata, Y. Aoki, M. Imai, O. Fujimori, M. Katashiro, and K. Matsumoto, A MEMS electromagnetic optical scanner for a commercial confocal laser scanning microscope, *J. Microelectromech. Syst.*, 12, 3, 243–251 (2003).

27. H.-J. Cho and E. Yoon, A low-voltage three-axis electromagnetically actuated micromirror for fine alignment among optical devices, *J. Micromech. Microeng.*, 19, 085007 (2009).

28. S. O. Isikman and H. Urey, Dynamic modeling of soft magnetic film actuated scanners, *IEEE. Trans. Mag.*, 45, 7, 2912–2919 (2009).

29. J. G. Smits and W. S. Choi, The constituent equations of piezo-electric heterogeneous bimorphs, *IEEE Trans. Ultrason. Ferroelectr. Freq. Control*, 38, 265–279 (1991).

30. T. Kobayashi, R. Maeda, and T. Itoh, A fatigue test method for Pb(Zr, Ti)O3 thin films by using MEMS-based self-sensitive piezoelectric microcantilevers, *J. Micromech. Microeng.*, 18, 11, 115007 (6pp) (2008).

31. S. S. Lee, L. Y. Lin, and M. C. Wu, Surface micromachined free space fiber-optic switches, *Electron. Lett.*, 31, 1481–1482 (1995).

32. R. A. Miller, Y. C. Tai, G. Xu, J. Bartha, and F. Lin, An electromagnetic MEMS 2×2 fiber optic bypass switch, *Proceedings of the International Conference on Solid-State Sensors and Actuators*, Chicago, IL, 1997, Paper 1A4, pp. 89–92.

33. H. Maekoba, P. Helin, G. Reyne, T. Bourouina, and H. Fujita, Self-aligned vertical mirror and v-grooves applied to an optical switch: Modeling and optimization of bi-stable operation by electromagnetic actuation, *Sens. Actuators A*, 87, 172–178 (2000).

34. W.-H. Juan and S. W. Pang, High-aspect-ratio si vertical micromirror arrays for optical switching, *IEEE J. Microelectromech. Syst.*, 7, 2, 207–213 (1998).

35. C. Marxer, C. Thio, M. A. Gretillat, N. F. de Rooij, R. Battig, O. Anthamatten, B. Valk, and P. Vogel, Vertical mirrors fabricated by deep reactive ion etching for fiber-optic switching applications, *J. Microelectromech. Syst.*, 6, 3, 277–285 (1997).

36. C. Marxer and N. F. de Rooij, Micro-opto-mechanical 2×2 switch for single-mode fibers based on plasma-etched silicon mirror and electrostatic actuation, *J. Lightwave Technol.*, 17, 1, 2–6 (1999).

37. B. Hichwa, M. Duelli, D. Friedrich, C. Iaconis, C. Marxer, M. Mao, and C. Olson, A unique latching 2×2 MEMS fiber optics switch, Presented at the *IEEE Conference on Optical MEMS*, Kauai, HI, 2000.

38. W. Noell, P.-A. Clerc, L. Dellmann, B. Guldimann, H.-P. Herzig, O. Manzardo, C. R. Marxer, K. J. Weible, R. Dändliker, and N. de Rooij, Applications of SOI-based optical MEMS, *IEEE J. Sel. Top. Quantum Electron.*, 8, 1, 148–154 (2002).

39. S. C. Lee, S. W. Kim, J. Park, J. Goo, I. Jung, and D. Cho, A fast optical switch using the SBM process, *Proceedings of IEEE/LEOS International Conference on Optical MEMS (Optical MEMS 2001)*, Okinawa, Japan, pp. 95–96, 2001.

40. D.-I. Cho and T. Chang, Micro-machined silicon on-off fiber optic switching systems. U.S. Patent 6520777 B2, 2003.

41. C. Lee, and C.-Y. Wu, Study of electrothermal V-beam actuators and latched mechanism for optical switch, *J. Micromech. Microeng.*, 15, 11–19 (2005).

42. L. Y. Lin, E. L. Goldstein, and R. W. Tkach, Free-space micromachined optical switches with submillisecond switching time for large-scale optical crossconnects, *IEEE Photon. Technol. Lett.*, 10, 4, 525–527 (1998).

43. L. Y. Lin, E. L. Goldstein, and R. W. Tkach, Free-space micromachined optical switches for optical networking, *IEEE J. Sel. Top. Quantum Electron.*, 5, 1, 4–9 (1999).

44. L. Fan, S. Gloeckner, P. D. Dobblelaere, S. Patra, D. Reiley, C. King, T. Yeh et al., Digital MEMS switch for planar photonic crossconnects, *OFC -Postconference Technical Digest* (IEEE Cat. No. 02CH37339), Washington, DC, 2002, Vol. 1, pp. 93–94.

45. L. Fan and M. C. Wu, Two-dimensional optical scanner with large angular rotation realized by self-assembled micro-elevator, in *Proceedings of the IEEE/LEOS Summer Topical Meeting on Digital Broadband Optical Network and Technologies: An Emerging Reality. Optical MEMS. Smart Pixels. Organic Optics and Optoelectronics* (Cat. No. 98TH8369), New York, 1998, pp. 107–108.

46. V. A. Aksyuk, F. Pardo, D. Carr, D. Greywall, H. B. Chan, M. E. Simon, A. Gasparyan et al., Beam-steering micromirrors for large optical crossconnects, *J. Lightwave Technol.*, 21, 3, 634–642 (2003).

47. D. S. Greywall, P. A. Busch, F. Pardo, D. W. Carr, G. Bogart, and H. T. Soh, Crystalline silicon tilting mirrors for optical cross-connect switches, *J. Microelectromech. Syst.*, 12, 5, 708–712 (2003).

48. M. Yano, F. Yamagishi, and T. Tsuda, Optical MEMS for photonic switching-compact and stable optical crossconnect switches for simple, fast, and flexible wavelength applications in recent photonic networks, *IEEE J. Sel. Top. Quantum Electron.*, 11, 2, 383–394 (2005).

49. T. D. Kudrle, C. C. Wang, M. G. Bancu, J. C. Hsiao, A. Pareek, M. Waelti, G. A. Kirkos, T. Shone, C. D. Fung, and C. H. Mastrangelo, Single-crystal silicon micromirror array with polysilicon flexures, *Sens. Actuators A*, 119, 2, 559–566 (2005).

50. J. Kim, C. J. Nuzman, B. Kumar, D. F. Lieuwen, J. S. Kraus, A. Weiss, C. P. Lichtenwalner et al., 1100×1100 port MEMS-based optical with 4-dB maximum loss, *IEEE Photonics Technol. Lett.*, 15, 11, 1537–1539 (2003).

51. R. Sawada, E. Higurashi, A. Shimizu, and T. Tohru Maruno, Single crystalline mirror actuated electrostatically by terraced electrodes with high-aspect ratio torsion spring, *Proceedings of the Optical MEMS 2001*, Okinawa, Japan, 2001, pp. 23–24.

52. R. A. Conant, J. T. Nee, K. Y. Lau, and R. S. Muller, A flat high frequency scanning micromirror, *Technical Digest of Solid-State Sensor and Actuator Workshop* (TRF Cat. No. 00TRF-0001), Cleveland, OH, 2000, pp. 6–9.

53. X. Mi, H. Soneda, H. Okuda, O. Tsuboi, N. Kouma, Y. Mizuno, S. Ueda, and I. Sawaki, A multi-chip directly mounted 512-MEMS-mirror array module with a hermetically sealed package for large optical cross-connects, *J. Opt. A: Pure Appl. Opt.*, 8, S341–S346 (2006).

54. J. E. Ford and J. A. Walker, Dynamic spectral power equalization using micro-opto-mechanics, *IEEE Photon. Technol. Lett.*, 10, 10, 1440–1442 (1998).

55. B. Barber, C. R. Giles, V. Askyuk, R. Ruel, L. Stulz, and D. Bishop, A fiber connectorized MEMS variable optical attenuator, *IEEE Photon. Technol. Lett.*, 10, 9, 1262–1264 (1998).

56. C. R. Giles, V. Askyuk, B. Barber, R. Ruel, L. Stulz, and D. Bishop, A silicon MEMS optical switch attenuator and its use in lightwave subsystems, *IEEE J. Sel. Top. Quantum Electron.*, 5, 1, 18–25 (1999).

57. C. Marxer, P. Griss, and N. F. de Rooij, A variable optical attenuator based on silicon micromechanics, *IEEE Photonics Technol. Lett.*, 11, 2, 233–235 (1999).

58. X. M. Zhang, A. Q. Liu, C. Lu, and D. Y. Tang, MEMS variable optical attenuator using low driving voltage for DWDM systems, *Electron. Lett.*, 38, 8, 382–383 (2002).

59. A. Q. Liu, X. M. Zhang, C. Lu, F. Wang, C. Lu, and Z. S. Liu, Optical and mechanical models for a variable optical attenuator using a micromirror drawbridge, *J. Micromech. Microeng.*, 13, 3, 400–411 (2003).

60. X. M. Zhang, A. Q. Liu, C. Lu, F. Wang, and Z. S. Liu, Polysilicon micromachined fiber-optical attenuator for DWDM applications, *Sens. Actuators A*, 108, 1–3, 28–35 (2003).

61. C. Lee, Y.-S. Lin, Y.-J. Lai, M. H. Tsai, C. Chen, and C.-Y. Wu, 3-V driven pop-up micromirror for reflecting light toward out-of-plane direction for VOA applications, *IEEE Photonics Technol. Lett.*, 16, 4, 1044–1046 (2004).

62. C. Lee and Y.-S. Lin, A new micromechanism for transformation of small displacements to large rotations for a VOA, *IEEE Sens. J.*, 4, 4, 503–509 (2004).

63. C.-H. Kim, N. Park, and Y.-K. Kim, MEMS reflective type variable optical attenuator using off-axis misalignment, *Proceedings of the IEEE/LEOS International Conference on Optical MEMS*, Lugano, Switzerland, 2002, pp. 55–56.

64. C. Chen, C. Lee, and Y.-J. Lai, Novel VOA using in-plane reflective micromirror and off-axis light attenuation, *IEEE Commun. Mag.*, 41, 8, S16–S20 (2003).

65. C. Lee, M. H. Tsai, C.-Y. Wu, S.-Y. Hung, C. Chen, Y.-J. Lai, M.-S. Lin, and J. Andrew Yeh, Characterization of MOEMS VOA based on various planar light attenuation configurations, *Proceedings of the IEEE/LEOS International Conference on Optical MEMS*, Kagawa, Japan, 2004, pp. 98–99.

66. C. Chen, C. Lee, and J. Andrew Yeh, Retro-reflection type MOEMS VOA, *IEEE Photonics Technol. Lett.*, 16, 10, 2290–2292 (2004).

67. T.-S. Lim, C.-H. Ji, C.-H. Oh, Y. Yee, and J. U. Bu, Electrostatic MEMS variable optical attenuator with folded micromirror, *Proceedings of the IEEE/LEOS International Conference on Optical MEMS*, Waikoloa, Hawaii, USA, 2003, pp. 143–144.

68. T.-S. Lim, C.-H. Ji, C.-H. Oh, H. Kwon, Y. Yee, and J. U. Bu, Electrostatic MEMS variable optical attenuator with rotating folded micromirror, *IEEE J. Sel. Top. Quantum Electron.*, 10, 3, 558–562 (2004).

69. J. A. Yeh, S.-S. Jiang, and C. Lee, MOEMS VOA using rotary comb drive actuators, *IEEE Photonics Technol. Lett.*, 18, 10, 1170–1172 (2006).

70. J. H. Lee, Y. Y. Kim, S. S. Yun, H. Kwon, Y. S. Hong, J. H. Lee, and S. C. Jung, Design and characteristics of a micromachined variable optical attenuator with a silicon optical wedge, *Opt. Commun.*, 221, 4–6, 323–330 (2003).

71. Y. Y. Kim, S. S. Yun, C. S. Park, J.-H. Lee, Y. G. Lee, H. K. Lee, S. K. Yoon, and J. S. Kang, Refractive variable optical attenuator fabricated by silicon deep reactive ion etching, *IEEE Photonics Technol. Lett.*, 16, 2, 485–487 (2004).

72. J. H. Lee, S. S. Yun, Y. Y. Kim, and K.-W. Jo, Optical characteristics of a refractive optical attenuator with respect to the wedge angles of a silicon optical leaker, *Appl. Opt.*, 43, 4, 877–882 (2004).

73. H. Cai, X. M. Zhang, C. Lu, A. Q. Liu, and E. H. Khoo, Linear MEMS variable optical attenuator using reflective elliptical mirror, *IEEE Photonics Technol. Lett.*, 17, 2, 402–404 (2005).

74. M. G. Kim and J.-H. Lee, A discrete positioning microactuator: linearity modeling and VOA application, *IEEE J. Microelectromech. Syst.*, 16, 1, 16–23 (2007).

75. X. M. Zhang, Q. W. Zhao, A. Q. Liu, J. Zhang, J. H. Lau and C. H. Kam, Asymmetric tuning schemes of MEMS dual-shutter VOA, *IEEE J. Lightwave Technol*, 26, 5, 569–579 (2008).

76. B. Glushko, S. Krylov, M. Medina, and D. Kin, Insertion type MEMS VOA with two transparent shutters, *Proceedings of the Asia-Pacific Conference on Transducers Micro-Nano Technology (APCOT)*, Marina Mandarin Hotel, Singapore, Paper 95-OMN-A0594, 2006.

77. J. C. Chiou and W. T. Lin, Variable optical attenuator using a thermal actuator array with dual shutters, *Opt. Commun.*, 237, 4–6, 341–350 (2004).

78. C. Lee, MOEMS variable optical attenuator with robust design for improved dynamic characteristics, *IEEE Photonics Technol. Lett.*, 18, 6, 773–775 (2006).

79. B. M. Andersen, S. Fairchild, N. Thorsten, and V. Aksyuk, MEMS variable optical amplifiers, *Proceedings of the Optical Fiber Communications Conference (OFC 2000)*, Baltimore, MD, 2, 260–262 (2000).

80. N. A. Riza and S. Sumriddetchkajorn, Digitally controlled fault-tolerant multiwavelength programmable fiber-optic attenuator using a two-dimensional digital micromirror device, *Opt. Lett.*, 24, 5, 282–284 (1999).

81. K. Isamoto, K. Kato, A. Morosawa, C. Chong, H. Fujita, and H. Toshiyoshi, Micromechanical VOA design for high shock-tolerance and low temperature-dependence, *Proceedings of the IEEE/LEOS International Conference on Optical MEMS*, Hawaii, USA, pp. 113–114 2003.

82. K. Isamoto, K. Kato, A. Morosawa, C. Chong, H. Fujita, and H. Toshiyoshi, A 5-Voperated MEMS variable optical attenuator by SOI bulk micromachining, *IEEE J. Sel. Top. Quantum Electron.*, 10, 3, 570–578 (2004).

83. K. Isamoto, A. Morosawa, M. Tei, H. Fujita, and H. Toshiyoshi, MEMS variable optical attenuator using asymmetrically driven parallel plate tilt mirror, *IEEJ. Trans. SM*, 124, 6, 213–218 (2004).

84. C. Lee, F.-L. Hsiao, T. Kobayashi, K. H. Koh, P. V. Ramana, W. Xiang, B. Yang, C. W. Tan, and D. Pinjala, A 1-V operated MEMS variable optical attenuator using piezoelectric PZT thin film actuators, *IEEE J. Sel. Top. Quantum Electron.*, 15, 5, 1529–1536 (2009).

85. A. Godil, Diffractive MEMS technology offers a new platform for optical networks, *Laser Focus World*, 38, 5, 181–185 (2002).

86. A. A. Godil and D. M. Bloom, Achromatic optical modulators, U.S. Patent 6169624 B1, 2001.

87. R. R. A. Syms, H. Zou, J. Stagg, and D. F. Moore, Multistate latching MEMS variable optical attenuator, *IEEE Photonics Technol. Lett.*, 16, 1, 191–193 (2004).

88. R. R. A. Syms, H. Zou, and P. Boyle, Mechanical stability of a latching MEMS variable optical attenuator, *IEEE J. Microelectromech. Syst.*, 14, 3, 529–538 (2005).

89. S.-S. Lee, J.-U. Bu, S.-Y. Lee, K.-C. Song, C.-G. Park, and T.-S. Kim, Low-power consumption polymeric attenuator using a micromachined membrane-type waveguide, *IEEE Photonics Technol. Lett.*, 12, 4, 407–409 (2000).

90. M.-C. Oh, S.-H. Cho, Y.-O. Noh, H.-J. Lee, J.-J. Joo, and M.-H. Lee, Variable optical attenuator based on large-core single-mode polymer waveguide, *IEEE Photonics Technol. Lett.*, 17, 9, 1890–1892 (2005).

91. S. K. Kim, Y.-C. Hung, K. Geary, W. Yuan, H. R. Fetterman, D. Jin, R. Dinu, and W. H. Steier, Metal-defined polymeric variable optical attenuator, *IEEE Photonics Technol. Lett.*, 18, 9, 1055–1057 (2006).

92. G.-D. J. Su, Y.-W. Yeh, C.-W. E. Chiu, C.-H. Li, and T.-Y. Chen, Fabrication and measurement of low-stress polyimide membrane for high resolution variable optical attenuator, *IEEE J. Sel. Top. Quantum Electron.*, 13, 2, 312–315 (2007).

93. H. Yu, G. Zhou, F. S. Chau, and F. Lee, A variable optical attenuator based on optofluidic technology, *J. Micromech. Microeng.*, 18, 11, 115016 (5pp) (2008).

94. J. B. Sampsell, An overview of Texas Instrument digital micromirror device (DMD) and its application to projection displays, *Proceedings of the International Symposium on Society of Information Display*, May 1993, Vol. 24, pp. 1012–1015.

95. L. J. Hornbeck, Current status of the digital micromirror device (DMD) for projection television applications (Invited Paper), *IEDM Tech. Dig.*, Washington, USA, 1993, 381–384.

96. F. Zimmer, M. Lapisa, T. Bakke, M. Bringe, G. Stemme, and F. Niklaus, One-megapixel monocrystalline-silicon micromirror array on CMOS driving electronics manufactured with very large-scale heterogeneous integration, *IEEE J. Microelectromech. Syst.*, 20, 3, 564–572 (2011).

97. O. Solgaard, F. S. A. Sandejas, and D. M. Bloom, Deformable grating optical modulator, *Opt. Lett.*, 17, 9, 688–690 (1992).

98. D. Bloom, The grating light valve: Revolutionizing display technology, *Proceedings of the SPIE Projection Displays Symposium V*, San Jose, USA, Vol. 3634, pp. 132–138, 1999.

99. T. S. Perry, A dark-horse technology—the grating light valve may join the competition to dethrone the CRT, *IEEE Spectr.* 41, 38–41 (2004).

100. W. Piyawattanametha, P. R. Patterson, D. Hah, H. Toshiyoshi, and M. C. Wu, Surface- and bulk-micromachined two-dimensional scanner driven by angular vertical comb actuators, *IEEE J. Microelectromech. Syst.*, 14, 6, 1329–1338 (2005).

101. Y. Du, G. Zhou, K. L. Cheo, Q. Zhang, H. Feng, and F. S. Chau, Double layered vibratory grating scanner for high-speed high-resolution laser scanning, *IEEE J. Microelectromech. Syst.*, 19, 5, 1186–1196 (2010).

102. K. Jia, S. Pal, and H. Xie, An electrothermal tip-tilt-piston micromirror based on folded dual S-shaped bimorphs, *IEEE J. Microelectromech. Syst.*, 18, 5, 1004–1015 (2009).

103. K. Jia, S. R. Samuelson, and H. Xie, High-fill-factor micromirror array with hidden bimorph actuators and tip-tilt-piston capability, *IEEE J. Microelectromech. Syst.*, 20, 3, 573–582 (2011).

104. Y. Eun and J. Kim, Thermally driven torsional micromirrors using pre-bent torsion bar for large static angular displacement, *J. Micromech. Microeng.*, 19, 4, 045009 (8pp) (2009).

105. S. O. Isikman, O. Ergeneman, A. D. Yalcinkaya, and H. Urey, Modeling and characterization of soft magnetic film actuated 2-D scanners, *IEEE J. Sel. Top. Quantum Electron.*, 12, 2, 283–289 (2007).

106. M. Tani, M. Akamatsu, Y. Yasuda, and H. Toshiyoshi, A two-axis piezoelectric tilting micromirror with a newly developed PZT-meandering actuator, *IEEE 20th International Conference on Micro Electro Mechanical Systems*, Kobe, Japan, January 21–25, 2007, pp. 699–702.

107. K. H. Koh, T. Kobayashi, F.-L. Hsiao, and C. Lee, Characterization of piezoelectric PZT beam actuators for driving 2D scanning micromirrors, *Sens. Actuators A*, 162, 336–347 (2010).

108. K. H. Koh, T. Kobayashi, J. Xie, A. Yu, and C. Lee, Novel piezoelectric actuation mechanism for gimbal-less mirror in 2-D raster scanning applications, *J. Micromech. Microeng.*, 21, 7, 075001 (9pp) (2011).

109. K. H. Koh, T. Kobayashi, and C. Lee, A 2-D MEMS scanning mirror based on dynamic mixed mode excitation of a piezoelectric PZT thin film S-shaped actuator, *Opt. Express*, 19, 15, 13812–13824 (2011).

110. H. Urey, D. Wine, and T. Osborn, Optical performance requirements for MEMS-scanner based microdisplays, *Conference on MOEMS and Miniaturized Systems*, Santa Clara, CA, 2000, SPIE Vol. 4178, pp. 176–185.

111. M. Tani, M. Akamatsu, Y. Yasuda, H. Fujita, and H. Toshiyoshi, A combination of fast resonant mode and slow static deflection of SOI-PZT actuators for MEMS image projection display, *IEEE Optical MEMS* 2006, 25–26, Big Sky, MT, August 2006.

112. H. M. Chu, T. Tokuda, M. Kimata, and K. Hane, Low-voltage operation micromirror based on vacuum seal technology using metal can, *IEEE J. Microelectromech. Syst.*, 19, 4, 927–935 (2010).

113. H. M. Chu and K. Hane, Design, fabrication and vacuum operation characteristics of two-dimensional comb-drive micro-scanner, *Sens. Actuators A*, 165, 422–430 (2011).

114. R. F. Wolffenbuttel, State-of-the-art in integrated optical microspectrometers, *IEEE Trans. Instrum. Meas.*, 55, 1, 197–202 (2004).

115. R. F. Wolffenbuttel, MEMS-based optical mini- and microspectrometer for the visible and infrared spectral range, *J. Micromech. Microeng.*, 15, 7, 145–152 (2005).

116. C.-K. Kim, M.-L. Lee, and C.-H. Jun, Electrothermally actuated Fabry–Pérot tunable filter with a high tuning efficiency, *IEEE Photonics Technol. Lett.*, 16, 8, 1894–1896 (2004).

117. N. Quack, S. Blunier, J. Dual, F. Felder, C. Ebneterm, M. Rahim, and H. Zogg, Tunable resonant cavity enhanced detectors using vertically actuated MEMS mirrors, *J. Opt. A: Pure Appl. Opt.*, 10, 4, 044015 (6pp) (2008).

118. H.-K. Lee, K.-S. Kim, and E. Yoon, A wide-range linearly tunable optical filter using Lorentz Force, *IEEE Photonics Technol. Lett.*, 16, 9, 2087–2089 (2004).

119. H.-K. Lee, K.-S. Kim, and E. Yoon, A wide range linearly-tunable optical filter using magnetic actuation, *IEEE International Conference on MEMS*, Maastricht, the Netherlands, 2004, pp. 93–96.

120. J. H. Correia, M. Bartek, and R. F. Wolffenbuttel, Bulk-micromachined tunable Fabry-Perot microinterferometer for the visible spectral range, *Sens. Actuators A*, 76, 191–196 (1999).

121. A. J. Keating, K. K. M. B. D. Silva, J. M. Dell, C. A. Musca, and L. Faraone, Optical characterization of Fabry–Pérot MEMS filter integrated on tunable short-wave IR detectors, *IEEE Photonics Technol. Lett.*, 18, 9, 1079–1081 (2004).

122. J. S. Milne, J. M. Dell, A. J. Keating, and L. Faraone, Widely tunable MEMS-based Fabry-Perot filter, *IEEE J. Microelectromech. Syst.*, 18, 4, 905–913 (2009).

123. N. Neumann, M. Ebermann, K. Hiller, and S. Kurth Tunable infrared detector with integrated micromachined Fabry-Perot filter, *Conference on MOEMS and Miniaturized Systems VII*, San Jose, CA, 2007, SPIE vol. 6466, 646606-1-646606-12.

124. N. Neumann, M. Ebermann, K. Hiller, and S. Kurth, Tunable infrared detector with integrated micromachined Fabry-Perot filter, *J. Micro/Nanolith. MEMS MOEMS*, 7, 2, 021004-1-021004-9 (2008).

125. N. Neumann, M. Ebermann, S. Kurth, and K. Hiller, Novel MWIR microspectrometer based on a tunable detector, *Conference on MOEMS and Miniaturized Systems VIII*, San Jose, CA, 2009, SPIE vol. 7208, 72080D.

126. M. Ebermann, N. Neumann, K. Hiller, E. Gittler, M. Meinig, and S. Kurth, Recent advances in expanding the spectral range of MEMS Fabry-Perot filter, *Conference on MOEMS and Miniaturized Systems IX*, San Jose, CA, 2010, SPIE vol. 7594, 75940V.

127. M. Meinig, S. Kurth, K. Harla, N. Neumann, E. Gittler, and T. Gessner, Tunable mid-infrared filter based on Fabry–Pérot interferometer with two movable reflectors, *MOEMS and Miniaturized Systems X*, San Jose, CA, 2011, SPIE vol. 7930, 79300K.

128. T. Yamanoi, T. Endo, and H. Toshiyoshi, A hybrid-assembled MEMS Fabry-Perot wavelength tunable filter, *The 14th International Conference on Solid-State Sensors, Actuators and Microsystems (Transducers'07)*, Lyon, France, June 10–14, 2007.

129. T. Yamanoi, T. Endo, and H. Toshiyoshi, A hybrid-assembled MEMS Fabry–Pérot wavelength tunable filter, *Sens. Actuators A*, 145–146, 116–122 (2008).

130. S. H. Kong, D. D. L. Wijingaards, and R. F. Wolffenbuttel, Infrared micro-spectrometer based on a diffraction grating, *Sens. Actuators A*, 92, 88–95 (2001).

131. C. Ataman, H. Urey, and A. Wolter, A Fourier transform spectrometer using resonant vertical comb actuators, *J. Micromech. Microeng.*, 16, 12, 2517–2523 (2006).

132. H. Yu, G. Zhou, F. S. Chau, F. Lee, S. Wang, and M. Zhang, An electromagnetically driven lamellar grating based Fourier transform microspectrometer, *J. Micromech. Microeng.*, 18, 5, 055016 (6pp) (2008).

133. F. S. Chau, Y. Du, and G. Zhou, A micromachined stationary lamellar grating interferometer for Fourier transform spectroscopy, *J. Micromech. Microeng.*, 18, 2, 025023 (7pp) (2008).

134. M.-C. M. Lee, D. Hah, E. K. Lau, H. Toshiyoshi, and M. Wu, MEMS-actuated photonic crystal switches, *IEEE Photonics Technol. Lett.*, 18, 2, 358–360 (2006).

135. J. Yao, D. Leuenberger, M.-C. M. Lee, and M. C. Wu, Silicon microtoroidal resonators with integrated MEMS tunable coupler, *IEEE J. Sel. Top. Quantum Electron.*, 13, 2, 202–208 (2007).

136. K. Takahashi, Y. Kanamori, Y. Kokubun, and K. Hane, A wavelength-selective add-drop switch using silicon microring resonator with a submicron-comb electrostatic actuator, *Opt. Express*, 16, 19, 14421–14428 (2008).

137. Y. Kanamori, K. Takahashi, and K. Hane, An ultrasmall wavelength-selective channel drop switch using a nanomechanical photonic crystal nanocavity, *Appl. Phys. Lett.*, 95, 17, 171911-1-171911-3 (2009).

138. T. Ikeda, K. Takahashi, Y. Kanamori, and K. Hane, Phase-shifter using submicron silicon waveguide couplers with ultra-small electro-mechanical actuator, *Opt. Express*, 18, 7, 7031–7037 (2010).

139. X. Chew, G. Zhou, H. Yu, F. S. Chau, J. Deng, Y. C. Loke, and X. Tang, An in-plane nano-mechanics approach to achieve reversible resonance control of photonic crystal nanocavities, *Opt. Express*, 18, 21, 22232–22244 (2010).

140. X. Chew, G. Zhou, H. Yu, F. S. Chau, J. Deng, X. Tang, and Y. C. Loke, Dynamic tuning of an optical resonator through MEMS-driven coupled photonic crystal nanocavities, *Opt. Lett.*, 35, 15, 2517–2519 (2010).

141. X. Chew, G. Zhou, F. S. Chau, and J. Deng, Enhanced resonance tuning of photonic crystal nanocavities by integration of optimized near-field multitip nanoprobes, *J. Nanophotonics*, 5, 1, 059503-1-059503-6 (2011).

142. X. Zhao, J. M. Tsai, H. Cai, X. M. Ji, J. Zhou, M. H. Bao, Y. P. Huang, D. L. Kwang, and A. Q.Liu, A nano⊠opto⊠mechanical actuator driven by optical radiation force, *The 16th International Conference on Solid⊠State Sensors, Actuators and Microsystems (Transducers'11)*, Beijing, China, June 5–9, 2011, pp. 1468–1471.

143. X. Zhao, J. M. Tsai, H. Cai, X. M. Ji, J. Zhou, M. H. Bao, Y. P. Huang, D. L. Kwang, and A. Q. Liu, A nano⊠opto⊠mechanical pressure sensor via ring resonator, *Opt. Express*, 20, 8, 8535⊠8542, (2012).

144. M. Ren, Y. F. Yu, J. M. Tsai, H. Caim W. M. Zhu, D. L. Kwang, and A. Q. Liu, Design and experiments of a nano⊠opto⊠mechanical switch using EIT⊠like effects of coupled⊠ring resonator, *The 16th International Conference on Solid⊠State Sensors, Actuators and Microsystems (Transducers'11)*, Beijing, China, June 5–9, 2011, pp. 1436–1439.

145. C. Lee, R. Radhakrishnan, C.-C Chen, J. Li, J. Thillaigovindan, and N. Balasubramanian, Design and Modeling of a nanomechanical sensor using silicon photonic crystals, *J. Lightwave Technol.*, 26, 7, 839–846 (2008).

146. C. Lee, J. Thillaigovindan, C.-C. Chen, X. T. Chen, Y.-T. Chao, S. Tao, W. Xiang, A. Yu, H. Feng, and G. Q. Lo, Si nanophotonics based cantilever sensor, *Appl. Phys. Lett.*, 93, 11, 113113-1-113113-3 (2008).

147. B. Li and C. Lee, NEMS diaphragm sensors integrated with triple nano ring resonator, *Sens. Actuators A*, 172, 1, 61-68 (2011).

148. W. Xiang and C. Lee, Nanophotonics sensor based on microcantilever for chemical analysis, *IEEE J. Sel. Top. Quantum Electron.*, 15, 5, 202–208 (2009).

149. F.-L. Hsiao and C. Lee, Computational study of photonic crystals nano-ring resonator for biochemical sensing, *IEEE Sens. J.*, 10, 7, 1185–1191 (2010).

150. F.-L. Hsiao and C. Lee, Nanophotonic biosensors using hexagonal nano-ring resonators—A computational study, *SPIE J. Micro/Nanolithogr., MEMS, MOEMS (JM3)*, 10, 1, 013001-1-013001-8 (2011).

15 Integrated Optofluidics and Optomechanical Devices Manufactured by Femtosecond Lasers

Yves Bellouard, Ali A. Said, Mark Dugan, and Philippe Bado

CONTENTS

15.1 INTRODUCTION

The unconventional laser–matter interaction resulting from the very high peak powers associated with femtosecond laser pulses provides novel ways to tailor material properties. Applied to dielectrics, such as fused silica, these localized changes of physical properties can be advantageously used to create miniaturized devices that combine optical, fluidics, and mechanical functions.

In this chapter, we discuss the opportunity of using femtosecond laser processes to form microsystems. The first part briefly introduces micro- and nanosystems with an emphasis on limitations of current fabrication practices. The second part reviews the work done to create miniaturized device with femtosecond lasers. The third part presents a generalized concept of integrated microdevices produced using femtosecond lasers.

15.2 MICRO- AND NANOSYSTEMS

15.2.1 Brief Overview of Microsystems

Micro- and nanotechnologies have emerged as key technologies for modern societies. These technologies are expected to significantly impact our industrialized economies as well as to generally contribute to the advancement of science.

Microsystems perform sophisticated tasks in a miniaturized volume. Shaping or analyzing light signals; mixing, processing, or analyzing ultrasmall volumes of chemicals; sensing mechanical signals; probing gas; and sequencing biomolecules are common operations that can be done by these small machines. Rationales for the use of microsystems are numerous. The reduction of consumables (reduced consumption of chemicals in Lab-on-a-Chip for example), a fast response time (critical in airbag sensors), an enhance portability Radio-Frequency Micro-Electro-Mechanical Systems (RF-MEMS), higher resolution (with Inkjet printer head), higher efficiency (using microchemical reactors), smaller footprint, etc. are typical benefits sought.

Starting from integrated circuits in the 1970s, followed by micromechanical systems in the 1980s–1990s and photonics and fluidics in the 1990s–2000s, and, recently, the addition of organic material and biomolecules, micro-/nanosystems are becoming complex machines performing sophisticated tasks (Figure 15.1).

15.2.2 Issues on Microsystems Integration and Fabrication

So far, the fabrication of microsystems [1] essentially relies on two technology platforms that are used separately or jointly.

The first technology platform is based on surface micromachining of substrates such as silicon. It relies on "clean-room" processes that, for the most part, were inherited from the microelectronics industry. Devices produced by surface micromachining are fabricated through successive steps of material deposition and selective material removal. By carefully selecting the proper combination of materials deposition and selective etching, one can fabricate, layer-by-layer, relatively complex devices. This approach produced elements that are planar or near-planar. Figure 15.2 shows an example of MEMS devices produced with this approach. The multilayer structure observed through sequential steps of etching/deposition is particularly visible on the left picture.

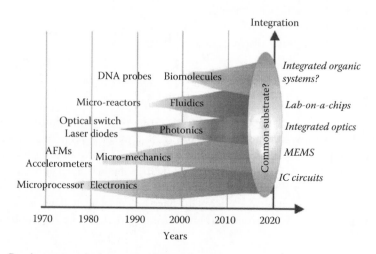

FIGURE 15.1   Roadmap toward microsystem integration.

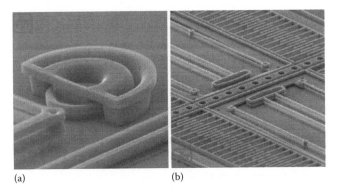

(a)   (b)

FIGURE 15.2 Illustration of a multilayer MEMS device produced by lithography methods. (Courtesy of Sandia National Laboratories, Albuquerque, NM.)

With the increasing number of desired functionalities, the surface micromachining approach faces numerous issues. The growing number of materials and processing steps introduce difficult material compatibility.

The second technology platform is the microassembly. Parts produced by surface micromachining or other single-process microfabrication techniques are put together by microassembly to form a device. There, the main challenges are to precisely position objects and to assemble them in a reliable manner through bonding or joining processes.

With the small size of the components of interest, the assembly process is often quite tedious and sometimes nearly impossible.

In addition to the technical limitations mentioned earlier, these two microfabrication approaches face economical and societal issues. The spectacular miniaturization trend ongoing for the past 30 years in various fields has not been paralleled by a similar miniaturization of the production means. Ironically, today, the microsystems industry uses very large pieces of sophisticated equipment to fabricate very small parts! Thus large capital investments are required to set up and operate microsystem production facilities. Consequently, only products with potential large markets are considered, and only a few large suppliers can make the necessary financial investments. Small and medium size enterprises are prevented from entering the field although they usually are a strong source of innovative ideas. This issue has a negative impact on our innovation capabilities as well as our abilities to rapidly adapt to new demands. Furthermore, foundries where surface micromachining takes place consume enormous amount of energy, most of it being wasted in operating the machinery and in the control of air temperature, humidity, and purity (which accounts for 90%–95% of the total energy budget!) as required for photolithography. As sustainable growth requirements become more prevalent, these power-hungry fabrication techniques will face increasing societal scrutiny.

As will be shown in the subsequent paragraphs, alternative production methods that can bypass some of these issues will be of increasing interest. One of these new approaches is the use of femtosecond lasers to fabricate glass-based microsystems and sensors. As we will see, this approach is highly flexible and provides a mean to create highly integrated three-dimensional elements.

15.3 MICROSYSTEMS FABRICATED USING FEMTOSECOND LASERS: REVIEW AND STATE OF THE ART

15.3.1 SPECIFICITIES OF FEMTOSECOND LASER–MATTER INTERACTION FROM THE VIEWPOINT OF MICROSYSTEMS DESIGN

Femtosecond lasers produce ultrahigh peak power pulses, leading to a fundamentally different laser–matter interaction than that associated with conventional long-pulse lasers: nonlinear absorption phenomena, such as multiphoton processes, are observed. This opens new and exciting opportunities to tailor the laser–matter, provides excellent spatial resolution, and furnishes a path to three-dimensional fabrication.

Although the peak power is enormous (in the GW/mm² or even TW/mm² ranges), the average power is small. For instance, the devices shown in the illustrations in this chapter were made with no more than 200-mW average power. The femtosecond lasers required for these types of applications are now rather small. The effect of femtosecond laser on dielectrics, although not completely understood, is well documented (see for instance Refs. [2,3]). Here, we just summarize some of the main features and, in particular, the effects on fused silica, a material that is of particular interest for microsystems. Others dielectrics, such as photoetchable glass [4], have also been considered for microsystems integration. See preceding chapter (by K. Sugioka) for details.

Synthetic amorphous silica (a-SiO_2) is a high-quality glass that has outstanding optical properties and is inert to almost all chemicals. Furthermore, although it may sound counter-intuitive, fused silica has excellent elastic properties making it a suitable material for micromechanics as it will be further emphasized. SiO_2 [5] is one of the most abundant material on earth and is used in a very large number of industrial applications including optics (telecommunications fibers and various optical elements), chemistry (catalyst, catalytic hosts, and absorbents), electronics (insulators and diffusion barrier), and biology (substrates for functionalization)—just to name a few examples. Furthermore, fused silica is biocompatible. In fact, amorphous silica forms the skeletons of most plankton (diatoms).

In fused silica, we observe three different regimes when the material is exposed to femtosecond lasers radiation (Figure 15.3):

- Above a first energy threshold E1 (which level depends on various parameters such as focusing optics), the material refractive index and etching susceptibility are locally increased (but no material ablation is observed). Although some aspects of this phenomenon are still debated, this effect is linked to a localized densification of the material [6].
- Above a second energy threshold E2, the formation of self-organized patterns is observed. These self-organized patterns have been reported by various authors [6–9]. Various

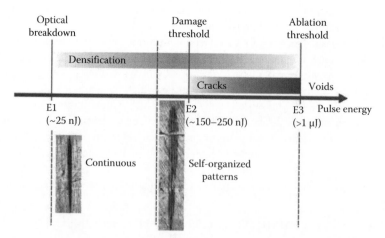

FIGURE 15.3 Effect on fused silica for different pulse energies. One can distinguish three typical regimes. The effective levels of pulse energies depend on the focusing optic (numerical aperture in particular). The images show continuous and self-organized patterns as viewed with a Scanning Thermal Microscope [6,10]. (The order of magnitude provided for the pulse energies correspond to pulse-length of 100 fs and for an objective with a numerical aperture [NA] of 0.55.)

interpretations for the formation of these patterns have been suggested. Note that cracks may be observed in these self-organized patterns [6,11].

- Finally, if the energy density is further increased to a given level E3, the formation of voids or the ablation of material is observed.

From the viewpoint of microsystems fabrication, the first two regimes are potentially the most attractive. In particular, the first regime provides a mean to locally increase the refractive index of the material (which is of particular interest for writing waveguides [12]) as well as a mean to locally increase the etching selectivity.

15.3.2 Integrated Optical Devices

To fabricate integrated optical devices, the laser is scanned through the glass structure in order to create continuous densified zones. Typically, the laser is stationary, and a moving stage is used to displace the specimen. This method has not only been used to create a variety of integrated optical devices in fused silica but also in other material systems including lithium niobate.

15.3.3 Optofluidics

The observation of an increased etching rate resulting from the laser exposure (first reported by A. Marcinkevicius et al. [13]) has triggered numerous developments toward the fabrication of fluidic channels and tunnels [14]. Combined with waveguides, these developments have led to novel concept of small-scale optofluidics devices ([15–19] for a variety of applications such as the detection and the screening of algae population [19].

15.3.4 Micromechanical Functionality

At about the same time, we suggested to further push down the integration path by inserting micromechanical functionalities into fused silica substrates. To illustrate this concept, a microsensor with an embedded position detection subsystem (waveguide array) was demonstrated. This development opens new avenues where mechanical and optical functions are deeply embedded and created using a single manufacturing process step.

15.4 MULTIFUNCTIONAL MONOLITHIC SYSTEM INTEGRATION

15.4.1 Concept

To reduce the fabrication complexity and to increase the performance and reliability of microsystems, we have focused our research on monolithic integration based on the concept of "system materials." Rather than building up a device by combining and assembling materials, this concept consists in turning a single piece of material into a system through spatially localized tailoring of its material properties. The material is no longer just an element of a device but becomes a device on its own. There are many advantages associated with this fabrication approach. It reduces microsystems assembly steps (a common source of significant cost, inaccuracy, and reliability issues), and it opens new design opportunities. This approach was first proposed to fabricate shape memory alloys used in microengineering [20]. There, a laser was used to locally introduce active and passive functions in a layer of material that was originally amorphous.

With regard to microsystems, the use of ultrafast lasers to process amorphous fused silica is of particular interest. Femtosecond lasers give a new dimension to the concept of system materials, when applied to dielectrics.

Femtosecond laser beam can locally increase the refractive index, enhance the etching rate [13], introduce sub-wavelength patterns [7], create voids [21], or change the thermal properties [6] of fused silica. By scanning the laser through the specimen volume, one can distribute, combine, and organize these material modifications to form complex patterns to be used, for instance, as waveguides or fluidic channels.

With this technique, instead of building up a device by combining layers of materials as common practice, the microdevice structure and function are directly "printed" into a monolithic piece material. Note that due to the nonlinear nature of the femtosecond laser–matter interaction, the material modifications can be introduced not only at the material surface but also anywhere in the bulk of the material.

15.4.2 TAXONOMY OF INDIVIDUAL ELEMENTS USED IN A MONOLITHIC DESIGN

15.4.2.1 Waveguides

The laser-affected zone (LAZ) shape and size can be determined using either a refractive index map technique or more recently a novel technique based on Scanning Thermal Imaging [6]. Typically, the LAZ has the shape of an ellipsoid stretched along the optical axis. This shape can be correlated to laser beam parameters (beam size, waist location, and energy). The stretching along the vertical direction depends on the chosen focusing optics. Noteworthy, as femtosecond laser matter interaction involved non-linear processes, the LAZ can be smaller than the laser spot-size itself. These observations are supported by near-field optical profilometry measurements that provide a refractive index map of the region of interest. A refractive map of a single line written in the glass using fs-laser is shown in Figure 15.4, left. To increase the mode-field diameter (MFD) of the waveguides so that to match a specific wavelength or to get particular waveguiding conditions, one can write multiple laser-written line next to another (Figure 15.4, right) [22].

Figure 15.5 illustrates a rather complex interleaver system written in a fused silica monolith using a femtosecond lasers. The interleaver is formed of an input short straight waveguide terminated by a 50–50 splitter. Two arms forming the main section of the interleaver follow this element. The two arms recombined at a coupler that is followed by a short straight output waveguide. All waveguides are single-mode at the design wavelength. The minimum waveguide curvature radius, approximately 15 mm, is a function of the change in the refractive index that can be introduced with the laser: the higher the change of refractive index, the smaller the radius of curvature.

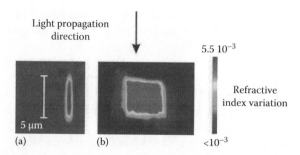

FIGURE 15.4 Refractive index map of a single and multiple lines patterns written in a fused silica glass. The center of the line has the highest index of refraction can be correlated with a local increase of glass density which in turn, can be related with the local increase of etching rate. By writing multiple lines adjacent one to another, one can form enlarged region of higher refractive index. (From P. Bado, A. A. Said, M. A. Dugan, and T. Sosnowski, "Waveguide fabrication methods and devices," U.S. patent 7391947, June 24, 2008.)

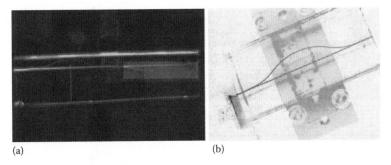

(a)　　　　　　　　　　　　　　　　　(b)

FIGURE 15.5　Dual stage (a) and single stage (b—negative image) interleaver. (Courtesy of Translume, Ann Arbor, MI.)

15.4.2.2　Channels, Grooves, etc.

To form three-dimensional structures, the following two-step procedure is applied (see Figure 15.6):

1. The material is selectively exposed by rasterizing a pattern according to a technique described in Ref. [14] and briefly outlined in the next paragraph. The laser used in our experiment is a Ti:Sapphire laser (RegA from Coherent) operating at 800 nm. The pulse width is typically 100 fs, and the repetition rate is set at 250 kHz. The average power ranges from 20 to 400 mW, which corresponds to pulse energies ranging from 55 nJ to 1.6 μJ. The linear spot size is approximately 1 μm in diameter at the focus. In our experiments, we used 20× and 50× long-working distance microscope objectives from Mittutoyo. With these objectives, one can work well below the surface of the glass substrate. Typical writing speeds are 0.5–2 mm/s. Affected regions are hit multiple times by the laser (typically 500–2500 times).

2. After laser exposure, the part is etched in a low-concentration Hydro-Fluoric acid (HF) bath. Concentrations between 2.5% and 5% are typically used. Etching time depends on pattern sizes and varies from 1 h to several hours for the deepest structures. Following etching, the part is rinsed in de-ionized water and dried.

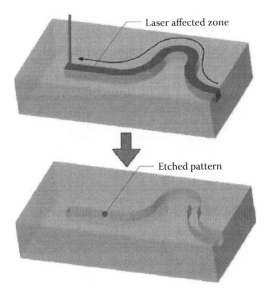

FIGURE 15.6　Process steps. (1) The material is exposed to femtosecond laser irradiation. (2) The part is etched with hydrofluoric acid. Exposed regions etch away much faster than unexposed regions.

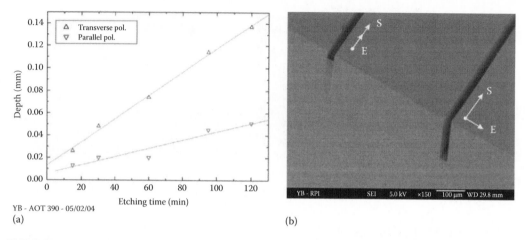

YB - AOT 390 - 05/02/04

(a)   (b)

FIGURE 15.7 Etching rate as a function of the polarization. (The convention used to define parallel and transverse polarization is shown in the right image featuring two etched channels under different polarization.)

Noticeably, the laser polarization has a strong effect on the etching efficiency as first reported by Hnatovsky et al. [23]. The explanation of this effect is still debated. One hypothesis rests on the presence of oriented nanocracks that would promote a rapid penetration of the HF etchant. However, this explanation fails to provide a satisfactory answer as to why index change tracks made at low energy (where no cracks are observed) also etched at a polarization-dependent rate.

Further investigations are needed to fully understand these observations. However, from a manufacturing point of view, this effect is easily controllable and has been quantified. In Figure 15.7, a typical etching time/etching depth curve is shown [24]. These results were achieved by scanning a 100-fs laser beam to form a rectangular pattern in the glass. The pattern was then etched, and the depth was observed at different time period.

The polarization state not only affects the etching rate but also the edge surface quality of microchannels.

As mentioned previously, by spatially arranging laser-affected tracks, one can form complex shapes and patterns of nearly arbitrarily dimensions. A collection of microchannels with various cross sections are shown in Figure 15.8. This ability to describe arbitrary three-dimensional volumes can also be exploited to create features such as micromixing chevrons.

15.4.2.3 Mechanical Component: Flexures

Flexures are mechanical elements used in micro- and precision engineering to precisely guide the motion of microparts. They consist of slender bodies that deform elastically upon the application of a force. As such, they can be considered as friction-less joint between two solid parts that constraint certain degree-of-freedoms in order to precisely guide the relative motion of two connected parts. Table 15.1 summarizes the main differences between a traditional mechanical joint and a flexure.

TABLE 15.1

Comparison of a Traditional Mechanical Joint and a Flexure

	Mechanical Joint	Flexure
Main characteristics	Assembled	Monolithic
Guiding mechanism	Geometrical surfaces	Elasticity of material
Pros	Large range of motion possible	No backlash/no play
Cons	Backlash/play	Limited range of motion

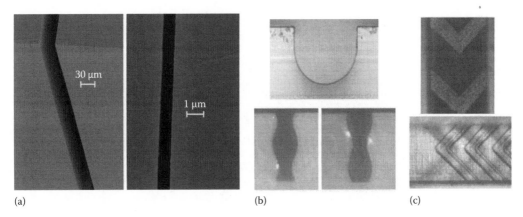

(a)   (b)   (c)

FIGURE 15.8 Collection of microfluidic channels made using the combination of femtosecond laser exposure and etching. (a) From several micron-wide deep trenches to submicron channel. (b) Front view of microchannels with various cross-section profiles. The channels are a few tens of microns deep. (c) Microfluidic channel with shaped walls and floor. Top, channel floor with chevrons. Bottom, sidewalls with chevrons (the back wall is out of focus).

One of the early examples of an elastic element to guide a motion is the clock that Christian Huygens, a prominent Dutch scientist, imagined in the seventeenth century. This clock revolutionized time keeping, offering a precision never obtained before.

While they were first confined to the niche of precision instruments, they are now much more broadly used with the general trend for miniaturization. They are extensively used in MEMS design and microrobotics [25]. In fact, the majority of MEMS with movable parts operate with (silicon) flexures.

We have investigated the use of fused silica as a material for miniaturized flexures and more specifically, the mechanical properties of femtosecond laser microfabricated flexures [26]. Using the same approach, we also investigated the strength of silica glass at the microscale level. We fabricated test structures as, shown in Figure 15.9 and Figure 15.10, in order to evaluate the elastic limits of our fused silica flexures.

The test structures consist of monolithic structures cut out of a 1 mm thick fused silica substrate. The structure is made of a slender body itself consisting of a thin part (the notch hinge) and a thicker beam. The test structure also has a frame to protect it during the fabrication process (in particular, the etching step). Additional features, such as reference surfaces and a mounting hole for fastening, are also built-in.

These test structures were loaded by pressing a contact pin against the thicker element of the slender beam and translating along the Y-axis (as shown in Figure 15.1). During the test, the contact pin

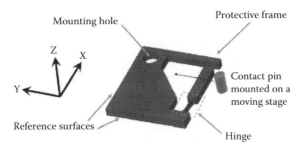

FIGURE 15.9 Schematic of a test structure used to evaluate fused silica flexures. The part is cut out from a fused silica wafer. A contact pin moving along the Y-axis is used to load the hinge in bending.

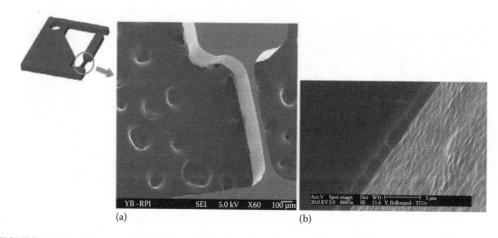

FIGURE 15.10 Micromachined flexure. Edge roughness resulting from the laser exposure typically varies from R_{tm} (Mean peak-to-valley) = 160–200 nm.

smoothly slides along the beam and applies a force that deforms elastically the structure. The pin-beam contact was dry (i.e., no lubricant was used). The pin diameter is sufficiently large compared to the glass roughness to prevent chattering or the application of an unwanted force along the X-axis.

During the experiments, a video of the deformed beam was captured and later analyzed using image processing techniques. Stationary features (such as the edge cut at 45° in the middle of the structure) were used as reference points to measure the angular deflection of the beam. We calculated the maximum stress in the beam using known equations for flexures on which a pure moment is applied (the loading case corresponds to a pure bending mode). Figure 15.10 shows a scanning electron microscope image of one of the micromachined hinges (left).

As shown in Figure 15.12, the surface roughness (R_{tm}) is in the few hundreds of nanometers, which is rather good for a laser-based machining process.

Our tests showed that our micromachined fused silica follows a brittle mechanical behavior. Our tests showed that the micromachined fused silica follows a brittle mechanical behavior but the presence of plastically deformed nanostructures [26] confirms that at the nanoscale, glass behaves like a metal [27]. Fused silica is characterized by an asymmetric mechanical load response: While it can be submitted to high compression stress level, it is weak while submitted to tensile stress.

Although glass materials have a high theoretical elastic limits (estimated at 15 GPa or higher), they tend to break at much lower stress levels (typically a few tens of MPa) due to presence of surface flaws that act as crack nucleation sites [28]. When a crack forms, it rapidly propagates through the structure and leads to catastrophic failure. In general, the tensile stress elastic limit dominates the flexure design as it defines the maximum possible excursion for a given hinge thickness and ultimately governs how small a flexure can be for a given displacement. As an example, 500 MPa is a typical value used as limit when designing flexures made of common steels. Figure 15.11 shows a sequence of photographs of a notch hinge (flexure) being loaded in bending [29]. The maximum bending angle achieved was 62°. The deformation shown in Figure 15.11 was reversible. This is quite remarkable result for a glass material that demonstrates the unusually high resistance to tensile stress of our micromachined hinges.

A preliminary analysis shows that the maximum strength of the flexure is function of the etching time. We observed that the ultimate tensile strength (UTS) limit linearly increases with the etching time. Similar observations have been made with macroscopic sample in the past [28]. We observed UTS as high as 2.5 GPa for specimens etched for a long period, which is quite remarkable for a non-trivial shape. Thus femtosecond laser combined with etching can be used to fabricate to fused silica devices of unusually high mechanical strength.

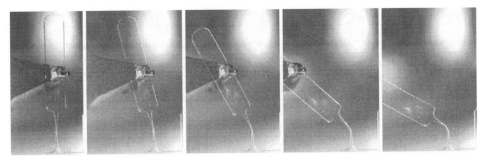

FIGURE 15.11   Flexure bending test sequence: The flexure is 40 μm wide at the neck. The bending is completely reversible. Note the severe deflection of the microhinge.

15.4.3   System Integration: Design Strategies and Interfacing

In the previous section, we described a taxonomy of simple elements that can be made using femtosecond laser processing of fused silica. The combination of these elemental structures opens numerous opportunities in term of system integration. As illustrated in Figure 15.12, through a combination of these various simple elements, one can create numerous complex devices such as lab-on-a-chip and optomechanical sensors.

Figure 15.13 illustrates the simultaneous integration of a waveguides and a fluidic channel [15] as well as waveguides and channels in a mechanical structure [29]. Figure 15.14 shows a

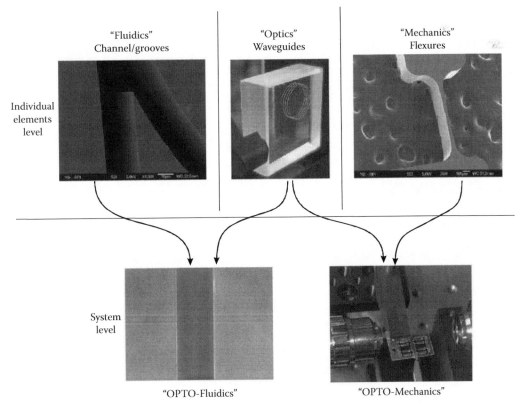

FIGURE 15.12   Monolithic system integration through the combination of individual elements.

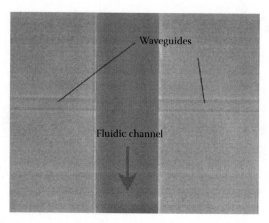

FIGURE 15.13 Optical microscope of a microchannel with two waveguides facing each other across the channel. The channel is 30 μm wide and 60 μm deep. The waveguides are 8 μm in diameter and are positioned 50 μm below the glass surface.

more complex assembly. In term of design, one has to take into consideration which element, if any, should be etched first. One may also want isolate some features, such as waveguides, from the etchant. This can be performed by carefully planning the etching pattern and the etchant progression through the microstructure.

15.4.4 ILLUSTRATION: MICRODISPLACEMENT SENSORS AND MICROFORCE SENSORS

To conclude, we illustrate the concept of monolithic integration with a presentation at a microforce instrument with an integrated optical microdisplacement sensor. This type of device, when fabricated using conventional practices, is made of many parts that require careful alignment and present difficult permanent fixturing/bonding challenges.

Our femtosecond laser-written microforce device is shown in Figure 15.15. A force applied to the sensor tip induces a linear motion of the mobile platform. The device has two key subsystems: a flexure-based micromechanism that accurately guides the motion of the platform along one axis and a waveguide-based element that senses the corresponding displacement. This displacement-sensing element consists of an array of optical waveguides embedded in the moving platform and two waveguide segments embedded in the stationary frame.

15.4.4.1 Sensor Kinematics

The kinematics model is based on two identical four-bar mechanisms serially connected, as shown in Figure 15.16c. This kinematics design is well known in precision engineering [30] and has the property to produce a well-defined linear motion (unlike a parallelogram four-bar mechanism as shown in Figure 15.16a without adding an internal mobility in the mechanism like a single compound would do Figure 15.16b. This strategy offers also the additional advantage that it is self-compensated for thermal expansion.

Figure 15.16 shows idealized rotational joints. In microengineering, due to scale and precision requirements, such a mechanical design is difficult to implement with multiple parts. Rather, a monolithic, flexure-type design is preferred. The principle is to replace traditional (i.e., multipart joints) by elastic hinges that provide the same kinematics. In our case, we use a notch-hinge to emulate the behavior of a rotational joint. The flexure was designed using both analytical and finite element modeling. The design procedure is detailed in Ref. [25]. The analytical model predicts that a force of 200 mN is required to reach the full 1-mm excursion.

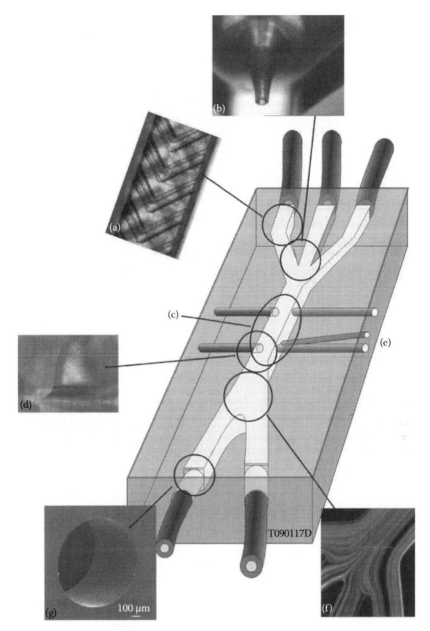

FIGURE 15.14 Schematic representation of a micro flow cytometer with an overlay of various relevant elements fabricated by femtosecond direct write.

As mentioned earlier, it is known that the elastic limit of fused silica depends on the presence or absence of surface flaws. Processes that eliminate these flaws, such as HF etching, can increase the elastic limit by several orders of magnitude. For this work, we used an elastic limit of 300 MPa. Experimentally, we found this value to be conservative.

To refine and optimize the hinge shape, a finite element analysis was conducted. Both static and dynamic analyses were investigated (details about this analysis can be found in Ref. [29]). Good agreement between analytical and Finite Element Analysis (FEA) model was found. From the finite

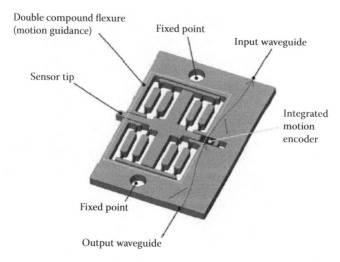

FIGURE 15.15 Drawing of a glass-based force-sensing device. The mobile "sensing" platform has a cross shape and is prolonged by a sensor tip.

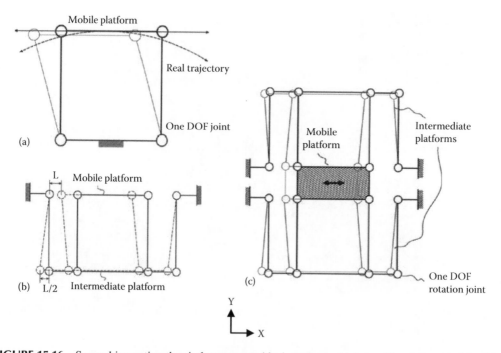

FIGURE 15.16 Sensor kinematics: the circle represents ideal mechanical joints with one degree of freedom (rotation in the plane). (a) shows a parallelogram four-bar mechanism, (b) represents a single compound design, and (c) a double-compound design.

element analysis, the force to get the full excursion is about 400 mN, and the maximum stress is 240 MPa. (Figure 15.17 illustrates an FEM simulation of the device in its deformed state).

The dynamic analysis indicates that the three first structural resonance modes are in-plane vibrations. The first natural mode is found at about 405 Hz, while the second and third modes are found around 1.2 kHz. Out-of-plane vibrations are activated at much higher frequencies (3.2–9.5 kHz).

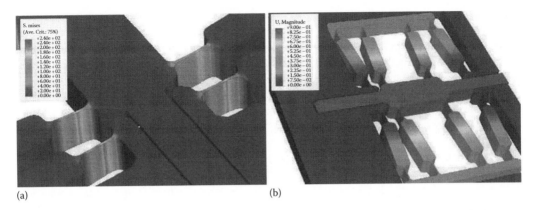

(a)   (b)

FIGURE 15.17   **(See color insert.)** FEM analysis—stress distribution in four hinges (a) and displacement distribution of the entire structure (b).

15.4.4.2   Integrated Linear Encoder

We used the variation of signal intensity induced by lateral misalignment between identical waveguides as the basic principle for the mobile platform displacement. In practice, a waveguide segment is incorporated in the mobile platform so that, at rest, it is aligned with two stationary frame waveguides used as transmitting and receiving waveguides for the integrated linear encoder (ILE) signals.

Using a single waveguide segment in the moving platform would limit the sensing range to approximately the width of the MFD. To extend the displacement sensing range, the platform contains an array of parallel waveguides. When a waveguide segment of the movable platform is aligned with the input and output stationary waveguides (points a and c in Figure 15.18), the intensity of the transmitted signal is maximized. Conversely, when the waveguides are misaligned (point b in Figure 15.18), the light is no longer guided through the platform (it is only guided in the input segment), which results in a severe loss of transmitted signal. The range of motion sensing can be extended indefinitely with this approach.

In practice, the ILE consists of a fixed 30 µm pitch waveguide array spanning the 1 mm end section of the movable platform. By design, a transmitting and a receiving waveguide in the stationary section of the flexure mount are in direct axial alignment with one of the array waveguides when the stage is unloaded and at rest. The optical signal crosses two identical free-space gaps (schematically

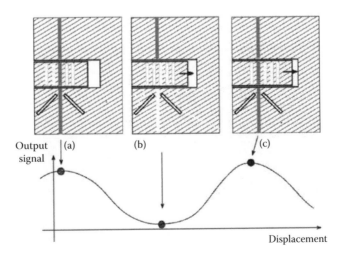

Output signal (a)   (b)   (c)

Displacement

FIGURE 15.18   Waveguide-based linear encoder principles.

represented by straight lines in Figure 15.18)—one located between the output of the transmitting waveguide and the input of the movable array waveguides and a second located between the output of the array waveguides and the input of the receiving waveguide. These gaps consist of a 30 μm air region sandwiched between two 20 μm glass regions where the light is unguided, as shown in Figure 15.19. The waveguides are 8 μm wide with an index difference (core – cladding) of ~5.25e−3. They are highly multimode at the test wavelength of 670 nm.

Light propagation through the structure was simulated using the finite difference method. The fundamental mode at 670 nm is launched into the input waveguide. After the propagation is completed, the output waveguide power is recorded, the array is displaced from the stationary waveguides by 0.6 μm, and the propagation is restarted. This is repeated through a couple of periods of the array. Figure 15.20 shows two of these sequences.

Beam expansion of the fundamental mode across the first gap results in a finite overlap with higher order modes at the input of the corresponding array waveguide even when the waveguides are perfectly aligned. From there on, the optical signal propagates in a multimode fashion as can be seen in Figure 15.20. Despite the multimode nature of the optical signal propagation, the roll-off of a single waveguide transmission with displacement is monotonic and can be scaled with distance.

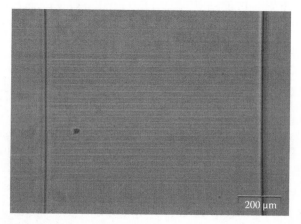

FIGURE 15.19 Magnified view of the actual ILE. The waveguide array is visible in the lower part of the image and the two input encoders in the upper part.

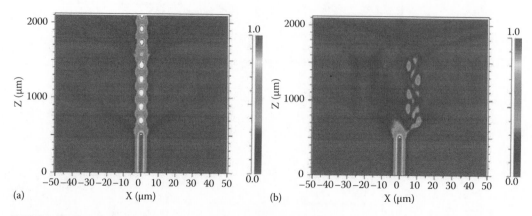

FIGURE 15.20 **(See color insert.)** Wave propagation in the ILE for two configurations as the array is moved from the right to the left. The horizontal line indicates the free-space gaps.

This type of signal response is maintained as long as the array pitch is large enough to prevent coupling between the parallel waveguides.

15.4.4.2.1 Experiments

A sketch and a partial view of the experimental test setup are shown in Figure 15.21. The device characterization was performed with a free-space optics setup. A mechanical finger is used to apply a displacement on the force sensor tip. The finger is stiff and is attached to a sub-micron accuracy piezo-actuated positioning stage whose stiffness is several orders of magnitude higher than that of the device under test. The finger position is measured using a triangulation measurement system made of a laser beam, a mirror, and a position-sensing device. This measurement system has a 50 nm resolution. As the finger applied a force on the sensor tip, the flexure is deformed and the waveguides embedded in the mobile platform moved laterally relative to the stationary waveguide pair located in the frame. A 670 nm laser diode is used as the light source and is injected into the stationary waveguide. At this wavelength, the waveguides are multimode. The light intensity transmitted through the sensor is measured by a photodetector. Acquired signals (transmitted intensity and finger position) are further processed. Details of the fabricated microsensor are shown in Figure 15.22: the two left pictures show an optical microscope view of half of the flexure (in deformed and non-deformed configurations); the right picture is a close view of the ILE.

Waveguides are placed 100 μm below the surface. There is a total of nine waveguides on the movable platform.

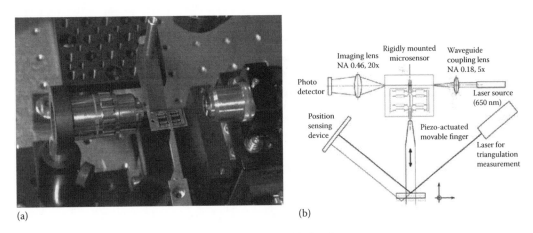

(a)   (b)

FIGURE 15.21 Experimental setup: (a) partial view and (b) sketch.

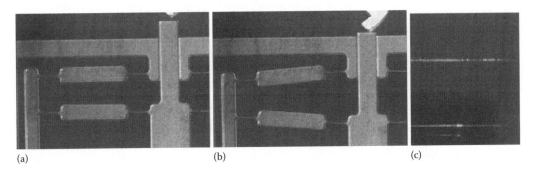

(a)   (b)   (c)

FIGURE 15.22 Optical microscope images of the sensor prototype: (a and b) hinges in deformed and non-deformed configuration, respectively (c) close view of the integrated linear encoder where waveguides are all aligned.

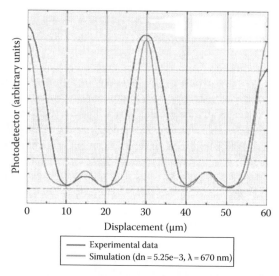

FIGURE 15.23 Experimental results (dark grey) compared with simulation results (light grey). The figure shows the intensity seen for the last three waveguides (going from left to right). The lowest intensity peaks (on the right) corresponds to the last waveguide.

The experimental results (shown in Figures 15.22 and 15.23) closely resemble the simulated results. The predicted intermediate peak is clearly visible in the experimental curve. We also notice some peak intensity variation for the largest peaks. This variation is typically within ±15% from waveguide to waveguide. These intensity fluctuations may arise from local changes in wall roughness (typ. 300 nm) associated with the two gaps. Inserting an index-matching liquid in the gaps tends to level the peak intensities.

The first resonant frequency was measured on a prototype having hinges of about 42 ± 2 μm as measured under an optical microscope. The measurement was done by imposing a sinusoidal mechanical vibration on the piezo-actuator that drives the moving finger main axis. We found the resonant frequency at 209 Hz. For this hinge thickness, the Finite Element Modeling simulation gives a first mode resonant frequency at 206 Hz showing a good agreement with the experimental results. Using the ILE signal, we know the microstage position with a resolution equal or better than 50 nm. (This positioning accuracy is presently limited by our experimental setup and not by the microstage itself.)

15.5 SUMMARY, BENEFITS, FUTURE PROSPECTS, AND CHALLENGES

Current microsystems technologies face multiple challenges. The increasing functional complexity of these tiny machines poses numerous issues with respect to packaging and reliability. To further increase function integration, new design and manufacturing techniques are needed. A promising approach is based on the concept of system materials where, rather than building a system through an assembling of parts, the material's intimate structure is locally modified so that the material can fulfill a specific function not present at first.

In that context, femtosecond lasers can be used to tailor the material properties of fused silica to create not only optical functions (such as waveguides) but also mechanical (flexures) or fluidics (microchannel) elements. These functions can be combined to form complex optofluidics or optomechanical devices. In this chapter, we presented a few illustrative femtosecond laser–machined microsystems. These examples demonstrated the viability of the concept of system materials. This novel microfabrication approach provides a means to produce fully integrated devices with advanced functionalities.

There are, however, still numerous challenges to address before this technology is widely accepted. So far, the increase of refractive index remains limited to a fraction of 1%. This imposes numerous design limitations, such as device footprint and ultimately affects the device integration. Machining time is another important limitation, which affects the commercialization of this technology. On-going efforts are targeting these issues.

REFERENCES

1. M. Madou, *Fundamentals of Microfabrication: The Science of Miniaturization*, Taylor & Francis Group, Boca Raton, FL, 2002, ISBN: 0849308267.
2. D. Du, X. Liu, G. Korn, J. Squier, G. Mourou, Laser-induced breakdown by impact ionization in SiO_2 with pulse widths from 7 ns to 150 fs, *Appl. Phys. Lett.*, 64, 233071 (1994).
3. S. S. Mao, F. Quéré, S. Guizard, X. Mao, R. E. Russo, G. Petite, P. Martin, Dynamics of femtosecond laser interactions with dielectrics, *Appl. Phys. A*, 79, 1695–1709 (2004).
4. Y. Cheng, K. Sugioka, K. Midorikawa, M. Masuda, K. Toyoda, M. Kawachi, and K. Shihoyama, Three-dimensional micro-optical components embedded in photosensitive glass by a femtosecond laser, *Opt. Lett.*, 28, 55–57 (2003).
5. L. W. Hobbs, C. E. Jesurum, V. Pulim, B. Berger, Local topology of silica networks, *Philos. Mag. A*, 78, 679–711 (1998).
6. Y. Bellouard, E. Barthel, A. A. Said, M. Dugan, P. Bado, Scanning thermal microscopy and Raman analysis of bulk fused silica exposed to low-energy femtosecond laser pulses, *Opt. Express*, 16, 19520–19534 (2008).
7. Y. Shimotsuma, P. G. Kazanksi, Q. Jiarong, K. Hirao, Self-Organized Nanogratings in Glass Irradiated by Ultrashort Light Pulses, *Phys. Rev. Lett.*, 91, 247405 (2003).
8. P. G. Kazansky, W. Yang, E. Bricchi, J. Bovatsek, A. Arai, Y. Shimotsuma, K. Miura, and K. Hirao, "Quill" writing with ultrashort light pulses in transparent materials, *Appl. Phys. Lett.*, 90, 151120 (2007).
9. V. R. Bardwaj, E. Simova, P. B. Corkum, D. M. Rayner, C. Hnatovsky, R. S. Taylor, B. Schreder, M. Kluge, and J. Zimmer, Femtosecond laser-induced refractive index modification in multicomponent glasses, *J. Appl. Phys.*, 97, 83102 (2005).
10. Y. Bellouard, M. Dugan, A. Said, P. Bado, Thermal conductivity contrast measurement of fused silica exposed to low-energy femtosecond laser pulses, *Appl. Phys. Lett.*, 89, 161911 (2006), DOI:10.1063/1.2363957.
11. R. Taylor, C. Hnatovsky, E. Simova, Applications of femtosecond laser induced self-organized planar nanocracks inside fused silica glass, *Laser Photonics Rev.*, 2, 26–46 (2008).
12. K. M. Davis, K. Miura, N. Sugimoto, and K. Hirao, Writing waveguides in glass with a femtosecond laser, *Opt. Lett.*, 21, 1729–1731 (1996).
13. A. Marcinkevičius, S. Juodkazis, M. Watanabe, M. Miwa, S. Matsuo, H. Misawa, J. Nishii, Femtosecond laser-assisted three-dimensional microfabrication in silica, *Opt. Lett.*, 26, 277–279 (2001).
14. Y. Bellouard, A. Said, M. Dugan, P. Bado, Fabrication of high-aspect ratio, micro-fluidic channels and tunnels using femtosecond laser pulses and chemical etching, *Opt. Express*, 12, 2120–2129 (2004).
15. A. Said, M. Dugan, P. Bado, Y. Bellouard, A. Scott, J. Mabesa, Manufacturing by laser direct-write of three-dimensional devices containing optical and microfluidic networks, *Proc. SPIE-The International Society for Optical Engineering* (2004), 5339, 94–204.
16. Y. Hanada, K. Sugioka, H. Kawano, I. Ishikawa, A. Miyawaki, K. Midorikawa, Nano-aquarium for dynamic observation of living cells fabricated by femtosecond laser direct writing of photostructurable glass, *Biomed. Microdevices*, 10, 403–410 (2008).
17. R.W. Applegate Jr, J. Squier, T. Vestad, J. Oakey, D. W. M. Marr, P. Bado, M. A. Dugan, and A. A. Said, "Microfluidic sorting system based on optical waveguide integration and diode laser bar trapping, *Lab. Chip*, 6, 422–426 (2006), DOI: 10.1039/b512576f.
18. R. M. Vazquez, R. Osellame, D. Nolli, C. Dongre, H. van den Vlekkert, R. Ramponi, M. Pollnau, and G. Cerullo, "Integration of femtosecond laser written optical waveguides in a lab-on-chip, *Lab. Chip*, 9, 91–96 (2009), DOI: 10.1039/b808360f.
19. A. Schaap, Y. Bellouard, and T. Rohrlack, Optofluidic lab-on-a-chip for rapid algae population screening, *Biomed. Opt. Express* 2, 658–664 (2011).
20. Y. Bellouard, T. Lehnert, J.-E. Bidaux, T. Sidler, R. Clavel, R. Gotthardt, Local annealing of complex mechanical devices: A new approach for developing monolithic micro-devices, *Mater. Sci. Eng. A*, 273–275, 795–798 (1999).

21. E. N. Glezer and E. Mazur, Ultrafast-laser driven micro-explosions in transparent materials, *Appl. Phys. Lett.*, 71, 882 (1997).
22. P. Bado, A. A. Said, M. A. Dugan, and T. Sosnowski, Waveguide fabrication methods and devices, U.S. patent 7391947 (June 24, 2008).
23. C. Hnatovsky, R. S. Taylor, E. Simova, V. R. Bhardwaj, D. M. Rayner, and P. B. Corkum, Polarization-selective etching in femtosecond laser-assisted microfluidic channel fabrication in fused silica, *Opt. Lett.*, 30, 1867–1869 (2005).
24. Y. Bellouard, A. A. Said, M. Dugan, P. Bado, All-optical, ultra-high accuracy displacement sensors with detection means, *Proc. SPIE*, 5989, 59890V–59890V–14 (2005).
25. Y. Bellouard, *Microrobotics: Methods and Applications*, Taylor & Francis Group, Boca Raton, FL, 2009.
26. Y. Bellouard, On the bending strength of fused silica flexures fabricated by ultrafast lasers [Invited], *Opt. Mater. Express,* 1, 816–831 (2011).
27. F. Celarie, S. Prades, D. Bonamy, L. Ferrero, E. Bauchoud, C. Guillot, C. Marliere, Glass Breaks like Metal, but at the Nanometer Scale, *Phys. Rev. Lett.*, 90, 075504 (2003).
28. M. Tomozawa, R.H. Doremus, *Treatise on Materials Science and Technology*, Academic Press, San Diego, CA, 1982, ISBN 0123418224, 9780123418227.
29. Y. Bellouard, A. Said, P. Bado, Integrating optics and micro-mechanics in a single substrate: A step toward monolithic integration in fused silica, *Opt. Express*, 13, 6635–6644 (2005).
30. R.V. Jones, An optical slit mechanism, *J. Sci. Instrum.*, 29, 345–350 (1952).

16 Multiscale, Hierarchical Integration of Soft Polymer Micro- and Nanostructures into Optical MEMS

Katsuo Kurabayashi, Nien-Tsu Huang, and Yi-Chung Tung

CONTENTS

16.1 INTRODUCTION

Modern technologies found in military, spacecraft, automotive, telecommunications, and biomedical applications highly demand reductions of the manufacturing cost, power consumption, size, and weight of integrated sensors and actuators. The research field of microelectromechanical systems (MEMSs) has seen quite a few significant innovations and advancements to meet this demand in the past two decades (Kovacs, 1998; Rebeiz, 2003; Grayson et al., 2004). Historically, MEMS technology has been seen as an offspring of silicon-based integrated circuit (IC) technology, which primarily relies on "top-down" photolithography techniques. In contrast, the technological promises that polymers hold in future micro/nanosystems have recently attracted much attention due to their cost effectiveness, manufacturability, various material properties, and compatibility with biological and chemical systems. A wide variety of polymer-based fabrication techniques, including those based on "bottom-up" self-assembly and soft printing approaches, meet the demand for forming nanometer-sized structures rapidly and economically, thus driving nanomanufacturing research and nanotechnology. Nanoscale polymer structures, such as nanopatterned polymeric films and self-organized polymeric materials, provide important building blocks in nanotechnology, nanoelectronics, and photonics (Muthukumar et al., 1997; Hamley, 2003; Yonzon et al., 2004; Ong et al., 2005; Oikawa et al., 2010; Slota et al., 2010). Some of these structures manifest material behavior radically different from that in micro/mesoscale domains (Fu and Yao, 2001). Integration of nanostructures into an MEMS across multiple dimensional scales ranging from a few nanometers to several micro/millimeters is expected to open up new unexplored MEMS research frontiers. The multiscale integration is the key to combine MEMS technology and emerging nanomanufacturing technology, making it possible to implement technological fruits offered by nanosciences and nanotechnology in sensor and actuator applications. The technology described here meets the need for manufacturing methods to integrate nanostructures within micro- and mesoscopic devices and systems in functionally scaled products, permitting hierarchical, continuous structuring across multiple scales.

Among existing polymer materials, an organic elastomer, polydimethylsiloxane (PDMS), has become one of the most attractive materials due to its unique material properties and moldability suited for low-cost rapid prototyping for micro/nanofabrication. In particular, PDMS has recently been used in many MEMS devices with integrated micro/nanofluidic channels for biological and chemical applications. These devices are primarily fabricated using soft lithography (Xia and Whitesides, 1998). Soft lithography is a collective name for fabrication techniques based on self-assembly and replica molding. Soft lithography provides a rapid and inexpensive way of forming and transferring patterns and structures with a feature size as small as 30 nm (Zhao et al., 1997), which are otherwise constructed using time-consuming and expensive nanofabrication methods, such as deep Ultraviolet (UV) light ($\lambda = 200–290$ nm) and extreme UV ($\lambda < 200$ nm) (Horiuchi et al., 2003) photolithography, phase-shift photolithography (Levenson, 1993), electron-beam writing (Chang et al., 1996), focused ion beam lithography (Matsui and Ochiai, 1996; Langford et al., 2002), x-ray lithography (Smith et al., 1996), and scanning probe lithography (Rolandi et al., 2002). However, PDMS micro/nanostructures find limited use since they are traditionally used as fully passive structural components separated from silicon-based control/sensing signal processing circuitry and fast-response actuators. The current gap between PDMS micro/nanofabrication techniques and silicon micromachining prohibits one to realize new MEMS devices potentially resulting from the simultaneous use of these two materials on the same device platform. The technology presented in this chapter combines silicon MEMS technology with the polymer micro/nanofabrication techniques—soft lithography and nanoimprint lithography, as illustrated in Figure 16.1. This chapter describes the development of a polymer/silicon hybrid technology fully compatible with silicon micromachining and complementary metal-oxide semiconductor (CMOS) technology. The technology allows for construction of polymer/silicon hybrid single-chip modules with multiple functionalities while taking advantage of unique material properties of both polymer and silicon.

The technological approach explores a fully new technological concept of "elastomer-on-silicon (EOS) microsystems." "EOS microsystems" refers to a new class of single-chip modules incorporating elastomeric polymer micro/nanostructures integrated onto silicon actuators and/or sensors. To develop new EOS MEMS devices, a new MEMS fabrication technique, namely "soft lithographic lift-off and grafting (SLLOG)," has been employed. The SLLOG process starts with soft lithography-based molding and release of a three-dimensional (3-D) PDMS microstructure. This is followed by assembling of the microstructure onto silicon MEMS devices with high accuracy. The SLLOG process is further extended to allow imprinting of nanoscale features on the surface of the 3-D PDMS microstructure. Based on micro/nanomolding, the developed fabrication method can permit transfer of a wide variety of nanoscale morphologies onto a MEMS device surface. Our device process could be robust enough to accommodate a wide variety of polymer self-assembly techniques to create the molding templates.

The proposed technological concept could lead to development of a new type of MEMS devices, including strain-tunable optical MEMS devices that integrate a PDMS microstructure containing nanoscale grating patterns and pillar arrays. These devices have a simple structural design, yet could exhibit very unique functions such as high-speed dynamic optical spectrum tuning and photonic

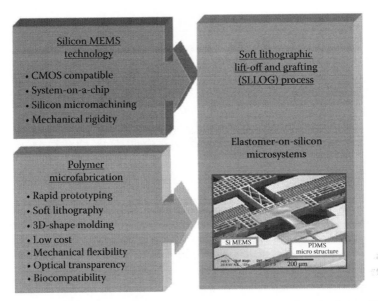

FIGURE 16.1 Concept of hybrid technology combining polymer micro/nanotechnology with silicon-based MEMS technology. The technology simultaneously employs the mechanical flexibility of polymer materials and the CMOS compatibility of silicon semiconducting materials.

band gap modulation. The functionalities of these devices originate from combining the high-strain elasticity and optical transparency of the PDMS micro/nanostructures and the fast dynamic response of silicon MEMS actuators. The devices can be constructed with reduced manufacturing cost and complexity, potentially facilitating their penetration into commercial applications and markets.

16.2 POLYMERS IN MICRO/NANOSYSTEMS

This section briefly reviews recent research on polymers and their material properties and processing methods and discusses the roles of polymers in emerging micro/nanosystems technology. The technological promises that polymers hold for future MEMS development are argued.

16.2.1 WHY POLYMERS?

Polymer is a collective name of macromolecular materials consisting of carbon-based molecular chains. In contrast to CMOS materials including silicon, nitride, and oxide, polymers can provide thousands of material species and material properties through engineering their chemical and/or crystal structures with varying synthesis and processing conditions (Matsumoto et al., 2000; Matsumoto, 2003). One of the most remarkable material properties of polymers is their mechanical flexibility with high strain strength. With an increasing demand for light-weight, low-cost, stretchable electronic modules, such as flexible displays (e.g., electronic paper) (Rogers et al., 2001; Chen et al., 2003), wearable electronics (Post et al., 2000; Khang et al., 2006), artificial skin pressure sensors (Boland, 2010), and polymers have recently gained widespread interest as promising active electronic materials and substrates that incorporate other inorganic nanomaterials (Bradley et al., 2003; Gray et al., 2003; McAlpine et al., 2005; Ju et al., 2007; Sun and Rogers, 2007). Some polymers are chemically and biologically compatible. Electronic circuits and devices made of both flexible and biocompatible polymers can be safely and directly used in contact with delicate biological tissues, thus allowing for electrical monitoring of biological systems with a bio/electronics interface. Certain polymers are suited for photonics and fiber optics communications applications due to their high optical transmission rate in the visible wavelength range. There are

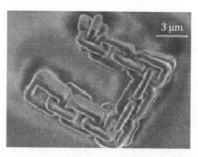

FIGURE 16.2 Three-dimensional microstructure fabricated using two-photon absorption. (Reprinted with permission from Sun, H.B. and Kawata, S., *J. Lightwave Technol.*, 21, 624–633. Copyright 2003, IEEE.)

some photosensitive polymers, which are solidified (cured) through photopolymerization, i.e., photo-induced cross-linking of molecular chains (Maruo et al., 1997).

The material properties described earlier lead to a wide variety of polymer-based micro/nanofabrication techniques that are not readily available in traditional silicon CMOS fabrication technology. Optical fabrication methods utilizing photopolymerization, such as microstereolithography and two-photon absorption, can create 3-D photonic devices and micro/nanodevices with feature sizes of a sub-diffraction limit resolution (~120 nm) (Maruo and Kawata, 1998; Zhang et al., 1999; Kawata et al., 2001; Maruo et al., 2003; Sun and Kawata, 2003). Figure 16.2 shows one example of such polymer 3-D structures.

The highly versatile construction of these truly 3-D shapes at the micro/nanometer scales can be achieved by using polymers. The highly stretchable mechanical property allows flexible fabrication and packaging techniques including soft lithography (Brittain et al., 1998; Xia and Whitesides, 1998), nanoimprinting (Chou et al., 1996; Guo, 2004), thermal embossing (Ornelas-Rodriguez et al., 2002), and injection molding (Gadegaard et al., 2003). These flexible fabrications significantly reduce the lead time to manufacture micro/nanodevices and components, allowing rapid prototyping in a highly economical and simple manner (Michel et al., 2001). In addition, emerging fields in nanosciences and nanotechnology highly benefit from the use of polymer-based fabrication methods. For example, a rich variety of nanoscale periodic patters formed by block copolymer self-assembly offers potential to fabricate high-density arrays for use in data storage, nanoelectronics, molecular separation, and combinatorial chemistry and DNA screening (Hamley, 2003). These patterns and structures can also be formed using surface-tension-driven self-assembly of polymer melts (Chou and Zhuang, 1999) and field-assisted alignment of polymers (Martin, 2002). With the availability of the rich material properties and fabrication methods, polymers inarguably play a critical role in many technological applications in the twenty-first century.

16.2.2 POLYMER MEMS—STATE OF THE ART

In the past two decades, the MEMS research community has recognized polymers as an important new class of materials for MEMS applications. An excellent review on this subject is provided in literature (Liu, 2007). The primary driving force for the use of polymers in MEMS is their greater mechanical yield strain, access to a wide variety of processing methods, unique chemical, structural, and biological functionalities, and low material cost compared to silicon, which is the traditionally predominant material in MEMS structures.

Recent work has introduced a significant number of polymer materials in MEMS applications. For example, Wang et al. (2003) demonstrated MEMS cantilever flow sensors and tactile sensors made of liquid crystal polymer (LCP). LCP is a thermoplastic polymer material consisting of well-aligned molecular chains and well suitable for mechanical sensing applications that require sufficient mechanical robustness.

Some studies indicate polymeric microstructures allow MEMS devices to operate in an environment where traditional silicon-based MEMS fail. Jager et al. (2000) and Smela et al. (1999) have reported several polymer-based MEMS actuators and manipulators for underwater applications. The actuation mechanisms of these devices are based on stress gradient in ionic conducting polymer films induced by ionic movement under electric fields or volumetric changes in polyaniline films caused by a reversible electrochemical oxidation–reduction reaction (Zhou et al., 2004a). These conjugated polymer actuators have a number of attractive features for use as an artificial muscle in biomedical applications (Smela, 2003). poly(acrylamide-ran-3-acrylamidophenylboronic acid) (PAA-ran-PAAPBA) films provide a chemical functionality promoting analyte surface binding and have been incorporated in a continuous glucose monitoring MEMS sensor (Huang et al., 2009a).

Tseng and Lin (2003) used a photosensitive polymer material, SU-8, in the form of a flexible micromirror structural layer in an optical fiber-based Faby-Perot shear stress sensor for use in an aqueous environment. Armani and Liu (2000) demonstrated biodegradable micromachined structures using polycaprolactone films. A research group led by Tai at Caltech has extensively investigated the material properties and processes of Parylene films for MEMS applications, including on-chip values (Yang et al., 1998) and electrostatic actuators (Yao et al., 2002). Using a pyrolysis process for a parylene precursor film, they have demonstrated the possibility to tailor the electrical and mechanical properties of a carbon film according to the desired MEMS device performance (Liger et al., 2004a,b).

The recent trend of migration of polymers in electronics and optoelectronics polymers open up a new device technology. Gokdel et al. (2010) have demonstrated integration of polymer light-emitting diode (PLED) arrays with a polymer composite-based electromagnetic MEMS actuator to construct a low-cost 2-D display. Each of these PLEDs has a thin-film structure consisting of indium tin oxide, semiconducting conjugated polymers, and aluminum. The authors fabricated the PLEDs on flexible polyethylene terephthalate sheets.

Regardless of their material type, the above-described polymeric MEMS components and devices are all fabricated using "top-down" fabrication techniques, which are based on photolithography and etching. In contrast, the demonstration of micromachining of MEMS devices using replica molding of PDMS, i.e., soft lithography, was a significant paradigm shift from these traditional microfabrication approaches, allowing for rapid prototyping of MEMS structures (Brittain et al., 1998). In particular, the use of the powerful and economical soft lithography technique for constructing 3-D microfluidic channels, valves, and pumps at low cost yielded a huge impact among the MEMS research community (Unger et al., 2000). In recent years, the conventional microfluidic device fabrication method following glass etching and glass-silicon bonding has been replaced in many cases by soft lithography-based rapid prototyping. Following the pioneering work by Whitesides (Duffy et al., 1998), Quake (Unger et al., 2000), and Beebe (Jo et al., 2000), a large number of PDMS-based microfluidic devices and biological MEMS (BioMEMS) have actively been reported in literature (McDonald et al., 2000; Ng et al., 2002; Sia and Whitesides, 2003; Quist et al., 2005). Figure 16.3 summarizes the characteristic features of PDMS as compared to those of other polymer materials to illustrate what has been discussed here.

16.2.3 PDMS MATERIAL PROPERTIES

Of particular interest here is the use of the flexible polymer fabrication techniques for development of a new class of MEMS devices. PDMS is a highly promising material for this purpose because it is inexpensive, easily molded, mechanically robust, disposable, chemically inert, nontoxic, and optically transparent (well into the UV). The surface chemical properties of PDMS can easily be modified using oxygen plasma treatment. Among commercially available polymers, PDMS has become highly common over the past decade in applications due to these highly desirable properties. These properties result from the presence of an inorganic siloxane backbone and organic methyl groups, which branch off of the silicon in the backbone (Clarson and Semlyen, 1993).

Material	Remarkable features	Device example	
Parylene	• Biocompatible • Taylor-made material properties by carbonization (pyrolysis) • Optical transparency	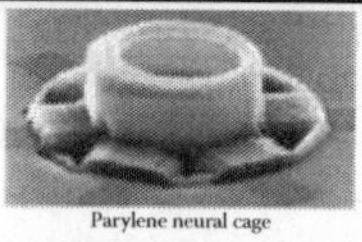 Parylene neural cage	Parylene neural cage (Tai, caltech)
LCP (liquid crystal polymer)	• Thermally bonded to various materials • IC process compatible • Low moisture absorption		LCP tactile MEMS sensor (Liu, U. Illinois)
Conjugated polymer	• Organic semiconductor • Mobility enhanced by embedding C60 or carbon nanotubes	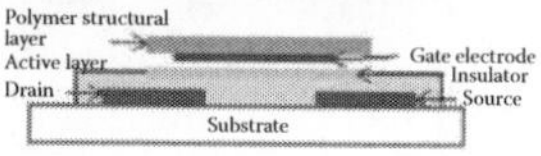 Polymer structural layer / Active layer / Drain / Gate electrode / Insulator / Source / Substrate	Organic TFT accelerometer (Varadan, PSU)
PDMS	• Biocompatible • Optical transparency • Soft lithography • Rapid prototyping closes to 3-D polymer machining processes	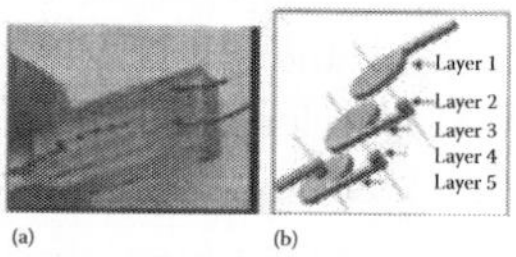 (a) Layer 1-5 (b)	Microfluidic devices (Beebe U. Wisconsin)

Left vertical arrows: Photolithography and etching; 3-D molding

FIGURE 16.3 Polymers for microsystems technology. (From He, Q. et al., Parylene neuro-cages for live neural networks study, *TRANSDUCERS* '03, Boston, MA, 2003, pp. 995–998; Varadan, V.K., *Proc. SPIE*, 5116, 149, 2003; Jo, B.-H. et al., *J. Microelectromech. Syst.*, 9, 76, 2000.)

Table 16.1 provides a comparison of the properties possessed by materials commonly used in MEMS devices (Lotters et al., 1996, 1997; Unger et al., 2000; Hoshino and Shimoyama, 2001; Shackelford and Alexander, 2002). The data show that PDMS has relatively low Young's modulus and high fracture strain, which are desirable for mechanical components such as complaint mechanical joints and membranes. Because of its transparency and refractive index similar to glass, PDMS is one of the most promising materials that can replace glass materials used in BioMEMS

TABLE 16.1
Comparisons of Material Properties

	PDMS	Boronsilicate Glass (PYREX®, 7740)	Silicon
Density	1.05 g/cm³	2.23 g/cm³	2.33 g/cm³
Young's modulus	~750 kPa	62.8 GPa	125–188 GPa
Poisson's ratio	~0.50	0.20	0.22–0.28
Tensile strength	—	—	6.9 GPa
Maximum strain	~40%	—	~1%
Thermal expansion ratio	3.10×10^{-4} K⁻¹	3.25×10^{-6} K⁻¹	2.6×10^{-6} K⁻¹
Thermal conductivity	0.17 W/m-K	—	168 W/m-K
Permittivity	2.75	4.60	11.7
Resistivity	1.2×10^{16} Ω-m	—	2.3×10^{7} Ω-m
Transparency (visible light)	Very good	Excellent	Opaque
Refractive index	~1.4	1.473	3.42
Breakdown voltage	~14 V/μm	—	30 V/μm

applications. Moreover, PDMS itself possesses a very good adhesion characteristic; it rarely creates sealing and bonding problems. This makes PDMS an excellent material for use in many microfluidic devices for fluorescence biological assays.

16.3 POLYMER/SILICON HYBRID SYSTEMS

Although an increasing number of research efforts have been directed toward constructing polymer-based devices, polymer technology is not yet matured enough to produce fully *organic* single-chip modules and microsystems that incorporate functional actuator components and signal-processing electronic circuits. Despite their economical and rapid manufacturability leading to the soft lithography-based fabrication, PDMS structures demonstrated in the previous studies only serve as passive substrate components, which contain micro/nanochannel patterns (Mello, 2002). The need for external off-chip valve/pump actuators usually prohibits the PDMS-based microfluidic devices to function literally as lab-on-a-chip systems. The application of the conjugated-polymer and hydrogel actuators is highly limited due to their slow dynamic response and low speed although they can yield large force, large displacement of motion, low voltage requirement, and biocompatibility (Smela, 2003). The charge-carrier mobility of conjugated polymers (typically less than 5 cm^2/Vs) (Dimitrakopoulos and Malenfant, 2002; Horowitz, 2004) is more than 2 orders of magnitude smaller than that of *n*-doped silicon, making it difficult to achieve satisfactory performance with organic ICs. Whereas, silicon-based CMOS transistors can achieve fast electronic circuits, which are not yet demonstrated by organic thin film transistors. Silicon-based MEMS can be fabricated using well-established CMOS compatible processing methods including bulk micromachining and surface micromachining. In addition, silicon is an excellent structural material for a wide variety of micromechanical devices due to its excellent mechanical properties, thus still playing a major role in MEMS technology. For these reasons, the need for silicon technology is still warranted in the development of "system-on-a-chip (SoC)" modules.

As discussed before, the emerging fields of nanosciences and nanotechnology employ polymers as key functional materials. Combining both polymer-based micro/nanotechnology and silicon-based MEMS technology holds remarkable promise to yield a new type of devices and modules. However, the nontraditional material processing techniques in these emerging fields have mostly been developed outside the scope of silicon MEMS/CMOS technology. Most of the current polymer-based nanofabrication studies are only focused on demonstration of the patterning of nanoscale polymeric features on a rigid substrate of silicon or glass. There is plenty of room for system-level integration of these polymer structures. There is a strong need for a new study that explores a polymer/silicon hybrid technology to achieve SoC with new functions enabled by integration of micro/nanometer-scale polymer structures.

16.4 OPTICAL MICRODEVICES WITH 3-D ELASTOMERIC STRUCTURES

As an example of the aforementioned polymer/silicon hybrid microsystem, monolithic integration of soft lithographically fabricated 3-D PDMS microstructures in silicon MEMS has been demonstrated (Tung and Kurabayashi, 2004, 2005a,c). The initial effort was focused on development of a multiaxis, single-layer MEMS actuator incorporating a 3-D PDMS microstructure placed on a silicon device layer. This effort resulted in a new fabrication technique called "SLLOG," which could be applicable to a wide range of polymer/silicon hybrid MEMS device fabrications.

16.4.1 PDMS-SILICON HYBRID MEMS ACTUATOR

Micrometer-scale actuation with multiple degrees of freedom plays an essential role in many MEMS applications, including microrobotics (Hollar et al., 2003), micromanipulation (Böhringer et al., 1996; Kanayama et al., 1997), and microoptics systems (Chiou and Lin, 1999; Kim and Kim, 1999;

Young and Shkel, 2001; Kwon et al., 2002; Tuantranont and Bright, 2002; Kwon and Lee, 2004). Significant research efforts have been made to develop electrostatic silicon MEMS actuators capable of multiaxis in-plane/out-of-plane motions (Kanayama et al., 1997; Chiou and Lin, 1999; Kim and Kim, 1999; Young and Shkel, 2001; Sun et al., 2002; Walraven and Bernhard Jokiel, 2003). Many of these electrostatic actuators are based on comb drives, which have advantages such as high-speed response, low power consumption, and a small device size. However, their fabrication is often challenging and time consuming, requiring multilayer structures (Chiou and Lin, 1999; Kwon et al., 2002) or relatively complex mechanisms (Walraven and Bernhard Jokiel, 2003). Furthermore, their assembly requires precise alignment and bonding of multiple silicon substrates, adding more complexity to the laborious fabrication processes.

For example, Kwon et al. (2002) constructed an asymmetric comb drive on a silicon-on-insulator (SOI) wafer. They used a three-step back and front side deep reactive ion etching (DRIE) process to allow the actuator to achieve out-of-plane piston motion. Using both surface (multiuser MEMS processes) and bulk micromachining (e.g., DRIE), Piyawattanametha et al. (2003) developed an angular vertical comb drive to generate out-of-plane tilting motion with one degree of freedom. Then, they integrated two sets of the actuators to generate out-of-plane tilting motion in two orthogonal axes. Xie et al. (2003) also exploited curled hinges, which resulted from residual stress between different materials, to construct a similar vertically angled offset comb drive that yields tilting motion. Walraven et al. (2003) constructed a spatial microstage that yields three degree-of-freedom motion (either XYZ translation or piston-tip-tilt) using the Sandia Ultraplanar MEMS Multilevel Technology silicon surface micromachining process. With the same process, Hah et al. (2004) developed a comb-drive actuated micromirror array that generates single degree-of-freedom motion (out-of-plane tilting).

Using a fundamentally different design/fabrication approach than the previous studies, a novel polymer/silicon hybrid actuator has been developed that allows multiaxis, out-of-plane vertical, and tilting actuation motions, only requiring a single silicon layer of comb drives, as shown in Figure 16.4 (Tung and Kurabayashi, 2004, 2005a,c).

The actuator is designed to translate the in-plane one-dimensional motion of comb drives to out-of-plane motion of a PDMS microstructure (more specifically, a PDMS microplatform) with multiple degrees of freedom (Tung and Kurabayashi, 2005b). The actuator structure consists of two major components: (1) four comb drives orthogonally fabricated on a SOI wafer and (2) a mechanically flexible PDMS micromotion translator connected to the comb drives. The PDMS micromotion translator is composed of a micrometer-scale platform and four connectors. Each of the PDMS microconnectors provides a mechanical connection between the platform and a comb-drive rotor beam via two thin flexural microjoints. Five different PDMS microplatform actuation modes are achieved by selectively activating some combinations among the four comb drives. Here, each comb drive laterally pulling the microconnector yields upward vertical motion as shown in Figure 16.5.

It is experimentally demonstrated that the actuator can achieve its motions in three independent axes with fast dynamic response reaching a bandwidth of about 5 kHz. The device yields a vertical displacement up to 5 μm and rotational motions with a 0.6° tilting angle at a 40 V peak-to-peak AC actuation voltage. This new hybrid actuator design shows how the integration of a 3-D PDMS microstructure can lead to high-degree-of-freedom microactuation with a single-layer silicon device structure. The performance of the actuator can be substantially enhanced by design optimization of the flexible polymer structure and the silicon comb drives.

16.4.2 SLLOG Process for PDMS–Silicon Hybrid MEMS Fabrication

In general, the SLLOG process involves releasing soft lithographically patterned 3-D PDMS structures from a micromold and attaching them onto a separate silicon-based MEMS structure with high accuracy. More specifically, the processes applied to fabricate the polymer-integrated MEMS actuator shown in Figure 16.4 are composed of three major steps: (1) fabrication of comb

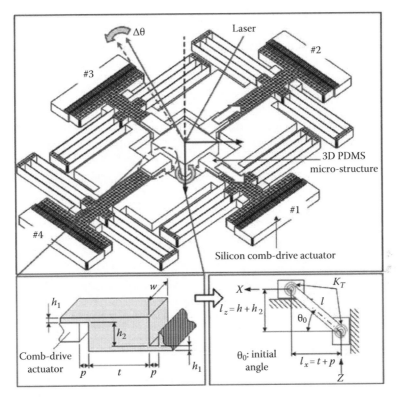

FIGURE 16.4　Schematic of Elastomer (PDMS)-on-Silicon hybrid microactuator. The double L-shaped flexible PDMS microstructure is modeled as a four-bar linkage.

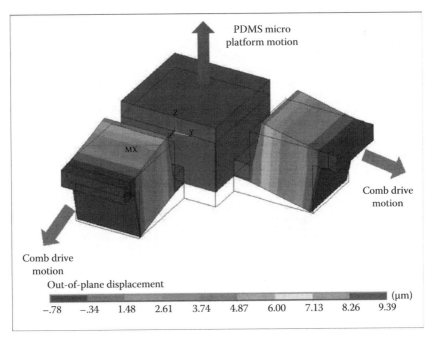

FIGURE 16.5　Finite element analysis (FEA) simulation of PDMS micromotion translator connected to comb drives.

drives on a SOI wafer, (2) soft lithography and lift-off of a 3-D PDMS microstructure, and (3) grafting of the PDMS microstructure onto the SOI patterns based on surface-tension-assisted fluidic microassembly. The entire process steps are summarized in Figure 16.6.

Two molds, namely, top and bottom molds, are typically needed to fabricate the 3-D PDMS microstructure. The top mold consists of a multilayer positive-tone photoresist AZ 9260 (Shipley, Marlborough, MA) microstructure constructed on a 4 in. transparent Pyrex glass wafer. The process first starts with spinning and hard baking of a 5 μm thick photoresist layer on the glass substrate, which later serves as a thin sacrificial layer. This process is followed by two-step spinning and patterning of two other photoresist layers on top of the first layer. It results in the construction of a photoresist micromold structure sitting on the thin positive photoresist sacrificial layer. The total thickness of the micromold structure can be varied between 10 and 100 μm, depending on the thickness of the aimed PDMS microstructure.

To fabricate the bottom mold, two-step silicon deep reactive ion etching (DRIE) is used to etch a 3-D mold shape into a single-side polished 4 in. silicon wafer. A special plasma treatment is

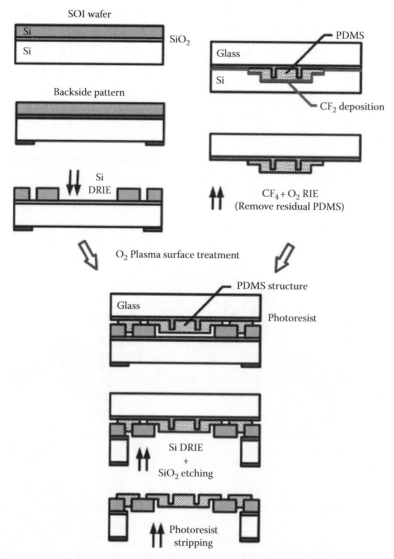

FIGURE 16.6   Steps of SLLOG process for fabrication of the PDMS/silicon hybrid microactuator.

performed for the bottom mold surface to promote release of a cured PDMS microstructure in the following lift-off process. A CF_4 passivation process using the DRIE tool (STS etch system, Surface Technology System) deposits a thin conformal layer of CF_x (~500 Å) on the entire surface of the bottom mold. The deposited CF_x, followed by 5 min annealing at 150°C, prevents adhesion of the PDMS microstructure to the bottom mold surface.

After finishing the preparation of the top and bottom molds, a few drops of PDMS precursor are deposited on the center of the bottom mold. Then, the top mold is pressed against the PDMS-covered bottom mold at room temperature with careful alignment under a stereo microscope. In this step, sufficient pressure is needed to squeeze out the excess PDMS and minimize the thickness of the residual PDMS layer. Subsequent PDMS curing is performed at 150°C for an hour in vacuum to prevent the entrapment of air bubbles. After fully cured, the microstructure is released from the bottom mold with its top attached to the photoresist layer on the top mold. The excess PDMS layer on the released side is removed by reactive ion etching (RIE). Directional etch is achieved using a gas mixture containing 25% O_2 and 75% CF_4, with 40mTorr chamber pressure and 250 W platen power (Semi-Group RIE). Figure 16.7 shows scanning electron microscope (SEM) images of a

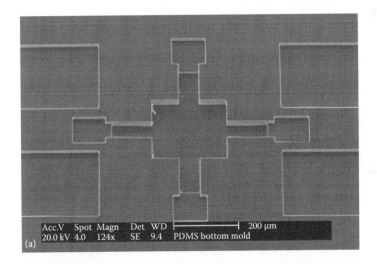

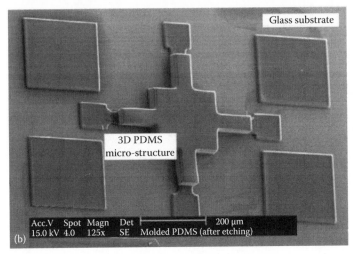

FIGURE 16.7 SEM images of (a) bottom mold etched in silicon and (b) 3-D PDMS microstructure fabricated by replica molding on the top mold. (Reprinted with permission from Tung, Y.C. and Kurabayashi, K., *J. Microelectromech. Syst.*, 14, 558–566. Copyright 2005, IEEE.)

sample bottom mold for PDMS fabrication, (Figure16.7a) and a 3-D molded PDMS microstructure attached onto a top mold after etching the residual PDMS (Figure16.7b).

The PDMS microstructure is monolithically integrated with silicon MEMS by a surface tension-assisted microassembly technique, as shown in Figure 16.8. A drop of deionized (DI) water is dispensed onto the surface of the DRIE-patterned comb drives, which assists the grafting of the PDMS microstructure onto the SOI substrate. The PDMS microstructure is roughly aligned through the optically transparent glass wafer and attached to the silicon structure. The surfaces of both the PDMS microstructure and the SOI comb-drive patterns are treated before the microfluidic assembly by oxygen plasma, which results in a permanent PDMS/silicon bonding. The oxygen-plasma-treated surfaces of the PDMS microstructure and SOI comb drives are both hydrophilic, thus tend to trap water between the top and bottom molds during the grafting process. The surface tension on the air/water interface pulls the two molds together.

To ensure the fine alignment process, microdimples are molded on the PDMS microstructure surface, and microgrooves are etched in the SOI patterns. Assisted by the surface tension, the PDMS microdimples are easily snapped into the grooves on the SOI wafer, while the glass wafer was manually tapped for shaking. The PDMS grafting process resulted in an alignment accuracy of ±2 μm. After the PDMS microstructure is grafted, the system is heated up to 120°C to evaporate the DI water between the top mold and SOI wafer. The PDMS and silicon patterns are well bonded together without applying any pressure after the DI water is totally dried out. It is found later that the resulting PDMS-silicon bond is strong enough to allow us to operate the fabricated device for more than million cycles without failure. The stress level required to break the silicon-PDMS bond is even higher than that for causing a fracture to the PDMS film itself. Thus, no mechanical reliability issue has been essentially observed with the fabricated PDMS/silicon hybrid actuator device.

As the last step, back-side silicon DRIE is performed to remove the handle silicon layer underneath the moving parts of the comb drives. Then, the buried oxide layer is etched by buffered hydrofluoric acid to release the comb-drive structures. Finally, the top photoresist mold is dissolved using a photoresist stripping solution PRS 2000 (J.T. Baker Inc., NJ), and the glass substrate is detached from the device surface. The processing temperature during the entire SLLOG process does not exceed 150°C. The non-silicon processing steps here can be implemented as post-CMOS processes. This makes SLLOG process to be CMOS compatible, allowing the EOS devices to be integrated with CMOS circuitry. It follows that the developed microactuator can be integrated with CMOS circuits, eventually permitting SoC incorporating a single-layer multiaxis actuation mechanism.

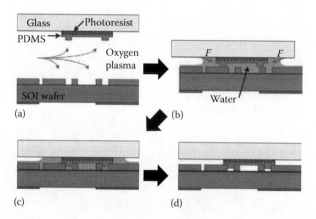

FIGURE 16.8 Microfluidic surface tension-assisted assembly and elastomer grafting onto silicon. (a) Oxygen plasma treatment of PDMS and silicon surfaces. (b) PDMS–silicon alignment assisted by surface tension and fluidic lubrication. (c) Precision snap-in of PDMS into silicon groove. (d) Permanent PMDS-silicon bonding after drying DI water.

Consequently, this device may find a wide variety of applications, including adaptive microoptical components and 3-D optical scanners (Tung and Kurabayashi, 2004, 2005a) for biological assays (Kwon and Lee, 2004).

16.4.3 HYBRID MEMS TECHNOLOGY FOR MICROOPTICS

Optical devices with variable-focus capability are critical platforms for dynamic-focus imaging (Kaneko et al., 1997; Kuiper and Hendriks, 2004), optical pickup data reading (Chao et al., 2005), optical information processing (Gorodetsky et al., 1994), optical interconnection (Ren et al., 2003), microlaser surgery (Lemberg and Black, 1996), and confocal microscopy (Oku and Ishikawa, 2003; Oku et al., 2004). The SLLOG process has been applied to construct a dynamic focus microlens capable of five degree-of-freedom motions using the SLLOG process (Tung and Kurabayashi, 2006), as shown in Figure 16.9.

The whole device structure consists of a SU-8 microlens with a numerical aperture of 0.18, a 3-D PDMS microstructure, and a single layer of silicon microactuators. The lens is formed on the top surface of the PDMS microstructure by surface tension-driven self-formation of SU-8 photoresist

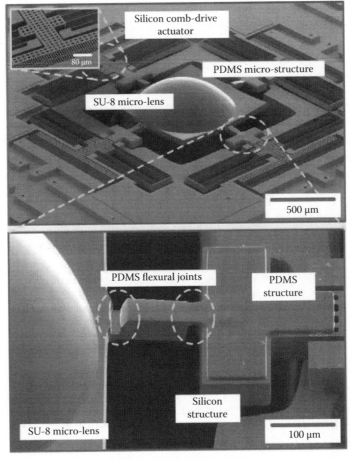

FIGURE 16.9 SEM images of the fabricated PDMS-on-Silicon multiaxis dynamic focus microlens. (Reprinted with permission from Tung, Y.C. and Kurabayashi, K., A single-layer multiple degree-of-freedom PDMS-on-silicon dynamic focus micro-lens, *MEMS 2006: Proceedings of the 19th IEEE International Conference on Micro Electro Mechanical Systems*, Technical Digest, Istanbul, Turkey, IEEE, New York, pp. 838–841. Copyright 2006, IEEE.)

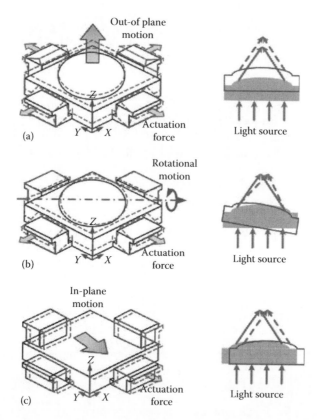

FIGURE 16.10 SEM images of the fabricated PDMS-on-Silicon multiaxis dynamic focus microlens. (Reprinted with permission from Tung, Y.C. and Kurabayashi, K., A single-layer multiple degree-of-freedom PDMS-on-silicon dynamic focus micro-lens, *MEMS 2006: Proceedings of the 19th IEEE International Conference on Micro Electro Mechanical Systems*, Technical Digest, Istanbul, Turkey, IEEE, New York, pp. 838–841. Copyright 2006, IEEE.)

pattern. As illustrated in Figure 16.10, the developed PDMS/silicon hybrid device translates the in-plane motion of silicon comb drives into dynamic focus motion of five degrees of freedom with fast response while taking advantage of the mechanical compliance of PDMS structures. The device allows the focal point to be varied in the vertical direction by approximately 20 μm at an actuation voltage of 140 V. The multiple degrees of freedom and simple structural design results in a high-performance dynamic focus microlens MEMS device with good structural reliability and manufacturability.

16.5 MULTISCALE INTEGRATION OF NANOIMPRINTED ELASTOMERIC STRUCTURES

This chapter has so far shown that the SLLOG process, which combines polymer soft lithography and silicon micromachining, is a promising technique to fabricate optical MEMS devices incorporating 3-D PDMS microstructures. This technique is now extended to allow for patterning or embedding of nanoimprinted PDMS structures in polymer-silicon hybrid MEMS. The processing technique derived from SLLOG is named "multi-scale SLLOG (MS-SLLOG)." As a representative optical MEMS device resulting from MS-SLLOG, this section presents a strain-tunable nanoimprinted grating device. This device is also electrostatically actuated using silicon comb drives to achieve voltage-controlled reconfigurability with large dynamic bandwidth and low power consumption.

By permitting real-time control over the unique optical properties of a nanofeatured PDMS microstructure, the grating device may serve as a new microoptics component for optical switching, filtering, routing, and spectroscopy.

16.5.1 MULTISCALE SOFT-LITHOGRAPHIC LIFT-OFF AND GRAFTING

The MS-SLLOG process can combine non-photolithographic nanofabrication methods, such as nanoimprint lithography (Chou et al., 1996; Guo, 2004), block copolymer self-assembly (Boker et al., 2001; Guarini et al., 2002; Hamley, 2003), and nanoscale pattern transfer (Zhao et al., 1997; Michel et al., 2001), with the above-described SLLOG process. MS-SLLOG allows for construction of a hierarchical device structure seamlessly incorporating feature sizes ranging from tens of nanometer to submillimeters on a single silicon chip. The general process steps of MS-SLLOG are illustrated in Figure 16.11.

The MS-SLLOG process starts with fabrication of two (top and bottom) micromolds and patterning of nanoscale features on the surfaces of one of the molds. These nanoscale features are later either imprinted on or embedded into a 3-D PDMS microstructure fabricated using the two-mold soft lithography technique. Here, a nanoimprint lithography approach (Guo, 2004, 2007) is employed to print these nanoscale features in a polymer mask layer covering the mold substrate. The mask layer material used here can be photoresist, polystyrene (PS), or poly(methylmethacrylate). Features of particular interest from the viewpoint of photonics applications include nanoscale grating patterns, concentric rings, dot arrays, pillar arrays, and cylindrical pores. These patterns are often found as important mask patterns (Guarini et al., 2002) or as building blocks in developing semiconductor devices such as nonvolatile memories (She et al., 2001), lasers (Zhukov et al., 2000), quantum computing (Zanardi and Rossi, 1998), and thermoelectric power generation (Khitun et al., 2001). These features can originally be formed on a silicon or polymer template by e-beam lithography or block copolymer self-assembly, and transferred into the polymer mask layer on the mold. Once the template is fabricated, it can *repeatedly* be used for patterning nanoscale structures many times no matter complex they may be. This will lead to significant reduction of manufacturing time and cost as well as simplification of the device fabrication process.

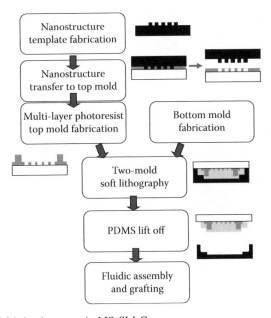

FIGURE 16.11 General fabrication steps in MS-SLLG.

The pattern transfer is achieved by physical embossing with sufficient pressure applied at elevated temperature (typically around 150°C). After these micromold preparation steps involving nanoimprint lithography, a 3-D PDMS microstructure is molded by pressing the two molds together at room temperature and then lifted off from the bottom mold after the curing process. This will be followed by the microfluidic assembly and grafting processes described in Section 16.4.2 to integrate the PDMS microstructure onto a silicon on-chip MEMS structure. Subsequently, the two-mold soft lithography process allows the nanoscale patterns to be transferred from the molds to the surface of the PDMS microstructure. The entire process steps results in a polymer/silicon hybrid MEMS device structure with a PDMS microstructure containing nanoscale patterns. This process has been employed to transfer submicron grating patters onto a PDMS microslab using the MS-SLLOG process. Figure 16.12 shows a SEM image of the integrated nanoimprinted structure. Figure 16.13 shows how exactly the nanoimprint process is incorporated in the SLLOG process to integrate the grating PDMS microbridge onto an on-chip silicon actuator.

Here, silicon templates are prepared by e-beam lithography, which allows us to precisely control the geometry and dimensions of nanoscale patterns. However, e-beam lithography is generally expensive and time consuming especially when the template area size exceeds 100×100 μm^2. It should be noted again that, in principle, the MS-SLLOG process may accommodate pattern transfer using self-assembled block copolymer templates as etching masks or nanoimprint stamps. Block copolymer self-assembly provides a promising route to unconventional, inexpensive, and robust nanofabrication (Hamley, 2003). Block copolymer films can easily be prepared by spin coating or dip coating, where a solution of the polymer in a volatile organic solvent is dispensed on a rigid substrate. The coated films typically show uniform flatness across the entire substrate, making the process suitable for large-area nanofabrication. Annealing of these films can induce nanophase separation of polymer blocks (Bates and Fredrickson, 1999). However, the arrangement, size, and shape of the resulting pattern are all highly affected by surface treatment, film thickness, annealing conditions, and the polymer blend type (Hamley, 2003). It is still unknown how one can construct a block copolymer self-assembled nanotemplate such that it provides sufficient flexibility, repeatability, and precision to meet the design of the EOS MEMS devices here. Fundamental study exploring the implementation of this approach could be a good future research topic.

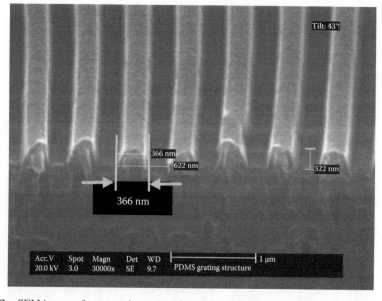

FIGURE 16.12   SEM image of nanograting arrays formed on PDMS surface using the proposed method.

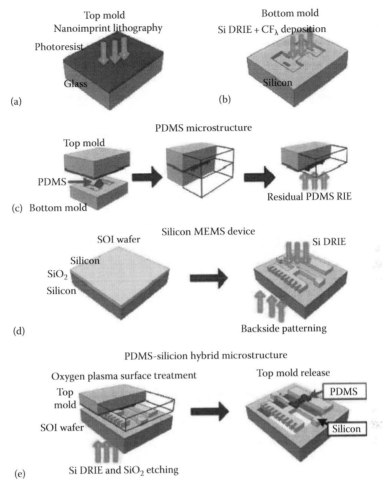

FIGURE 16.13 MS-SLLOG process steps to integrate a PDMS microstructure with nanoimprinted surface features onto silicon microactuators suspended on a silicon chip. (a) Photoresist is spun onto glass and nanoimprinted for the top mold. (b) The bottom mold is a two-step DRIE of silicon followed by CF_λ deposition. (c) The top and bottom molds shape the PDMS microstructure with the submicron grating on the top surface. (d) An SOI wafer is patterned and etched to make the MEMS structures. (e) The top mold and SOI wafer are treated with oxygen plasma, aligned, and bonded. The backside and oxide layer are etched, and the top mold is released by dissolving the photoresist.

16.5.2 Demonstration of Strain Tunable Optical Grating MEMS

A diffraction grating serves as a critical component in spectroscopy, display technology, and laser tuning, and has been employed in various microoptical devices (Bloom, 1997; Burns and Bright, 1998; Schueller et al., 1999; Sumetsky et al., 2004; Zhou et al., 2004b). However, gratings fabricated using conventional techniques offer only limited controllability and versatility in their profiles (Sumetsky et al., 2004; Xie et al., 2004). Once the gratings are formed, they cannot be adjusted easily to compensate for process errors or changing application needs. As one of devices following the concept of EOS microsystems constructed by the MS-SLLOG process, a new *stretchable* microoptical MEMS device has been developed. The device incorporates a nanoimprinted polymer diffraction grating with its spatial period varied under mechanical strain introduced by silicon comb drives. Figure 16.14 shows a transmission optical grating with a spatial period, *a*, which lies in the

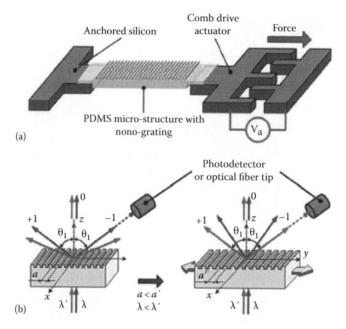

FIGURE 16.14 Schematic of strain-reconfigurable elastomeric optical grating coupled with silicon comb-drive actuator. (a) Concept of device design. (b) Device operation principle.

x-y-plane, and a laser beam with a wavelength λ incident to the z direction. The grating equation indicates that the angle of the first order diffraction light θ_1 can be given by

$$\theta_1 = \sin^{-1}\left(\frac{\lambda}{a}\right). \tag{16.1}$$

When introducing a mechanical strain to elongate the spatial period of the grating to a', the angle of the first-order diffraction light becomes θ_1'. The corresponding mechanical strain ε is defined as

$$\varepsilon = \frac{a'-a}{a} = \frac{a'}{a} - 1. \tag{16.2}$$

Therefore, the relationship between the diffraction angle change and the mechanical strain is given by

$$\theta_1' - \theta_1 = \Delta\theta_1 = \sin^{-1}\left(\frac{\lambda}{a}\right) - \sin^{-1}\left[\frac{\lambda}{a(\varepsilon+1)}\right]. \tag{16.3}$$

The wavelength of the first-order diffraction at the same angle θ_1 shifts to λ', with the spatial period of the grating changed to a'. The relation between the wavelength shift $\lambda' - \lambda$ and the mechanical strain is given by

$$\lambda' - \lambda = a\varepsilon \sin\theta_1 = \lambda\varepsilon. \tag{16.4}$$

This relation indicates that the wavelength shift of the first-order diffraction at the angle θ_1 is linearly proportional to the mechanical strain introduced to the grating structure. The peak wavelength of the diffracted light detected at a specific angle can be tuned by controlling the mechanical strain while maintaining the linear strain-diffraction angle relation. One can tune the wavelength of the light collected by a photodetector or an optical fiber placed at the angle θ_1 through varying the actuation voltage. It follows that plots of emission spectra can be obtained by mapping the detected intensity and the actuation voltage following:

$$\lambda' - \lambda \propto \varepsilon \propto F = \frac{\varepsilon_o t N_g V_a^2}{2g}, \tag{16.5}$$

where

    t is the in-depth thickness of the comb drive
    N_g is the total number of the gaps
    g is the gap between the comb electrodes
    ε_o is the permittivity of free space
    V_a is the actuation voltage

Optical spectroscopy employing this principle requires no extensive and complex spectrum-plot construction and signal processing algorithms.

PDMS is used as the grating material in this EOS hybrid MEMS device. The optical transparency (~92% for visible light) and mechanical compliance (~750 kPa Young's modulus and ~40% maximum elongation) described in Section 16.2.3 are expected to provide two technological advantages: (1) incident light can be transmitted through the grating component itself, making it easy to integrate the device with other on-chip solid-state optoelectronic components and (2) varying the actuation voltage allows for tuning the directions of diffracted light, enabling voltage-controlled reconfiguration of grating patterns. The center-to-center distance and dimensions of the grating can be 50–350 nm, accounting for the diffraction angle range, the strain-voltage relation requirement, and the light wavelength. In this case, the nominal spatial period, a, is set to be comparable with, or smaller than, the wavelength $\lambda \sim 650$ nm.

To fabricate the grating device, a silicon template with designed nanoscale grating patterns is first constructed using e-beam lithography and RIE. Then, the template is pressed into a photoresist layer deposited on a transparent glass top mold substrate, and the grating patters are translated into the photoresist layer. This photoresist layer pattern serves as a nanoscale mold pattern in the subsequent soft lithography of PDMS as shown in Figure 16.13a. The EOS hybrid MEMS device incorporating a soft polymer nanograting structure is shown to be excellent at easily testing the polymer's both static and dynamic mechanical response to a large variety of stress conditions and can also serve as a vehicle for microscale material characterization (Truxal et al., 2008a). The integrated PDMS structure yields a high level of microscale actuated strain (>13%) at a large actuation bandwidth of 2 kHz. With the elasticity, transparency, and soft lithographic nanopatterning of PDMS with silicon MEMS actuators, our technology permitting the on-chip hybrid PDMS integration leads to the development of a new photospectroscopic technique with both a high detection limit (~10 pW) and a short time window of 250 µs (Truxal et al., 2008b).

16.5.3 BIOPHOTONICS APPLICATIONS OF STRAIN TUNABLE OPTICAL GRATING MEMS

In biological imaging, wavelength tuning in the visible band often facilitates optical detection of multicolor fluorescence-labeled cells or biomarkers for elucidating complex biological phenomena. Current laser wavelength tuning methods used in fluorescent imaging and spectroscopy normally incorporate acoustic-optical tunable filters and liquid crystal tunable filters (Gat, 2000). However, the bulky size of these filters and their peripheral electronics could limit their use for portable point-of-care applications. As an alternative to these optical filters, the nanoimprinted grating device was

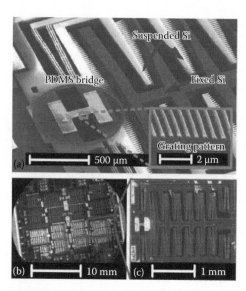

FIGURE 16.15   (a) SEM image of nanoimprinted grating EOS MEMS device. (b) Optical images of 16 optical filter device array. (c) Optical image of the whole device unit in a 3×3 mm^2 area. (Reprinted with permission from Huang, N.T. et al., *Appl. Phys. Lett.*, 95, 211106. Copyright 2009b, American Institute of Physics.)

implemented as a tunable optical filter that achieves high-speed two-color laser switching in an on-chip setting (Huang et al., 2009b). The SEM and optical images of the device(s) are shown in Figure 16.15a and b. With a large strain (~18%) introduced to the PDMS grating microbridge by the comb drive actuators, the optical filter device can yield large wavelength tunability ($\Delta\lambda > 40$ nm) in the visible band for such a small device size.

Using the tunable grating optical filter, bandpass (FWHM = 4–8 nm) wavelength switching between two wavelengths ($\lambda = 473$ and $\lambda = 532$ nm) of laser light passing through the PDMS grating within 0.5 ms has been demonstrated. Here, the optical filtering is achieved by spatially blocking light other than the one diffracted to the targeted optical path leading to a photodetector or an optical fiber coupler. A sinusoidal actuation voltage of 20–180 V at 1 KHz was applied to the comb drives of the optical filter device. Figure 16.16a shows the excitation spectrum as a function of the actuation voltage V_a, measured by commercial spectrometer. At $V_a = 20 \sim 80$ V, the blue excitation ($\lambda = 473$ nm) is guided onto the detector, while the green excitation ($\lambda = 532$ nm) is blocked. With V_a increased to 120 V, the blue excitation intensity significantly drops more than 3 orders of magnitude. At $V_a = 175$ V, the green excitation intensity reaches the maximum value with a signal-to-noise ratio at nearly 4×10^3.

Tuning the actuation voltage adjusts the ratio between the blue and green excitation laser intensities irradiated onto a specimen in a programmable manner. This would enable dual-excitation ratiometric dye detection (Fukano et al., 2006). Tunable dual-excitation of PC-3 prostate cancer cells stained by LIVE/DEAD Viability/Cytotoxicity Kit has been by optical coupling between excitation light diffracted by the grating optical filter and a slit-projected optical fiber. The cell specimen is exposed to the light propagating through the optical fiber. Figure 16.16b through e constitute an image sequence that shows the fluorescent color variation resulting from the excitation laser wavelength tuning at varying V_a. The intensity of a single cell with the live (Calcein AM, green emission) or dead (Ethidium homodimer-1, red emission) stain is obtained by ImageJ software (National Institute of Health). With this voltage-controlled wavelength tuning method, the wavelength of an excitation light source can be adjusted in situ during imaging of multicolor fluorescent labels.

Coupled with a photomultiplier tube (PMT), the nanoimprinted grating EOS MEMS device can serve as a voltage-controlled strain-tunable dispersive optical component for optical spectroscopy

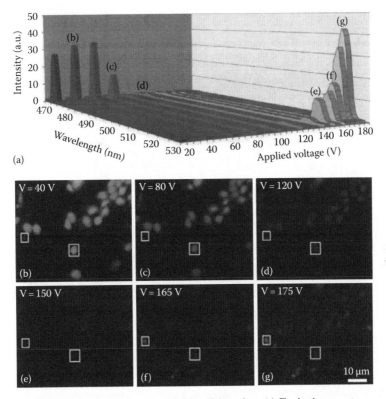

(a)

FIGURE 16.16   Results for two-color live/dead PC3 cell imaging. (a) Excitation spectrum as a function of optical filter actuation voltage. (b–g) Sequence of images corresponding to the excitation spectrum in (a). (Reprinted with permission from Huang, N.T. et al., *Appl. Phys. Lett.*, 95, 211106. Copyright 2009b, American Institute of Physics.)

(Truxal et al., 2008b). Coupling the grating device-based monochromator system with a microfluidic flow channel via an optical fiber waveguide enables a new microscale optofluidic detection technique named "microfluidic multispectral flow cytometry (MMFC)" (Huang et al., 2010). MMFC permits quantitative enumeration of the spectral signatures of various fluorescence color-coded bioparticles in flow. Figure 16.17 shows a schematic of the MMFC setup, which is constructed with three units: (1) a fluorescent microscope with an excitation laser ($\lambda = 473$ nm), (2) a microfluidic flow chamber, and (3) a spectral detection system consisting of the strain-tunable nanoimprinted grating MEMS device and a PMT with a slit. The sample flow is hydrodynamically focused by two sheath flows of buffer solution within the microfluidic channel. A 60× objective and an embedded optical fiber together enable fluorescence excitation and detection with a highly focused interrogation zone. The PMT provides high signal sensitivity and a fast acquisition rate. Adjusting the angular position of the PMT allows one to set up a wavelength range to cover emission spectra that are to be detected.

MMFC permits dynamic in situ spectral detection for four types of in-flow PS microspheres with different colors. The tunable grating strain ratio and the PMT angle are adjusted to achieve a 518–558 nm wavelength tuning range, where these microspheres exhibit notable spectral overlaps (Figure 16.18c). From the continuous spectrum measurement, one can map the maximum intensity and wavelength of maximum intensity of each microsphere color type (Figure 16.18b). The histograms (Figure 16.18a and d) show the particle population distributions for each of these spectral parameters. By using the combination of the two parameters, four fluorescent color bands of the PS microspheres can be well discriminated even in such a narrow wavelength bandwidth as 40 nm. The use of particles coded with spectrally similar colors enables our system to carry out

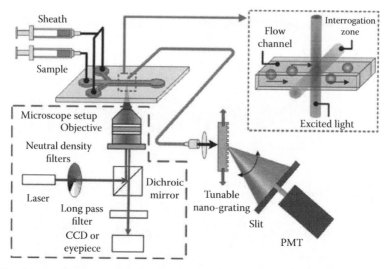

FIGURE 16.17 Schematic of the microfluidic multispectral flow cytometry (MMFC) setup. (Reprinted with permission from Huang, N.T. et al., *Anal. Chem.*, 82, 9506. Copyright 2010. American Chemical Society.)

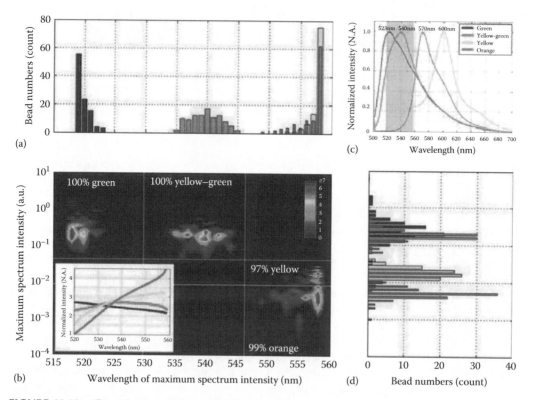

FIGURE 16.18 **(See color insert.)** Four color discrimination of polystyrene microspheres in a 518–558 nm wavelength (λ) range (shadowed region in (c)) by MMFC. (a) A univariate histogram of the wavelength of maximum spectrum intensity. (b) A population density contour plot for the maximum spectrum intensity and the wavelength of maximum spectrum intensity. (c) Spectral data of the microspheres measured by a commercial spectrometer. (d) A univariate histogram of the maximum spectrum intensity. (Reprinted with permission from Huang, N.T. et al., *Anal. Chem.*, 82, 9506. Copyright 2010. American Chemical Society.)

high-speed, multiplexed detection in a *single* excitation source and a *single* photodetector. This represents a capability unique to MMFC that has not been demonstrated with conventional flow cytometry settings.

The nanoimprinted grating EOS MEMS device has shown various functionalities in biophotonics applications. Integrated as a light-source modulating component in a microfluidic system, this device could potentially allow for high-speed alternating laser excitation or spectroscopy for capturing dynamic biological interactions (Kapanidis et al., 2004; Hesch et al., 2008) in a lab-on-a-chip setting. When coupled with a microfluidic cell culture system, the multiwavelength detection platform based on the strain-tunable nanoimprinted grating device may allow dynamic in situ analysis of cell groups emitting nearly identical fluorescence spectra. This could lead to on-chip quantification of protein binding on cell surfaces and monitoring of intracelluar Ca^{2+} concentration and pH (Palmgren, 1991; Krasnowska et al., 1998; Zhu et al., 2010). With this spectral resolving power, researchers could also combine and indentify more fluorescent labeled cell or biomarkers for disease diagnosis and drug discovery in microfluidic environments.

16.6   SUMMARY AND RECOMMENDATIONS

Enabling multiscale, hierarchical integration of PDMS elastomer micro/nanostructures in silicon MEMS, the technology discussed in this chapter holds great promise to guide development of a new type of optical MEMS devices. This technology may allow one to incorporate the unique material properties and physical phenomena of polymers occurring at the nanometer scale into a MEMS structure. This could open the door for polymer-based nanofabrication and nanotechnology to find a wider variety of commercial, military, and biomedical applications. Furthermore, adopting polymer nanofabrication methods in MEMS technology may help create a niche market for MEMS products benefitting from the combination of polymer and silicon materials. The technological concept discussed in this chapter is expected to stimulate further development of other single-chip modules that are difficult to achieve without simultaneously integrating PDMS micro/nanostructures with high-speed, CMOS-compatible silicon sensors and actuators.

The EOS hybrid MEMS device fabrication technique, i.e., the SLLOG process, permits low-cost, high-yield manufacturing of highly functional MEMS devices. This is possible by replacing silicon-based structural components with flexible polymeric ones to eliminate design complexity, as demonstrated in the previous study (Tung and Kurabayashi, 2005b,c). The study also demonstrated a manufacturing yield higher than 90% of 30 device units. If using nontraditional laser machining to fabricate molds and templates in SLLOG, highly complex 3-D micro/nanosoft structures could be integrated in optical MEMS. The EOS optical MEMS devices provide a new way of signal modulation useful in optical communications, optical scanning, optical data storage, optical sensing, and optical computing. They also provide an opportunity for us to study fundamental photonic phenomena affected by mechanical strain. This may contribute to yield new scientific knowledge useful for the emerging field namely "stretchable photonics."

Demonstration of the devices incorporating the unique physical properties of polymer nanostructures could additionally promote future multiscale modeling research for full prediction of their performance. Micro/mesoscopic device-level phenomena that result from polymer nanostructural behavior in the electrical, mechanical, thermal, and optical domains are very intriguing and essential research subjects for future system integration and packaging of polymer–silicon hybrid devices.

REFERENCES

Armani, D. and Liu, C. (2000) Microfabrication technology for polycaprolactone, a biodegradable polymer. *Journal of Micromechanics and Microengineering*, 10, 80–84.

Bates, F. S. and Fredrickson, G. (1999) Block copolymers—Designer soft materials. *Physics Today*, 52, 32–38.

Bloom, D. (1997) The grating light valve: Revolutionizing display technology. *SPIE Proceedings*, 3013, 165–171.

Böhringer, K.-F., Donald, B. R., and Macdonald, N. C. (1996) Single-crystal silicon actuator arrays for micro manipulation tasks. *9th IEEE International Workshop on Microelectromechanical Systems*, San Diego, CA, pp. 7–12.

Boker, A., Muller, A. H. E., and Krausch, G. (2001) Nanoscopic surface patterns from functional ABC triblock copolymers. *Macromolecules*, 34, 7477–7488.

Boland, J. J. (2010) Flexible electronics: Within touch of artificial skin. *Nature Materials*, 9, 790–792.

Bradley, K., Gabriel, J. C. P., and Gruner, G. (2003) Flexible nanotube electronics. *Nano Letters*, 3, 1353–1355.

Brittain, S., Paul, K., Zhao, X. M., and Whitesides, G. (1998) Soft lithography and microfabrication. *Physics World*, 11, 31–33.

Burns, D. M. and Bright, V. M. (1998) Development of microelectromechanical variable blaze gratings. *Sensors and Actuators A: Physical*, 64, 7–15.

Chang, T. H. P., Thomson, M. G. R., Yu, M. L., Kratschmer, E., Kim, H. S., Lee, K. Y., Rishton, S. A., and Zolgharnain, S. (1996) Electron beam technology—SEM to microcolumn. *Microelectronic Engineering*, 32, 113–130.

Chao, P. C.-P., Lai, C.-L., and Huang, J.-S. (2005) Intelligent actuation strategy for a three-DOF four-wire type optical pickup. *Sensors and Actuators A: Physical*, 117, 28–40.

Chen, Y., Au, J., Kazlas, P., Ritenour, A., Gates, H., and Mccreary, M. (2003) Flexible active-matrix electronic ink display. *Nature*, 23, 136.

Chiou, J. C. and Lin, Y.-C. (1999) Micromirror device with tilt and piston motions. *Design, Characterization, and Packaging for MEMS and Microelectronics*, Gold Coast, Australia, pp. 298–303.

Chou, S. Y., Krauss, P. R., and Renstrom, P. J. (1996) Nanoimprint lithography. *Journal of Vacuum Science and Technology, Part B: Microelectronics and Nanometer Structures: Processing, Measurement, and Phenomena*, 14, 4129–4133.

Chou, S. Y. and Zhuang, L. (1999) Lithographically induced self-assembly of periodic polymer micropillar arrays. *Journal of Vacuum Science and Technology B*, 17, 3197–3202.

Clarson, S. J. and Semlyen, J. A. (1993) *Siloxane Polymers*, Prentice Hall, Englewood Cliffs, NJ.

Dimitrakopoulos, C. D. and Malenfant, P. R. L. (2002) Organic thin film transistors for large area electronic. *Advanced Materials*, 14, 99–117.

Duffy, D. C., Mcdonald, J. C., Schueller, O. J. A., and Whitesides, G. M. (1998) Rapid prototyping of microfluidic systems in poly(dimethylsiloxane). *Analytical Chemistry*, 70, 4974–4984.

Fu, H. B. and Yao, J. N. (2001) Size effects on the optical properties of organic nanoparticles. *Journal of the American Chemical Society*, 123, 1434–1439.

Fukano, T., Shimozono, S., and Miyawaki, A. (2006) Fast dual-excitation ratiometry with light-emitting diodes and high-speed liquid crystal shutters. *Biochemical and Biophysical Research Communications*, 340, 250–255.

Gadegaard, N., Mosler, S., and Larsen, N. B. (2003) Biomimetic polymer nanostructures by injection molding. *Macromolecular Materials and Engineering*, 288, 76–83.

Gat, N. (2000) Imaging spectroscopy using tunable filters: A review. *Proceedings of SPIE*, 4056, 50–64.

Gokdel, Y. D., Sevim, A. O., Mutlu, S. and Yalcinkaya, A. D. (2010) Polymer-MEMS-Based Optoelectronic Display. *IEEE Transactions on Electron Devices*, 57, 145–152.

Gorodetsky, A. E., Kompan, M., and Sergeyev, A. G. (1994) Variable focus electro-optical lens for analog-to-digital optical converting. *SPIE Proceedings*, 2169, 192–194.

Gray, D. S., Tien, J., and Chen, C. S. (2003) High-conductivity elastomeric electronics. *Advanced Materials*, 16, 393–397.

Grayson, A. C. R., Shawgo, R. S., Johnson, A. M., Flynn, N. T., Li, Y. W., Cima, M., and Langer, R. (2004) A BioMEMS review: MEMS technology for physiologically integrated devices. *Proceedings of IEEE*, 92, 6–21.

Guarini, K. W., Black, C. T., Zhang, Y., Kim, H., Sikorski, E. M., and Babich, I. V. (2002) Process integration of self-assembled polymer templates into silicon nanofabrication. *Journal of Vacuum Science and Technology B*, 20, 2788–2792.

Guo, L. J. (2004) Recent progress in nanoimprint technology and its applications. *Journal of Physics D: Applied Physics*, 37, R123–R141.

Guo, L. J. (2007) Nanoimprint lithography: Methods and material requirements. *Advanced Materials*, 19, 495–513.

Hah, D., Huang, S. T.-Y., Tsai, J.-C., Toshiyoshi, H., and Wu, M. C. (2004) Low-voltage, large scan angle MEMS analog micromirror arrays with hidden vertical comb-drive actuators. *Journal of Microelectromechanical Systems*, 13, 279–289.

Hamley, I. W. (2003a) *Developments in Block Copolymer Science and Technology*, Wiley, New York.

He, Q., Meng, E., Tai, Y-C., Rutherglen, C.M., Erickson, J., and Pine, J. (2003) Parylene neuro-cages for live neural networks study. *TRANSDUCERS '03*, Boston, MA, pp. 995–998.

Hesch, C., Hesse, J., and Schutz, G. J. (2008) Implementation of alternating excitation schemes in a biochip-reader for quasi-simultaneous multi-color single-molecule detection. *Biosensors and Bioelectronics*, 23, 1891–1895.

Hollar, S., Flynn, A., Bellew, C., and Pister, K. S. J. (2003) Solar powered 10 mg silicon robot. *16th IEEE International Conference on Microelectromechanical Systems*, Kyoto, Japan, pp. 706–711.

Horiuchi, S., Fujita, T., Hayakawa, T., and Nakao, Y. (2003) Micropatterning of metal nanoparticles via UV photolithography. *Advanced Materials*, 15, 1449–1452.

Horowitz, G. (2004) Organic thin film transistors: From theory to real devices. *Journal of Materials Research*, 19, 1946–1962.

Hoshino, K. and Shimoyama, I. (2001) An elastic thin-film microlens array with a pneumatic actuator. *Proceedings of the 14th IEEE International Conference on Micro Electro Mechanical Systems*, Interlaken, Switzerland, pp. 321–324.

Huang, X., Li, S. Q., Schultz, J. S., Wang, Q., and Lin, Q. (2009a) A MEMS affinity glucose sensor using a biocompatible glucose-responsive polymer. *Sensors and Actuators B-Chemical*, 140, 603–609.

Huang, N.-T., Truxal, S. C., Tung, Y.-C., Hsiao, A. Y., Luker, G. D., Takayama, S., and Kurabayashi, K. (2010) Multiplexed spectral signature detection for microfluidic color-coded bioparticle flow. *Analytical Chemistry*, 82, 9506–9512.

Huang, N. T., Truxal, S. C., Tung, Y. C., Hsiao, A., Takayama, S., and Kurabayashi, K. (2009b) High-speed tuning of visible laser wavelength using a nanoimprinted grating optical tunable filter. *Applied Physics Letters*, 95, 211106.

Jager, E. W. H., Inganas, O., and Lundstrom, I. (2000) Microrobots for micro-sized objects in aqueous media: Potential tools for single-cell manipulation. *Science*, 288, 2235–2238.

Jo, B.-H., Lerberghe, L. M. V., Motsegood, K. M., and Beebe, D. J. (2000) Three-dimensional micro-channel fabrication in polydimethylsiloxane (PDMS) elastomer. *Journal of Microelectromechanical Systems*, 9, 76–81.

Ju, S. Y., Facchetti, A., Xuan, Y., Liu, J., Ishikawa, F., Ye, P. D., Zhou, C. W., Marks, T. J., and Janes, D. B. (2007) Fabrication of fully transparent nanowire transistors for transparent and flexible electronics. *Nature Nanotechnology*, 2, 378–384.

Kanayama, H., Tsuruzawa, T., Mitsumoto, N., Idogaki, T., and Hattori, T. (1997) Micromanipulator utilizing a bending and expanding motion actuator. *10th IEEE International Workshop on Microelectromechanical Systems*, Nagoya, Japan.

Kaneko, T., Ohmi, T., Ohya, N., Kawahara, N., and Hattori, T. (1997) A new, compact and quick-response dynamic focusing lens. *Transducers '97*, Chicago, IL, pp. 63–66.

Kapanidis, A. N., Lee, N. K., Laurence, T. A., Doose, S., Margeat, E., and Weiss, S. (2004) Fluorescence-aided molecule sorting: Analysis of structure and interactions by alternating-laser excitation of single molecules. *Proceedings of the National Academy of Sciences of the United States of America*, 101, 8936–8941.

Kawata, S., Sun, H., Tanaka, T., and Takada, K. (2001) Finer features for functional microdevices—Micromachines can be created with higher resolution using two-photon absorption. *Nature*, 412, 697–698.

Khang, D. Y., Jiang, H. Q., Huang, Y., and Rogers, J. A. (2006) A stretchable form of single-crystal silicon for high-performance electronics on rubber substrates. *Science*, 311, 208–212.

Khitun, A., Balandin, A., Liu, J. L., and Wang, K. L. (2001) The effect of the long-range order in a quantum dot array on the in-plane lattice thermal conductivity. *Superlattices and Microstructures*, 30, 1–8.

Kim, C.-H. and Kim, Y.-K. (1999) Integration of a microlens on a micro XY-stage. *Device and Process Technologies for MEMS and Microelectronics*, Gold Coast, Australia, SPIE, pp. 109–117.

Kovacs, G. T. (1998) *Micromachined Transducers Sourcebook*, McGraw-Hill, New York.

Krasnowska, E. K., Gratton, E., and Parasassi, T. (1998) Prodan as a membrane surface fluorescence probe: Partitioning between water and phospholipid phases. *Biophysical Journal*, 74, 1984–1993.

Kuiper, S. and Hendriks, B. H. W. (2004) Variable-focus liquid lens for miniature cameras. *Applied Physics Letters*, 85, 1128–1130.

Kwon, S. and Lee, L. P. (2004) Micromachined transmissive scanning confocal microscope. *Optics Letters*, 29, 706–708.

Kwon, S., Milanovic, V., and Lee, L. P. (2002) Vertical microlens scanner for 3D imaging. *Solid-State Sensor and Actuator Workshop*, Hilton Head Island, SC, pp. 227–230.

Langford, R. M., Petford-Long, A. K., Rommeswinkle, M., and Egelkamp, S. (2002) Application of a focused ion beam system to micro and nanoengineering. *Materials Science and Technology*, 18, 743–748.

Lemberg, V. G. and Black, M. (1996) Variable-focus side-firing endoscopic device. *SPIE Proceedings*, 2671, 398–402.

Levenson, M. D. (1993) Wave-front engineering for photolithography. *Physics Today*, 46, 28–36.

Liger, M., Harder, T. A., Tai, Y.-C., and Konishi, S. (2004a) Parylene-pyrolyzed carbon for MEMS applications. *Proceedings of 17th IEEE International Conference on MEMS*, Maastricht, the Netherlands, pp. 161–164.

Liger, M., Konishi, S., and Tai, Y. C. (2004b) Uncooled all-parylene. *Proceedings of the 17th IEEE International Conference on MEMS,* Maastricht, Netherlands pp. 593–596.

Liu, C. (2007) Recent developments in polymer MEMS. *Advanced Materials*, 19, 3783–3790.

Lotters, J. C., Olthuis, W., Veltink, P. H., and Bergveld, P. (1996) Polydimethylsiloxane as an elastic material applied in a capacitive accelerometer. *Journal of Micromechanics and Microengineering*, 6, 52–54.

Lotters, J. C., Olthuis, W., Veltink, P. H., and Bergveld, P. (1997) The mechanical properties of the rubber elastic polymer polydimethylsiloxane for sensor applications. *Journal of Micromechanics and Microengineering*, 7, 145–147.

Martin, D. C. (2002) Controlled local organization of lyotropic liquid crystalline polymer thin films with electric fields. *Polymer*, 43, 4421–4436.

Maruo, S., Ikuta, K., and Korogi, H. (2003) Force-controllable, optically driven micromachines fabricated by single-step two-photon micro stereolithography. *Journal of Microelectromechanical Systems*, 12, 533–539.

Maruo, S. and Kawata, S. (1998) Two-photon-absorbed near-infrared photopolymerization for three-dimensional microfabrication. *Journal of Microelectromechanical Systems*, 7, 411–415.

Maruo, S., Nakamura, O., and Kawata, S. (1997) Three-dimensional microfabrication with two-photon-absorbed photopolymerization. *Optics Letters*, 22, 132–134.

Matsui, S. and Ochiai, Y. (1996) Focused ion beam applications to solid state devices. *Nanotechnology*, 7, 247–258.

Matsumoto, A. (2003) Polymer structure control based on crystal engineering for materials design. *Polymer Journal*, 35, 93–121.

Matsumoto, A., Nagahama, S., and Odani, T. (2000) Molecular design and polymer structure control based on polymer crystal engineering. Topochemical polymerization of 1,3-diene mono- and dicarboxylic acid derivatives bearing a naphthylmethylammonium group as the counteraction. *Journal of the American Chemical Society*, 122, 9109–9119.

Mcalpine, M. C., Friedman, R. S., and Lieber, C. M. (2005) High-performance nanowire electronics and photonics and nanoscale patterning on flexible plastic substrates. *Proceedings of the IEEE*, 93, 1357–1363.

Mcdonald, J. C., Duffy, D. C., Anderson, J. R., Chiu, D. T., Wu, H. K., Schueller, O. J. A., and Whitesides, G. M. (2000) Fabrication of microfluidic systems in poly(dimethylsiloxane). *Electrophoresis*, 21, 27–40.

Mello, A. D. (2002) Plastic fantastic? *Lab on a chip*, 2, 31N–36N.

Michel, B., Bernard, A., Bietsch, A., Delamarche, E., Geissler, M., Juncker, D., Kind, H. et al. (2001) Printing meets lithography: Soft approaches to high-resolution printing. *IBM Journal of Research and Development*, 45, 697–719.

Muthukumar, M., Ober, C. K., and Thomas, E. L. (1997) Competing interactions and levels of ordering in self-organizing polymeric materials. *Science*, 277, 1225–1232.

Ng, J. M. K., Gitlin, I., Stroock, A. D., and Whitesides, G. M. (2002) Components for integrated poly(dimethylsiloxane) microfluidic systems. *Electrophoresis*, 23, 3461–3473.

Oikawa, H., Onodera, T., Masuhara, A., Kasai, H., and Nakanishi, H. (2010) New class materials of organic-inorganic hybridized nanocrystals/nanoparticles, and their assembled micro- and nano-structure toward photonics. *Polymer Materials: Block-Copolymers, Nanocomposites, Organic/Inorganic Hybrids, Polymethylenes*, Springer-Verlag, Berlin, Germany.

Oku, H., Hashimoto, K., and Ishikawa, M. (2004) Variable-focus lens with 1-kHz bandwidth. *Optics Express*, 12, 2138–2149.

Oku, H. and Ishikawa, M. (2003) A variable-focus lens with 1 kHz bandwidth applied to axial-scan of a confocal scanning microscope. *16th Annual Meeting of the IEEE Lasers and Electro-Optics Society (LEOS)*, Tucson, AZ, pp. 309–310.

Ong, B. S., Wu, Y. L., Liu, P., and Gardner, S. (2005) Structurally ordered polythiophene nanoparticles for high-performance organic thin-film transistors. *Advanced Materials*, 17, 1141.

Ornelas-Rodriguez, M., Calixto, S., Sheng, Y. L., and Turck, C. (2002) Thermal embossing of mid-infrared diffractive optical elements by use of a self-processing photopolymer master. *Applied Optics*, 41, 4590–4595.

Palmgren, M. G. (1991) Acridine orange as a probe for measuring pH gradients across membranes: Mechanism and limitations. *Analytical Biochemistry*, 192, 316–321.

Piyawattanametha, W., Patterson, P. R., Hah, D., Toshiyoshi, H., and Wu, M. C. (2003) A 2D scanner by surface and bulk micromachined angular vertical comb actuator. *2003 IEEE/LEOS International Conference on Optical MEMS*, Waikoloa, HI, pp. 93–94.

Post, E. R., Orth, M., Russo, P. R., and Gershenfeld, N. (2000) E-broidery: Design and fabrication of textile-based computing. *IBM Systems Journal*, 39, 840–860.

Quist, A. P., Pavlovic, E., and Oscarsson, S. (2005) Recent advances in microcontact printing. *Analytical and Bioanalytical Chemistry*, 381, 591–600.

Rebeiz, G. M. (2003) *RF MEMS: Theory, Design, and Technology*, John Wiley & Sons, Inc., Hoboken, NJ.

Ren, H., Fan, Y.-H., and Wu, S.-T. (2003) Tunable Fresnel lens using nanoscale polymer-dispersed liquid crystals. *Applied Physics Letters*, 83, 1515–1517.

Rogers, J. A., Bao, Z., Baldwin, K., Dodabalapur, A., Crone, B., Raju, V. R., Kuck, V., et al. (2001) Paper-like electronic displays: Large-area rubber-stamped plastic sheets of electronics and microencapsulated electrophoretic inks. *Proceedings of the National Academy of Sciences of the United States of America*, 98, 4835–4840.

Rolandi, M., Quate, C. F., and Dai, H. J. (2002) A new scanning probe lithography scheme with a novel metal resist. *Advanced Materials*, 14, 191–194.

Schueller, O. J. A., Duffy, D. C., Rogers, J. A., Brittain, S. T., and Whitesides, G. M. (1999) Reconfigurable diffraction gratings based on elastomeric microfluidic devices. *Sensors and Actuators A: Physical*, 78, 149–159.

Shackelford, J. F. and Alexander, W. (2002) *CRC Materials Science and Engineering Handbook, 3rd edn*, CRC Press, Boca Raton, FL.

She, M., King, Y.-C., King, T.-J., and Hu, C. (2001) Modeling and design study of nanocrystal memory devices. *Annual Device Research Conference Digest*, Notre Dame, Indiana, pp. 139–140.

Sia, S. K. and Whitesides, G. M. (2003) Microfluidic devices fabricated in poly(dimethylsiloxane) for biological studies. *Electrophoresis*, 24, 3563–3576.

Slota, J. E., He, X. M., and Huck, W. T. S. (2010) Controlling nanoscale morphology in polymer photovoltaic devices. *Nano Today*, 5, 231–242.

Smela, E. (2003) Conjugated polymer actuators for biomedical applications. *Advanced Materials*, 15, 481–494.

Smela, E., Kallenbach, M., and Holdenried, J. (1999) Electrochemically driven polypyrrole bilayers for moving and positioning bulk micromachined silicon plates. *Journal of Microelectromechanical Systems*, 8, 373–383.

Smith, H. I., Schattenburg, M. L., Hector, S. D., Ferrera, J., Moon, E. E., Yang, I. Y., and Burkhardt, M. (1996) X-ray nanolithography: Extension to the limits of the lithographic process. *Microelectronic Engineering*, 32, 143–158.

Sumetsky, M., Dulashko, Y., Fleming, J. W., Kortan, A., Reyes, P. I., and Westbrook, P. S. (2004) Thermomechanical modification of diffraction gratings. *Optics Letters*, 29, 1315–1317.

Sun, H.-B. and Kawata, S. (2003) Two-photon laser precision microfabrication and its applications to micro-nano devices and systems. *IEEE Journal of Lightwave Technology*, 21, 624–633.

Sun, Y., Piyabongkarn, D., Sezen, A., Nelson, B. J., and Rajamani, R. (2002) A high-aspect-ratio two-axis electrostatic microactuator with extended travel range. *Sensors and Actuators A: Physical*, 102, 49–60.

Sun, Y. G. and Rogers, J. A. (2007) Inorganic semiconductors for flexible electronics. *Advanced Materials*, 19, 1897–1916.

Truxal, S. C., Tung, Y. C., and Kurabayashi, K. (2008a) A flexible nanograting integrated onto silicon micromachines by soft lithographic replica molding and assembly. *Journal of Microelectromechanical Systems*, 17, 393–401.

Truxal, S. C., Tung, Y. C., and Kurabayashi, K. (2008b) High-speed deformation of soft lithographic nanograting patterns for ultrasensitive optical spectroscopy. *Applied Physics Letters*, 92, 3.

Tseng, F. G. and Lin, C. J. (2003) Polymer MEMS-based Fabry-Perot shear stress sensor. *IEEE Sensors Journal*, 3, 812–817

Tuantranont, A. and Bright, V. M. (2002) Segmented silicon-micromachined microelectromechanical deformable mirrors for adaptive optics. *Journal on Selected Topics in Quantum Electronics*, 8, 33–45.

Tung, Y.-C. and Kurabayashi, K. (2004) Multi-axis single-layer PDMS-on-silicon micro optical reflector. *SPIE Proceedings*, 5604, 126–135.

Tung, Y.-C. and Kurabayashi, K. (2005a) A metal-coated polymer micromirror for strain-driven high-speed multi-axis optical scanning *IEEE Photonics Technology Letters*, 17, 1193–1195.

Tung, Y.-C. and Kurabayashi, K. (2005b) A single-layer PDMS-on-silicon hybrid micro actuator with multi-axis out-of-plane motion capabilities: Part I: Modeling and design. *Journal of Microelectromechanical Systems*, 14, 548–557.

Tung, Y.-C. and Kurabayashi, K. (2005c) A single-layer PDMS-on-silicon hybrid micro actuator with multi-axis out-of-plane motion capabilities: Part II: Fabrication and characterization. *Journal of Microelectromechanical Systems*, 14, 558–566.

Tung, Y. C. and Kurabayashi, K. (2006) A single-layer multiple degree-of-freedom PDMS-on-silicon dynamic focus micro-lens. *MEMS 2006: 19th IEEE International Conference on Micro Electro Mechanical Systems, Technical Digest*, Istanbul, Turkey, IEEE, New York, pp. 838–841.

Unger, M. A., Chou, H., Thorsen, T., Scherer, A., and Quake, S. R. (2000) Monolithic microfabricated valves and pumps by multilayer soft lithography. *Science*, 288, 113–116.

Varadan, V. K. (2003) Polymer-based MEMS accelerometer with modified organic electronics and thin film transistor. *Proceedings of SPIE*, 5116, 149–159.

Walraven, J. A. and Bernhard Jokiel, J. (2003) Failure analysis of a multi-degree-of-freedom spatial microstage. *Reliability, Testing, and Characterization of MEMS/MOEMS II*, SPIE, San Jose, CA, 4980, 87–96.

Wang, X., Engel, J., and Liu, C. (2003) Liquid crystal polymer (LCP) for MEMS: Processes and applications. *Journal of Micromechanics and Microengineering*, 13, 628–633.

Xia, Y. and Whitesides, G. M. (1998) Soft lithography. *Annual Review of Materials Science*, 28, 153–184.

Xie, H., Pan, Y., and Fedder, G. K. (2003) A CMOS-MEMS with curled-hinge comb drives. *Journal of Microelectromechanical Systems*, 12, 450–457.

Xie, Y., Xu, X., Hong, Y., and Fu, S. (2004) Fabrication of varied-line-spacing grating by elastic medium. *Optics Express*, 12, 3894–3899.

Yang, X., Grosjean, C., and Tai, Y. C. (1998) A low power MEMS silicone/parylene valve. *Technical Digest, Solid-State Sensor and Actuator Workshop*, Hilton Head Island, SC, pp. 316–319.

Yao, T. J., Yang, X., and Tai, Y. C. (2002) BrF3 dry release technology for large freestanding parylene micro-structures and electrostatic actuators. *Sensors and Actuators A: Physical*, 97–8, 771–775.

Yonzon, C. R., Jeoungf, E., Zou, S. L., Schatz, G. C., Mrksich, M., and Van Duyne, R. P. (2004) A comparative analysis of localized and propagating surface plasmon resonance sensors: The binding of concanavalin A to a monosaccharide functionalized self-assembled monolayer. *Journal of the American Chemical Society*, 126, 12669–12676.

Young, J. I. and Shkel, A. M. (2001) Comparative study of 2-DOF micromirrors for precision light manipulation. *Smart Structures and Materials 2001: Smart Electronics and MEMS*, SPIE, Newport Beach, CA, pp. 328–335.

Zanardi, P. and Rossi, F. (1998) Quantum information in semiconductors: Noiseless encoding in a quantum-dot array. *Physical Review Letters*, 81, 4752–4755.

Zhang, X., Jiang, X. N., and Sun, C. (1999) Micro-stereolithography of polymeric and ceramic microstructures. *Sensors and Actuators A: Physical*, 77, 149–156.

Zhao, X. M., Xia, Y. N., and Whitesides, G. M. (1997) Soft lithographic methods for nano-fabrication. *Journal of Materials Chemistry*, 7, 1069–1074.

Zhou, J. W. L., Chan, H. Y., To, T. K. H., Lai, K. W. C., and Li, W. J. (2004a) Polymer MEMS actuators for underwater micromanipulation. *IEEE-ASME Transactions of Mechatronics*, 9, 334–342.

Zhou, G., Vj, L., Tay, F. E. H., and Chau, F. S. (2004b) Diffraction grating scanner using a micromachined resonator. *Proceedings of the 17th IEEE International Conference on MEMS*, Nagoya, Japan, pp. 45–48.

Zhu, B. C., Jia, H. Y., Zhang, X. L., Chen, Y., Liu, H. P., and Tan, W. H. (2010) Engineering a subcellular targetable, red-emitting, and ratiometric fluorescent probe for Ca2+ and its bioimaging applications. *Analytical and Bioanalytical Chemistry*, 397, 1245–1250.

Zhukov, A. E., Kovsh, A. R., Ustinov, V. M., Livshits, D. A., Kop'ev, P. S., Alferov, Z. I., Ledentsov, N. N., and Bimberg, D. (2000) 3.5 W continuous wave operation from quantum dot laser. *Materials Science and Engineering B*, B74, 70–74.

Part V

Applications of Optical Actuation

Although optically driven technologies have been investigated since the 1970s, they have not made the impact on medicine or product design as originally anticipated. In recent years, however, optical nano- and microactuation systems have been incorporated into a broad spectrum of applications. In Chapter 17, Peipei Jia and Jun Yang discuss how optical actuation has played a critical role in the advancement of BioMEMS and biophotonics. Specifically, the authors focus on applications with significant biological importance such as optical manipulation and actuation of biological objects and, in particular, for single-molecule and single-cell manipulation, sorting, and detection. Lab-on-a-chip/BioMEMS technologies have numerous advantages over traditional experimental approaches such as less operating time, less sample consumption, and portability. Particularly in biology, lab-on-a-chip/BioMEMS provides all-in-one, automated, compacted, high-throughput, real-time, and on-site solutions for the entire experimental procedure, including sample preparation, reagent delivery, reactant mixing, biochemical reactions, and signal detection. Optical manipulation and actuation play a major role in all these functions.

Hiroshi Toshiyoshi describes in detail a particular light-driven optical scanner for a fiber-optic medical endoscope in Chapter 18. A medical endoscope is a miniaturized version of photogastroscope used to visualize the tissue in the body such as esophagus, trachea, and blood vessel with minimum invasion. To build a safe medical instrument, Toshiyoshi's research group has developed an all-optical-type fiber endoscope, where the actuation energy for a MEMS scanner is transferred by an infrared light traveling in a single mode fiber; at the same time, a probe light at a different wavelength is superposed onto the identical fiber for optical inspection, based on the principle of wavelength division multiplexing (WDM). The author discusses the architecture of the developed all-optical endoscope and provieds a detailed view of the MEMS optical scanner.

The remote control and delivery of power to microrobotic devices is a critical step in realizing miniaturized machines. In light-driven actuators (LDAs), the control and the energy supply are performed simultaneously by light, thereby eliminating the need for bulky conductive wiring. Wiring and on-board electronics are critical design constraints in developing microrobots and micro-machines because these devices require multiple actuators to be integrated in very small spaces. In Chapter 19, Hideki Okamura introduces energy-efficient light-activated and -powered shape memory alloy (SMA) actuators. The author provides a detailed discussion on the application of NiTi SMA to a tweezer type of light-driven actuator. The relatively high energy conversion efficiency of the SMA LDA means smaller heat production; this allows integration of the optically driven device into small spaces. Yukitoshi Otani further explores the photothermal effects on metal alloys and optical fibers for creating microrobots in Chapter 20.

In a very broad sense, electromagnetic radiation exerts a very small pressure upon any exposed surface. This radiation pressure and optical gradient forces have proven to be useful for driving nano- and microscale devices on Earth. In space, however, these same forces of light can propel very large solar sails and even specially designed spacecraft to nearby stars. Bernd Dachwald from FH Aachen University of Applied Sciences (Germany) introduces the notion of light propulsion systems for spacecraft in Chapter 21. Utilizing solar radiation pressure (SRP) for propulsion, solar sails can be propelled by reflecting the solar photons off large and lightweight mirroring surfaces, thereby transforming their momentum into a propulsive force. The author describes in detail the basic SRP force models used to describe the thrust force that is acting on a solar sail and the orbital dynamics essential for control. The performance parameters necessary to evaluate different designs are discussed and applications are summarized. To complete the presentation, both photon propulsion and laser propulsion systems are introduced.

The book concludes with Chapter 22, an opportunity for the contributing authors to provide their viewpoints as to the future prospects of optical nano- and microactuation technology and innovative applications. It is hoped that this more speculative discussion will spur discussion and provide young researchers with insight on future directions. This presentation may also be of interest to individuals who wish to see a snapshot of views of a rapidly changing technology in the early part of the twenty-first century.

17 Biophotonics and Its Applications in Lab-on-a-Chip or BioMEMS Platforms

Peipei Jia and Jun Yang

CONTENTS

17.1 INTRODUCTION

Following its success in information technology, photonics has significantly facilitated research in life science and biology in recent years. As a result, a new area, biophotonics, has emerged and is under fast development. The term biophotonics refers to an interdisciplinary subject at the interface between biology and photonics. This emerging subject mainly deals with the interaction between light and biological quantities with a particular focus on the biomedical applications of featured optical technologies such as optical microscopy, spectroscopy, biosensing or diagnostics, monitoring, and manipulation. Among the wide range of impacts that photonics have had in the last several decades, one of the most significant advances is the applications of optical forces such as optical trapping, manipulation, and sorting of biological objects. These applications enable us to explore the dynamics of single molecules, kinetics of molecular biochemical reactions, and cell mechanics, which have extensively advanced our fundamental understandings of many biological events at the single-molecule level and/or at the single-cell level. In addition to the fundamental research, many practical applications can be exploited by integration of optical trapping or optical sorting into biomedical devices such as lab-on-a-chip and biomicroelectromechanical systems (BioMEMS).

Historically, there are several key discoveries that make optical manipulation of single molecules and cells possible. Back to the seventeenth century, Kepler (1619) introduced the concept of radiation pressure or light pressure. The basis for the idea that light could generate a force was proposed for the first time. Radiation pressure was further elaborated by Maxwell (1873), who thought radiation with much higher energy exerted on a thin metallic disk might produce an observable mechanical effect like movement or deformation.

Applications of biophotonics are largely dependent on the development of optical microscope. The invention and application of optical microscope is probably one of the most important events for research in the history of science. Jansen, a Dutch spectacle-maker, built the first compound microscope around 1595. Leeuwenhoek described his detailed observations of bacteria, erythrocyte, and spermatozoa through a high-quality magnification glass in 1670. After centuries of evolution in the theory, development and application of microscopes, introducing a laser into an optical microscope system first implemented by Maiman (1960) enabled experiments of high precision and analytical values. Integration of laser into optical microscope systems laid the foundation for manipulating mesoscopic particles through radiation pressure. Ashkin, the key pioneer in this field, first demonstrated the particle confinement with single-beam optical traps, popularly known as optical tweezers (Ashkin 1970).

For several reasons, there has been a considerable requirement for manipulation of biological objects under the microscope without physical contact. Optical tweezers have opened up many new experimental opportunities for biologists that now could operate biological objects at the single-cell or the single-molecule level with the merit of noninvasive, full 3D control over the trapped object. In addition, the concept of optical manipulation and actuation also inspires micro-/nanosystem engineers to adopt this versatile tool into lab-on-a-chip and BioMEMS devices for realizing various functions instead of traditional mechanical or electrical approaches.

Biomedical microdevices represented by two typical categories of devices, lab-on-a-chip and BioMEMS, are based on the general concept of system miniaturization. As one of the most important technologies in the twenty-first century, lab-on-a-chip and BioMEMS exhibit an unprecedented perspective in studying biology. Lab-on-a-chip and BioMEMS devices provide unique functions that allow researchers to manipulate and probe individual cells and even single molecules with precisely defined microenvironments. Thus, lab-on-a-chip and BioMEMS technologies enable quantitative measurements with high biological/chemical selectivity and sensitivity, as well as high temporal and spatial resolution. In general, lab-on-a-chip/BioMEMS devices consist of microactuating and microsensing components, which are fabricated and assembled by micro-/nanotechnology. Lab-on-a-chip/BioMEMS technologies are of many advantages over traditional experimental approaches such as less operating time, less sample consumption, and portability. Particularly to biology, lab-on-a-chip/BioMEMS provides all-in-one, automated, compacted, high-throughput, real-time, and on-site solutions for the entire experimental procedure including sample preparation, reagent delivery, reactant mixing, biochemical reactions, and signal detection. Optical manipulation and actuation can play a major role in all these functions.

This chapter will present a review of the state-of-the-art biophotonics, not the entire area of the subject but the topics related to optical driving, actuation, and manipulation, in order to align with the major theme of the book. Although the general idea of manipulating matter with light has been applied to many fields, the focus will be given to the applications of significant biological importance such as optical manipulation and actuation of biological objects, and in particular, for the single-molecule and single-cell manipulation, sorting, and detection. Again, this overview is not to give a comprehensive list of experiments of these topics but rather to highlight some of the major studies with particular emphases on cases in the emerging fields of lab-on-a-chip and BioMEMS.

17.2 THEORY OF OPTICAL TRAPPING

Most optical manipulation or actuation is implemented by focusing a laser beam through a high numerical aperture (NA) microscope objective to create a focusing region where a dielectric particle will experience an optical force. The light-pressure force consists of two basic components, a scattering force in the direction of incident light propagation and a gradient force produced by a gradient of field intensity, as shown in Figure 17.1a. The scattering force always acts in the direction of the laser beam and tends to push the particle along the propagation direction of the light (Neuman and Block 2004). The scattering force is the cause of a net momentum transferred from incident photons to the particle. Regarding the gradient force component, since vibrating dipoles in the dielectric particle

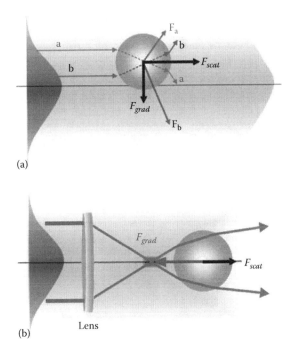

FIGURE 17.1   The principle of optical trapping with optical tweezers. (a) This shows the origin of the scattering (F_{scat}) and gradient (F_{grad}) components of the optical force on a dielectric sphere. (b) This shows the axial stability of a dielectric sphere on a focused laser beam, which is established by the gradient force along the centerline of the beam, toward the focal region. (Reprinted from Hormeño, S. and Arias-Gonzalez, J.R., *Biol. Cell*, 98, 679, 2006. With permission from Portland Press Ltd.)

are induced by the laser, each dipole existing in an inhomogeneous electromagnetic field experiences a net force in the direction of the field gradient of the laser beam. As a whole of the force on each dipole, the overall gradient force exerted on the particle is always toward the focal point (Jackson 1975). By designing the configuration or setup and the pattern of laser beams, one particle or a group of particles can be trapped, rotated, and even moved in an accelerated or decelerated manner.

The proportion between the size of trapped particles and the wavelength of the trapping laser source determines the light-particle interaction. If the radius of a trapped dielectric particle is much smaller than the light wavelength, the conditions of Rayleigh scattering can be applied (Ashkin et al. 1986). When the sphere size is significantly larger than the wavelength of the trapping laser, the criteria for stability are governed by the Mie regime (Ashkin 1992).

In the Rayleigh regime, the trapped particle can be considered as an induced point dipole, and the electromagnetic field of the laser beam pulls the particle toward the area of the highest light intensity of the "Gaussian intensity profile" of the laser beam due to the principle of minimum energy of a system. The function of trapping can be attributed to the effects of the both force components, the scattering force and the gradient force. The scattering force is proportional to the light intensity of the incident laser beam and points toward the focal point in the direction of the incident laser beam, a restoring force perpendicular to the focal plane. On the other hand, the gradient force is proportional to the gradient intensity of the "Gaussian beam" and points to the center of the beam, a restoring force within the focal plane. Together, both forces attract the particle into a close vicinity of the focal point, forming a stable trapping (Ashkin et al. 1986).

In the case of Mie scattering, since the size of the particle is much greater than the light wavelength, optical trapping can be described by simple ray optics, and the conservation of momentum model is applicable to this case. If the refraction index of the particle is greater than that of the surrounding medium, light refracted through the particle will transfer part of their momentum to the particle, causing an optical force in the direction of the intensity gradient or the direction toward the focal point. For example, refraction of light by a spherical particle corresponds to a change in the momentum carried by the light. When the particle is located at the centre of the trap, individual light rays are refracted through the particle symmetrically as shown in Figure 17.1b, resulting in zero net transverse force and compensating the scattering force of the laser light (Ashkin 1992). The scattering component of the optical force is attributed to both the absorption and reflection of the light by the trapped particle. Due to the use of a high-NA objective, the gradient force overcomes the scattering force and dominates the resultant 3D optical trapping.

When the size of trapped particles is comparable to the wavelength of the trapping laser, the majority of cases of experiments in practice, classical electrodynamics is required to supply an accurate description (Barton 1995). For example, the sizes of most biological cells, such as bacteria, yeast, and organelles of larger cells, lie in the range of dimension $0.1 \lambda – 10 \lambda$, where λ denotes the wavelength of the light. In this case, the net optical force exerted on a particle of an arbitrary shape may be calculated by the Maxwell stress tensor approach (Jackson 1999). The optical force expressed in this approach is valid for a non-deformable shape-arbitrary object and is completely determined by the electric and magnetic fields at the surface. Detailed explanation of this approach can be found in the work by Barton et al. (1989). A more detailed description of the entire theory of optical forces can be found in other chapters of this book.

17.3 OPTICAL MANIPULATION IN BIOLOGY

The manipulation of single cells and subcellular components in biology is achieved by optical trapping or optical tweezers. Optical tweezers are one of the most popular tools for actively gripping or trapping, positioning, and manipulating biological samples with the micro- or nanoscale spatial resolution and up to 100 aN force resolution (Zhang and Liu 2008). Compared to other methods of manipulating cells or proteins like atomic force microscope, optical tweezers have softer spring constant at the order of fN/nm. Typical applications of optical tweezers for biomanipulation include confining, organizing, assembling, locating, sorting, and modifying single cells, DNAs, proteins, and molecules. For the first time, Ashkin et al. demonstrated optical trap and manipulation of individual biological specimens such as viruses, bacteria, yeasts, and eukaryotic cells (Ashkin and Dziedzic 1987; Ashkin et al. 1987) and even noninvasive optical manipulation of subcellular organelles in living cells (Ashkin and Dziedzic 1989). Another featured application is to select cells; optical tweezers were harnessed to select single archaea from a highly mixed culture including multispecies of coexisting microorganisms in an anaerobic environment for subsequent cloning experiments by Huber et al. (1995). Since optical tweezers are a noninvasive technique, they have the incomparable advantage for an isolated and sealed working space with limited accessibility (e.g. an anaerobic environment). It is probably also a universal advantage of any optical technology. For example, this capability of optical tweezers has been also applied to capture a single cell without physical contact and thus to further explore the structural and elastic properties of red blood cells (Svoboda et al. 1992; Bronkhorst et al. 1995). Another case of manipulating subcellular organelles is that optical tweezers have been applied to monitor, track, and control the chromosome movement during cell division (Berns et al. 1989; Vorobjev et al. 1993).

Optical tweezers are also undoubtedly one of the most powerful tools to study the foundation of many biomolecular events. Biomolecules are often too small to be handled directly by optical tweezers. A typical case of application is to attach a microsphere to the molecule of interest and then observe and manipulate movement of the microsphere that keeps the same pace of movement with the single biomolecule. The spatial resolution for manipulation of biomolecule is around 0.1 nm, and a well-defined force of 0.1 pN resolution can be applied to the single biomolecule. Through observation and analysis of the position signals of the microsphere, one can indirectly discover dynamics, kinetics, and conformational changes of the bound single molecule when it performs a function or interacts with another molecule. This optical tweezers-based single-molecule technique can be

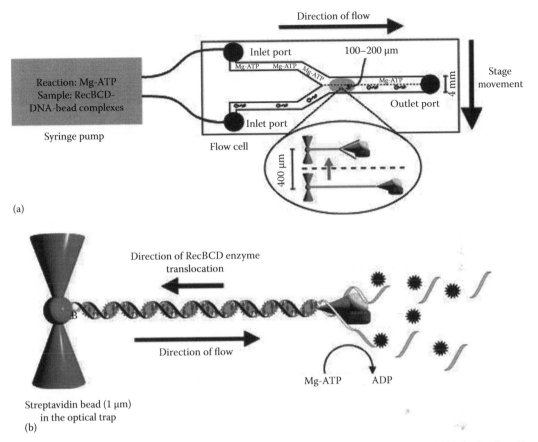

FIGURE 17.2 (See color insert.) Visualization of DNA helicase action on individual DNA molecules. (a) Syringe pump and flow cell. The red arrow indicates movement of the trapped DNA–bead complex across the boundary between solutions. (b) A trapped and stretched fluorescent DNA molecule is shown. As RecBCD enzyme translocates, it both unwinds and degrades the DNA, simultaneously displacing dye molecules (black stars). (Reprinted by permission from Macmillan Publishers Ltd. *Nature*, Bianco, P.R., Brewer, L.R., Corzett, M. et al., Processive translocation and DNA unwinding by individual RecBCD enzyme molecules, 409, 374–378, Copyright 2001.)

applied to a wide variety of different reagents such as DNAs, RNAs, proteins, and so on. Bianco et al. (2001) employed optical tweezers for actively manipulating forward and reverse translocation of single-RECbcd enzymes and DNA unwinding. In this experiment, a microsphere was bound to the end of a fluorescently labeled DNA and an inactive RECbcd was bound to the other end of the DNA molecule. Once a RecBCD enzyme-DNA-bead complex is trapped and the RECbcd is activated, both helicase movement and DNA unwinding can be monitored. As a processive DNA helicase, the RECbcd moves along the double-stranded DNA and simultaneously splits and unwinds it. The shortening of the fluorescently labeled DNA molecule was observed in real time as the RECbcd translocated or the reaction proceeded. Figure 17.2 schematically shows the experimental design and the biochemical reaction. A similar experiment was performed by Handa et al. (2005) through monitoring the displacement of a fluorescent nanoparticle attached to the RECbcd. Maier et al. (2004) utilized a different experimental design to investigate DNA transport through the cell envelope in *Bacillus subtilis* where a biotinylated DNA molecule was linked to both an optically trapped bead and a protein of an immobilized bacterium.

In the past two decades, various experimental configurations of optical tweezers have been developed, and new innovative optical tweezers have been emerging continuously (Stevenson et al. 2010).

The optical tweezers setup has evolved from the old version of single-beam optical tweezers to multiple optical traps that can be realized by time-sharing a single laser beam or by dynamic holographic optical trapping. Meanwhile, different research objectives such as movement or conformational changes of different specimen have guided new experimental designs including the optics configurations and the sample preparation of reagents and operating environments.

17.3.1 OPTICAL SCALPELS

Optical tweezers combined with the lasers, which function as optical scalpels, can be employed to perform very precise laser surgeries including controlled transiently cutting and noninvasively perforating with submicron resolution and at the cellular level. With accurate positioning and holding of surgical targets by optical tweezers, optical scalpels, therefore, enable cell fusion and molecular injection. Steubing et al. (1991) first demonstrated that two selected cells brought into contact using an optical trap can be fused by cutting the common wall of the two cells using pulsed ultraviolet (UV) laser microbeams. This innovative method effectively enabled the creation of hybrid cells without losing cell function. For fertilization procedures, optical manipulation has been adopted for selection and delivery of single human sperm into intimate contact with a selected oocyte (Tadir et al. 1991). Schutze et al. (1994) further advanced the optics-assisted fertilization procedure. They first combined optical tweezers with laser scalpels to hold a cell and drill channels in the zona pellucida of oocytes to facilitate sperm penetration. A safe and efficient reproduction technology will soon become clinically available, thanks to the assistance of noninvasive optical scalpels that ensure nonradiative damage (Ebner et al. 2005). Similar trapping–cutting technical assembly was also applied to the chromosome dissection (Greulich 1992) and gene cloning with polymerase chain reaction techniques to amplify microdissected chromosomes (He et al. 1997). Recently, such methods have been employed for separating individual cells from cell clusters, for example, for separating individual Frankia strains from root nodules of two Frankia–Alnus symbioses (Leitz et al. 2003). Moreover, transient permeabilization of the cell plasma membrane can also be realized by applying pulsed light, resulting in cell transfection (Stevenson et al. 2009).

Some assays in biology require that specific organelles are handled in a controlled manner on condition that other components should still maintain intact. However, pulsed UV or near-infrared laser light of certain intensity can irreversibly destroy the functions of organelles while shooting into cells. Thus, optical scalpels must be well designed in terms of light dose and/or irradiation time in the order of nanosecond according to different parts of a cell in order to ensure precise operation at a well-defined position. Grigaravicius et al. (2009) ingeniously applied this technique to aging research, where optical tweezers were utilized to exert a vertical pressure on human umbilical vein endothelial cells, mimicking the pathological condition of high blood pressure, to study the resultant changes in cell morphology. DNA repair system protects human body from various DNA damages and keeps the normal physiological functions of all levels of living systems. To investigate the repairing mechanism of how to recruit DNA repair molecules to the sites of damage of DNAs, optical scalpels provides sub-micrometer spatial and sub-second temporal resolution for purposely inducing DNA damage at defined locations. Precisely and controllably inducing DNA damage in time and space is pivotal. Such molecular-level DNA surgeries can be realized by tuning a variety of adjustable beam parameters. The Combination of optical scalpels with fluorescence microscopy opens a new avenue for the study of U2OS human osteosarcoma cells, they finally found that DNA repair arose near the non-homologous end joining and then turned to a homologous recombination repair pathway.

17.3.2 OPTICAL STRETCHING

Study of cell's capability to deform is of significant biological relevance such as blood cells pass through capillaries smaller than the diameters of cells. The cell cytoskeleton is a cellular protein

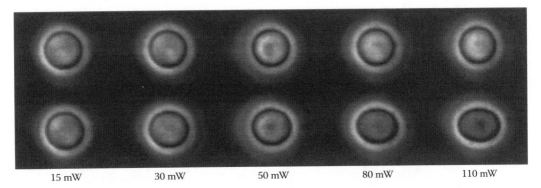

15 mW   30 mW   50 mW   80 mW   110 mW

FIGURE 17.3 Typical sequence of the stretching of one erythrocyte for increasing light powers. The top row shows the RBC trapped at 5 mW in each beam. The power was then increased to the higher powers given in the following, which lead to the stretching shown underneath. (Reprinted from *Biophys. J.*, 81, Guck, J., Ananthakrishnan, R., Mahmood, H., Moon, T.J., Cunningham, C.C., and Käs, J., The optical stretcher: A novel laser tool to micromanipulate cells, pp. 767–784, Copyright 2001, with permission from Elsevier.)

scaffold that can withstand deforming stress and plays important roles in support for both integrity and division of cells. The morphological change of cell shape is a representation of the state of the cytoskeleton reorganization. Optical stretching, exerting optical forces on cell membrane at the single-cell level, has been developed to measure viscoelastic properties of the cytoskeleton so that one can infer the characterization of cells. When a ray of light is reflected or refracted during passing a trapped cell, this deformable cell will bulge as a consequence of the transfer of part momentum carried by the light. Guck et al. (2001) constructed an optical stretcher with two opposed, non-focused, identical laser beams to trap the dielectric object, resulting in an additive surface force to stretch the object along the axis of beams. They utilized this optical stretcher to change the ellipticity of erythrocytes while applying different powers of beams (Figure 17.3). It is obvious that the radius along the beam axis clearly increases with increasing light power. Because no focusing of light is necessary in optical stretching, one obvious advantage of optical stretching is the minimal radiation damage.

In theory, due to changes of cellular function in certain diseases, disordered cells exhibit weaker cytoskeletons than those of normal cells on average. As a result, the cytoskeletons of diseased cells are more deformable than the counterparts of healthy cells. Hence, the statistical distribution of the ellipticity of cell morphology reflects the stage of some diseases and is significant in diagnosis and therapy of these diseases, for example, sickle-shaped red blood cells in sickle-cell anemia. In Guck group's subsequent work, optical stretching was performed to analyze mechanical properties for cancer cells. A microfluidic optical stretcher was demonstrated to analyze deformability of cancer cells. By comparing elasticity of a large number of healthy and cancerous cells, they found that the normal cells appear less deformable than the cancerous cells. Therefore, this group confirmed that optical deformability can be used as an inherent cell marker to monitor the minute change in terms of cell states from the normal to the cancerous (Guck et al. 2005). In another instance of optical stretching with single-beam optical tweezers, erythrocytes were attached to two adhesive microspheres, which were optically stretched to induce a tension on the two counter ends along the diameter of cells. This technique was developed by Sleep et al. (1999) to study the elasticity of erythrocyte membranes after chemical modifications of the membrane cytoskeletons where protein networks composed of various cytoskeletal proteins such as spectrin, ankyrin, and actin are disturbed (Figure 17.4).

Besides features of whole cells, the mechanical characteristics of some intracellular or extracellular organelles also were investigated. Wei et al. (2008) employed an oscillatory optical tweezers-based microrheometer to measure micromechanical properties of human lung epithelial cells.

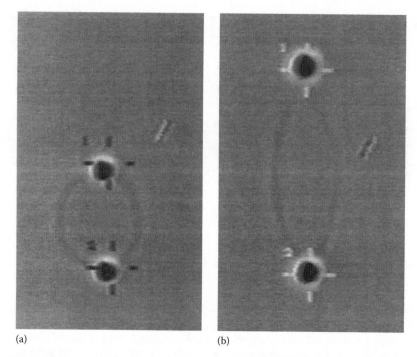

FIGURE 17.4   An erythrocyte ghost modified with *N*-ethylmaleimide, (a) before and (b) after optical stretching. (Reprinted from *Biophys. J.* 77, Sleep, J., Wilson, D., Simmons, R., and Gratzer, W., Elasticity of the red cell membrane and its relation to hemolytic disorders: An optical tweezers study, pp. 3085–3095, Copyright 1999, with permission from Elsevier.)

They got comparable intercellular viscoelasticity by forced oscillations of either an anti-integrin conjugated silica bead extracellularly bound to the cytoskeleton or an endogenous intracellular organelle.

17.4   OPTICAL MANIPULATION IN LAB-ON-A-CHIP OR BIOMEMS SYSTEMS

Since being integrated into lab-on-a-chip or BioMEMS platforms, the technique of optical manipulation has immediately made significantly greater impacts in a wide variety of disciplines such as biology and biochemistry. Microfluidic systems, also under the name of lab-on-a-chip or micrototal analysis systems, are typically a series of channels with dimensions of cross section ranging from several microns up to 100 μm, plus other microfluidic components. Many types of flows in microchannels are completely laminar, thereby mixing mainly through diffusion. By exposing cells to various microenvironments with well-defined biochemistry and biophysics, the systems can mimic in-vivo physiological conditions. In addition, the systems are usually made of transparent materials that facilitate the utility of optical tweezers in combination with microscopes. Thus, the real-time responses of cells to different chemical, physical, and biomechanical environment can be observed in these microfluidic platforms and be analyzed with other techniques like fluorescent and Raman simultaneously.

17.4.1   CELL TRANSPORTATION

The ability to move single-cell in a microenvironment filled with different reagents is critical for studying cell responses to external stimuli. Optical tweezers, in combination with microfluidic systems, have emerged as a robust technique in this area because of their capability of transporting a

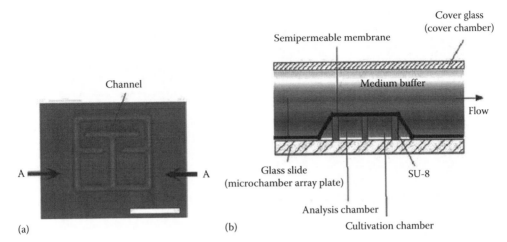

FIGURE 17.5 Microchambers of an on-chip microculture system. On the microchamber array plate, there are several microchambers, including an analysis chamber and a cultivation chamber, connected by a channel. Buffer medium in the chambers is exchanged through a semipermeable membrane. Cells in the microchambers are transported by means of optical tweezers. (a) Optical micrograph. (b) Schematic diagram of A–A cross-sectional view of microchamber (a). (Reprinted with kind permission from Springer Science + Business Media: *Fresen. J. Anal. Chem.*, Analysis of single-cell differences by use of an on-chip microculture system and optical trapping, 371, 2001, 276–281, Wakamoto, Y., Inoue, I., Moriguchi, H., and Yasuda, K.)

single cell in a controllable and non-interfering way. For instance, cells can be optically manipulated and transferred between different microfluidic reservoirs or between distinct laminar flow streams in the same channel in order to expose them to different reagents. Wakamoto et al. (2001) demonstrated a microculture system with two microchambers made of thick photoresist on the glass slide (Figure 17.5) for comparing genetically identical cells. In that setup, optical tweezers were employed to capture and transport a single *Escherichia coli* cell through a microchannel connecting the cultivation chamber and the analysis chamber. When a cell divided into two daughter cells, they separated and isolated one descendant cell from the other and further delivered the trapped cell into the cultivation chamber by optical tweezers to monitor its growth for in-depth analysis. They repeated such differentiation and isolation processes many times for comparison to draw a statistical conclusion. This technique will enlighten us to conduct the investigations of unequal phenomena in cell division in an environment without any contamination and/or disturbance. Eriksson et al. (2006) developed a microfluidic platform with a Y-shaped channel combined with epifluorescence microscopy and optical tweezers to reveal rapid responses of cells in different reagent streams with different osmolarities. In their experiments, as shown in Figure 17.6, the Yap1p transcription factor fused with a green fluorescent protein (GFP) accumulated in the nucleus when the yeast cell was moved from a normal cell culture environment to an environment containing the tert-butyl hydroperoxide (t-BOOH) that leads to oxidative stress. The accumulation of Yap1p transcription factor was explained as a proper transcriptional response to counteract the increase of peroxide stress. Because manipulation by optical tweezers and observation by microscopy were performed simultaneously, this approach of monitoring cellular responses in gene expression to environment changes shows a great promise of application in getting insight into different cellular processes and various properties of cells in real time.

Moreover, the ability to shape the intensity profile and phase of the light implies a variety of profound consequences in optical manipulation. The harness of novel-shaped light in optical tweezers has great potential in optical manipulation due to its flexibility and controllability in the application of optical transportation. Baumgartl et al. (2009) redistributed microparticles and mammalian cells among prepatterned and isolated microwells using Airy and parabolic laser beams (Figure 17.7),

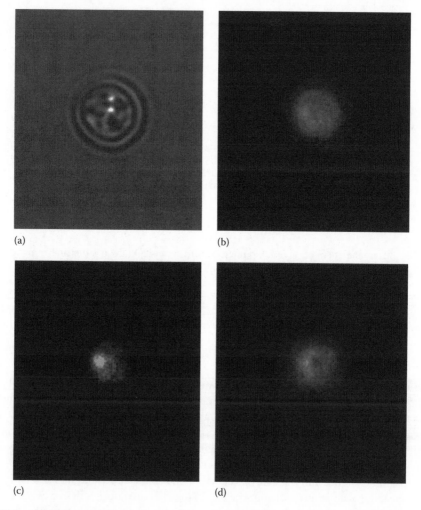

FIGURE 17.6 t-BOOH stress of a yeast cell. A transmission image of the cell is shown in image (a). The Yap1p-Gfp proteins are distributed over the entire cytosol, shown in image (b). The cell was then moved to the t-BOOH environment upon which the Yap1p-GFP accumulated in the nucleus, which can be seen in image (c). The intensity in the white spot is much higher than in the other parts. The cell was then moved back to the "neutral" environment, where the spatial GFP distribution returned to normal (d). (Eriksson, E., Enger, J., Nordlander, B. et al., A microfluidic system in combination with optical tweezers for analyzing rapid and reversible cytological alterations in single cells upon environmental changes, *Lab Chip*, 7, 71–76, 2006. Reprinted by permission of The Royal Society of Chemistry.)

where the traditional microflow for transportation became unnecessary. Their method may bring new opportunities that will allow for the transportation, redistribution, and exposure of any cell types to different environments in a BioMEMS platform.

17.4.2 CELL SORTING

The cell sorting to distinguish cellular types is another promising application of optical manipulation in the field of lab-on-a-chip and BioMEMS systems. Although flow cytometry is the standard setup for cell sorting that has been widely used in biology, microfluidic-based cell sorting provides some special advantages over conventional methods. In particular, microfluidic sorting systems

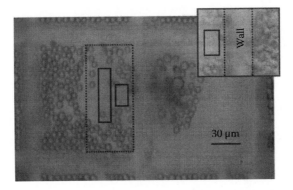

FIGURE 17.7 Particle and cell redistribution between microwells. Particles are guided up and over the 20 μm PDMS wall into the neighboring microwell. Insert: clearing of red blood cells with an Airy beam (black rectangle) oriented to the right. (Baumgartl, J., Hannappel, G.M., Stevenson, D.J., Day, D., Gu, M., and Dholakia, K., Optical redistribution of microparticles and cells between microwells, *Lab Chip*, 9, 1334–1336, 2009. Reprinted by permission of The Royal Society of Chemistry.)

require considerably less sample volume of cells compared to typical flow cytometers. Meanwhile, miniaturization of operation platforms also reduces reagent consumption without losing accuracy of the assay. A powerful method to sort cells is fluorescence-activated cell sorting (FACS; Fulwyler 1965). In this technique, optical tweezers can drag biologically tagged cells on-demand based on the detected fluorescence signal from the detection region. In this way, tagged single cells can be dragged and thus sorted from a group of cells moving in a flow stream and finally stored into an isolated part of the microfluidic system. Wang et al. (2005) utilized an all-optical control switch in a high-throughput microfluidic FACS for sorting mammalian cells. After flowing into the microfluidic system, cells were first aligned to the center of the channel and optically switched into different branch channels based on their detectable fluorescence signal. The target cells were transported to the collection output, while all others were directed to the waste channel. This type of optical switch facilitates the reduction of chip complexity and connectivity by displacing the cell within the laminar flow. In a similar work, Perroud et al. (2008) utilized an infrared laser to actively deflect green-labeled and hydrodynamically focused cells from a mixed population into a collection channel.

Cells can also be separated through passive sorting according to optical potential energy landscapes created by an array of optical traps, which now has become a key technique in the field of optical sorting and separation. Cells that have different shapes, sizes, or refractive indices may experience different optical forces in the light field, which can be utilized for passive sorting with a prominent advantage that no specific marker is necessary. On the other hand, motion of cells across such an optical potential energy landscape depends not only on the beam power but also on its optical field pattern. For instance, Paterson et al. (2005) applied a Bessel beam in the passive separation of lymphocytes from erythrocytes with no need of using fluid flow. When the two types of cells were exposed to the Bessel beam, the erythrocytes moved toward the core of the beam and then were trapped in the outer rings of the beam due to the optical force, whereas the lymphocytes migrated directly to the beam center. MacDonald et al. (2003) demonstrated a 3D and interlinked optical lattice for cell sorting, which is dynamically reconfigurable and can be readily integrated into microfluidic platforms. Figure 17.8 illustrates the basic concept for a sorting system based on a specifically designed optical lattice, which provides an effective way of removing biological matters according to the selection criteria from another laminar stream. When a flow of mixed cells or particles passed through the fractionation chamber and interacted with the optical lattice, targeted cells/particles deviated from their original trajectories due to strong optical forces, whereas others moved straight through the deflection region without any distortion.

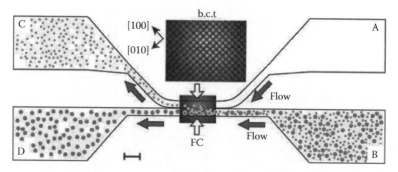

FIGURE 17.8 The concept of optical fractionation. By introducing a 3D optical lattice into the fractionation chamber, one species of particle is selectively pushed into the upper flow field. The reconfigurability of the optical lattice allows for dynamic updating of selection criteria. (Reprinted by permission from Macmillan Publishers Ltd. *Nature*, MacDonald, M.P., Spalding, G.C., and Dholakia, K., Microfluidic sorting in an optical lattice, 426, 421–424, Copyright 2003.)

17.5   CONCLUSIONS AND FUTURE DIRECTIONS

As one of the most important tools in biophotonics, optical manipulation has been making extensive impacts in many aspects of biology, particularly showing incomparable advantages in research at the single-molecule and the single-cell levels. Although it is difficult to foresee the future direction of technique advancement, further endeavor should be first focused on developing highly integrated lab-on-a-chip and/or BioMEMS platforms with biophotonic means. This will create numerous opportunities in cell manipulation and operation including cell sorting and cell surgery. By means of the integration with lab-on-a-chip/BioMEMS devices, appropriately shaped lights enable us to watch, control, and interfere in motions of single molecules intracellularly or extracellularly and even transcellularly. These techniques certainly have facilitated and will continue facilitating observation and analysis of fundamental biological processes under various microenvironments and ever in vivo for further understanding their essence. Moreover, these advances will no doubt accelerate the development of new methods for molecular surgery, intracellular organelle surgery, fertility clinics, and drug screening as well as many other potential applications.

From another perspective, it is prospective to noninvasively transform and assemble biological objects with nanometer accuracy and precision and to do all these things in parallel at many different locations in one single instrument with properly designed optical configurations. Meanwhile, the automatization of lab-on-a-chip/BioMEMS functions, such as optically driven pumping and valving, is expected to implement built-in optical manipulation. As a result, optical manipulation, assays, and detection could be realized in an all-in-one single chip or system that offers multiple functions such as pumping, flow control like valving, sorting, reaction, surgery, organization, synthesis, assembly, and so on. By doing so, optomechanical approaches could expand the applications of lab-on-a-chip/BioMEMS technologies in the field of biology and medicine to provide a broad range of services in pathology, diagnostics, and drug discovery. To conclude, biophotonics combined with lab-on-a-chip/BioMEMS systems has been evolving into a powerful interdisciplinary field to get a real insight into the nature of life.

REFERENCES

Ashkin, A. 1970. Acceleration and trapping of particles by radiation pressure. *Phys. Rev. Lett.* 24, 156–159.
Ashkin, A. 1992. Forces of a single-beam gradient laser trap on a dielectric sphere in the ray optics regime. *Biophys. J.* 61, 569–582.
Ashkin, A. and Dziedzic, J. M. 1987. Optical trapping and manipulation of viruses and bacteria. *Science* 235, 1517–1520.

Ashkin, A. and Dziedzic, J. M. 1989. Internal cell manipulation using infrared laser traps. *Proc. Natl Acad. Sci. USA*. 86, 7914–7918.

Ashkin, A., Dziedzic, J. M., Bjorkholm, J. E., and Chu, S. 1986. Observation of a single-beam gradient force optical trap for dielectric particles. *Opt. Lett.* 11, 288–290.

Ashkin, A., Dziedzic, J. M., and Yamane, T. 1987. Optical trapping and manipulation of single cells using infrared laser beams. *Nature* 330, 769–771.

Barton, J. P. 1995. Internal and near-surface electromagnetic fields for a spheroidal particle with arbitrary illumination. *Appl. Opt.* 34, 5542–5551.

Barton, J. P., Alexander, D. R., and Schaub, S. A. 1989. Theoretical determination of net radiation force and torque for a spherical particle illuminated by a focused laser beam. *J. Appl. Phys.* 66, 4594–4602.

Baumgartl, J., Hannappel, G. M., Stevenson, D. J., Day, D., Gu, M., and Dholakia, K. 2009. Optical redistribution of microparticles and cells between microwells. *Lab Chip* 9, 1334–1336.

Berns, M. W., Wright, W. H., Tromberg, B. J., Profeta, G. A., Andrews, J. J., and Walter, R. J. 1989. Use of a laser-induced optical force trap to study chromosome movement on the mitotic spindle. *Proc. Natl Acad. Sci. USA*. 86, 4539–4543.

Bianco, P. R., Brewer, L. R., Corzett, M. et al. 2001. Processive translocation and DNA unwinding by individual RecBCD enzyme molecules. *Nature* 409, 374–378.

Bronkhorst, P. J., Streekstra, G. J., Grimbergen, J., Nijhof, E. J., Sixma, J. J., and Brakenhoff, G. J. 1995. A new method to study shape recovery of red blood cells using multiple optical trapping. *Biophys. J.* 69, 1666–1673.

Ebner, T., Moser, M., and Tews, G. 2005. Possible applications of a non-contact 1.48 micron wavelength diode laser in assisted reproduction technologies. *Hum. Reprod. Update* 11, 425–435.

Eriksson, E., Enger, J., Nordlander, B. et al. 2006. A microfluidic system in combination with optical tweezers for analyzing rapid and reversible cytological alterations in single cells upon environmental changes. *Lab Chip* 7, 71–76.

Fulwyler, M. J. 1965. Electronic separation of biological cells by volume. *Science* 150, 910–911.

Greulich, K. O. 1992. Chromosome microtechnology: Microdissection and microcloning. *Trends Biotechnol.* 10, 48–51.

Grigaravicius, P., Greulich, K. O., and Monajembashi, S. 2009. Laser microbeams and optical tweezers in ageing research. *Chem. Phys. Chem.* 10, 79.

Guck, J., Ananthakrishnan, R., Mahmood, H., Moon, T. J., Cunningham, C. C., and Käs, J. 2001. The optical stretcher: A novel laser tool to micromanipulate cells. *Biophys. J.* 81, 767–784.

Guck, J., Schinkinger, S., Lincoln, B. et al. 2005. Optical deformability as an inherent cell marker for testing malignant transformation and metastatic competence. *Biophys. J.* 88, 3689–3698.

Handa, N., Bianco, P. R., Baskin, R. J., and Kowalczykowski, S. C. 2005. Direct visualization of RecBCD movement reveals cotranslocation of the RecD motor after χ recognition. *Mol. Cell*. 17, 745–750.

He, W., Liu, Y. G., Smith, M., and Berns, M. W. 1997. Laser microdissection for generation of a human chromosome region-specific library. *Microsc. Microanal.* 3, 47–52.

Hormeño, S. and Arias-Gonzalez, J. R. 2006. Exploring mechanochemical processes in the cell with optical tweezers. *Biol. Cell* 98, 679–695.

Huber, R., Burggraf, S., Mayer, T., Barns, S. M., Rossnagel, P., and Stetter, K. O. 1995. Isolation of a hyperthermophilic archaeum predicted by in situ RNA analysis. *Nature* 376, 57–58.

Jackson, J. D. 1975. *Classical Electrodynamics*, 2nd edn. New York: Wiley.

Jackson, J. D. 1999. *Classical Electrodynamics*, 3rd edn. Hoboken, NJ: Wiley.

Kepler, J. 1619. De cometis libelli tres, typis Andreae Apergeri, sumptibus Sebastiani Mylii bibliopolae augustani, Avgvstae Vindelicorum.

Leitz, G., Lundberg, C., Fällman, E., Axner, O., and Sellstedt, A. 2003. Laser-based micromanipulation for separation and identification of individual Frankia vesicles. *FEMS Microbiol. Lett.* 224, 97–100.

MacDonald, M. P., Spalding, G. C., and Dholakia, K. 2003. Microfluidic sorting in an optical lattice. *Nature* 426, 421–424.

Maier, B., Chen, I., Dubnau, D., and Sheetz, M. P. 2004. DNA transport into Bacillus subtilis requires proton motive force to generate large molecular forces. *Nat. Struct. Mol. Biol.* 11, 643–649.

Maiman, T. 1960. Stimulated optical radiation in ruby. *Nature* 187, 493.

Maxwell, J. C. 1873. *Treatise on Electricity and Magnetism*. Oxford, U.K.: Clarendon Press.

Neuman, K. C. and Block, S. M. 2004. Optical trapping. *Rev. Sci. Instrum.* 75, 2787.

Paterson, L., Papagiakoumou, E., Milne, G. et al. 2005. Light-induced cell separation in a tailored optical landscape. *Appl. Phys. Lett.* 87, 123901.

Perroud, T. D., Kaiser, J. N., Sy, J. C. et al. 2008. Microfluidic-based cell sorting of *Francisella tularensis* infected macrophages using optical forces. *Anal. Chem.* 80, 6365–6372.

Schutze, K., Clement-Sengewald, A., and Ashkin, A. 1994. Zona drilling and sperm insertion with combined laser microbeam and optical tweezers. *Fertil. Steril.* 61, 783–786.

Sleep, J., Wilson, D., Simmons, R., and Gratzer, W. 1999. Elasticity of the red cell membrane and its relation to hemolytic disorders: An optical tweezers study. *Biophys. J.* 77, 3085–3095.

Steubing, R. W., Cheng, S., Wright, W. H., Numajiri, Y., and Berns, M. W. 1991. Laser induced cell fusion in combination with optical tweezers: The laser cell fusion trap. *Cytometry* 12, 505–510.

Stevenson, D. J., Gunn-Moore, F., Campbell, P., and Dholakia, K. 2009. Single cell optical transfection. *J. R. Soc., Interface* 7, 863–871.

Stevenson, D. J., Gunn-Moore, F., and Dholakia, K. 2010. Light forces the pace: Optical manipulation for biophotonics. *J. Biomed. Opt.* 15, 041503.

Svoboda, K., Schmidt, C. F., Branton, D., and Block, S. M. 1992. Conformation and elasticity of the isolated red blood cell membrane skeleton. *Biophys. J.* 63, 784–793.

Tadir, Y., Wright, W. H., Vafa, O., Liaw, L. H., Asch, R., and Berns, M. W. 1991. Micromanipulation of gametes using laser microbeams. *Hum. Reprod.* 6, 1011–1016.

Vorobjev, I. A., Liang, H., Wright, W. H., and Berns, M. W. 1993. Optical trapping for chromosome manipulation: a wavelength dependence of induced chromosome bridges. *Biophys. J.* 64, 533–538.

Wakamoto, Y., Inoue, I., Moriguchi, H., and Yasuda, K. 2001. Analysis of single-cell differences by use of an on-chip microculture system and optical trapping. *Fresen. J. Anal. Chem.* 371, 276–281.

Wang, M. M., Tu, E., Raymond, D. E. et al. 2005. Microfluidic sorting of mammalian cells by optical force switching. *Nat. Biotechnol.* 23, 83–87.

Wei, M. T., Zaorski, A., Yalcin, H. C. et al. 2008. A comparative study of living cell micromechanical properties by oscillatory optical tweezers. *Opt. Express* 16, 8594–8603.

Zhang, H. and Liu, K. 2008. Optical tweezers for single cells. *J. R. Soc. Interface* 5, 671–690.

18 Light-Driven Optical Scanner for Fiber-Optic Endoscope

Hiroshi Toshiyoshi

CONTENTS

18.1 INTRODUCTION

Medical endoscope is a miniaturized version of photogastroscope used to visualize the tissue in the body such as esophagus, trachea, and blood vessel with a minimum invasion [1]. In the early stage of endoscope development, a fiber scope, which was a bundle of optical fibers as illustrated in Figure 18.1a, was used to transfer the image by pressing the fiber bottom directly onto the tissue [2]. However, the fiber bundle could not be bent at a hard angle due to the mechanical rigidity, and the access is limited within a rather short straight range from the body surface. In addition to this, the image resolution of a fiber bundle endoscope was poor due to the small number of fiber assembly. Endoscope for practical medical use today is called videoscope, which is equipped with a tiny illumination device and a chip of electronic imaging element based on the Change Coupled Device (CCD) or Complementry Metal Oxide Semiconductor (CMOS) technology. Videoscope is usually used in combination with other medical equipment inserted into a multitube catheter.

Besides the surface image observation, recent progress in the optical coherence tomography (OCT) has enabled us to see the cross-sectional images of tissue such as retina membrane by optically probing the deep tissue and by reconstructing the sectional image using the computer-controlled optical interference system [3,4]. A use of a microelectromechanical systems (MEMS) optical scanner attached at the end of an optical fiber as schematically illustrated in Figure 18.1b is, therefore, thought to be a solution to build a thin and flexible endoscope with the OCT function; the surface tissue profile is converted into time-sequential signal of back-scattered light intensity from the optical probe scanned over the sample.

Research and development of MEMS endoscopes has become active in past 10 years, and various types of spatial light modulation have been reported including electrostatic, electromagnetic, and electrothermally driven mechanisms [5–10]. In those cases, the actuation energy for the MEMS scanner is transferred from the outside by electrical current (a few mA to several tens of mA) or voltage (a few volts to several tens of volts) through a pair of metallic wires assembled in parallel

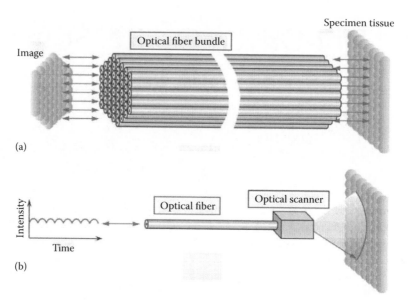

FIGURE 18.1 Fiber-optic endoscope architectures. (a) Image transfer by fiber bundle and (b) image scanner by MEMS optical scanner.

with the optical fiber. A lot of cautions, therefore, have to be paid to ensure the electrical insulation on the endoscope packaging to avoid internal electrification due to the leakage of current, because the ventricular fibrillation of heart may occur at a current level as low as 0.1 mA, particularly when it is directly exposed to the electrocution in the body. A safety guideline has been set for electronic medical instruments to keep the leak current as low as 10 microampere level [11]. Nonetheless, endoscopes with wired cables are not free from such a risk, because it could still be coupled with other medical instruments through the electromagnetic induction.

To build a safe medical instrument, we have recently developed an all-optical type fiber endoscope, where the actuation energy for a MEMS scanner is transferred by an infrared light traveling in a single mode fiber; at the same time, a probe light at a different wavelength is superposed onto the identical fiber for optical inspection, based on the principle of wavelength division multiplexing (WDM) [12,13]. This chapter deals with the architecture of the developed all-optical endoscope along with the detail view of the MEMS optical scanner and measurement results of OCT.

18.2 ENDOSCOPE ARCHITECTURE

Figure 18.2 shows the schematic view of the all-optical medical endoscope equipped with a MEMS optical scanner. The entire probe housing is 6 mm in the outer diameter and 20 mm in length, excluding the butt-coupled single-mode optical fiber. A 10 mW infrared light of 1.5 μm in wavelength is supplied through the optical fiber and selectively reflected by a dichroic beam splitter; the reflected light is projected onto a photovoltaic cell, where the drive voltage for the MEMS optical scanner is generated. The voltage is transferred to the electrostatic optical scanner through the pair of metallic interconnection patterns made on the submount substrate. At the same time, another light of 1.3 μm in wavelength is superposed on the fiber. This light goes through the beam splitter and the focal lens before hitting the MEMS scanner and is spatially scanned over the tissue appressed on the endoscope window. The probe light reflected from the tissue surface as well as the deep tissue is collected by the identical optics and transferred back to the OCT measurement system located at the other end of the optical fiber. Table 18.1 summarizes the target specifications for the MEMS optical scanner.

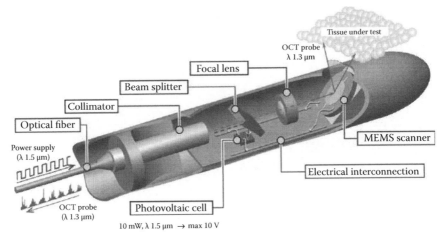

FIGURE 18.2   Structure of all-optical endoscope.

TABLE 18.1
Specifications of MEMS Optical Scanner for Endoscope

Parameter	Specification	Note
Probe housing	φ 6 mm	Pyrex glass tube of 0.5 mm wall thickness
Size of chip	2 mm×3 mm	Placed at a 45° titled angle
Mirror size	1 mm×2 mm	Long axis 1.5 mm for titled incident light
Drive voltage	Max 5 V	
Drive current	Max 1 mA	Limited by the photovoltaic cell's current capacity
Scan angle	±5°	Optical angle
Scan rate	250 Hz	

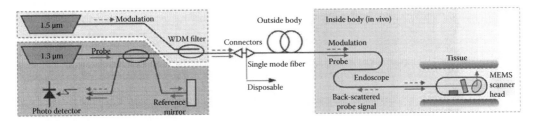

FIGURE 18.3   Optical apparatus of all-optical endoscope based on the WDM fiber telecom technology.

The entire system of the OCT is shown in Figure 18.3. Two laser diodes for powering (λ 1.5 μm) and probing (λ 1.3 μm) are accommodated. The intensity of the powering light is chopped into a square wave signal at the mechanical resonant frequency of the MEMS scanner. On the other hand, the probe light wavelength is tuned at 1270–1380 nm to scan the penetration depth into the tissue, which is speculated by the OCT interference measurement [14]. A simplified principle of time domain OCT is shown in Figure 18.4. The back-scattered light and the reference light of the probe wavelength are guided onto a balanced detector, where the constructive/deconstructive interference

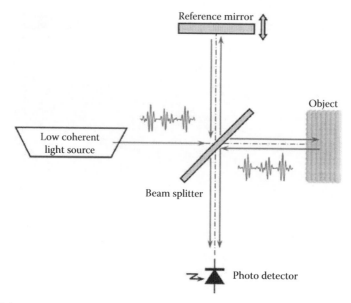

FIGURE 18.4 Principle of optical coherence tomography.

of light takes place depending upon the difference of the optical path lengths, i.e., the depth in the tissue. A cross-sectional image is reconstructed by mapping the dot brightness at the (X and Y) coordinates; the lateral X position corresponds to the MEMS scanner angle, while the vertical Y position is the depth in the tissue. In the practical implementation, we used frequency domain OCT, and the fast Fourier transform to tell the point of depth at a fast frame rate of 500 fps [14].

Figure 18.5 illustrates the arrangement of the MEMS optical scanner chip in the endoscope tube. The scanner is used to redirect the probe light at the right angle and spatially scan it in the transverse directions. This work used a tube of 6 mm in the outer diameter and 5 mm in the inner diameter. Considering the dimension of the sub-mount substrate for the MEMS chip and the focal length, maximum 3 mm is allocated for the longer axis of the MEMS chip that is set at 45°. The effective mirror size of 1.5 mm × 1.5 mm or larger has been set such that the incoming light is not clipped off the mirror edges.

As an actuation principle for the MEMS optical scanner, we chose electrostatic actuation [15], which was thought to consume less power due to the higher electrical impedance than other approaches

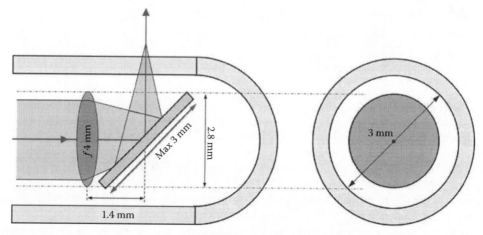

FIGURE 18.5 Geometrical arrangement of MEMS scanner chip in endoscope tube.

such as electromagnetic or electrothermal operation. Presuming a use of an electrostatic microactuator made by the silicon-on-insulator (SOI with a 2 μm-thick silicon oxide) bulk micromachining, the most significant contribution to the electrical capacitance comes from the buried oxide (BOX), which lies in between the top silicon layer and the bottom handle wafer. Assuming a MEMS chip of $S = 2$ mm $\times 3$ mm, the maximum load capacitance of the MEMS is, therefore, estimated to be $C_{max} = \varepsilon_r \varepsilon_0 S/t$, where $\varepsilon_r = 4.2$ is the relative permittivity of silicon oxide, $\varepsilon_r = 8.85 \times 10^{-12}$ F/m the dielectric constant, and t the oxide thickness. Substituting the parameters, one would obtain $C_{max} \sim 10^{-10}$, implying that a power needed to charge and discharge the capacitance is in the order of $\omega C_{max} V^2$ at maximum, which is no greater than 20 μW when the scanner is operated at 10 V and 300 Hz.

We used an Indium Gallium Phosphide (InGaP) super-lattice type photovoltaic cell [16] as a power source, shown in Figure 18.6; it had a total energy conversion efficiency of 0.44% (including the clipping loss of the Gaussian beam and the insertion loss of a lens). When exposed under an infrared light of 10 mW at 1.5 μm wavelength, the cell was found to generate maximum power of 44 μW (=7.6 V × 5.7 μA), which gave a good power budget to operate an electrostatic MEMS optical scanner. For a CW light of 4 mW or higher, the cell generated maximum 12 V as shown in Figure 18.7a under the open condition of circuit, where no load resistance was attached. When the incident light intensity was chopped to a periodic square waveform, the photovoltaic cell was found to follow frequency upward of 1 kHz, as shown in Figure 18.7b, which was higher than the operation frequency of the MEMS optical scanner.

Near-infrared wavelengths at 665 or 880 nm are usually preferable for OCT measurement because the optical absorption loss by hemoglobin is low in this wavelength range [17], as shown in Figure 18.8. Nonetheless, we used the wavelengths of 1.3 and 1.5 μm due to the benefit that we could use optical components originally developed for the fiber telecom applications. For instance,

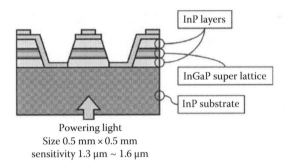

FIGURE 18.6 Structure of photovoltaic cell.

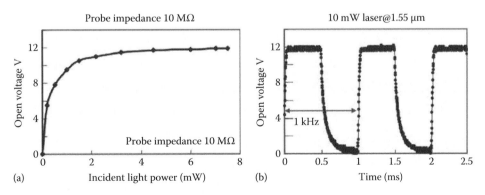

FIGURE 18.7 Response of photovoltaic cell. (a) DC voltage output under CW incident light and (b) transient response under chopped light.

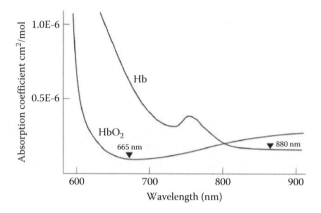

FIGURE 18.8 Optical absorption by hemoglobin.

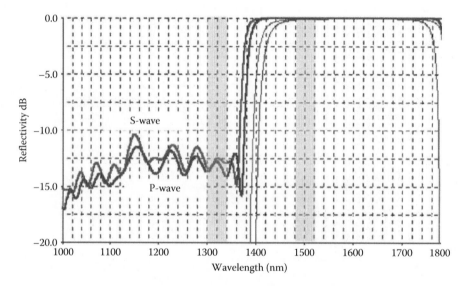

FIGURE 18.9 Reflection selectivity of infrared beam splitter.

a beam splitter of superior performance is available from the standard component sets of WDM; as shown in Figure 18.9, optical contrast of more than 10 dB was obtained to selectively reflect the 1.5 μm wavelength while transmitting the 1.3 μm wavelength in both S- and P-polarization. In our all-optical system, we used a 10 mW infrared light at 1.5 μm wavelength for powering and low power 1.3 μm for the OCT measurement; the latter had lower optical absorption in hemoglobin.

18.3 ELECTROMECHANICAL DESIGN OF OPTICAL SCANNER

The optical scanner in this work has been modeled by using a quadratic oscillation system as schematically illustrated in Figure 18.10a and b. For simplicity of modeling, the mirror plate is represented by a rectangular plate of length l_m (in parallel with the rotation axis) and a total width w_m (perpendicular to the rotation axis), which is supported by a pair of straight torsion bars of width w_s and length l_s. The suspensions are connected and fixed to the stationary anchors on the substrate. Comb-drive electrodes of width w_c, length l_c, and the mutual gap g are arranged on the edges of

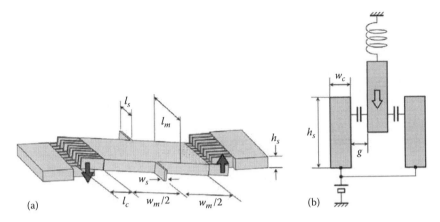

FIGURE 18.10   Electrostatic actuator model for MEMS optical scanner.

the mirror; counter electrodes for the movable combs are also arranged on the substrate to induce electrostatic torque. The structures have a uniform height of h_s.

The moment of inertia of the mirror plate is written

$$I = \rho \int_{-w_m/2}^{w_m/2} l_m h_s x^2 \, dx = \frac{\rho}{12} l_m w_m^3 h_s, \qquad (18.1)$$

where ρ is the density of silicon (2.33×10^3 kg/m³). The pair of torsion bars has a suspension rigidity of [18]

$$k = 2 \times \frac{G h_s w_s^3}{3 \, l_s} \left\{ 1 - \frac{192}{\pi^5} \cdot \frac{w_s}{h_s} \cdot \tanh\left(\frac{\pi h_s}{2 \, w_s} \right) \right\}, \qquad (18.2)$$

where G is the shear stress of rigidity (73 GPa for single crystalline silicon). From these two equations, the fundamental resonant mode of the scanner is written as $f_0 = (1/2\pi)\sqrt{(k/I)}$ at this mode, the mirror plate vibrates about the torsion axis to cause the out-of-plane motion.

Static tilt angle of the mirror can be calculated by considering the electrostatic torque generated in between the vertical comb electrodes. Ignoring the fringing effect of the electrical field near the electrode edges, the electrical capacitance between two parallel plates separated by g is written $C(\xi) = \varepsilon_0 (\xi - \xi_0)/g$ for the unit length, where ε_0 is the dielectric constant of vacuum (8.85×10^{-12} F/m), ξ the vertical displacement of the movable electrode, and ξ_0 the initial overlap between the plates [15]. Such parallel plate capacitance is formed on the both sides of the movable plate, and the movable comb is pulled into the gap as the voltage is applied. The electrostatic force acting on the movable comb is calculated by

$$dF = 2 \times \frac{1}{2} \frac{\partial C(\xi)}{\partial \xi} V^2 \, d\psi = \frac{\varepsilon_0}{g} V^2 \, d\psi, \qquad (18.3)$$

where $d\psi$ is the length of a small section along the comb length. Considering the distance between the section and the rotation axis, $(w_m/2) + \psi$, we write the electrostatic torque as

$$dT = \left(\frac{w_m}{2} + \psi\right) dF = \frac{\varepsilon_0}{g} V^2 \left(\frac{w_m}{2} + \psi\right) d\psi. \tag{18.4}$$

We integrate the torque along the comb length to obtain the total electrostatic torque acting on the mirror plate as

$$T(V) = 2N \times \int_0^{l_c} \frac{\varepsilon_0}{g} V^2 \left(\frac{w_m}{2} + \psi\right) d\psi = \frac{\varepsilon_0 \, NV^2}{g} \left(w_m + l_c\right) l_c, \tag{18.5}$$

where N is the number of the comb teeth on one side.

Optical scanner in this work is expected to operate at optical angle of ±5°, which corresponds to a mechanical angle of ±3.5° for an incident angle of 45° with respect to the rotation axis [19]. To lower the drive voltage, the scanner is excited at its fundamental resonant frequency. Amplitude at resonance is calculated by the product of the quality factor Q and the static tilt angle δ at the identical dc voltage. Estimation of quality factor requires the contribution of the air-damping loss, which is usually difficult to predict for an oscillator in the out-of-plane motion. Therefore, we used an empirical value of 8 for the quality factor from our own experience [13] and set a target static tilt angle of ±0.4 at 5 V. Figure 18.11 shows the static tilt angle numerically simulated as a function of drive voltage; three different suspension widths (2.0, 2.5, and 3.0 µm) are used to investigate the effect of the suspension rigidity. For an optical scanner compatible with a 5 V operation, the plot indicates that a suspension width smaller than 2.5 µm should be used.

Another condition for the scanner design is the resonant frequency. Due to the maximum speed of the analog-to-digital converter used in the OCT data acquisition system, the resonant frequency of the MEMS scanner should be made to be lower than 300 Hz. We used the values listed in Table 18.2 as default design parameters and swept the suspension width as a tailoring parameter to adjust the resonant frequency. Figure 18.12 shows the theoretical simulation results of the resonant frequency as a function of suspension width. Resonant frequency lower than 300 Hz was obtained by designing the suspension width of 2.3 µm or thinner, which was a feasible value for our silicon micromachining process.

From these two design considerations, we used a suspension width of 2.5 µm for on the computer-aided design layout, expecting a slight reduction of width through the photolithography steps and the dry etching of silicon.

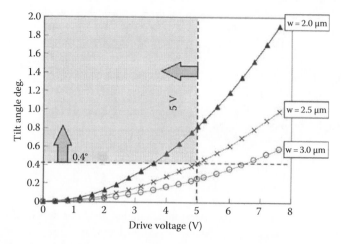

FIGURE 18.11 Static tilt angle as a function of suspension width (simulation).

TABLE 18.2
Design Parameters for MEMS Optical Scanner

Parameter	Symbol	Dimension
Mirror and suspension thickness	h_s	30 µm
Chip size	—	2 mm × 3 mm
Mirror size (full width and length)	$w_m \times l_m$	1.5 mm × 1.5 mm
Suspension width	w_s	2.3 µm
Suspension length	l_s	250 µm
Radius of suspension support	r_s	50 µm
Comb width	w_c	8 µm
Comb length	l_c	150 µm
Comb gap	g	5 µm
Number of inner comb teeth	$2N$	132 pairs
Comb root position	$p = w_m/2$	750 µm

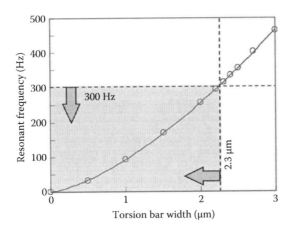

FIGURE 18.12   Resonant frequency as a function of suspension width (simulation).

18.4 MEMS OPTICAL SCANNER

A MEMS optical scanner for the endoscope has been developed by using the SOI bulk micromachining as shown in Figure 18.13. An eclipse shape was used for the chip to fit in a glass tube at the 45° inclination. The total chip dimensions were 3 mm × 2 mm in the longer and shorter axes, respectively, and 0.6 mm in thickness. A flat mirror of 1.5 mm × 1.5 mm in effective area and 30 µm in thickness was supported by a pair of torsion bars, around which the mirror rotates in the out-of-plane direction to control the direction of the reflected light. A through hole was made in a part of the handle wafer behind the mirror plate to let it rotates with minimum damping effect due to the air viscosity. Vertical comb drive electrodes were arranged on the edges (root location at 750 µm from the rotation axis) of the mirror to induce electrostatic torque. The torsion bar was rectangular in cross-section of only 2.3 µm in width, 30 µm in height, and 250 µm in length. The comb electrodes were 8 µm wide, 150 µm with, and 5 µm gap; the scanner dimensions were designed by considering the resonant frequency as well as the mechanical scan angle under a given voltage. Figure 18.14 shows the detailed close-up view of the vertical comb-drive electrodes. An initial tilt offset was built in the mirror in order to smoothly start the oscillation by the electrostatic operation.

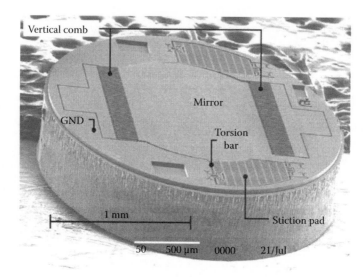

FIGURE 18.13   SEM image of electrostatic MEMS optical scanner.

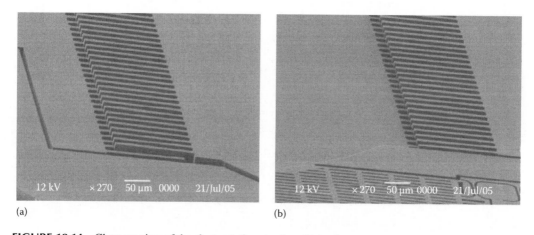

(a)   (b)

FIGURE 18.14   Close-up view of the electrostatic actuation electrodes.

18.5   FABRICATION PROCESS

The fabrication process of the MEMS optical scanner has been originally developed for another application of fiber-optic variable optical attenuator scanner reported elsewhere [20]. The process uses only two photolithography steps on a bonded SOI wafer (30-μm-thick SOI, 2-μm-thick BOX, and 500-μm-thick handle wafer) and has good compatibility with a MEMS foundry process.

Figure 18.15 shows the simplified steps of the fabrication process. After stripping the native oxide from the top and bottom surfaces in Step 1, an aluminum layer of 100 nm in thickness was deposited by the vacuum evaporation and was protected under a passivation photoresist. On the front surface, we put a 1.8-μm-thick positive-type photoresist and photolithographically form the mask pattern of the optical scanner. Deep reactive-ion etching (DRIE) was used to transfer the mask pattern into the 30-μm-thick SOI layer; the etching process was automatically stopped at the interface to the BOX. In Step 2, the DRIE-processed SOI patterns were protected under a passivation photoresist, and the back side aluminum was photolithographically shaped into the through hole; the double-side

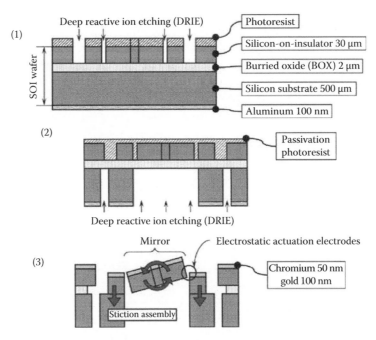

FIGURE 18.15 Silicon micromachining process for MEMS optical scanner.

mask-alignment was used to bring the reverse side patterns to match with the front side. The back side of the silicon substrate was etched by DRIE to make a through hole behind the mirror plate; the hole became a room for the mirror plate to tilt in the out-of-plane direction. At the same time, the MEMS chip was separated from the substrate frame by forming trenches surrounding the chip. In Step 3, the passivation photoresist on the front side was removed in oxygen plasma ashing, and the BOX layer was selectively removed in a 47% hydrofluoric acid; the etching time was controlled to minimize the under-cut etching (typically 10 min at room temperature) such that the mirror plate was released but the anchors and electrode pads remain fixed on the substrate. Finally, the mirror surface was finished with a 50-nm-thick chromium and a 100-nm-thick gold by the vacuum evaporation to make a highly reflective mirror for infrared light.

Surface stiction is a problem commonly seen in most MEMS processes that use the sacrificial release in wet etching; a released structure is unexpectedly brought into contact with the neighborhood surfaces by the surface tension force of liquid that is used at the final rinsing step. Various processing and designing techniques have been developed to avoid the surface stiction [21]. In our process, however, we intentionally used the surface stiction force to produce the initial height offset of the vertical comb-drive mechanism, by which the scanner oscillation could be swiftly triggered by the electrostatic force [22].

Figure 18.16 illustrates the steps for the surface stiction self-assembly. A meshed pattern (called "stiction pad") attached to the torsion bar near the anchoring place is fully released by the under-cut etching of the BOX; after the sacrificial etching, the wafer is picked up from the rinsing water and left in the ambient air for natural dry. As the water evaporates from the surface, the stiction pad is brought downward onto the substrate surface due to the surface tension force and permanently bonded. The vertical motion of the stiction pad is transferred to the rotation of the torsion bars to yield the initial offset of the mirror plate. The comb electrodes on the mirror edges thereby have a vertical offset, which generates electrostatic torque in the out-of-plane direction.

Close-up scanning electron microscope images of the stiction pad and the torsion bars are shown in Figure 18.17. The mesh opening was 10-μm wide and 50-μm long, and the remaining frame area was 50% of the total pad outline. The tiny hinges connecting the stiction pad and the torsion bar

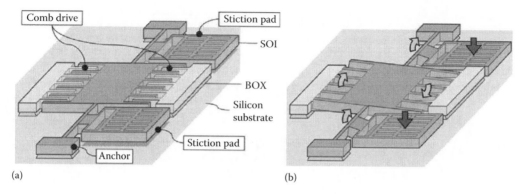

FIGURE 18.16   Self-assembly technique for vertical comb electrodes using surface stiction force.

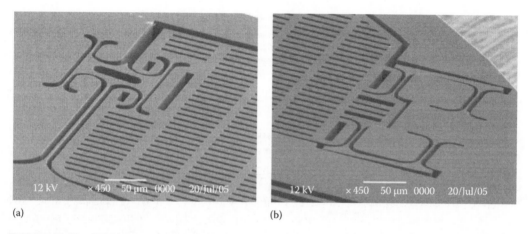

FIGURE 18.17   SEM images of vertical comb electrodes at rest position with initial offset.

were 5 area wide and 100-μm long; the structure height was equal to the SOI thickness, 30 μm. The dimensions of the stiction pad were found after trial-and-error steps, and the process yield for successful self-assembly is more than 90%.

18.6   ELECTROMECHANICAL OPERATION

The scanner was electrostatically operated by applying voltages to the drive electrodes on the both sides of the mirror plate; the mirror plate and the substrate were electrically grounded. Figure 18.18 shows the mirror angle as a function of applied voltage determined by the laser Doppler vibrometer. As theoretically designed, a static displacement greater than 0.4° was experimentally obtained at a drive voltage of 5 Vdc. The mirror's motion was directed to lower the offset angle due to the electrostatic attractive torque, and hence the maximum tilt angle was limited by the initial offset angle. Nonetheless, we did not see the saturation because the in-plane instability of the mirror occurred before reaching such range. Figure 18.19 shows the frequency responses (phase and gain) of the scanner under ac voltage operation. The out-of-plane motion around the torsion bar was the fundamental resonant mode found at 294 Hz. Thanks to the leverage effect through the Q-factor; optical scan angle of 8° was obtained at 5 V operation.

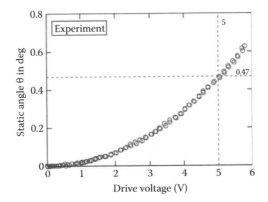

FIGURE 18.18   Static tilt angle of mirror as a function of drive voltage (experiment).

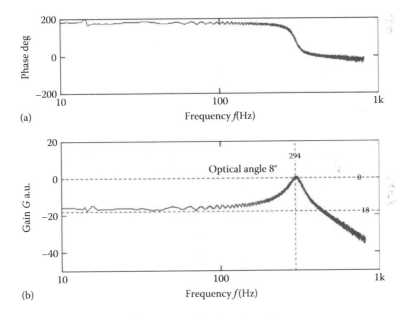

FIGURE 18.19   Frequency response of scanner motion (experiment).

18.7   ENDOSCOPE ASSEMBLY AND OCT MEASUREMENT

To construct an endoscope probe, we used a 45°-tilted submount, as shown in Figure 18.20, to set the mirror chip. A glass substrate with gold patterns was used to place the submount, and the electrical interconnection to the chip was made by wire-bonding of gold threads. Under a guided laser beam of visible wavelength, we placed the beam splitter and the photovoltaic cell and then adjust the position and attitudes such that the powering light could be coupled into. In the same manner, a focal lens was placed between the beam splitter and the MEMS scanner to make straight inline alignment. Focal position was brought to the outer surface of the housing tube by adjusting the mutual distance between the Gradient-index (GRIN) lens fiber and the submount module. Finished endoscope assembly is shown in Figure 18.21. The total dimensions were 6 mm in outer diameter and 20 mm in length. Diameter could be further reduced to 4 mm by using a smaller focus lens. We have recently developed a more convenient jig for optical alignment by using a pair of metallic guiding pins [23].

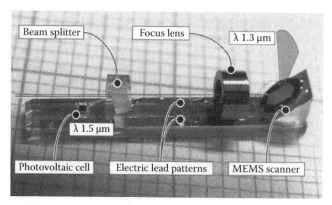

FIGURE 18.20   Optical assembly of endoscope probe.

FIGURE 18.21   Endoscope probe assembled in a 5 mm glass tube.

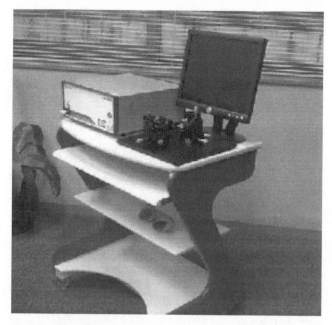

FIGURE 18.22   OCT measurement system.

We used the developed endoscope head to visualize the cross section of human finger print. The total measurement system is shown in Figure 18.22. A skin surface was brought into contact with the endoscope housing shell and optically scanned over the finger print profile as shown in Figure 18.23. The penetration depth of optical probe was 2.3 mm, and the lateral scan range by the MEMS scanner was 1.6 mm containing two periods of finger print profile. The skin surface was

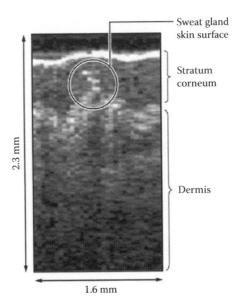

FIGURE 18.23 OCT image obtained by the all-optical endoscope.

seen in a thin bright band at the top in the figure, and the boundary between the dermis and the stratum corneum could also be seen in a burred region. Above the boundary, a spiral object could be spotted, which was thought to be a sweat gland. From the pixel size of the screen, the lateral and vertical resolutions were 40 and 8 μm, respectively.

18.8 INTEGRATED OPTOMECHANISM

As an extension of the optically actuated device, we also have developed a fully integrated version by using a series of photodiodes and an electrostatically driven actuator as schematically illustrated in Figure 18.24. P-I-N layers of silicon were epitaxially grown on the 30-μm-thick SOI layer and etched by DRIE to make the isolation trenches to separate the cells. After making the surface insulation of silicon oxide, the N-type silicon was partially exposed to open a contact window. Electrical series connection was made by cascading the P- and N-contact with metallic interconnection. A cantilever made of metallic layers (40-nm-thick chromium, 50-nm-thick gold, and 120-nm-thick

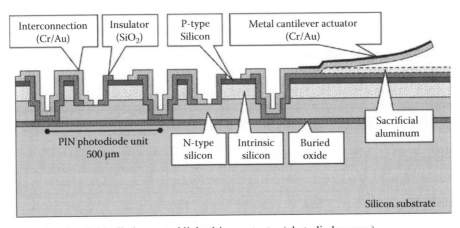

FIGURE 18.24 Monolithically integrated light-driven actuator (photodiodes array).

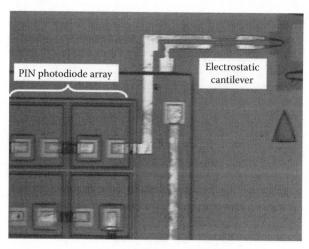

FIGURE 18.25   Microscope image of photodiode array and metal cantilever actuator.

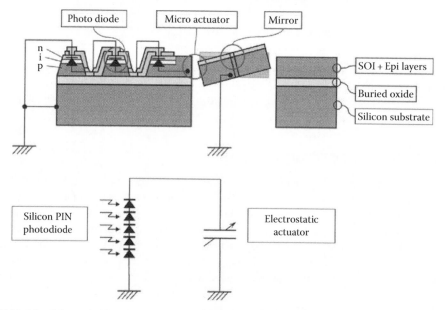

FIGURE 18.26   Schematic illustration of monolithically integrated light-driven MEMS optical scanner.

chromium) was also integrated at the output port of the photovoltaic cell. Figure 18.25 shows a microphotograph of the developed chip. The photovoltaic cell size was 400 μm × 400 μm, and total 4 × 4 elements were arranged in an area of 1.5 mm × 1.5 mm. Maximum output voltage of 4.8 V was generated under a 1 mW He-Ne light of 632 nm, which was supplied to the actuator to show electrostatic motion at the tip of a 300 μm long cantilever. By using the SOI part as a mechanical material, a bulk-micromachined optical scanner with integrated photovoltaic cells can be designed as illustrated in Figure 18.26, which is under development.

18.9   SUMMARY

As an example of optically driven microactuator, we reported an all-optical type fiber-optic endoscope with a SOI bulk micromachined electrostatic micromirror scanner. Apart from the conventional optical scanners that used electrical wires for electrostatic, electromagnetic, or electrothermal actuation, we used the wavelength division multiplex system to transfer the actuation energy as well as the endoscope probe signal by the superposition of lights of different wavelengths. Due to the simplicity of implementation, we employed the optical components such as collimator lens and beam splitter used in the 1.3–1.5 μm infrared wavelength for fiber telecommunications. An infrared light of 10 mW at 1.5 μm was intensity modulated at 294 Hz and was converted into the ac voltage by a photovoltaic cell to drive the electrostatic optical scanner. Probe light at 1.3 μm was superposed in the identical optical fiber to scan the living cell tissue appressed onto the window of the endoscope hull, and the cross-sectional image of the tissue was successfully retrieved by means of the OCT measurement. Thanks to the optical actuation scheme, the developed fiber endoscope is thought to be safe to human body without a risk of internal electrocution or electromagnetic coupling with other medical instruments. Miniaturized version of the endoscope less than 2 mm in outer diameter is under development by removing the dead volume

REFERENCES

1. T. Uji, M. Sugiura, M. Fukami, Camera for taking photographs of inner wall of cavity of human or animal bodies, U.S. Patent 2,641,977, 1953.
2. B. Hirschowitz, Flexile light transmitting tube, U.S. Patent 3,010,357, 1961.
3. G. J. Tearney et al., Scanning single-mode fiber optics catheter-endoscope for optical coherence tomography, *Opt. Lett.*, 21, 543–545, 1996.
4. G. J. Tearney et al., In vivo endoscopic optical biopsy with optical coherence tomography, *Science*, 276, 2037, 1997.
5. T. Xie, H. Xie, G. Fedder, and Y. Pan, Endoscopic optical coherence tomography with a modified microelectromechanical systems mirror for detection of bladder cancers, *Appl. Opt.*, 42(31), 6422–6426, 2003.
6. J. T. W. Yeow, V. X. D. Yang, A. Chahwan, M. L. Gordon, B. Qi, I. A. Vitkin, B. C. Wilson, and A. A. Goldenberg, Micromachined 2-D scanner for 3-D optical coherence tomography, *Sens. Actuators A*, 117, 331–340, 2005.
7. H. Miyajima, MEMS optical scanners for microscope, *IEEE J. Sel. Top. Quantum Electron.*, 10(3), 514–527, 2004.
8. B. A. Flusberg, E. D. Cocker, W. Piyawattanametha, J. C. Jung, E. L. M. Cheung, and M. J. Schnitzer, Fiber-optic fluorescence imaging, *Nat. Methods*, 2(12), 941–950, 2005.
9. D. T. McCormick, W. Jung, Z. Cheng, and N. C. Tien, 3-D MEMS based minimally invasive optical coherence tomography, in *Proceedings of the 13th International Conference on Solid-State Sensors, Actuators and Microsystems (Transducers 05)*, Seoul, Korea, June 5–9, 2005, pp. 1644–1648.
10. H. Ra, Y. Taguchi, D. Lee, W. Piyawattanametha, and O. Solgaard, Two-dimensional MEMS scanner for dual-axes confocal in vivo microscopy, in *Proceedings of the IEEE International Conference on Micro Electro Mechanical Systems (MEMS 2006)*, Istanbul, Turkey, January 22–26, 2006, pp. 862–865.
11. C. F. Dalzidel, Effects of electric shock on man, *IRE Trans. Med. Electron*, 5, 44–62, 1956.
12. C. Chong, K. Isamoto, and H. Toshiyoshi, Optically modulated MEMS scanning endoscope, *Photon. Technol. Lett.*, 18(1), 133–135, 2006.
13. M. Nakada, C. Chong, A. Morosawa, K. Isamoto, T. Suzuki, H. Fujita, and H. Toshiyoshi, Optical coherence tomography by all-optical MEMS fiber endoscope, *IEICE Electron. Express*, 7(6), 428–433, 2010.
14. K. Isamoto, K. Totsuka, T. Sakai, T. Suzuki, A. Morosawa, C. Chong, and H. Toshiyoshi, High speed MEMS scanner-based swept source laser for SS-OCT, *IEEJ Sensors and Micromachine Society, The 27th SENSOR SYMPOSIUM on Sensors, Micromachines and Application Systems*, October 14–15, 2010, Matsue, Japan (presented in Japanese), paper id C-3-5, p. 160 in Proceeding.
15. H. Toshiyoshi, Electrostatic actuation, in Y. B. Gianchandani, O. Tabata, and H. Zappe, eds., *Comprehensive Microsystems*, Vol. 2, pp. 1–38, Elsevier, Amsterdam, the Netherlands, 2008.
16. T. Hirata, I. Mitama, M. Abe, K. Makita, K. Shibata, K. Hane, and M. Sasaki, Development of MEMS-based optical surge suppressor, in *Technical Digest of Optical Fiber Communication Conference (OFC/NFOEC)*, Anaheim, CA, Vol. 4, p. 3, 2005.
17. W. G. Zijlstra, A. Buursma, and O. W. van Assendelft, *Visible and Near Infrared Absorption Spectra of Human and Animal Hemoglobin*, VSP Publishing, Utrecht, the Netherlands, 2000.

18. S. P. Timoshenko and J. N. Goodier, *Theory of Elasticity*, 3rd edn., Mc-Graw-Hill, New York, 1970.
19. H. Toshiyoshi, W. Piyawattanametha, C. Ta Chan, and M. C. Wu, Linearization of electrostatically actuated surface micromachined 2D optical scanner, *IEEE/ASME J. Microelectromech. Syst.*, 10, 205–214, 2001.
20. K. Isamoto, A. Morosawa, M. Tei, H. Fujita, and H. Toshiyoshi, A 5-volt operated MEMS variable optical attenuator by SOI bulk micromachining, *IEEE J. Sel. Top. Quantum Electron. (JSTQE)*, 10(3), 570–578, 2004.
21. R. Maboudian and R. T. Howe Critical review: Adhesion in surface micromechanical structures, *J. Vac. Sci. Technol.*, B 15(1), 1–20, 1997.
22. K. Isamoto, T. Makino, A. Morosawa, C. Chong, H. Fujita, and H. Toshiyoshi, Self-assembly technique for MEMS vertical comb electrostatic actuators, *IEICE Electron. Express*, 2(9), 311–315, 2005.
23. H. Toshiyoshi, Lateral spread of MEMS WDM technologies, *SPIE Photonic West (MOEMS-MEMS)*, January 24, 2011, Program p. 22 (SPIE#7930).

19 Light Activated and Powered Shape Memory Alloy

Hideki Okamura

CONTENTS

19.1 INTRODUCTION

The scheme of manipulating objects by light realizes a remote, noncontact actuation of objects. Well-known examples of this scheme are optical tweezers (Ashkin 1970) and laser cooling (Chu et al. 1986). In the case of microactuators, the components are usually in contact with some fixture so this feature might not seem particularly beneficial; however, in reality, this has a profound meaning. In light-driven actuators (LDAs), the control and the energy supply are performed simultaneously by light, which means all the wires for control and power supply can be eliminated. This is very important because one of the restraints in using conventional actuators for microscopic scale arises from the need for wiring. In particular, when multiple actuators are integrated into small spaces, increased number of interconnections becomes a serious problem. In contrast to conductive wires, light passes through each other without interaction; therefore, overlap of beams for interconnection does not cause any problem. Also, light can be focused to a micrometer size and can access those small structures quite easily. For these reasons, LDAs are considered to have certain potential advantages as microscale actuators. Also, by using LDAs there are added benefits of robustness toward electric and magnetic noises, and simple and lightweight structure. This allows these actuators to be used in extreme environments.

Various physical phenomena have been applied to LDAs so far, including radiation pressure (Ashkin 1970), photothermal effect (Inaba et al. 1995; Hahtela and Tittonen 2005; Okamura et al. 2009, 2011; Okamura 2011), photostrictive effect (Poosanaas et al. 2000), photochemical reaction (Yu et al. 2003; Kobatake et al. 2007), and photo-induced phase transition (Ikehara et al. 2004).

The material bends, shrinks, or stretches, and, subsequently, these changes are converted into a usable form of mechanical motion.

In general, forces produced by LDAs are small (typically pN or nN in the case of optical tweezers), and consequently, most of the LDAs reported so far are intended for small-scale applications; however, it does not necessarily mean they can be applied to microactuators readily. Energy conversion efficiency of an LDA, defined as the ratio of the mechanical output work to the input light energy, is typically in the order of $10^{-5}\%–10^{-10}\%$ (Okamura 2006). For instance, the efficiency of the polymer polyvinylidene fluoride (PVDF) film LDA was $8.3 \times 10^{-5}\%$ (Sarkisov et al. 2006) and that of a cantilever LDA made from polydiacetylene crystal was evaluated to be $2 \times 10^{-6}\%$ (Ikehara et al. 2004). The efficiency of photothermal optical fiber core LDA (Inaba et al. 1995) can be roughly estimated to be $10^{-6}\%$ (Yamaguchi et al. 2009). Currently, single-crystal photoisomeric material, which reacts in millisecond or microsecond timescale, has been actively studied (Al-Kaysi et al. 2006; Kobatake et al. 2007); however, at present, this material requires intense light for reaction and, therefore, seems to need further development to exceed the currently available technology in terms of efficiency.

With energy conversion efficiency far smaller than unity, a large amount of light energy has to be injected into a small region, which can lead to damage by photo irradiation or resultant increased temperature. As a matter of fact, in optical tweezers the temperature increase is a problem in the manipulation of living cells. It can be anticipated that heat dissipation will become a bottleneck for the integration of microactuators in small size.

Currently, it has been reported that the energy conversion efficiency of a shape memory alloy (SMA)-based LDA can reach 1% (Okamura et al. 2009). An energy conversion efficiency of 1% means that, in essence, the required light energy can be reduced by a factor or 10^5 or more to produce the same amount of work when compared with other type of LDAs. This leads to a significantly smaller amount of heat production, and thus allows a larger scale of integration into small dimensions possible. There are few actuating mechanisms, which produce more useful work per unit volume than SMA (Otsuka and Wayman 1999), and the above-mentioned work evidently demonstrates the possibility of SMA-based LDAs. The details of the application of SMAs for LDAs are discussed in the following.

19.2   PROPERTIES OF SHAPE MEMORY ALLOYS

Shape memory alloys are known to have peculiar characteristics of remembering their original shapes. Figure 19.1 illustrates the memory effect of SMA. Below the transition temperature, the material is in martensitic state and is flexible. When the stress is small it reacts elastically. The yield stress is relatively small (~100 MPa for NiTi), and above this point, SMA undergoes plastic

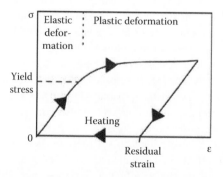

FIGURE 19.1   Explanation of the shape memory effect. σ and ε denote stress and strain, respectively. Below the transition temperature, some amount of strain remains after complete unloading. This residual strain may be recovered by heating the sample.

deformation. The deformation remains after the stress is removed, which is recovered by heating the alloy above the transition temperature, thus it returns to its original shape.

The alloys that have been confirmed to have shape memory effect (SME) include AgCd, AuCd, CuZn, CuZnX (X = Si, Sn, Al, Ga), CuAlNi, CuSn, CuAuZn, NiAl, TiNi, TiNiCu, TiPdNi, InTl, InCd, MnCd, MnCu, FePt, FeNiCoTi, FeNiC, FeNiNb, FeMnSi, FeCrNiMnSi, FeMnSiC, FePd, and FePt (Suzuki 1988; Otsuka and Wayman 1999). Many of them exhibit perfect SME only in single crystals and are difficult to be used in industry; consequently, very few of them, such as NiTi (nickel–titanium), CuZnAl (copper–zinc–aluminum), and CuAlNi (copper–aluminum–nickel), have been put into practical use. In particular, nickel–titanium alloys have been found to be the most useful of all SMAs, and we mainly focus on this material in this chapter. NiTi SMA can recover strains of up to 8% and when physically restrained, stress up to 700 MPa can be generated (Otsuka and Wayman 1999). A NiTi SMA wire of 1 mm diameter is capable of lifting up a human (Suzuki 1988). When used as an actuator, NiTi SMA can induce a displacement 10 times larger than that of the bimetal for the same temperature difference. The cycle strength is also superior: Properly prepared NiTi SMAs show no degradation in SME after one million cycles. They have superb mechanical properties such as high strength, high corrosion, and abrasion resistance. Also, NiTi is one of the rare SMA materials that exhibit SME in polycrystal and is easier to manufacture when compared with materials that exhibit SME only in single crystals.

The transformation temperature of NiTi SMAs can be adjusted by changing the composition: Change in the composition of 0.1% shifts the transformation temperature by 10 K (Suzuki 1988). The range of transition temperature is roughly 50°C–100°C (Suzuki 1988), with large variances depending on the literature (Otsuka and Wayman 1999). Most commercially available NiTi SMAs have transition temperatures around 60°C–70°C. The transition has a hysteresis and their separation is typically 10–30 K. It is also possible to obtain different temperature ranges by using variants of NiTi such as NiTiPd or NiTiZr. For applications such as cellular phone antennas and eyeglass frames, the transformation temperature is made below the ambient temperature. In this case, the alloy is in austenite throughout without a load and transforms into martensite as stress is applied, exhibiting pseudoelasticity.

The shape recovery during heating can occur against an opposing force, and it is possible to produce work. Based on this principle, various types of SMA-based actuators have been built and used in many fields (Raynaerts and Van Brussel 1991), including space applications such as Mars Pathfinder Sojourner Rover and Hubble Space Telescope. One of the most successful applications of SMA actuators is that these actuators react to changes in the ambient temperature and automatically actuate the components in a desired manner. Since they are actuators that own the function of sensors as well, they are a type of smart materials. Some examples of such actuators are ventilation doors for desiccators, automatic shield of automobile fog lamps, and moving air flap for directing airflow in air conditioners. Another type of SMA-based actuators works in the same manner as the conventional actuators, and is actively controlled by applying heat. Most commonly used method to apply heat is Joule heating by passing an electrical current directly to the alloy. Other methods include passing warm water or hot air, radio-frequency heating, and infrared or visible light heating. One of the most important applications of SMA actuators are robotic actuators (Raynaerts and Van Brussel 1991), in particular, robotic hands (Cho et al. 2007; Kheirikhah et al. 2010). As for microactuators (Galhotra et al. 2000; Kuribayashi 2000), Ikuta demonstrated a microgripper using two SMA springs (Ikuta 1990), and several other reports have been made by other authors (Cho et al. 2007). Also, a monolithic microgripper and translation stage has been reported (Bellouard et al. 1998). This reduces the number of components and will help reduce the size of the device.

As another application of SMAs, heat engines have been extensively studied since 1970s as an environment-friendly alternative to extract mechanical energy from low-grade energies such as warm wastewater. Various types of SMA heat engines have been proposed (Banks 1986), and they can usually be classified into three main types—offset crank type, turbine type, or field type— although there are some proposals that cannot be categorized in any of the aforementioned groups.

In offset crank type, the reciprocal motion of SMAs is converted into a rotary motion, just like in automobile engines. Typically, they have several SMA components and each of them undergoes cycles with different phases to each other. The heated SMA component deforms and rotates the engine by a small amount, which exposes another SMA component to the high-temperature heat bath and each component takes turns. For the turbine type, the restoring force of SMAs is used continuously to deform the low temperature portion of SMAs. The belt type engines are typical examples of this category, and, generally speaking, they are faster than crank type engines. The belt type SMA heat engines can produce output power of 0.2–2.0 W with frequency of rotation up to 400 rpm (Tanaka and Saito 1993). For the field type engines, the restoring force of SMA is used to lift up a mass, and subsequently, its potential energy is converted into rotational motion.

The energy efficiency of transforming thermal energy to mechanical energy of SMAs has been extensively studied for the cases of material itself and for the system of actual actuator or engine. Theoretically, it has been predicted that the maximum efficiency can be up to 20% in an ideal situation (Wayman and Tong 1975). Mercier and Melton measured the chemical and mechanical parameters on an actual NiTi sample and calculated a theoretical efficiency of 13% (Mercier and Melton 1981). In their experiment, using the same sample they obtained a conversion efficiency of 1%–7%. The discrepancy between the calculation and the measurement was ascribed to a nonideal martensitic transformation. Tanaka and Saito estimated an energy conversion efficiency of up to 2.5% using the material parameters obtained from a sample (Tanaka and Saito 1993). The overall efficiency of a SMA heat engine system depends largely on its designs, and also other effects such as friction affects the efficiency in actual situations. Zhu et al. calculated a conversion efficiency of 2%–3% for the twin crank heat engine (Zhu et al. 2001a). Simulation models for SMA heat engines were also developed (Govindjee and Hall 2000; Zhu et al. 2001b).

The energy conversion efficiency of SMA actuators is subjected to the restrictions imposed by thermodynamics. They can be regarded as heat engines, because their motion is cyclic and they return to their original state after one cycle without leaving any change in the system other than heat. Carnot's theory dictates that the maximum efficiency is given by $e = 1 - T_C/T_H$, where T_C and T_H are the temperatures of the low-temperature and high-temperature heat baths, respectively. For our case, T_C and T_H are usually the ambient temperature and the temperature that resulted from heat laser irradiations, respectively. The minimum temperature difference required for the NiTi SMA actuator to undergo one full cycle is typically 40 K for commercially available SMA materials, and if we take this temperature difference it gives a maximum conversion efficiency of about 12%.

19.3 SMA AS LDA: THEORETICAL TREATMENT

19.3.1 THERMAL PROCESSES

The scheme of activating SMAs by photoinduced heat has been used in real applications for quite a long time. One example is the arms of the solar panel for the Hubble Space Telescope, which was launched in 1990. The arms of the solar panel are made of SMAs. Once in space, the sun rapidly warmed up the telescope, heating the arms and hence causing the solar panels to fold out automatically. Despite this, application of SMA to LDA has started to be investigated relatively recently. The fundamental difference between these applications is that a cyclic operation is required for an LDA. To give some examples, Sutapun et al. built a valve that opens upon irradiation of light by using SMA film of 3 μm thickness (Sutapun et al. 1998). SMA film of 30 μm thickness was used to investigate the possibility of inducing two-dimensional mechanical displacement using laser beam (Furuya and Tokura 2006). Bellouard et al. developed a mobile micro-robot that is driven by light (van den Broek et al. 2007). Okamura et al. developed an LDA using an SMA wire (Okamura et al. 2009) and an SMA-based actuator that provides a rotational motion (light-driven motor) (Okamura2011).

19.3.1.1 Light Reception

In the scheme of applying SMA to LDA by the way of photothermal effect, several processes are involved. The first process is the photo-irradiation. The portion of SMA needs to be exposed to receive light. The important point associated with laser irradiation is the shape of the irradiation area: Laser beam typically has a Gaussian beam profile, and if the irradiation area does not match the beam profile, the unused portion of light is wasted and degrades the overall efficiency. This is particularly problematic when an SMA wire is used as the active element: The stroke of a wire-based actuator will be directly proportional to the length of the wire, so a longer wire is preferable; however, it is difficult to illuminate a long wire evenly and focusing light onto a wire along its length is virtually impossible. One solution to this problem is to line up the wires to form a desired shape. In the SMA wire actuator (Okamura et al. 2009), the SMA wire was wound around two shafts, forming a rectangular plane of SMA wire. Light was then irradiated at an angle so that the projected area becomes a square, which is much more compatible with the shape of the laser beam.

An alternative idea for increasing the amount of light reception is to receive the light energy elsewhere and pass the heat to the actuator. The light receiver should have as little heat capacity as possible and should be made of a material with large thermal conductivity.

19.3.1.2 Light Absorption

The second process is the light absorption. Only the portion of light energy that was absorbed turns into heat and contributes to output work; therefore, absorption coefficient directly links to the energy conversion efficiency. The NiTi SMAs that we tested in our laboratory are from several different companies and some were shiny and some had much darker shines, resulting in largely different absorption coefficients. This is due to the presence of oxidization layer on the surface, whose composition is largely dependent on the manufacture process.

The bare NiTi alloy has a reflective shine. The optical property of NiTi alloy has been extensively studied both theoretically and experimentally for the purpose of studying the electronic structures of the material. For this purpose, the oxidization layer on the surface of NiTi is removed by polishing, spattering, or annealing in ultravacuum. Rhee et al. measured the optical conductivity from 1.5 to 5.4 eV (230–827 nm) on a NiTi thin film that has been annealed in ultravacuum (Rhee et al. 1996), and obtained results that are consistent with theoretical calculations (Wang et al. 1998; Kulkova et al. 2001). Sutapun et al. measured the reflectivity spectrum in the range of 550–850 nm on a NiTi film of about 2–3 μm thickness annealed under vacuum (Sutapun et al. 1998). The reflectivity of NiTi surface changes depending on its state. It looks opaque and cloudy at room temperature, and becomes shiny with a metallic luster as it is heated. The reflection of the austenitic phase was higher than that of the martensitic phase by a factor ranging from 15% at 750–800 nm to 45% at 500–600 nm. Note that it does not necessarily mean the absorption has changed by the same amount. The reflectivity is determined by surface geometry (surface relief), which causes light scattering, and surface electrical properties, which is connected to light absorption (Danilov et al. 2004).

The oxidization layer on the surface of NiTi alloy consists mostly of TiO_2-based oxides with a little amount of Ni (0%–5%) in them (Shabalovskaya 1995; Wever et al. 1998). The composition varies with temperature and oxygen pressure (Chan et al. 1990). For atmospheric oxidation at room temperature, NiO was formed in the first minute of air exposure. Under low oxygen pressure (10^{-4} Torr), TiO, Ti_2O_3, and TiO_2 are developed. TiO_2 single crystal is transparent for the wavelength range of 0.4–5 μm. Ti_2O_3, on the other hand is black, and TiO has a bronze color in powder form. TiO is naturally formed in the atmosphere on the surface of Ti metal as a passive layer and gives it a corrosion resistance, as well as beautiful interference color. NiO has greenish color in powder form. The reflectivity for pressed powder samples of NiO was as low as 0.1 around a wavelength of 700 nm and increases as the wavelength becomes smaller, and reaches 0.25 at around 530 nm (Musella 2001). It has also been found that the crystal structure of the surface oxidized layer mostly follows that of the underlying substance. It changes as the structure underneath it changes; however, sometimes the process is irreversible (Danilov et al. 2004).

With a metallic surface of polished NiTi, a large part of the incident light is reflected. Oxidization layer effectively increases the absorption coefficient of NiTi SMA and improves the energy conversion efficiency when applied for an LDA. The same goal might be achieved by coating the surface with some black substance; however, there is a possibility of inducing unwanted stress as well as additional heat capacity. Oxidization layer does not cause such problems, so it will be a preferred way of increasing the light absorption of NiTi alloy. By controlling the oxygen pressure and temperature, the surface oxidization layer can be developed in a controlled manner.

19.3.1.3 Cooling

The final process is cooling. LDA needs to return to its original state after one cycle of operation in order to prepare for the next cycle; therefore, when laser irradiation is ceased the heat has to be removed promptly. In general, cooling process takes more time than the photothermal process, and it determines the overall response speed. The response speed is important also with respect to the output power: The output power is the work produced in a unit time, so if a larger number of cycles can be achieved within a unit time, the output power becomes larger accordingly.

For LDA, heat is usually taken away from the surface either by radiation or conduction through the surrounding substance such as air. In either case, larger surface area and smaller heat capacity help rapid temperature decrease. One criterion would be the ratio of the surface area, S, to the volume, V. If we consider an example of an SMA wire of diameter, d, it will be given by

$$\frac{S}{V} = \frac{\pi d}{\left(\pi d^2 / 4\right)} = \frac{4}{d} \tag{19.1}$$

The ratio becomes larger for smaller d, which means a faster operation can be expected with smaller d.

One very important note in using SMA for LDA is the control of the maximum temperature. If a laser beam irradiates a specific portion of SMA for a prolonged time, it can drive that portion of SMA to exceed the maximum operating temperature. If this happens, the SMA can forget the trained form, and it will result in a loss of functionality of LDA. Conversely, this property might be used to actively train the SMA to attain a desired function.

19.3.2 Mechanism of Actuation

In this chapter, we introduce formulations of the basic mechanism of SMA-based LDA. We consider the case where an SMA material is deformed in martensitic state due to externally applied stress and then heated by means of light irradiation, transforming to austenitic state. The deformed portion restores its original shape and produces work. In Figure 19.2, a stress–strain curve of SMA is shown. The material is deformed in martensite under external stress σ_1 and a strain ε is induced (state B in

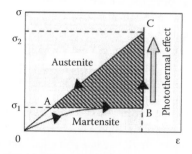

FIGURE 19.2 A stress–strain curve of SMA for one cycle of operation of SMA-LDA. The material is deformed in martensitic state and then transforms into austenitic state by the way of photothermal effect. The hatched area corresponds to the work produced in one cycle.

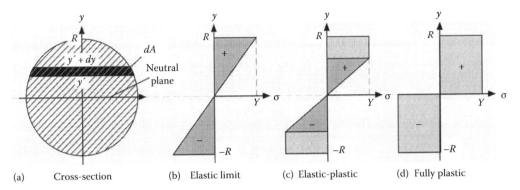

FIGURE 19.3 (a) The cross section of the SMA wire/rod and its stress distributions for (b) elastic limit, (c) elasto-plastic region, and (d) fully plastic state.

Figure 19.2). The stress σ_2 in austenitic state is larger than σ_1 for the same strain ε. Therefore, when the material is heated and transforms to austenitic state (state C in Figure 19.2), if the external stress is maintained at σ_1, the material can deform against the external stress until the stress relaxes at σ_1 (state A in Figure 19.2). After the heat is removed, the material transforms to martensite and relaxes to point B. Now the system is ready for the next operation and the process can be repeated. The net work produced in one cycle is the hatched area in Figure 19.2.

Note that the deformation in martensite can be either elastic or plastic, so there is no lower limit in the stress σ_1. In contrast, if the load is excessive and exceeds σ_2 the LDA cannot restore to its original shape and is unable to operate.

In the actual LDA, the material is formed in a particular form to achieve the desired function, and various forms of deformation are possible depending on the design. In the following, we will consider the case of bending a wire/rod. For any type of deformation a similar discussion can be made. The deformation by bending takes three regimes (Figure 19.3), which are discussed in the following.

19.3.2.1 Elastic Regime

When bending moment is small, all portions of the material react in elastic manner (elastic regime). In this scheme, the bending moment $M = EI/\rho$ holds, where E is Young's modulus, I is the area moment of inertia, and ρ is the radius of curvature. The elastic limit is the point where the stress at the skin of the material (skin stress), $\sigma(R) = M/z$, starts to exceed the yield strength Y of the material, where z is the section modulus. For the case of circular cross section, the area moment of inertia $I = (\pi/4)R^4 = \pi d^4/64$, and the section modulus $z = (\pi/4)R^3 = \pi d^3/32$, where $d = 2R$ is the diameter. For this case, the bending moment at elastic limit (yield bending moment) is given by $M_y = (\pi/4)R^3 Y = (\pi d^3/32)Y$.

19.3.2.2 Fully Plastic Regime

When stress is larger than yield strength, the material no longer reacts in an elastic manner and it undergoes a plastic deformation. If all portions of the material are in the plastic state, the bending moment of a material with a circular cross section is calculated as

$$M_p = 2\int_0^R YydA = 2Y\int_0^{\pi/2} R\sin\theta(2R^2\cos^2\theta)\,d\theta = \frac{4R^3Y}{3} = \frac{d^3Y}{6} \tag{19.2}$$

where
  A is the area
  y is the distance from the neutral plane

M_p is larger than the yield bending moment by a factor of $(4/3)/(\pi/4) = 1.698$

19.3.2.3 Elasto-Plastic Regime

The transitional regime between the elastic limit and the fully plastic regimes is the elasto-plastic regime. In this regime, part of the material is in elastic state and the remaining is in plastic state. If the dividing plane between the two regimes is located at distance y' from the neutral plane, then the stress distribution in the cross section is

$$\sigma(y) = \begin{cases} Y & \text{if } y > y' \\ \dfrac{y}{y'} Y & \text{otherwise} \end{cases} \tag{19.3}$$

Using this expression, the bending moment is obtained as

$$M_{ep}(y') = 2\frac{Y}{y'}\int_0^{y'} y^2\, dA + 2Y\int_{y'}^{R} y\, dA \tag{19.4}$$

where A is the area of integration. We introduce a parameter $s' = 2y'/d$ to denote the degree of elastic deformation. The parameter s' takes a value between 0 and 1, with $s' = 1$ meaning fully elastic, and $s' \to 0$ meaning fully plastic state. Similarly, we define $s = 2y/d$. Using the relation $dA = (d^2/2)\sqrt{1 - s^2}\, ds$, Equation 19.4 reduces to

$$M_{ep}(s') = \frac{d^3 Y}{2}\left[I_1(s') + \frac{1}{s'} I_2(s') \right] \tag{19.5}$$

where

$$I_1(s') \equiv \int_{s'}^{1} s\sqrt{1 - s^2}\, ds, \quad I_2(s') \equiv \int_{0}^{s'} s^2\sqrt{1 - s^2}\, ds \tag{19.6}$$

I_1 and I_2 represent the geometrical factors of the plastic and the elastic deformations, respectively.

The parameter s' represents the degree of elastic deformation. It decreases as the material is bent and approaches the fully plastic state. The strain distribution in a bent material is determined geometrically as, $\varepsilon(y) = y/\rho$. From Hooke's law, for the elastic portion of the material $\sigma(y) = E\varepsilon(y) = Ey/\rho$ holds. Therefore, we obtain the relation between the parameter s' and the radius of curvature ρ as

$$\rho = \frac{Ey'}{Y} = \frac{Eds'}{2Y} \tag{19.7}$$

For $s' = 1$, the material is at the elastic limit, and the corresponding radius of curvature is $\rho_c = Ed/2Y$, which we will call the critical radius of curvature in the following. A plot of critical radius of curvature is shown in Figure 19.4 for Young's modulus (martensite) of 28 GPa and yield strength (martensite) of 100 MPa (TiNi-Aerospace-inc 2001).

In Figure 19.5, the geometrical part of the bending moment for an SMA material with a circular cross section is plotted as a function of inverse radius of curvature normalized by ρ_c. The lower curve in Figure 19.5 represents the contribution from the plastic component. For $\rho_c/\rho < 1$, the material is in elastic regime and there is no plastic component. When the material is bent further and $\tilde{n}_c/\tilde{n}$ reaches unity, the plastic component starts to emerge, and the elastic component decreases monotonically thereafter. As ρ_c/ρ becomes larger, the plastic component increases and finally saturates.

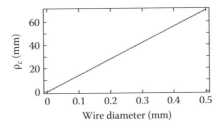

FIGURE 19.4 Critical radius of curvature as a function of wire diameter for the case of $E_m = 28$ GPa.

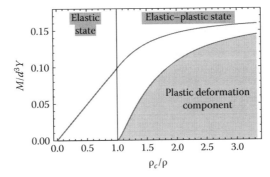

FIGURE 19.5 Plot of the geometrical part of bending moment of an SMA material with a circular cross section as a function of the inverse radius of curvature normalized by the critical radius of curvature ρ_c. The lower curve represents the plastic component of the bending moment.

The changeover of elastic and plastic components is important for designing an efficient SMA LDA engine, and it will be further discussed later.

19.3.3 THERMODYNAMICS

SMA-based LDAs are basically heat engines, because incident light is converted into heat upon absorption. Also, LDAs are typically designed to return to their initial states after each cycle without leaving any change in the system other than heat, thus the thermodynamics of heat engines can readily be applied. Thermodynamics dictates that the theoretical limit of the energy conversion efficiency of a heat engine is given by $\eta_{th} = 1 - T_C/T_H$, where T_C and T_H are the absolute temperatures of the cold and hot reservoirs, respectively, and in our case, they respectively correspond to the ambient temperature and the temperature of the actuating part that is heated due to light irradiation. The hysteresis in transition for NiTi is usually 10–30 K. For $T_C = 300$ K and $T_H = 310$ K, the maximum efficiency is $\eta_{th} = 3.2\%$. For $T_C = 300$ K and $T_H = 330$ K, the maximum efficiency is $\eta_{th} = 9\%$. These figures give some idea of maximum conversion efficiencies of SMA.

It is worth discussing the effect of thermal loss here. The thermal loss at the illuminated surface reduces the hot reservoir temperature, T_H, and directly degrades the energy efficiency in accordance with the above-mentioned formula. Thermal loss is caused by radiation and conduction through air and/or neighboring object. The rate of thermal conduction is proportional to the temperature difference, while the rate of radiation is proportional to T^4 (Stefan–Boltzmann law), where T is the absolute temperature of the object. For SMA-based LDAs, conduction usually plays a major role in heat dissipation. As is mentioned in the previous section, in order to obtain a fast response speed, heat has to be removed promptly after light is turned off, so there is a trade-off between the efficiency and the response speed. It has been proposed to use pulsed lasers to overcome this limitation and increase the efficiency (Okamura 2008). Further, a detailed discussion on the reconciliation of heat dissipation and photothermal effect has been made in terms of the thermal diffusion length (Okamura et al. 2011).

As another example, the temperature distribution in a semi-infinite object under illumination of laser beam of uniform intensity of radius a is given by (Mori 1984)

$$T_{z,t} = \frac{2P}{\pi a^2} \frac{\sqrt{\kappa t}}{K} \left(\text{ierfc} \frac{z}{2\sqrt{\kappa t}} - \text{ierfc} \frac{\sqrt{z^2 + a^2}}{2\sqrt{\kappa t}} \right) \tag{19.8}$$

$$\text{ierfc}(x) = \frac{1}{\sqrt{\pi}} e^{-x^2} - x \cdot \text{erfc}(x) \tag{19.9}$$

$$\text{erfc}(x) = 1 - \frac{2}{\sqrt{\pi}} \int_0^x e^{-y^2} dy \tag{19.10}$$

where
  P is the absorbed laser power
  κ is the thermal diffusion length
  K is the thermal conductivity
  t is the irradiation time
  a is the beam radius.

The thermal distribution at the surface is given by

$$T_{0,t} = \frac{2P}{\pi a^2} \frac{\sqrt{\kappa t}}{K} \left(\frac{1}{\sqrt{\pi}} - \text{ierfc} \frac{a}{2\sqrt{\kappa t}} \right) \tag{19.11}$$

The temperature monotonically increases with time, and then saturates. The saturation temperature can be obtained by taking a limit of $t \to \infty$, and is given by

$$T_{0,\infty} = \frac{P}{\pi K a} \tag{19.12}$$

Saturation of temperature is caused by heat dissipation. The high reservoir temperature, T_H, is limited and the maximum efficiency is limited accordingly.

19.4 RECIPROCAL MOTION

Light-driven actuators typically provide reciprocal motions, such as pull/push, pinch, or twist motions. In Figure 19.6, an SMA-based LDA developed in our laboratory is shown. It consists of a torsion spring and two arms, and SMA wires are wound around the arms near the spring. It operates in a tweezer-like motion and closes the tip upon irradiation of light. The response speed is roughly 0.1 s for closing and 0.5 s for opening. The SMA we used is an anisotropically organized NiTi-based SMA wire actuator (Toki Corporation BMF series). This wire remembers only its length: When heated from 338 to 348 K it shrinks over 4% of its original length. The torsion spring gives a pretension to the SMA wire. The stress is evenly applied to the wire, and it can handle a large load. The wire has a diameter of 50 µm. By winding SMA wire the applicable load increases accordingly.

Based on the same principle, another prototype was built to evaluate its performance (Figures 19.7 and 19.8). It consists of two aluminum channels joined by a hinge and operates in a scissors-like motion. The wire is wound 35 times, thereby multiplying the force by a factor of 70 compared with a single wire.

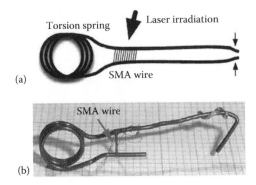

FIGURE 19.6 (a) Schematics and (b) actual photograph of an SMA tweezers LDA. The overall length is 2 cm.

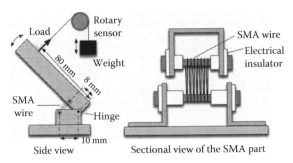

FIGURE 19.7 Schematic for the light-driven actuator and experimental setup.

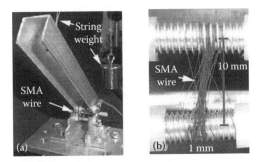

FIGURE 19.8 (a) The photograph of the light-driven actuator. The SMA wire is indicated with a circle. (b) The close-up view of the circled area.

In the experiment, a continuous wave (cw) argon-ion laser was used as the light source. When shined by a laser beam, the upper arm of LDA immediately went down by a few millimeters (Figure 19.9). When the laser beam is cut off it returned to its original position. With 1.0 W argon-ion laser beam, this actuator produced work of 3.7 mJ with 0.5 s response time. The efficiency was 0.97% when a load of 0.98 N was applied. This is an improvement of over a factor of 10^4 compared with the highest efficiency reported so far for LDAs, which is in the order of 10^{-5}% (Sarkisov et al. 2006). Assuming T_C and T_H of 338 K and 348 K, respectively, the theoretical limit of this LDA's conversion efficiency is calculated to be 2.9%; therefore, the obtained efficiency is fairly close to the theoretical limit. This was achieved by pre-tensioned wire design, as well as a small temperature difference of SMA in operation, which leads to a small heat loss.

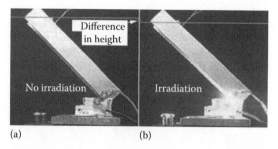

FIGURE 19.9 Photographs of LDA (a) before and (b) under irradiation. The difference in height between the two photographs is shown with a thick line.

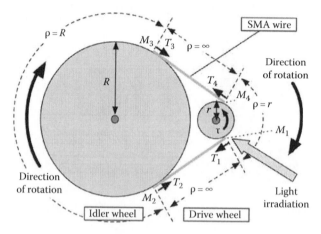

FIGURE 19.10 Principle of the SMA-based light-driven motor.

19.5 ROTATIONAL MOTION

Although reciprocal motions have the most important applications, for some applications rotational motion is a preferred type of motion. The most obvious case is when the device itself translates over a large distance. Also, it can take advantage of established mechanisms using conventional motors. Recently, an SMA-based light-driven motor has been demonstrated (Okamura 2011). In Figure 19.10, the structure of this device is shown. The system consists of two wheels of different sizes, and a looped SMA wire, which is initially trained straight shape. Light is irradiated at one side of the smaller wheel (drive wheel) where the wire undergoes a change in the radius of curvature from $\tilde{n} = r$ to infinity, where r is the radius of the drive wheel. As it is heated by light irradiation, the wire straightens and produces a torque. The part of the wire in the austenitic state is cooled when it leaves the drive wheel and advances to the straight part between the wheels. When it makes contact with the larger wheel (idler wheel), it is further cooled and goes back to martensite. It then becomes flexible again and can follow the curvature of the idler wheel. Further, a larger wheel helps cool the SMA wire while it is in contact with the wheel. In this system, the discussion in Section 19.3.2.3 plays an important role. We introduced the critical radius of curvature ρ_c. If SMA is bent further than this, a plastic deformation starts. In light-driven motors, the wire is repeatedly deformed as it arrives and leaves the idler wheel. If the radius of idler wheel is smaller than ρ_c, the wire undergoes plastic deformation for each deformation and a drag is induced.

The light-driven motors have some similarities to the belt-type SMA engine; however, there is a fundamental difference that the phase transition is localized in the case of light-driven motors. Due to this property, it starts spontaneously upon irradiation of light, while for the belt-type SMA engine it is usually necessary to provide an initial rotation to start the engine. This is an indispensable feature for remotely operated devices. Another advantage is that one can change the direction of rotation simply by irradiating the opposite side of the wheel. For the belt-type SMA engine, the direction of rotation is dependent on the direction of initial rotation.

Theoretical formulation of the system makes use of the bending moment discussed in Section 19.3.2. As the light-driven motor rotates, the wire repeats bending and straightening, changing its radius of curvature from infinity to r, and then to infinity again and then to R, and back to infinity. We represent each of the bending moments at these points as M_i ($i = 1 - 4$) as is shown in Figure 19.10. The steady-state equation of motion for the drive and idler wheels are $(T_1 - T_4)r - M_1 + M_4 - \tau = 0$ and $(T_3 - T_2)R - M_3 + M_2 = 0$, respectively. We assumed that a load is given on the drive wheel, asserting a torque ô. Imposing $T_1 = T_2$ and $T_3 = T_4$, we obtain the output torque as $-\tau = (M_1 - M_4) + (r/R)(M_3 - M_2)$. Since the wire is in austenitic and martensitic states at positions 1 and 4, respectively, in the first term $M_1 > M_4$ holds, and this term produces a positive contribution. The latter term of the expression vanishes if the wire deformation at idler wheel is in elastic regime and $M_2 = M_3$ holds.

The elastic energy stored in a bent wire per unit length is given by $dU = M^2 dx/2EI$, where the energy stored by shearing force is ignored and x-axis was taken along the length of the wire. Using the relations used in the previous discussion, it reduces to

$$dU = \frac{EI dx}{2\rho^2} = \left(\frac{\pi E d^4}{128 \rho^2} \right) dx \qquad (19.13)$$

The net energy produced per unit length is given by

$$dU = (E_a - E_m) \left(\frac{\pi d^4}{128 \rho^2} \right) dx \qquad (19.14)$$

where E_a and E_m are Young's modulus for the austenitic and martensitic states, respectively. Figure 19.12 shows the numerical plot of dU/dx as a function of the wire diameter. For this plot we used $E_a = 75$ GPa and $E_m = 28$ GPa (TiNi-Aerospace-inc 2001).

In the experiment, we used aluminum wheels with $R = 27.76$ mm and $r = 3.93$ mm at the place where a wire runs. A photograph of the apparatus is shown in Figure 19.11. The samples were made from NiTi SMA wires with a transition temperature of 60°C initially trained straight shape (Nilaco corp.). We prepared samples of various lengths and a number of loops out of wires of diameters 0.1, 0.3, and 0.5 mm. The laser beam from a diode-pumped solid state cw laser (Spectra Physics, BL-106C, wavelength 1064 nm) was expanded with a beam expander (Newport, T81–3X) and was directed by a mirror to the location of the inflection point of the SMA wire.

The experiment revealed that wire diameter is the most critical parameter. Samples made from 0.3 mm wire rotated immediately upon irradiation of light. The rotational speed of the drive wheel ranged from 23 to 140 rpm. As for samples made from 0.1 mm wire, a 27 times loop sample rotated (<10 rpm), but, a single and a double loop samples did not rotate. Samples made from 0.5 mm wire exhibited an opposed tendency: A single loop 0.5 mm sample rotated (<10 rpm), but a double loop sample barely rotated. These results are consistent with the theory as is explained in Figure 19.10. Rotation takes place when the energy exceeds the threshold, and in our experiment the threshold was estimated to be ~0.2 mJ/mm. For a 0.5 mm wire, the bending at the idler wheel was well over the elastic limit and a substantial drag was produced, which is indicated by the plastic component of the bending moment, $M_{plastic}$, in Figure 19.12. Another observation was that a larger drive wheel (diameter 11.97 mm) slowed down the rotation. This is also consistent with the expression for the output power given earlier. A steady and continuous rotation was achieved with a bundled loop of 0.3 mm diameter wire. Rotational speed of 60 rpm was obtained at 3.6 W of light power. The rotational speed was proportional to the incident light power, indicating the process as purely photothermal (Figure 19.13). Figure 19.14 shows the photograph of light-driven motor in operation.

FIGURE 19.11 A photograph of the light-driven motor.

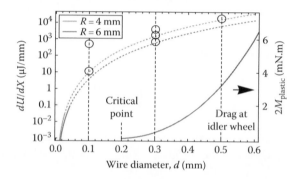

FIGURE 19.12 (Left-hand set of curves) Calculated output energy per unit length at the drive wheel versus wire diameter. The broken part of each curve indicates the elastic–plastic bending in martensitic phase, and thus the theoretical curve is not accurate in this region. (right-hand curve) Plastic component of the bending moment at the idler wheel that becomes a drag associated with rotation. This drag is present only when the radius of curvature exceeds the critical radius of curvature for that wire diameter. The open circles indicate the experimental conditions.

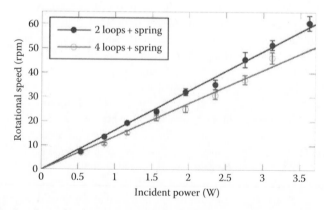

FIGURE 19.13 Rotational speeds of a two-loop (filled circles) and a four-loop (open circles) 0.3 mm spring-bundled samples as a function of the incident light power.

FIGURE 19.14 Captured images of the rotating drive wheel. The sample is the four-loop 0.3 mm sample in a closed coil spring. Light power was 0.5 W at the sample. To visualize the rotation, stickers were applied on the top of both wheels. Direction of rotation is indicated by arrows.

19.6 SUMMARY

The details of application of SMAs, in particular NiTi SMA, to LDAs were explained from a practical view. Conversion efficiency of 1% has been obtained for tweezer type SMA LDA, and a rotational speed of 60 rpm for continuous operation, and up to 140 rpm for noncontinuous operation has been obtained for SMA light-driven motors. These works clearly demonstrate the usefulness of SMAs as LDAs. High energy conversion efficiency means smaller heat production and it allows integration of LDAs into small space possible, which is an essential feature for microactuators.

REFERENCES

Al-Kaysi, R. O., A. M. Müller, and C. J. Bardeen. 2006. Photochemically driven shape changes of crystalline organic nanorods. *J. Am. Chem. Soc.* 128(50):15938–15939.

Ashkin, A. 1970. Acceleration and trapping of particles by radiation pressure. *Phys. Rev. Lett.* 24:156–159.

Banks, R. 1986. Phase transitions in shape memory alloys and implications for thermal energy conversion. *Phase Transform.* 1986:77–96.

Bellouard, Y., R. Clavel, R. Gotthardt, J. E. Bidaux, and T. Sidler. 1998. A new concept of monolithic shape memory alloy micro-devices used in micro-robotics. Paper read at *Proceedings of Actuator 98. 6th International Conference on New Actuators*, Bremen, Germany, June 17–19, 1998.

van den Broek, P. J., B. Potsaid, Y. Bellouard, and J. T. Wen. 2007. Laser actuated shape memory alloy mobile micro-robot: Initial results. Paper read at *Optomechatronic Actuators and Manipulation III*, Lausanne, Switzerland, October 8, 2007.

Chan, C. M., S. Trigwell, and T. Duerig. 1990. Oxidation of an NiTi alloy. *Surf. Interface Anal.* 15(6):349–354.

Cho, K-J., J. Rosmarin, and H. Asada. 2007. SBC Hand: A lightweight robotic hand with an SMA actuator array implementing C-segmentation. Paper read at *2007 IEEE International Conference on Robotics and Automation*, Roma, Italy, April 10–14, 2007, IEEE Robotics and Automation Society, Roma, Italy.

Chu, S., J. E. Bjorkholm, A. Ashkin, and A. Cable. 1986. Experimental observation of optically trapped atoms. *Phys. Rev. Lett.* 57(3):314–317.

Danilov, A., T. Tuukkanen, J. Tuukkanen, and T. Jamsa. 2004. Effect of strain on NiTi surface-optical reflectivity. *EDP Sciences. Journal de Physique IV (Proceedings)* 115:287–295.

Furuya, S. and H. Tokura. 2006. Local heating and local transformation of a shape memory alloy foil by laser irradiation. *J. Jpn. Soc. Precis. Eng.* 72(2):203–207. Supplement. Contributed papers.

Galhotra, V., V. Gupta, V. Martynov, and A. D. Johnson. 2000. Shape memory alloy based micro actuators. Paper read at *Proceedings of 7th International Conference on New Actuators—ACTUATOR 2000*, Bremen, Germany, June 19–21, 2000.

Govindjee, S. and G. J. Hall. 2000. A computational model for shape memory alloys. *Int. J. Solids Struct.* 37(5):735–760.

Hahtela, O. and I. Tittonen. 2005. Optical actuation of a macroscopic mechanical oscillator. *Appl. Phys. B* B81(5):589–596.

Ikehara, T., S. Shimada, H. Matsuda, and M. Tanaka. 2004. Mechanical strain generation in a polydiacetylene crystal due to photoinduced phase transition. *Jpn. J. Appl. Phys.* 43(2):654–658.

Ikuta, K. 1990. Micro/miniature shape memory alloy actuator. Paper read at *1990 IEEE International Conference on Robotics and Automation*, Cincinnati, OH, May 13–18, 1990, IEEE, Cincinnati, OH, USA.

Inaba, S., H. Kumazaki, and K. Hane. 1995. Photothermal vibration of fiber core for vibration-type sensor. *Jpn. J. Appl. Phys. Part 1 (Regular Papers & Short Notes)* 34(4A):2018–2021.

Kheirikhah, M. M., A. Khodayari, and M. Tatlari. 2010. Design a new model for artificial finger by using SMA actuators. Paper read at *2010 IEEE International Conference on Robotics and Biomimetics (ROBIO)*, Tianjin, China, December 14–18, 2010.

Kobatake, S., S. Takami, H. Muto, T. Ishikawa, and M. Irie. 2007. Rapid and reversible shape changes of molecular crystals on photo irradiation. *Nature* 446(7137):778–781.

Kulkova, S. E., D. V. Valujsky, K.J. Sam, G. Lee, and Y. M. Koo. 2001. Optical properties of TiNi, TiCo and TiFe thin films. *Physica B* 304(1–4):186–192.

Kuribayashi, K. 2000. Micro SMA actuator and motion control. Paper read at *MHS2000. Proceedings of 2000 International Symposium on Micromechatronics and Human Science*, Nagoya, Japan, October 22–25, 2000, Nagoya Junior Chamber, Nagoya, Japan.

Mercier, O. and K. N. Melton. 1981. Theoretical and experimental efficiency of the conversion of heat into mechanical energy using shape-memory alloys. *J. Appl. Phys.* 52(2):1030–1037.

Mori, M. 1984. In R. Kyokai (ed.), *Reza Ouyo Handobukku (Laser Application Technology Handbook)*, Asakura Publishing Co., Ltd., Tokyo, Japan, p. 71 (in Japanese).

Musella, M. 2001. Spectral reflectivity of Fe_2O_3 and NiO at high temperature. Paul Scherrer Institut Scientific Report, Annex V, pp. 29–30.

Okamura, H. 2006. Laser motor. *Proc. SPIE—Int. Soc. Opt. Eng.* 6374:637401-1-10.

Okamura, H. 2008. On the efficiency of heat engines by pulsed laser. *Proc. SPIE—Int. Soc. Opt. Eng.* 7266:726609-1-7.

Okamura, H. 2011. Light driven motor. *Opt. Eng. Lett.* 50(2):020503-1-3.

Okamura, H., K. Kawakami, and A. Mitani. 2011. Optimization of photothermal oscillators based on thermal diffusion analysis. *Int. J. Optomech.* 5(1):1.

Okamura, H., K. Yamaguchi, and R. Ono. 2009. Light-driven actuator with shape memory alloy for manipulation of macroscopic objects. *Int. J. Optomech.* 3(4):277–288.

Otsuka, K. and C. M. Wayman. 1999. *Shape Memory Materials*, Cambridge University Press, Cambridge, U.K.

Poosanaas, P., K. Tonooka, and K. Uchino. 2000. Photostrictive actuators. *Machanics* 10(4–5):467–487.

Raynaerts, D. and H. van Brussel. 1991. Development of a SMA high performance robotic actuator. Paper read at *91 ICAR. Fifth International Conference on Advanced Robotics. Robots in Unstructured Environments*, Pisa, Italy, June 19–22, 1991.

Rhee, J. Y., B. N. Harmon, and D. W. Lynch. 1996. Optical properties and electronic structures of equiatomic XTi (X = Fe, Co and Ni) alloys. *Phys. Rev. B* 54(24):17385–17391.

Sarkisov, S. S., M. J. Curley, L. Huey, A. Fields, and G. Adamovsky. 2006. Light-driven actuators based on polymer films. *Opt. Eng.* 45(3):34302-1-10.

Shabalovskaya, S. A. 1995. Biological aspects of TiNi alloy surfaces. *Journal de Physique IV* 5 (C8, pt.2): 1199–1204.

Sutapun, B., M. Tabib-Azar, and M.A. Huff. 1998. Applications of shape memory alloys in optics. *Appl. Opt.* 37(28):6811.

Suzuki, Y. 1988. *Keijokiokugokin no Hanashi (About Shape Memory Alloys)*, Nikkan Kogyo, Tokyo, Japan (in Japanese).

Tanaka, M. and K. Saito. 1993. Fundamental study on energy conversion system using shape memory alloy. *J. Mech. Eng. Lab.* 47(6):257–271.

TiNi-Aerospace-inc. 2001. *Shape Memory Alloys*, San Leandro, CA http://www.tiniaerospace.com

Wang, X., Y. Y. Ye, C. T. Chan, K. M. Ho, and B. N. Harmon. 1998. Calculated structural-dependent optical properties of alloys TiNi, TiPd, and TiPt. *Phys. Rev. B* 58(6):2964–2968.

Wayman, C. M. and H. C. Tong. 1975. On the efficiency of the shape memory effect for energy conversion. *Scr. Metall.* 9(7):757–760.

Wever, D. J., A. G. Veldhuizen, J. de Vries, H. J. Busscher, D. R. A. Uges, and J. R. van Horn. 1998. Electrochemical and surface characterization of a nickel-titanium alloy. *Biomaterials* 19(7–9):761–769.

Yamaguchi, K., R. Ono, and H. Okamura. 2009. Light-driven actuator with energy conversion efficiency in the order of 1%. *Appl. Phys. Express* 2(3):034502(3).

Yu, Y., M. Nakano, and T. Ikeda. 2003. Directed bending of a polymer film by light. *Nature* 425:145.

Zhu, J. J., N. G. Liang, W. M. Huang, and K. M. Liew. 2001a. Energy conversion in shape memory alloy heat engine. II. Simulation. *J. Intell. Mater. Syst. Struct.* 12(2):133–140.

Zhu, J. J., N. G. Liang, K. M. Liew, and W. M. Huang. 2001b. Energy conversion in shape memory alloy heat engine. I. Theory. *J. Intell. Mater. Syst. Struct.* 12(2):127–132.

20 Optically Driven Microrobotics

Yukitoshi Otani

CONTENTS

20.1 INTRODUCTION

Optically driven microrobotics has some interesting features, such as no generation of magnetic noise and remote receipt of the energy. Many researchers want to make optical robots move with the speed of light, but it might be dream right now. The force created by photon energy is about a pico Newton, but it works to move particles with micrometer size. This chapter shows novel optically driven micromanipulators at the beginning style of optically driven microrobotics. First, we compare different manipulators, and then a photothermal manipulator is shown for practical applications.

20.2 CLASSIFICATION OF VARIOUS MANIPULATORS

Quite a few applications of light energy have been reported comparing with possibilities for information optics. In recent years, there have been new attempts at actuating and/or driving to small objects with nano to micrometer size because they are too small to be controlled by electric motors. A novel optical actuator is utilized for a miniaturized inspection robot in an extreme environment, for example, inside of nuclear reactors and the human body. A microgripper and a manipulator have been required for applications ranging from industries to the biomedical area. Samples for manipulation, handling, and carriage spread out not only solid materials as mechanical parts but also soft ones as biocells and noncontact and remote control. Moreover, there are different sizes, shapes, and weights (Higuchi et al., 2010).

Figure 20.1 shows the function of actuators compared in different methods with various sizes of samples and required minimum gripping power. The actuator is defined as a machine that converts various forms of energy to mechanical energy. We focus only on a manipulator or gripper in this figure. The horizontal axis indicates the size of the sample, and the vertical axis indicates the required gripping power. The conventional methods, such as using electric motors, ultrasonic motors, air actuators, and piezo actuators, have their own niche. The electrical motor that generates power from magnetic power from electricity is a powerful tool for the industry. Most robot arms or manipulators employ it for joint actuators. The recent development of an ultrasonic motor that generates ultrasonic vibration has been quick. It has a big advantage of precise positioning in comparison with electric motors. Its size becomes smaller and it has more handling power, so it is applied in practical instruments. An air pressure that is controlled an airflow under regulation are used the same way as electrical motors for the robot arms. It has a feature for power control by regulating air pressure.

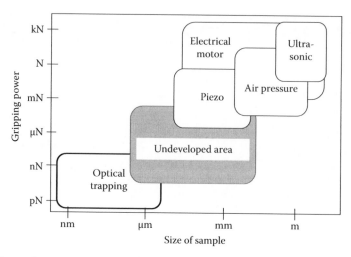

FIGURE 20.1 Comparison of sample size and gripping power for different actuation methods.

These methods are applicable in a wide range of products for our life from submillimeters to several meters of sample size with mN to kN of handling power. Especially, actuators for focusing of lens in small cameras have a big market for this method.

A piezo actuator made of a piezoelectric element (piezo ceramic), for example lead titanate zirconate (PZT), moves with precise positioning and less than nanometers. There are some drawbacks for small objects and soft materials because their methods are contact type and wired connections. On the other hand, optical trapping is a powerful tool in sample sizes with nanometers, as shown in Figure 12.1 (Ashkin, 1970). It has already been used commercially in biomedical manipulators. It generates a trapping power with nano to micro Newtons and can handle objects with the size of nano to micrometers. However, there is an undeveloped area as shown in Figure 20.1 because the weight of the wire is one of the problems in electric manipulators. An optical manipulator has an advantage for noncontact method for this area. A micromanipulator by photothermal effect which is made of optical fiver cantilever (Inaba, 1995) is demonstrated for a three-dimensional fabrication for its applications (Matsuba, 2002).

20.3 OPTICALLY DRIVEN CHOPSTICKS

Figure 20.2 shows optically driven chopsticks made of two optical fiber cantilevers (OFCs) shown in Figure 20.3. One of the fibers is cut for a bevel as shown in Figure 20.3a. The surface at the end of the fiber is painted in black for photothermal effect after illumination by light. It can easily absorb light to convert to heat in Figure 20.1b. A thermal expansion occurs by photothermal effect in response to incident light. This effect makes the end of the fiber stretch to the deformation direction shown in Figure 20.3c. In case the light is on, the optical fiber is bent because of thermal expansion. If the light is off, it returns to the initial position. As the irradiation area for photothermal effect is so small that its frequency response is high because it can take a short time for heat exchange. We studied the theoretical analysis of the OFC by using the finite element method (FEM). We used a plastic fiber with 250 μm of diameter. The front edge size of the optical fiber is 10 mm of bevel and has a He–Ne laser with 35 mW as a light source. The highest temperature is 60° and it bends at 60 μm. The result of the generative force of the front edge of OFC is shown by changing the diameters and lengths as shown in Figure 20.4. The best condition of the OFC is chosen to be 0.5 mm of diameter and 10 mm of bevel. By using these parameters, the circles in Figure 20.4 are compared with experimental results.

The optically driven chopsticks work to pinch a small object. The OFCs is mounted on the top of gripper cases. A laser diode is also mounted on the other side of the OFC with focusing lasers.

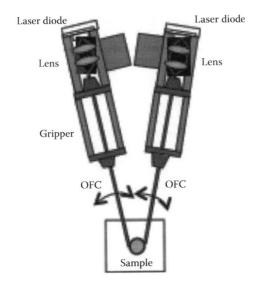

FIGURE 20.2 Optically driven chopsticks (Otani, 2003).

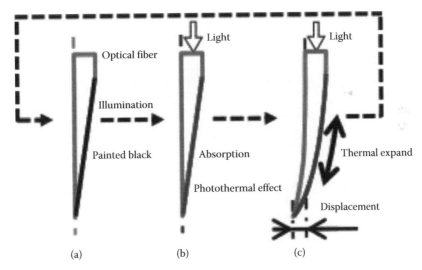

FIGURE 20.3 Optical fiber cantilevers (OFCs) for optically driven chopsticks. (a) Initial position (b) After illumination through optical fiber (c) Displacement by photothermal effect (Otani, 2001).

The angle between two OFCs can be easily changed along the arrow's direction after illuminating the laser light shown in Figure 20.2. We can easily pinch and release a sample by switching laser diodes. Moreover, we can carry the sample by moving the optical driven chopsticks. The demonstration of the optical chopsticks is shown to manipulate a screw with 1 mm diameter in Figure 20.5. This manipulator can also be utilized in the water. The glass beads are aligned to make the characters "TUAT" as in Figure 20.6.

20.4 OPTICALLY DRIVEN MICROMANIPULATOR

Figure 20.7 shows the construction of an optical micromanipulator made of the OFCs. Figure 20.7a shows three fingers made of the OFC are mounted on a moving stage. Figure 20.7b is a photograph

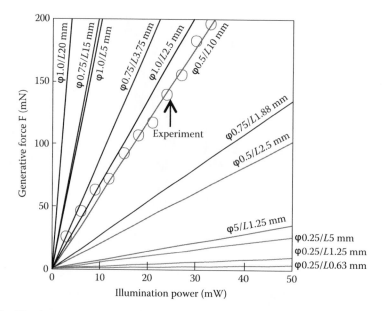

FIGURE 20.4 The forces generated by the optical fiber cantilevers (OFCs) with varying diameters and lengths. The circles represent experimental data.

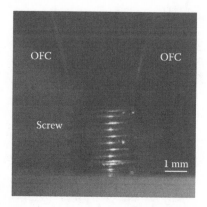

FIGURE 20.5 Optical chopsticks are shown to manipulate a screw with 1 mm diameter.

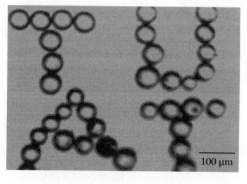

FIGURE 20.6 Two-dimensional alignment of glass beads.

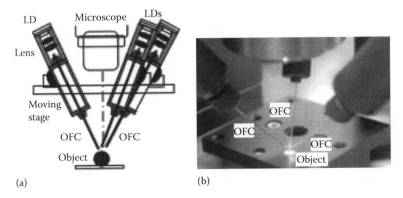

(a)   (b)

FIGURE 20.7 Experimental setup for optical manipulator. (a) Illustration of construction (b) Photograph of experimental setup (Otani, 2006).

of the end part. The OFC is made of the optical fiber with 0.5 mm of diameter and 10 mm of bevel. The optically driven micromanipulator is mounted on the xyz-stage. A laser diode with 810 nm of wavelength is used as a light source. Its maximum power is 500 mW, but it is usually use less than 50 mW. The light comes in one of the optical fiber by lens. A material of the optical fiber is selected as plastic. The deformation of plastic was examined and found to be 30 times that of quartz glass. The size of the micromanipulator is 10 mm of length, 3 mm of width, and a diameter of 500 μm. Three optically driven pawls are driven by each laser diode. The optical manipulator is set on an optical microscope. The xyz moving stage was controlled to position the manipulator and switch the laser diode to move the OFC for gripping and releasing by a personal computer.

Figure 20.8 shows experimental results for handling a small plastic particle with a diameter of 100 μm. The microscope was used to capture the base plate from the top. Figure 20.8a indicates the status before illumination and (b) is just after illumination. After illuminating the OFC, it manipulates a microparticle and it takes 0.3 s while it manipulates in case intensity of 35 mW. Figure 20.8c

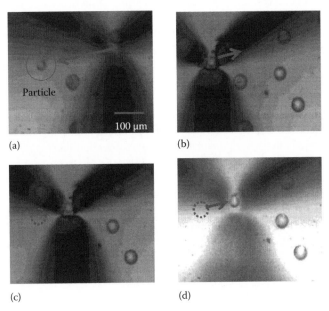

(a)   (b)

(c)   (d)

FIGURE 20.8 Manipulation process of glass bead sample. (a) Before manipulation (b) Manipulation (c) Moving sample (d) After manipulation.

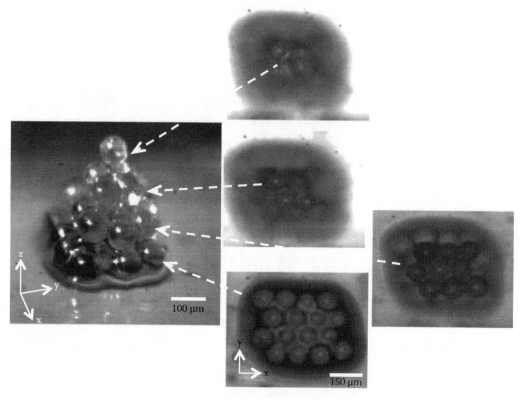

FIGURE 20.9   Glass beads manipulated in three dimensions.

shows the status after the movement to x direction and (d) indicates it just after the release. We were successful in picking the microparticle.

Finally, the experimental result was obtained by manipulation of a three-dimensional construction shown in Figure 20.9. After repeating the two-dimensional fabrication process, the three-dimensional structure was constructed. Figure 20.9 demonstrates three-dimensional fabrication of a microparticle with 100 μm diameter with four layers. A UV cure adhesive on each layer was used to protect the breakdown. The 3D structure was fabricated with a width of 300 μm.

20.5   CONCLUSION

An optical manipulator consisting of optical fiber cantilevers was demonstrated. We succeeded in moving a new type of micromanipulator that is driven by photothermal effect. It consists of an optical fiber cantilever with a size of 3 × 3 × 10 mm. It is possible to move at a speed less than 1 s. Moreover, deformation of optical fiber cantilevers by FEM was analyzed. Particle movement and alignment of glass beads were demonstrated even if in water condition. Finally, the manipulator can be demonstrated for three-dimensional structures of glass particles. We hope to expand to an optically driven micro robot for micro-optomechatronic machines. This manipulator seems to be a powerful tool to handle biomedical samples.

REFERENCES

Ashkin, A. 1970, Acceleration and trapping of particles by radiation pressure, *Phys. Rev. Lett.* 24:156–159.
Higuchi, T., Suzumori, K., Tadokoro, T. 2010, *Next-Generation Actuators Leading Breakthroughs*, Springer, London, U.K.

Inaba, S., Kumazaki, H., Hane, K. 1995, Photothermal vibration of fiber core for vibration type sensor, *Jpn. J. Appl. Phys.* 34:2018–2021.

Matsuba, Y., Otani, Y., Yoshizawa, T. 2002, Fiber type actuator using photothermal effect—Experiment of two dimensional movement, *Proc. SPIE*, 4902:78–82.

Otani, Y., Hirai, Y., Mizutani, Y., Umeda, N., 2006, Light-driven micromanipulator and its application for 3D fabrications, *Proc. SPIE*, 6374, 63740N.

Otani, Y., Matsuba, Y., Chimura, S., Umeda, N., Yoshizawa, T. 2003, Micromanipulator by photothermal effect, *Proc. SPIE*, 5264:150–153.

Otani, Y., Matsuba, Y., Yoshizawa, T. 2001, Photothermal actuator composed of optical fibers, *Proc. SPIE*, 4564:216–219.

21 Light Propulsion Systems for Spacecraft

Bernd Dachwald

CONTENTS

Symbols

A	Sail area
$\vec{a}$	Thrust acceleration
a_1, a_2, a_3	Optical solar radiation pressure force coefficients
a_c	Characteristic acceleration
B_b	Non-Lambertian coefficient of the sail's back side
B_f	Non-Lambertian coefficient of the sail's front side
c	Speed of light in vacuum
$\vec{d}$	Unit vector along the desired thrust direction
E	Emissive power

$\vec{e}_f$	Unit vector that is perpendicular to $\vec{e}_r$ and along the projection of $\vec{f}$ in the $\vec{e}_t$-$\vec{e}_h$-plane
$\vec{e}_h$	Coordinate frame unit vector in orbit normal direction
$\vec{e}_r$	Coordinate frame unit vector in radial direction
$\vec{e}_t$	Coordinate frame unit vector in transversal direction
$\vec{f}$	Thrust unit vector
$\vec{F}$	Thrust force
$\vec{F}_f$	Thrust force along $\vec{e}_f$
$\vec{F}_n$	Thrust force along $\vec{n}$
$\vec{F}_r$	Thrust force along $\vec{e}_r$
$\vec{F}_t$	Thrust force along $\vec{t}$
h	Sail film's surface roughness
m	Spacecraft mass
m_0	Initial spacecraft mass
m_f	Final spacecraft mass
$\vec{n}$	Sail normal (unit) vector
$\mathcal{O}$	Heliocentric coordinate frame
P	Solar radiation pressure
P_0	Solar radiation pressure at Earth's solar distance
$\mathcal{P}$	Set of optical sail coefficients
r	Solar distance
r_0	Earth's (mean) solar distance, 1 astronomical unit (1 AU)
S	Solar radiation flux
s	Specular reflection factor
T	(Absolute) sail temperature
$\vec{t}$	Sail tangential (unit) vector
$\vec{v}$	Spacecraft velocity vector
v_{esc}	Solar system escape velocity
V_e	Propellant exhaust velocity
ΔV	Required velocity increment that has to be provided by the spacecraft propulsion system
α	Sail pitch angle
β	Solar sail lightness number
ε	Emission coefficient
ε_b	Emission coefficient of the sail's back side
ε_f	Emission coefficient of the sail's front side
$\tilde{\alpha}$	Reflection coefficient
δ	Sail clock angle
θ	Thrust cone angle
η	Sail efficiency parameter
ρ	Reflection coefficient
ρ_b	Back reflection coefficient
ρ_d	Diffuse reflection coefficient
ρ_s	Specular reflection coefficient
τ	Transmission coefficient
σ	Stefan–Boltzmann constant
Ψ	Non-perfectly reflecting sail function
ϕ	Centerline angle
μ	Gravitational parameter of the Sun
Σ	Solar radiation dose
$*$	Optimal value

21.1 INTRODUCTION AND OVERVIEW OF LIGHT PROPULSION SYSTEMS FOR SPACECRAFT

While being useful to drive nano- and micro-scale actuators down here on Earth, in space, the radiation pressure of light can be used to propel even macroscopic devices, i.e., spacecraft. Systems that are propelled by sunlight allow—already with current and near-term technology—innovative space missions. In the far future, the radiation pressure of laser light may propel spacecraft even to nearby stars. This chapter focuses on the current to near-term technology of solar sails as light propulsion systems for spacecraft. The more futuristic concept of laser sails and other concepts are also introduced briefly.

The feasibility of ambitious and innovative space missions crucially depends on the capability of the propulsion system to achieve the required orbital energy change, which is typically expressed as a velocity increment, ΔV.* Challenging missions with high ΔV-demands comprise, e.g., fast missions to the outer solar system and into near-interstellar space, missions to close and highly inclined solar orbits, comet rendezvous missions, multiple asteroid rendezvous (and sample return) missions, missions to deflect asteroids from an impact with the Earth, and missions that maintain spacecraft in "exotic" non-Keplerian orbits. Such missions require ever more demanding propulsion capabilities that exceed the capabilities of the well-established chemical propulsion systems, which are presently the most commonly used in-space propulsion systems, and even the higher capabilities of electrical propulsion systems (like ion engines). The so-called rocket equation gives the total ΔV that a spacecraft can gain as

$$\Delta V = V_e \ln\left(-\frac{m_f}{m_0}\right) \tag{21.1}$$

where
   V_e is the exhaust velocity of the propellant
   m_0 is the initial spacecraft mass (with propellant)
   m_f is the final spacecraft mass (without propellant left)

Due to the energy barrier inherent in chemical combustion, chemical propulsion systems (rocket engines) have a limited V_e and thus a limited ΔV-capability. The exhaust velocity of chemical rocket engines is typically limited to $V_e \approx 4.5$ km/s for liquid hydrogen and liquid oxygen, as on the space shuttle or on the Ariane 5, and certainly to $V_e < 5$ km/s. Assuming a minimum mass ratio of $m_f/m_0 < 0.05$ (at least 5% of the total mass is required for structure, tanks, engines, etc.), the maximum velocity increment is limited to $\Delta V < 3V_e$, i.e., $\Delta V < 15$ km/s for a chemical propulsion system. The typical exhaust velocity of current ion engines is $V_e \approx 30$ km/s, but their thrust is low.

Utilizing solely the freely available solar radiation pressure (SRP) for propulsion, solar sails are propelled by reflecting the solar photons off large and lightweight mirroring surfaces, thereby transforming their momentum into a propulsive force. Consequently, solar sails do not consume any propellant, so that their ΔV-capability is theoretically unlimited. Practically, it is only limited by the lifetime of the solar sail in the harsh space environment and its distance from the Sun (because the SRP decreases with $1/r^2$, where r is the solar distance). Despite this constraint, even fast missions into the outer solar system and into near-interstellar space are feasible [4].

---

* The ΔVs are always counted positive, no matter whether they are used for increasing or decreasing the spacecraft's orbital energy.

This chapter is organized as follows: in Section 21.2, the basic SRP force models that are used to describe the thrust force that is acting on a solar sail are specified. Then, some advanced models to describe the SRP force more realistically are briefly sketched in Section 21.3, which might be skipped by the expeditious reader. Afterward, the orbital dynamics and control of solar sails are described in Section 21.4. The orbit control of solar sails requires changes in the orientation (termed *attitude* in the space engineering language) of the solar sail. The potential concepts that can be used for solar sail attitude control are illustrated in Section 21.5. To compare different solar sail designs, adequate performance parameters are required, as they are described in Section 21.6. After the reader has learned *how* solar sails work, the *why*, i.e., the mission applications, where solar sails can outperform traditional propulsion concepts, are described in Section 21.7. Section 21.8 briefly describes an option for the far future of sailing to other stars, laser sails, which areextremely challenging from an engineering point of view, but physically feasible. In the last section, Section 21.9, two other (non-actuator) spacecraft propulsion concepts that are based on light are briefly mentioned for the sake of completeness, photon propulsion, and laser propulsion.

21.2 BASIC SOLAR RADIATION PRESSURE FORCE MODELS

Solar sails obtain their propulsive force from the momentum of solar photons. When being absorbed, their momentum is transferred to the absorbing body, and when being specularly reflected, this momentum transfer is doubled. According to Newton's second law, this change in momentum results in a force on the absorbing or reflecting body. The force per unit surface area is called *solar radiation pressure* (SRP) and the total force on the body is called *SRP force*. The SRP at a distance r from the Sun is

$$P = \frac{S_0}{c}\left(\frac{r_0}{r}\right)^2 = P_0\left(\frac{r_0}{r}\right)^2 = 4.563\frac{\mu N}{m^2}\left(\frac{r_0}{r}\right)^2 \qquad (21.2)$$

where $S_0 = 1368\,W/m^2$ is the (mean) solar radiation flux at a solar distance of $r_0 = 1$ AU (1 astronomical unit),* the so-called solar constant, and c is the speed of light in vacuum. Because the SRP is very low, solar sails must be very large and lightweight to experience a significant acceleration.

The SRP force exerted on a solar sail, $\vec{F}$, is conveniently described with two unit vectors. The first one is the *sail normal (unit) vector $\vec{n}$*, which is perpendicular to the sail surface and always directed away from the Sun. The second one is the *thrust unit vector $\vec{f}$*, which points along the direction of the SRP force (in the very simplifying case of ideal reflection, $\vec{n}$ and $\vec{f}$ are identical). Let $\mathcal{O} = \{\vec{e}_r, \vec{e}_t, \vec{e}_h\}$ be a heliocentric orthogonal right-handed polar coordinate frame, where $\vec{e}_r$ points along the Sun-spacecraft line, $\vec{e}_h$ is normal to the orbit plane (pointing along the spacecraft's orbital angular momentum vector), and $\vec{e}_t$ completes the right-handed coordinate system ($\vec{e}_r \times \vec{e}_t = \vec{e}_h$). In $\mathcal{O}$, the direction of the sail normal vector $\vec{n}$, which describes the sail attitude, is expressed through the *sail pitch angle α* and the *sail clock angle δ*, while the direction of the thrust unit vector $\vec{f}$, which describes the thrust force ($\vec{F} = F\vec{f}$), is expressed by the *thrust cone angle θ* and by δ (see Figure 21.1).

The thin-film sails that are projected for solar sails are covered with a highly reflective coating on the front side (e.g., aluminum), and, typically, with a highly emissive thermal coating on the back side (e.g., chromium). All basic SRP force models assume that the solar sail is a "flat plate." Different levels of simplification for the optical characteristics of the sail film result in different models for the magnitude and direction of the SRP force, the most simple one being the ideal solar sail model.

---

* 1 AU = 149.5879×10^6 km is Earth's mean distance from the Sun.

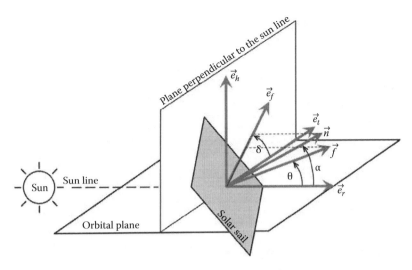

FIGURE 21.1   Definition of the sail normal vector and the thrust unit vector.

21.2.1   Ideal Solar Sail Model

In the ideal solar sail model, which will here be referred to as model IR (ideal reflection), an ideally reflecting sail surface is assumed. The SRP force exerted on an ideal sail of area A is (see [23] for derivation)

$$\vec{F} = 2PA\cos^2\alpha\,\vec{n} \tag{21.3}$$

In the solar sail-related literature, a similar SRP force model is sometimes employed, which uses an overall *sail efficiency parameter* η (therefore, it will here be referred to as model ηIR). Using this parameter, the SRP force acting on the sail is described similarly to Equation 21.3 by

$$\vec{F} = 2\eta PA\cos^2\alpha\,\vec{n} \tag{21.4}$$

Model ηIR is widely used because—as for model IR—the thrust is always along the sail normal vector ($f \equiv \vec{n}$), which allows an easy analytical treatment of solar sail dynamics. This model, however, as it will be shown later, provides only a rough approximation of the SRP force and is therefore not recommended except for very preliminary mission feasibility analysis. A better description for the SRP force can be obtained with the so-called optical solar sail model, which will be detailed in the next section.

21.2.2   Optical Solar Sail Model

For a thorough mission analysis, a more sophisticated SRP force model than IR or ηIR must be employed, which takes the (thermo-)optical coefficients of the real sail film into account. This model will be referred to as model OR (optical reflection). It was proposed in the 1970s for solar sail trajectory optimization by Sauer [29] and further studied by Forward [14]. In model OR, the optical characteristics of the sail film are parameterized through the *absorption coefficient* $\tilde{\alpha}$, the *reflection coefficient* ρ, the *transmission coefficient* τ, and the *emission coefficient* ε, with the constraint $\tilde{\alpha} + \rho + \tau = 1$. Assuming $\tau = 0$ for the reflectingside of the solar sail, the absorption coefficient is $\tilde{\alpha} = 1 - \rho$. Because not all photons are reflected specularly, the reflection coefficient can

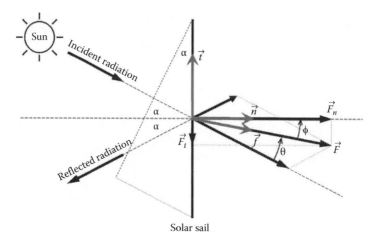

FIGURE 21.2   SRP force on a non-perfectly reflecting sail.

be further divided into a *specular reflection coefficient* ρ_s, a *diffuse reflection coefficient* ρ_d, and a *back reflection coefficient* ρ_b, with the constraint $\rho_s + \rho_d + \rho_b = \rho$. Assuming $\rho_b = 0$, the latter constraint can equivalently be expressed by introducing a *specular reflection factor* s, such that $\rho_s = s\rho$ and $\rho_d = (1-s)\rho$. The emission coefficient ε describes the emissive power E from a surface of area A at absolute temperature T, $E = \varepsilon\, \sigma T^4 A$, where $\sigma = 5.67051 \times 10^{-8}$ W m^{-2} K^{-4} is the Stefan–Boltzmann constant. The emission coefficients of the sail's front and back side are ε_f and ε_b, respectively. The angular distribution of the emitted and diffusely reflected photons is described by the *non-Lambertian coefficients* of the sail's front and back side, B_f and B_b, respectively. As a result, model OR parameterizes the optical characteristics of the sail film by the following set of six optical coefficients: $\mathcal{P} = \{\rho,\, s,\, \varepsilon_f,\, \varepsilon_b,\, B_f,\, B_b\}$. According to Wright [37], the optical coefficients for a sail with a highly reflective aluminum-coated front side and a highly emissive chromium-coated back side are $\mathcal{P}_{AllCr} = \{\rho = 0.88,\, s = 0.94,\, \varepsilon_f = 0.05,\, \varepsilon_b = 0.55,\, B_f = 0.79,\, B_b = 0.55\}$. By using these optical coefficients, the SRP force exerted on the solar sail can be decomposed into a normal component $\vec{F}_n$, along $\vec{n}$, and a tangential component $\vec{F}_t$, along $\vec{t}$ (see Figure 21.2), where

$$\vec{F}_n = 2PA\cos\alpha\,(a_1\cos\alpha + a_2)\vec{n} \tag{21.5a}$$

$$\vec{F}_t = -2PA\cos\alpha\, a_3\sin\alpha\,\vec{t} \tag{21.5b}$$

with the *optical SRP force coefficients*

$$a_1 \triangleq \frac{1}{2}(1 + s\rho) \tag{21.6a}$$

$$a_2 \triangleq \frac{1}{2}\left[B_f(1-s)\rho + (1-\rho)\frac{\varepsilon_f B_f - \varepsilon_b B_b}{\varepsilon_f + \varepsilon_b} \right] \tag{21.6b}$$

$$a_3 \triangleq \frac{1}{2}(1 - s\rho) \tag{21.6c}$$

The total SRP force can then be written as

$$\vec{F} = 2PA \cos \alpha \, \Psi \, \vec{f} \qquad (21.7)$$

where

$$\Psi \triangleq \left[(a_1 \cos \alpha + a_2)^2 + (a_3 \sin \alpha)^2 \right]^{1/2} \qquad (21.8)$$

depends only on the pitch angle α and the optical coefficients of the sail film. Note the symmetry in Equations 21.3 and 21.7: one $\cos \alpha$ is replaced by Ψ and $\vec{n}$ is replaced by $\vec{f}$ (the other $\cos \alpha$ remains unchanged because it describes the light-collecting projected sail area). Recall that the angle between $\vec{e}_r$ and $\vec{f}$ is the thrust cone angle θ, while the angle between $\vec{f}$ and $\vec{n}$ is referred to as the *centerline angle* ϕ. It can be calculated via

$$\phi = \arctan \left(\frac{a_3 \sin \alpha}{a_1 \cos \alpha + a_2} \right) \qquad (21.9)$$

The thrust cone angle is then obtained as $\theta = \alpha - \phi$.

More important for orbital dynamics, the SRP force can also be written in terms of components along the radial unit vector $\vec{e}_r$ and a unit vector $\vec{e}_f$ that is perpendicular to $\vec{e}_r$ and along the projection of $\vec{f}$ in the $\vec{e}_t$–$\vec{e}_h$-plane (see Figure 21.1). The components of $\vec{F}$ along $\vec{e}_r$ and $\vec{e}_f$ can be obtained from F_n and F_t via

$$\begin{pmatrix} F_r \\ F_f \end{pmatrix} = \begin{bmatrix} \cos \alpha & -\sin \alpha \\ \sin \alpha & \cos \alpha \end{bmatrix} \begin{pmatrix} F_n \\ F_t \end{pmatrix} \qquad (21.10)$$

so that

$$\vec{F}_r = 2PA \cos \alpha (a_1 \cos^2 \alpha + a_2 \cos \alpha + a_3 \sin^2 \alpha) \vec{e}_r \qquad (21.11a)$$

$$\vec{F}_f = 2PA \cos \alpha \sin \alpha (a_1 \cos \alpha + a_2 - a_3 \cos \alpha) \vec{e}_f \qquad (21.11b)$$

As already noted, the acceleration capability of a solar sail increases with $1/r^2$ when going closer to the Sun. The minimum solar distance, however, is constrained by the temperature limit of the sail film (and the rest of the spacecraft). The equilibrium temperature of the sail film is (see [23] for derivation)

$$T = \left(\frac{S_0}{\sigma} \frac{1 - \rho}{\varepsilon_f + \varepsilon_b} \left(\frac{r_0}{r} \right)^2 \cos \alpha \right)^{1/4} \qquad (21.12)$$

Thus, the sail temperature does not only depend on the solar distance, but also on the sail attitude, $T = T(r, \alpha)$ (and on the optical parameters $\mathcal{P}$).

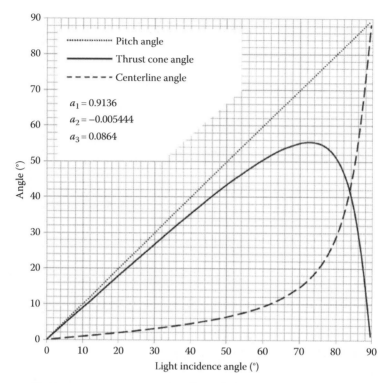

FIGURE 21.3 Sail pitch angle, cone angle, and centerline angle for model OR.

21.2.3 SOLAR SAIL MODEL COMPARISON

For model IR and ηIR, the SRP force is always perpendicular to the sail surface, i.e., $\vec{f} \equiv \vec{n}$. This way, both models allow an easy analytical treatment of solar sail dynamics, but misrepresent the sail normal SRP force component $\vec{F}_n$ and completely ignore the sail tangential SRP force component $\vec{F}_t$. In doing so, they ignore the deviation of the thrust cone angle from the pitch angle. Figure 21.3 shows how the deviation becomes larger as the light incidence angle increases. As a consequence, the SRP force in model OR is not only smaller than model IR (which may also be taken into account by model ηIR), but also much more constrained in its direction (as Figure 21.3 shows, there is a maximum thrust cone angle of 55.5° for a pitch angle of 72.6°). The set of possible SRP force vectors can be illustrated by a so-called SRP force bubble. Figure 21.4 shows for each SRP force model the SRP force bubble on whose surface the SRP force vector tip is constrained to lie (vector tail at origin). From the perspective of solar sail dynamics, model μIR is equivalent to model IR because the *shape* of both bubbles is identical. A decrease in sail efficiency η can be offset with a proportional increase in sail area, so that both bubbles have the same shape *and* size. This equivalency is not the case for model OR. Even if the bubbles are scaled to the same size, their shape is different. In the following sections, only the more general model OR will be used. The equations for model IR can be obtained easily by simplifying the equations for model OR (from $a_1 = 1$ and $a_2 = a_3 = 0$ follows $\Psi = \cos \alpha$ and $\phi = 0$).

21.3 ADVANCED SOLAR RADIATION PRESSURE FORCE MODELS

For the high-fidelity description of solar sail dynamics, the SRP force model OR, as described in Section 21.2.2, may still not be sufficiently sophisticated. It assumes that the sail film is a "flat plate"

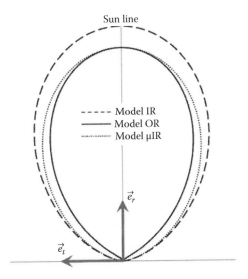

FIGURE 21.4   SRP force bubbles for the three different SRP force models.

with constant (thermo-)optical properties. In this section, some recently developed SRP models are introduced that overcome this limitation.

21.3.1   Generalized Solar Sail Force Model

Rios-Reyes and Scheeres [27] have developed a method for the analytic description of the force and torque generated by a solar sail of arbitrary shape and optical surface properties. In this *generalized solar sail force model*, Equation 21.5 are applied to differential solar sail area elements dA and the total force on the solar sail is then found by integration of the $d\vec{F}_n$ and $d\vec{F}_t$ over the sail surface. This yields a completely analytic description for the SRP force acting on a solar sail with arbitrary shape. With this formalism, it is also possible to accommodate solar sails with varying optical properties. Also, it is possible to add additional solar sail elements (e.g., attitude control flaps, see Section 21.5) by adding the corresponding terms for those additional sail elements.

21.3.2   Refined Solar Sail Force Model

Supported by experimental data, Mengali et al. [26] have developed a *refined solar sail force model*. In this model, the reflection coefficient is assumed to depend on the light incidence (or pitch) angle α, i.e., $\rho = \rho(\alpha)$, the specular reflection coefficient is assumed to depend on the pitch angle and the sail film's surface roughness h, i.e., $s = s(\alpha, h)$, and the emission coefficients of the sail's front and back side are assumed to depend on the sail film's temperature T, i.e., $\varepsilon = \varepsilon(T)$. Polynomial dependencies are used to approximate the measured variation of those optical coefficients. Due to their complexity, however, they are not given here. Because of those dependencies, $a_1 = a_1(\alpha, h)$, $a_2 = a_2(\alpha, h, T)$, and $a_3 = a_3(\alpha, h)$. Solar sail trajectory optimization has shown that this model might yield reduced mission durations ($\approx 5\%–10\%$) w.r.t. model OR and that the surface roughness of the sail film significantly affects the control of the sail [26].

21.3.3   Optical Solar Sail Degradation Model

The optical properties of thin metalized polymer solar sail films are likely to be affected by the damaging effects of the space environment. Their real degradation behavior, however, is still—to a great extent—unknown. The previously described SRP force models do not take such *optical solar sail degradation* into account. Therefore, Dachwald et al. [7] have developed a parametric SRP force model that allows to describe the optical degradation of the sail film. In their model, they assume that the only source of optical degradation are the solar photons and particles (a simplification that is reasonable at least in the inner solar system and far from planetary atmospheres and magnetic fields) and that the fluxes do not depend on time (average Sun without solar events). With those assumptions, the optical degradation depends on the sail's environmental history through the *solar radiation dose* $\Sigma(t)$ accepted by the solar sail during its mission (until time t). $\Sigma(t)$ depends on the solar distance history, $r[t]$, and the attitude history, $\alpha[t]$, of the solar sail. With increasing optical degradation, the SRP force bubble shrinks and the performance of the solar sail decreases. The potential effects of optical degradation on solar sail mission performance are described in [6,7].

21.4 SOLAR SAIL ORBITAL DYNAMICS AND CONTROL

The orbital dynamics of solar sails is in many respects similar to the orbital dynamics of other low-thrust spacecraft (e.g., with ion engines). The latter, however, may orient its thrust vector into any desired direction, whereas the thrust vector of solar sails is constrained to lie on the surface of the SRP force bubble, which is always directed away from the Sun. Due to this circumstance, it is a widespread misconception that solar sails can only fly away from the Sun. But as a matter of fact, by controlling the sail orientation relative to the Sun, solar sails can gain orbital energy (when $\vec{F} \cdot \vec{v} > 0$) or lose orbital energy (when $\vec{F} \cdot \vec{v} < 0$). In the heliocentric case, they spiral toward the Sun when losing orbital energy and they spiral away from the Sun when gaining orbital energy, as shown in Figure 21.5. The attitude of the solar sail, as described by the sail normal vector $\vec{n}$, or alternatively by the pitch angle α and the clock angle δ, provides control over the thrust direction, as given by the thrust unit vector $\vec{f}$, i.e., $\vec{F} = \vec{F}(\vec{n}) = \vec{F}(\alpha, \delta)$. In general, if $\vec{d}$ denotes the unit vector along the desired thrust direction, $\vec{f}$ must point into the direction that maximizes $\vec{F} \cdot \vec{d}$, the SRP force along $\vec{d}$. For models IR and μIR, the optimal attitude $\vec{n}^*$ (or α^* and δ^*) can be derived analytically from $\vec{d}$ (see [23]). For model OR, this is not possible. The problem of solar sail steering is then to determine

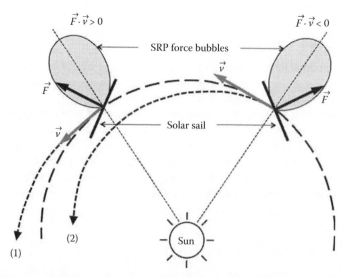

FIGURE 21.5 Solar sail steering. (1) Gaining orbital energy and spiraling away from the sun. (2) Losing orbital energy and spiraling towards the sun.

the optimal sail attitude $\vec{n}^*$ that yields the optimal thrust unit vector $\vec{f}^*$. Because the optimal thrust direction typically changes with time, the solar sail must also change its attitude with time. The potential concepts to do this are illustrated in the next section.

21.5 SOLAR SAIL ATTITUDE DYNAMICS AND CONTROL

The translational motion of a solar sail is coupled to its rotational motion because both the magnitude and the direction of the thrust force $\vec{F}$ depend on the sail attitude $\vec{n}$. Therefore, accurate thrust vector control requires accurate attitude control. In contrast to conventional spacecraft, solar sails have large moments of inertia and they are made of soft and flexible structures (membranes). As a result, their eigenfrequencies are very low and their amplitudes and periods of oscillation are very high due to low damping rates [30]. Consequently, attitude changes of the central spacecraft bus do not immediately change the attitude of the sail. The sail will follow the commanded attitude of the bus only with a large time delay and will oscillate around the new attitude for a long time. Therefore, the rigidity of the solar sail structure is a main driver for the achievable agility. This circumstance renders solar sailing in planetary orbits, where fast attitude maneuvers are required [5], much more difficult than solar sailing in interplanetary space, where attitude changes of a few degrees per day suffice. Solar sail attitude dynamics and control is a complex, challenging, and lively field of research, a treatment of which can be found in [32,35].

Attitude determination and control requires sensors, a control logic, and actuators. For attitude determination, the same sensors can be used as for other spacecraft, given that they are sufficiently lightweight. So far, many potential methods have been devised for solar sail actuation. Those attitude control methods might be categorized as follows:

1. "Traditional" attitude control methods:
   a. Thrusters at the sail boom tips or at the tips of dedicated booms
   b. Reaction wheels, etc.
2. Attitude control methods that change the center of mass w.r.t. the center of pressure:
   a. Gimballed central mast
   b. Displacement of masses inside the spacecraft bus, along the sail booms, or along dedicated booms
3. Attitude control methods that change the center of pressure w.r.t. the center of mass:
   a. Reflective flaps at the boom tips
   b. Roll stabilizer bars (also called spreader bars) at the boom tips
   c. Local retraction of sail segments
   d. Local changes of sail reflectivity

Figure 21.6 sketches those methods (although not in an advisable combination). All attitude control methods have specific advantages and disadvantages and some combinations are possible. For example, Wie and Murphy [35] advise for a first solar sail flight validation mission movable masses along the booms together with roll stabilizer bars for primary attitude control and pulsed plasma thrusters at the boom tips for backup attitude control in case of off-nominal conditions.

The optimal choice of the attitude control system (ACS) depends considerably on the required total changes of angular momentum and the required control torques about the three axes. In interplanetary space, only slow attitude maneuvers are required and the required total change in angular momentum is consequently low. Close to planets, however, fast attitude maneuvers and large changes in total angular momentum are required [5]. In this case, solar sail control is much more demanding, so that an ACS with a higher control authority is required. Additional criteria for the optimal ACS design are system mass, structural loads, reliability and redundancy, reusability (for different missions), complexity (structure, deployment, control), available know-how, and cost. Because all criteria are highly interdependent and the ACS has a strong influence on the overall

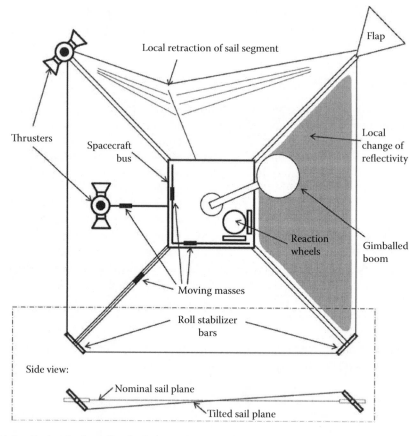

FIGURE 21.6 Optional solar sail attitude control methods (in a not advisable combination, size of spacecraft bus and sail are not to scale).

solar sail design, the mission-optimal choice of the ACS is a vital and difficult task that should be assessed early in the project.

21.6   SOLAR SAIL PERFORMANCE PARAMETERS

The performance of a solar sail is usually quantified through the values taken by the characteristic acceleration or the lightness number. The *characteristic acceleration*, a_c, is defined as the SRP acceleration acting on a solar sail that is oriented perpendicular to the Sun line ($\vec{n} \equiv \vec{e}_r$) at $r_0 = 1$ AU. It follows from Equations 21.7 and 21.8 that the characteristic acceleration for a solar sail with mass m is

$$a_c = \frac{2P_0 A}{m}(a_1 + a_2) \tag{21.13}$$

The *lightness number*, β, is defined as the ratio of the SRP acceleration acting on a solar sail that is oriented perpendicular to the Sun line and the gravitational acceleration of the Sun, μ/r^2:

$$\beta = \frac{a_c(r_0/r)^2}{\mu/r^2} = \frac{a_c}{\mu/r_0^2} \tag{21.14}$$

where $\mu/r_0^2 = 5.930\,mm/s^2$ is the Sun's gravitational acceleration at 1 AU solar distance. Note that the lightness number is independent of solar distance. Inserting Equations 21.13 and 21.14, respectively, into Equation 21.7 gives the SRP acceleration in terms of the characteristic acceleration or the lightness number as

$$\vec{a} = \frac{\vec{F}}{m} = a_c \left(\frac{r_0}{r}\right)^2 \cos\alpha\, \Psi\, \vec{f} = \beta \frac{\mu}{r^2} \cos\alpha\, \Psi\, \vec{f} \qquad (21.15)$$

21.7 SOLAR SAIL MISSION APPLICATIONS

Being propelled solely by SRP, solar sails have in principal unlimited propulsive capability (as long as they do not fly too far away from the Sun). This way, they offer the potential to make challenging high-ΔV missions feasible. Such missions comprise, e.g., missions to close and highly inclined solar orbits [10,21], fast missions to the outer solar system and into near-interstellar space [4,25], comet rendezvous missions [17] and multiple asteroid rendezvous missions [9], asteroid deflection missions [24,33], and missions that maintain spacecraft in "exotic" non-Keplerian orbits [22]. Some of these mission applications will be outlined in the following sections.

21.7.1 MISSIONS TO CLOSE AND HIGHLY INCLINED SOLAR ORBITS

Because the SRP and consequently the thrust force increases proportional to $1/r^2$ when approaching the Sun, solar sails are especially effective for missions that go close to the Sun because they can attain such orbits rapidly. A mission that goes into a very close solar orbit ($\lesssim 8$ Solar radii) would be very attractive for testing current gravitational theories [10]. For the required high-precision measurements, the solar sail would be jettisoned after the final orbit is reached. The ΔV that is required for a conventional Hohmann transfer from a low Earth orbit to the final orbit is $\Delta V \approx 84\,km/s$. Thus, using Equation 21.1 and $V_e = 4.5\,km/s$, thepayload ratio of a chemical rocket would be as small as $m_f/m_0 = \exp(-\Delta V/V_e) \approx 10^{-8}$, which is clearly infeasible. Besides the solar sail, only an advanced electric propulsion system with an exhaust velocity larger than about 50 km/s would make such a mission feasible. Figure 21.7a shows the solar sail trajectory to such a close solar orbit within a transfer time of only 6.6 years. The required characteristic acceleration for such a mission is only 0.1326 mm/s², being near-term technology. Nevertheless, the solar radiation flux would be tremendous at very close solar distances. Even if the temperature of the solar sail can be kept below 300°C by orienting it near-edge-on to theincoming radiation, the thermal control of the whole spacecraft would be extremely challenging. A more detailed mission design for this mission can be found in [10].

Another mission application that would profit from the high effectiveness of solar sails close to the Sun would be a solar observation mission out of the ecliptic and over the solar poles, to investigate the global structure and dynamics of the solar corona and to reveal the secrets of the solar cycle [2]. The Solar Polar Imager (SPI) mission was one of several solar sail roadmap missions envisioned by NASA. A similar solar sail mission, called Solar Polar Orbiter (SPO), was studied by ESA [21]. One SPI reference mission design is based on a $160\,m \times 160\,m$, 150 kg square solar sail assembly with a 250 kg spacecraft bus and a scientific payload of 50 kg (450 kg total mass) [36]. This yields a characteristic acceleration of $a_c = 0.35\,mm/s^2$. The SPI target orbit is a heliocentric circular orbit at 0.48 AU solar distance with an inclination of 75° out of the ecliptic. Figure 21.7b shows the SPI trajectory with a transfer time of only 4.7 years. Along the trajectory, the temperature of the solar sail is kept below 240°C. The ΔV that is required for a conventional Hohmann transfer from a circular low Earth orbit to the final orbit, including the inclination change maneuver, is $\Delta V \approx 42\,km/s$. A more detailed mission design for the SPI mission can be found in [8].

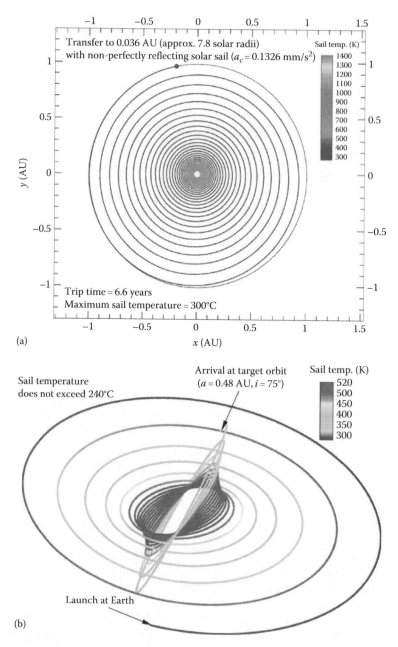

FIGURE 21.7 **(See color insert.)** Trajectory to (a) a very close solar orbit and (b) to the Solar Polar Imager (SPI) orbit.

21.7.2 FAST MISSIONS TO THE OUTER SOLAR SYSTEM AND INTO NEAR-INTERSTELLAR SPACE

Although the SRP decreases proportional to $1/r^2$ when going away from the Sun, solar sails enable also missions that go very far away from the Sun and leave the solar system with a high velocity. Sauer [28] observed that a solar sail may gain a large amount of energy by making a close approach to the Sun that turns the trajectory hyperbolic, a maneuver for which Leipold coined the term "solar photonic assist" (SPA) [18,19]. The objective of such missions would be the investigation of the heliopause, the boundary between the heliosphere and interstellar space, which is expected to lie

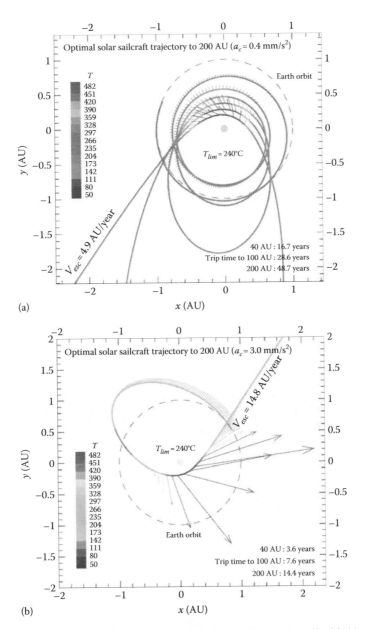

FIGURE 21.8 (See color insert.) Solar system escape trajectory for a solar sail with (a) moderate performance and (b) high performance. The arrows indicate the SRP acceleration.

at a solar distance between 100 and 200 AU, to study near-interstellar space itself, and to measure weak gravitational fields on a large scale for testing current gravitational theories [10]. After the last SPA, the solar sail would be jettisoned at about 5 AU to allow high-precision measurements. For this mission type, however, a more advanced solar sail technology with $a_c \geq 0.4 \, mm/s^2$ is required. This could be achieved by using larger solar sails, thinner sail films, more advanced materials for the films and the booms, and by reducing the mass of the deployment mechanism. Figure 21.8 shows the trajectories for optimal transfers to 200 AU for two different characteristic accelerations, 0.4 and 3.0 mm/s^2, the solar sail film temperature being limited to 240°C in both cases. One can see that more SPAs are required to reach 200 AU in minimum time as the characteristic acceleration of the sail decreases because a larger fraction of flight time must be spent in the inner solar system for gaining orbital energy [4]. For $a_c = 3.0\,\text{mm/s}^2$, the solar sail reaches 200 AU in less than 15 years and leaves the solar system with a velocity of $\approx 70\,\text{km/s}$ or $\approx 250,000\,\text{km/h}$.

21.7.3 MISSIONS FOR ASTEROID DEFLECTION

Near-Earth objects (NEOs) are asteroids and short-period comets with orbits that intersect or pass near the orbit of Earth. More than a thousand near-Earth asteroids (NEAs) with a diameter $\gtrsim 1\,\text{km}$ are currently known [1]. All NEAs that can approach the Earth closer than 0.05 km (and with a diameter $\gtrsim 200\,\text{m}$) are defined as potentially hazardous asteroids (PHAs). According to the latest discovery statistics (February 2011), there are currently 186 known PHAs with a diameter $\gtrsim 1\,\text{km}$, 7 of them having a diameter $\gtrsim 5\,\text{km}$. In the long run, they pose a significant hazard to human civilization and to life on Earth. Today, it is widely accepted that NEO impacts have caused at least one mass extinction (65 Myr ago), and a few NEOs will continue to do so in the future, if they are not deflected prior to impacting the Earth. Even NEAs that do not intersect Earth's orbit may evolve into Earth-crossers because their orbits are chaotic, having a relatively short dynamical lifetime in the order of several million years [3,16].

One approach to deflect an NEA is to impact it with a Kinetic Energy Impactor (KEI), a massive projectile at a high relative velocity, as shown in Figure 21.9a. The highest impact velocity can be

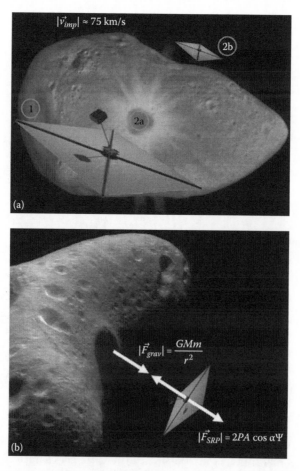

FIGURE 21.9 Two options for asteroid deflection using a solar sail: (a) the Kinetic Energy Impactor (KEI) concept (the solar sail brings the KEI onto a collision course with target (1) and releases the KEI, which impacts target (2a) while the solar sail monitors the impact (2b)) and (b) the Gravity Tractor concept.

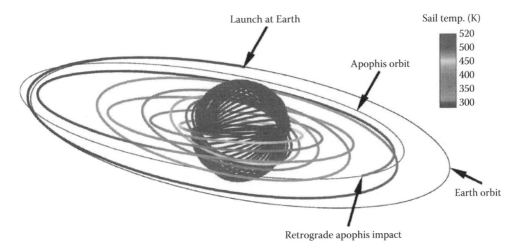

FIGURE 21.10 (See color insert.) Trajectory to bring a solar sail onto a retrograde intercept orbit with asteroid Apophis.

achieved from a trajectory that is retrograde to thetarget's orbit, i.e., from an orbit that rotates in the opposite direction around the Sun. Again, the large ΔV that is required to make the orbit retrograde is prohibitive for chemical propulsion systems and extremely difficult for electric propulsion systems. Figure 21.10 shows an exemplary trajectory to bring a solar sail onto a retrograde intercept orbit with asteroid Apophis. For this target, the impact velocity will be between 75 and 81 km/s, depending on the launch date and the impact geometry. The effective impulse imparted to the asteroid will be the sum of the pure kinetic impulse (linear momentum) of the impactor plus the impulse due to the "thrust" of material being ejected from the impact crater. The last term can be very significant and even dominant [11]. The velocity change of the target depends on the impact velocity and the masses of the target and the impactor, but will generally be small, in the order of 1 mm/s. The change in the target's Earth-miss distance due to the impact, however, might be sufficient, depending on the lead time, i.e., the time between the KEI's impact and the target's Earth impact. A more detailed mission design for an asteroid deflection mission with a KEI can be found in [11,33].

Another approach to deflect an asteroid from a collision course with the Earth is to tow it with a Gravity Tractor (GT). In the GT concept, invented quite recently by Lu and Love [20], a spacecraft hovers close to the asteroid, as shown in Figure 21.9b. Because—according to Newton's third law—not only the asteroid attracts the spacecraft, but also vice versa, momentum transfer can be achieved by keeping the relative position of the GT via some propulsive means. While conventional spacecraft require propellant to keep the GT in position, a solar sail can provide the required continuous thrust without using any propellant. A more detailed mission design for an asteroid deflection mission with a GT can be found in [34].

21.7.4 MISSIONS INTO NON-KEPLERIAN ORBITS

Up to now, only Keplerian orbits have been considered. Keplerian orbits are a sufficient approximation for most space missions, where a spacecraft is flying in the gravitational field of a much heavier central body (two-body problem). In this case, the spacecraft flies always in a plane that contains the central body. Non-Keplerian orbits are orbits where the central body does not lie in the orbital plane of the spacecraft.

Between the Earth and the Sun, at a distance of about 1.5 million km from the Earth, the so-called L_1-point (Lagrange point 1) is located, where the gravitational attraction of the Sun and the Earth add up in a way that the orbital period of a spacecraft that is located there exactly equals the

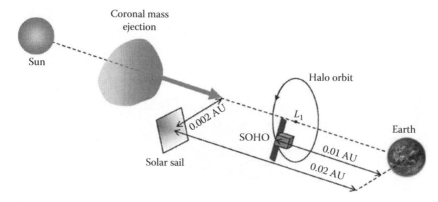

FIGURE 21.11   Solar sail on a non-Keplerian orbit for enhanced solar storm warning (not to scale).

orbital period of the Earth, so that it is fixed in a rotating frame of reference. Because this point is unstable (errors in the spacecraft position increase over time), spacecraft typically go into a so-called halo orbit around it. This situation is sketched in Figure 21.11. One spacecraft that is currently located there is SOHO (Solar and Heliospheric Observatory), a NASA/ESA mission that monitors the Sun and the solar wind that reaches the Earth about 30–60 min later. Sometimes, the Sun expels so-called coronal mass ejections into the direction of the Earth, which may contain billions of tons of high-energy particles, which endanger space assets and power grids on the Earth. The warning time by SOHO could approximately be doubled by using a solar sail spacecraft, where the continuous thrust of the solar sail is used to move the natural L_1-point another 1.5 million km closer toward the Sun (to an artificial Lagrange point). This solar sail mission application was studied by NASA and called Geostorm Warning Mission [31]. The required characteristic acceleration for such a mission would be in the order of $a_c \approx 0.2\,\text{mm/s}^2$. This mission concept would be very attractive for a first solar sailing mission because a solar sail deployment failure would not result in a loss of the mission, but only in the degradation to a "regular" L_1-point mission.

Another interesting application of solar sails in highly non-Keplerian orbits is the PoleSitter mission, as it was proposed by Forward [15] (and originally named "Statite"), based on previous work by McInnes. In this concept, the solar sails thrust vector is directed in a way that the orbital plane of the spacecraft is above the orbital plane of Earth and that the spacecraft's angular velocity is exactly Earth's angular velocity, so that—as seen from the Earth—the spacecraft "sits" on the pole (Figure 21.12). Those missions, however, require quite large characteristic accelerations in the order of $5\,\text{mm/s}^2$.

21.8   LASER SAILS AND INTERSTELLAR TRAVEL

In the far future, laser sails may be able to propel spacecraft to nearby stars—within a human lifetime—and even to return them to Earth. While such a system—in contrast to other "advanced" future space propulsion systems—does not require yet undiscovered physical laws, the technical challenge is so immense that it will probably not be built within this millennium. It is to note, however, that in contrast to the predominate educated opinion, propelling spacecraft to other stars is not extremely difficult per se, just reaching them in a moderate time, i.e., in less than several thousands of years or so, is extremely difficult.

The following discussion mainly follows the seminal paper by Forward [13]. A laser sail is in principle not much different to a solar sail; however, the light does not come from the Sun but from a high-power laser in space. If the laser light would be ideally parallel, the acceleration capability of the sail would not depend on its distance from the laser. The maximum laser power on the sail is

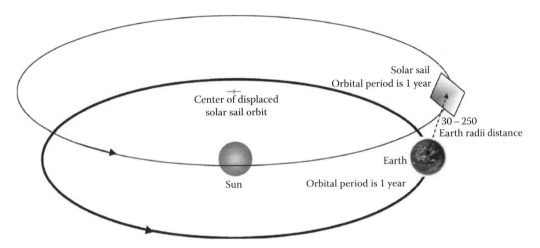

FIGURE 21.12   Solar sail on a non-Keplerian orbit for enhanced polar communications (not to scale).

just limited by the optical parameters, which have to be almost ideal, and the temperature limit of the sail film material. For a real laser, however, a huge Fresnel zone lens with a diameter of about 1000 km is required in the optical path of the laser, ideally being located somewhere between Saturn and Uranus and being levitated in a way that the laser pressure and the solar gravitation, both acting on the lens, cancel out. This lens would have 111,410 rings with a radial tolerance of 56 cm, and, having a thickness of 1 µm, the mass of the lens would be 560,000 metric tons. Already those numbers arouse our suspicion that such a mission might be technically challenging. A stellar flyby at Alpha Centauri, being the closest star at a distance of only 4.3 light years, would require a 1000 kg 3.6 km diameter laser sail and a 65 GW laser to accelerate the sail to a velocity of $0.11c$, yielding a stellar flyby after 40 years. A stellar rendezvous mission requires a 71,000 kg 30 km diameter payload sail that is surrounded by a 710,000 kg ring-shaped decelerator sail with a 100 km outer diameter. The two sails would be launched together from the solar system by a 7.2 TW laser until they have reached a coasting velocity of $0.21c$. As they approach Alpha Centauri, the inner payload sail detaches from the outer deceleration sail. An until then upgraded 26 TW laser from the solar system strikes the heavier ring sail and accelerates it past Alpha Centauri. The curved surface of the outer ring sail focuses the light back onto the inner payload sail, slowing it to halt in the Alpha Centauri system after a mission time of 41 years. Using a three-stage sail, a round trip would also be possible, but even more difficult.

21.9   OTHER LIGHT PROPULSION SYSTEMS FOR SPACECRAFT

For the sake of completeness, two other potential light propulsion concepts for spacecraft should be briefly mentioned here: photon propulsion and laser propulsion.

21.9.1   PHOTON PROPULSION

Photon propulsion uses an energy source, e.g., a nuclear reactor, to heat up a radiator to an extremely high temperature. The momentum carried away by the emitted photons is then transferred to the spacecraft. Therefore, such a spacecraft is sometimes also called "photon rocket." The disadvantage is that it takes a lot of power to generate only a small amount of thrust in this way, so that the acceleration is very poor. Although photon propulsion is technologically feasible, it is rather impractical with current technology.

21.9.2 LASER PROPULSION

Laser propulsion uses a high-power laser on the ground to heat up the solid propellant of a small ascent spacecraft, tracking the spacecraft with the laser beam during its ascent into orbit. This way, the energy source is separated from the spacecraft. With this propulsion concept, reasonably small spacecraft might be launched into space in the future [12].

GLOSSARY

Astronomical unit (AU): Earth's mean solar distance.

Attitude: Spacecraft orientation.

Characteristic acceleration: Maximum thrust acceleration of a solar sail at Earth's solar distance (performance parameter).

Hohmann transfer: ΔV-optimal orbit transfer between two circular coplanar orbits (under certain conditions).

Interstellar space: Space outside the Sun's heliosphere (beginning at about 150 AU solar distance) that is dominated by interstellar particles and fields.

Ion engine: State-of-the-art propulsion technology that accelerates ions (typically Xenon) in an electrostatic field to very high velocities.

Laser sail: Extremely large, highly reflective, and extremely lightweight structure that uses laser radiation pressure for propulsion in space.

Lightness number: Ratio of the thrust acceleration of a solar sail and the gravitational acceleration of the Sun (performance parameter).

Non-Keplerian orbit: Orbit that does not follow Kepler's laws of two-body planetary motion due to a thrust acceleration that is larger than the disturbing accelerations (in the three-body problem) or comparable to the gravitational acceleration of the central body (in the two-body problem).

Orbital energy: Sum of a spacecraft's kinetic energy and potential energy w.r.t. the central body.

Solar constant: Mean solar radiation flux at Earth's mean solar distance.

Solar sail: Large, highly reflective, and very lightweight structure that uses solar radiation pressure for propulsion in space.

Three-body problem: The problem of describing the motion of a body (e.g., spacecraft) in the gravitational field of two much larger masses (e.g., Sun and Earth).

Two-body problem: The problem of describing the motion of a body (e.g., spacecraft) in the gravitational field of a single larger mass (e.g., Sun).

REFERENCES

1. European Asteroid Research Node website: http://earn.dlr.de
2. Sun-Solar System Connection. Science and technology roadmap 2005–2035. NASA Publication, NASA, 2005.
3. W. F. Bottke, M. C. Nolan, R. Greenberg, and R. A. Kolvoord. Collisional lifetimes and impact statistics of near-Earth asteroids. In T. Gehrels (ed.). *Hazards due to Comets and Asteroids*, pp. 337–357. The University of Arizona Press, Tucson, AZ, 1994.
4. B. Dachwald. Optimal solar-sail trajectories for missions to the outer solar system. *Journal of Guidance, Control, and Dynamics*, 28(6):1187–1193, 2005.
5. B. Dachwald, R. Blockey, and W. Shyy (eds.). Solar sail dynamics and control. In *Encyclopedia of Aerospace Engineering*, pp. 3139–3152. Wiley, Hoboken, NJ, 2010.
6. B. Dachwald, M. Macdonald, C. R. McInnes, G. Mengali, and A. A. Quarta. Impact of optical degradation on solar sail mission performance. *Journal of Spacecraft and Rockets*, 44(4):740–749, 2007.
7. B. Dachwald, G. Mengali, A. A. Quarta, and M. Macdonald. Parametric model and optimal control of solar sails with optical degradation. *Journal of Guidance, Control, and Dynamics*, 29(5):1170–1178, 2006.

8. B. Dachwald, A. Ohndorf, and B. Wie. Solar sail trajectory optimization for the Solar Polar Imager (SPI) mission. *AIAA/AAS Astrodynamics Specialist Conference*, Keystone, CO, August 2006, AIAA Paper 2006-6177.

9. B. Dachwald and W. Seboldt. Multiple near-Earth asteroid rendezvous and sample return using first generation solar sailcraft. *Acta Astronautica*, 57(11):864–875, 2005.

10. B. Dachwald, W. Seboldt, and C. Lammerzahl. Solar sail propulsion: An enabling technology for fundamental physics missions. In H. Dittus, C. Lämmerzahl, and S.G. Turyshev (eds.). *Lasers Clocks, and Drag-Free Control: Exploration of Relativistic Gravity in Space*, pp. 379–398. Springer, Berlin, Germany, 2007.

11. B. Dachwald and B. Wie. Solar sail kinetic energy impactor trajectory optimization for an asteroid deflection mission. *Journal of Spacecraft and Rockets*, 44(4):755–764, 2007.

12. H.-A. Eckel, W. Schall, C. Bruno, and A. Accettura (eds.). Laser propulsion systems. In *Advanced Propulsion Systems and Technologies, Today to 2020*, pp. 357–406. American Institute of Aeronautics and Astronautics, Reston, VA, 2008.

13. R. L. Forward. Roundtrip interstellar travel using laser-pushed lightsails. *Journal of Spacecraft and Rockets*, 21(2):187–195, 1984.

14. R. L. Forward. Grey solar sails. *The Journal of the Astronautical Sciences*, 38(2):161–185, 1990.

15. R. L. Forward. Statite: A spacecraft that does not orbit. *Journal of Spacecraft and Rockets*, 28(5):606–611, 1991.

16. B. Gladman, P. Michel, and C. Froeschle. The near-Earth object population. *Icarus*, 146:176–189, 2000.

17. G. W. Hughes and C. R. McInnes. Small-body encounters using solar sail propulsion. *Journal of Spacecraft and Rockets*, 41(1):140–150, 2004.

18. M. Leipold. To the Sun and Pluto with solar sails and micro-sciencecraft. *Acta Astronautica*, 45(4–9):549–555, 1999.

19. M. Leipold and O. Wagner. 'Solar photonic assist' trajectory design for solar sail missions to the outer solar system and beyond. In T. H. Stengle (ed.). *Spaceflight Dynamics 1998*, Vol 100 Part 2 of Advances in the Astronautical Sciences, pp. 1035–1045. Univelt, Inc., Escondido, CA, 1998.

20. E. T. Lu and S. G. Love. Gravitational tractor for towing asteroids. *Nature*, 438:177–178, 2005.

21. M. Macdonald, G. W. Hughes, C. R. McInnes, A. Lyngvi, P. Falkner, and A. Atzei. Solar polar orbiter: A solar sail technology reference study. *Journal of Spacecraft and Rockets*, 43(5):960–972, 2006.

22. C. R. McInnes. Dynamics, stability, and control of displaced non-Keplerian orbits. *Journal of Guidance, Control, and Dynamics*, 21(5):799–805, 1998.

23. C. R. McInnes. *Solar Sailing. Technology, Dynamics and Mission Applications*, Springer–Praxis Series in Space Science and Technology. Springer–Praxis, Berlin, Germany, 1999.

24. C. R. McInnes. Deflection of near-Earth asteroids by kinetic energy impacts from retrograde orbits. *Planetary and Space Science*, 52(7):587–590, 2004.

25. C. R. McInnes. Delivering fast and capable missions to the outer solar system. *Advances in Space Research*, 34(1):184–191, 2004.

26. G. Mengali, A. A. Quarta, C. Circi, and B. Dachwald. Refined solar sail force model with mission application. *Journal of Guidance, Control, and Dynamics*, 30(2):512–520, 2007.

27. L. Rios-Reyes and D. L. Scheeres. Generalized model for solar sails. *Journal of Spacecraft and Rockets*, 42(1):182–185, 2005.

28. C. G. Sauer. Optimum solar-sail interplanetary trajectories. *AIAA/AAS Astrodynamics Conference*, San Diego, CA, August 1976, AIAA Paper 76-792.

29. C. G. Sauer. A comparison of solar sail and ion drive trajectories for a Halley's comet rendezvous mission. *AAS/AIAA Astrodynamics Conference*, Jackson, MS, September 1997, AAS Paper 77-104.

30. C. Sickinger, L. Herbeck, and E. Breitbach. Structural engineering on deployable CFRP booms for a solar propelled sailcraft. *Acta Astronautica*, 58(4):185–196, 2006.

31. J. L. West. The Geostorm warning mission: Enhanced opportunities based on new technology. *AAS/AIAA Space Flight Mechanics Meeting*, Maui, HI, February 2004, AAS Paper 04-102.

32. B. Wie. Solar sail attitude control and dynamics. *Journal of Guidance, Control, and Dynamics*, 27(4):526–544, 2004.

33. B. Wie. Solar sailing Kinetic Energy Interceptor (KEI) mission for impacting and deflecting near-Earth asteroids. *AIAA Guidance, Navigation, and Control Conference*, San Francisco, CA, August 2005, AIAA Paper 2005-6175.

34. B. Wie. Dynamics and control of gravity tractor spacecraft for asteroid deflection. *Journal of Guidance, Control, and Dynamics*, 31(5):1413–1423, 2008.

35. B. Wie and D. Murphy. Solar-sail attitude control design for a sail flight validation mission. *Journal of Spacecraft and Rockets*, 44(4):809–821, 2007.

36. B. Wie, S. Thomas, M. Paluszek, and D. Murphy. Propellantless AOCS design for a 160-m, 450-kg sail-craft of the Solar Polar Imager mission. *41st AIAA/ASME/SAE/ASEE Joint Propulsion Conference and Exhibit*, Tuscon, AZ, July 2005, AIAA Paper 2005-3928.

37. J. L. Wright. *Space Sailing*. Gordon & Breach Science Publishers, Philadelphia, PA, 1992.

22 Perspectives on Light-Driven Actuators

George K. Knopf and Yukitoshi Otani

CONTENTS

22.1 INTRODUCTION

Light-driven and optically controlled actuators have evolved rapidly in the past decade because of unprecedented advances in innovative materials, new fabrication processes, and a multidisciplinary approach to product and process design. These devices are typically small remotely activated transducers that transform the spectral, intensity, or phase properties of light into very small structural displacements and forces. Radiation pressure and optical gradient forces have been used to manipulate minuscule objects, while the spectral properties of light have changed the mechanical behavior of various stimulus responsive polymers. Alternatively, optically driven photothermal effects have been used to heat liquids and gases in an effort to increase the fluid pressure acting on a flexible diaphragm. Many of these induced effects are not readily observable with the human eye and may not appear to be significant in the larger more familiar world, but within the micro- and nanodomain of the *very very small*, these light-driven mechanisms become viable solutions.

As predicted by Richard Feynman in his 1959 lecture to the American Physical Society at Caltech, "There's Plenty of Room at the Bottom" (Feynman 1960), the world of the very small will play a significant role in future technology innovations. The future, he believed, lies in "manipulating and controlling things on a small scale" (Feynman 1960). With respect to light, a tightly focused beam provides a highly versatile precision tool for reaching out and manipulating tiny cells, or providing a feathered touch to rearrange molecules, or creating the gentle forces necessary to continually drive a submicron motorotor. Specific wavelengths of electromagnetic radiation can also induce physical changes in the shape of preformed photoresponsive polymers and gels. The smaller the polymer structure, the faster and more pronounced the observed reaction to the incident light. These light-induced shape changing materials can be exploited by creative engineers to develop embedded mechanisms for regulating microfluidic flow or manipulating molecules in Feynman's world of the *very small*. An interesting illustration of exploiting the unique characteristics of photoresponsive materials is the light-driven single-molecule DNA hairpin structured nanomotor introduced by Kang et al. (2009). The nanomotor incorporates photoisomerizable azobenzene molecules that enable the DNA structure to undergo a reversible light-controlled switching operation.

Three main objectives were set out at the beginning of this book. The first was to present the scientific language and fundamental principles of light-driven technologies helping to shape the future of nano- and microsystems. The second was to provide readers with a holistic view of optical nano- and microactuator systems, thereby enabling them to begin the process of understanding how light can be used to physically manipulate material properties and mechanical structures. The third objective was to help readers realize the practical applications of light-driven systems. By addressing these goals, the various authors hoped to advance the underlying science and provide inspiration for the next generation of engineers as they move forward to develop innovative solutions far beyond anything that we can imagine today.

Through their chapter contributions and personal communications, a number of the book's coauthors have provided a brief glimpse into areas of future research that will advance the state of the art in optical actuation technology. These perspectives are provided by an international group of researchers such as Khaled Al-Aribe, Christopher Barrett, Yves Bellouard, Roman Boulatov, Bernd Dachwald, Katsuo Kurabayashi, Shoji Maruo, Hideki Okamura, Balaji Panchapakesan, Halina Rubinsztein-Dunlop, Atsushi Suzuki, Kenji Uchino, and Jun Yang. It is clear from the discussion that the future of nano- and micro-optically driven technologies will depend largely on the design of customized materials, intensified exploitation of leading-edge microfabrication technologies, and development of functional nano- and microstructures that simplify the construction of sophisticated light-driven micromachines. It is hoped that the following, and often more speculative comments, will spur vigorous discussion and provide young researchers with a path to future research success. In addition, these words and thoughts may also be of interest to individuals who wish to see a snapshot of views expressed by experts and leading-edge researchers in the early part of the twenty-first century.

22.2 ADVANCING THE STATE OF THE ART

Future advances in optically driven actuators are dependent upon the discovery and exploitation of new photoresponsive materials that enable designers to create very small optically driven structures, mechanisms, and machines that have never been previously envisioned. Christopher Barrett (Chapter 4, personal communication, 2011) of McGill University believes that the key advances in optically driven technologies will occur with the introduction of custom-designed materials. Barrett and his collaborators have explored the reversible changes in shape that can be induced with material systems that incorporate azo compounds. The photomechanical effect of this class of materials provides a reversible change in shape induced by the adsorption of light, which results in a significant macroscopic mechanical deformation of the host material. Azobenzene materials exhibit a wide variety of switching behavior, from altering optical properties to surface energy changes to eliciting bulk material phase changes. However, a deep understanding of these azobenzene polymers and their potential applications is still at an early stage.

Soft azobenzene polymers may be one of the most promising materials for the next-generation photomechanical devices because of their efficient and robust photochemistry. In particular, liquid crystalline elastomers with azobenene may enable future engineers to create soft actuators that mimic the behavior of human muscles. Barrett stresses that recent studies have shown that these soft polymer systems can generate both two- and three-dimensional (3D) movements. With all this promise, unfortunately, there are critical material properties such as fatigue resistance and biocompatibility that still need to be understood in greater detail. He also points out that azobenzene materials still appear to be a "solution in need of a problem to solve." The material science and theoretical understanding of these light responsive soft polymers have far outpaced the work on practical applications. Although a few recent "proof-of-principle" devices can be found in the published literature, these engineered azobenzene-based systems still lack sufficient unifying problems or application areas where the material can fully display its inherent advantages. Clearly, creative and inspired engineering must still be employed to develop the technology beyond the current basic science.

Kenji Uchino (Chapter 5, personal communication, 2011), from Pennsylvania State University, restates the importance of thoroughly exploiting the unique properties of these new photomechanical materials. Uchino draws on his groundbreaking work on photostrictive materials (Uchino 1990, Uchino et al. 1983) to identify engineering challenges in developing photoresponsive materials and light-driven actuator technology. Photostrictive materials exhibit light-induced strains that arise from the superposition of the "bulk" photovoltaic effect (voltage generated from light irradiation) and the converse-piezoelectric effect (material expansion or contraction under applied voltage). Through advances in material design the response speed of these photostrictive bimorph actuators have improved by three orders of magnitude (10^3 times) since its initial discovery in the early 1980s. Further advances led to $(Pb,La)(Zr,Ti)O_3$ (PLZT) ceramics doped with WO_3 that have exhibited even larger photostriction effects under uniform illumination of near-ultraviolet light. Simple photo-driven relays and a micro walking "machine" have demonstrated the functionality of this biomorph configuration. Critical engineering challenges must still be addressed if practical micromachines are to be made from these bimorph structures. For example, these photostrictive actuators being very small devices make it difficult, if not nearly impossible, to attach any lead wires to increase the functionality of these microactuators.

The importance of searching for nontraditional functional materials is further stressed by Yokohama University's Atsushi Suzuki. Among the various kinds of smart actuators being investigated, Suzuki believes that light sensitive hydrogels hold a particular promise for future development. The gels consist of covalently cross-linked copolymer networks of a thermoresponsive polymer and chromophore. By making use of the gel's hysteresis phenomenon in response to the changes in external stimuli, the phase of the gel can be successfully activated by visible light ("switched-on") and deactivated ("switched-off") by altering the local environmental conditions. This switching phenomenon can be exploited by engineers and used for a variety of applications such as actuators, sensors, and display units. Another interesting property is that local heating of the polymer network will change the light transmission properties of optically transparent gel at the exposed location.

Suzuki (Chapter 7, personal communication, 2011) points out that there is another, equally important, aspect to using hydrogels when designing BioMEMS devices, Lab-on-Chip (LoC) platforms, and micro-Total Analysis Systems (µTAS). Hydrogels are biocompatible with human tissues and cells and are being currently used in medicine to repair tissues, organs, and vascular systems. Although hydrogels have comparable characteristics with tissues, it is important to understand that they are not identical and lack several important biological properties associated with the living tissue. In the future, however, it may be possible to externally control the mechanical, chemical, and chemomechanical functions of these gels in order to make these biocompatible structures more closely mimic natural biological tissue.

As a scientist, Suzuki also believes that two of the most significant technological problems that must be addressed in the early twenty-first century concerns global warming and our rapidly depleting natural resources. He feels that a greater understanding about the physics and chemistry of gel science will no doubt play greater role in creating more human-friendly and environmentally conscious products. Optically driven actuators can contribute to a reduction in the raw materials and energy necessary to meet the growing technology demands of the future. It is conceivable that in the coming decades very small clean "factories" that produce a variety of high-quality consumer products and customized medication will be created on single chips.

Balaji Panchapakesan, from the University of Louisville, has spent years investigating carbon nanotube–based photomechanical actuators. The light-induced actuation properties of these nanotube-polymer systems depend on the physical structure of the actuator, nanotube alignment and entanglement, and the presence of prestrains in the sample. Looking into the future, Panchapakesan (Chapter 6) sees graphene-based photomechanical devices making a measureable impact. Recent research shows outstanding mechanical properties for infrared-triggered graphene nanocomposites (Huang et al. 2009) including a change in stress of 50 MPa and strains over 100%. In addition, heat-treated graphene papers have recently demonstrated superior hardness, an yield strength higher than carbon steel, and an extremely high modulus of elasticity during bending. Mixing graphene with polymeric materials could lead to a broad spectrum of high performance nanogrippers, rotational actuators, and nanocantilevers. Hybrid materials such as liquid crystal elastomers mixed with both nanotubes and graphene could result in new types of light-driven actuators that absorb sunlight to create mechanical work. Finally, light absorbing and low weight polymer composites would have extraordinary strength and could be a suitable material to construct advanced spacecraft suitable for extremely long interplanetary missions.

Khaled Al-Aribe (personal communication, 2011), from the University of Western Ontario, encourages future scientists and engineers to explore biological molecules in their search for new photoresponsive materials. Among the biologically derived light-responsive materials is bacteriorhodopsin (bR), the protein found in the salt marsh archaebacteria *Halobacterium salinarum*. In nature, the bR molecules use the sun's energy to transport hydrogen ions across the *Halobacterium* cell wall to generate the necessary potential difference for synthesizing adenosine triphosphate (ATP) from adenosine diphosphate (ADP) (Hampp 2000, Wang et al. 2005). Under low oxygen conditions, the *Halobacterium* cell grows planar purple membrane (PM) patches in the form of a hexagonal 2-D crystalline lattice of bR trimers. When exposed to light in the visible spectrum, each bR protein acts as an independent proton pump that pumps hydrogen ions from the cytoplasmic to the extracellular side through a transmembrane ion channel that connects both sides of the membrane. Because of its crystalline structure, the bR in the PM patches exhibits unique chemical and thermal stability properties when exposed to long periods of sunlight. The stability and functional capability of bR have made it one of the most studied biologically photosensitive materials. The PM even preserves its photochemical and photoelectric activities under dry conditions and can withstand relatively high temperatures of up to 140°C and can continue functioning as a proton pump under very harsh chemical conditions such as extreme acidic and alkaline environments that are often considered corrosive for semiconductor technology (Hampp 2000).

The bR protein has been used to create a variety of bioelectronic imaging arrays (Wang et al. 2008) and color sensors (Lensu et al. 2004). bR-coated microcavities have also been introduced as a mechanism for developing all-optical switches (Roy et al. 2010). Current work has also shown that bR-based photoelectric structures can induce volumetric phase transitions in pH-sensitive polymer gels. The flow of ions from the photon activated bR changes the pH value of the ionic solution that surrounds the hydrogel actuator or microvalve (Al-Aribe et al. 2011, 2012). The chargeable polymeric network undergoes a measureable geometric change when the pH of the ionic solution is shifted to the phase transition point pKa (Liu et al. 2002). These light-responsive devices require the bR to be properly immobilized on the surface of an electrode in order for the PM to act as a directional proton pump. The fact that bR can function as a thin film for microapplications, or as molecules for nanoapplications, makes it a remarkable material.

The increasing complexity of these tiny microsystems poses numerous technical challenges in component design, system integration, packaging, and part fabrication. Microelectromechanical systems (MEMS) are often viewed as an off-shoot of silicon-based integrated circuit (IC) technology, which relies on "top-down" photolithography techniques for part fabrication. Although photolithography is fairly well developed, the processes involved restrict the types of materials, microfeatures, and component packaging possible. Yves Bellouard from Eindhoven University of Technology discusses how new micromanufacturing processes are necessary to locally alter a material's structure to create innovative mechanism designs and better realize full functional integration (Chapter 15, personal communication, 2011). The unconventional laser–matter interaction arising from the very high-peak powers associated with femtosecond laser pulses is one approach to tailoring material properties. It is possible, for example, to locally change the properties of fused silica to create miniaturized devices that combine optical functions (such as waveguides), mechanical (flexures) features, and fluidic (microchannels) elements.

Manufacturing integrated microsystems from polymers also introduces challenges for traditional IC fabrication processes. Katsuo Kurabayashi, from the University of Michigan, reminds us that polymers have attracted significant attention in recent years due to their versatile material properties, compatibility with biological and chemical systems, cost-effectiveness, and ease of manufacturability. Nano-patterned polymeric films and self-organized polymeric materials provide important building blocks in nanotechnology, nanoelectronics, and photonics. Unlike photolithography, polymer-based fabrication techniques often involve "bottom-up" self-assembly and soft printing approaches. Kurabayashi (Chapter 16) believes that it will be necessary to combine the emerging area of nanomanufacturing with more established MEMS technology to develop new types of micro-optical devices. The multiscale hierarchical integration of polydimethylsiloxane (PDMS) elastomer micro- and nanostructures in silicon MEMS would enable engineers to incorporate the unique material properties and physical phenomena of polymers. As Kurabayashi points out, this could open the door for polymer-based nanofabrication and nanotechnology to find a wider variety of commercial, military, and biomedical applications.

Richard Feynman anticipated the idea of micromachines in his 1959 talk (Feynman 1960) by suggesting a micromechanical "surgeon," which could be swallowed and could operate inside a faulty blood vessel. At the end of his talk, he challenged researchers to build a microscale electric motor that could fit in a 1/64 in.[3] Not long after his talk in November 1960, the first micromachine was built by an electrical engineer, James McLellan. However, the first true MEMS devices were silicon electrical micromotors with a diameter of 100 μm (Fan et al. 1989, Mehregany et al. 1990) built in the late 1980s using IC fabrication methods. Today, MEMS technology has been incorporated in a variety of sensor and actuator applications including accelerometers that trigger airbags in cars, inkjet printers, and optical switches for data communications. Although less common, MEMS technology is beginning to make an impact in the field of medicine with new types of blood pressure sensors (Benzel et al. 2004, Goh and Krishnan 1999, Ishiyama et al. 2002).

Conventional silicon-based MEMS do not represent the only approach to creating sophisticated integrated micromachines. An alternative approach takes advantage of electronic printing and rapid prototyping technologies such as stereolithography (Deitz 1990). Halina Rubinsztein-Dunlop and her colleagues at the University of Queensland (Chapter 9) have shown how the functional mechanism for an optically driven micromachine can be fabricated using two-photon photopolymerization (2PP) of UV curing resins. The 2PP method is successfully used to create arbitrarily shaped 3D structures including a functional micromotor. Rubinsztein-Dunlop and her colleagues have also shown that these microstructures can be functionalized with biomolecules making them great candidates for biological experiments in optical tweezers.

Shoji Maruo and his colleagues, from Yokohama University, have developed several types of optically driven micromachines including microtweezers and micropumps using the two-photon microfabrication approach. Maruo (Chapter 10) believes that microfabrication technology will lead to the development of an all-optically controlled LoC. LoC devices play an important role in microfluidic systems for analyzing DNA and RNA, medical screening, monitoring the environment, and chemical analysis. Often, these chips must be disposable because they can only be used once to avoid contamination of samples. To address this practical need, it is necessary to develop low-cost manufacturing methods that can produce large numbers of microfluidic systems inexpensively. Maruo points out that by creating the constituent components of the LoC using the two-photon microfabrication techniques it is possible to manufacture complex mechanically moving systems that are readily disposed of in an environmentally friendly manner. In the near future, he expects that advanced manufacturing techniques will enable all-optically controlled biochips to be fully integrated with light sources and open pathways for future medical care and diagnosis.

Jun Yang (Chapter 17, personal communication, 2011), from the University of Western Ontario, reminds us that one of the most important tools in biophotonics is the ability to manipulate single molecules and cells using optical trapping and tweezer techniques. Although he feels that it may be difficult to foresee the future, it is clear that seamlessly integrating biophonics with LoC and BioMEMS platforms will be an important step toward significant advances in the not too distant future. Yang states that this will create numerous new opportunities in cell manipulation including cell sorting and cell surgery. Appropriately shaped light beams would enable researchers to watch, control, and intentionally interfere with the intracellular, extracellular, or even transcellular movement of single molecules. Further, properly designed optical configurations could be used to non-invasively transform and assemble biological objects, with nanometer accuracy and precision. This would enable optical manipulations, assays, and detections to be performed in an all-in-one single chip or microsystem for diagnostics and drug discovery.

Hideki Okamura (Chapter 19, personal communication, 2011), with the International Christian University in Japan, echoes one of the main themes of the book by reminding us that the opportunities for exploiting optically driven technologies are immense. The key is to fully exploit the nature of light and photon–material interactions. An important advantage of optical technology, often overlooked, is the elimination of the need for conductive wiring in the micromachine design. Electrical wiring for power delivery and transmitting information signals is a limiting factor in assembling and packaging complex microsystems. Optically driven systems are free from undesirable side effects produced by electrical current losses and resistive heat dissipation, and through clever designs hundreds of actuation points can be addressed simultaneously. This is possible because the intersecting light beams do not interfere with each other. By taking advantage of this characteristic of light, it may be possible to independently control the shape of numerous thin film actuators. At present, no alternative conventional actuating mechanism can be found.

Roman Boulatov (Chapter 3, personal communication, 2011), from the University of Illinois, hopes that basic research into light-driven materials will lead to previously unimagined technologies that can be used to propel small (100 nm or less) swimming or flying micromachines. In the future, he foresees researchers being able to purchase off-the-shelf optical nano- and microactuators (similar to buying an electrical motor) and combine them to create whatever submicron machine the design engineer requires. Boultatov looks forward to having access to a modular microtechnology that would enable him to build hundreds of sub-microswimmers powered by light who instinctively swim toward, or away, from the light source, exhibit complex interactions, and demonstrate emergent collective of a school of fish or a swarm of insects.

Finally, Bernd Dachwald, from FH Aachen University of Applied Sciences, looks into the *far far future* where he sees interstellar spacecraft being driven by light. Dachwald (Chapter 21, personal communication, 2011) comments that while the physical laws for this type of propulsion are well known, the technical challenges in creating light-driven spacecraft are so immense that it is unlikely that humanity shall see the concept realized for several hundred years (Forward, 1984). With current laser technology and sophisticated optics, it is possible to generate the radiation pressure and optical gradient forces necessary to propel giant lightweight mirrored surfaces, or laser sails, through the vacuum of space. Unfortunately, a trip to our closest neighboring star Alpha Centauri, approximately 4.3 light years away, would require a 1000 kg, 3.6 km diameter laser sail and a 65 GW laser to accelerate the craft through its 40 year journey. The technology may capture the imagination of scientists and engineers, but the costs and overwhelming technical challenges make this dream impractical in our lifetimes.

22.3 CONCLUDING REMARKS

Although optically activated nano- and microactuators are probably the least developed of all force-generating structures, the emerging technology offers a number of interesting and beneficial design features. All-optical circuits and light-driven devices have advantages over conventional electronic components because they are activated by streams of photons instead of currents and voltages. In many of these designs, the photons provide both the energy into the system and the control signal used to initiate the desired mechanical response. Furthermore, optical systems are free from current losses, resistive heat dissipation, and mechanical friction forces that can greatly diminish the performance and efficiency of conventional electronic or electromechanical systems. The negative effects of current leakage and power loss are greatly amplified as design engineers strive for product miniaturization through the exploitation of micro- and nanotechnology. Even the radiation pressure arising from a beam of photons becomes a viable force for driving mechanisms that have a picogram mass or exist in a nanometer size. Optical actuators are also ideal components for smart structures because they are immune from electromagnetic interference, safe in hazardous or explosive environments, and exhibit low signal attenuation.

The fundamental and unique characteristics of light-activated optical actuators are explored in this book. The primary means of actuation is to project light onto a nano- or microactuator shell in an effort to generate mechanical deformation that, in turn, produces the desired displacement or force. Light is used to both initiate movement (i.e., power) and control the actuation mechanism that performs the work. Current researchers have exploited optical gradient forces to manipulate tiny objects and biological cells, and have used the spectral properties of light to change the shape of various stimulus responsive polymers or gels. In addition, optically driven photothermal effects were used to heat small quantities of liquids and gases to increase the pressure acting on mechanically flexible structures. The opportunities for advancement and future innovation are based on the development of new high-performance photomechanical materials, leading-edge microfabrication technologies that enable molecular self-assembly and efficient integration of polymers and traditional MEMS, and microproduct designs that take advantage of the unique light and material interactions that occur in the world of the *very very* small.

REFERENCES

Al-Aribe, K., Knopf, G.K., and Bassi, A.S. 2011. Photoelectric monolayers based on self-assembled and oriented purple membrane patches. *IEEE/ASME Journal of Microelectromechanical Systems* 20(4): 800–810.

Al-Aribe, K., Knopf, G.K., and Bassi, A.S. 2012. Fabrication of an optically driven pH-gradient generator based on self-assembled proton pumps. *Microfluidics and Nanofluidics* 12: 325–335.

Benzel, E., Ferrara, L., Roy, S., and Fleishman, A. 2004. Micromachines in spine surgery. *Spine* 29: 601–606.

Deitz, D. 1990. Stereolithography automates prototyping. *Mechanical Engineering* 112(2): 34–39.

Fan, L.S., Tai, Y.C., and Müller, R.S. 1989. IC-processed electrostatic micromotors. *Sensors and Actuators* 20: 41–47.

Feynman, R.P. 1960. There's plenty of room at the bottom. California Institute of Technology. *Journal of Engineering and Science* 4: 23–36.

Forward, R.L. 1984. Roundtrip interstellar travel using laser-pushed lightsails. *Journal of Spacecraft and Rockets* 21(2): 187–195.

Goh, P. and Krishnan, S.M. 1999. Micromachines in endoscopy. *Bailliere's Clinical Gastroenterology* 13: 49–58.

Hampp, N. 2000. Bacteriorhodopsin as a photochromic retinal protein for optical memories. *Chemical Reviews* 100: 1755–1776.

Huang, Y., Zhang, L., Wang, Y., Ma, Y., Li, F., Guo, T., and Chen, Y. 2009. Infrared-triggered actuators from graphene-based nanocomposites. *Journal of Physical Chemistry C* 113(22): 9921–9927.

Ishiyama, K., Sendoh, M., and Arai, K.I. 2002. Magnetic micromachines for medical applications. *Journal of Magnetism and Magnetic Materials* 242: 41–46.

Kang, H., Liu, H., Phillips, J.A., Cao, Z., Kim, Y., Chen, Y., Yang, Z., Li, J., and Tan, W. 2009. Single-DNA molecule nanomotor regulated by photons. *Nano Letters* 9(7): 2690–2696.

Lensu, L., Frydrych, M., Parkkinen, J., Parkkinen, S., and Jaaskelainen, T. 2004. Photoelectric properties of bacteriorhodopsin analogs for color-sensitive optoelectronic devices. *Optical Materials* 27: 57–62.

Liu, R.H., Yu, Q., and Beebe, D. 2002. Fabrication and characterization of hydrogel-based microvalves. *Journal of Microelectromechanical Systems* 11(1): 45–53.

Mehregany, M., Bart, S.F., Tavrow, L.S., Lang, J.H., and Senturia, S.D. 1990. Principles in design and microfabrication of variable-capacitance side-drive motors. *Journal of Vacuum Science and Technology A* 8: 3614–3624.

Roy, S., Prasad, M., Topolancik, J., and Vollmer, F. 2010. All-optical switching with bacteriorhodopsin protein coated microcavities and its application to low power computing circuits. *Journal of Applied Physics* 107: 053115–053124.

Uchino, K. 1990. Photostrictive actuator. In *IEEE Ultrasonics Symposium*. Honolulu, HI, December 4–7, 1990, pp. 721–723, DOI: 10.1109/ULTSYM.1990.171457.

Uchino, K., Miyazawa, Y., and S. Nomura 1983. Photovoltaic effect in ferroelectric ceramics and its applications. *Japanese Journal of Applied Physics* 22: 102.

Wang, W.W., Knopf, G.K., and Bassi, A.S. 2005. Photoelectric properties of a detector based on dried bacteriorhodopsin film. *Biosensors and Bioelectronics* 21: 1309–1319.

Wang, W.W., Knopf, G.K., and Bassi, A.S. 2008. Bioelectronic imaging array based on bacteriorhodopsin film. *IEEE Transactions on Nanobioscience* 7(4): 249–256.

Index